Laboratory Manual for Anatomy and Physiology

Laboratory Manual for Anatomy and Physiology

With Cat Dissections

THIRD EDITION

Patricia J. Donnelly
George A. Wistreich

East Los Angeles College

HarperCollins*Publishers*

Grateful acknowledgement is made for permission to use the following illustrations: From Gerard J. Tortora and Nicholas P. Anagnostakos, Principles of Anatomy and Physiology, Third, Fourth, Fifth, and Sixth Editions HarperCollins*Publishers*, Inc., 1981, 1984, 1987, and 1990): 7-2a, 7-5, 7-6, 7-7, 7-8, 7-9, 7-11, 7-12, 7-14, 7-15, 10-1, 10-2, 10-3, 11-1, 11-2, 11-3, 11-6, 12-1, 12-2, 12-3, 13-1, 13-5, 13-6, 14-3, 14-6c, 14-6d, 14-9a, 14-9b, 14-10a, 14-10b, 14-11a, 14-11b, 14-13a, 14-13b, 15-1, 15-2, 16-5, 16-6, 17-4, 17-5, 18-3, 18-4, 19-3, 19-4, 19-5, 19-6, 21-5, 21-6, 22-3, 31-4, 31-6, 31-7, 31-8, Q38-1, 39-3, 39-8, 39-9a, 39-9b, 39-10a, 39-10b, 39-11a, 39-11b, 39-12, 48-4, 51-4, 54-10, and 54-11.
From Gerard J. Tortora, Principles of Human Anatomy, Third, Fourth, and Fifth Editions HarperCollins *Publishers*, Inc., 1983, 1986, and 1989): 7-1, 7-2b, 7-3, 7-4, 7-9, 7-10, 7-13, 7-16, 7-19, 7-20, 7-21, 7-22, 11-4, 13-2, 13-4b, 13-7, 14-1, 14-2, 14-5, 14-6a, 14-6b, 14-8a, 14-8b, 15-3, 19-9, 20-5, 20-6, Q20-1, 31-5, 31-9, and Q54-1.

Sponsoring Editor: Elizabeth A. Dollinger
Project Editor: Thomas R. Farrell
Art Direction: Teresa J. Delgado
Text Design and Cover Coordinator: Heather A. Ziegler
Cover Design: Circa '86, Inc.
Cover Photo: From *The Dance Workshop.* Courtesy Gaia Books Ltd.
Production: Kewal K. Sharma

LABORATORY MANUAL FOR ANATOMY AND PHYSIOLOGY: With Cat Dissections,
Third Edition

Library of Congress Cataloging-in-Publication Data

Donnelly, Patricia J.
Laboratory manual for anatomy and physiology : with cat dissections / Patricia J. Donnelly, George A Wistreich. --3rd ed.
p. cm.
Includes bibliographical references.
ISBN 0-06-041708-0
1. Physiology, Experimental—Laboratory manuals. 2. Cats—Physiology—Laboratory manuals. 3. Cats—Anatomy—Laboratory manuals. 4. Dissection—Laboratory manuals.
I. Wistreich, George A. II. Title.
QP44.D66 1990 89-19971
599'01'078—dc20 CIP

90 91 92 9 8 7 6 5 4 3 2

Contents

SECTION IX

The Senses 409

SECTION X

Cat Dissection: An Introduction to the Visceral Organs 461

SECTION XI

The Cardiovascular System 471

SECTION XII

Blood and the Immune System 547

SECTION XIII

The Respiratory System 603

SECTION XIV

The Digestive System 623

SECTION XV

The Urinary and Reproductive Systems 667

Preface

The purpose of *Laboratory Manual for Anatomy and Physiology: With Cat Dissections* is to provide students with functional and effective experiences that will enable them not only to observe but also to understand fundamental principles and concepts of human anatomy and physiology. The exercises in this manual are organized into complete units of instruction so that students study each organ system in its entirety and observe and recognize fundamental similarities and differences among the systems.

As the manual's subtitle indicates, we have used the cat as the primary dissection animal. Tables in the exercises on the muscular system and certain other systems relate information learned in the dissection procedure on the cat to human anatomy. The choice of the cat as a dissection specimen was made for two reasons: first, human cadavers are difficult to obtain and maintain; second, we feel that a student's own dissection has more instructional value than mere observation of demonstration material. In designing the physiology experiments we kept in mind the need to provide modern and functional approaches to demonstrate physiological concepts and techniques. The experiments and studies use relatively new, inexpensive, and easily obtainable specimens and laboratory setups. It is our feeling that basic principles of physiology can be demonstrated with simple techniques and a minimum of complex equipment.

Organization of the Manual

The 59 laboratory exercises in this manual include a variety of special features that are designed to stimulate interest in the subject matter, to involve students in the learning process, and to guide them through specifically planned activities.

The manual offers a broad range of exercises from which instructors may choose. It may be used for a one- or two-semester course in anatomy and physiology at the undergraduate level. Some of the exercises presume a knowledge of basic high school biology and chemistry. Although the manual was designed for courses related to allied health curricula, it can be used equally well in courses for general mammalian anatomy and physiology.

The 59 exercises are arranged in 15 sections as follows:

Section I emphasizes chemical and cellular measurements. Major components of this section are techniques for making scientific measurements, handling chemicals, and the care and use of the microscope.

Section II serves as an introduction to the cellular and tissue levels of organization. Exercises are designed to show the features of cellular structure and function and the chemical and physical aspects of cellular permeability.

Section III is a survey of anatomical orientation and terminology and serves as an introduction to many terms used in subsequent exercises.

Section IV demonstrates the general features of the skin, nails, and hair. Particular attention is given to fingerprints and the general appearance of blood vessels in response to different temperatures.

Section V familiarizes students with the structure and functions of the skeletal system. Exercises in this section provide a thorough treatment of the support system in the human body.

Section VI contains numerous exercises designed to demonstrate the properties of mammalian muscle. The cat is the laboratory animal, and there are black and white photographs, color photographs, and line drawings to assist the students in their dissections. Other exercises cover the structure of muscle and the chemistry of muscular contraction.

Section VII provides students with a comparative and functional treatment of nervous system coordination and control. The anatomy of the human system is stressed, and the frog is used to demonstrate cranial and spinal reflex responses.

Section VIII provides an overall view of the human endocrine system. The interrelationships, locations, and functions of the glands are covered in sufficient detail to emphasize the importance and special properties of the endocrine system.

Section IX contains numerous simple experiments that emphasize the importance and special properties of the endocrine system.

Section X familiarizes students with the dissection procedure. It emphasizes and demonstrates the human organ systems and indicates their counterparts in the cat.

Section XI presents the major anatomy and physiology of the cardiovascular system. Several exercises are included that deal with the effects of physical and chemical factors on the heart, along with an exercise demonstrating clinical features of coronary circulation.

Section XII presents coverage of the components of blood, the immune system, and various aspects of diagnostic hematology, including factors involved with coagulation, hemoglobin determinations, blood typing, and cross-matching.

Section XIII familiarizes students with the basic anatomical and physiological features of the respiratory system. Procedures for conducting pulmonary function tests are incorporated in this section.

Section XIV treats the structure of the digestive system. Exercises dealing with enzymatic action, digestion chemistry, teeth, dental caries, liver function tests, and the demonstration of the normal flora are important parts of this section.

Section XV deals with the anatomy and physiological features of both the urinary and reproductive systems. It includes an examination of the urogenital organs and demonstrates the principles associated with dialysis, fluid and electrolyte balance as well as acid–base balance. An exercise dealing with routine diagnostic urinalysis is contained within this portion of the manual. In addition, attention is given to sex cell formation, pregnancy testing, and karyotyping.

Learning Aids

A number of learning aids have been incorporated to make this manual an effective tool for students. These include the following:

Section overviews. Each major section of the laboratory manual is preceded by a description of the range of topics, exercises, and experiments to be covered. A list of general references is included to provide background information and a basis for further study.

Illustrations. Photographs and diagrams have been included to provide a greater understanding of anatomical relationships and physiological principles, techniques, and test results. Photographs showing important structures, reactions, and results are used throughout the manual. Special sections containing color photographs of cat muscles and histology are presented to simplify identification.

Procedure diagrams. Exercises that introduce basic techniques contain step-by-step illustrations of the procedure to show the proper handling of equipment, animals for dissection, and related items.

Questions for review. At the end of each exercise there are selected questions to test students' comprehension of the major principles and results of the exercises performed. Several of these sections also contain a photograph quiz.

Index. An index has been included in the manual. Students should consult this index to facilitate location of anatomical parts, tests, and procedures.

Instructor's Manual

An instructor's manual to accompany the laboratory manual is available. It contains specific directions for the preparation of laboratory materials and descriptions of alternative procedures and the use of supplementary aids. There are suggested sources for audiovisual aids, laboratory equipment and supplies, and questions for laboratory examinations and quizzes.

Acknowledgments

We thank the following colleagues who read and reviewed the manuscript versions of our work and whose comments we found constructive: Susan Athwal, Monmouth College; Elizabeth Hager, Trenton State College; Bonnie Kalison, Mesa Community College; Brian O'Connor, University of Massachusetts–Amherst; John Marshall Wakeman, Louisiana Tech University; Carol Haspel, Fiorello H. LaGuardia Community College; Charles Achenbach, Northhampton County Area Community College; John Mackenzie, Boricua College.

The authors also gratefully acknowledge the HDC Corporation, SmithKline Diagnostics, Sunnyvale, California, and Wampole Laboratories, a Division of Carter-Wallace, Inc., Cranbury, New Jersey, for providing samples of their products for use in developing this manual.

Special thanks to College Biological Supply Company, Escondido, California, for supplying the dissection specimens: Dr. Joel Schechter of the Department of Anatomy, University of Southern California Medical School, and Leonard D. Dank, President of Medical Illustrators, Inc., for art work; and Dr. Daniel Fertig and Mr. Richard Holdredge of East Los Angeles College for photographic assistance.

We are also deeply grateful to the members of the editorial and production staffs of Harper & Row, Publishers, Inc.

P. J. Donnelly
G. A. Wistreich

Note to the Student

This laboratory manual is written for the student who wants to obtain a broad foundation in anatomy and physiology and some insight into the roles body structure and function play in health and disease.

The manual is divided into 15 major sections; each section contains one or more laboratory exercises. To make the manual an effective tool for you, we have incorporated into each exercise a number of learning aids. The *results and observations section* not only provides organized space to record important results and observations but also can serve as an excellent review of specific anatomical and physiological concepts. Answering the *questions for review* after you complete the laboratory work will help you to judge how well you have learned the material and to identify areas in which you need more work.

In the exercises relating to bones and muscles, you will find *phonetic pronunciations* for some anatomical terms. Taking time to sound out new terms and to say them aloud once or twice will help you master one of the tasks of the anatomy and physiology course—learning its specialized vocabulary. The following key explains the system used for the prounciations.

1. The strongest accented syllable appears in capital letters (e.g., bī-LAT-er-al, dī-ag-NŌ-sis). A syllable that has a secondary accent is marked by a single quote mark (e.g., kon'-sti-TOO-shun, fiz'-ē-OL-ō-jē).
2. Vowels pronounced with long sounds are indicated by a line above the vowel and are pronounced as in the following common words.

   ā as in *māke*
   ē as in *bē*
   ī as in *īvy*
   ō as in *pōle*
3. Vowels not marked for long sounds are pronounced with the short sound, as in the following words.

   e as in *bet*
   i as in *sip*
   o as in *not*
   u as in *bud*
4. Other phonetic symbols are used to indicate the following sounds.

   a as in *above*
   soo as in *sue*
   kyoo as in *cute*
   oy as in *oil*

Through the use of this laboratory manual you should gain a deeper understanding of human anatomy and physiology. Such knowledge will demand not only a genuine effort but also an organized coordination of text reading and laboratory experimentation. A solid foundation in the basic principles of anatomy and physiology will make the effort well worthwhile.

P.J.D.
G.A.W.

SECTION I

INTRODUCTION TO BIOCHEMISTRY, MICROSCOPY, AND MEASUREMENT

The world around us consists of energy and matter in several different forms. To study them effectively, a clear understanding of basic units is necessary. In dealing with matter, the basic unit is the *atom,* groups of which make up *molecules.* Molecules, in turn, group together and comprise the basic units of life, *cells.* To remain alive, grow, and reproduce, all living systems require chemical substances and undergo chemical changes. These systems also share in common the fact that carbon-containing organic molecules form their essential components. Carbon atoms are able to establish strong bonds with hydrogen, nitrogen, oxygen, and other carbon atoms and are well adapted for uniting the atoms of living systems into stable macromolecules. Cells are composed of a large variety of organic molecules representing four major classes of biochemicals: carbohydrates, lipids, proteins, and nucleic acids. Biochemical reactions involving these same classes of molecules are essential for the metabolic activities of living systems and serve as the bases for cellular growth and reproduction. Physiology requires an understanding of the biochemical macromolecules that form cellular structures and the biochemical reactions involved in the metabolic activities of cells. Examination of molecular level activity is necessary to understand the structure and function of living systems. Since most cells and their structures are too small to be seen with the naked eye, various instruments and techniques are used to reveal them. Certain exercises in this and later sections will serve as the bases for such instrumentation, thereby providing the necessary background for physiological exercises and studies.

One goal of investigations in the biological sciences is to separate and identify chemical compounds associated with cells and their activities. Separations are made on the basis that the chemical of interest reacts differently from the other chemicals in the mixture in which it is found. For example, some chemicals are more soluble in one type of solvent than another or more readily adsorbed to a particular porous surface, such as filter paper, when treated with a particular solvent.

Detecting and identifying small quantities of organic and inorganic substances that have to be separated can be done in a variety of ways. Some substances normally have color and can be seen easily. For colorless chemicals, specific procedures are used to make them visible. Exercise 1 presents the technique of *paper chromatography.* When this analytical technique is performed under controlled conditions of temperature and solvent purity, it can be used not only to separate various chemicals in a mixture, but to identify them as well.

Among the different tools essential for studying cells and their activities are microscopes. The primary function of a microscope is to enable the observer to distinguish structures and points that are separated by short distances. This function is called *resolution* or *resolving power.*

Resolution is more important than magnification: it is not always desirable to obtain the largest image possible, but it is necessary to obtain sharp detail. Resolution is dependent on many factors, including illumination and the nature of the specimens. The resolving power of a microscope is inversely related to the wavelength of light used, that is, the shorter the wavelength, the greater the resolving power.

Two categories of microscopes in general use are the *bright-field instrument,* which contains optical lenses and uses visible light as the source of illumination, and the *electron microscope,* which contains electromagnetic lenses and uses a beam of electrons as the source of illumination. These electron microscopes provide higher magnifications and greater detail than the bright-field instruments. Since several of the exercises in this manual will require the use of the compound bright-field microscope, Exercise 2 will deal with microscope construction, application, care, and basic concepts.

The application of various analytical instruments and techniques has played an enormous role in acquiring information about the biochemical processes occurring in cells and the structure and properties of both macromolecules and cellular organelles. Exercises 1 and 2 will show how some methods, such as chromatography, spectrophotometry, and microscopy, are used to study and obtain physical and physiochemical measurements of macromolecules and cells.

GENERAL REFERENCES

Burrells, W. *Microscope Techniques. A Comprehensive Handbook for General and Applied Microscopy.* New York: Halsted Press, 1976.

Heettman, E. *Chromatography.* New York: Van Nostrand Reinhold, 1975.

Cargille, J. J. *Immersion Oil and the Microscope.* Technical Reprint 10-51. Cedar Grove, NJ: R. R. Cargil Laboratories, 1975.

Mollring, F. K. *Microscopy from the Very Beginning.* Oberkochem, Federal Republic of Germany: Carl Zeiss.

Sackheim, G. I. and D. D. Lehman. *Chemistry for the Health Sciences,* 5th ed. New York: Macmillan Publishing Company, 1985.

SPECIFIC REFERENCES

De Robertis, E. D. P., and E. M. F. De Robertis, Jr. *Cellular and Molecular Biology,* 8th ed. Philadelphia: Lea & Febiger, 1987.

Stock, R., and C. D. Rice. *Chromatographic Methods.* London: Chapman and Hall, 1978.

Strong, F. M., and G. H. Koch. *Biochemistry Laboratory Manual.* Dubuque, IA: Wm. C. Brown, 1981.

EXERCISE 1

THE CHEMICAL BASIS OF LIFE

The principal atoms commonly found in cells are carbon, hydrogen, oxygen, nitrogen, phosphorus, and sulfur. The molecules formed from these atoms are of four major classes: *carbohydrates, lipids, proteins,* and *nucleic acids.* These molecules, referred to as *macromolecules* because of their large size, can associate with each other and even become integral parts of cellular structure.

Carbohydrates play a major role in the energy requirements of organisms. Sugars, such as glucose, and starches, which are complex sugars called polysaccharides, are the principal sources of energy in the ordinary diet of most forms of life. In addition, carbohydrates provide important structural skeletons—carbon atoms linked together with hydrogen and oxygen that are used in the construction of cell parts.

Lipids are a general group of organic substances that, unlike carbohydrates, are hydrophobic; that is, they do not mix with water. The basic chemical structure of many lipids consists of a molecule of glycerol to which three fatty acid molecules have been attached. These organic substances have several important functions such as providing fuel for metabolic reactions, forming structural components of cellular membranes, and serving as biologically active hormones.

Proteins are large molecules. They make up a significant portion of cells and are associated with the structures, functions, and metabolic activities of cells or tissues. Various proteins serve as catalytic agents called *enzymes.* Such proteins are associated intimately with cellular chemical processes. Proteins consist of basic building blocks called *amino acids* (Figure 1-1). These units are connected to one another by chemical connections known as peptide bonds. Proteins are formed by hundreds of amino acids connected to one another by these bonds.

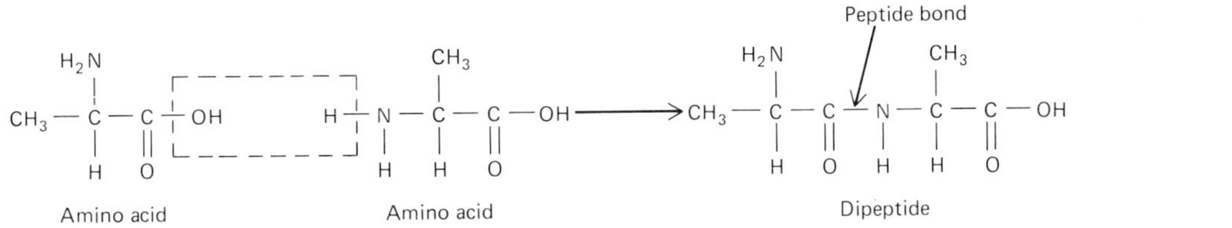

Figure 1-1 Protein formation. Amino acids serve as building blocks for macromolecules. Two molecules of the amino acid alanine are shown being joined in the formation of a peptide bond.

Two general kinds of nucleic acids are known, ribonucleic acid (RNA) and deoxyribonucleic acid (DNA). Both of the macromolecules are composed of basic units called *nucleotides.* Nucleotides consist of combinations of five-carbon sugars, phosphates, and either a pyrimidine or a purine, which are nitrogen-containing ring molecules. The ribonucleic acids are associated principally with the manufacture of proteins, while deoxyribonucleic acid contains the genetic or hereditary information that is passed from cell to cell and from generation to generation.

This exercise will demonstrate the types of procedures used to separate and to identify amino acids from biological materials. In addition, a simple chemical test to show the presence of peptide bonds will be considered.

A. PEPTIDE BONDS: THE BIURET TEST

The Biuret reaction will detect the presence of two or more peptide bonds in a protein molecule. A positive test is indicated by the appearance of a pink to violet color after the addition of test reagents (see Color Plate 1).

Materials

1. The following solutions, along with Pasteur pipettes and rubber bulbs, should be provided for class use.
   a. 1 percent albumin (egg white)
   b. Skim milk
   c. 1 percent starch
   d. 1 percent glycine (amino acid)
   e. Distilled water
   f. Biuret test reagents—10 percent sodium hydroxide (NaOH) and 1 percent copper sulfate ($CuSO_4$)
2. The following materials should be provided per four students.
   a. Ten test tubes
   b. One test tube rack
   c. Four Pasteur pipettes with rubber bulbs
   d. One marking pencil or pen
3. One unknown preparation per student

Procedure

This procedure should be performed by students in groups of four.

1. Place six test tubes in a test tube rack and label one tube for each of the test solutions. Specify the last tube as a reagent control.
2. Using the individual Pasteur pipettes provided for each solution, place a small, measured amount of each test solution in its labeled tube. Figure 1-2 shows how to use a Pasteur pipette.
3. Add an equal volume of the 10 percent NaOH and 4 drops of $CuSO_4$ to each tube.
4. Gently mix the contents of each preparation.
5. Observe for the presence of a pink to violet color. Save all tubes showing a positive reaction. (Refer to Color Plate 1.)
6. Enter your findings for each test solution in the appropriate table in the results and observations section.
7. Obtain an unknown test solution from your instructor. Place a small amount of it in a clean test tube and add the Biuret test reagents as you did with the known solution. Were

peptide bonds present? Enter your findings in the appropriate table. Check with your instructor to see if your observations were correct.

8. After all tests have been performed, empty the contents of the test tubes as indicated by your instructor and clean all glassware.

B. CHROMATOGRAPHY

Chromatography is a procedure commonly used in studying the chemistry of cells. While there are many forms of chromatography, the principles are similar in all. By means of this procedure a liquid mixture *(mobile phase)* of organic compounds can be separated into individual components. Such separation is based on the differing rates at which different substances migrate through porous materials *(stationary phase)* such as filter paper. Color reactions produced by treatment with chemical sprays also can be used for identification.

In this exercise, solutions of several different known and unknown amino acid preparations (mobile phase) will be run simultaneously on a single sheet of filter paper (stationary phase). The test paper is placed in a chamber, sealed, and after a specified period sprayed with ninhydrin reagent. Various amino acids form characteristic color spots (Table 1-1). Unknown samples of amino acids can be identified by comparing spots formed with those of known amino acids on prepared *chromatograms.*

Table 1-1 Color Reactions of Amino Acid Standards

Amino Acid	Color Results
Alanine	Dark purple
Aspartic acid	Generally light blue*
Glutamic acid	Purple
Glycine	Orange
Serine	Greenish brown

*If the chromatogram is removed too soon from the drying oven, a bright green color usually develops.

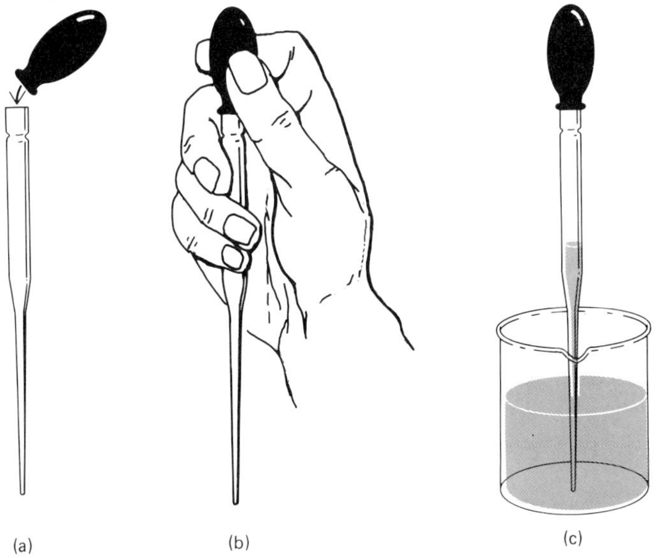

Figure 1-2 The use of a Pasteur pipette. (a) Attach a rubber bulb to the end of the Pasteur pipette. It should fit snugly. (b) Test the drawing power of the pipette by depressing the rubber bulb with your index finger and thumb and then releasing it. (c) To get a sample of solution, depress the rubber bulb, insert the pipette into the solution, and then slowly release the bulb, allowing the fluid to begin to fill the pipette. Leave the tip of the pipette in the fluid until the bulb has reached its normal size. Be careful not to allow the solution to pass into the bulb by tapping the sides of each tube.

Materials

1. The following materials should be provided for class use.
   a. 0.5 percent solutions of the amino acids: alanine, aspartic acid, glycine, glutamic acid, and serine
   b. Chromatography solvent—*n*-butanol, water, and glacial acetic acid in the proportion 4:5:1
   c. Indicator solution—2 percent ninhydrin–cupric nitrate
   d. Millimeter rulers
   e. Wooden applicator sticks
   f. Large sheets of wax paper
   g. Plastic wrap
   h. Drying oven set at 105°C
   i. Stapler
   j. Hair dryer
   k. Paper towels
2. The following materials should be provided per two students.
   a. Two Whatman #1 chromatography papers (10 × 15 cm)
   b. Two metal clamps (to hold chromatography paper)
   c. One 600 ml beaker
   d. One pair of forceps
3. One unknown preparation per student

Procedure

This procedure is to be performed by students in pairs.

1. Put a large sheet of wax paper on your laboratory table. Place one piece of Whatman #1 chromatography paper on the wax paper. Handle the chromatography paper only by its edges to prevent contamination by the amino acids on your hands.
2. On the Whatman paper draw a pencil line 1.5 cm from and parallel to the bottom edge. Mark a dot along this starting line 1.5 cm from the left-hand edge of the paper; moving to the right, mark successive dots every 2 cm for a total of seven dots (Figure 1-3).
3. Label one dot for each of the known amino acid solutions and for each unknown to be identified. Use the following symbols: alanine, al; aspartic acid, as; glutamic acid, gl; glycine, gy; serine, se; unknowns, x_1, x_2.
4. Draw a second pencil line 2 cm from and parallel to the top edge of the Whatman paper.
5. Using separate applicator sticks for each, place a small amount of each amino acid and unknown preparation on their respective labeled positions. Make three applications of each preparation, drying each application completely with the aid of a hair dryer or similar device. *It is extremely important to keep each spot area as small as possible.*
6. Carefully form a cylinder with the chromatography paper and staple it once at the top and bottom (Figure 1-4).
7. Fill the 600 ml beaker with 20 ml of chromatography solvent.
8. Place the paper cylinder in the beaker. (Do not allow it to touch the sides of the beaker.)
9. Quickly cover the beaker with plastic wrap. Place the chromatography system where it will not be disturbed while the solvent begins to move up the chromatogram preparation.
10. Observe the system periodically.
11. When the solvent reaches the pencil line near the top of the paper, take the chromatogram out of the beaker, remove the staples, and place the Whatman paper on a paper towel in a hood with ventilation.

EXERCISE 2

THE MICROSCOPE AND SPECIMEN PREPARATION

The compound microscope is a precision instrument. It is essentially a system of accurately ground lenses arranged to give sharp, clear, magnified images of minute objects. Magnification which may be obtained with different lens systems ranges from approximately 10 to 2000 times (×) the diameter of the specimen being observed. It should be noted that for each lens system the maximum magnification usually is obtained by highly trained microscopists using auxiliary equipment under ideal conditions.

The optical lenses of the microscope include the ocular or eyepiece, usually three objectives with different magnifying powers, and the substage condenser. All other parts of the microscope are accessory to the main purpose of magnification by these lenses. They are designed to hold the object to be examined steadily in place, to regulate the amount of light necessary for clear viewing, and to aid focusing by adjusting the space relationship between the ocular, the objective, and the specimen on the side.

Microscopes differ greatly from laboratory to laboratory. Your instructor will demonstrate and explain the general construction and manipulation of the particular instrument you will use. There is, however, a set of basic rules that should always be observed for all instruments to ensure maximum efficiency and proper maintenance.

USE AND CARE OF THE MICROSCOPE

1. Always carefully remove the microscope from its place of storage by holding it firmly by the arm with one hand and supporting it at the base with the other. Keep the instrument in an upright position.
2. Place the microscope at least 6 inches from the edge of your laboratory table.
3. Do not tamper with or remove any parts. If the microscope does not seem to be functioning properly, call your instructor immediately. Other laboratory sections may use the same instrument, and any part damaged but not reported by someone else might be charged to you.
4. Do not handle the lenses with your fingers. Perspiration contains fatty acids and other substances that can mar the lens glass. Always use the specified type of lens paper for cleaning the optical system. *Never use paper towels or cloths.* These materials will scratch the delicate lens surfaces.
5. Wipe the lenses of the instrument before and after laboratory use.
6. Use a clean, soft, linen cloth to remove dirt, dust, or other materials from parts other than the lenses.
7. Do not allow liquids, particularly acids and alcohol, to come in contact with any part of the microscope.

8. Always use a cover glass when examining objects or organisms mounted in water or other fluids.
9. Remove immersion oil from the microscope with lens paper.
10. Always raise the body tube sufficiently before placing a slide on or removing a slide from the microscope stage.
11. Before putting the microscope away, always put the low power objective in working position (i.e., in line with the body tube), open the diaphragm, and raise the condenser to its highest fixed position.
12. When disconnecting the light source, do not pull on the wire, but grasp the plug firmly to remove it from the socket.
13. If a cover is provided for the microscope, be certain that it covers the entire instrument.

PRINCIPLES OF MICROSCOPY

To serve its purpose a microscope must provide adequate magnifying power so that the finest details of a specimen are separated enough to be visible to the eye. The magnifying power of microscope lenses usually is indicated by markings on objectives and eyepieces (e.g. 45×, 100×). The total magnification obtainable is the product of the magnifications produced by the lenses individually.

While magnification is an important microscope property, the limiting factor that determines the usefulness of the light microscope is its resolving power. The resolving power of an optical system is its ability to produce separate images of small parts of a specimen that are only a short distance apart, that is, the ability to distinguish fine detail. In practice, the limit of resolution obtainable with a compound light microscope is about 0.2 μm, or roughly 1000 times the resolving power of the unaided human eye.

The quality of the image obtained or its definition depends on the degree of contrast between the specimen and its surroundings and between parts of a specimen. Such contrast is due in part to differences in light absorption, and various techniques, which affect absorption, are used to improve contrast. For example, the use of chemical stains results in color differences. Reducing the intensity of the light illuminating a specimen makes it easier to distinguish cells that have a refractive index similar to the fluid in which they are suspended.

Direct examination of living specimens can be extremely useful in determining size and shape relationships, motility, and reactions to various chemicals. One of the methods in general use for studies of this type is the wet mount technique. This method maintains the natural shape of cells and reduces the distorted effects that can occur when specimens are dried and preserved.

In this exercise the wet mount technique is used to emphasize certain properties of the light microscope and to introduce features of living cells.

A. IDENTIFICATION OF MICROSCOPE COMPONENTS

Although compound microscopes from different manufacturers vary in appearance or design, they consist of the same basic components. The parts of a typical microscope and their functions are described in Table 2-1. Use this information as you examine and locate specific parts of the microscope.

Table 2-1 Microscope Components and Their Functions

Microscope Component	Location or Description	Functions
Ocular (eyepiece)	Uppermost series of lenses	Magnification
Body tube (barrel)	Main cylindrical part	Holds oculars, conducts light rays from specimen
Nosepiece	Movable, usually circular plate at bottom of body tube	Holds objectives
Low power objective	Shortest objective, usually magnifies 10 ×	Magnification
High (dry) power objective	Intermediate-sized objective, usually magnifies 43 × to 45 ×	Magnification
Oil immersion objective	Longest objective; marked with either etched, red, or black circles; usually magnifies 100 ×	Magnification
Condenser	Lens system located below central stage opening	Concentrates and directs light beam through specimen
Stage	Platform on which specimens for examination are placed	Specimen support
Mechanical stage	Control devices that permit movement of slides from left to right and forward and backward on stage	Used to move specimen
Iris diaphragm	Located beneath stage in association with condenser unit, controlled with a lever	Regulates brightness or intensity of light passing through lenses
Coarse adjustment knob*	Generally located below stage	Used for preliminary and coarse focusing by raising or lowering body tube
Fine adjustment knob*	Generally located below stage	Used for final or fine focusing by raising or lowering body tube
Condenser adjustment knob	Control knob located below stage	Used to obtain full illumination by raising or lowering condenser
Base	Heavy, bottom portion on which instrument rests	Microscope support
Arm	Somewhat curved portion of microscope used in carrying instrument	Microscope support
Mirror**	Double-faced mirror for reflecting light through specimen	Reflects light from source through lens systems

vertical 10 power X10

*On several instruments the coarse and fine adjustments are combined in one knob device.
**Found on older instruments.

Materials

1. Compound microscope with an appropriate light source (one per student).
2. Large cut-away diagram of a representative microscope for class use.

Procedure

This procedure is to be performed by students individually.

1. Carefully remove the microscope assigned to you from the cabinet by firmly grasping the

microscope's arm with one hand and placing the other hand under the instrument's base for support.
2. Place the microscope approximately 6 inches from the edge of the laboratory table with the microscope arm facing you.
3. Examine your microscope and with the aid of Figure 2-1 locate and learn the functions of each component listed in Table 2-1.
4. Answer items pertaining to this portion of the exercise in the questions for review section.

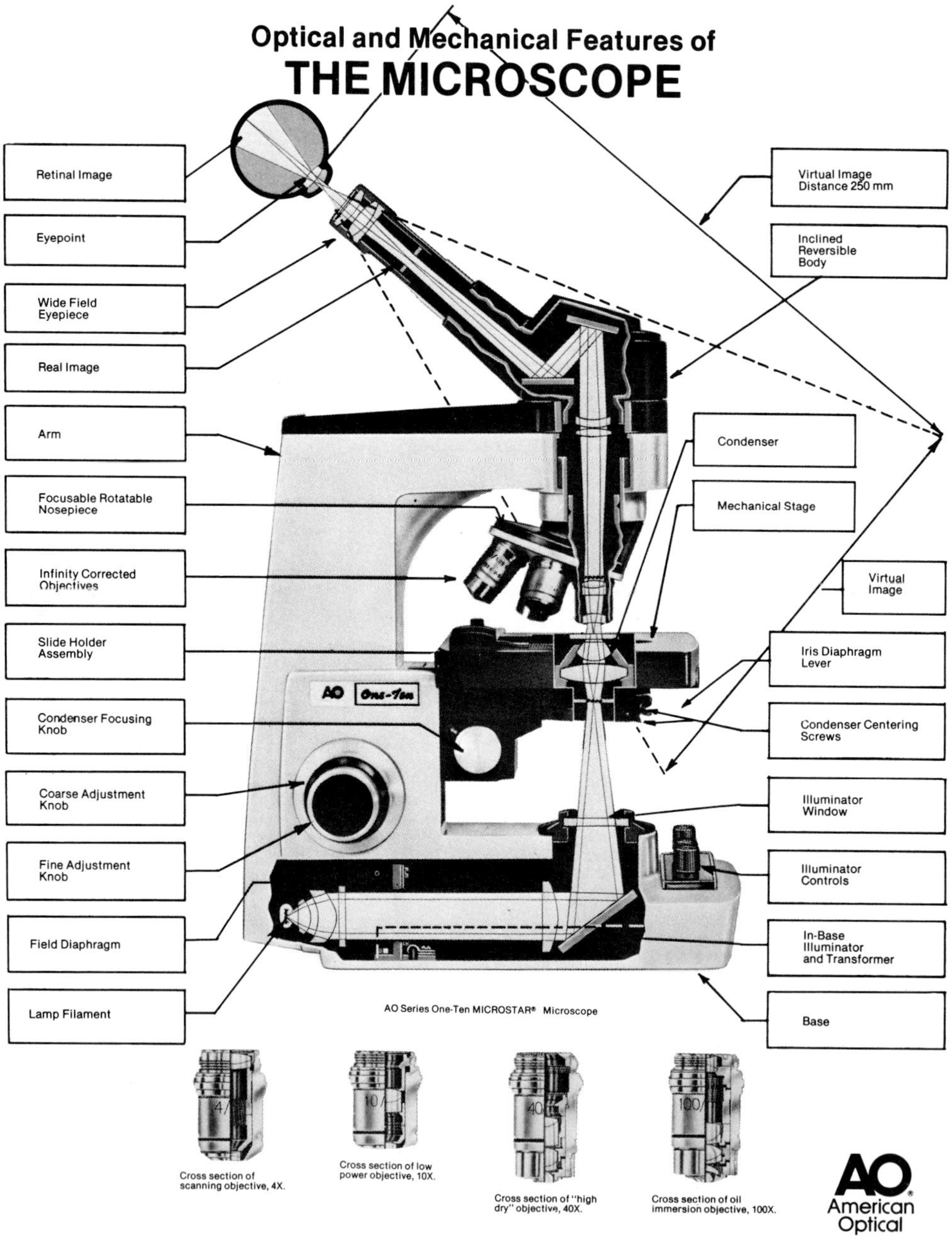

Figure 2-1 A cut-away diagram of a compound monocular microscope showing major components. Many instrument components are combined. Note that the fine adjustment and coarse adjustment knobs are combined and that the condenser adjustment is permanently fixed in position. (Courtesy of Reichert Scientific Instruments.)

B. TEMPORARY WET MOUNT SPECIMEN PREPARATION

Specimens for microscopic examination can be prepared in several ways. The temporary wet mount method described here is sufficient for materials to be examined for a short time period. (*Note:* The specimens prepared in the following two procedures will be used for viewing in Part C of this exercise.)

Materials

1. The following materials should be provided per four students.
   a. One eyedropper
   b. One small beaker of water
   c. One pair of scissors
   d. One dissecting needle
   e. One bottle of methylene blue stain
   f. Four wooden applicator sticks
   g. Container with disinfectant for applicator sticks
2. The following materials should be provided for class use.
   a. Microscope slides and cover slips
   b. Newspaper
   c. Lens paper

Procedure 1: Letter "e"

This procedure is to be performed by students individually.

1. Carefully clean a glass slide and a cover slip as indicated by the instructor. Place a drop of water on the center of the microscope slide (Figure 2-2).
2. Cut a row of a few letters from newspaper. Be certain that at least one of the letters is a small "e."
3. Place the string of letters in the drop of water on the slide and make certain that the liquid covers them.
4. Holding the cover slip at a 45° angle, place one edge on the slide in contact with the drop of water and then slowly lower it into place over the letters (Figure 2-2).
5. Save this preparation for later viewing.

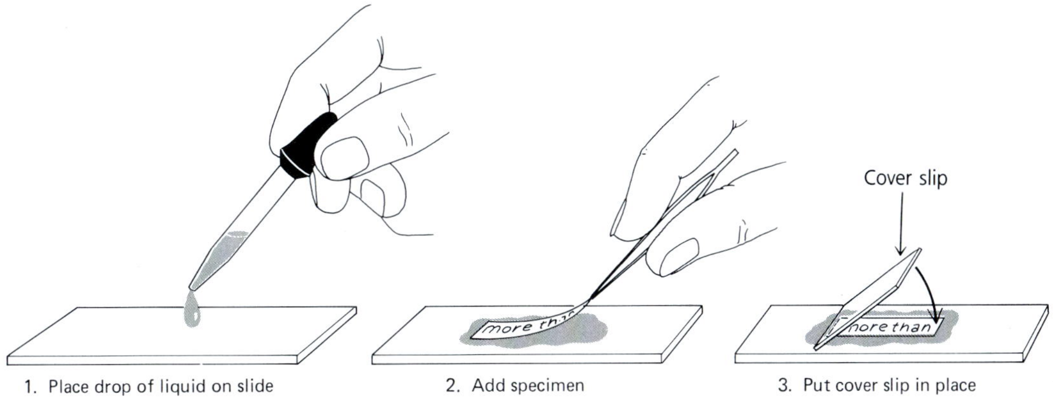

Figure 2-2 Steps in the preparation of a temporary wet mount.

Procedure 2: Cheek Cells

This procedure is to be performed by students individually.

The squamous epithelium lining of the human cheek provides a readily available source of cells for microscopic examination. These cells will serve as an introduction to general properties of animal cells.

1. Place 1 drop of water and 1 drop of methylene blue on the center of a clean glass slide.
2. Gently scrape the inside of your cheek with the end of a clean wooden applicator stick. (*Note:* Force is not necessary to dislodge cells.)
3. Mix the scraped material with the drops of fluid on the slide.
4. Apply the cover slip as described earlier and shown in Figure 2-2.
5. Save this preparation for viewing in the next part of the exercise.
6. Dispose of the applicator stick as indicated by your instructor.

C. USE OF THE MICROSCOPE

This portion of the exercise is designed to familiarize you with basic features of microscope parts and operation. The procedures described are for instruments with movable body tubes.

Materials

1. The following materials should be provided for class use.
   a. Temporary wet mounts of the letter "e" and of squamous epithelial cells prepared in the previous procedures
   b. Lens paper
2. Prepared slides of crossed colored silk threads (one per four students).

Procedure 1: Microscope Operation

This procedure is to be performed by students individually.

1. Wipe all lenses with lens paper.
2. Move the low power objective into position under the body tube. You will feel a click when it is correctly in place.
3. Connect your light source and turn it on.
4. Place the letter "e" slide on the mechanical stage and secure it with the slide holder or other means provided. The finger control lever for the slide holder is generally located at the rear left-hand corner of the mechanical stage. Forward pressure on the lever will cause the slide holder to open, and the slide may be inserted into the holding area. The bottom of the slide should rest on the microscope stage and not on the mechanical stage apparatus. Release the finger control lever to allow the slide holder to close and to hold the slide within the mechanical stage.

5. Using the control knobs on the mechanical stage, move the slide until the specimen on the slide is over the center of the condenser opening in the stage. One control knob moves the slide left to right and the other moves it front to back.
6. Use the coarse adjustment knob to lower the low power objective to approximately ¼ inch above the cover slip.
7. Use the condenser adjustment knob to raise the condenser as far as it will go.
8. While looking into the ocular, move the iris diaphragm lever back and forth. Note the change in illumination. The intensity or brightness of illumination is controlled by adjusting the iris diaphragm.
9. Locate your specimen. Focus *upward,* that is, move the objectives upward, with the fine adjustment knob until the specimen comes into clear view. In a microscope with a fixed body tube the objectives remain in a fixed position. Focus is controlled by moving the stage *downward* until the specimen comes into view. If you have difficulty in locating your specimen, look for a region with a color or intensity different from other portions of the slide.
10. In order to change objectives (switch from low power to high power) do not raise the body tube; simply turn the nosepiece to bring the objective desired into place. The specimen will generally stay in focus with any objective, but a slight turn of the fine adjustment knob may be necessary. Most microscopes are *parfocal,* that is, once the specimen is in focus it will remain so when the objective lens is changed.
11. When looking through the microscope, keep both eyes open at all times. This may be difficult at first with a monocular microscope, but it will become easier with practice. Keeping both eyes open prevents eye strain and enables you to observe and draw microscopic views of specimens without moving your head.
12. When the microscope is no longer needed, the following steps should be performed.
    a. Remove any slide from the stage.
    b. Clean the instrument including the lenses to remove any fluid from specimens.
    c. Open the diaphragm.
    d. Raise the condenser to its highest position.
    e. Move the low power objective into position over the stage.
    f. Wrap the electrical cord around the base.
    g. Return the microscope to its proper storage place.

Procedure 2: Examination of Temporary Wet Mounts

This procedure is to be performed by students individually.

LETTER "e"

1. Examine the letter "e" as it appears on the slide to the unaided eye in a normal reading position. Make a sketch of the letter in the results and observations section. All that is necessary is an outline of the "e" because we are interested only in the orientation or position of the letter rather than detail. (Refer to Color Plate 2.)
2. Focus and examine the preparation under the low power objective following the steps listed in procedure 1.
3. Slowly move the slide first side to side and then forward (up) and backward (down) with the mechanical stage control knobs. Compare this view of the orientation of the "e" with the view with the unaided eye.

4. Sketch a representative microscopic field in the results and observations section.
5. Change to the high power objective. Adjust the light and carefully focus with the fine adjustment knob. A portion of the letter "e" will be off center.
6. Center the portion of the letter while you have it under high power and then turn to low power to determine how far off center the "e" is in your particular microscope. This is important and should aid you in all your future microscope work.
7. Practice shifting from low to high power and back again. (Note that only a small amount of manipulation with the fine adjustment knob is necessary to bring the letter into proper focus whenever the other objective is swung into position.)
8. Sketch the representative high power microscopic field in the results and observations section.
9. On your sketch of the "e" as observed under low power, draw a circle around the portion of the letter seen under high power.
10. Remove, clean, and return the slide to a tray or other location indicated by your instructor.
11. Answer the questions for review pertaining to this portion of the exercise.

SQUAMOUS EPITHELIAL CELLS

Note that these cells are thin and flat and somewhat irregular in shape. The edges of the cells often become folded over in the process of preparation and the cells tend to clump together in groups. Certain cells may be covered by small, almost dotlike structures. These are usually bacteria from the oral region.

1. Examine your cheek cell preparation under the low power objective. Make the necessary adjustments for light intensity.
2. Select a microscopic field for study in which the cells are separate from each other and exhibit somewhat centrally located and darker staining nuclei. Refer to Color Plate 3.
3. Using first low power and then high power, draw several squamous cells in the space provided in the results and observations section. Label the following: cytoplasmic (cell) membrane, cytoplasm, and nucleus. If bacteria (small spherical or sticklike forms) are present indicate them also.

Procedure 3: Depth of Field

This procedure is to be performed by students individually.

To focus any microscope, the specimen being examined must be brought to the proper distance from the objective lens. This is done as indicated in procedure 1 by the vertical movement of the body tube. In this portion of the exercise you will consider the depth of field (focus) or the thickness of a specimen that is in focus at any one time. By use of this technique determinations can be made as to what structures and location details are in a specimen. In addition, knowledge of the width, length, and depth of a specimen can be used to understand the three-dimensional feature of objects.

1. Place a prepared slide containing three different colored threads on the microscope stage so that the point where the threads cross is over the center of the light opening in the stage.
2. Using the low power objective, find the location on the slide where the three threads cross.

3. Use the coarse adjustment to raise the body tube slowly until the place where the threads cross is barely in focus. The colors should be visible, but the image will be fuzzy and blurred.
4. Slowly lower the body tube until the edge of the first thread appears clear. What is the color of the first thread to appear? red
5. Use the fine adjustment knob to focus on the second and third threads. Determine the color of the threads in the middle and on the bottom. Enter your findings in the results and observations section.
6. Obtain an unknown thread slide from your instructor. Examine it the same way you examined the known slide. Determine the color on top, in the middle, and on the bottom, and enter your findings in the results and observations section. Check with the instructor to see if your observations are correct.

RESULTS AND OBSERVATIONS

LETTER "e"

1. Sketches

Unaided eye view | Low power | High power

2. List three ways the microscope changes the appearance of the letter "e."

a. makes it larger
b. turns it upside down
c. makes it backwards

SQUAMOUS EPITHELIAL CELLS

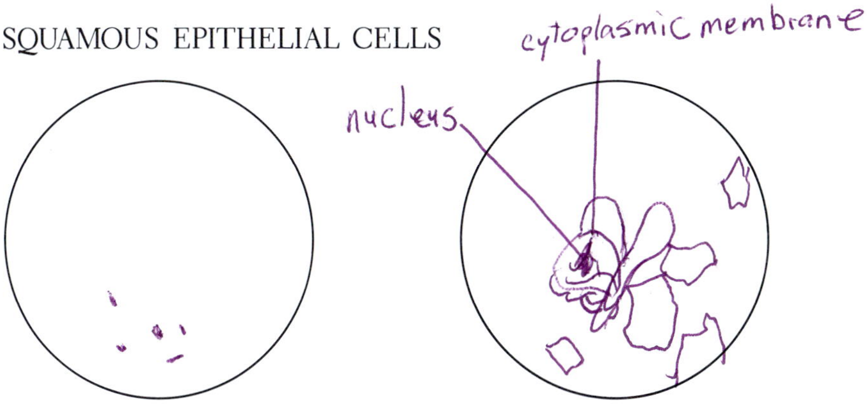

Low power | High power

DEPTH OF FOCUS

1. Unknown thread slide number: 2

2. Complete Table 2-2.

Table 2-2

Thread Layer	Color of Thread Layer in Known Slide	Color of Thread Layer in Unknown Slide
Top	Red	orange
Middle	Blue	raspberry
Bottom	yellow	purple

D. WORKING DISTANCE AND THE OIL IMMERSION LENS

The distance between a specimen and an objective lens is known as the *working distance.* Magnification increases as this distance decreases and more light is needed for better and more accurate viewing of a specimen. Figure 2-3 shows the relationship between the working distances of objectives and the iris diaphragm adjustments needed to provide adequate illumination. In addition, the more a specimen is magnified, the greater is the chance for the specimen and the lens of the objective to come into contact. Great care should be exercised when the high power or the oil immersion objective is in position. Improper use of objectives can result in broken slides and damaged objectives.

The objective of highest power on compound light microscopes is usually the oil immersion objective. The objective must always be immersed in cedar oil or some other substance with about the same index of refraction as glass. The refraction index of a substance determines the angle at which light is bent when it passes through that substance. If a glass slide has air above and below it, as when the high or low power objective is used, light is bent upon entering the glass and again when it emerges into air. Because of this refraction, image details become blurred and some light is lost as it passes by rather than enters the objective (Figure 2-4). With the oil immersion lens, light passing through the slide proceeds in a straight path through the immersion oil. Immersion oil thus increases the resolving power of the objective so that finer details can be seen and more of the incident light is used.

In this portion of the exercise a prepared blood smear will be used to demonstrate the properties of the oil immersion objective. The general microscopic features of human blood cells are given in Table 2-3, Figure 2-5, and Color Plate 4.

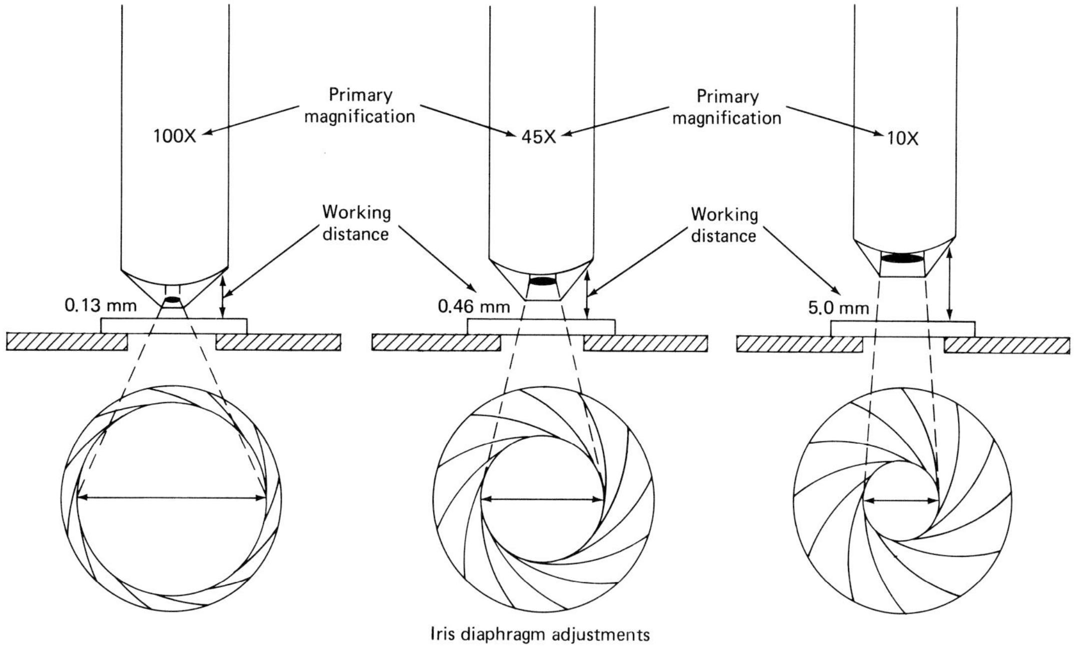

Figure 2-3 The relationship between the working distances of the low (10x), high dry (45x), and oil immersion (100x) objectives, and the iris diaphragm adjustments necessary to obtain satisfactory specimen illumination. The primary magnification and working distance (in millimeters) are given for each objective.

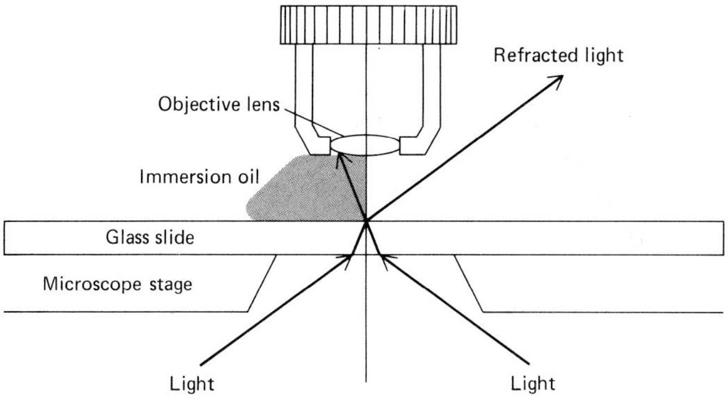

Figure 2-4 A diagrammatic comparison of the effects of immersion oil and air on the path of light leaving a microscope glass slide.

Table 2-3 General Microscopic Features of Mammalian Blood Cells

Blood Cell	General Description
Erythrocyte (red blood cell)	Circular, biconcave disc appearance No nucleus in mature state Most numerous of all types of cells found in the peripheral blood (bloodstream) Usually measures approximately 7 μm in length and 2 μm in thickness
Leucocytes (white blood cells)	
Granulocyte (granular leucocyte)	Contains distinct cytoplasmic granules which, upon staining, react with dyes Granulocytes also have irregular and multi-lobed (lobose) nuclei
Eosinophil	Granules react with dyes to yield characteristic color
Basophil	Granules stain blue
Neutrophil	Granules stain neutral or orange
Agranulocyte	Does not contain granules Has rounded, rather than lobose, nucleus
Lymphocyte	Has individual nucleus that occupies major portion of cell
Monocyte	Has kidney-shaped nucleus generally larger than lymphocyte
Platelet (thrombocyte)	A circulating, non-nucleated fragment of a large bone marrow cell (megakaryocyte) that participates in blood clotting.

Materials

The following materials should be provided for class use.

1. Prepared human blood smears
2. Charts and other reference materials showing blood cell types
3. Immersion oil
4. Lens paper and xylene

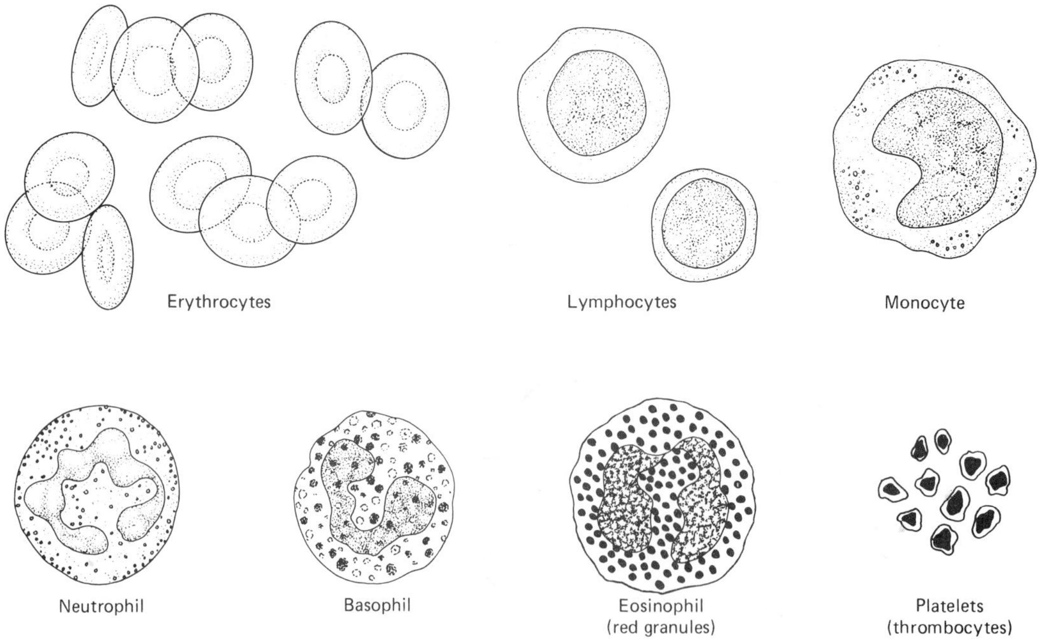

Figure 2-5 General appearance of blood cell types.

Procedure

This procedure is to be performed by students individually.

1. Place a blood smear slide in position on the microscope stage.
2. Rotate the low power objective into position.
3. Focus and locate a field showing both red and white blood cells.
4. Swing the low power objective aside.
5. Put one drop of immersion oil on the slide region over the light opening in the stage.
6. Move the oil immersion objective into position over the slide. The objective should touch the oil and be just short of touching the slide.
7. Look through the ocular and move the iris diaphragm lever to make any necessary light adjustments. If the light is too bright, it will be difficult to distinguish the cells from the background.
8. Focus upward with the fine adjustment knob until the blood cells come clearly into view. Refer to Color Plate 4.
9. Examine the slide and identify all the blood cell types described in Table 2-3.
10. Identify the cells indicated in the blood smear in Figure 2-6 in the results and observations section.
11. Answer items in questions for review pertaining to this portion of the exercise.
12. After completing this procedure, remove the slide and clean it with the xylene and lens paper.
13. Wipe the oil from the immersion oil objective, and swing the low power objective into position.

RESULTS AND OBSERVATIONS

1. Identify all labeled cells in Figure 2-6.

a. Lymphocytes d. platelets

b. Monocyte e. erythocytes

c. Lymphocytes f. erythocytes

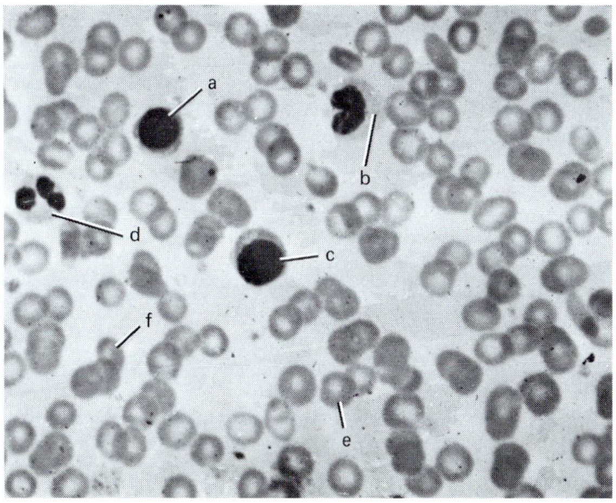

Figure 2-6 Representative human blood smear preparation.

Questions for Review

EXERCISE **2**

THE MICROSCOPE AND SPECIMEN PREPARATION

Name Kathy Martin

Laboratory Section A&P Lab Mon. 8:00AM Date 1-2-91

1. Label the microscope parts in Figure Q2-1.

a. wide field eyepiece
b. infinity corrected objectives
c. slide holder assembly
d. mechanical stage
e. coarse adjustment knob
f. fine adjustment knob
g. condenser focusing knob.
h. illuminator window
i. Base
j. iris diagram lever
k. arm

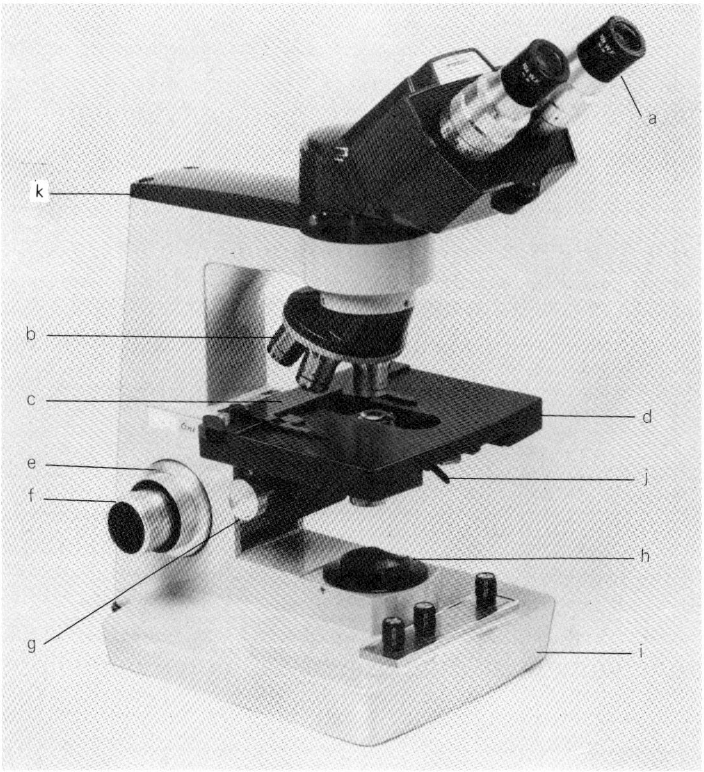

Figure Q2-1 (Courtesy, American Optical Corporation, Buffalo, N.Y.)

2. Complete Table Q2-1.

Table Q2-1

Microscope Part	Functions
Ocular	a. Magnification
Low power objective	b. Magnification
High power objective	c. Magnification
Oil immersion objective	d. Magnification
e. Condenser	Lens, located beneath the stage opening, that controls the light-beam size
f. iris diaphram	Controls intensity of light
g. stage	Platform on which slides or other materials to be examined are studied
Coarse adjustment knob	h. Used for preliminary and coarse focusing by raising or lowering body tube
i. Fine adjustment knob	Adjustment device used for final focusing
Condenser adjustment knob	j. Used to obtain full illumination by raising or lowering condenser
k. Base	Support during transport

3. What is the magnification of the following objectives without an ocular?

a. Low power 10X c. Oil immersion 100 X

b. High power 43X to 45X

4. Which of the objectives comes closest to a specimen and is most likely to break a slide if proper precautions are not taken? Oil immersion

5. What materials should be used for the cleaning of microscope lenses? lense paper

6. a. Distinguish between resolving power and magnification. Resolving power distinguishes fine detail + magnifying make larger.

b. Which of these is the more important property of a microscope? The resolving power.

c. List the factors that contribute to resolving power. Light absorption, various techniques

7. What is parfocal? once the speciman is in focus it well remain so even when the objective lense is changed.

8. Table Q2-2 lists some common problems encountered in microscope usage. Indicate the proper procedure for correcting each of them.

Table Q2-2

Problem	Correction Procedure
Oil immersion lens does not enter immersion oil on slide	a. adjust the microscope stage.
Microscope field is too bright	b. adjust the iris diaphragm
Specimen appears blurred after proper light and focusing adjustments are made	c. adjust the fine adjustment knob.
Microscope view does not change as slide is moved	d. Change facusable rotatable nosepiece
Only half of the field appears to be illuminated	e. adjust illinator controls

9. Indicate the appropriate microscope part in Figure Q2-1 that is used in the situations in Table Q2-3.

Table Q2-3

Situation	Microscope Part
Fine focusing of a specimen	a. Fine adjustment knob
Securing a slide for viewing	b. Slide holder assemble
Adjusting the brightness of the light	c. Iris diaphragm
Providing full illumination for viewing	d. Condensor adjustment knob
Preliminary focusing of a specimen	e. Coarse adjustment knob.
Changing objectives	f. Focusable rotatable nosepiece

SECTION II

CELLULAR STRUCTURE, ORGANIZATION, AND FUNCTION

Physiologic systems usually exhibit two characteristic properties: *transduction* and *translocation.* The first of these properties involves the changing of energy from one form to another, while the other property, translocation, is associated with the movement of either material or information from one location to another. Since the cell is considered to be the most basic component of a physiologic system, the properties of such units and the way in which they are organized into higher systems, such as tissues and organs, determine overall function. Furthermore, the performance of cellular structures, as well as the coordination of their activities, helps to maintain the state of dynamic equilibrium in the body's internal environment; this is known as *homeostasis.*

As mentioned in Section I, microscopes are among the different tools essential to the study of cells and their activities. A variety of such instruments (Table II-1) and associated techniques allow a more detailed examination of cells and reveal the finer aspects of cellular structure, function, and activities. Exercise 2 presented the features of the bright-field instrument and emphasized that it incorporates optical lenses and visible light as the source of illumination. With electron microscopes, a beam of electrons is used as the source of illumination and is focused by electromagnetic lenses instead of the optical lenses characteristic of other microscopes. Since the electron beam has a considerably shorter wavelength than ordinary light, the resolving power and magnification obtainable with electron microscopes far exceeds those obtained with other optical systems (Table II-1). Two general types of electron microscope are currently in use: the transmission and the scanning instruments. Another instrument, the scanning transmission electron microscope, combines many of the properties of both instruments.

Table II-1 Comparison of Microscope Types

Type	Magnification Range	Resolution (in nm)	Applications Include
Compound bright-field	10–2,000	300	Used extensively for the examination of various cell types, either stained or unstained
Fluorescence	10–3,000	200	Employs fluorescent dyes for staining; useful in many medical diagnostic procedures for identifying cells, including cancer and microorganisms and/or their parts
Phase contrast	10–1,500	200	Used to examine and study cellular structures; staining of specimens is not required
Transmission electron (TEM)	200–1,000,000	1	Used to examine and study the ultrastructure of cells and viruses and certain biochemicals
Scanning electron (SEM)	200–200,000	5.0–10.0	Used to examine and study surface features of cells and nonbiological materials; images are presented in three-dimensional form
Scanning transmission electron (STEM)	Combination of TEM and SEM	Combination of TEM and SEM	Used for applications similar to those listed for TEM and SEM; also used for analysis of chemical elements in specimens; equipped with computer for measurements and calculations

Several elaborate techniques are used in electron microscopy in preparing specimens and providing specimen contrast. Such techniques include ultrathin sectioning (slicing) and staining (Figure II-1a), the use of electron stains, and depositing the atoms of heavy metals in a vacuum onto the surfaces of specimens. The microscope used is a deciding factor in the selection of the preparation technique. The scanning electron microscope is a particularly useful instrument, since it provides three-dimensional views of specimens without the need for extensive preparation techniques (Figure II-lb). Such views provide a basis for understanding complex relationships within cells, tissues and organs, and enables observers to gain a perspective as to the true locations and organization of cellular structures.

Specimens examined by electron microscopy must be viewed in a vacuum. Furthermore, specimens viewed or photomicrographs taken of specimens lack color, which is a major means of obtaining contrast in light microscopes.

Sizes and shapes are important identifying properties of cells. In the biological sciences several common metric units are used to express the dimensions of cells and their parts. The common metric units of measurement include the millimeter (mm), the micrometer (μm), and the nanometer (nm). The relationships of the units and their application to cells are shown in Figure II-2.

Cells from different forms of life and from different regions of a single multicellular organism differ greatly in various features including shape, size, rate of metabolic activity, and functions. While all cells share certain basic structural features, some have specialized structures called *organelles*. The exercises in this section not only emphasize numerous properties of cells and their organization but demonstrate that they are efficient basic units of life.

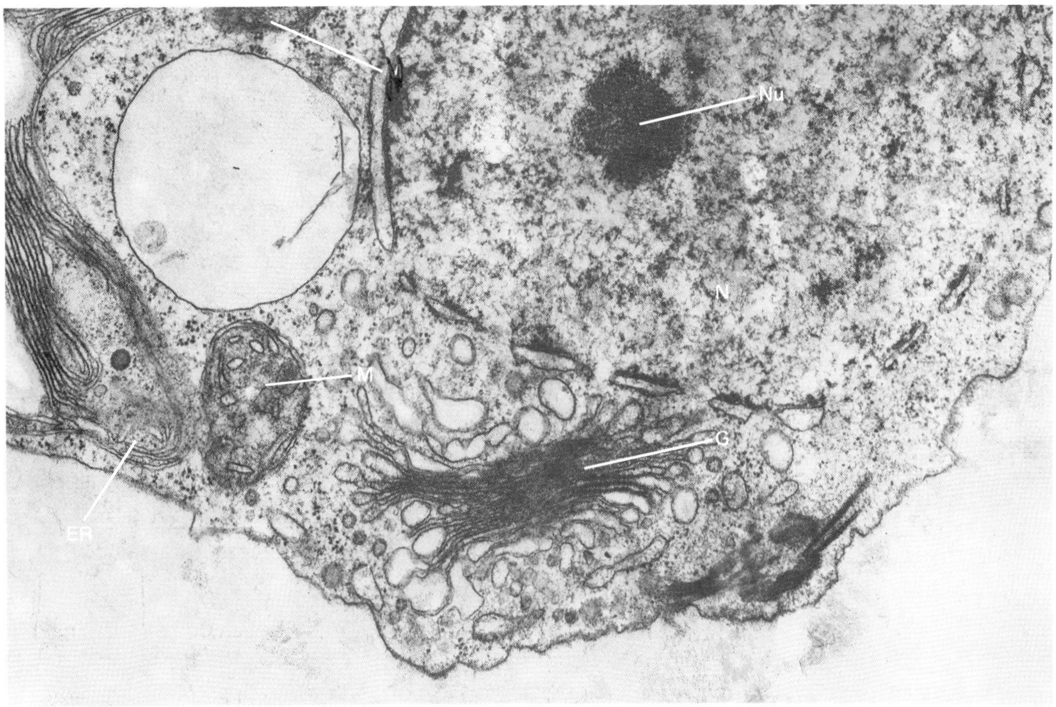

(a)

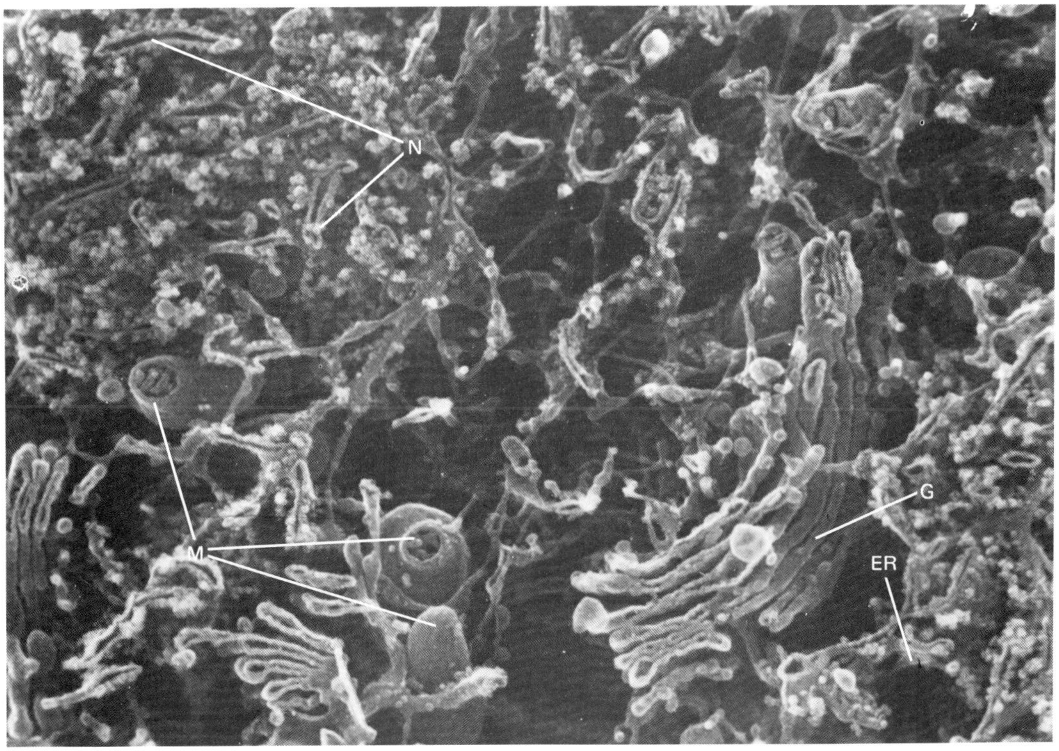

(b)

Figure II-1 Electron micrographs. (a) A transmission micrograph showing Golgi apparatus (G), mitochondrion (M), endoplasmic reticulum (ER), nucleus (N), and nucleolus (Nu). (b) A scanning micrograph providing an internal three-dimensional view of an eukaryotic cell. Cellular organelles that can be seen include endoplasmic reticulum (ER) with ribosomes attached, Golgi apparatus (G), nucleus (N), and individual mitochondria (M). The original magnification is 26,000×. (From K. Tanaka and A. Mitsushima. *J. Microscopy, 133:* 213-222, 1984.)

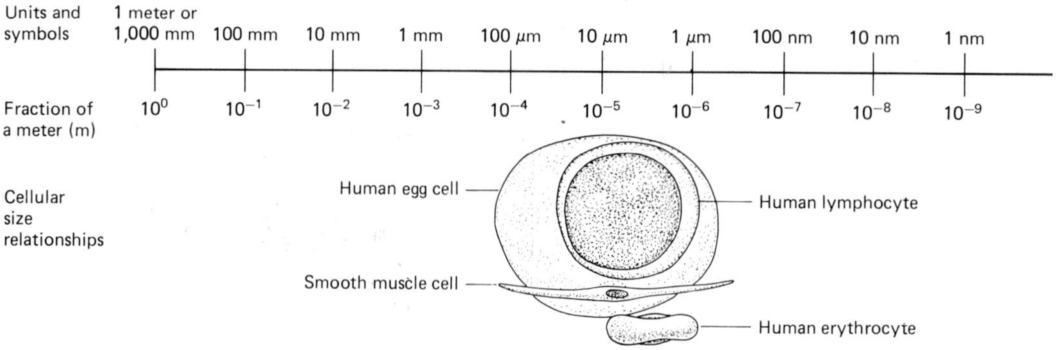

Figure II-2 Metric units of length commonly used in the biological sciences: 1 meter = 1,000 millimeters (mm); 1 mm = 1,000 micrometers (μm); 1 μm = 1,000 nanometer (nm). The relative sizes of various types of cells are shown (Refer to Color Plate 3).

GENERAL REFERENCES

DeRobertis, E. D. P., and E. M. F. DeRobertis, Jr. *Cell and Molecular Biology,* 8th ed., Philadelphia: Lea & Febiger, 1987.

Gartner, L. P., and J. L. Hiatt. *Atlas of Histology.* Baltimore: Williams & Wilkins, 1987.

Tortora, G. J., and N. P. Anagnostakos. *Principles of Anatomy and Physiology,* 6th ed. New York: Harper & Row, 1990.

SPECIFIC REFERENCES

Dautry-Varsat, A., and H. F. Lodish. "How Receptors Bring Proteins into Cells," *Scientific American,* 250: 52–58, 1984.

Nomura, M. "The Control of Ribosome Synthesis," *Scientific American,* 250: 102–114, 1984.

EXERCISE 3

THE CELLULAR LEVEL OF ORGANIZATION

Living organisms are composed of basic units of structure and function called *cells.* Furthermore, cells contain subcellular parts, referred to as *organelles* (or little organs), that exhibit distinctive structural and functional properties.

The use of electron microscopes and associated techniques for the preparation of biological specimens has demonstrated clearly the existence of two fundamentally different types of cellular organization, namely, the *prokaryotic* (primitive nucleus) and *eukaryotic* (true nucleus).

The prokaryotic cell, which is the unit of cellular organization for bacteria, is distinguished by its small size and the absence of many of the organelles normally found in eukaryotic cells (Figure 3-1). All living organisms other than bacteria have a eukaryotic type of cellular organization (Table 3-1).

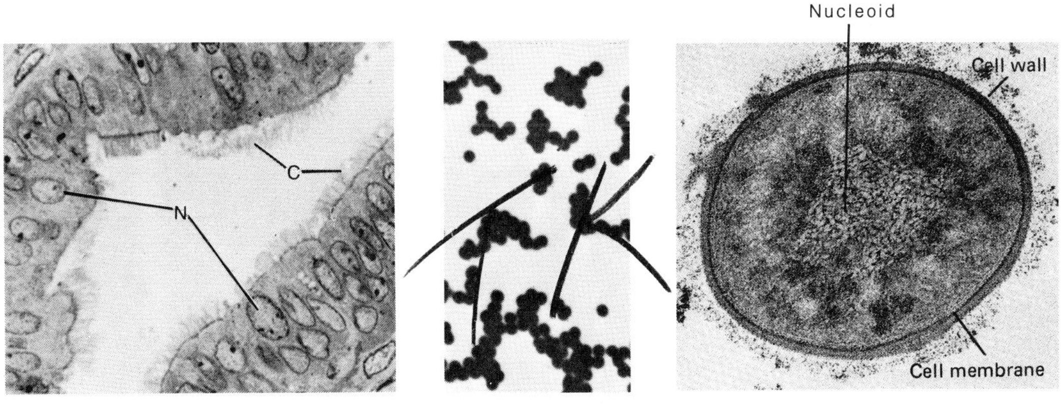

(a) (b) (c)

Figure 3-1 Prokaryotic and eukaryotic cells. (a) Eukaryotic cells taken from a human oviduct. Note the well-defined nuclei (N) and cilia (C). Such details are obtainable with properly prepared specimens. (From I. A. Brosens and G. Vasquez. *J. of Reproductive Medicine* 16: 171, 1976.) (b) The appearance of the much smaller prokaryotic cells. The forms are called cocci (spherical shape). (c) An electron micrograph of a prokaryotic cell clearly showing the absence of membrane-bound internal structures (organelles) and a well-defined nucleus (nucleoid). (Figure (b) and (c) from C. M. Ward, Jr., and G. W. Claus. *Journal of Bacteriology,* 114:378-389, 1973.)

Table 3-1 Eukaryotic Organelles and Their Functions and Activities

Organelle	Associated Functions and Activities
Cell plasma membrane	Transport of substances into and out of cells Phagocytosis (engulfment of foreign material) in some cells Pinocytosis (uptake of fluid)
Cell wall (found only in plants and certain microorganisms)	Imparts shape and strength to the cell Protection against certain osmotic imbalances
Chloroplast	Photosynthesis
Cilium	Motion or movement of substances past the ciliated cell
Endoplasmic reticulum	Protein synthesis
Flagellum	Propulsion
Golgi apparatus	Storage and sorting structure for cell products Transport of proteins, lipids, and other cell products to the cell surface
Microbody or peroxisome	Enzymatic activities
Microtubule	Transport of materials within cell Development and maintenance of cell shape Cell division Part of internal structure of cilia and flagella
Mitochondrion	Synthesis of the energy-rich compound adenosine triphosphate (ATP)
Nucleolus	Major site for the formation of ribosomal components
Nucleus	Control of cellular physiological process Transfer of hereditary factors to subsequent generations
Ribosome	Protein synthesis
Vacuole	Storage site for water; certain amino acids, carbohydrates, and proteins; cell products; or cell wastes

Dyes (coloring agents) termed "biologic stains" frequently are used to distinguish organelles and/or chemical substances found in and between cells. Several categories of biologic stains are recognized; however, tissue stains are the most commonly used in microscopic work. In procedures involving these stains, one, two, and occasionally three dyes are used to differentiate (distinguish) the nucleus from the cytoplasm of cells and to permit the differentiation among different tissue types. Stains may be used to color both living and preserved tissue. In addition, special techniques employing enzymes or acids can be used to identify specific chemicals in cellular structures. For example, the nucleic acids deoxyribonucleic acid (DNA) and ribonucleic acid (RNA) are identified with the use of the enzymes deoxyribonuclease (DNase) and ribonuclease (RNase) and acids such as trichloroacetic acid and perchloric acid.

In this exercise major attention is given to the basic parts of eukaryotic cells that are readily visible with the compound bright-field microscope. Specimens for the exercise include stained representatives of animal eukaryotic cells, such as blood and cheek cells, and several of their organelles. Examples of enzyme-treated cells also are considered.

A. AN INTRODUCTION TO CELLULAR ORGANELLES

Materials

1. Six demonstration microscopes, one for each of the following prepared, stained slides:
   a. Cross section of frog *(Rana pipiens)* liver
   b. Frog or fish blood smear
   c. Smooth muscle cells (teased preparation or a cross section of intestinal tissue)
   d. Human blood smear
   e. Squamous epithelial cheek cells
   f. Ciliated epithelial cells
2. Unknown slide preparation, one per student
3. Lens paper

Procedure

This procedure is to be performed by students individually.

1. Examine the preparations provided under low and high powers. Refer to Figure 3-2 and Color Plates 3, 4, 5, 6, 7, 8, and 9 as guides.
2. Look for the various organelles associated with eukaryotic cells.
3. Pay particular attention to the size and shape of the various cell types.
4. Complete the table and questions in the results and observations section.
5. After you have examined the six demonstration slides, obtain an unknown slide from your instructor. Examine the slide and identify the specific type of cell it contains. The preparation will be similar to one of those studied in the exercise. Enter your findings on the report form in the results and observations section. Check with your instructor to see if your identification is correct.

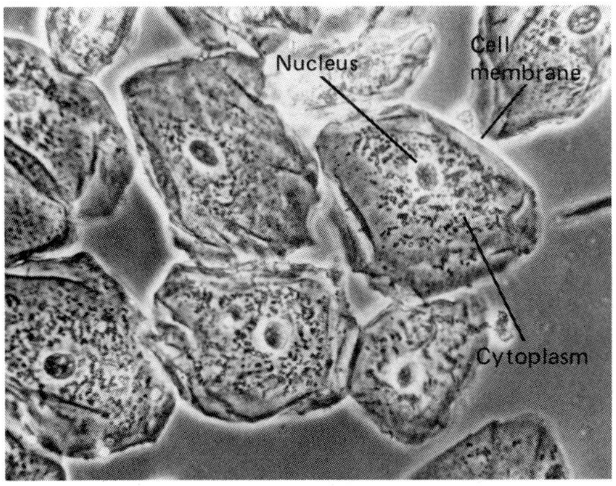

Figure 3-2 Squamous epithelial animal cells. Note that animal cells show variation in size and shape. (The original magnification is 450 ×.)

RESULTS AND OBSERVATIONS

INTRODUCTION TO CELLULAR ORGANELLES

1. List four organelles of animal cells.

a. Nucleolus c. mitochondrian

b. Nucleous d. endoplasmic aparadom

2. Compare the various cell types provided in this exercise as to the presence of organelles listed in Table 3-2. Indicate the presence of an organelle by a "+" and its absence by a "−", and briefly describe the shape of the cell under study, for example, round, oval, or spindle.

Table 3-2

Organelle and Cell Shape	Cell Type						
	Frog Blood		Human Blood		Frog Liver	Smooth Muscle	Squamous Epithelium
	Red	White	Red	White			
Cell membrane	+	+	+	+	+	+	+
Nucleolus (Dark center; DNA RNA)	−	+	−	+	+	+	−
Nucleus (Darker center)	+	+	−	+	−	+	+
Vacuole (gas bubbles)	−	−	−	−	+	−	−
Shape of cell	round	oval	round	round	oval	spindle	oval

3. Give the type of cell the unknown slide preparation contains.

B. DEMONSTRATION OF ORGANELLES BY SPECIAL STAINS AND TREATMENT

Materials

+ 1. *Ascaris* egg in mitosis showing centriole (iron hematoxylin and orange G stain)
2. *Amphiuma* liver cells showing mitochondria (iron hematoxylin stain)
+ 3. Golgi apparatus preparation (silver stain)
4. *Necturus* liver cells showing distinctive nucleus and nucleoli (azure B stain)
5. Liver tissue stained with iron hematoxylin and eosin
6. Liver tissue stained for DNA with Feulgen reaction
7. Deoxyribonuclease-treated liver tissue and stained with Feulgen reaction
+ 8. Ribonuclease-treated liver tissue

Procedure

This procedure is to be performed by students individually.

1. Examine the preparations provided under the high power and oil immersion objectives. Refer to Figure 3-2 and Color Plates 10 and 11 as guides.

2. Look for the various organelles or reactions indicated on each slide. Use the following descriptions to aid you in finding specific organelles or interpreting reactions.
   a. *Asters*—strands extending from and around a central body.
   b. *Mitochondria*—small, individual, threadlike strands generally blue within the cytoplasm; some are short and sticklike (rod) in appearance.
   c. *Golgi apparatus*—brown or black strands forming a network around a cell's nucleus.
   d. *Nucleolus*—small, darkly stained bodies within a cell's nucleus.
3. Sketch a representative view of each specially stained organelle as seen under oil immersion in the spaces provided in the results and observations section. Include all organelles observed and the general outline of a cell.
4. Next, examine the liver cells stained by the Feulgen reaction under low and high power. This staining procedure shows the presence of DNA. Refer to Color Plates 10 and 11. Use these color plates for comparison purposes.
5. Now examine the enzyme-treated preparations. Based on your observations of the stained cells, determine what portion(s) of these cells was affected by the respective enzyme treatments.
6. Complete Table 3-3 and answer the questions in the results and observations section.

ORGANELLES BY SPECIAL STAINS

1. Sketch the organelles and their locations.

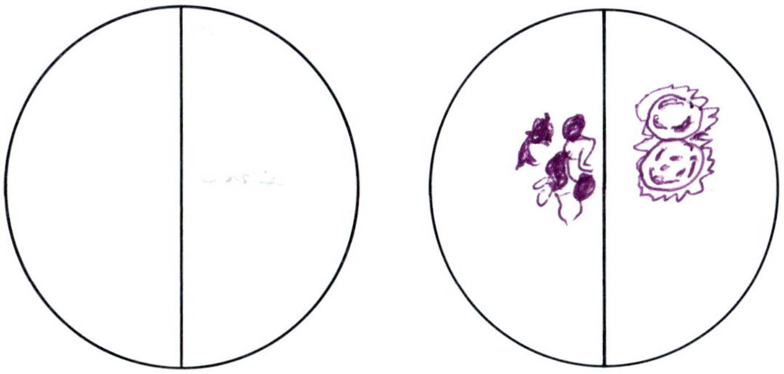

Aster Mitochondria Golgi apparatus Centriole

2. Compare the differently enzyme-treated cells as to the presence of the cellular components listed in Table 3-3. Indicate the presence or absence of a component by a "+" or "−" sign, respectively.

Table 3-3

Treatment	Cellular Components			
	Cell Membrane	Nucleus	Nucleolus	Cytoplasm
No treatment				
RNase treatment	+	+	−	+ +
DNase treatment	+	+	+	+

Questions for Review

EXERCISE **3**

THE CELLULAR LEVEL OF ORGANIZATION

Name Kathy Martin

Laboratory Section Mon, 8:00 AM Date 2-1-91

1. Identify the cell parts shown in Figure Q3-1.

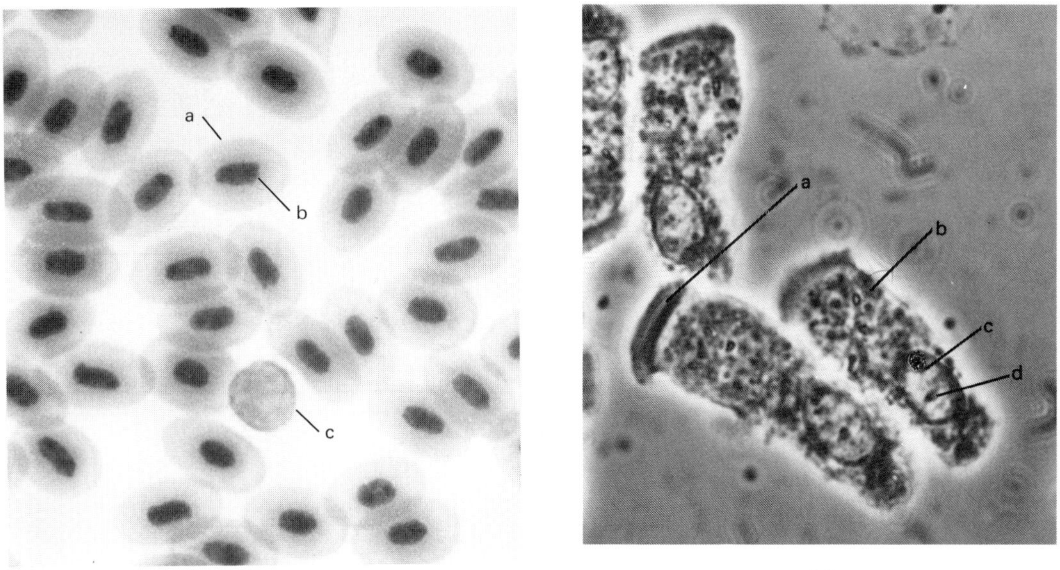

(a) (b)

Figure Q3-1 (a) Frog blood cells. (b) Enterocytes from the small intestine. (From S. Knutton et al. *Infect. Immun., 44:*514-518, 1984.)

(a) Frog blood cells

a. cell membrane

b. nucleus

c. red blood cell — Erythrocytes

(b) Enterocytes

a. cilia

b. mitochondria vacuolas

c. nucleus

d. Nucleolus

2. Using your textbook or other references, complete Table Q3-1 by giving the functions of the organelles listed.

Table Q3-1

Organelle	Functions
Nucleus	a. transfer of heridity factors to future generations, Control to cellular process
Nucleolus	b. Major cite for the formation of ribosomal components.
Endoplasmic reticulum	c. protein synthesis
Golgi apparatus	d. transports parts of proteins, lipids and other cell products to the cell service, also storage structure for c[illegible] pro[illegible]
Mitochondrion	e. Synthesis of the Energy rich compound (ATP)
Centriole	f. Helps organize spindle fibers and distribute chromosomes during mitosis.
Cilia	g. Moves particles along cell or moves cell
Ribosome	h. Synthesis proteins
Plasma membrane	i. Gives form to cell + allows passage of materials in + out of cell

3. Is the cell in electron micrograph Figure Q3-2 *prokaryotic* or *eukaryotic*? prokaryotic

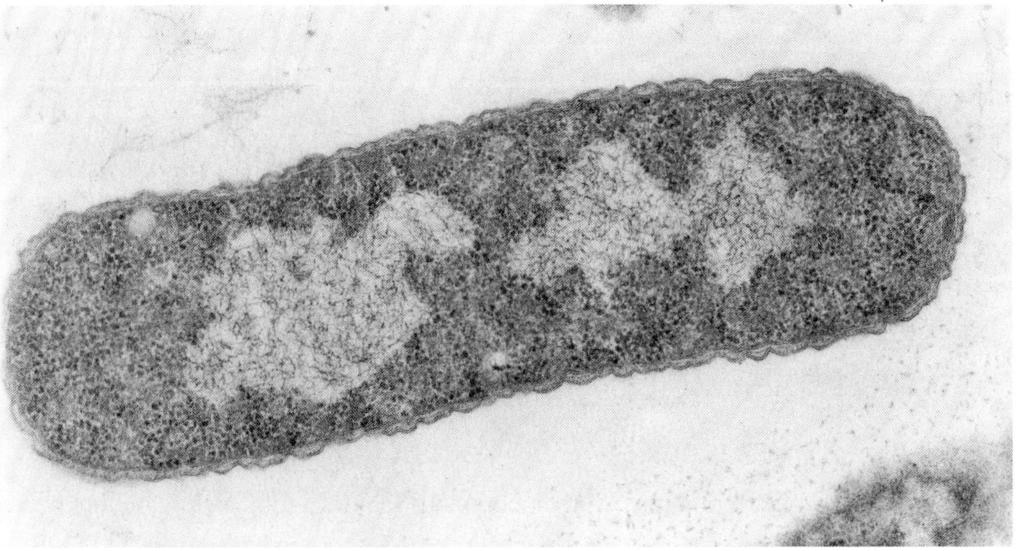

Figure Q3-2

4. Identify the organelles in Figure Q3-3.

a. nucleolus d. mitochondrion

b. nucleus e. cytoplasm

c. golgi apparatus f. Lysosome

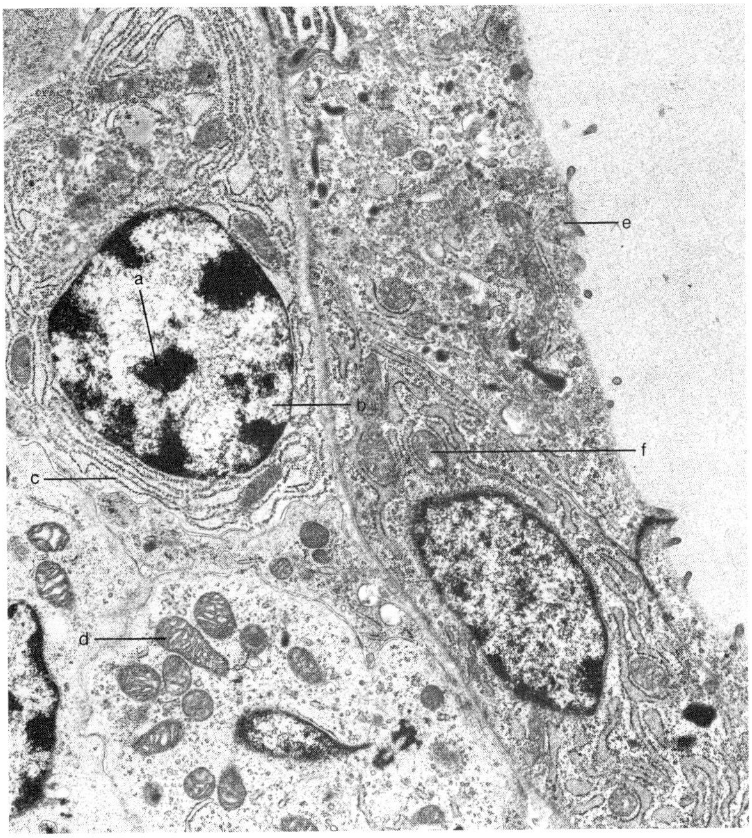

Figure Q3-3

5. Give the appropriate metric equivalent units for the following.

a. 11 millimeters = 11,000 μm d. 17μm = .017 mm

b. 100 nanometers = .0001 mm e. 10 nm = .00001 mm

c. 1 μm = .001 mm f. 3 μm = 3000 nm

6. Give the function or purpose of the following in the examination of cells.

a. Biologic stains are used to distinguish organelles and/or chemical substances found in + between cells.

b. DNase Idenitifies the nucleic acid (DNA)

c. RNase Idenitifies the nucleic acid (RNA)

EXERCISE 4

CELLULAR REPRODUCTION (MITOSIS AND CYTOKINESIS)

As indicated in Exercise 3, a eukaryotic cell's internal organization is a complicated assembly of chemical systems designed to perform a number of basic activities and functions to maintain its life on a day-to-day basis. However, cells under various conditions become injured, diseased, or worn and then die. Thus, new cells are needed for purposes of replacement and growth. The nucleus of each cell generally contains a precise set of genetic information built into more or less elongated structures known as chromosomes (Figure 4-1). A cell's genetic information is vital to its proper functioning and to determining the properties of the next generation of cells. Nuclear division resulting in the precise distribution of duplicated chromosomes into two newly formed cells is called *mitosis. Meiosis,* a specialized type of cell division that occurs during the formation of gametes (mature sex cells), is described in Exercise 58.

During a cell's existence, it passes through a regular, organized sequence of physiological events known as the *cell cycle* (Figure 4-2). The cycle is complex and is divided into four phases: G_1 (gap), S (synthesis), G_2 (gap), and mitosis. As Figure 4-2 shows, G_1, S, and G_2 comprise interphase. During the G_1 phase, active protein synthesis occurs, as well as the formation of new mitochondria, Golgi apparatus, endoplasmic reticulum, and other cytoplasmic materials. Cell size also increases in the G_1 phase. In the S phase, DNA synthesis takes place and chromosomes are duplicated. The deoxyribonucleic acid in a eukaryotic nucleus together with a large amount of protein and a small portion of ribonucleic acid form chromatin strands. In the G_2 phase, structures directly related to mitosis are produced. Mitosis follows the G_2 phase. During mitosis, chromatin strands in the nucleus condense by coiling and folding to form compact chromosomes (Figure 4-1a). Microscopically, each chromosome has a characteristic shape and appears to consist of two chromatin strands called *chromatids* (see Table 4-1). Those chromatids are attached firmly to each other at a specific region called a centromere (Figure 4-1b). Mitosis usually occurs in close association with the division of the cytoplasm *(cytokinesis).* During this event, the various duplicated cytoplasmic organelles in a parent cell are divided between two newly formed cells.

Mitosis is an extremely efficient process that provides for the even distribution of hereditary material during cell division. In describing the events taking place in mitosis, it is convenient to divide the process into four stages: *prophase, metaphase, anaphase,* and *telophase.* The period between successive stages is called *interphase.* It should be noted that mitosis is actually a continuous process and does not occur in discrete steps. It is not always readily apparent when one stage ends and another begins. Table 4-1 summarizes the main events of mitosis.

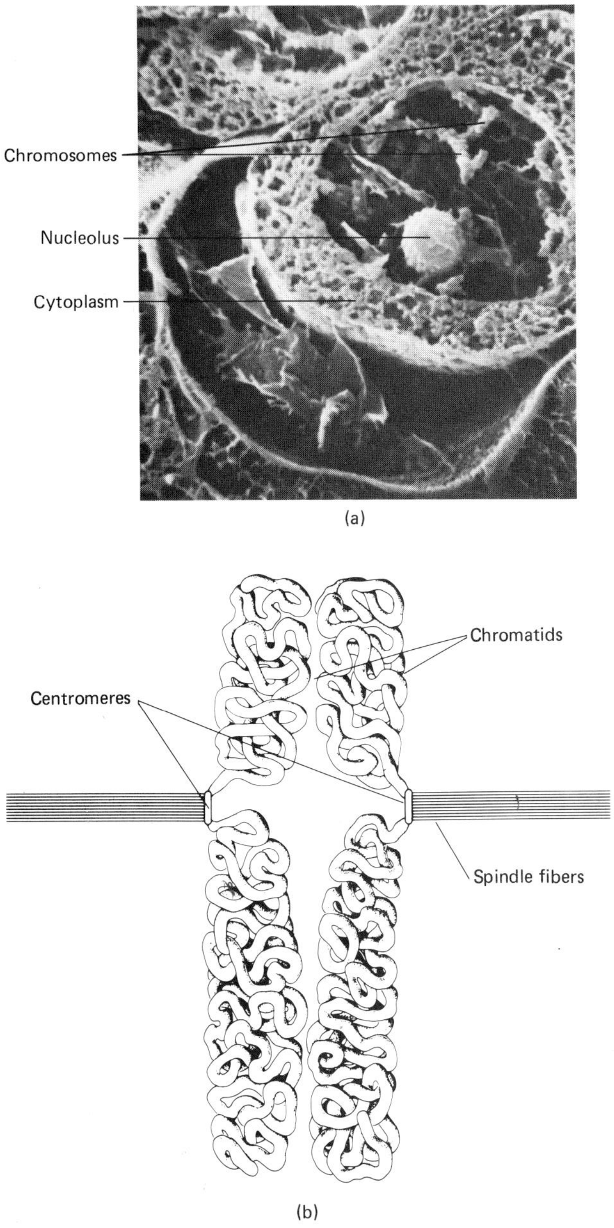

Figure 4-1 Cell during division. (a) Scanning micrograph showing chromosomes and a nucleolus within a nucleus. (Reproduced by permission of the National Research Council of Canada from E. D. P. Whelan and others. *Canadian Journal of Botany, 52:*1438-1440, 1974.) (b) Chromatids attached to spindle fibers at centromeres.

Normal mitosis during the development of a human from a fertilized egg cell (ovum) produces only cells with the number of chromosomes characteristic of a human. Thus, all cells of the body should contain the same number. There are, however, situations in which abnormalities occur. These include development and mutation.

The stages of mitosis are not limited to normal cells. Abnormal types of cell growth, spoken of as cancers, occur when growth is no longer regulated by the processes that control normal cell growth (Figure 4-3). Any cell in the human body has the potential of becoming a cancer cell. Cancer cells grow and divide more rapidly than normal cells and form a mass called a malignant tumor. Such growths tend to invade surrounding tissues and disrupt the normal structure and function of body organs.

Materials

1. The following materials should be provided for class use.
   a. Prepared slides of whitefish *(Coregonus)* blastulas showing mitosis
   b. Models or charts showing stages of mitosis

Procedure

This procedure is to be performed by students individually.

1. Use both low and high power objectives to examine the prepared slide showing mitotic stages in the whitefish cells. The slide should contain several embryonic cells exhibiting different stages of mitosis. (Refer to Color Plates 12 and 13.)
2. With the aid of Table 4-1 identify the stages of mitosis and sketch each one in the results and observations section. Label all significant parts of your sketches.
3. Find a clear (good) cell in metaphase and count the number of chromosomes present. Do the same for a cell in anaphase.
4. Answer the questions pertaining to this portion of the exercise in the results and observations section.

RESULTS AND OBSERVATIONS

MITOSIS

1. Sketch and label the stages of mitosis.

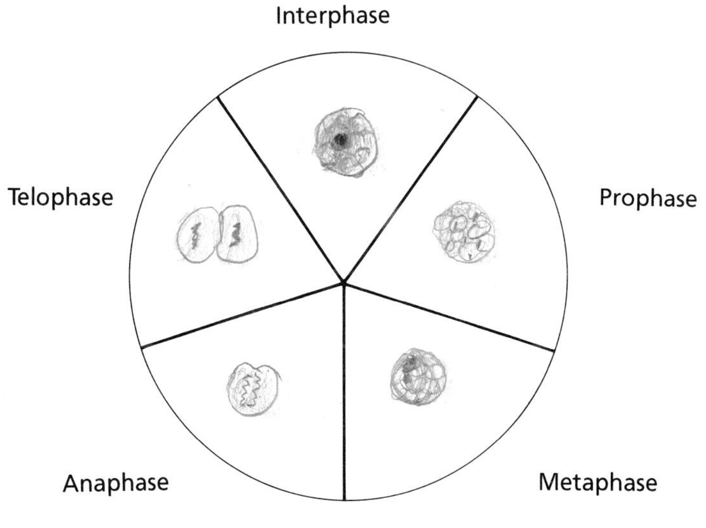

2. In which stage of mitosis is an equatorial plate evident? metaphase

3. How many chromosomes were present during metaphase? 4

4. Were the number of chromosomes present during anaphase the same as the number during metaphase? Explain your answer. No because they separate and become chromatids

Table 4-1 Cell Division and Mitosis

Stage	General Events	Diagrammatic Representation
Interphase	1. Chromosomes are uncoiled. 2. Intense cellular metabolic activity includes DNA synthesis (replication).	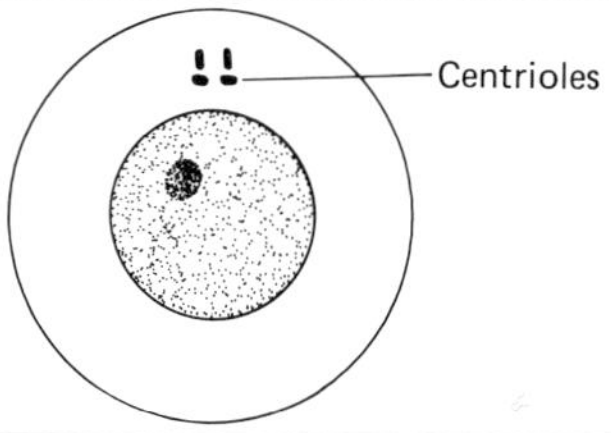
Prophase	1. Individual chromosome structure becomes visible and recognizable as identical chromatids held together by centromeres. 2. Centrioles of cells separate and migrate toward opposite poles of cells. 3. Nucleolus and nuclear membrane disappear. 4. Spindle appears. 5. Asters are present in animal cells only.	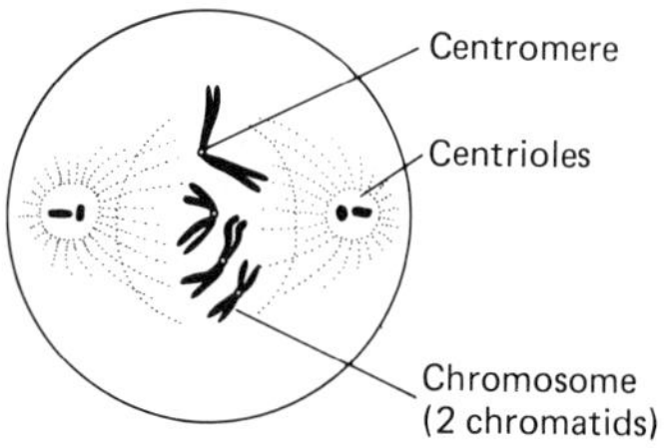
Metaphase	1. Chromosomes line up by their centromeres at the equator or spindle.	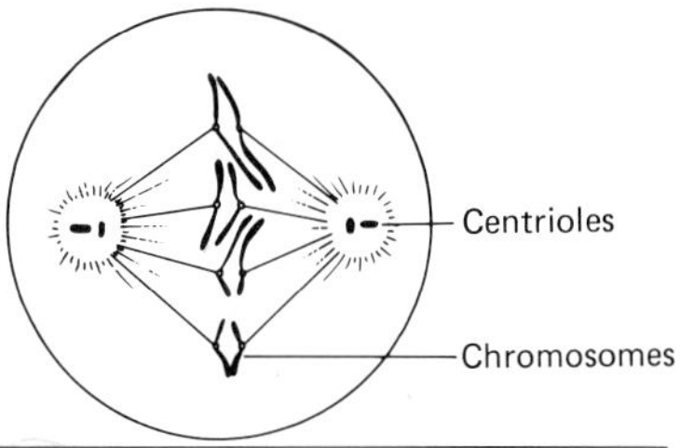
Anaphase	1. Chromatids separate and move to opposite poles of the cell. 2. Each chromatid becomes an independent chromosome. 3. Cytokinesis begins.	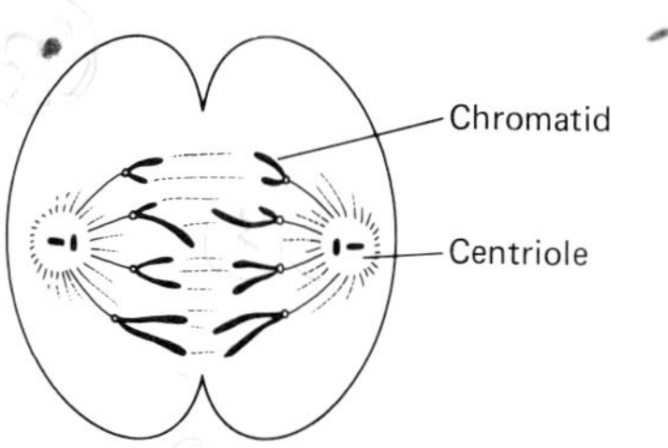
Telophase	1. Chromosome movement is completed. 2. Chromosomes assume the appearance they had in interphase. 3. New nucleolus and nucleus appear. 4. Cell plate occurs only in plants.	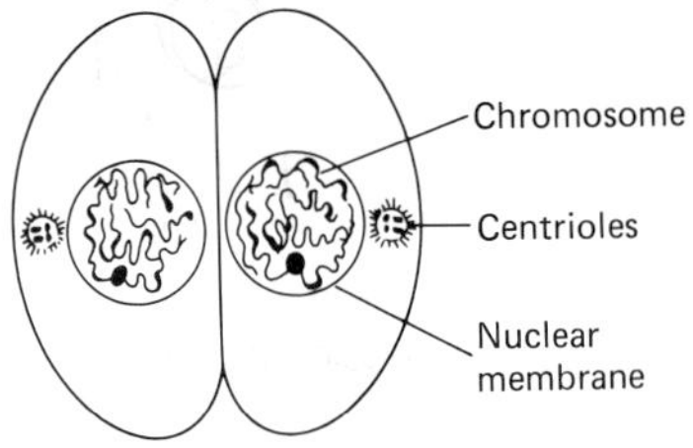

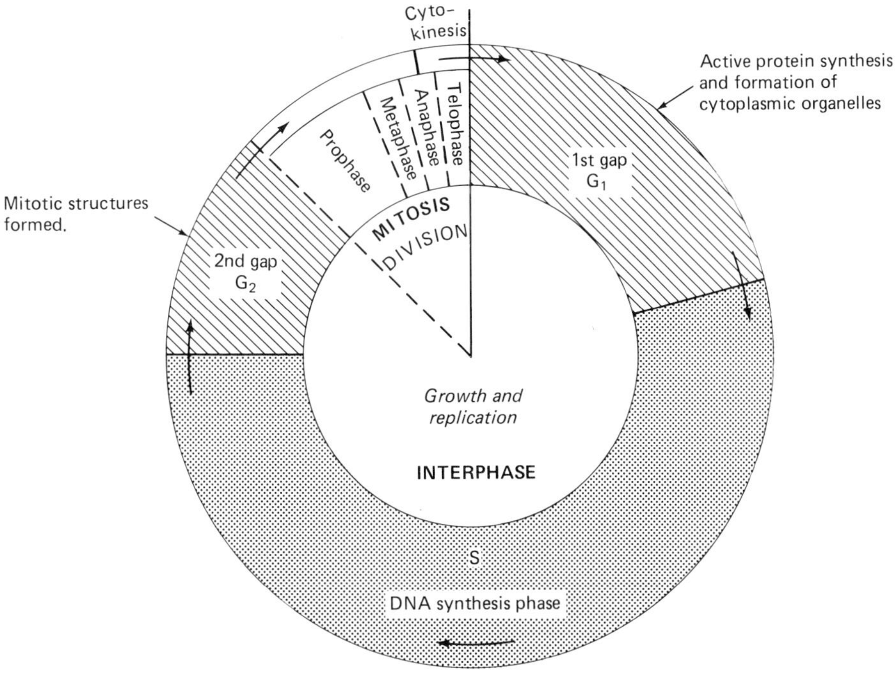

Figure 4-2 The phases of a cell cycle. The relationship of the four phases (G_1, S, G_2 and mitosis) of the cycle are shown. Events such as cytoplasmic organelle production, DNA synthesis, and cellular and nuclear division are related to cell cycle phases.

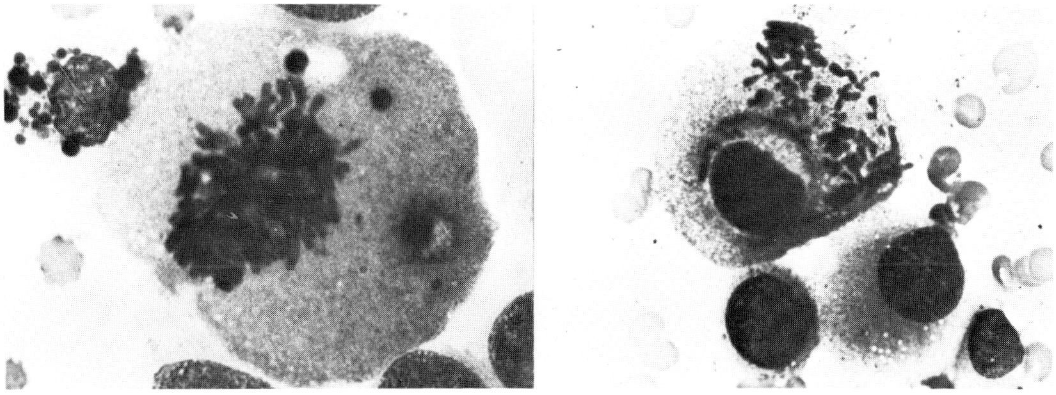

(a) (b)

Figure 4-3 Micrographs of malignant cells dividing. The stages of division are different visually and physiologically from those of normal cells. Note the general appearance of chromosomes. (Reproduced with permission from Friedman and others. "Needle Aspiration of Metastatic Melanoma," *Acta Cytologica,* 24:715, 1980.)

Questions for Review

EXERCISE **4**

CELLULAR REPRODUCTION (MITOSIS AND CYTOKINESIS)

Name Kathy Martin

Laboratory Section Mon 8:00 AM Date 2-7-91

1. List and explain the four phases of the cell cycle.
   a. G1 or gap 1 protein synthesis + cytoplasmic organelles form
   b. S or synthesis - DNA synthesis takes place
   c. G2 or gap 2 - mitotic structures are formed
   d. Mitosis - the separation of cells which include four phases

2. Draw and label all parts of an animal cell undergoing metaphase in mitosis.

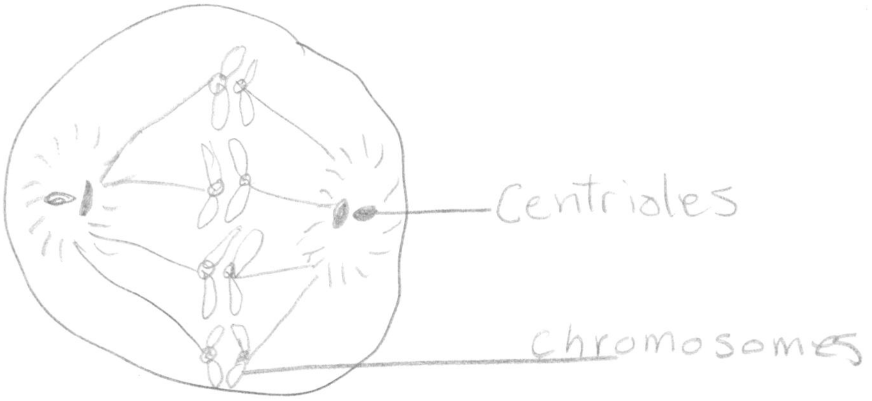

3. List and briefly describe the stages of mitosis in their correct sequence.
   a. interphase - chromosomes are uncoiled
   b. prophase - spindle appears
   c. metaphase - chromosomes line up at the equator
   d. anaphase - chromosomes become chromatids and separate
   e. Telophase - New Nucleolus & Nucleus appears cytokinesis

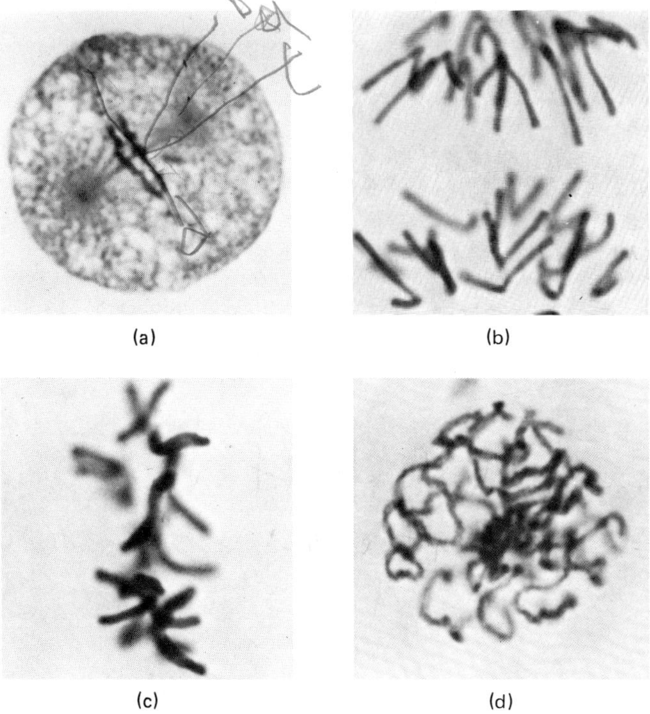

Figure Q4-1

4. Identify the stages of mitosis shown in Figure Q4-1.

a. metaphase c. prophase

b. anaphase d. Telaphase

5. Identify the labeled structures shown in Q4-1(a).

a. centrioles c. centromeres

b. chromosomes d. equator

6. Define or explain the following.

a. Nucleolus dark areas within the nucleus composed of RNA + protein molecule

b. DNA Deoxyribonucleic acid serves as a basis for the genetic code.

c. Meiosis a cell devision that occures in sex cells

d. Centromere The center of the chromosomes that pull out on it.

e. Chromatid is ½ of a chromosome in anaphase

EXERCISE 5

FLUID TRANSPORT AND MOLECULAR ACTIVITY: BROWNIAN MOVEMENT, FILTRATION, DIFFUSION, AND OSMOSIS

All the chemical reactions in a cellular environment take place in *solution.* Before a substance can enter a cell it must dissolve. The term solution can refer to the situation in which a substance is dissolved more or less evenly in a liquid. If, for example, the carbohydrate glucose is mixed thoroughly with water, the solid crystals of the substance dissolve and are dispersed throughout the liquid. The resulting mixture—a solution—is composed of the liquid *solvent* and the dissolved *solute.* Substances that dissolve are *soluble,* while those that do not dissolve are *insoluble.* The digestive process in humans and other animals generally involves turning a variety of large compounds into small compounds that are soluble in water and can be carried in and out of cells.

Substances move into and out of cells by several physical processes including: (1) *diffusion,* resulting from a concentration gradient (movement of molecules from an area of higher concentration to a lower one); (2) *osmosis,* the diffusion of water molecules across a semipermeable membrane; and (3) *filtration,* produced by hydrostatic pressure differences.

Within cells numerous minute particles appear to move in an erratic, zigzag motion known as *Brownian movement.* The movement occurs as small particles suspended within the cytoplasm are bumped about by the molecules of the cell's internal environment.

A cell membrane consists mainly of protein and lipid molecules compactly arranged into a delicate, thin layer. Most of its functions are associated with regulating the passage of materials into and out of cells. A membrane that allows molecules to pass through it is called *permeable.* Since cell membranes allow certain ions and certain molecules to pass through them, but do not allow the passage of others, they are said to be *selectively permeable, differentially permeable,* or *semipermeable.*

Movement of dissolved inorganic and organic substances within solutions or through the cell membrane involves a natural physical process called *diffusion.* Diffusion can be demonstrated easily by placing a small dye crystal (the solute) in a volume of water (the solvent). The solute moves from an area of higher solute concentration to an area of lower concentration.

In *osmosis,* a type of diffusion, a solvent (the water in which substances are dissolved) moves from a solution of lower solute concentration to a solution of higher solute concentration when the solutions are separated by a semipermeable membrane.

Filtration is described as the movement of solvents and dissolved substances across a semipermeable membrane under the influence of mechanical pressure. Such movement is always from an area of higher pressure to an area of lower pressure and continues as long as a difference in pressure exists.

This exercise will demonstrate these various phenomena and processes. (Dialysis, a process involving the separation of small molecules from larger ones by diffusion of the smaller molecules through a semipermeable membrane, will be presented in Section XV.)

A. BROWNIAN MOVEMENT

Materials

The following materials should be provided for class use.

1. Dilute solution of India ink
2. Eyedropper and rubber bulb
3. Glass slides and cover slips
4. Microscopes

Procedure

This procedure is to be performed by students individually.

1. Place a small drop of the dilute India ink solution on a glass slide and add a cover slip.
2. Observe the preparation under the high power objective of your microscope.
3. Reduce the light intensity and note the direction of movement of the India ink particles. Indicate the type of movement seen in the results and observations section.

B. FILTRATION

Chemical compounds in solution may behave in one of two ways. The molecules may remain intact, that is, undissociated, or they may dissociate, that is, separate into components called *ions.* This process is called *ionization.* Compounds that behave in this manner are known as *electrolytes;* compounds made up of molecules that do not dissociate are called *nonelectrolytes.*

Ions carry an electrical charge. For example, when the electrolyte sodium chloride is dissolved in water, it provides positively charged sodium ions (Na^+) and negatively charged chloride ions (Cl^-).

Both nonelectrolytes and electrolytes are used in this exercise to show how dissolved substances can be forced through physical barriers.

Materials

1. The following materials should be provided per four students.
   a. One funnel (65 mm diameter)
   b. One ring stand and clamp
   c. One 50 ml beaker
   d. One sheet of filter paper of appropriate diameter
   e. Two test tubes

2. The following materials should be provided for class use.
   a. 5 percent starch solution
   b. 10 percent sodium chloride (NaCl) solution
   c. Iodine reagent
   d. Silver nitrate ($AgNO_3$) solution (0.1 molar)
   e. Powdered wood charcoal
   f. 10 ml graduated cylinders
   g. Wooden tongue depressors

Procedure

This procedure is to be performed by students in groups of four.

1. Use a graduated cylinder to measure 2 ml of the 10 percent sodium chloride (NaCl) solution. Pour the solution into a test tube. Add 2 drops of the silver nitrate ($AgNO_3$). Observe and record your findings in the results and observations section. This is a test for the presence of chloride ions (Cl^-).
2. Put 3 ml of the 5 percent starch solution into the other test tube. Add 3 drops of the iodine reagent. Observe and record your findings in the results and observations section. This is a test for the presence of starch.
3. Rinse out both test tubes well and save for a later portion of this exercise.
4. Fold one piece of filter paper in quarters, and open the folded paper to form a cone. Place the cone in a funnel.
5. Set up a filtration system similar to the one shown in Figure 5-1.
6. Prepare a mixture for filtering by mixing the following items in a small beaker.
   a. 4.5 ml starch solution
   b. 2 ml sodium chloride solution
   c. A small amount of the powdered charcoal
   d. 10 ml water
7. Place a test tube under the funnel and slowly pour 10 ml of water through the system. Collect the water and discard it. The filter paper should now be firmly in place.
8. Place another test tube temporarily under the funnel, and pour the mixture through the system.
9. Collect the *filtrate* (the solution that passes through) in a beaker.
10. Divide the filtrate between two tubes. Perform the test for starch with one tube and the test for chloride ions with the other.
11. Record your findings and answer the questions in the results and observations section.

C. DIFFUSION

methods Brownian Motion
Kidney operation

Materials

1. The following materials should be provided per four students.
   a. One plastic Petri plate containing a layer of agar
   b. One pair of forceps
   c. One millimeter ruler
2. The following materials should be provided for class use.
   a. Filter paper discs approximately 5 mm in diameter
   b. 1 percent crystal violet solution
   c. 1 percent methylene blue solution
   d. 1 percent safranin solution
   e. Wax pencil or other marking device
   f. Paper towels

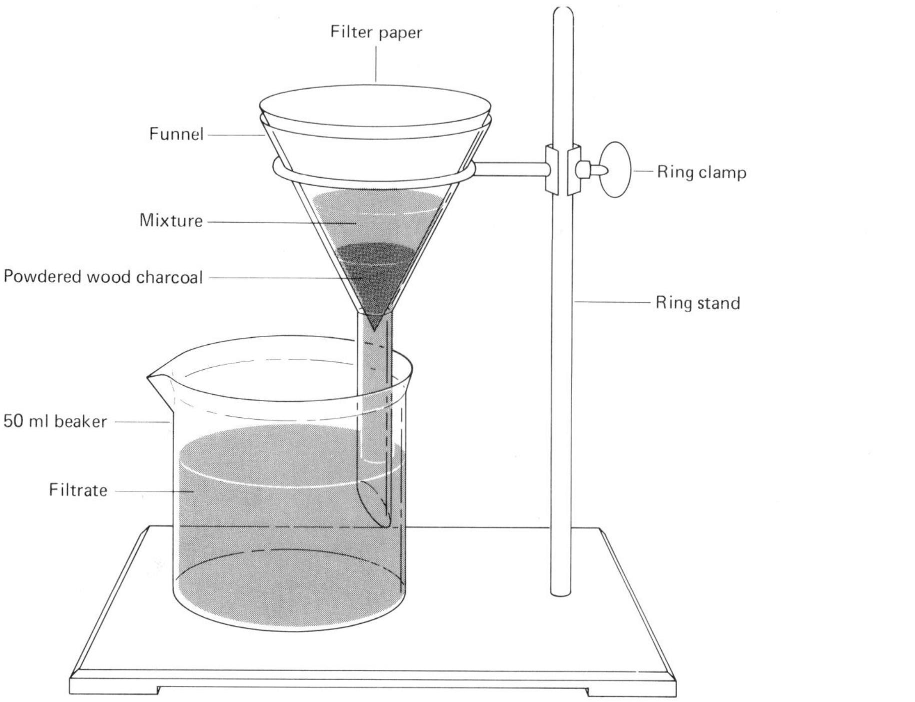

Figure 5-1 A filtration system. Be certain that all parts of the system are firmly in place.

Procedure

This procedure should be performed by students in groups of four.

1. Measure the diameter of the filter paper discs. Enter your measurement in the results and observations section.
2. Obtain a Petri plate containing a thin layer of jelled agar. Mark the bottom of the plate with a wax pencil to divide the agar preparation into three sectors.
3. Holding a filter paper disc with the forceps, touch it to the surface of the crystal violet solution until the dye penetrates throughout the disc.
4. Remove the Petri plate top and place the disc in the center of one of the marked sectors. Press the disc gently with the forceps so that a good contact is made with the agar surface. Close the plate.
5. Wipe the forceps free of the dye solution. Repeat steps 3 and 4 with the methylene blue, and safranin solutions placing the discs in the centers of their respective sectors of the Petri plate.
6. Examine the discs on the plate at the beginning of the experiment (0 time) and at 15, 30, 45, and 60 minutes. Measure the diameters of the developing diffusion zones. (See Color Plate 14.)
7. Enter your findings and answer the questions in the results and observations section.

D. OSMOSIS

Brownian Motion Kidney glomerulus

The membrane used in this portion of the exercise is a piece of synthetic material often used as a substitute for a true biological membrane. Such synthetic material contains pores or holes that are approximately the same size as those in biological membranes but more uniform.

Materials

1. The following materials should be provided per four students.
   a. One thistle tube (bell-shaped funnel tube)
   b. One dialysis membrane square (about 10 mm square)
   c. One rubber band
   d. One 250 ml beaker and one 500 ml beaker
   e. One ring stand and clamp or similar support system
   f. One millimeter ruler
   g. One marking pen
2. The following materials should be provided for class use.
   a. Four sucrose (cane sugar) solutions having concentrations of 1, 10, 20, and 40 percent sucrose by weight, respectively
   b. 1 percent methylene blue solution in dropper bottle
   c. Marking pen or similar device
   d. Transparent tape

Procedure

This procedure is to be performed by students in groups of four.

An osmometer system will be constructed with which to determine the rates at which water diffuses through a synthetic membrane separating solutions with differing sucrose concentrations.

1. The instructor will assign a specific sucrose concentration to each student group so that all concentrations will be used. Be certain to examine all other systems and the results obtained by other members of the laboratory, so that the significance and influence of different sucrose concentrations can be noted.
2. Using the sucrose concentration assigned, prepare an osmometer system as follows (Figure 5-2).
   a. Moisten one dialysis membrane square thoroughly.
   b. Put about 50 ml of the assigned sucrose solution and enough methylene blue to color the preparation into the 250 ml beaker.
   c. While one laboratory partner places a finger firmly over the stem opening of the thistle tube, pour the sucrose solution slowly into the tube until the head (funnel portion) is full.
   d. Allow a small amount of the solution to escape from the stem opening by releasing the finger at the stem opening.
   e. Place the wet membrane square over the funnel end and wind the rubber band around it several times to be sure that the system is secure, that is, that it will not leak.
3. With the funnel portion facing down, attach the thistle tube to a ring stand and clamp or similar support.
4. Fill the 500 ml beaker with water, and place it under the osmometer.
5. Lower the osmometer into the water to the depth as shown in Figure 5-2.
6. Allow 5 minutes for the system to equilibrate. Then locate the top level (meniscus) of the sucrose solution in the tube and mark it.
7. Using the ruler, measure the level of the meniscus in terms of millimeters moved from the starting point. Measure it again at 15, 30, 45, 60, and 90 minutes.
8. Record your findings and answer the questions in the results and observations section.
9. Determine the rate of diffusion for each of the sucrose solutions used in the class. (This will require getting the results of others.) To calculate the rate, select a specific time interval, for example, 0–15 minutes, and note the distance covered by the fluid. The rate of diffusion, expressed in millimeters/minute, is equal to the distance divided by the time.

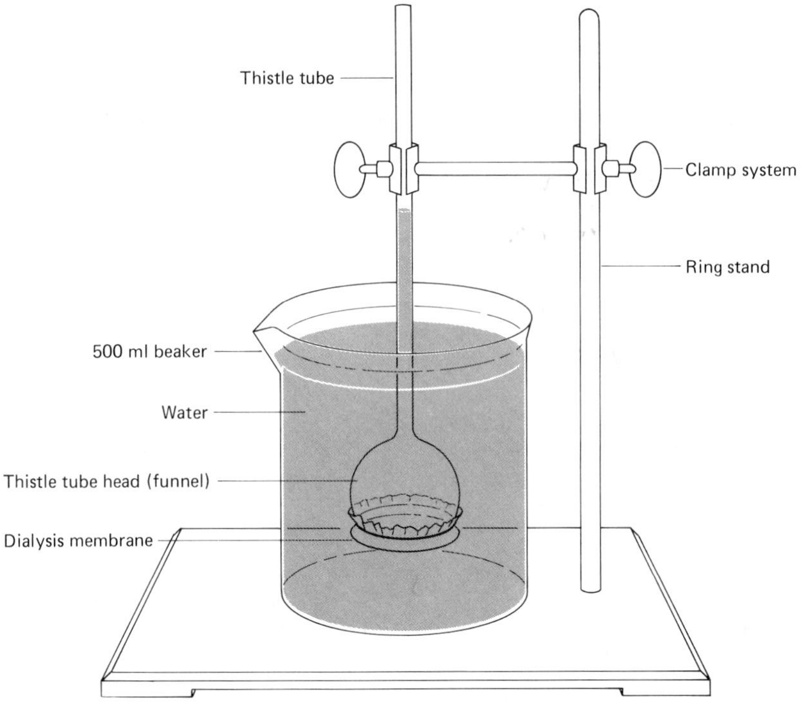

Figure 5-2 An osmometer system.

RESULTS AND OBSERVATIONS

BROWNIAN MOVEMENT

1. Use arrows to indicate the type of movement of India ink particles.

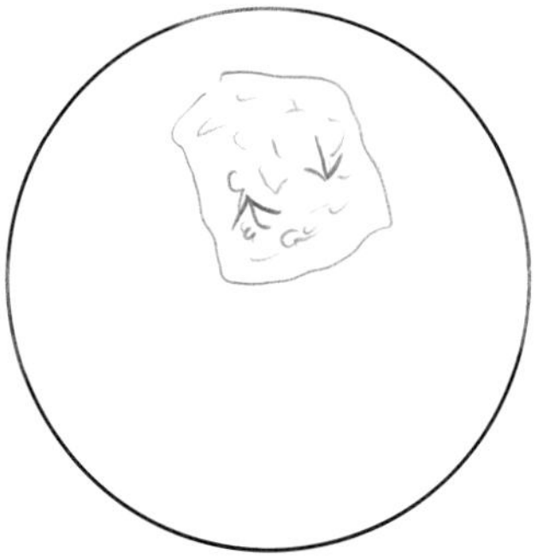

FILTRATION

1. Indicate the appearance of the solutions after the reagents were added.

   Iodine added to starch: precipitated into Black then to purple

   Silver nitrate added to sodium chloride: precipitated instantly

2. Which substances were present in the filtrate? starch + sodium chloride

3. What happened to the charcoal? It stayed in the filter paper

Margie + Todd 686-0010

DIFFUSION

1. Enter your measurements in Table 5-1. 6 mm

Table 5-1

Time (in min)	Zone Diameters (in mm) Crystal Violet	Methylene Blue	Safranin
0	6 mm	6 mm.	6 mm
15	10 mm	8 mm	10 mm
30	12	9	11
45	12	9	11
60	12	9	12

2. Were the diffusion zones clear and sharp? they were in the methylene Blue and safranin but not the crystal violet

3. Did the dyes diffuse down through the agar? yes

Scott 682-2238

pg 938+939

Becky 682-2618

Ernie

pg Table 27.4 Mosis

OSMOSIS

1. Enter your measurements in Table 5-2.

Table 5-2

Time	Meniscus Level 10%	20%	30%	40%
0 5	0	0	0	0
15	5	7	10	14
30	10	12	20	24
45	15	21	29	41
60	19	28	40	55
90	28	40	59	79

diaphram leaked on ours

2. Enter the diffusion rates of the different solutions in Table 5-3.

Table 5-3

Sucrose Concentration	Rate of Diffusion (in mm/min)
30	.65 mm/min
10	.31 mm/min
20	.44 mm/min
40	.87 mm/min

3. Which of the solutions had a faster rate of diffusion? 40%

Explain your answer. Because osmosis happens when a lower solute concentration moves to a higher solute concentration through a semipermeable membrane

Margie & Todd 686-0010

8. Keeping the finger in a downward position, place the bleeding site over the open end of the vial containing saline. Hold the vial firmly against the fingertip and invert it so that the formed blood drop is washed off the finger into the solution.
9. Wipe the finger dry and repeat step 8 with each of the other solutions.
10. Examine the vials and note if each blood-containing solution is cloudy or clear. H_2O cloudy NaCl cloudy Phys cloudy 8.4
11. With a wooden applicator stick put a *small* amount of petroleum jelly on each corner of three cover slips.
12. Transfer 1 drop of the suspension from the saline vial to the center of one of the prepared cover slips.
13. Place a depression slide face down over the cover slip so that the depression portion is directly over the drop of sample material. Press down gently on the slide so that it will stick firmly to the vaseline on the corners of the cover slip.
14. Turn the slide over in one quick motion so that the drop is suspended in the depression.
15. Examine this preparation first under low power and then under high power.
16. Sketch a number of cells showing their general appearance in the results and observations section. Note whether the cells are intact, round, or show any unusual feature.
17. Repeat steps 12–16 with the other solutions. (Avoid contamination with the pipette.)
18. Answer the questions in the results and observations section.

RESULTS AND OBSERVATIONS

SELECTIVE MEMBRANE PERMEABILITY

1. Which of the preparations contained stained yeast cells? ________________
2. Did living or dead yeast cells absorb the dye? ________________
3. On the basis of this exercise how are cell death and cell membrane control functions related? By homeostasis inside & out of cell.

STRUCTURAL INTEGRITY OF CELL MEMBRANES

1. Sketch representative fields.

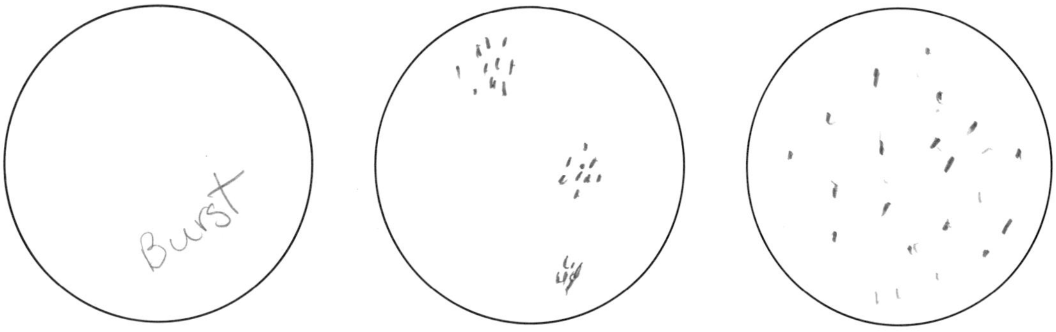

Cells in distilled water — Hypotonic N osmosis

Cells in 0.85 percent saline — isotonic

Cells in 2.0 percent sodium chloride — Hypertonic No Osmosis

2. Which of the preparations used was hypotonic to the red blood cells? Explain. cells in distilled water. Because it had more H_2O into cell than o

3. Which of the preparations used was hypertonic to the red blood cells? Explain. NaCl 4 solution Because it had more H_2O outside than inside of cell.

4. Which of the preparations used was isotonic to the red blood cells? Explain. saline solution because it had same H_2O in & outside of cell

5. How would the different preparations used affect plant cells? Explain. (Diagrams can be used.) the solutions used is used cell wall with cellulose instead of membrane & has vacules to disburse fluids in + out of plant cells

6. Which of the systems exhibited crenation? Hypertonic

Questions for Review

EXERCISE **6**

SELECTIVE PERMEABILITY AND THE STRUCTURAL INTEGRITY OF CELL MEMBRANES

Name Kathy Martin

Laboratory Section Mon 8:00 AM Date 2-9-91

1. a. Do all types of cell have cell membranes? no

   b. What types of cell lack a cell membrane? prokaryotic

   c. What is the chemical composition of a cell membrane? 42% phospholipids pg 59
   55% proteins + carbohydrates

2. Briefly explain how the exposure to the following factors would affect the functioning of a cell membrane.

   a. Heat ? ask

   b. Lipid solvents

3. Figure Q6-1 shows red blood cells suspended in a 2 percent salt solution.

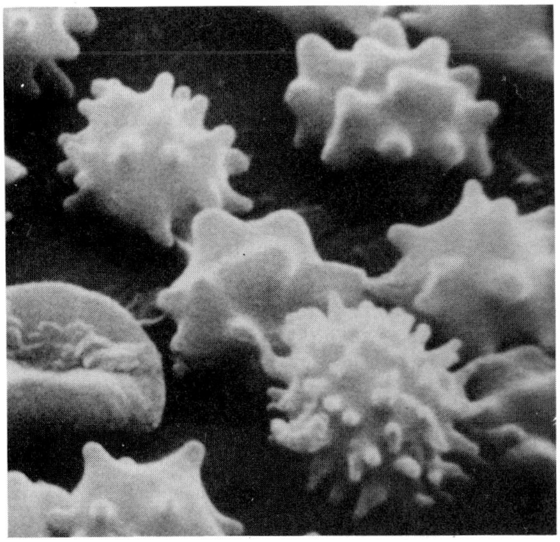

Figure Q6-1

a. Describe the effect produced. Hypertonic

b. Sketch the effect produced when blood cells are placed in a hypotonic solution.

4. Figure Q6-2 shows red blood cells suspended in an isotonic solution.

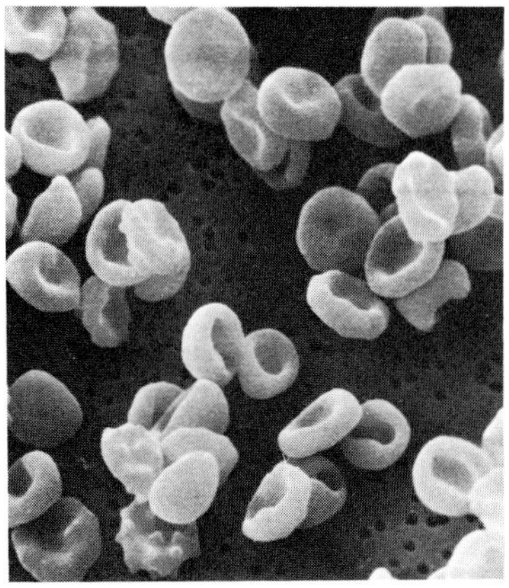

Figure Q6-2

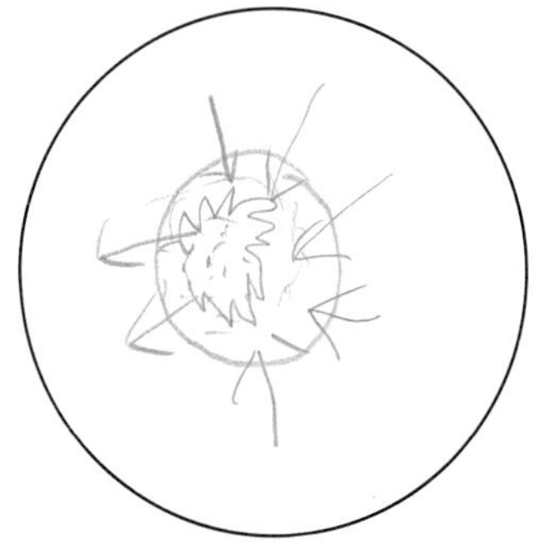

Hypotonic solution

a. Describe the effect produced. cell is unchanged

b. Sketch the effect produced when blood cells are placed in a hypertonic solution.

5. Define or explain the following.

a. Osmotic pressure the force made by the water molecules as they flow across the semipermeable membrane

b. Crenation The shrinking and wrinkling of red blood cells in a hypertonic solution

c. Hemolysis a rupture of red blood cells

d. Isotonic the cells remain the same in other solution

6. In the system shown in Figure Q6-3, will water molecules pass from A to B or from B to A? Explain your answer. A to B. Because Osmosis goes from lower concentration to higher concentration separated by the semipermeable membrane

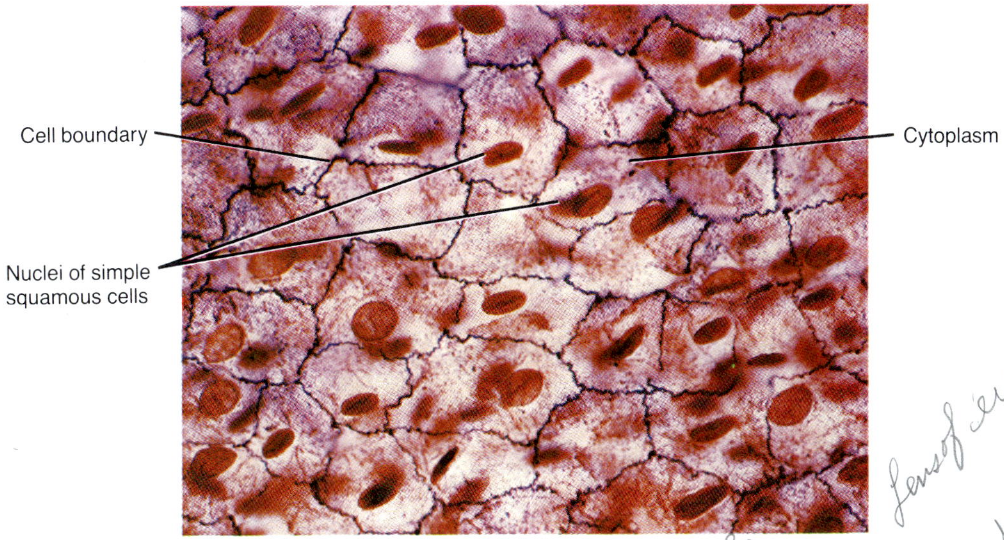

Figure 7-1 Squamous epithelium.

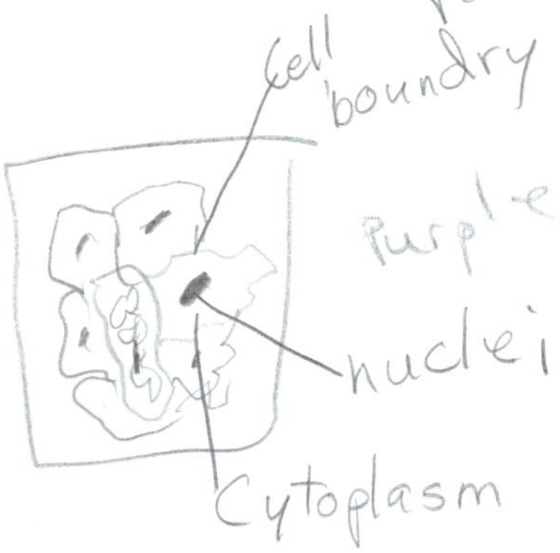

Student drawing

Location in the body of simple squamous epithelium:

1. Peritoneum (lining of the body cavity and its visceral organs)
2. Blood vessel lining (called endothelium)
3. Pleural membranes that surround the lungs
4. Alveoli of lungs
5. Glomerulus of kidney
6. Lens of eye
7. Eardrum (tympanic membrane) layers

SIMPLE CUBOIDAL EPITHELIUM

These cells are cube shaped (Figure 7-2a) or, if pinched together at one end as they are in a duct with a small lumen, they are pyramidal (triangular) (Figure 7-2b).

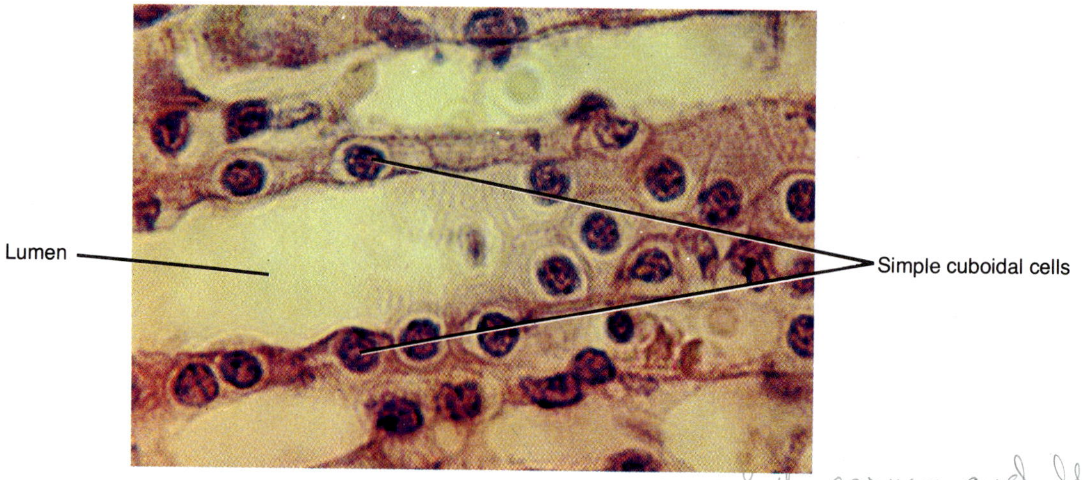

Figure 7-2a Cuboidal epithelium.

Student drawing

Location in the body of cuboidal epithelium:

1. Lines ducts of kidney tubules
2. Lines small ducts of exocrine and endocrine glands
3. Germinal epithelium (outer layer) of ovary
4. Inner surfaces of the cornea and lens of the eye

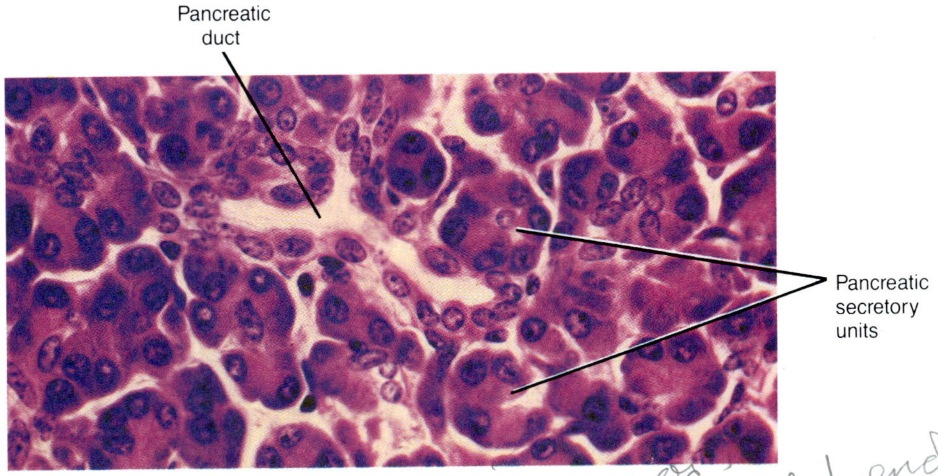

Figure 7-2b Pyramidal epithelium — lines exocrine + endocrine glands

No

Student drawing

SIMPLE COLUMNAR EPITHELIUM

These cells are elongated (Figure 7-3). The shape of the nucleus usually conforms to the cell shape by becoming somewhat flattened, and it is located at the basal end of the cell (away from the free end). *Goblet cells* are columnar cells that produce mucus. The cells in Figure 7-3 have a great deal of mucus, which is released from the cells' free ends into the digestive tract.

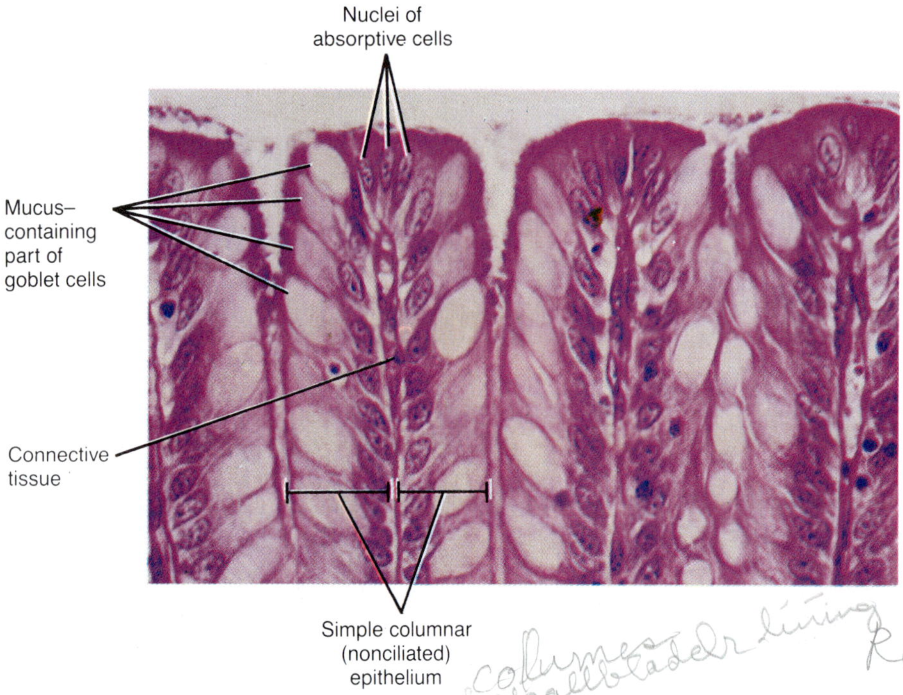

Figure 7-3 Columnar epithelium.

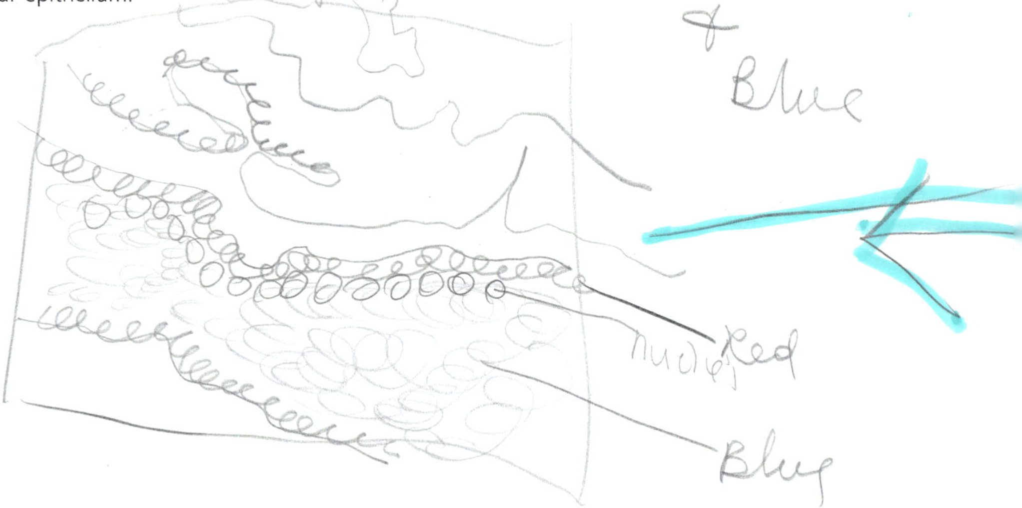

Student drawing

Location in the body of simple columnar epithelium:

1. Lining of the digestive tract from stomach to anus
2. Gallbladder lining
3. Ducts of many glands

CILIATED COLUMNAR EPITHELIUM

Often the columnar cells in the body have cilia projecting from their free ends (Figure 7-4). Their purpose is to pass materials along in a specific direction. Goblet cells are also present in this tissue.

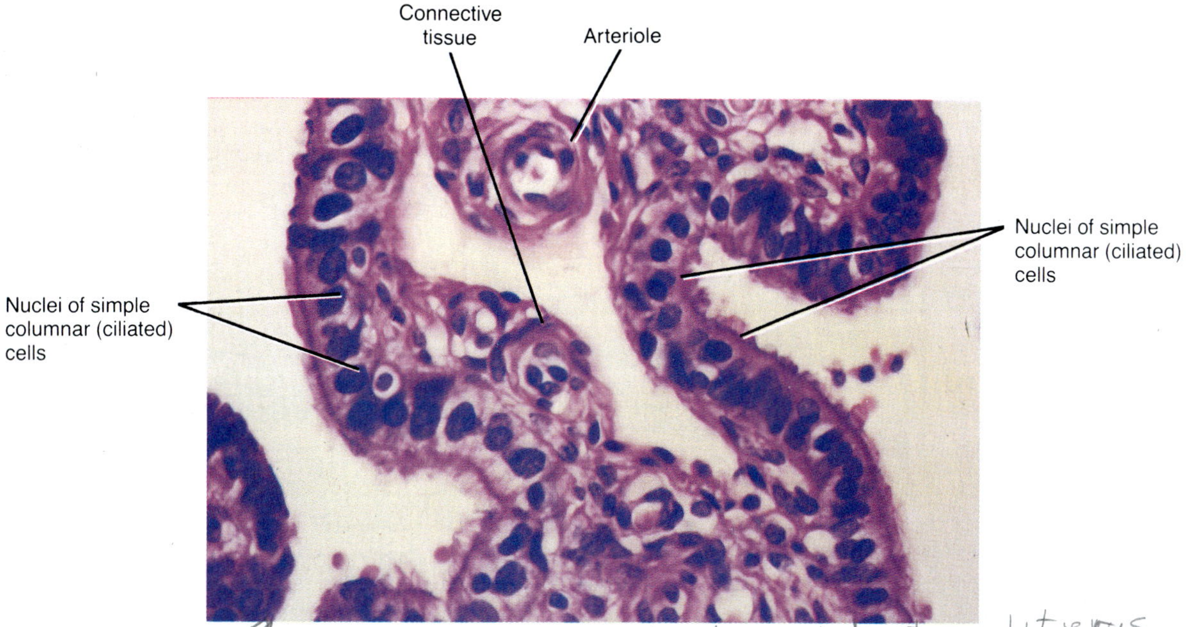

Figure 7-4 Ciliated columnar epithelium.

Student drawing

Location in the body of simple columnar (ciliated) epithelium:

1. Upper respiratory tract lining
2. Fallopian tubes or oviducts
3. Uterus

STRATIFIED EPITHELIUM

This type of epithelium consists of more than one layer of cells, often combining the three cell types we have just examined: squamous epithelium, cuboidal epithelium, and columnar epithelium. Here, however, all these cells are considered as *one* type of epithelium. The cheek cells previously studied are part of this tissue.

STRATIFIED SQUAMOUS EPITHELIUM

The uppermost layer of this many-layered tissue consists of squamous cells. In Figure 7-5, the cells shown are of the cuboidal type, but they may range into the columnar variety in some tissues.

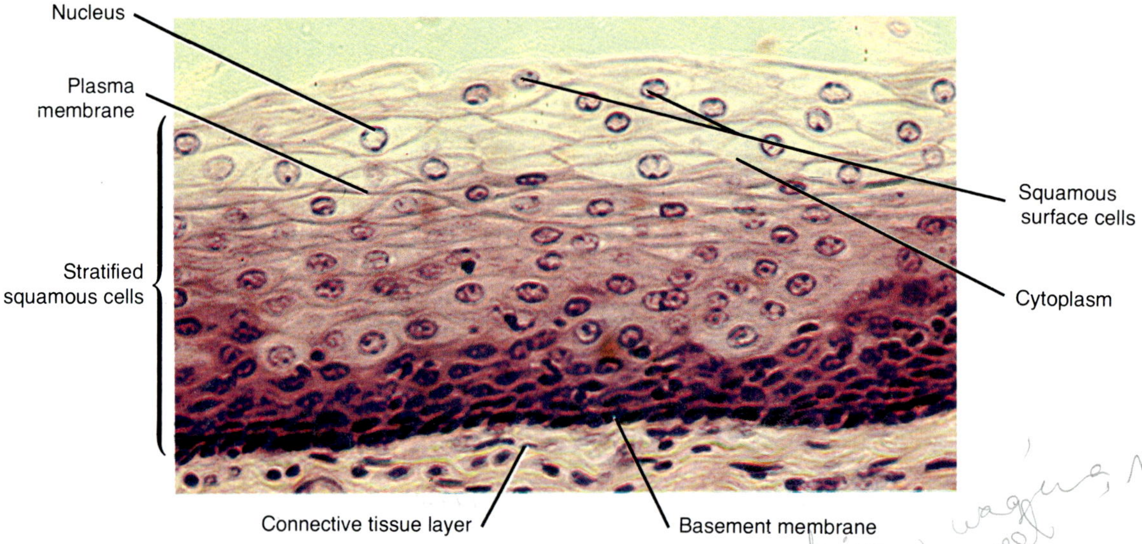

Figure 7-5 Stratified squamous epithelium. (Human cheek cells—430X)

Student drawing

Location in the body of stratified squamous epithelium:

1. Epidermis of the skin
2. Lining of the mouth and esophagus
3. Lining of the vagina

STRATIFIED CUBOIDAL EPITHELIUM

This type of tissue has two or more layers of cuboidal cells (Figure 7-6).

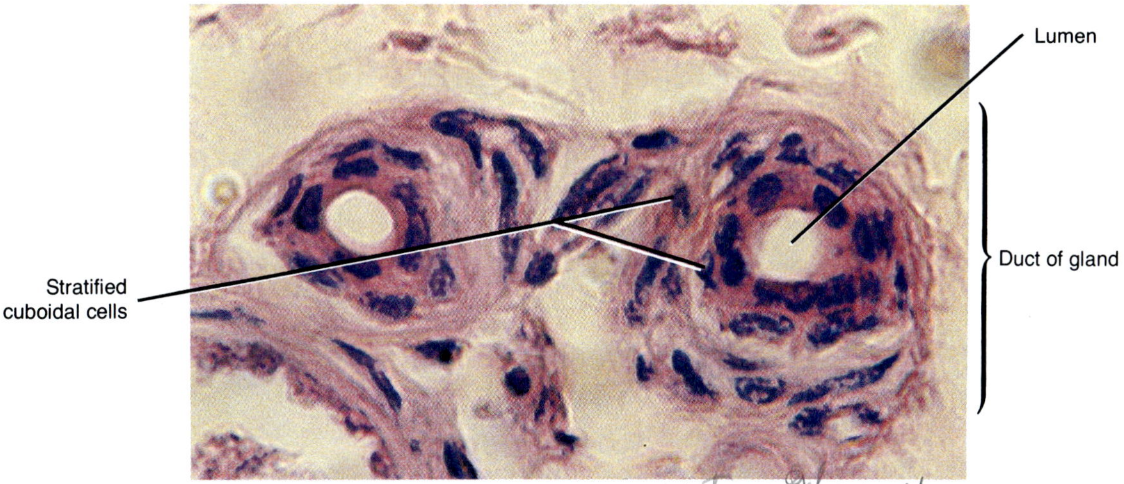

Figure 7-6 Stratified cuboidal epithelium.

sweet gland

No

Student drawing

Location of stratified cuboidal epithelium in the body:

Ducts of the sweat glands

STRATIFIED COLUMNAR EPITHELIUM

This tissue is composed of two or more layers of many-sided cells at its base and a single layer of columnar cells at the free surface (Figure 7-7).

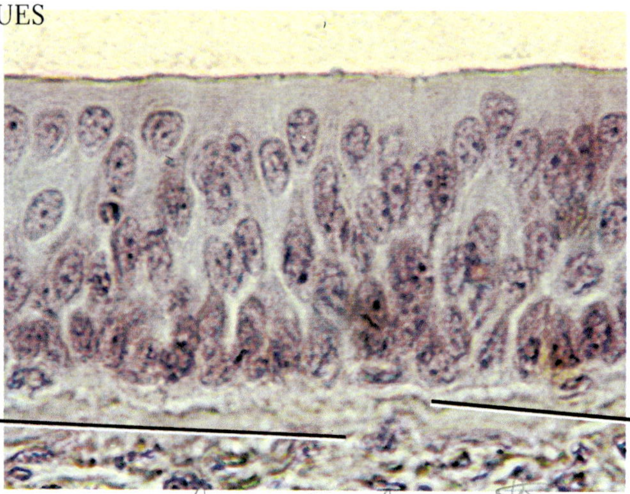

Figure 7-7 Stratified columnar epithelium.

Student drawing

Location of stratified columnar epithelium in the body:

1. Lining of male urethra
2. Lining of some excretory ducts

STRATIFIED TRANSITIONAL EPITHELIUM

This type of tissue is similar to stratified squamous except that the layer of cells at the free end is more cube-shaped than squamous (Figure 7-8).

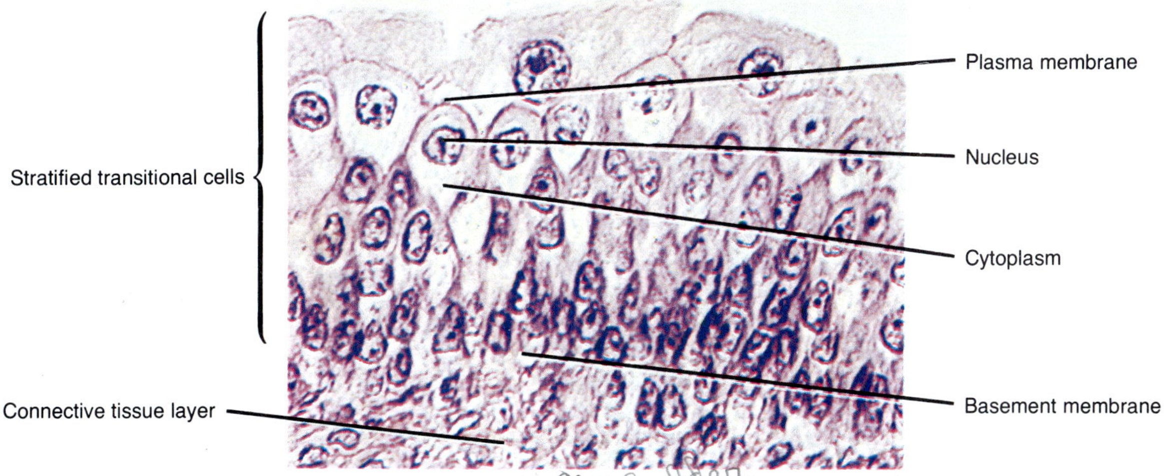

Figure 7-8 Stratified transitional epithelium.

lining of urinary bladder purpliest pinkest

just like this

Student drawing

Location of stratified transitional epithelium in the body:

Lining of urinary bladder

PSEUDOSTRATIFIED EPITHELIUM

Pseudostratified epithelium appears to be layered, but in fact it is not. All cells are attached to the basement membrane (Figure 7-9). Two types of cell are present: the columnar cell, whose nucleus is near the free end, and the basal cell, whose nucleus is near the basement membrane.

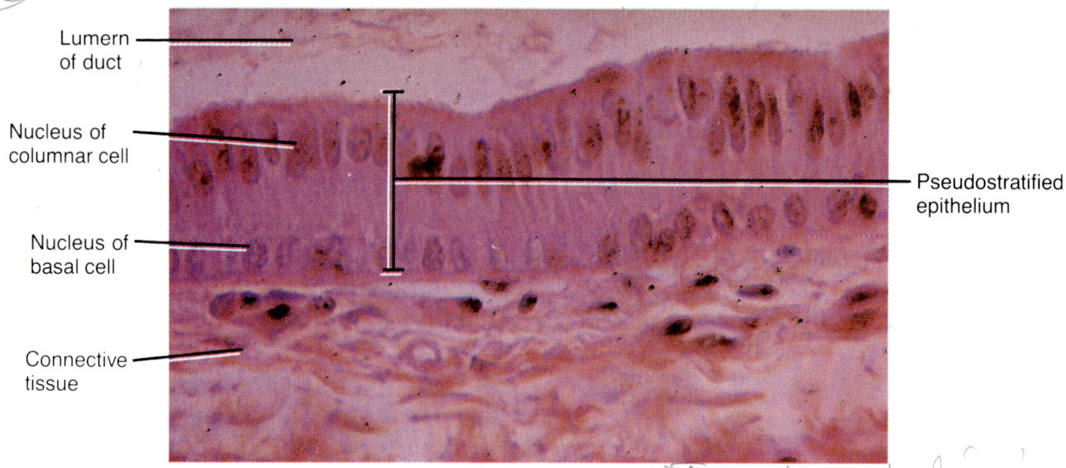

Figure 7-9 Pseudostratified epithelium.

Student drawing

Location in the body of pseudostratified epithelium:

1. Many excretory ducts
2. Male urethra
3. Lining of eustachian tube
4. Upper respiratory tract lining

EPITHELIUM OF EXOCRINE GLANDS

Glands usually are composed of epithelial cells that have grown inward from the free surface. They are classified structurally as follows and as shown in Figure 7-10.

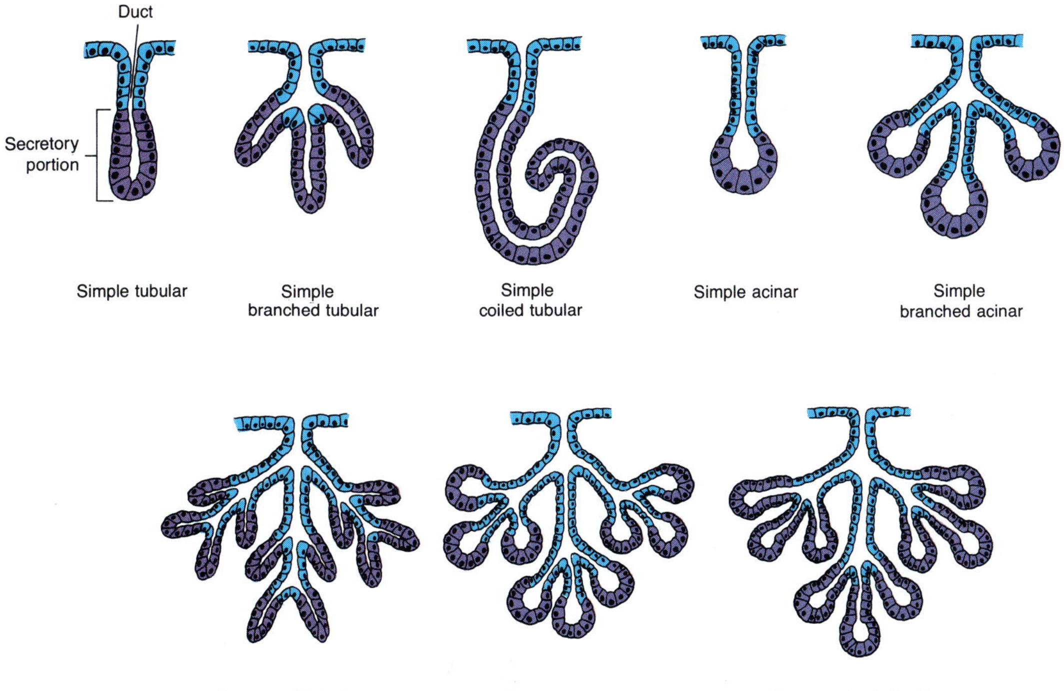

Figure 7-10 Structural types of multicellular exocrine glands. The secretory portions of the glands are indicated in purple. The blue areas represent the ducts of the glands.

Unicellular—a one-celled gland that produces mucus. An example has already been seen in the *goblet cell* of columnar epithelium.

Multicellular—glands composed of many cells.

Simple—has a single, nonbranching duct

1. Tubular
2. Branched tubular
3. Coiled tubular
4. Acinar
5. Branched acinar

Compound—has a branched duct

1. Tubular
2. Acinar

B. CONNECTIVE TISSUE

Materials

The following prepared slides should be provided for class use.

CONNECTIVE TISSUE

1. Trachea cross section (hyaline cartilage)
2. External ear (elastic cartilage)
3. Long bone cross section (compact bone)
4. Subcutaneous tissue (adipose or fat cells)
5. Human blood smear (red and white blood cells)
6. Subcutaneous tissue (areolar tissue)
7. Bone marrow (hemopoietic tissue)
8. Tendon (dense connective tissue)
9. Liver (reticular connective tissue)
10. Blood smear (formal blood elements)

This large group of diverse tissues has two determining characteristics:

1. The cells produce a nonliving intercellular matrix.
2. The cells are derived from embryonic mesenchyme cells.

There are some exceptions to the production of an intercellular matrix, but in a short classification such as this exceptions are made for two of the connective tissues. The special pecularities of these tissues will be noted as they are encountered in this exercise. The connective tissue types you will study here are areolar, adipose, dense, elastic, reticular, cartilage, bone, and vascular (blood).

CONNECTIVE TISSUE PROPER

There are many tissues in this category and they are found in all parts of the body. They bind cells of an organ together, surround and protect major organs, and serve as important links between muscle and bone and between bone and bone.

AREOLAR OR LOOSE CONNECTIVE TISSUE

The living cells of this tissue (fibroblasts) produce both collagenous and elastic tissue arranged loosely (Figure 7-11).

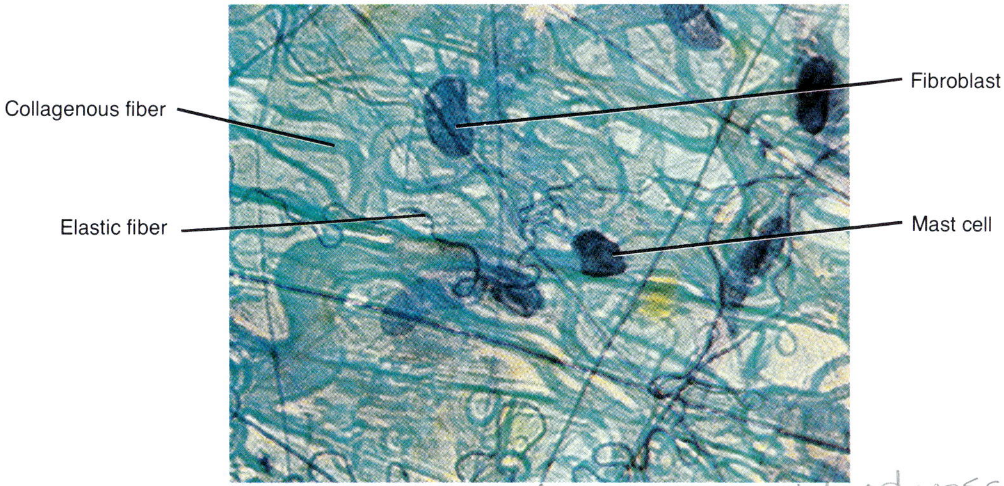

Figure 7-11 Areolar connective tissue.

Binds nerves, blood vessels, etc.
hairy
?
Subcutaneous
layer of the skin
tissue

Student drawing

Location in the body of areolar connective tissue:

1. Binds nerves, blood vessels, and body organs
2. Subcutaneous layer of the skin

ADIPOSE TISSUE

These cells are specialized for fat storage intracellularly (within the cells) (Figure 7-12). This is an exception to the usual intercellular matrix.

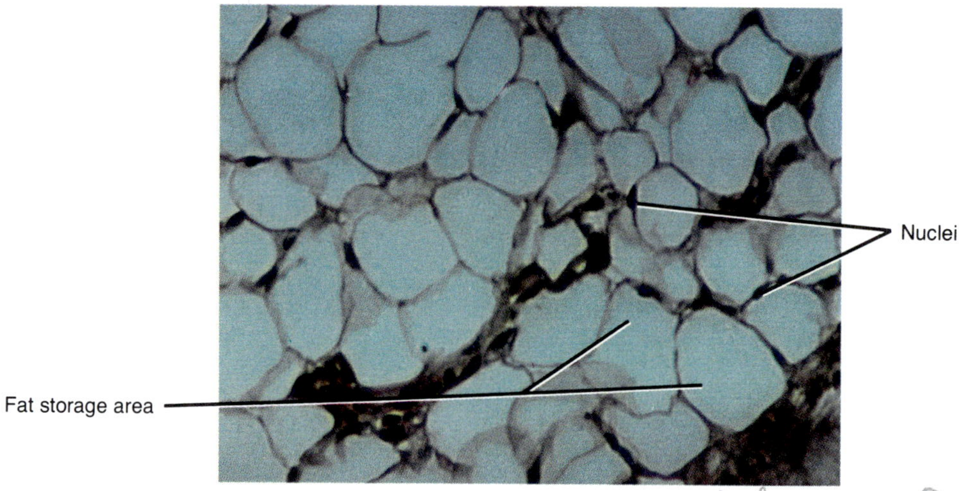

Figure 7-12 Adipose tissue.

yellow bone marrow, fatty depots, Just like on Blue

Left

Student drawing

Location of adipose tissue in the body:

1. Subcutaneous layer of the skin
2. Surrounding heart, kidneys, and major arteries
3. Yellow bone marrow
4. Various other "fatty depots"

DENSE (COLLAGENOUS) CONNECTIVE TISSUE

The fibroblasts produce strong collagenous (sticky) fibers (Figure 7-13), which are often arranged in parallel fashion forming tough capsules around organs and forming tendons, ligaments, and aponeuroses (a-po-nyoo-RO-sees).

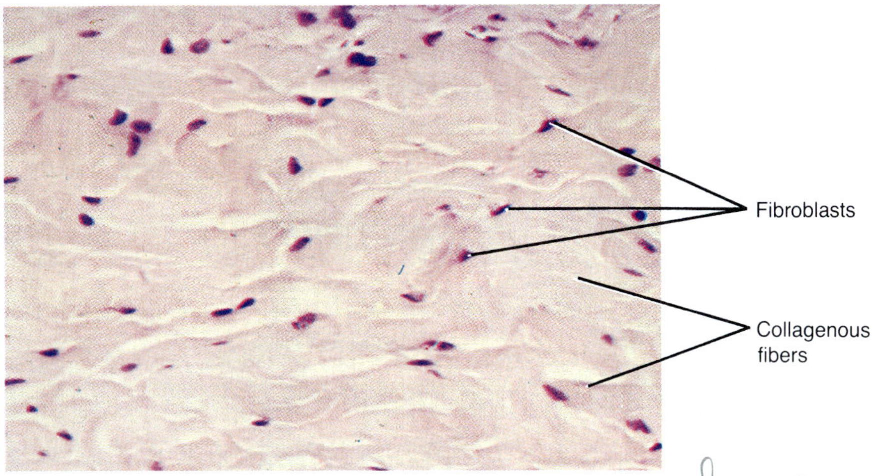

Figure 7-13 Dense (collagenous) connective tissue.

tendons - fascia

Student drawing

Location in the body of dense connective tissue:

1. Tendons
2. Ligaments
3. Aponeuroses
4. Capsule surrounding testis
5. Periosteum
6. Perichondrium
7. Fascia

ELASTIC CONNECTIVE TISSUE

The fibroblasts of this tissue produce both collagenous and elastic fibers (Figure 7-14). The latter lend elasticity to organs.

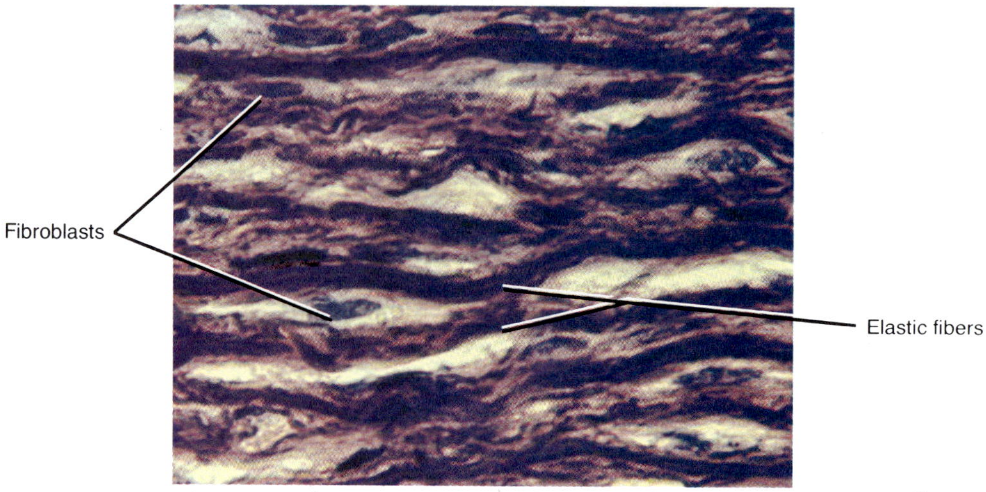

Figure 7-14 Elastic connective tissue.

lungs, larynx, trachea, bronchi, vocal cords

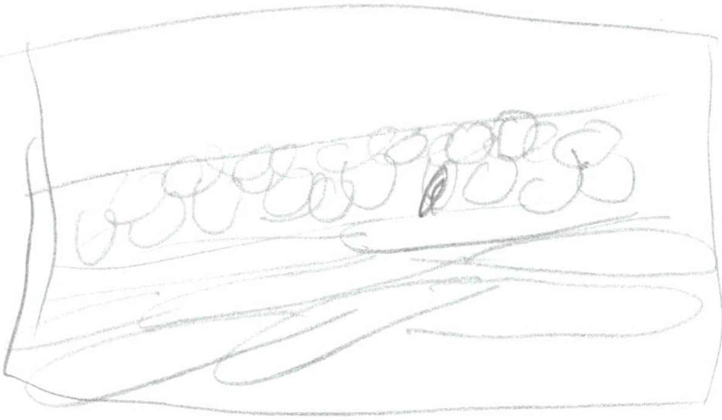

Purple

Student drawing

Location of elastic connective tissue in the body:

1. Lungs
2. Larynx, trachea, and bronchi
3. Vocal cords

RETICULAR CONNECTIVE TISSUE

Very delicate fibers are produced by the fibroblasts, which tend to interlace, forming a weblike structure (Figure 7-15).

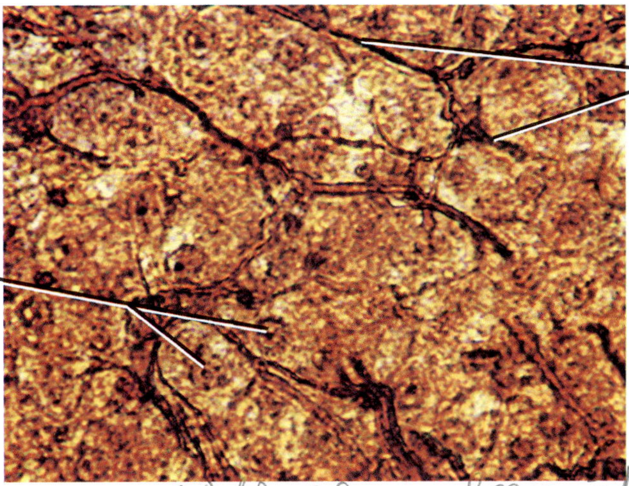

Figure 7-15 Reticular connective tissue.

Student drawing

Location of reticular connective tissue in the body:

The infrastructure of many organs such as the liver and spleen

CARTILAGE

This type of connective tissue is a skeletal tissue. Cells (chondrocytes) in cartilage produce a nonliving matrix, which surrounds and traps them within it. There are three types of cartilage: hyaline cartilage, fibrocartilage, and elastic cartilage. Only one, hyaline cartilage, is depicted here.

Hyaline Cartilage

The matrix surrounding the cells is clear because the fibers within it are so small and fine they cannot be seen (Figure 7-16).

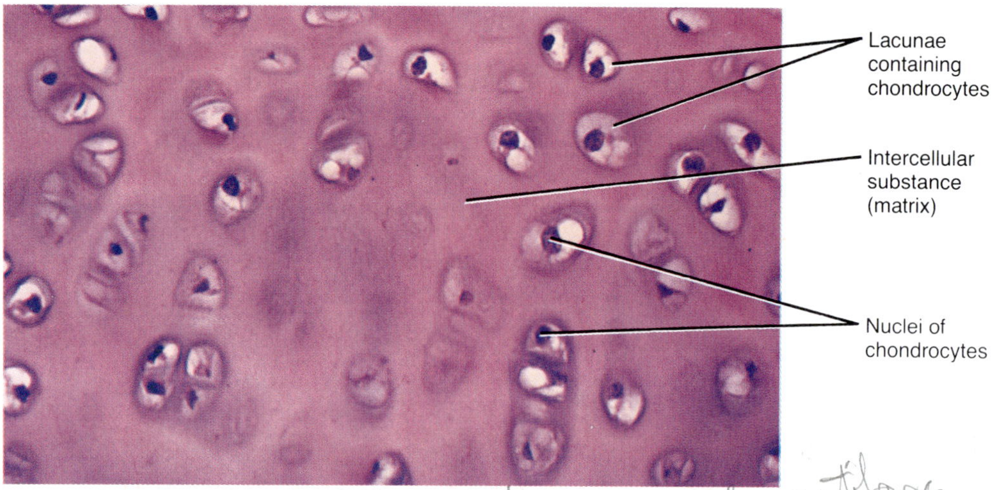

Figure 7-16 Hyaline cartilage. long bones, rib cartilage

Looks like this
More Blue

Student drawing

Location of hyaline cartilage in the body:

1. Articular cartilages of long bones
2. Earlobe cartilages
3. Larynx, trachea, and bronchi
4. Costal (rib) cartilages

BONE

The cells that produce the collagenous fibers in this skeletal tissue (osteoblasts) become embedded in a hard matrix but retain connections with each other and the blood vessels by means of cytoplasmic extensions (canaliculi) (Figure 7-17).

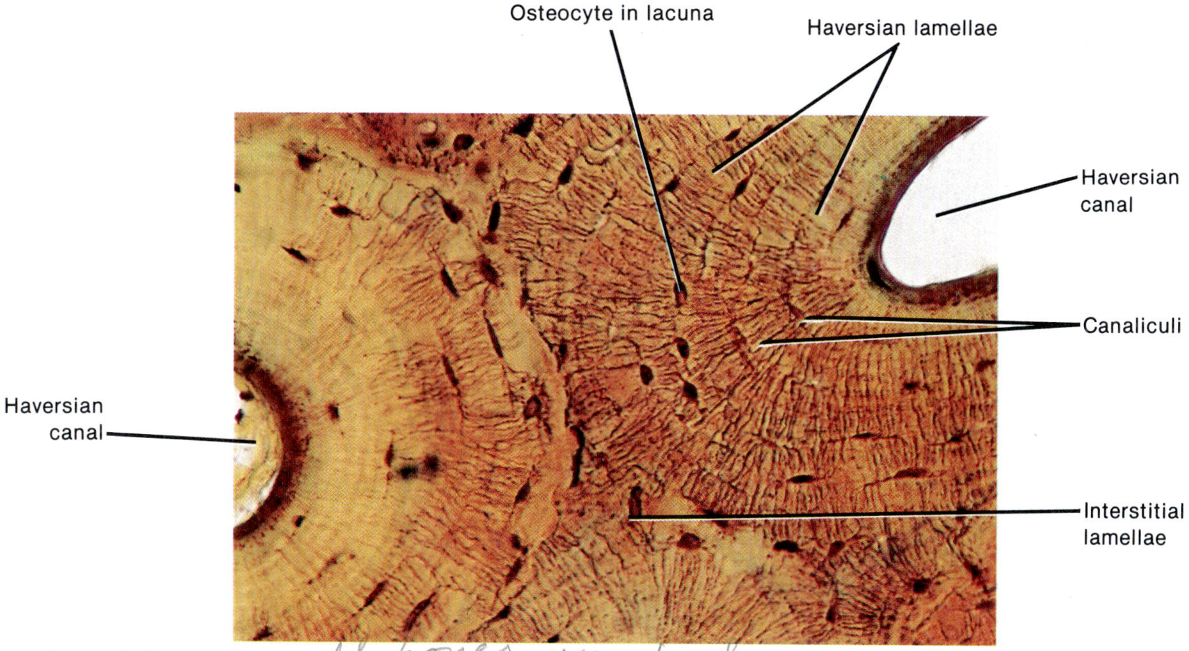

Figure 7-17 Bone.

all bones in body
tree —
brownish gray
No

Student drawing

Location of bone in the body:

The entire framework of bones that makes up the body

VASCULAR TISSUE (BLOOD)

This tissue consists of formed elements (erythrocytes, leukocytes, and platelets) and a liquid plasma (Figure 7-18). A stained blood slide will show the pink, disc-shaped enucleated (no nucleus) red blood cells and the blue-stained white cells. Examine a blood slide and pick out as many different white blood cells as you can. (Blood will be studied in detail in a later exercise).

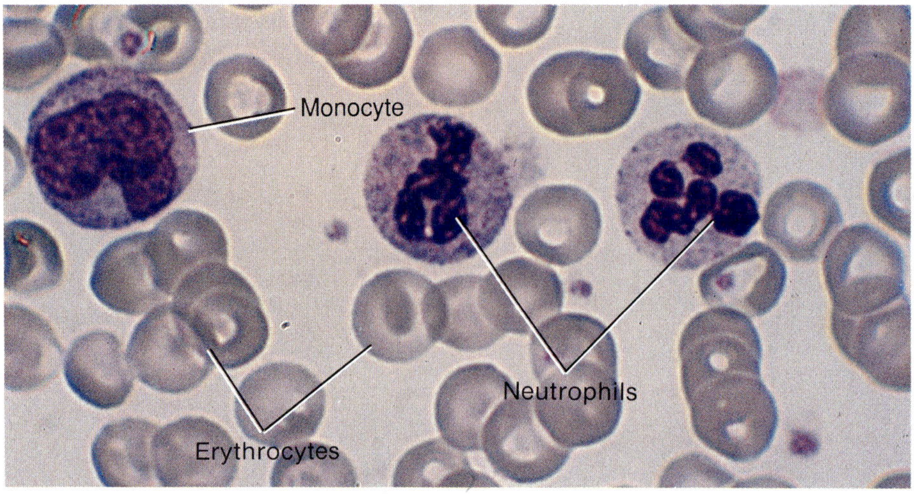

Figure 7-18 Human blood cells. – veins

Student drawing

Location of blood in the body:

Found circulating within arteries, veins, and capillaries of the body's closed circulatory system

C. MUSCULAR TISSUE

Materials

The following prepared slides should be provided for class use.

MUSCLE TISSUE

1. Visceral teased smooth muscle
2. Skeletal muscle
3. Cardiac (heart) muscle

This highly specialized tissue contains cells that are able to shorten and thicken (contract) upon stimulation. Muscle is classified in three different ways:

Based on Location in the Body

1. *Skeletal*—attached to parts of the skeleton
2. *Cardiac*—found only in the heart
3. *Visceral*—found in many visceral organs

Based on Function

1. *Voluntary*—skeletal muscle that is under the control of the will
2. *Involuntary*—muscle that is not under the control of the will. Cardiac and visceral muscle are involuntary.

Based on Microscopic Appearance

1. *Striated Skeletal*—cross striations are seen under the light microscope. (See text for explanations of this phenomenon.)
2. *Striated Cardiac*—found only in the heart. Cross striations are also visible here.
3. *Nonstriated or Smooth*—found in visceral organs. Usually layered. No cross striations present.

Examine the three types of muscle cells (fibers) based on location.

STRIATED SKELETAL MUSCLE

In a good preparation one can see that these muscle fibers are *multinucleate* (contain more than one nucleus). This is an unusual characteristic because most cells in the body are *uninucleate* (one nucleus). Seen within the cells are longitudinal muscle fibrils or *myofibrils* and dark-staining masses that appear to be lined up in a row transversely giving the impression of a line. The myofibrils run longitudinally, and the dark-staining masses at right angles to them are the cross striations. A highly specialized cell membrane is called the *sarcolemma* and the cytoplasm is referred to as *sarcoplasm*. Find all parts labeled in Figure 7-19 and in the slide given you for laboratory study.

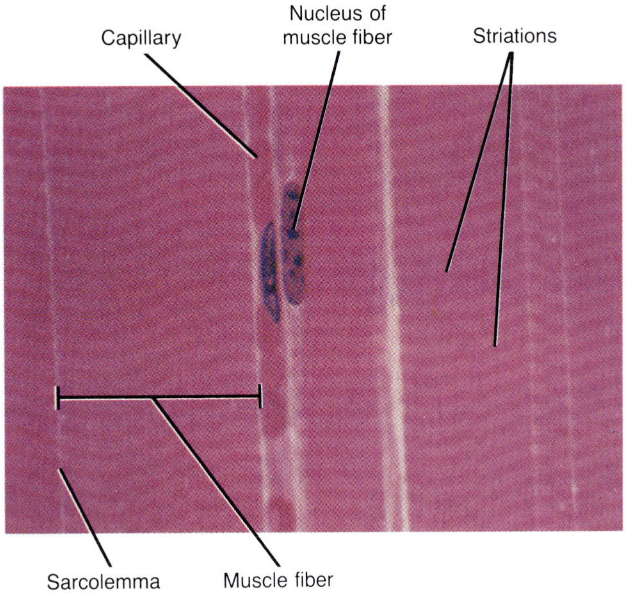

Figure 7-19 Striated skeletal muscle.

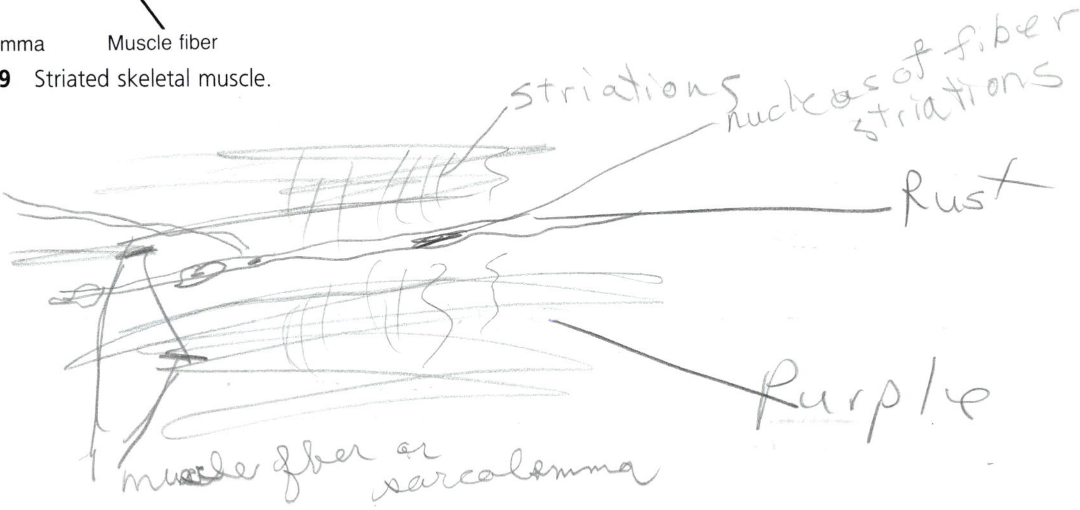

Student drawing

STRIATED CARDIAC MUSCLE

The fibers in this type of muscle are all united, the cells are much smaller, and it is difficult to see the myofibrils (Figure 7-20). Notice the dark-staining bars (not seen in striated skeletal muscle) that appear at regular intervals. These bars are called *intercalated discs*. Although both cardiac and skeletal muscle are striated, they differ in the following ways:

Striated Skeletal	Striated Cardiac
1. Multinucleate	1. Uninucleate
2. Nuclei peripherally located	2. Nucleus centrally located
3. Muscle fibers do not branch	3. Muscle fibers are branched, leaving prominent spaces
4. There are no intercalated discs	4. Intercalated discs present

Ucinda

Nervous	o. Neurons – storing memory, thinking, regulating the activity of other organs or glands
	p. Neuroglia – support and bind neurons together.

3. With the aid of your laboratory manual and text determine the locations in the body of the following types of tissue.

a. Smooth muscle walls of hollow internal organs

b. Elastic cartilage Framework of external ear, auditory canal, portions of larynx

c. Compact bone hard outer layer of all bones in body

d. Squamous epithelium depends on simple or stratified: stratified – skin; simple – linings of body cavities

e. Hyaline cartilage walls of respitory passages, fetal skeleton, nose.

f. Adipose Subcutaneous layer of the skin, surround heart kidneys, yellow bone marrow

g. Cardiac muscle heart muscle

h. Columnar epithelium lining of organs, ducts, uterus etc.

i. Hemopoietic red bone marrow, spleen, lymph nodes, tonsil

Hemopoiesis

Read

4. Place a check (√) in the appropriate space of Table Q7-2 to indicate which muscle type(s) exhibits the property listed.

Table Q7-2

	Cardiac	Skeletal	Visceral (Smooth)
a. Attached to parts of the skeleton		√	
b. Found only in the heart	√		
c. Found in many visceral organs			√
d. No cross striations present			√
e. Shows cross striations	√	√	
f. Has intercalated discs	√		
g. Multinucleated		√	
h. Uninucleated	√		
i. Tapered cells			√
j. Voluntarily controlled		√	
k. Involuntarily controlled	√		√

10

SECTION III

AN INTRODUCTION TO THE ORGAN AND SYSTEM LEVELS OF ORGANIZATION

A division of labor occurs early in the development of higher animals. As Exercise 7 demonstrated, such cells become specialized to perform specific functions and are arranged to form tissues, a process known as histogenesis. As the animal body continues to enlarge and develop further, the tissues become organized to produce organs. Later the organs become clustered to form organ systems as their functions become more complex.

The exercises of this section emphasize the terms used to specify body position and direction.

GENERAL REFERENCES

Tortora, G. J., and N. P. Anagnostakos. *Principles of Anatomy and Physiology,* 6th ed. New York: Harper & Row, 1990.

Witherspoon, J. D. *Human Physiology.* New York: Harper & Row, 1984.

EXERCISE 8

ANATOMICAL ORIENTATION AND TERMINOLOGY

Materials

The following materials should be provided for class use:

1. Human skeleton
2. Human torso model
3. Skeleton of a four-legged animal

TERMS OF LOCATION AND POSITION

Terminology for the location and position of body parts of the two-legged human varies somewhat from that used for four-legged animals. In four-legged animals the head end is forward in locomotion and the belly surface is parallel to the ground, whereas in humans the head is above the body and the belly surface is not parallel to the ground. The differences in terminology are minor, but they can present some difficulty for the beginning student. There is no alternative but to learn all the variations since they are used (often interchangeably) by different authors. These terms are important and you should make them a part of your vocabulary as soon as possible (see Figure 8-1).

1. Terms referring to the head end and the rear end fall into three groups:
   SUPERIOR and *INFERIOR*
   a. **Superior**—the head or upper end; used for humans but not for four-legged animals
   b. **Inferior**—the rear or lower end, away from the head; used for humans, but not for four-legged animals
   CRANIAL and *CAUDAL*
   a. **Cranial**—the head end; used for all animals that have a brain case or cranium
   b. **Caudal**—the rear or tail end; used for all animals that have a tail
   ANTERIOR and *POSTERIOR*
   a. **Anterior**—the part of the body that goes forward in locomotion
      i. The head end as in four-legged animals
      ii. The front or belly side of humans
   b. **Posterior**—the part of the body that is not forward in locomotion
      i. The rear end of four-legged animals
      ii. The back side of the human
2. Terms referring to the back side and the belly side of a four-legged animal:
   DORSAL and *VENTRAL*
   a. **Dorsal**—the back side of four-legged animals; sometimes also applied to the human back, for example, the dorsal body cavity
   b. **Ventral**—the belly side or underside of four-legged animals; sometimes also applied to the human belly side, for example, the ventral body cavity

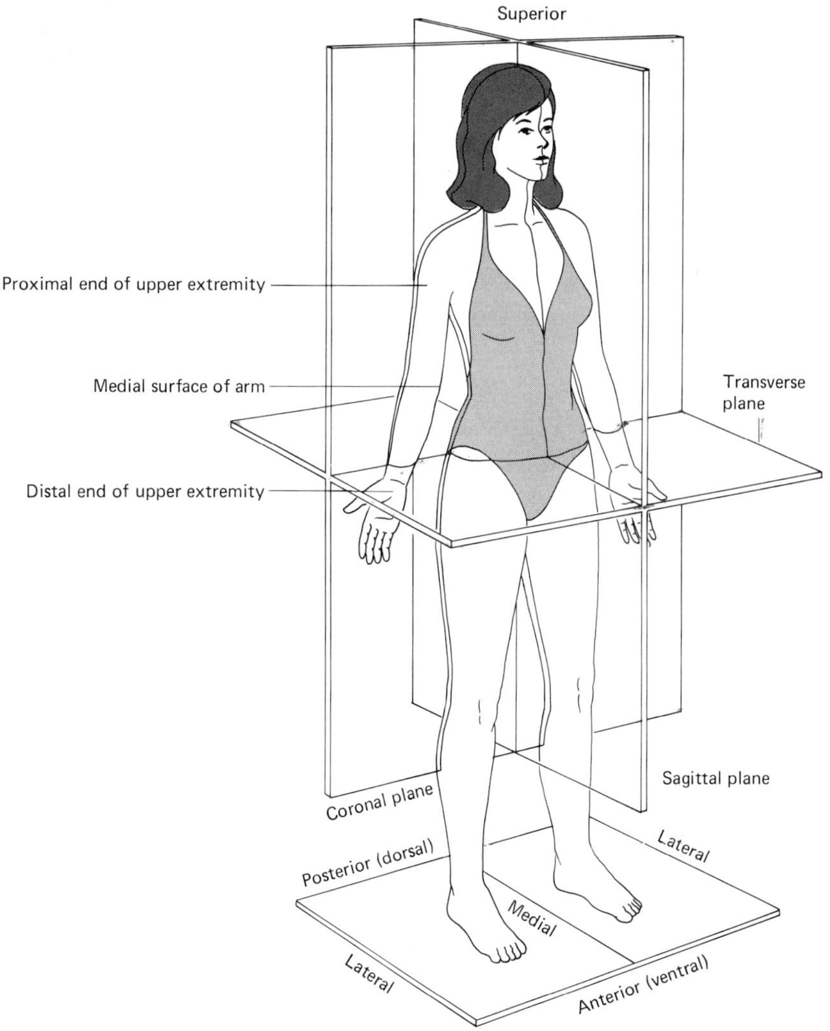

Figure 8-1 Anterior view of the human body, including terms of location, position, and body planes.

3. Terms referring to points on or away from the midline or center:
   <u>*MEDIAL*</u> *and* <u>*LATERAL*</u>
   a. **Medial**—toward or on the midline of the body or an organ
   b. **Lateral**—away from or to the side of the midline of the body or an organ
   <u>*CENTRAL*</u> *and* <u>*PERIPHERAL*</u>
   a. **Central**—the center of a spheroidal part or organ ?
   b. **Peripheral**—along the outside of a spheroidal part or organ
   <u>*SUPERFICIAL*</u> *and* <u>*DEEP*</u>
   a. **Superficial** (external)—on the outside
   b. **Deep** (internal)—on the inside
4. Terms describing appendages or parts of appendages in reference to the point of attachment to the main axis of the body or to some specific point of reference inside the body:

<u>*PROXIMAL*</u> *and* <u>*DISTAL*</u>

a. **Proximal**—nearer to the point of attachment to the body axis or trunk in the case of extremities; nearer to the point of reference on an internal organ
b. **Distal**—farther from the point of attachment to the body axis or trunk in the case of extremities; farther from the point of reference on an internal organ

TERMS DESCRIBING BODY PLANES

Sections cut through a body or through an organ are very helpful in understanding anatomy. A section or plane is a cut passing through the body that divides it into two parts.

1. **Transverse or cross section**—a cut at right angles to the longitudinal axis of the body or organ, dividing it into superior and inferior parts (in the human)
2. **Sagittal section**—a cut parallel to the longitudinal axis of the body or organ, dividing it into right and left sides
   a. **Midsagittal**—a sagittal section through the midline, resulting in "equal" right and left sides
   b. **Parasagittal**—a sagittal section to the right or left of the midline, resulting in unequal right and left sides
3. **Coronal or frontal section**—a cut parallel to the longitudinal axis of the body or organ, dividing it into dorsal and ventral (anterior and posterior in the case of humans) parts

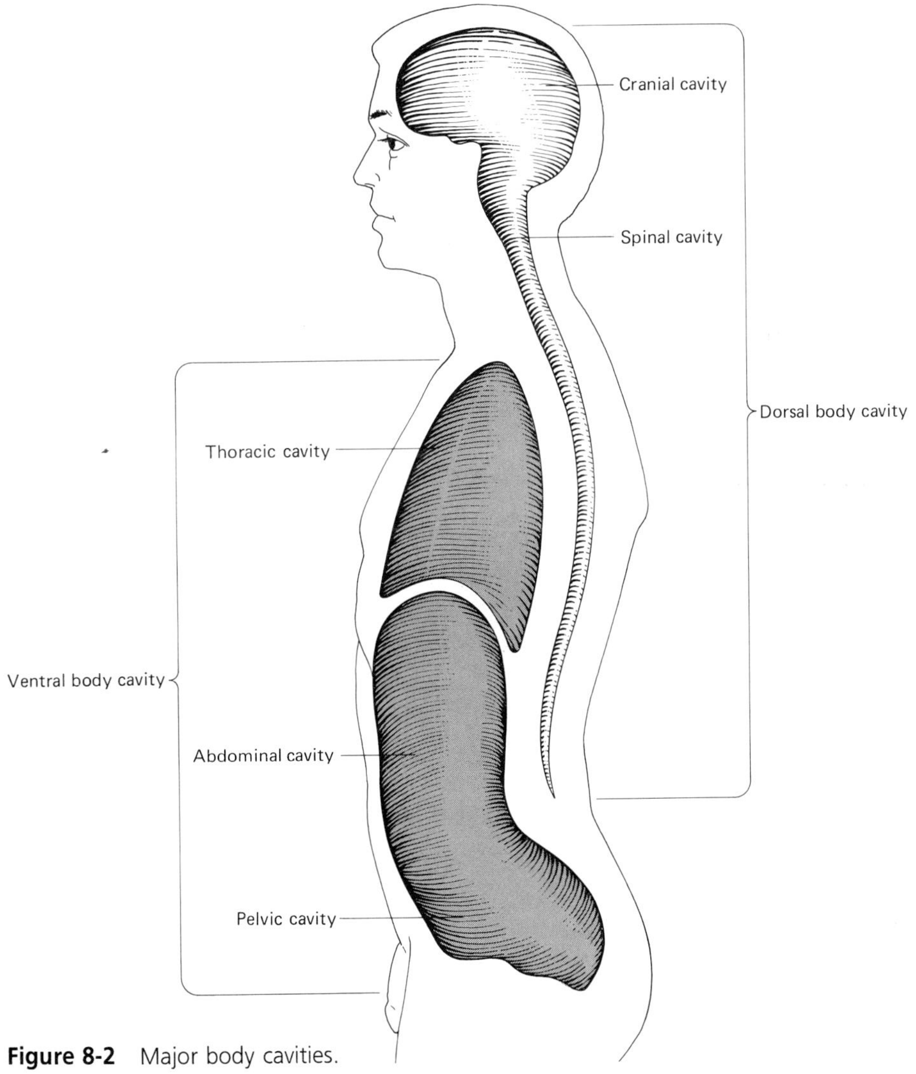

Figure 8-2 Major body cavities.

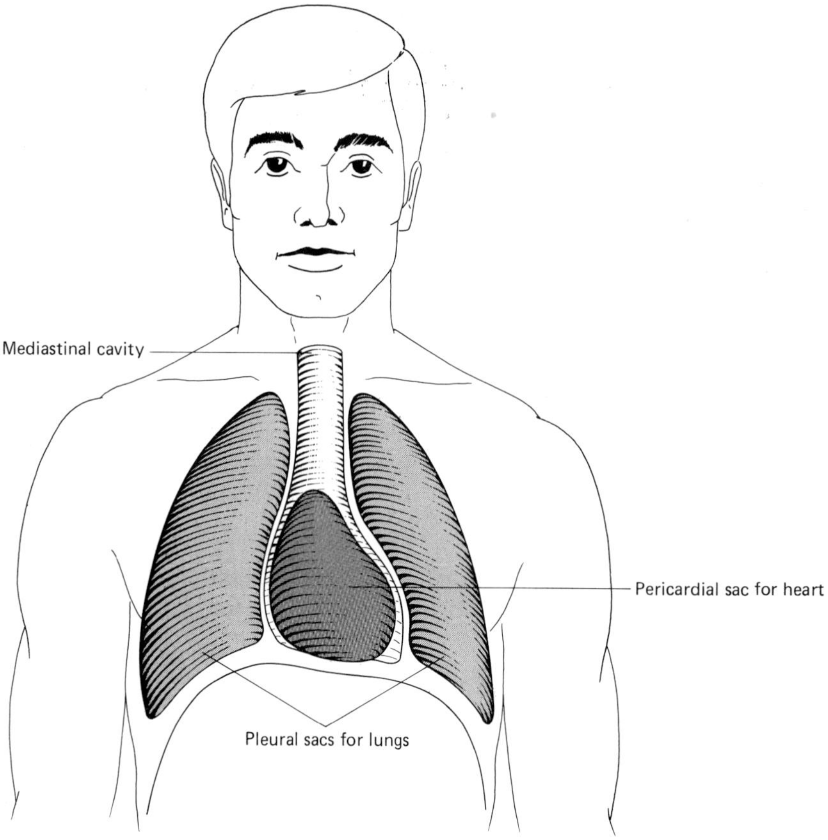

Figure 8-3 Subdivision of thoracic cavity, anterior view.

BODY CAVITIES

The spaces within the body that contain the internal organs are divided into two principal cavities, both of which contain further subdivisions (Figures 8-2 and 8-3).

1. *DORSAL BODY CAVITY*—cavity that has its embryonic origin along the back side of the animal (notice that the word dorsal is used and not posterior for the human as well as the cat)
   a. **Cranial cavity**—cavity formed by the skull, which contains the brain
   b. **Spinal cavity**—cavity formed by the vertebrae, which contains the spinal cord
2. *VENTRAL BODY CAVITY*—cavity that is located in the belly region in both two-legged and four-legged animals
   a. **Thoracic cavity**—located above the diaphragm; the chest cavity
      i. **Pleural cavities**—cavities between the pleural membranes, which surround the lung
      ii. **Pericardial cavity**—cavity between the pericardial membranes, which surround the heart
      iii. **Mediastinal cavity**—the cavity between the lungs, containing the heart, trachea, esophagus, and thymus gland
   b. **Abdominal (abdominopelvic) cavity**—located below the diaphragm; contains the viscera
      i. **Abdominal cavity proper**—contains stomach, spleen, liver, gallbladder, pancreas, small intestines, and most of large intestine
      ii. **Pelvic cavity**—contains urinary bladder, sigmoid colon, rectum, and internal reproductive organs

Note: The abdominal cavity, pleural cavities, and pericardial cavity are located between two layers of peritoneum: the outer layer is called the **parietal peritoneum,** and the inner layer that surrounds the organs is called the **visceral peritoneum.** A large cavity, called the abdominal cavity, occupies the space between the parietal and visceral peritoneum below the diaphragm. In the case of both pericardial and pleural cavities, however, the space is very small and the two layers of peritoneum are closely applied to each other and to their respective organs with only a small fluid-filled space between them.

SUBDIVISIONS OF THE ABDOMINAL REGION

The outer surface of the body is marked off into regions for the purpose of locating and specifying internal parts. The regions are shown in Figure 8-4.

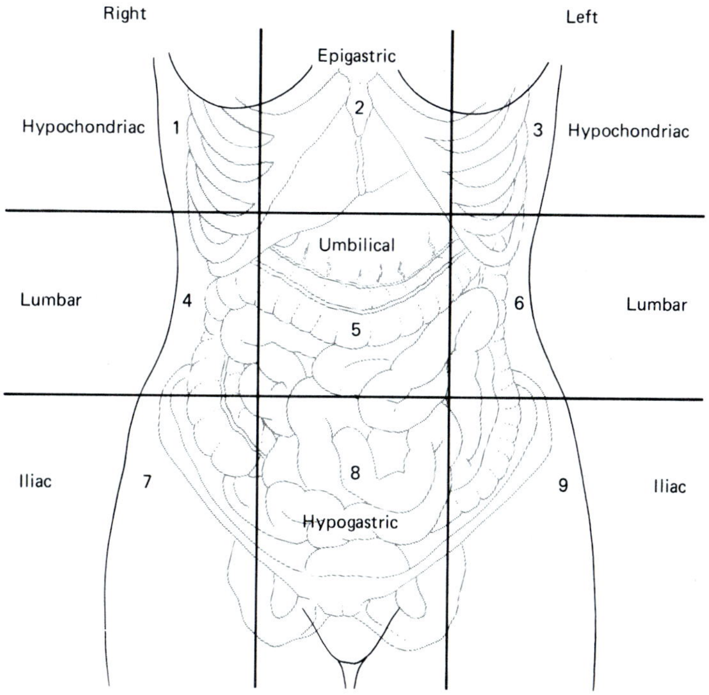

Figure 8-4 Anatomical regions of the ventral body surface.

TERMS NAMING BODY PARTS

Identify the following body parts on Figure 8-5. The common terms and some anatomical terms are listed below. These terms will be used in future laboratory exercises in which more detailed descriptions will be given. Insert the proper numbers from the list below on the diagrams of the human body.

Head (1)
- Skull, cranium (2)
- Skull, face (3)

Neck (4)

Trunk (5)
- Thorax (chest) (6)
- Breast (7)
- Axilla (armpit) (8)
- Shoulder (9)
- Abdomen (10)
- Pelvis (11)
- Back (12)
- Lumbar region (13)
- Navel (14)

Upper extremity or arm (15)
- Upper arm (brachium) (16)

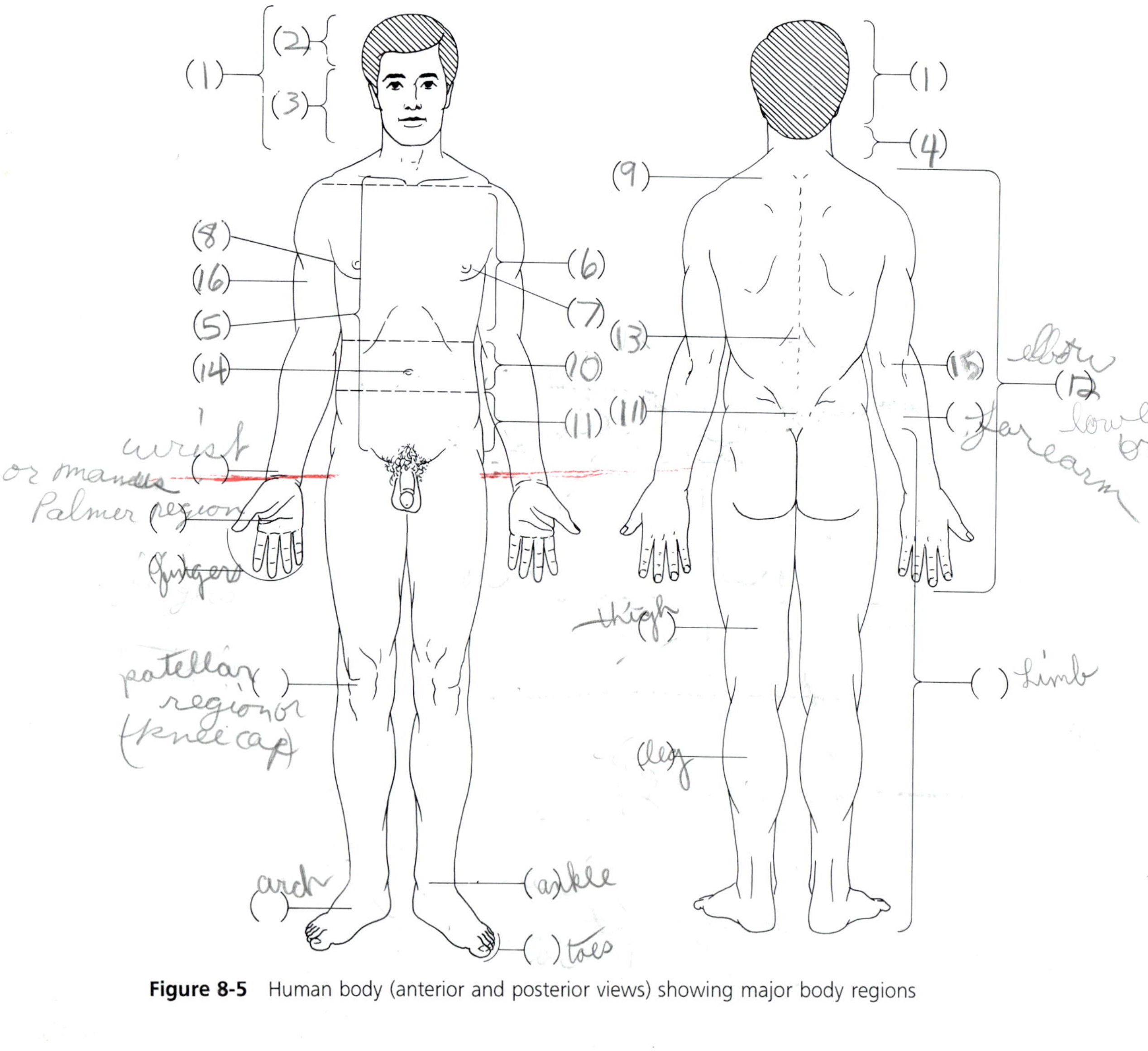

Figure 8-5 Human body (anterior and posterior views) showing major body regions

Questions for Review

EXERCISE **8**

ANATOMICAL ORIENTATION AND TERMINOLOGY

Name Kathy Martin

Laboratory Section ______ Date 1-22-91

1. Match the terms on the right with the definitions on the left.

(6) A longitudinal cut, dividing body into right and left sides	1. Cranial cavity
(8) The belly or underside	2. Coronal section
(1) Cavity containing brain	3. Posterior
(10) Transverse cut dividing body into anterior and posterior	4. Lateral
(5) Space between membranes that surround the heart	5. Pericardial cavity
(9) Term meaning on the outside of an organ	6. Sagittal section
(2) Longitudinal cut dividing body into dorsal and ventral parts	7. Mediastinum
(4) Away from or to the side of the midregion	8. Ventral
(7) Cavity between the lungs	9. Peripheral
(3) Rear end of an animal	10. Cross section

2. In which specific body cavities would the organs listed below be found? Use your textbook or a chart to locate the organs.

a. Brain cranial cavity

b. Stomach abdominal cavity

c. Heart paricardial cavity

d. Thymus gland Mediastenal cavity

e. Spinal cord spinal cavity

f. Pancreas abdominal cavity proper

g. Lungs pleural cavities

h. Esophagus mediastinal cavity

i. Uterus pelvic cavity

j. Liver abdominal cavity proper

3. Fill in the correct term for location and position.

a. The human head is superior to the neck.

b. The cat head is cranial to the neck.

c. The human neck is ___inferior___ to the head.

d. The cat neck is ___caudal___ to the head.

e. The elbow is ___proximal___ to the wrist.

f. The eye is ___lateral___ to the nose.

g. The ankle is ___distal___ to the knee.

h. The mouth is ___medial___ to the ears.

i. The tail of the cat is ___caudial___ to the trunk.

j. The vertebral column of the cat is ___horizontal___ in position.

4. On Figure Q8-1 draw the following lines.

   a. A red line to indicate a transverse section

   b. A blue line to indicate a midsagittal section

   c. A green line to indicate a parasagittal section

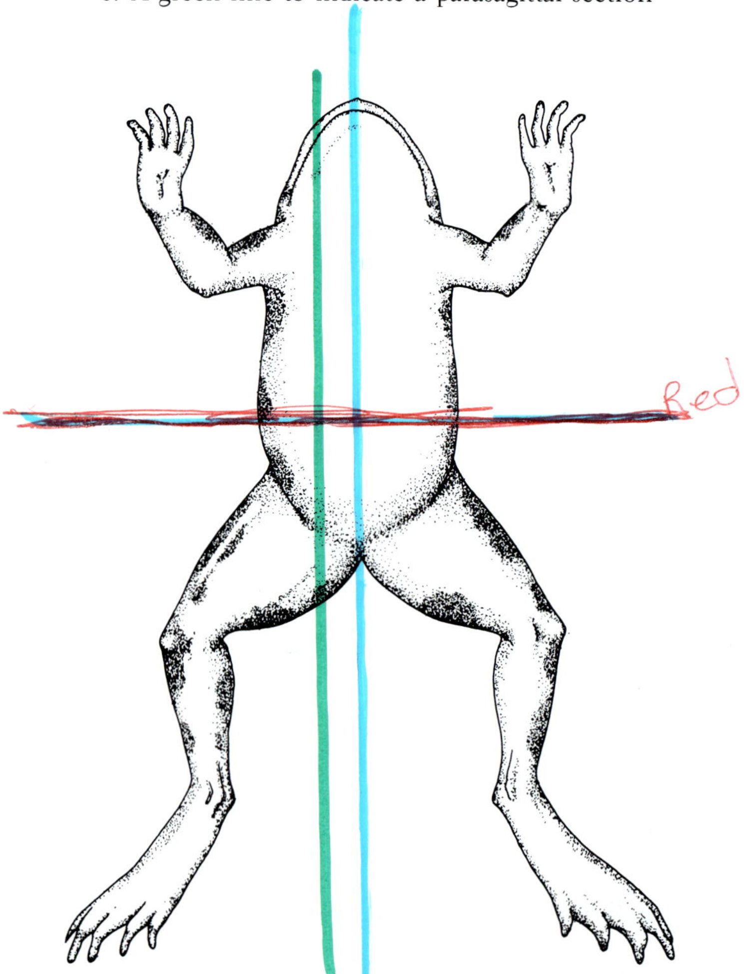

Figure Q8-1

5. To which of the major parts of the body (head, neck, trunk, arm, leg) do the following belong?

a. Navel Trunk

b. Knee leg

c. Eye head

d. Wrist arm

e. Lumbar area Trunk

f. Breast trunk

g. Ankle leg

h. Chin head

i. Pelvis trunk

j. Thigh leg

10

SECTION IV

THE INTEGUMENTARY SYSTEM

The skin and certain derivatives of the skin, such as hair, nails, and various glands, make up the integument of the human body. The skin is only a few millimeters thick but it is the body's largest organ. Within it a variety of highly specialized cells are organized into intricate structures and subsystems to perform many complex functions critical for an individual's survival. For example: the skin: (1) provides protection against infection and injury; (2) protects against too little or too much heat loss; (3) serves as the chief means of obtaining vitamin D; (4) plays a vital role in fluid and electrolyte balance; and (5) functions as an indicator of changes in the body's external environment. In addition, this organ has been recently recognized to be an integral and active component of the body's immune system.

The organization, and the structure and functions of the skin together with those of its associated derivatives are presented in this section. The experiments emphasize various properties of the skin, nails, and hair.

GENERAL REFERENCES

Edelson, T. L., and J. M. Fink, "The Immunologic Function of the Skin," *Scientific American,* 252: 46–53, 1985.

Goldsmith, L. A. (Ed.). *Biochemistry and Physiology of the Skin.* New York: Oxford University Press, 1983.

Kapit, W., R. I. Macey, and E. Meisami. *The Physiology Coloring Book.* New York: Harper & Row, 1987.

Schaumann, B., and M. Alter. *Dermatoglyphics in Medical Disorders.* New York: Springer-Verlag, 1976.

Tortora, G. J., and N. P. Anagnostakos. *Principles of Anatomy and Physiology,* 6th ed. New York: Harper & Row, 1990.

SPECIFIC REFERENCES

Beutner, E. H., T. P. Chorzelski, and S. F. Bean (Eds.). *Immunopathology of the Skin.* New York: Wiley, 1979.

Koss, L. G. *Diagnostic Cytology and Its Histopathologic Bases,* 3rd ed. Philadelphia: Lippincott, 1979.

Streilein, J. W. "Skin-Associated Lymphoid Tissue (SALT): Origins and Functions," *Journal of Investigative Dermatology,* 80: 12–16, 1983.

EXERCISE 9

THE SKIN AND EPIDERMAL DERIVATIVES

The outer surface of the human body is covered with a protective layer of skin that includes such modified derivatives as nails and hair. The skin is composed of two distinct layers—the superficial layer, known as the *epidermis*, and the deeper layer, called the *dermis* or *corium*. A subcutaneous layer of tissues is located below the dermis.

The dermis is composed of connective tissue interlaced with blood vessels, fat, glands, hair follicles, and nerves. Fibers from the dermis serve to attach the skin to the subcutaneous layer, which is firmly attached to underlying structures. The subcutaneous layer nourishes, supports, and cushions the skin. The general structure and organization of the skin are shown in Figure 9-1.

The thickness of the skin and of the individual layers varies in different parts of the body. For example, it is thickest on the back, palms, and soles. The thinner skin areas of the body occur on the tympanic (eardrum) membrane and over the eyelids.

In the areas where the epidermis is thickest it is made up of five *strata* or regions. The outermost region, the *stratum corneum*, consists of numerous layers of dead, scalelike cells composed of proteins called *keratins* (Figure 9-1b). Cells of this region are constantly being sloughed or lost by a process called *exfoliation*. The deepest epidermal region known as the *stratum basale* or *stratum germinativum* consists of a single layer of columnar cells undergoing division. Newly formed cells are pushed upward into each successive layer and are eventually sloughed. The epidermis does not contain blood vessels and so depends on the dermis for nutrients.

Skin color depends on the presence of the yellow pigment *carotene* in the surface layer, the varying concentrations of the brown pigment *melanin* throughout epidermal layers, and the color of the circulating blood in the dermis. The so-called flush appearance is influenced by the degree of oxygenation of the blood and the extent to which tiny blood vessels open up or close (peripheral blood flow). When the skin is exposed to heat, numerous small blood vessels open up and more blood flows through the tissue. This blood is slightly warmer than usual because it picks up more heat from the warmed tissue. The cooling of the skin causes several arterioles and capillaries to constrict, which reduces the net amount of blood flowing through the tissue.

The entire surface of the skin of humans, in contrast to that of most other mammals, is pitted and grooved by many intersecting lines that form geometric patterns characteristic of each specific area. No area is absolutely free of such markings. These lines, craters, and various microscopic features are stamped indelibly on the skin surface. The palm (palmar) and sole (plantar) surfaces of an individual characteristically consist of alternating ridges and grooves, deeper and closer together and more clearly seen than those elsewhere on the body. These markings, referred to as the true *dermatoglyphics* (from *derma*, "skin," and *glyphics*, "carvings"), are nearly indestructible and are different in every individual. The ridge patterns on fingertips are divided into three basic digital groups: *arches, loops,* and *whorls* (Figure 9-2). The simplest pattern found on the fingertips is the arch, while the most common is the loop. These markings increase the surface area of the body, which is an adaptation for a more efficient dissipation of heat. In addition, dermatoglyphics are obviously adapted to give the hand and foot a better grip.

Several rules, principles, and definitions have been proposed that allow reliable analyses of dermatoglyphics. In criminology investigation the intricate details of ridge patterns serve as the bases for fingerprint identifications. In addition, relationships have been reported between unusual dermatoglyphic features and certain genetic abnormalities. Unusual ridge configurations have been shown to exist in individuals with certain chromosomal defects, in persons with single-gene disorders, and in some in whom the genetic basis of disorder is unclear.

Normally every person has hair on most areas of the skin surface except the palms and soles. Each individual hair is the product of the epidermis and develops from hair *follicles.* The free portion of a hair is referred to as the *shaft* and the part within the follicle is called

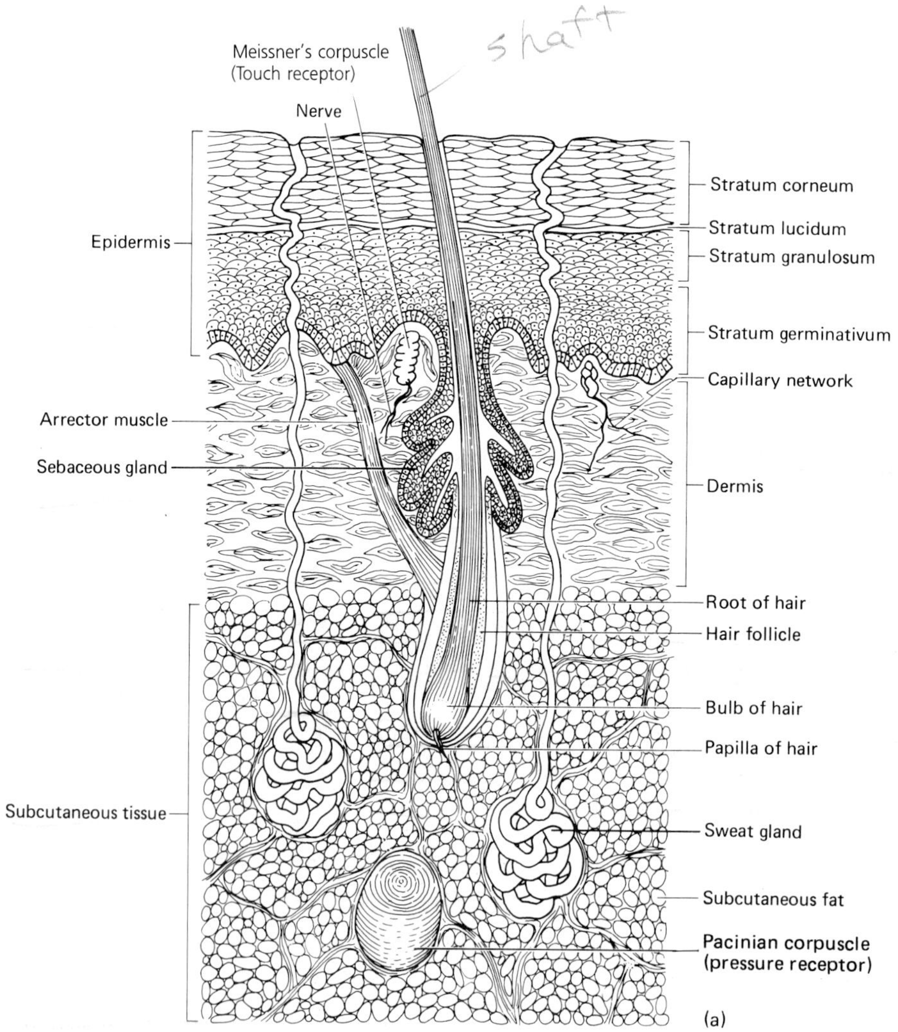

Figure 9-1a Microscopic features of human skin showing its most important components. The layers of skin showing a longitudinal section through a hair shaft.

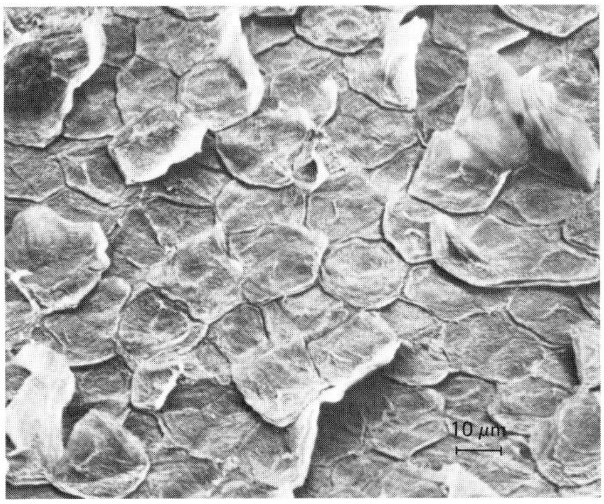

(b)

Figure 9-1b A scanning micrograph of the skin surface showing the uniform appearance and size of stratum corneum cells. (From F. C. Tring and R. I. Jolly, *Acta Dermatovener* (Stockholm), 53:343–346, 1973.)

the *root* (see Figure 9-1). A hair shaft consists of (1) an inner layer, the *medulla*, composed of loosely arranged, soft, horny cells and several air spaces; (2) a layer surrounding the medulla, the *cortex*, composed of hard, closely placed, horny cells; and (3) an outer layer, the cuticle, composed of hard horny cells tightly joined (Figure 9-3). Individual hairs have one or several oil (sebaceous) glands whose ducts empty into the follicle. Smooth muscles, called *arrector pili*, are attached to the follicle below the sebaceous glands and can move the shaft in response to various emotions and temperatures. "Goose pimples" result from hair shaft movements.

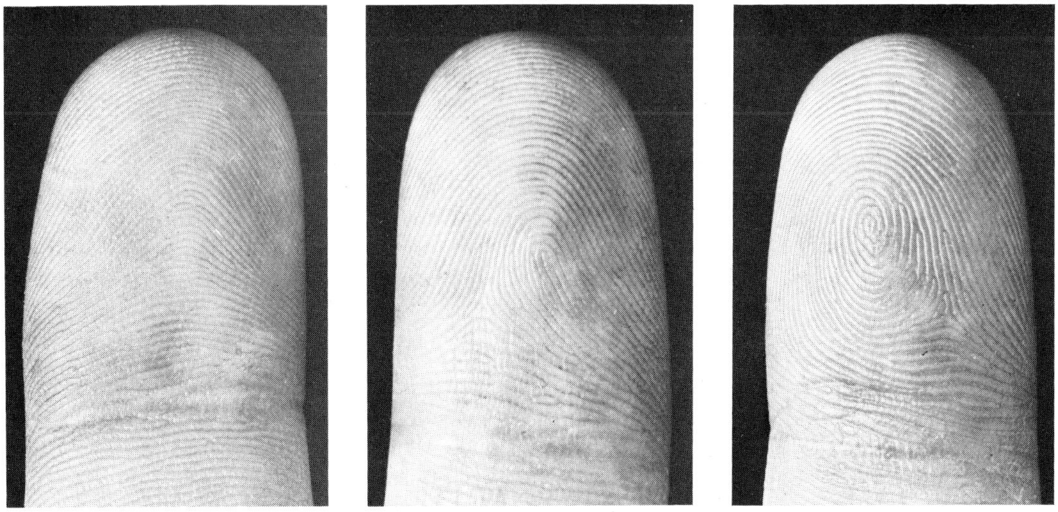

(a) (b) (c)

Figure 9-2 Dermatoglyphics. Photographs demonstrating the three basic digital patterns: (a) arch, (b) loop, and (c) whorl. (From J. R. Miller and J. Giroux, *Journal of Pediatrics*, 69:302–312, 1966.)

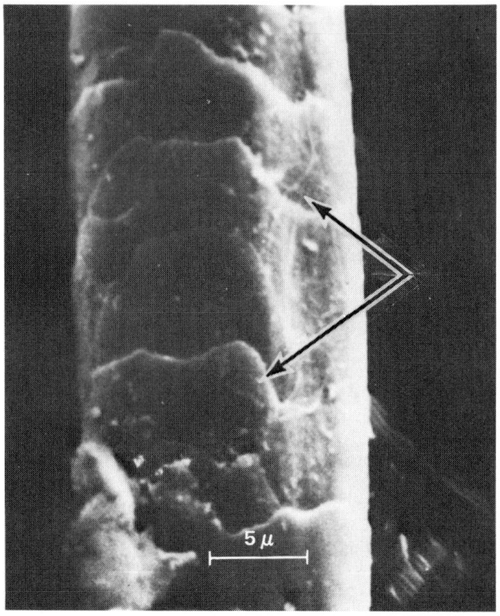

Figure 9-3 The surface appearance of hair. A scanning electron micrograph of the hair shaft showing the cuticular scales (arrows) on its surface. (from D. M. Singh, W. B. Green, R. A. Osborne, J. Upshur, and G. R. Hennigar, *Journal of Invest. Dermatology,* 63:334–336, 1974.)

Adult nails consist of a plate of horny keratinized cells situated on a *nail bed* over the terminal phalanges (bones) of the fingers and toes (Figure 9-4a). The visible portion of the nail *(nail body)* is bordered laterally and proximally by a fold of skin called the *nail fold.* A *nail groove* lies between this fold and the nail. The portion of the nail hidden under the nail fold is known as the *root.*

Other noticeable features include the *lunula,* a white semicircle at the base of each nail, and the cuticle or *eponychium,* which is the region where the base of the nail connects to the finger or toe. The distal end of the finger or toe under the free edge of the nail is the *hyponychium.* The characteristic pink color of the nail is due to the blood vessels that show through the semitransparent nail structure.

The procedures in this exercise emphasize the external structure of nails, the dermatoglyphics of fingertips, and the microscopic properties of skin and hair.

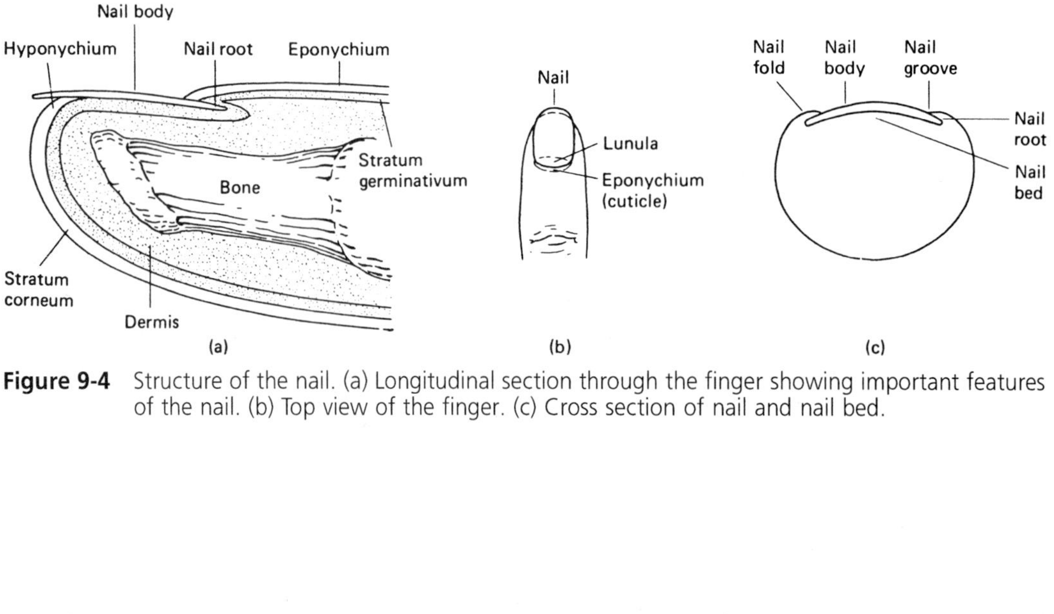

Figure 9-4 Structure of the nail. (a) Longitudinal section through the finger showing important features of the nail. (b) Top view of the finger. (c) Cross section of nail and nail bed.

A. EXAMINATION OF FINGERS AND FINGERPRINTS

Materials

The following materials should be provided for class use.

1. Printer's ink
2. One ink roller
3. Two glass or metal slabs or ordinary stamp pads
4. Soap (cleanser and paper towels)

Procedure 1: External Structure of Nails

This procedure is to be performed by students in pairs.

1. Examine the fingers of your partner's right hand.
2. Locate all parts of the nail shown in Figure 9-4.

Procedure 2: Fingerprints

This procedure is to be performed by students in pairs.

1. Put a small amount of ink on an inking slab. Spread it with a roller into a thin, even film.
2. Place one edge of the thumb of your right hand against the inked slab. Allow your thumb to roll naturally across the slab to its opposite edge, so that the entire area to be printed is covered with ink.
3. Repeat the movement with thumb on the space provided in the results and observations section. The sheet of paper on which the inked finger is pressed must rest on a firm surface.
4. One at a time, place the remaining fingers of your hand directly down on the inked slab and press them directly onto the spaces provided in the results and observations section.
5. Repeat steps 2–4 with your partner's right hand.
6. Compare the fingerprints obtained and answer the questions in the results and observations section.

B. MICROSCOPIC IDENTIFICATION OF THE SKIN AND HAIR

Materials

The following materials should be provided for class use.

1. Prepared slides of
   a. Human skin showing both epidermis and dermis
   b. Human scalp showing hair follicles
2. Glass slides and cover slips
3. Eyedroppers

Procedure

This procedure is to be performed by students individually.

1. Examine the prepared slides of human skin.
2. With the aid of Figure 9-1, Color Plates 15 and 16, and your textbook identify the specific layers of the epidermis and the various glands typically found in the dermis.
3. Examine the preparation of human scalp and identify the components of a hair follicle (see Color Plate 16).

4. Remove a hair from your scalp and one from your partner's scalp. Prepare a wet mount. (Refer to Exercise 2 for directions.)
5. Examine the hair preparations under both low power and high power.
6. Sketch the appearance of both hairs in the space provided in the results and observations section.
7. Look for any unusual surface features, such as scales and patterns.
8. Enter your findings in the results and observations section and answer items in the questions for review pertaining to this exercise.

RESULTS AND OBSERVATIONS

FINGERPRINTS

1. Student's name: Kathy

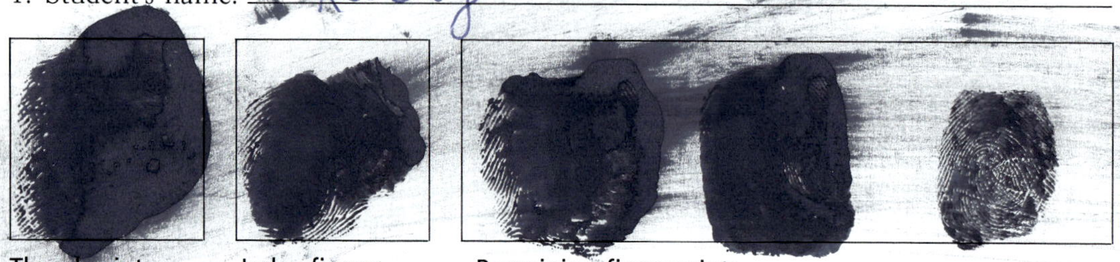

Thumbprint Index finger Remaining fingerprints

2. Partner's name: Becky

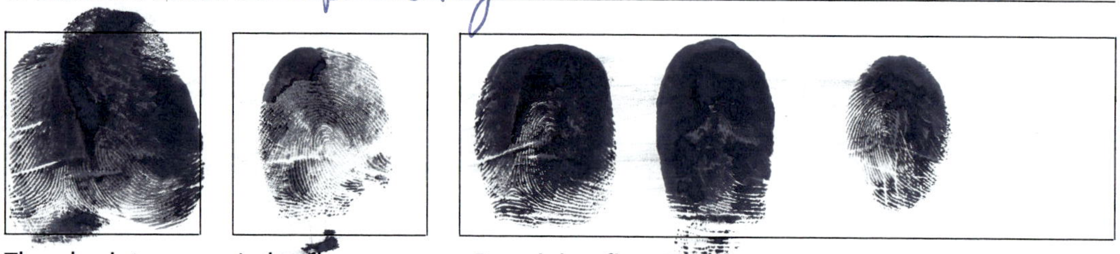

Thumbprint Index finger Remaining fingerprints

3. What type of fingerprint pattern did your fingers exhibit? (Refer to Figure 9-2.)

whorl whorl, loop whorl whorl

4. What type of fingerprint pattern did your partner's fingers exhibit?

loop, loop, whorl, loop loop

5. Were the patterns the same for both individuals?

No

6. Were there any unusual features to the patterns obtained? If so, encircle them on the specific fingerprint.

MICROSCOPIC EXAMINATION OF THE HAIR

1. Hair sketch

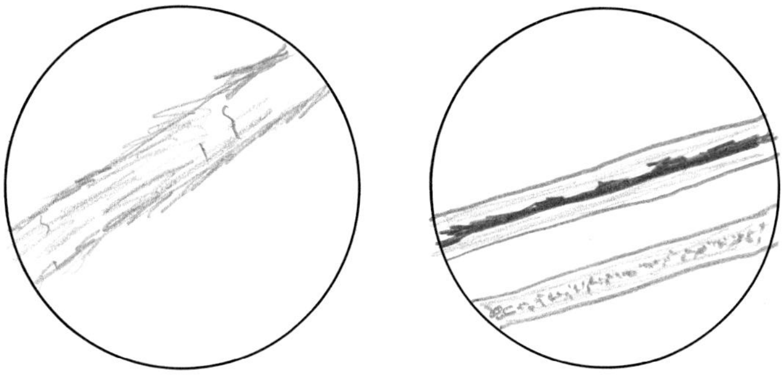

Your hair　　　　Partner's hair

2. Describe any unusual surface features.

having split, dk line in middle

Scalpi

Questions for Review

EXERCISE **9**

THE SKIN AND EPIDERMAL DERIVATIVES

Name Kathy Martin

Laboratory Section Mon. 8:00 Am Date 2-29-91

1. Identify the parts of the skin in Figure Q9-1.

a. hair shaft h. sweat gland

b. sebaceous gland i. dermis

c. stratum corneum j. subcutanious tissue

d. stratum lucidum k. arrector muscle

e. epidermis l. hair follicle

f. Meissners corpuscle m. bulb of hair

g. capillary network n. hair follicle

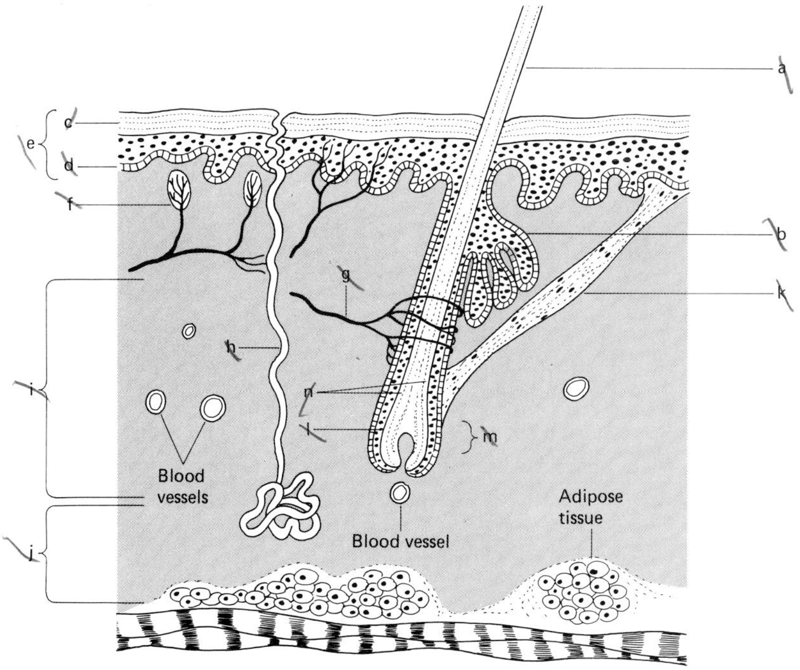

Figure Q9-1

2. Identify the parts of the hair follicle in Figure Q9-2.

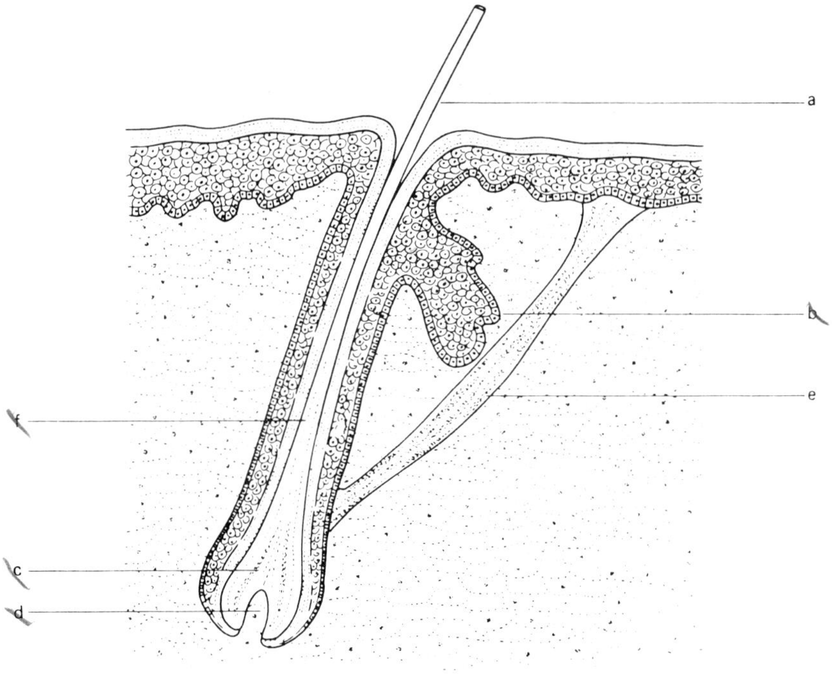

Figure Q9-2

a. shaft
b. sebaceous gland
c. bulb of hair
d. papilla of hair
e. hair follicle
f. root of hair

3. The fingerprint in Figure Q9-3a was found on a stolen object. The fingerprints of three suspects (b, c, d) were taken and compared.

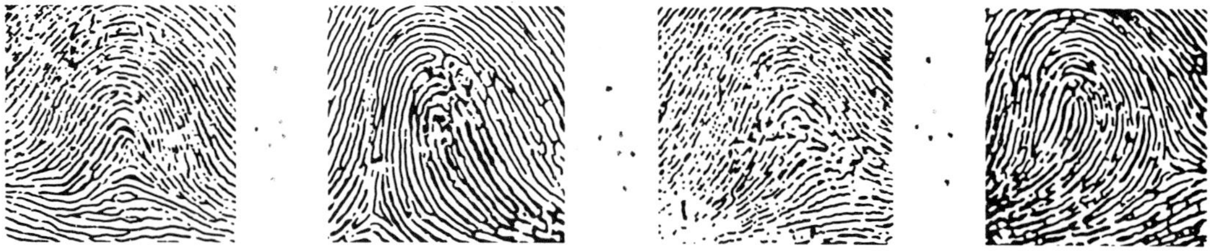

(a) (b) (c) (d)

Figure Q9-3

a. On the basis of this evidence which of the three individuals would you suspect?

b

b. What are the digital patterns shown for each fingerprint?

a. arch
b. loop
c. arch
d. whorl

EXERCISE 10

INTRODUCTION TO THE STUDY OF THE SKELETON

MICROSCOPIC STRUCTURE OF BONE

Examine a demonstration slide showing a section through human compact bone. You will see that bone is made up of numerous **Haversian systems** (see Figure 10-1 and review the work done in Exercise 7). These systems are very active. They develop quickly, pushing other older Haversian systems out of the way. As a result you see many Haversian system fragments on the slide. These are called **interstitial lamellae.** In the center of each system is the **Haversian canal**, containing blood vessels and nerves, and surrounding the canal are concentric rings of bone called **lamellae.** Trapped between the lamellae are the bone cells or **osteocytes** in their **lacunae.** Their long cytoplasmic extensions named **canaliculi** extend toward other bone cells or the Haversian canal ensuring an intricate connecting mechanism.

MACROSCOPIC APPEARANCE OF BONE

If a long bone that has been split lengthwise is available, study the parts. If no specimen is available, use Figure 10-2 to locate the following:

1. **Head** or **epiphysis** (ē-PIF-i-sis)—the end of the bone beyond the epiphyseal (ep′-i-FIZ-ē-al) line sometimes visible on the outside of the bone. A layer of compact bone covers the inner spongy bone. The spongy (cancellous) bone contains the *red bone marrow,* which is the site of blood cell formation. A layer of *articular cartilage* covers the head where it is part of a joint.
2. **Shaft** or **diaphysis** (dī-AF-i-sis)—the part of the long bone between the two heads. The bone shaft is composed of compact bone and a hollow center, called the **marrow cavity** or **medullary cavity**, which contains **yellow bone marrow** in the living specimen.
3. **Periosteum** (pair-ē-OS-tē-um)—a double layered membrane surrounding the outside of a bone (silvery in a freshly killed specimen, but absent in the dried bone). This membrane is active in producing more bone on the outer surface.
4. **Endosteum** (end-OS-tē-um)—a membrane lining the medullary cavity. It is similar to the periosteum and produces bone on the inner surface. (It, too, cannot be seen in the dried specimen.)

MAJOR DIVISIONS OF THE SKELETON

The bones of the human skeleton are grouped into two major divisions, as shown in Figure 10-3.

1. **Axial skeleton** (shaded)—the bones located along the midline axis of the body
2. **Appendicular skeleton** (unshaded)—the bones that make up the appendages and the girdles

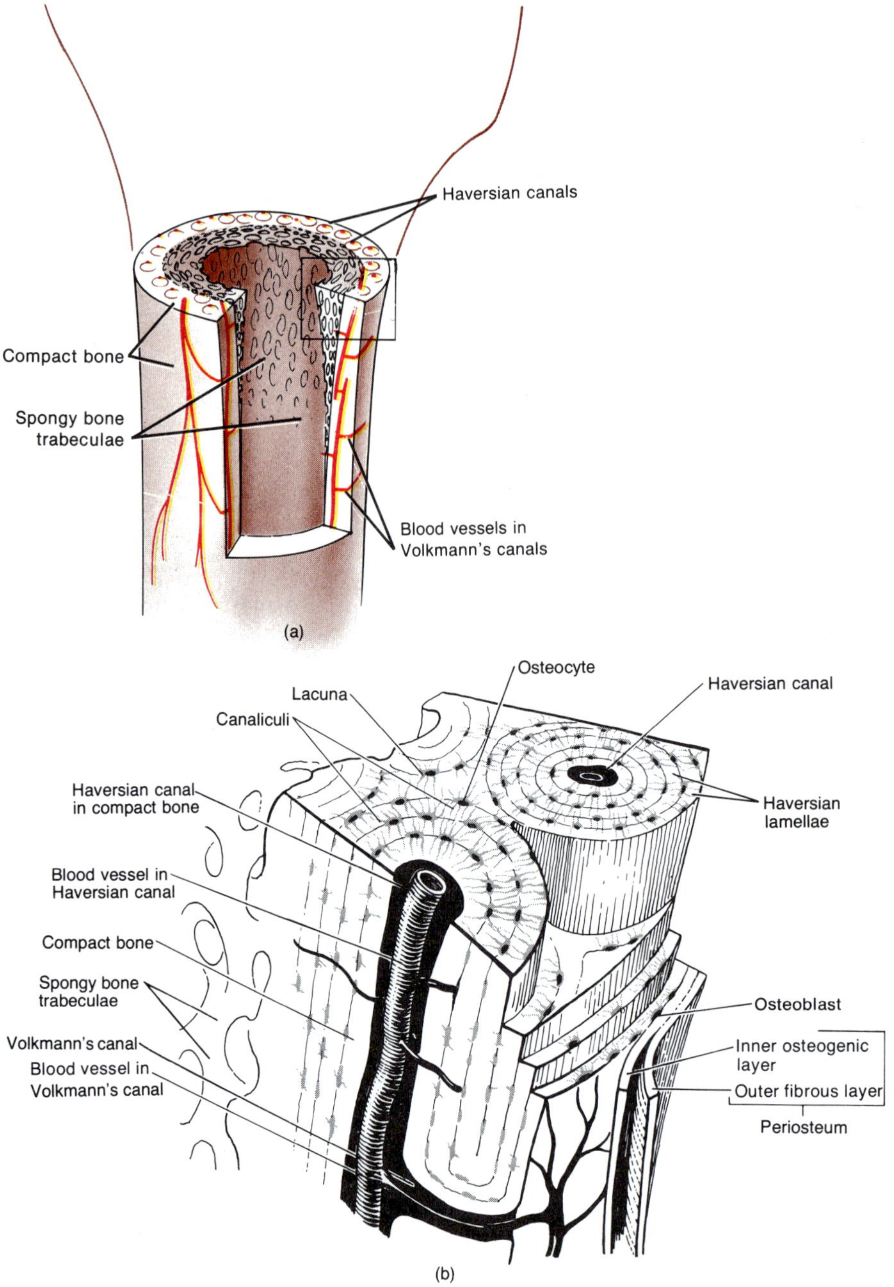

Figure 10-1 Haversian systems in compact bone. (a) Structure of bone. (b) enlarged view of Haversian systems.

Each major division is further subdivided into regions. With the help of Figure 10-3 look over the mounted skeleton and identify as many bones as you can. Be sure you have a clear idea of the bones in each subdivision before beginning detailed study in Exercises 11–14.

AXIAL SKELETON

1. **Skull**—*cranial* and *facial bones*
2. **Visceral skeleton**—derivatives of the embryonic visceral or branchial (gill) arches
3. **Vertebral column**—the vertebrae, series of bones in the mid-dorsal region inferior to the skull

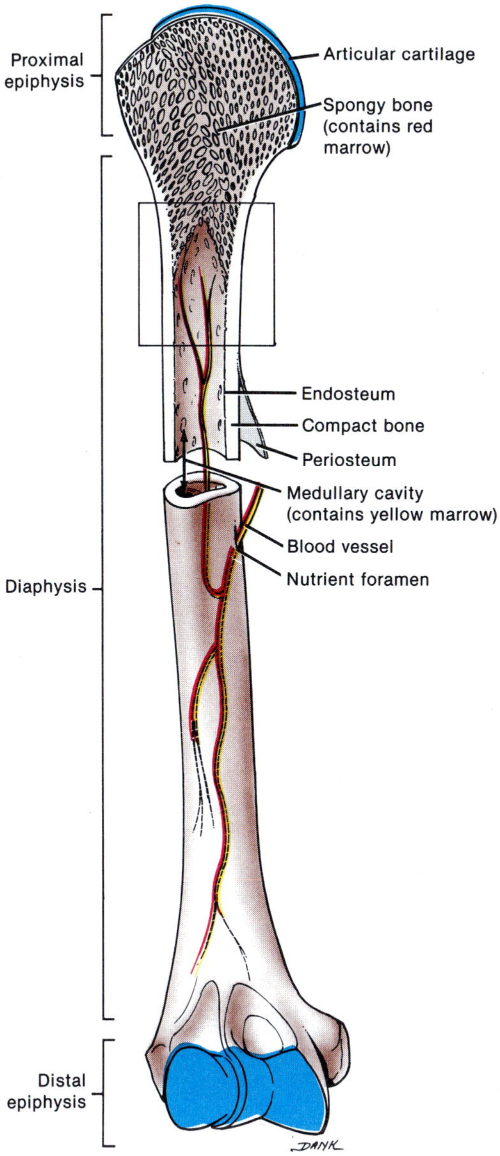

Figure 10-2 Osseous tissue. Macroscopic appearance of a long bone that has been partially sectioned.

4. **Sternum**—commonly known as the "breast bone," located in the midventral or anterior thoracic region
5. **Ribs**—12 pairs of widely separated bones in the thoracic region

APPENDICULAR SKELETON

1. **Pectoral (shoulder) girdle**—two bones located in the shoulder that form a bony socket for articulation with the arm
2. **Upper extremity**—series of bones beginning at the pectoral girdle that make up the arm
3. **Pelvic girdle** *(os coxae)*—fused bones that form a socket for articulation with the leg
4. **Lower extremity**—series of bones beginning at the pelvic girdle that make up the leg

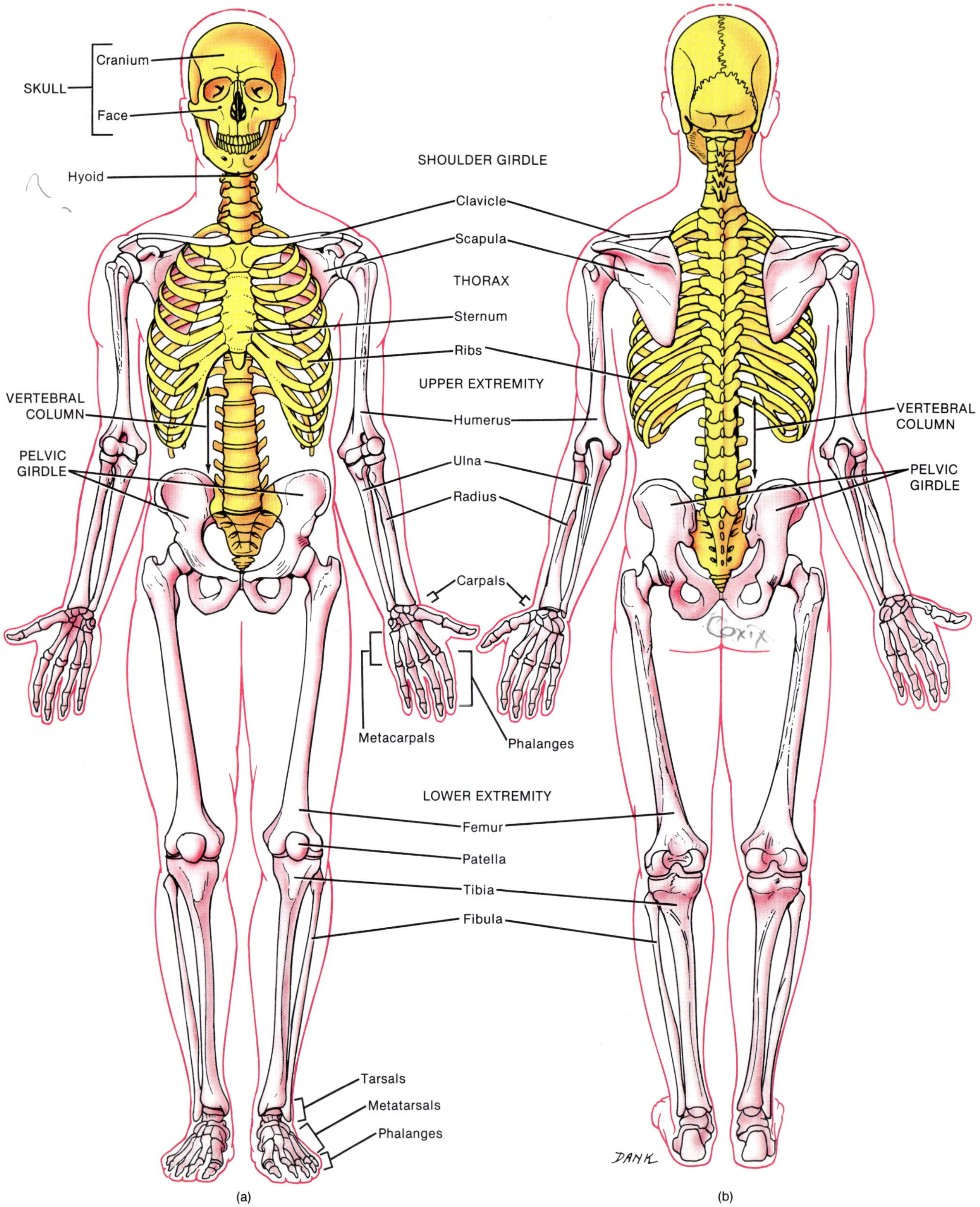

Figure 10-3 Division of the skeletal system. The axial skeleton is indicated in yellow. (a) Anterior view. (b) Posterior view.

Questions for Review

EXERCISE **10**

INTRODUCTION TO THE STUDY OF THE SKELETON

Name Kathy Martin

Laboratory Section Monday AM Date 3/18/91

1. Name the divisions of the axial skeleton.
   a. sKull
   b. Visceral skeleton - Hyoid
   c. Vertebral column
   d. Sternum
   e. Ribs
2. Name the divisions of the appendicular skeleton.
   a. Pectoral girdle
   b. Upper Extremity
   c. Pelvic girdle
   d. Lower Extremity
3. What characteristic must a bone have to be placed in the axial skeleton?

   the bones located along the midline axis of the body
4. What characteristic must a bone have to be placed in the appendicular skeleton?

   the bones that make up the appendages and girdles.
5. Pronounce the names of the bones described in this exercise at least two times or until you do not have any difficulty with them. Use this method throughout the laboratory exercises whenever large numbers of unfamiliar words are encountered.

EXERCISE 11

THE SKULL (PART I)

Materials

The following materials should be provided for class use.

1. Mounted human skeleton
2. Human skull
3. Mounted disarticulated skull (Beauchene skull)
4. Disarticulated skull bones
5. Colored pencils

The bones of the skull are subdivided into *cranial bones* and *facial bones.* Figures 11-1 through 11-6 provide various views and labels and are color coded. Use them along with the laboratory specimens so that you will be able to identify the actual bones and not just pictures of bones.

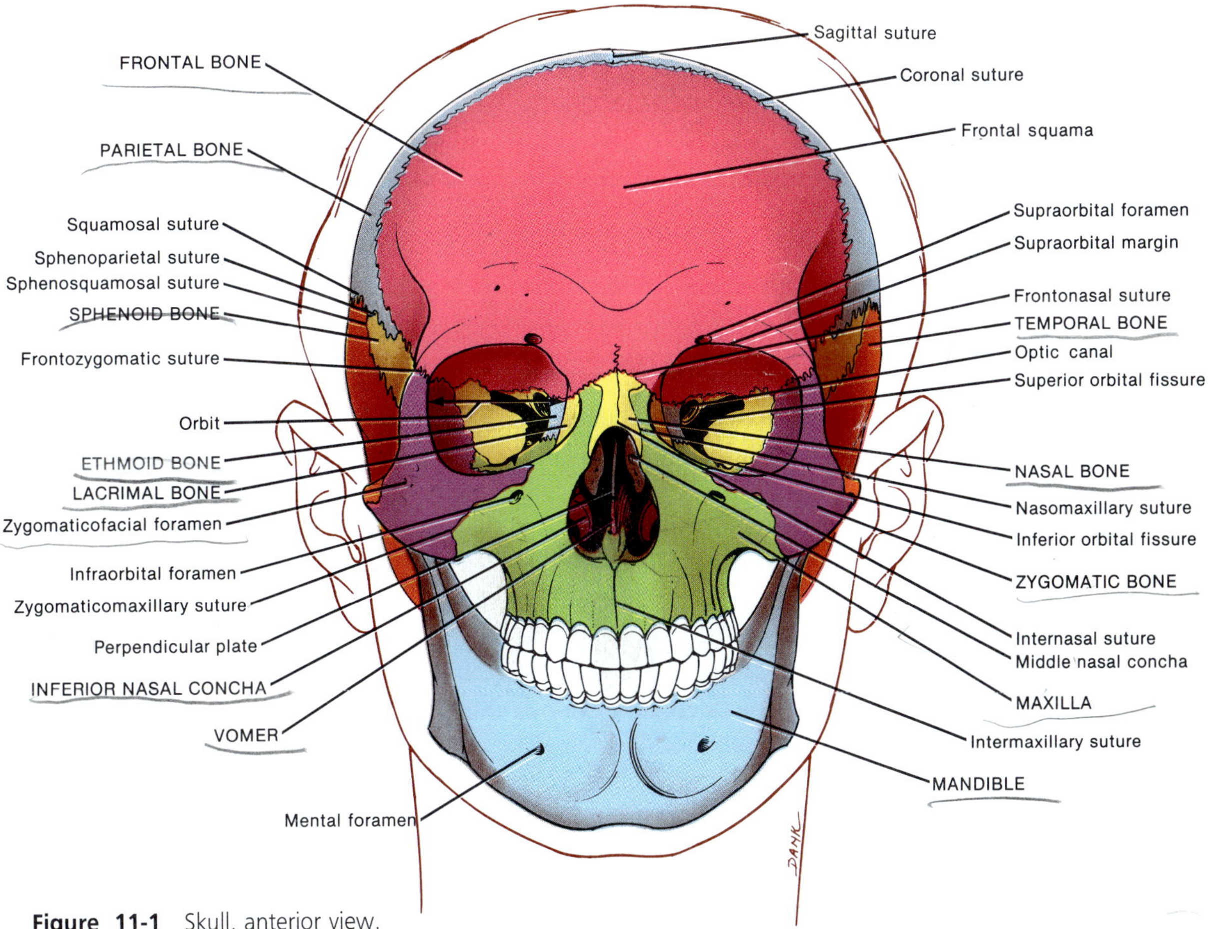

Figure 11-1 Skull, anterior view.

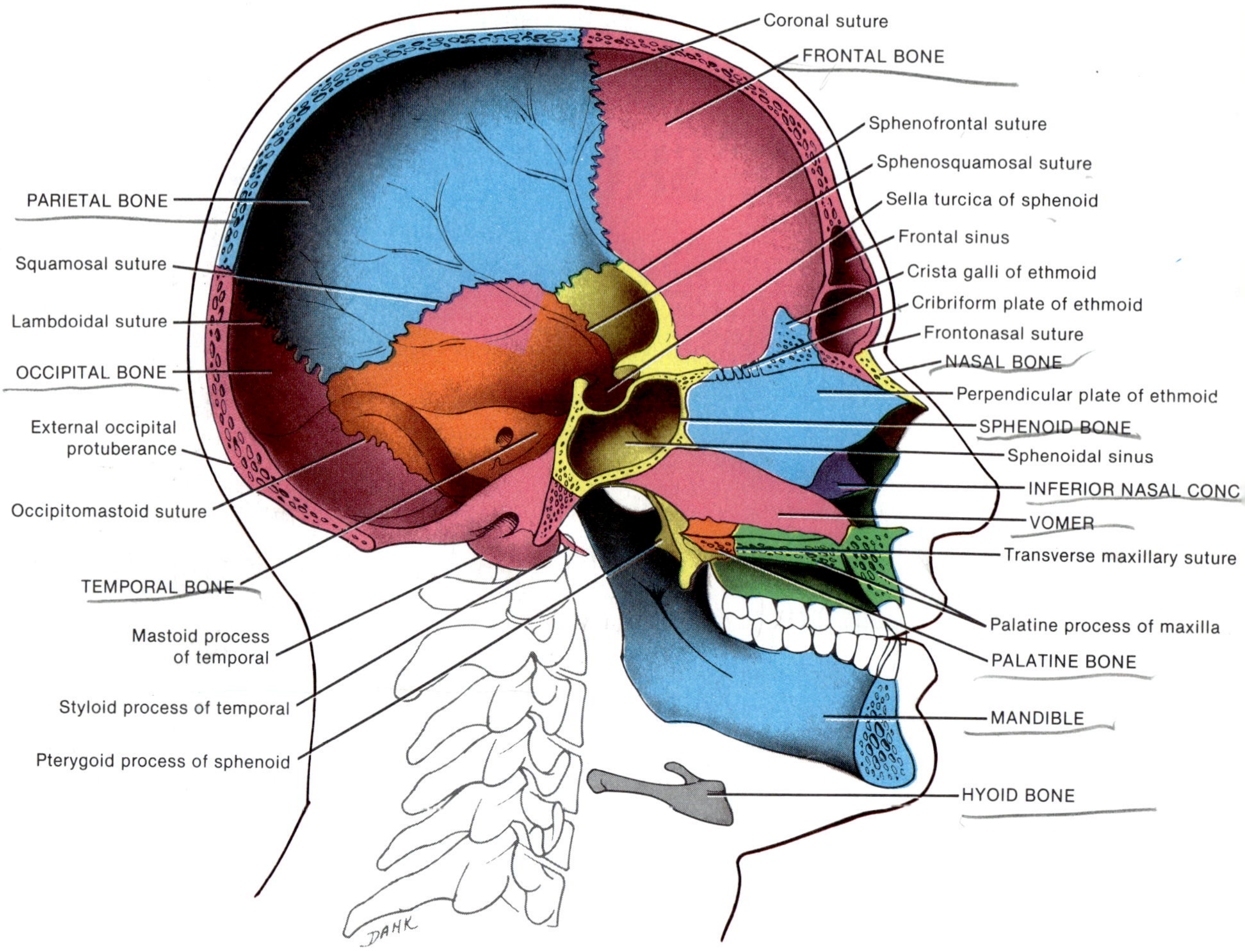

Figure 11-2 Skull, median view

A. THE CRANIAL BONES

The eight bones that surround and protect the brain form the brain case or *cranium.* Be able to identify these bones on the skulls in the laboratory. Always learn the correct pronunciation and spelling.

1. *Frontal bone* (FRUN-tal)—large, single bone that forms the forehead
2. *Parietal bones* (pa-RĪ-i-tal)—paired bones that make up most of the top of the head posterior to the frontal bone
3. *Occipital bone* (ok-SIP-i-tal)—large, single bone at the back of the head posterior to the parietals
4. *Temporal bones* (TEM-por-al)—paired bones at the sides of the skull inferior to the parietals
5. *Sphenoid bone* (SFĒ-noyd)—large, single bone that forms most of the floor of the cranium
6. *Ethmoid bone* (ETH-moyd)—single bone on the inner surface of the cranium anterior to the sphenoid

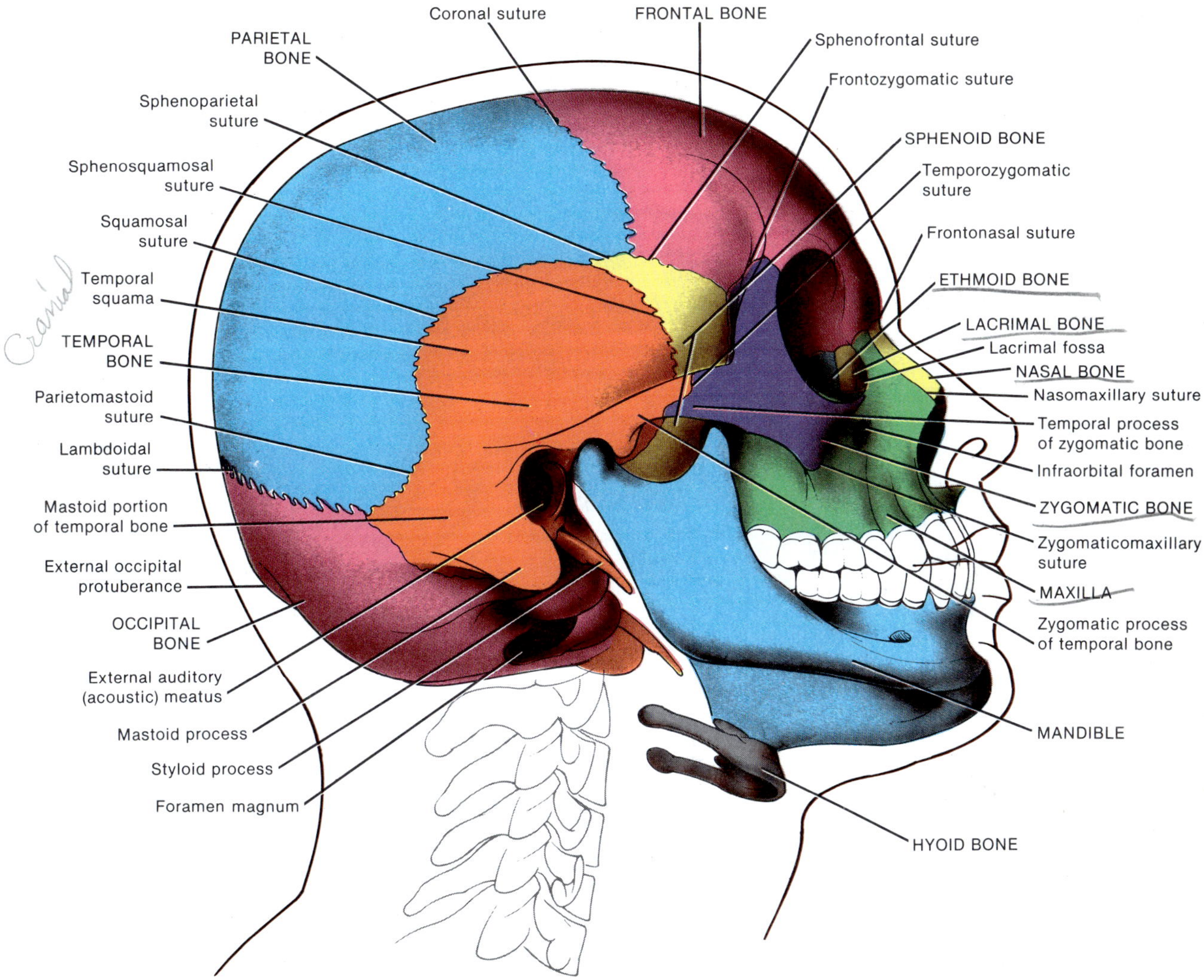

Figure 11-3 Skull, right lateral view.

B. THE FACIAL BONES

The remainder of the skull consists of the 14 bones we call the face.

1. *Nasal bones* (NĀ-zal)—small, paired bones that constitute the bridge of the nose (Other skeletal structure within the nose is cartilage and is not preserved on this skeleton.)
2. *Maxillary bones* (MAK-si-lair-ē)—large, paired bones that form the upper jaw
3. *Lacrimal bones* (LAK-ri-mal)—small, paired bones located within the orbit of the eye and posterior to the uppermost portion of the maxillaries
4. *Zygomatic bones* (zī-gō-MAT-ik)—paired bones that articulate with the maxillaries, commonly known as the cheek bones
5. *Palatine bones* (PAL-a-tīn)—paired bones located in the mouth cavity posterior to the maxillaries and forming part of the hard palate
6. *Vomer bone* (VŌ-mer)—single, dagger-shaped bone located in the center of the nasal cavity superior to the maxillaries
7. *Inferior nasal concha bones* (KONG-ka)—paired, flat bones that form the lateral walls of the nasal cavity

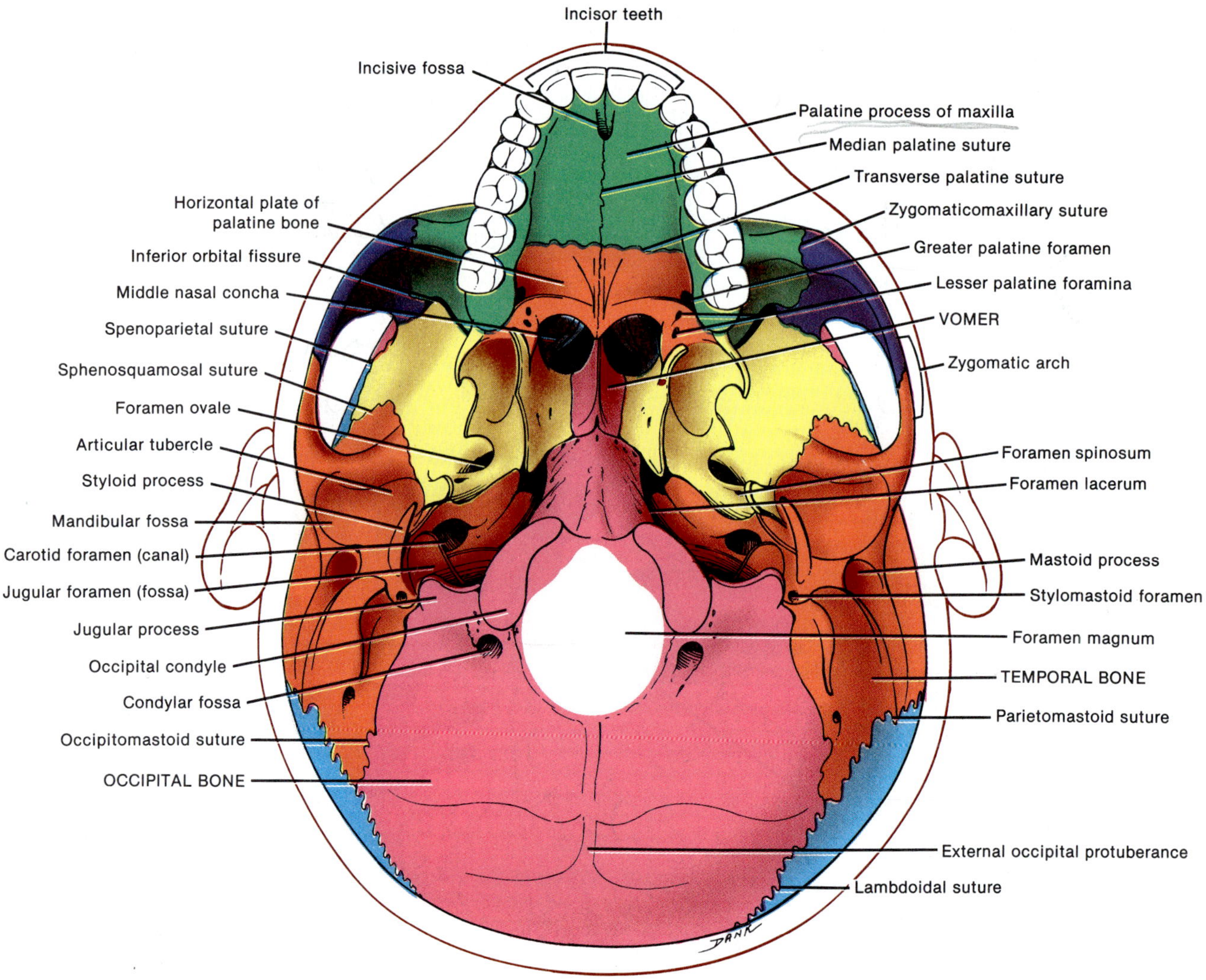

Figure 11-4 Skull, inferior view.

8. *Mandible bone* (MAN-di-bul)—single bone that forms the lower jaw, bearing all the teeth in the lower jaw

C. THE SKULL SUTURES

Continue your examination of the skull. Bones come together at lines that are called *sutures* (SOO-churs). A suture is an immovable joint in which bones with serrated edges fit together, separated by a thin layer of fibrous connective tissue. Joints, ar articulations, will be taken up in a later exercise, but it is appropriate to name the major skull sutures at this time.

1. *Coronal* (kō-RŌ-nal)—suture between the frontal and parietal bones
2. *Sagittal* (SAJ-i-tal)—suture extending from the coronal suture in the midline and separating the two parietal bones
3. *Squamosal* (skwa-MŌ-sal)—suture extending around the temporal bone and separating it from the parietal and sphenoid bones
4. *Lambdoidal* (lam-DOY-dal)—suture separating the occipital bone from the parietals

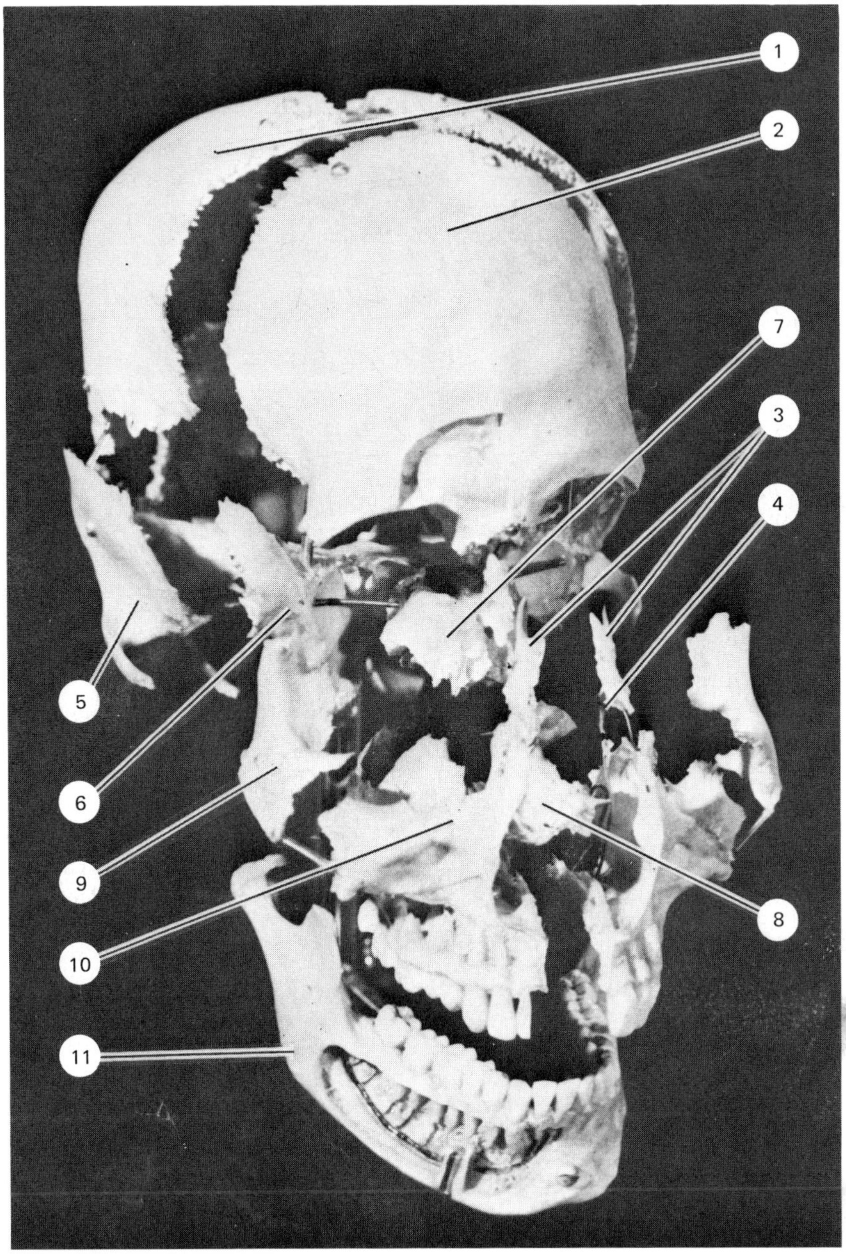

Figure 11-5 Beauchene skull.
1. Parietal
2. Frontal
3. Nasal
4. Lacrimal
5. Temporal
6. Sphenoid
7. Ethmoid
8. Vomer
9. Zygomatic
10. Maxillary
11. Mandible

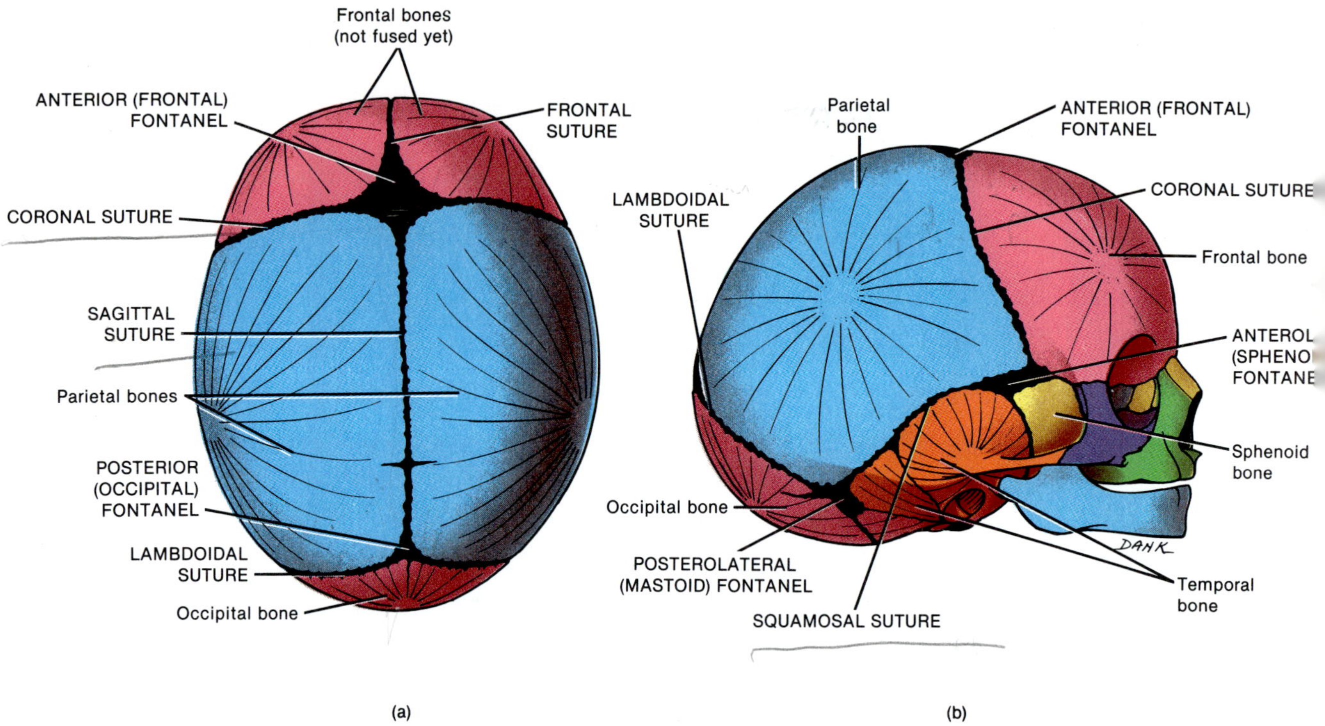

Figure 11-6 Infant Skull. (a) Superior view. (b) Lateral view.

D. THE INFANT SKULL

Examine Figure 11-6 and other demonstration materials. Note the absence of teeth and general fragility of the bones. It is obvious that the infant skull is still in the process of developing and maturing. Look for the following features.

1. Locate the suture separating the paired frontal bones. In the fully developed skull, the bones have fused to form a single frontal bone and the suture has disappeared.
2. The *fontanels* or "soft spots" are places where the bone has not yet ossified over the membrane that gives rise to it. Locate the *anterior fontanel,* the *posterior fontanel,* and the smaller, paired posterolateral and anterolateral fontanels.

Questions for Review

EXERCISE **11**

THE SKULL (PART I)

Name Kathy Martin

Laboratory Section Mon AM Date 3/18-91

1. Match the phrase on the left with the numbered choices on the right.

a. 9 Total number of cranial bones

b. 13 Dagger-shaped bone in nasal cavity

c. 14 Bone that forms the forehead

d. 11 The cheek bone

e. 2 Total number of facial bones

f. 15 Bone that forms bridge of the nose

g. 4 Bone that forms back of skull

h. 10 One bone that makes up the hard palate

i. 6 Another bone that makes up the hard palate

j. 3 Bone that makes up the lower jaw

1. Lacrimal
2. 14
3. Mandible
4. Occipital
5. 20
6. Maxillary
7. Sphenoid
8. Parietal
9. 8
10. Palatine
11. Zygomatic
12. Inferior nasal concha
13. Vomer
14. Frontal
15. Nasal

Foramen

2. Identify the bones in Figure Q11-1. You may color them if you wish. (If you do use color, be sure to use the same color for each bone in all the drawings.)

a. optical	h. vomar
b. parietal	i. lacrimal bone
c. Frontal	j. zygomatic bone
d. temporal	k. coronal suture
e. ethmoid	l. maxilla
f. sphenoid bone	m. mandible
g. nasal bone	n. inferior nasal concha

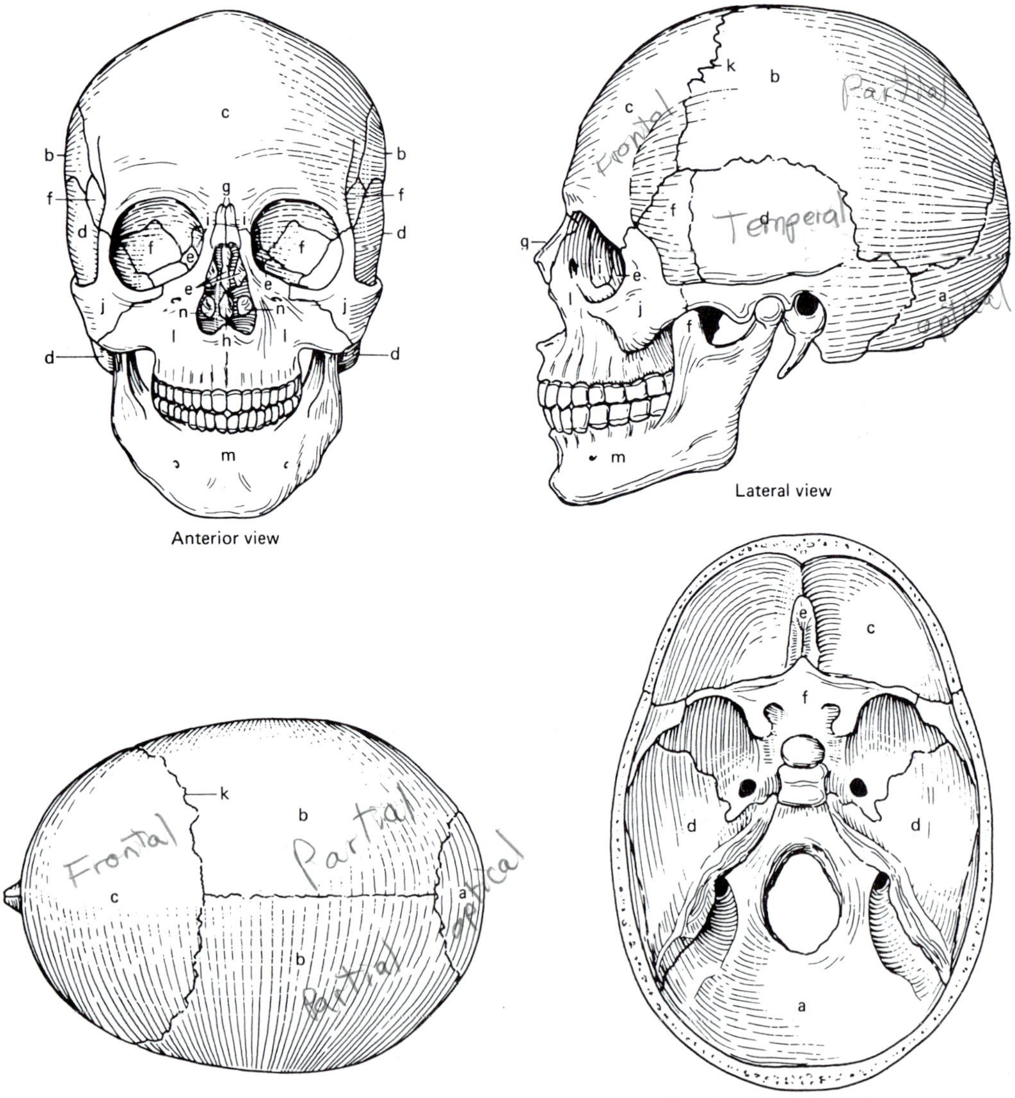

Figure Q11-1

10

EXERCISE 12

THE SKULL (PART II) AND THE VISCERAL SKELETON

Materials

The following materials should be provided for class use.

1. Mounted human skeleton
2. Human skull
3. Beauchene skull
4. Disarticulated skull bones
5. Demonstration of ear ossicles

A. SPECIAL FEATURES OF SKULL BONES

Special surface features of bones are called markings. They include prominent or projecting parts (e.g. processes, facets, crests, and spines); depressions, or *fossae;* openings, or *foramina;* grooves, or *sulci;* and enclosed cavities, or *sinuses.* Some of the more important markings of the skull bones are listed below.

SPHENOID BONE

Situated at the base of the skull and forming most of the base of the cranium, this very complex bone articulates with the temporals, frontal, occipital, and ethmoid. Locate the following on a separate sphenoid bone or on the Beauchene skull. Consult Figures 12-1 and 11-3.

1. **Sphenoidal sinuses**—two large air cavities that drain into the nasal cavity
2. **Sella turcica** (SEL-a TUR-si-ka)—depression in center of bone which contains pituitary gland
3. **Greater wings**—large, lateral, paired projections that can be seen on the outer surface of the skull
4. **Lesser wings**—small, paired projections superior to the greater wings
5. **Pterygoid processes** (TER-i-goyd)—large, paired projections on the under surface of the bone
6. **Optic foramina** (OP-tik fō-RĀ-men)—paired holes through which the optic nerves pass
7. **Superior orbital fissures**—paired openings through which cranial nerves III, IV, and parts of V and VI pass
8. **Foramina ovales** (fō-RĀ-men ō-VA-lē)—paired openings through which part of cranial nerve V passes

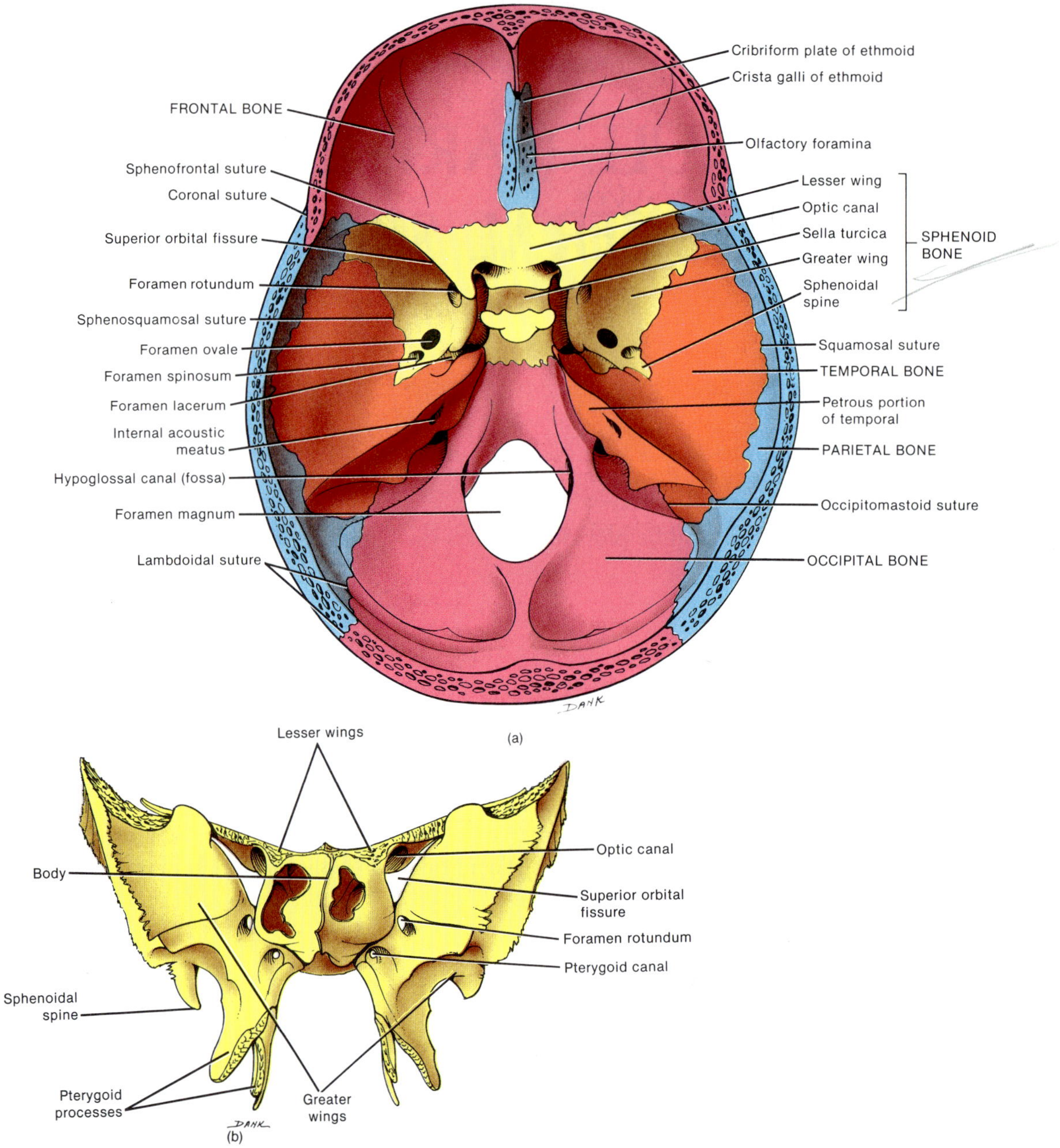

Figure 12-1 Sphenoid bone. (a) Viewed in the floor of the cranium from above. (b) Anterior view.

9. **Foramen spinosum** (spī-NŌ-sum)—An opening through which medial meningeal vessel passes.
10. **Foramen lacerum** (la-SER-um)—A slit between the sphenoid and temporal bones through which the internal carotid artery passes.
11. **Foramen rotundum** (rō-TUN-dum)—An opening through which part of cranial nerve V passes.

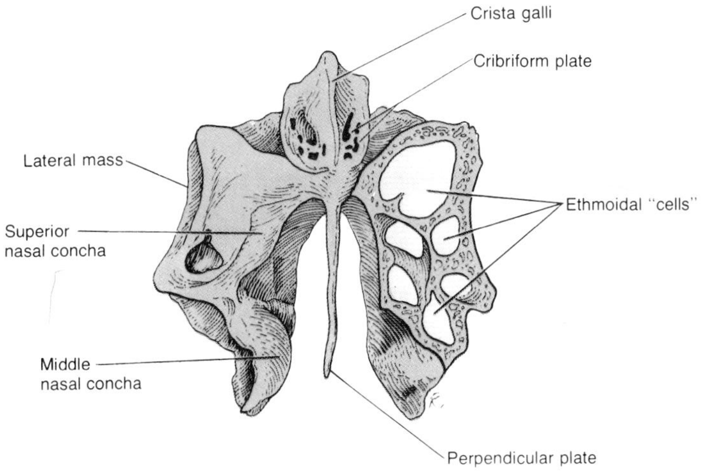

Figure 12-2 Ethmoid bone, anterior view, with left side of bone shown in section.

ETHMOID BONE

The ethmoid bone is a large, but very light, spongy bone with a honeycombed structure (Figure 12-2). It is the main support of the nasal cavity and forms part of the cranial floor and orbits.

1. **Perpendicular plate**—forms nasal septum with the vomer bone
2. **Cribriform plate** (KRIB-ri-form)—porous plate that forms part of the floor of the cranium and contains openings for the olfactory nerve
3. **Crista galli** (KRIS-ta GAL-lē)—triangular process projecting upward from the cribriform plate and forming part of the floor of the cranium
4. **Superior nasal conchae** (KON-kē)—paired lateral folds that create air flow for efficient circulation of inhaled air
5. **Middle nasal conchae**—paired lateral folds below the superior nasal conchae, with the same function

MAXILLARY BONE

The paired maxillary bones form the upper jaw and most of the hard palate. They also form a part of the orbits and nasal cavity. (See Figures 11-1 and 11-2.)

1. **Maxillary sinus**—air cavity emptying into the nasal cavity
2. **Alveolar process** (al-VĒ-ō-lar)—part of the bone that contains sockets for the teeth
3. **Infraorbital foramina** (in′-fra-OR-bi-tal)—paired holes through which the infraorbital nerve and artery pass

TEMPORAL BONE

These paired bones form the inferior lateral portion of the cranium. They contain parts of the ear. (See Figures 11-1–11-3.)

1. **Zygomatic process**—projection that articulates with the temporal process of the zygomatic bone and forms most of the **zygomatic arch**
2. **Mandibular fossa** (man-DIB-yoo-ler)—depression into which the lower jaw inserts
3. **Mastoid process** (MAS-toyd)—portion of the bone behind the earlobe which contains mastoid air cells and serves as the point of attachment for several neck muscles
4. **External auditory meatus** (AW-di-tor-ē mē-Ā-tus)—canal that leads toward the middle ear deep within the temporal bone
5. **Styloid process** (STĪ-loyd)—long, downward-projecting process to which muscles of the tongue and neck are attached

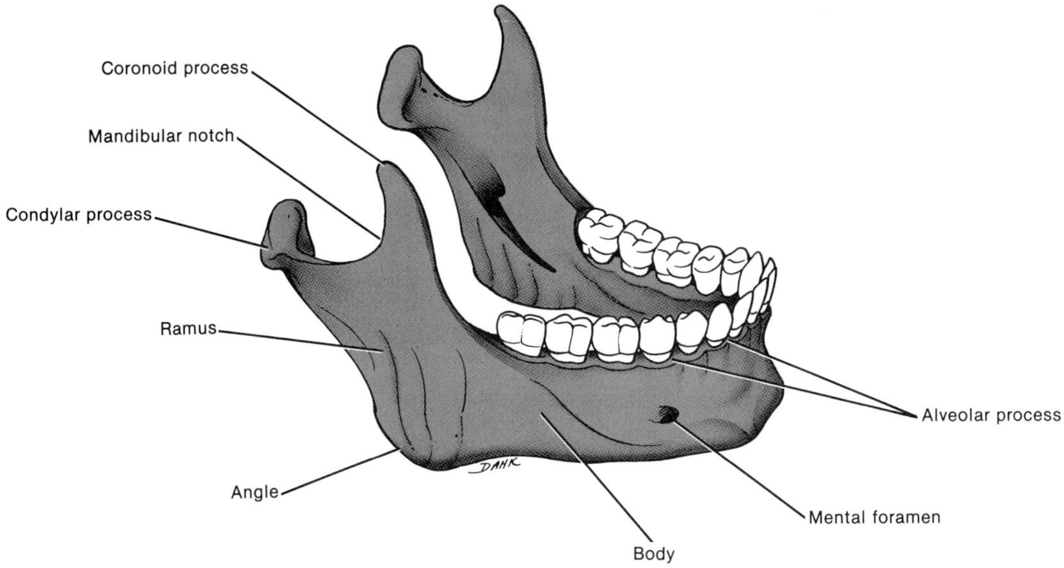

Figure 12-3 Mandible in right lateral view.

FRONTAL BONE

The frontal bone is a bone of both the cranium and the face. It forms the forehead and parts of the orbit and the cranial floor. (See Figures 11-1–11-3.)

1. **Glabella** (gla-BELL-a)—flattened area above the nose
2. **Supraorbital foramina**—paired openings through which supraorbital nerve and supraorbital artery pass
3. **Frontal sinus**—air cavity that communicates with the nasal cavity

MANDIBLE

The mandible is the large bone that forms the entire lower jaw and contains all the teeth of the lower jaw (Figure 12-3).

1. **Condylar process** (KON-di-lar)—projection that articulates with the mandibular fossa of the temporal bone
2. **Coronoid process** (KOR-ō-noyd)—projection to which temporal muscle attaches
3. **Mandibular notch**—depression between coronoid and condylar processes
4. **Mental foramen**—opening through which mental nerve and mental blood vessels pass
5. **Alveolar process** (al-VĒ-ō-lar)—part of the bone that contains sockets for the teeth

B. THE VISCERAL SKELETON

These skeletal elements (some bone and some cartilage) arise from the embryonic gill arch (visceral arch) region.

HYOID BONE

The hyoid (HĪ-oyd) is a U-shaped bone suspended from the styloid process of the temporal bone by ligaments. It serves as point of origin (attachment) for some of the muscles of the tongue. Observe it on the mounted skeleton just inside the mandibular area and on Figure 11-2.

EAR OSSICLES

Three pairs of tiny bones span the air cavity of the middle ear and serve to transmit sound waves to the inner ear. They are not preserved on the laboratory specimens, but you may have a special preparation to study. Diagrams can be found in the text in the exercise on the special senses. The names of the bones are: *malleus* (MAL-ē-us), meaning "hammer"; *incus* (ING-kus), meaning "anvil"; and *stapes* (STĀ-pēz), meaning "stirrup."

CARTILAGES OF THE LARYNX

These highly specialized skeletal elements of a visceral organ will be observed when the dissection of the cat is made. They may be seen on a diagram of the larynx in the text in the exercise on the respiratory system.

Questions for Review

EXERCISE **12**

THE SKULL (PART II) AND THE VISCERAL SKELETON

Name Kathy Martin

Laboratory Section ____________ Date 3/18/91

1. Match the terms on the left with the bones listed on the right. This question reviews the markings of the bones of the skull.

a. 5 Sella turcica
b. 4 Mastoid process
c. 4 Zygomatic process
d. 14 Mandibular notch
e. 6 Cribriform plate
f. 5 Greater wing
g. 14 Mental foramen
h. 6 Middle nasal concha
i. 8 Maxillary sinus
j. 5 Pterygoid processes
k. 5 Foramen spinosum
l. 1 Supraorbital foramen
m. 8 Infraorbital foramen
n. 6 Perpendicular plate
o. 14 Condylar process
p. 5 Foramen lacerum
q. 1 Glabella
r. 14 Coronoid process
s. 4 Styloid process

1. Frontal
2. Parietal
3. Occipital
4. Temporal
5. Sphenoid
6. Ethmoid
7. Nasal
8. Maxillary
9. Lacrimal
10. Zygomatic
11. Palatine
12. Vomer
13. Inferior nasal concha
14. Mandible

t. 6 Superior nasal concha

u. 5 Foramen ovale

v. 6 Crista galli

w. 5 Sphenoidal sinus

x. 5 Optic foramen

y. 4 External auditory meatus

2. Anwer the following questions briefly. 210 - 229 pg

a. Why is the "visceral skeleton" so named? because they have some bone and some cartiledge. Not really connected.

b. What is the function of the laryngeal cartilages? they close the glottis during swallowing and speach

c. What is the function of the middle ear ossicles? they Transmit sounds to the inner ear.

d. Which skeletal muscle originates on the hyoid bone and what is the function of this muscle? Ligaments it serves as a point of origin for the tongue.

e. Distinguish between the terms *visceral skeleton* and *viscera*. viscera means the organs within the abdominal or thoracic cavities and visceral skeleton is the bones around the abdominal or thoracic cavity.

EXERCISE 13

THE VERTEBRAL COLUMN, STERNUM, AND RIBS

Road

Materials

The following materials should be provided for class use.

1. Mounted human skeleton
2. Disarticulated skeleton
3. Strung vertebral columns
4. Individual vertebrae

A. THE VERTEBRAL COLUMN

The bones of the vertebral column surround and protect the spinal cord in the living organism. They are named for their location along the body axis. Study them on the laboratory specimens and compare with Figure 13-1.

Continue your observations by locating the parts of a lumbar vertebra shown in Figure 13-2 or a thoracic vertebra shown in Fig. 13-4a.

1. **Spinous process**—posterior projection that serves as point of attachment for muscles (It can be felt on your own body in the midback region.)
2. **Vertebral foramen**—central canal through which the spinal cord passes
3. **Body** or **centrum**—large circular area anterior to the vertebral foramen that forms the joints at the intervertebral discs
4. **Transverse processes**—lateral projections that serve as points of attachment for muscles
5. **Articular processes**—smooth surfaces that articulate with other vertebrae
   a. **Superior articular processes**—pair of smooth articular surfaces facing toward the spinous process
   b. **Inferior articular processes**—pair of smooth articular surfaces facing away from the spinous process
6. **Vertebral arch**—arch over the vertebral foramen between the centrum and the spinous process
   a. **Pedicles**—lateral portions of arch
   b. **Laminae** or **roof plates**—posterior portion of arch
7. **Facets** and **demifacets**—found on thoracic vertebrae for articulation with ribs.

CERVICAL VERTEBRAE

The seven vertebrae in the neck region are called the cervical vertebrae. They differ from other vertebrae in that each transverse process contains a lateral opening called the **transverse**

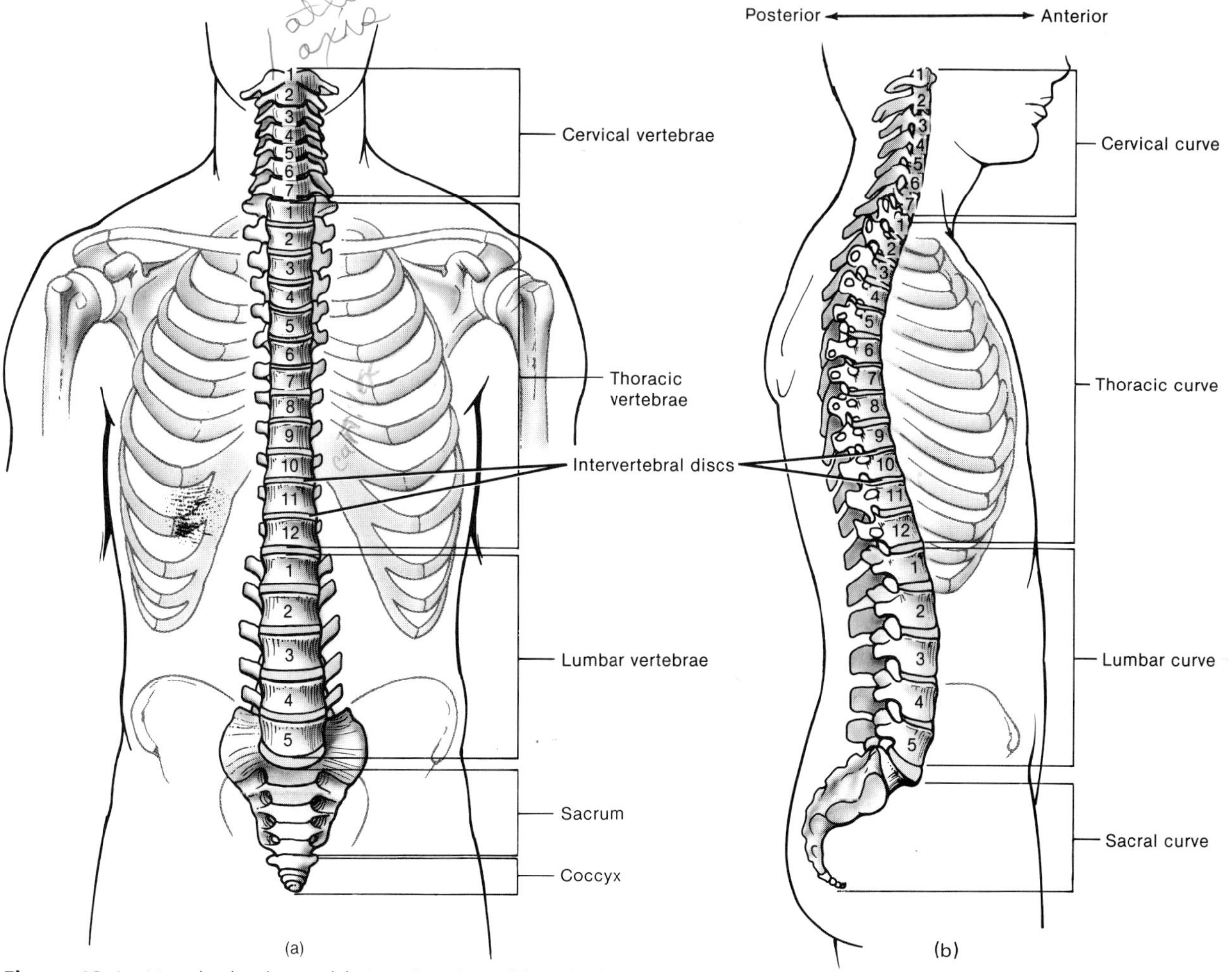

Figure 13-1 Vertebral column. (a) Anterior view. (b) Right lateral view.

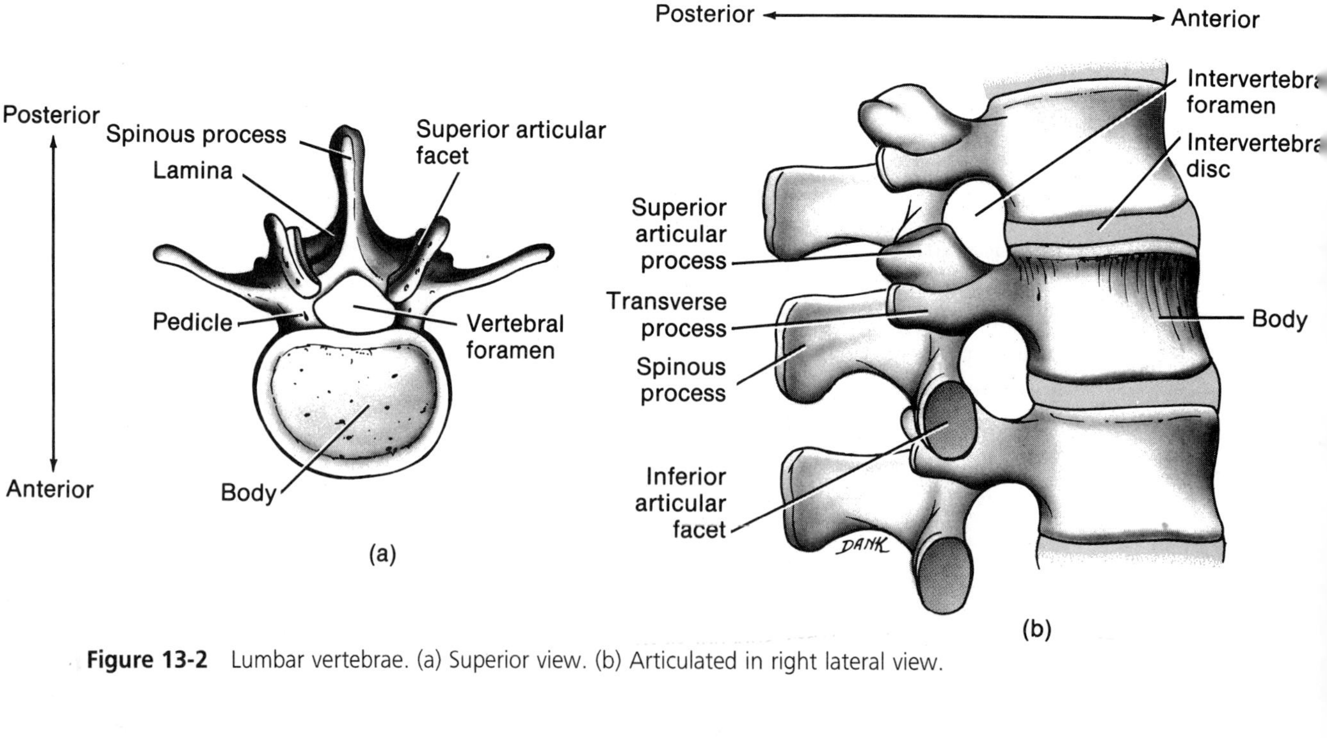

Figure 13-2 Lumbar vertebrae. (a) Superior view. (b) Articulated in right lateral view.

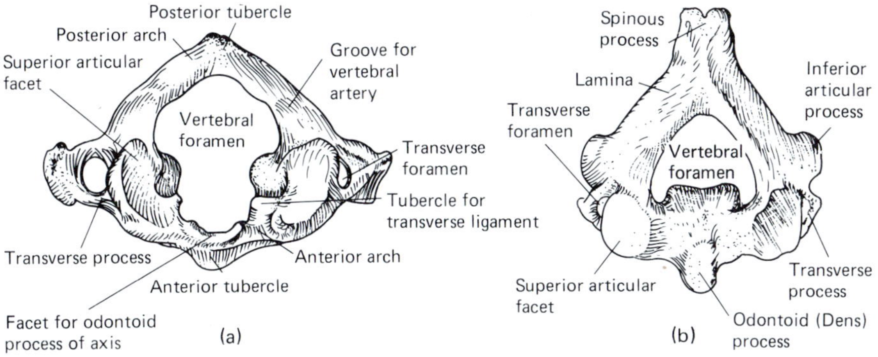

Figure 13-3 Cervical vertebrae. (a) Superior view of the atlas. (b) Superior view of the axis.

foramen. The first two are highly specialized to form the pivotal joint of the head and neck. The first cervical vertebra is called the **atlas** and the second is the **axis** (Figure 13-3). Study the atlas and axis. Demonstrate how the **dens** of the axis fits into the atlas to form a pivotal joint.

THORACIC VERTEBRAE

The 12 thoracic vertebrae are characterized by long spinous processes. Each vertebra articulates with a pair of ribs that have whole *facets* or *demifacets* on the sides of the vertebral centra. Thoracic vertebrae 2 through 8 have two *demifacets* on each side so that each rib head articulates with 2 vertebrae. Examine the mounted human skeleton for these features. The ninth thoracic vertebra has only one *demifacet* on each side. In addition to the *facets* and *demifacets* on the centra, thoracic vertebrae 1 through 10 have *facets* located superiorly on the transverse processes for articulation with the rib tubercles.

LUMBAR VERTEBRAE

The five lumbar vertebrae are generally larger and heavier than the other vertebrae and are located in an area of the back which bears a considerable amount of the body's weight. Their spines, which project backward almost horizontally, are not as long and pointed as those of the thoracic group (see Figure 13-2).

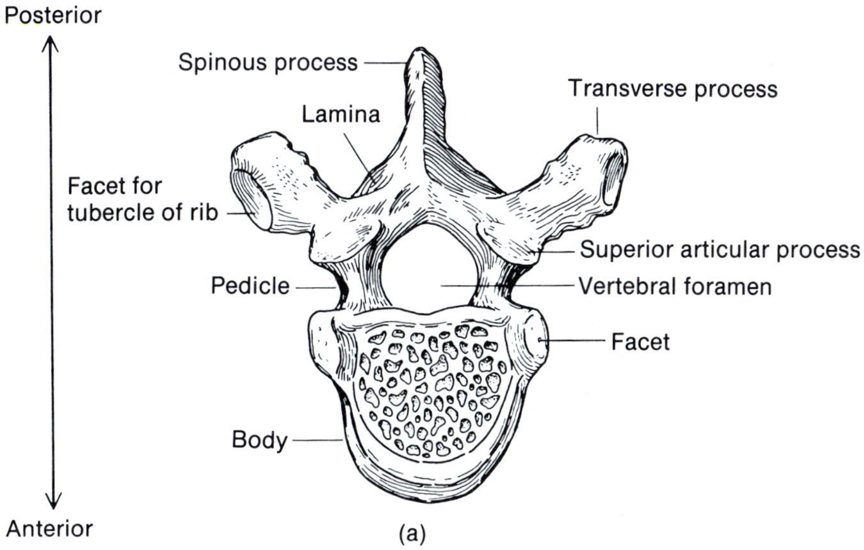

Figure 13-4a Thoracic vertebra, superior view.

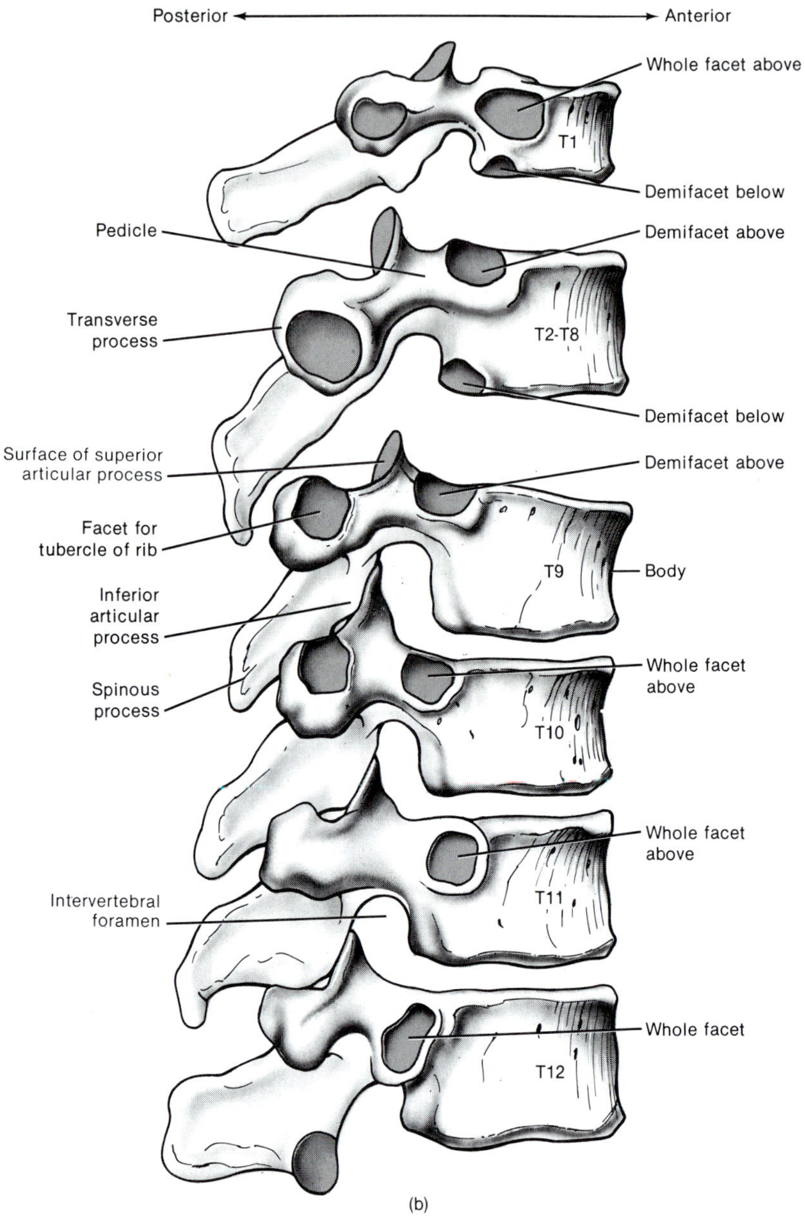

Figure 13-4b Vertebrae articulated in right lateral view.

SACRAL VERTEBRAE OR SACRUM

At this point in the vertebral column five vertebrae have fused (for added strength) to form one large triangular bone called the **sacrum** (Figures 13–1 and 13–5). It supports the pelvic girdle at the **sacroiliac joint.**

COCCYGEAL VERTEBRAE OR COCCYX

This is a vestigial (nonfunctional) part of the vertebral column which represents the tail (Figure 13–5). There may be three, four, or five of these small bones, and they may or may not be fused.

B. THE STERNUM AND RIBS

STERNUM

Identify the three bones that make up the sternum (Figure 13-6).

1. *Manubrium* (ma-NOO-brē-um)—most superior in position
2. *Body* or *gladiolus* (glad′-ē-Ō-lus)—middle and longest bone
3. *Xiphoid process* (ZĪ-foyd)—most inferior and smallest

With the aid of the diagram, locate the *jugular notch* and the *clavicular notch* on the manubrium.

RIBS

The ribs are divided into two groups. The first seven pairs articulate with the sternum by means of their own costal cartilages. They are called *true ribs.*

The last 5 pairs of ribs do not articulate directly with the sternum by means of their own costal cartilages, but indirectly, if at all, through the cartilages of the seventh true rib. These are called *false ribs.* Study Fig. 13-6 and the mounted human skeleton. The cartilages of the eighth, ninth, and tenth false ribs join each other and then articulate with the cartilage of the seventh true rib.

The last two pairs of the five pairs of false ribs do not have costal cartilages and do not attach to the sternum at all. They are called *floating ribs.*

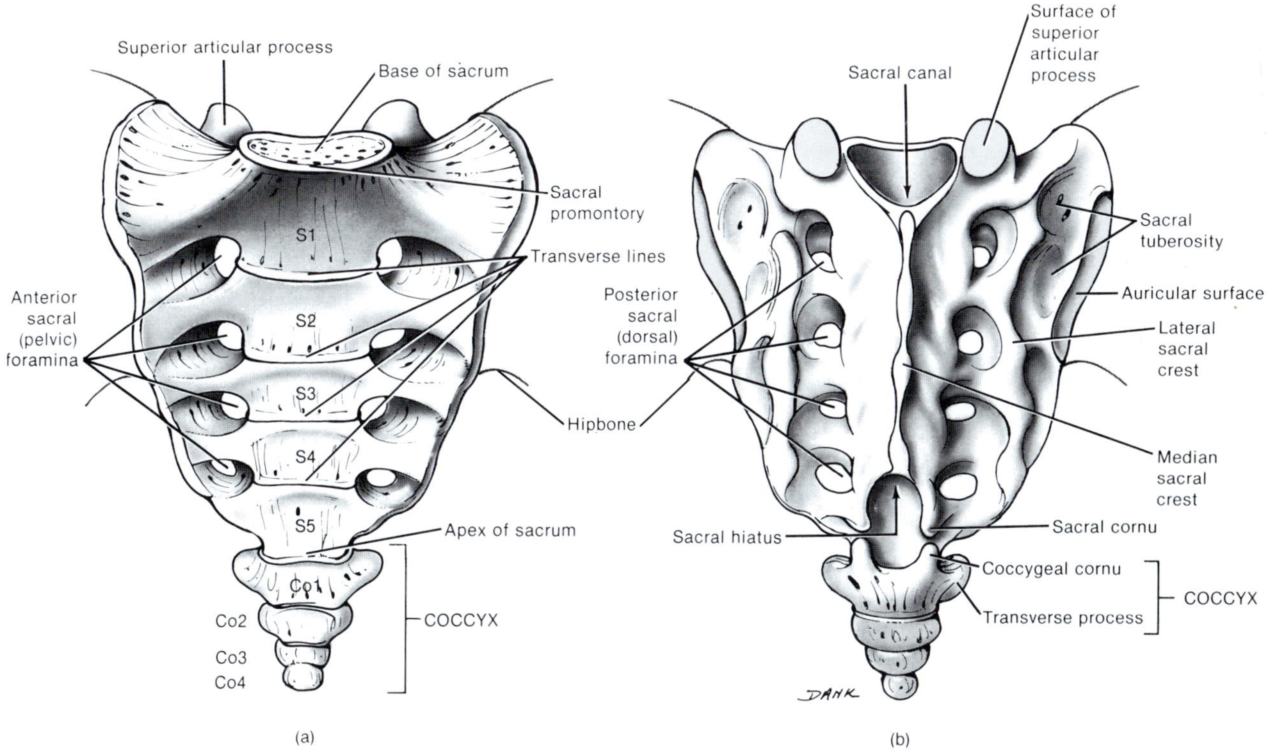

Figure 13-5 Sacrum and coccyx. (a) Anterior view. (b) Posterior view.

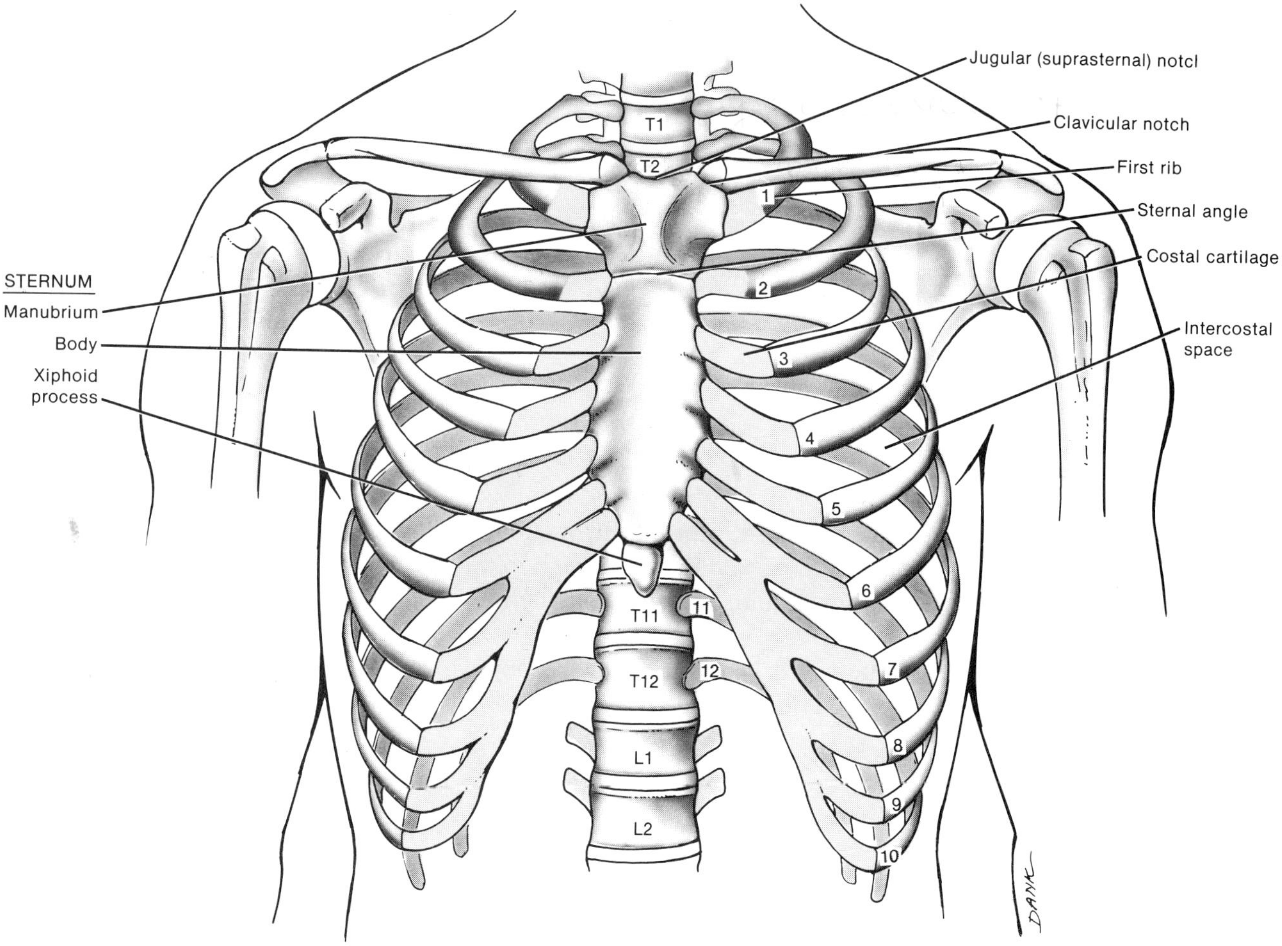

Figure 13-6 Sternum and ribs, anterior view.

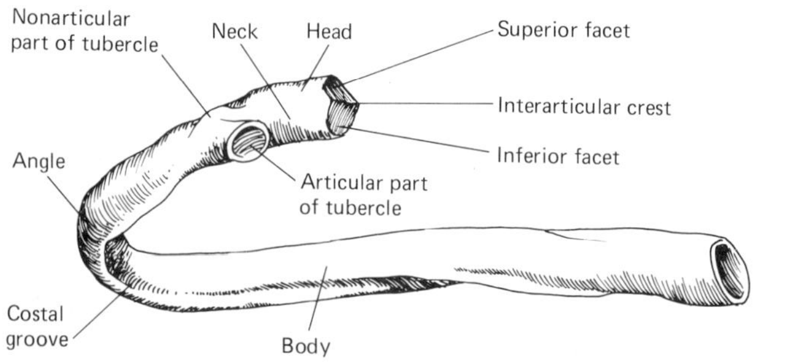

Figure 13-7 Typical rib. A left rib viewed from below and behind.

Questions for Review

EXERCISE **13**

THE VERTEBRAL COLUMN, STERNUM, AND RIBS

Name Kathy Martin

Laboratory Section 8:00 AM Mon Date 3/23-91

1. Fill in the blanks.

a. Facet — Name the part of a vertebra that forms the point of articulation for a rib.

b. seven — How many cervical vertebrae does the human have?

c. atlas — What is the special name given to the first cervical vertebra?

d. axis — What is the special name given to the second cervical vertebra?

e. sacrum — Which bone is formed from five fused sacral vertebrae?

f. Coccyx — Which part of the human vertebral column is considered vestigial?

g. vertebral foramen — What is the name for the cavity in a vertebra through which the spinal cord passes?

h. lumbar — Which vertebrae are the largest and heaviest?

i. lumbar — On which part of the vertebral column does walking place the most stress?

j. points of attachment — What is the function of the transverse processes?

k. between — Do spinal nerves exit through vertebrae or between them?

l. manubrium — Name the superior portion of the sternum.

m. body — Name the middle portion of the sternum.

n. xiphoid process — Name the inferior portion of the sternum.

lacrimal = tear duct

o. spine Which part of a vertebra is most posterior in position?

2. What criterion makes a rib a "true" rib? they articulate + have their own costal cartilage

3. Name and locate the spinal curvatures.

a. Cervical curve

b. Thoracic

c. Lumbar curve

d. sacral curve

4. Match the terms on the left with the part of the skeleton to which they belong.

a. 1 Transverse process
b. 2 Tubercle
c. 1 Atlas
d. 3 Gladiolus
e. 1 Centrum
f. 3 Manubrium
g. 2 Inferior articular process
h. 1 Spinous process
i. 3 Xiphoid process
j. 3 Jugular notch
k. 1 Coccyx

1. Vertebral column
2. Ribs
3. Sternum

EXERCISE 14

THE APPENDICULAR SKELETON

Materials

The following materials should be provided for class use.

1. Mounted human skeleton
2. Disarticulated skeleton
3. Individual pelvis and scapula bones
4. Articulated hand–wrist and foot–ankle bones

A. SUBDIVISIONS OF THE APPENDICULAR SKELETON

The bones of the appendicular skeleton form the appendages (arms and legs) and the girdles (pectoral and pelvic). The girdles are bony structures to which the arms and legs attach and join the *torso.* The bones are grouped in four subdivisions.

1. *Upper extremity* or *arm* (Figure 14-1)
   a. **Humerus** (HYOO-mer-us)—bone of the upper arm
   b. **Radius** (RĀ-dē-us)—lateral bone of the lower arm or forearm
   c. **Ulna** (UL-na)—medial bone of the lower arm or forearm
   d. **Carpals** (KAR-puls)—bones of the wrist
   e. **Metacarpals**—bones of the hand
   f. **Phalanges** (fa-LAN-jēz), sing., *phalanx* (FA-lanks)—bones of the fingers
2. *Pectoral girdle* (Figure 14-1)
   a. **Scapula** (SKAP-yoo-la)—posterior component of girdle, commonly known as the shoulder blade
   b. **Clavicle** (KLAV-i-kul)—anterior component, commonly known as the collarbone
3. *Lower extremity* or *leg* (Figure 14-2)
   a. **Femur** (FĒ-mer)—bone of the thigh
   b. **Patella** (pa-TEL-a)—sesamoid bone (develops in a tendon) that forms the kneecap
   c. **Tibia** (TIB-ē-a)—medial bone of the lower leg
   d. **Fibula** (FIB-yoo-la)—lateral bone of the lower leg
   e. **Tarsals** (TAR-sals)—bones of the ankle
   f. **Metatarsals—**bones of the foot
   g. **Phalanges—**bones of the toes
4. *Pelvic girdle* (Figure 14-2)
   a. *Coxal bone*—commonly known as the hipbone, made of three fused bones
      i. **Ilium** (ILL-ē-um)
      ii. **Ischium** (IS-kē-um)
      iii. **Pubis** (PYOO-bis)

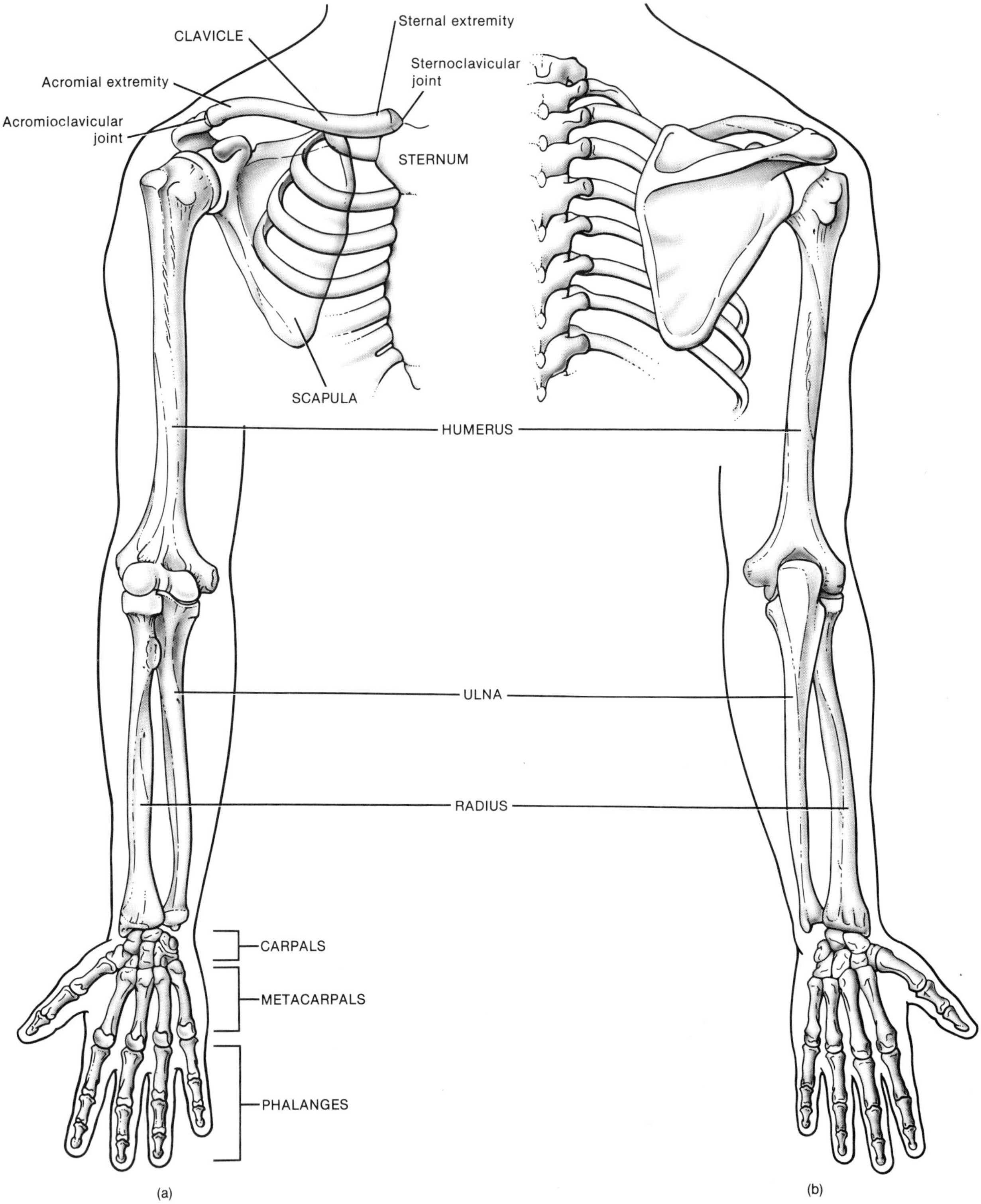

Figure 14-1 Right pectoral (shoulder) girdle and upper extremity. (a) Anterior view. (b) Posterior view.

B. BONES OF THE UPPER EXTREMITY

Having learned the names of the bones of the appendicular skeleton, the student should now examine each bone for some of the more important markings. We will consider the bones of the upper extremity first.

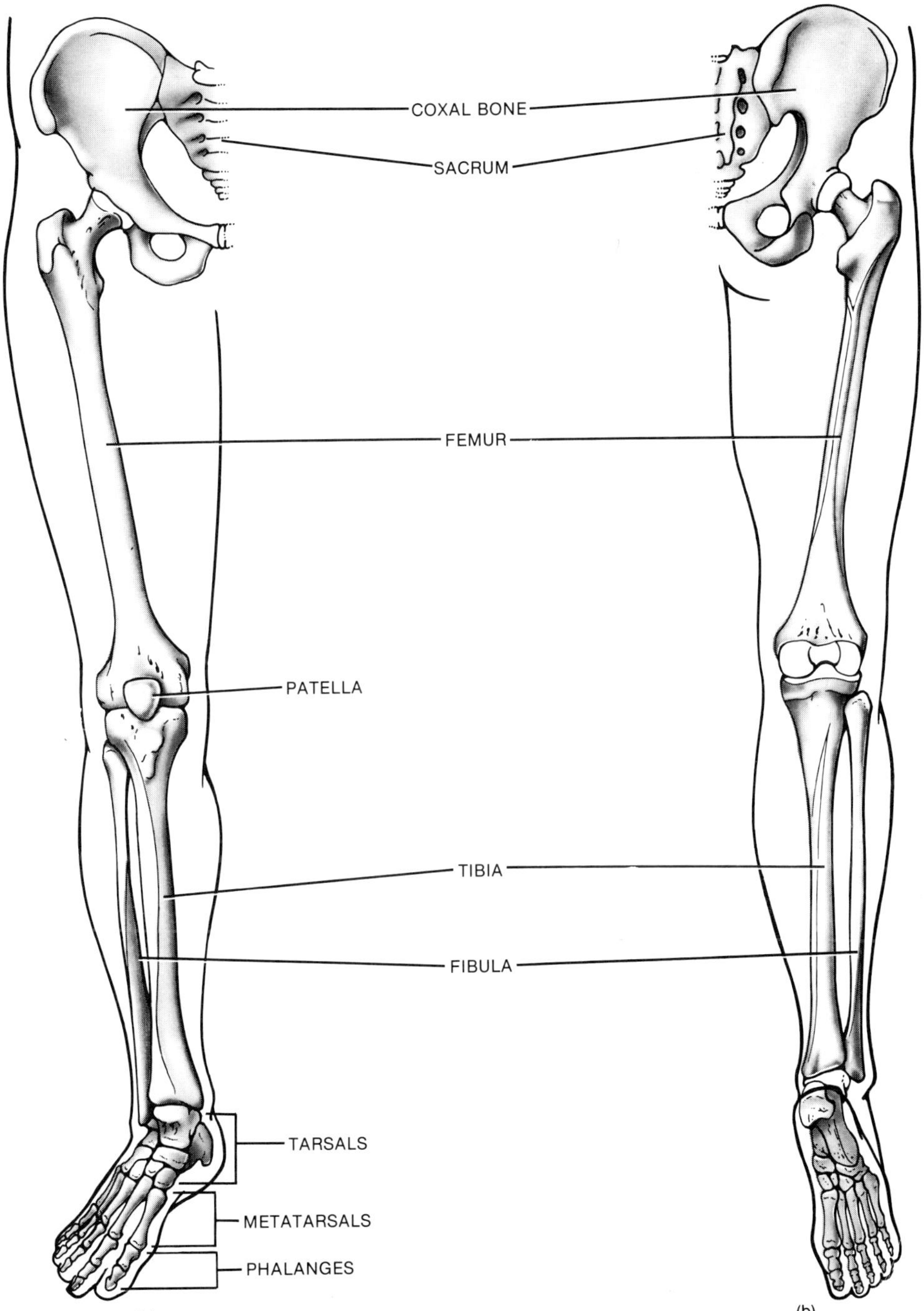

Figure 14-2 Right pelvic girdle and lower extremity. (a) Anterior view. (b) Posterior view.

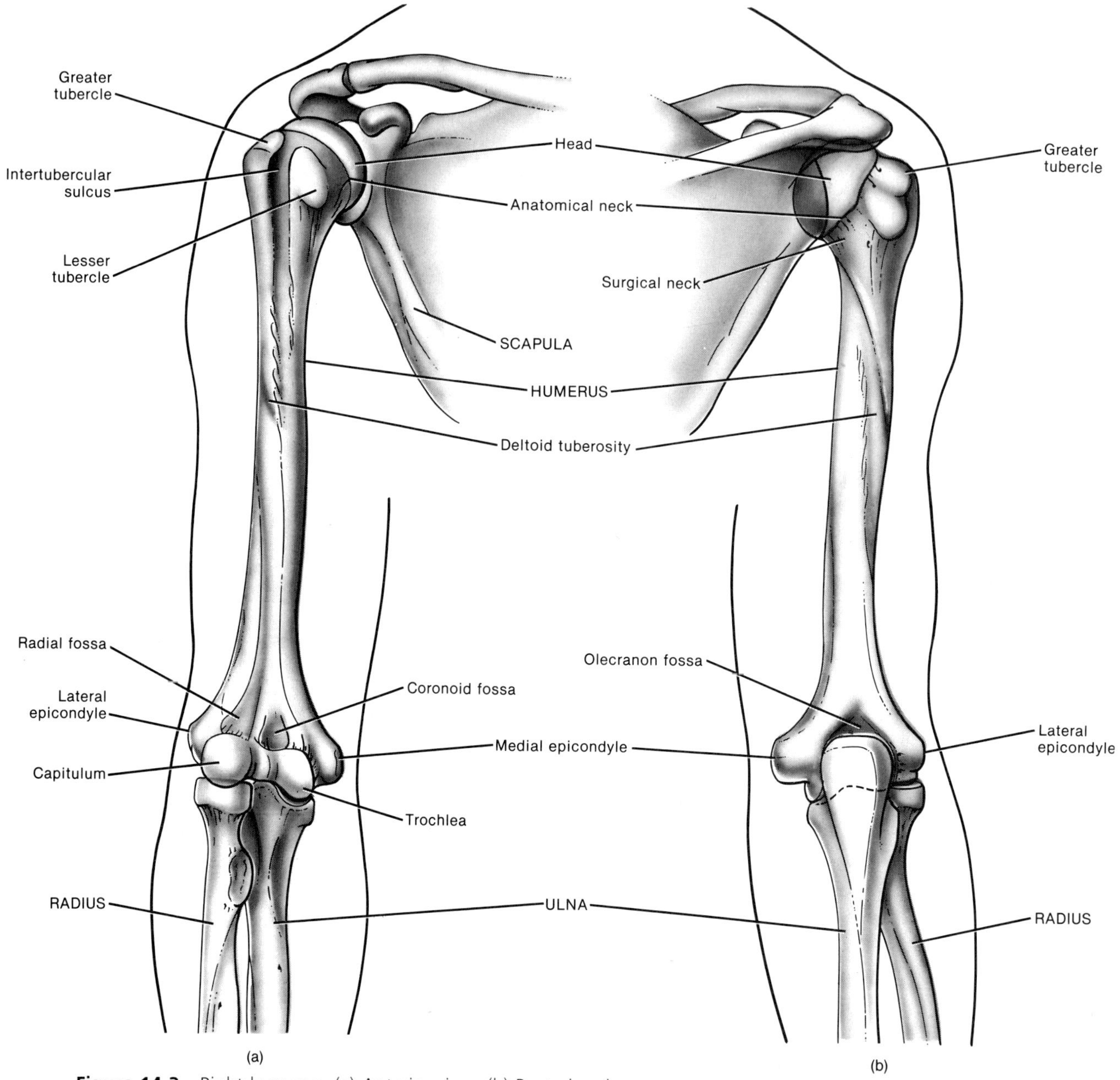

Figure 14-3 Right humerus. (a) Anterior view. (b) Posterior view.

HUMERUS (Figures 14-3 and 14-4)

The longest bone of the arm, the humerus, articulates with the pectoral girdle (scapula) proximally and with the radius and ulna distally at the elbow. Locate the following parts on a laboratory specimen.

1. **Greater tubercle** (TOO-ber-kul)—lateral projection distal to the neck of the humerus
2. **Lesser tubercle**—anterior projection distal to the neck of the humerus
3. **Intertubercular sulcus** or **bicipital groove** (in′-ter-too-BER-kyoo-lar SUL-kus) (bī-SIP-i-tal)—furrow between the two tubercles
4. **Deltoid tuberosity** (too-ber-OS-i-tē)—roughened area in the middle of the shaft
5. **Capitulum** (ka-PIT-yoo-lum)—knoblike process articulating with the radius
6. **Radial fossa**—depression for the head of the radius
7. **Trochlea** (TRŌK-lē-a)—articular surface for the ulna
8. **Coronoid fossa**—anterior depression for the ulna when forearm is flexed
9. **Olecranon fossa** (ō-LEK-ra-non)—posterior depression for the olecranon process of the ulna when forearm is extended
10. **Medial epicondyle** (ep′-i-KON-dīl)—roughened projection on the medial distal surface
11. **Lateral epicondyle**—roughened projection on the lateral distal surface

RADIUS AND ULNA (Figures 14-1 and 14-5)

The *radius* is the lateral bone of the forearm. It forms a radial or pivotal joint with the ulna at the proximal end of each bone. Locate the following parts.

1. **Radial tuberosity**—roughened area on the medial surface just below the head
2. **Styloid process** (STĪ-loyd)—sharp lateral projection at the distal end
3. **Ulnar notch**—distal concavity for articulation with the ulna

The *ulna* is the medial bone of the forearm. Locate the following parts.

1. **Olecranon process**—large projection of the proximal end of the ulna that forms the elbow
2. **Coronoid process**—anterior projection at the proximal end that articulates with the trochlea of the humerus
3. **Trochlear** (TRŌK-lē-ar) **notch** or **semilunar notch**—curved area that receives the trochlea of the humerus
4. **Radial notch**—lateral depression, inferior to the trochlear notch, that receives the head of the radius
5. **Styloid process**—sharp lateral projection at the distal end

CARPALS, METACARPALS, AND PHALANGES (Figures 14-5 and 14-6)

The *carpus* or wrist consists of eight small bones closely held together by ligaments. They are arranged in two rows, each with four bones. Using Figure 14-6 and a human skeleton or separate hand–wrist preparation, locate all the carpal bones.

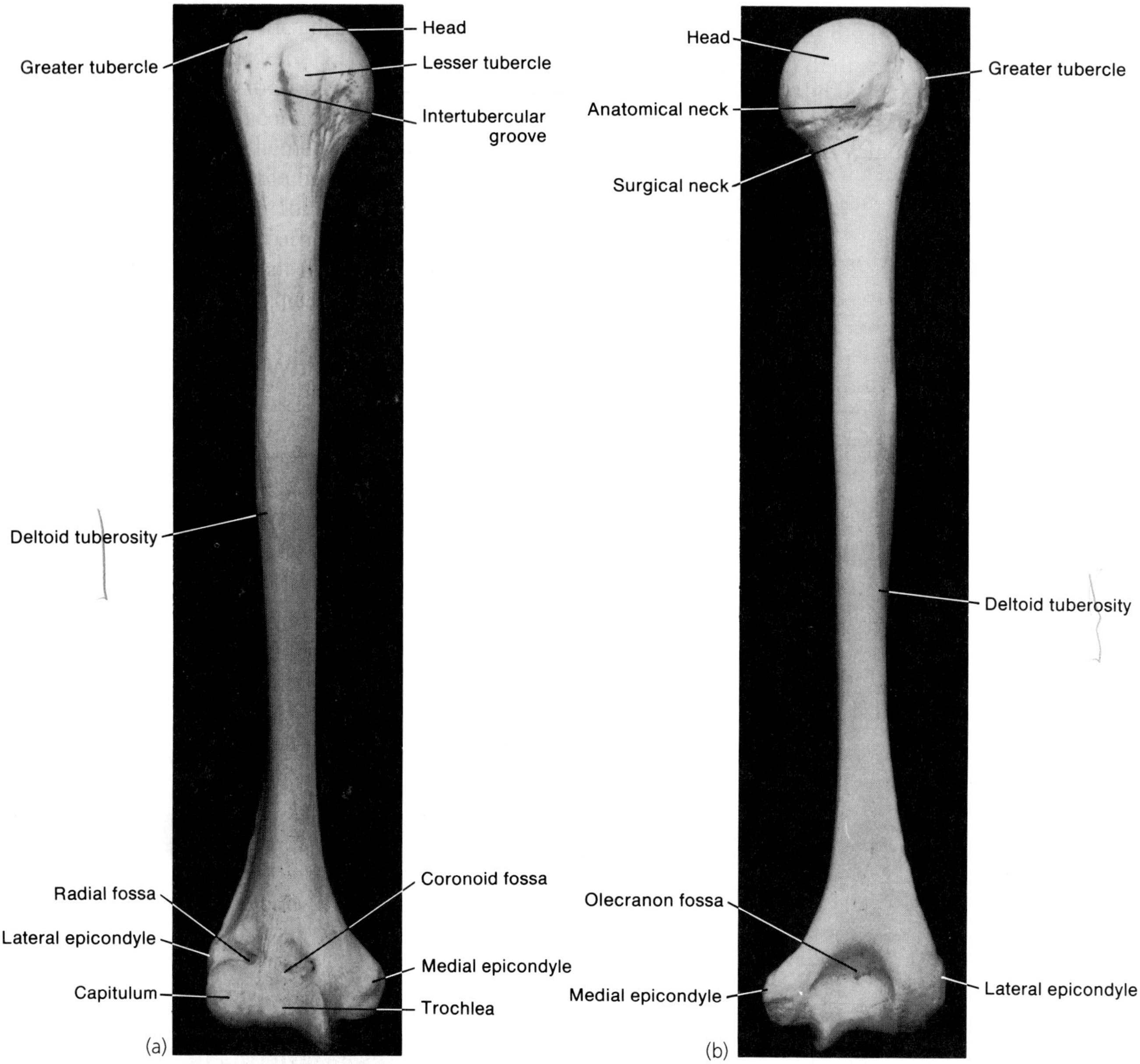

Figure 14-4 Right humerus. (a) Anterior view. (b) Posterior view. (Photographs courtesy of Lenny Patti.)

The four bones of the proximal row, from medial to lateral, are:

1. **Pisiform** (PĪ-si-form)
2. **Triquetral** (trī-KWĒ-tral) or *triangular*
3. **Lunate** (LOO-nāt)
4. **Scaphoid** (SKAF-oyd) or *navicular* (na-VIK-yoo-lar)

The four bones of the distal row, from medial to lateral, are:

1. **Hamate** (HAM-āt)
2. **Capitate** (KAP-i-tāt)
3. **Trapezoid** (TRAP-i-zoyd) or *lesser multangular* (mul-TANG-gyoo-lar)
4. **Trapezium** (tra-PĒ-zē-um) or *greater multangular*

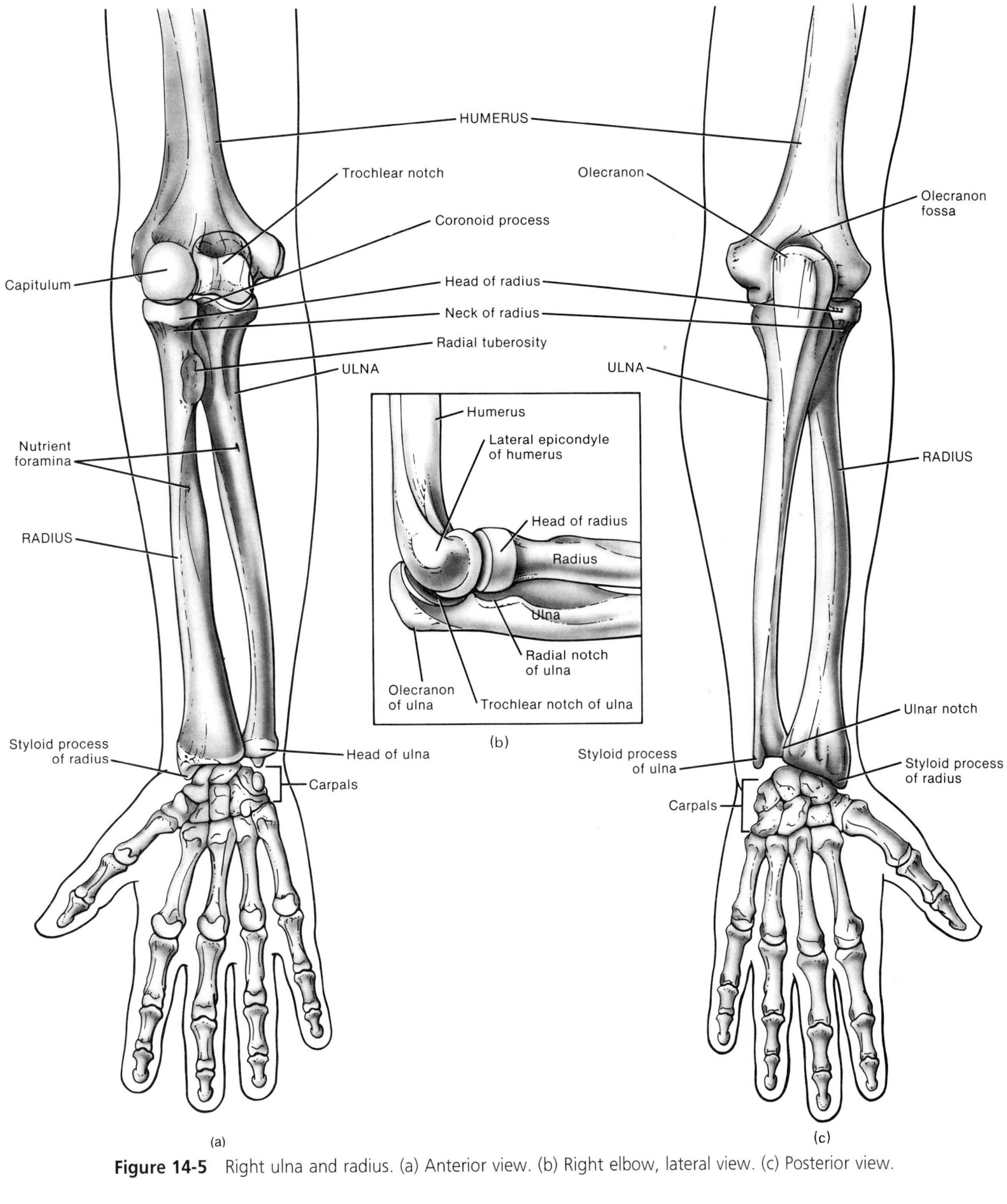

Figure 14-5 Right ulna and radius. (a) Anterior view. (b) Right elbow, lateral view. (c) Posterior view.

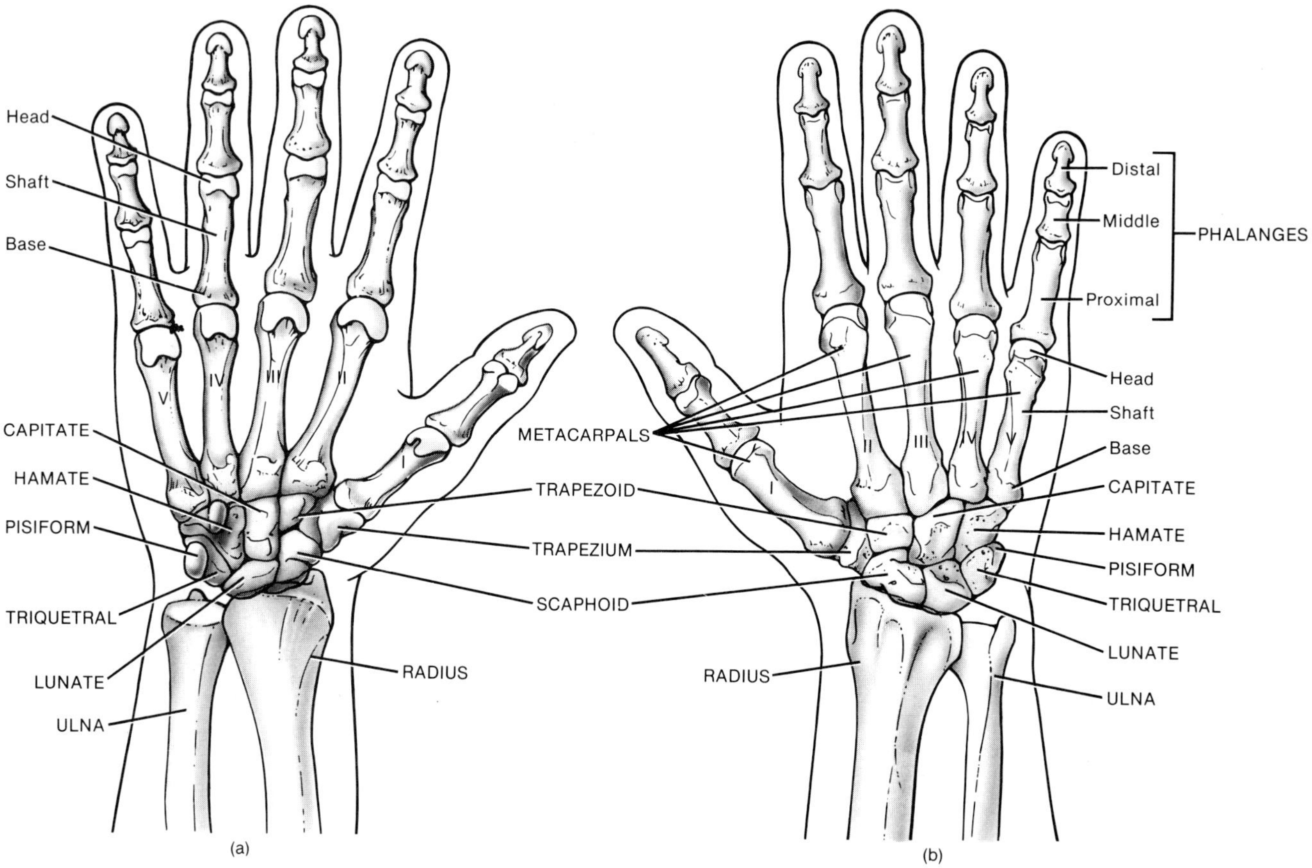

Figure 14-6 Right wrist and hand. (a) Anterior view. (b) Posterior view.

The *metacarpus,* the region between the wrist and the fingers, consists of five bones that form the palm of the hand. They are numbered from I to V, beginning with the metacarpal that articulates with the thumb. Assign the proper numbers to the metacarpals on your lab specimens.

The *phalanges* are the bones of the digits, or fingers. There are 14 in each hand and 3 in each finger except the thumb, which has only 2.

C. BONES OF THE PECTORAL GIRDLE

The pectoral girdle is made up of the bones which join the arm to the torso (Figures 14-7 and 14-8).

Clavicle—The anterior bone of the girdle which is long, slender and horizontally placed or oriented.

Scapula—The posterior bone of the girdle is flat and triangular with many markings for muscle attachment.

1. **Spine**—large, sharp ridge that runs transversely across the posterior surface
2. **Acromion** (a-KRŌ-mē-on)—large projection at the end of the spine which articulates with the clavicle
3. **Coracoid process** (KOR-a-koyd)—a somewhat smaller projection on the anterior surface
4. **Glenoid fossa** or *cavity*—depression below the acromion and coracoid process which articulates with the proximal head of the humerus to form the shoulder joint

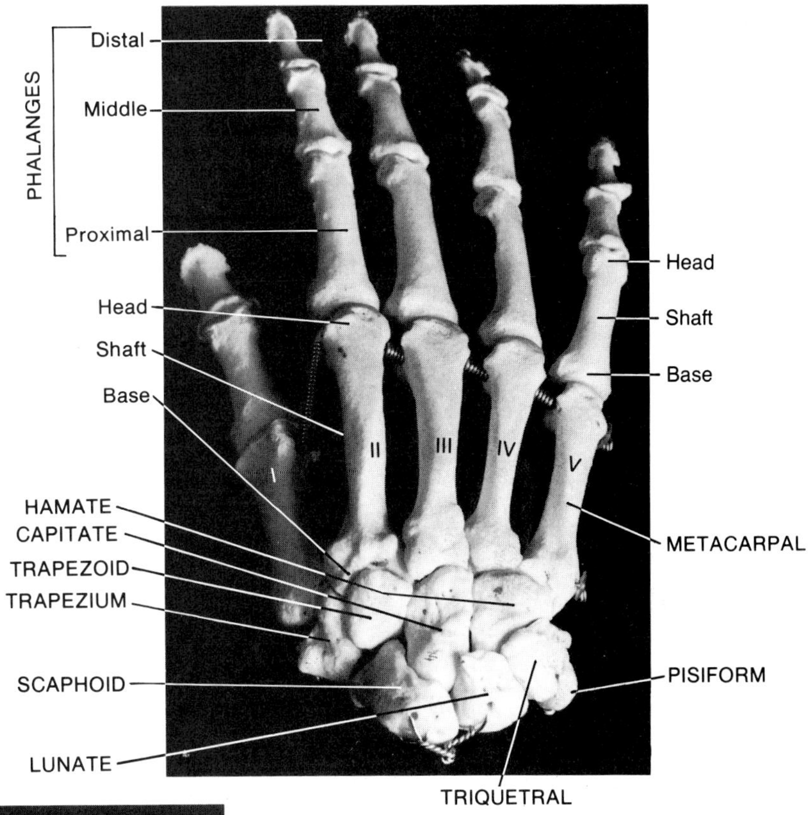

(c)

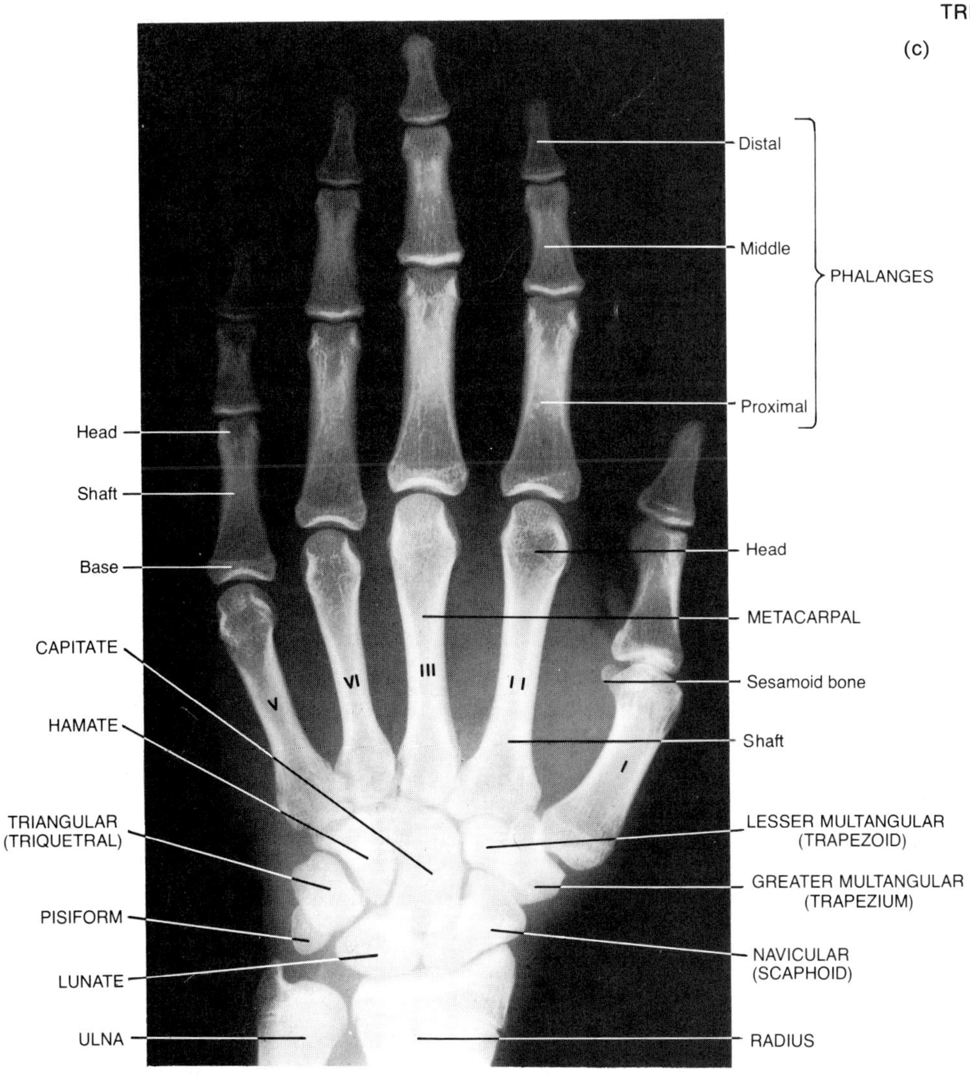

(d)

Figure 14-6 *(Continued)* (c) Posterior view. (Courtesy of Lenny Patti.) (d) Anteroposterior roentgenogram. (Courtesy of Daniel Sorrentino.)

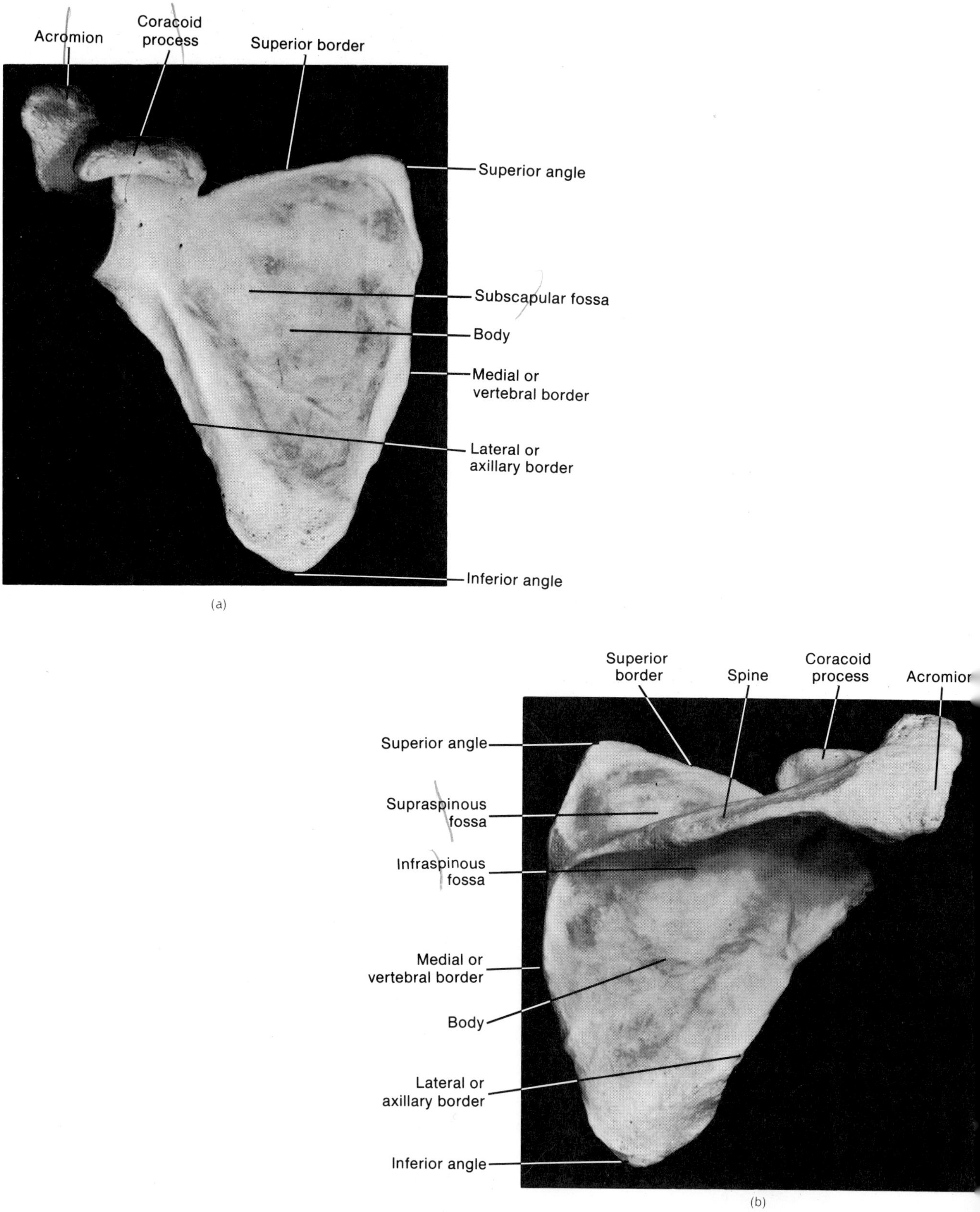

Figure 14-7 Right scapula. (a) Anterior view. (b) Posterior view.

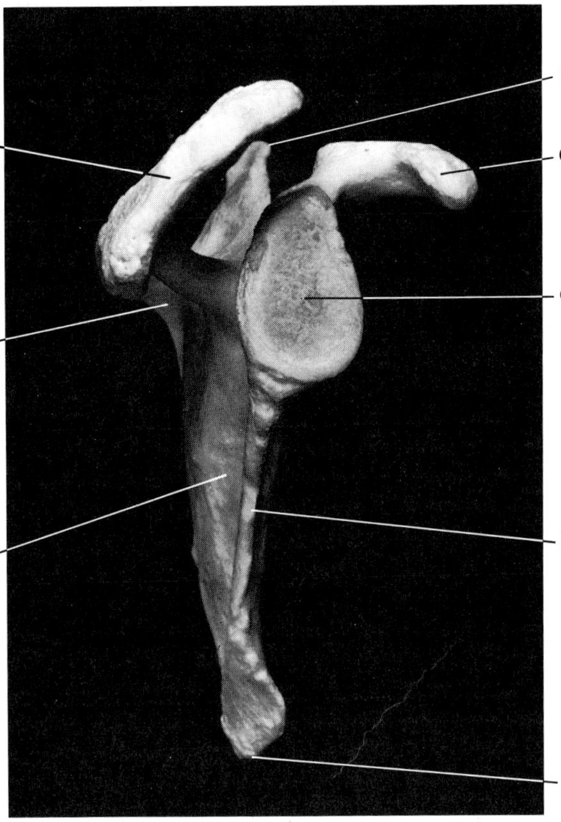

(c)

Figure 14-7 (continued) (c) Medial view. (Courtesy of Lenny Patti.)

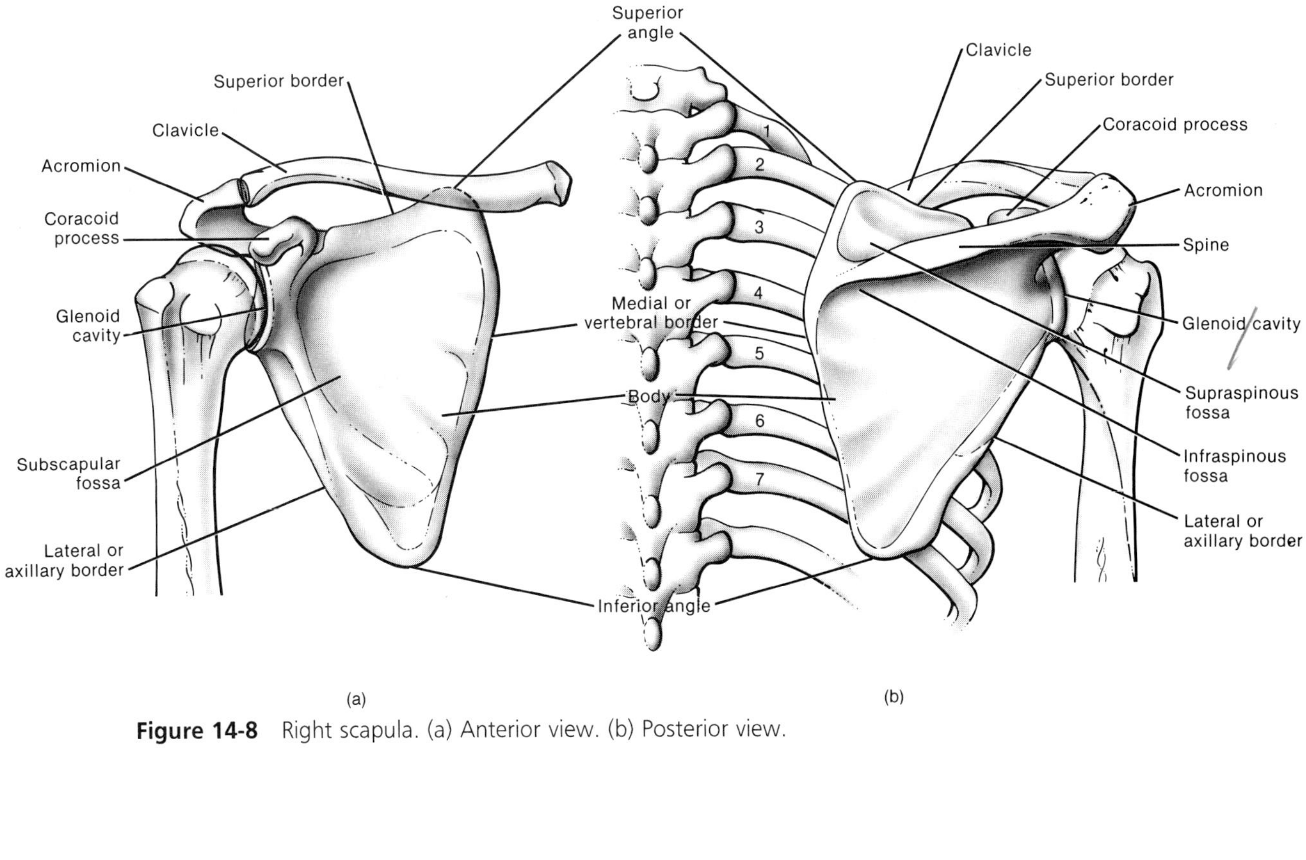

(a) (b)

Figure 14-8 Right scapula. (a) Anterior view. (b) Posterior view.

5. **Supraspinous fossa** (soo′-pra-SPĪ-nous)—depression above the spine on the posterior surface, point of attachment for supraspinatus muscle
6. **Infraspinous fossa** (in-fra-SPI-nous)—large concavity below the spine on the posterior surface, point of attachment for infraspinatus muscle
7. **Subscapular fossa** (sub-SKAP-yoo-lar)—slight concavity occupying most of the anterior surface, point of attachment for subscapularis muscle
8. **Vertebral** or **medial border**—thin edge of the body near the vertebral column
9. **Axillary** or **lateral border**—thickened edge of the body near the arm
10. **Superior border**—superior edge of the body
11. **Inferior angle**—angle formed where medial and lateral borders meet, most inferior part of the bone
12. **Superior angle**—angle formed where superior and medial borders meet

D. BONES OF THE LOWER EXTREMITY

Review the names of the bones of the leg, and then study their important markings.

FEMUR

The femur, or thigh bone, is the longest and heaviest bone in the body. Using Figure 14-9 and a laboratory specimen, locate the following parts.

1. **Head**—rounded proximal end that articulates with the hip socket in the pelvic girdle
2. **Neck**—constricted area distal to the head
3. **Greater trochanter** (trō-KAN-ter)—large, roughened lateral projection at the proximal end
4. **Lesser trochanter**—smaller, roughened medial projection
5. **Linea aspera** (LIN-ē-a AS-per-a)—roughened vertical ridge on the shaft
6. **Medial condyle** (KON-dīl)—projection at the distal end for articulation with the tibia
7. **Lateral condyle**—projection at the distal end for articulation with the tibia
8. **Intercondylar fossa**—depression between the medial and lateral condyles
9. **Medial epicondyle**—projection superior to the medial condyle
10. **Lateral epicondyle**—projection superior to the lateral condyle

TIBIA AND FIBULA

The *tibia* is the larger of the two bones of the lower leg. It is located medially. Identify the following parts using Figure 14-10 and the laboratory specimen.

1. **Lateral condyle**—lateral projection at the proximal end which articulates with the femur
2. **Medial condyle**—medial projection at the proximal end which articulates with the femur
3. **Intercondylar eminence** (in′-ter-KON-di-lar)—small projection between the condyles
4. **Tibial tuberosity**—roughened projection on the anterior surface just below the head
5. **Medial malleolus** (mal-LĒ-ō-lus)—medial prominence at the distal end
6. **Fibular notch**—lateral surface at the distal end which articulates with the fibula

The *fibula* is the lateral bone of the lower leg. It is much smaller than the tibia. Locate the following parts.

1. **Head**—proximal end that articulates with the tibia
2. **Lateral malleolus**—long projection at the distal end which articulates with the talus of the ankle

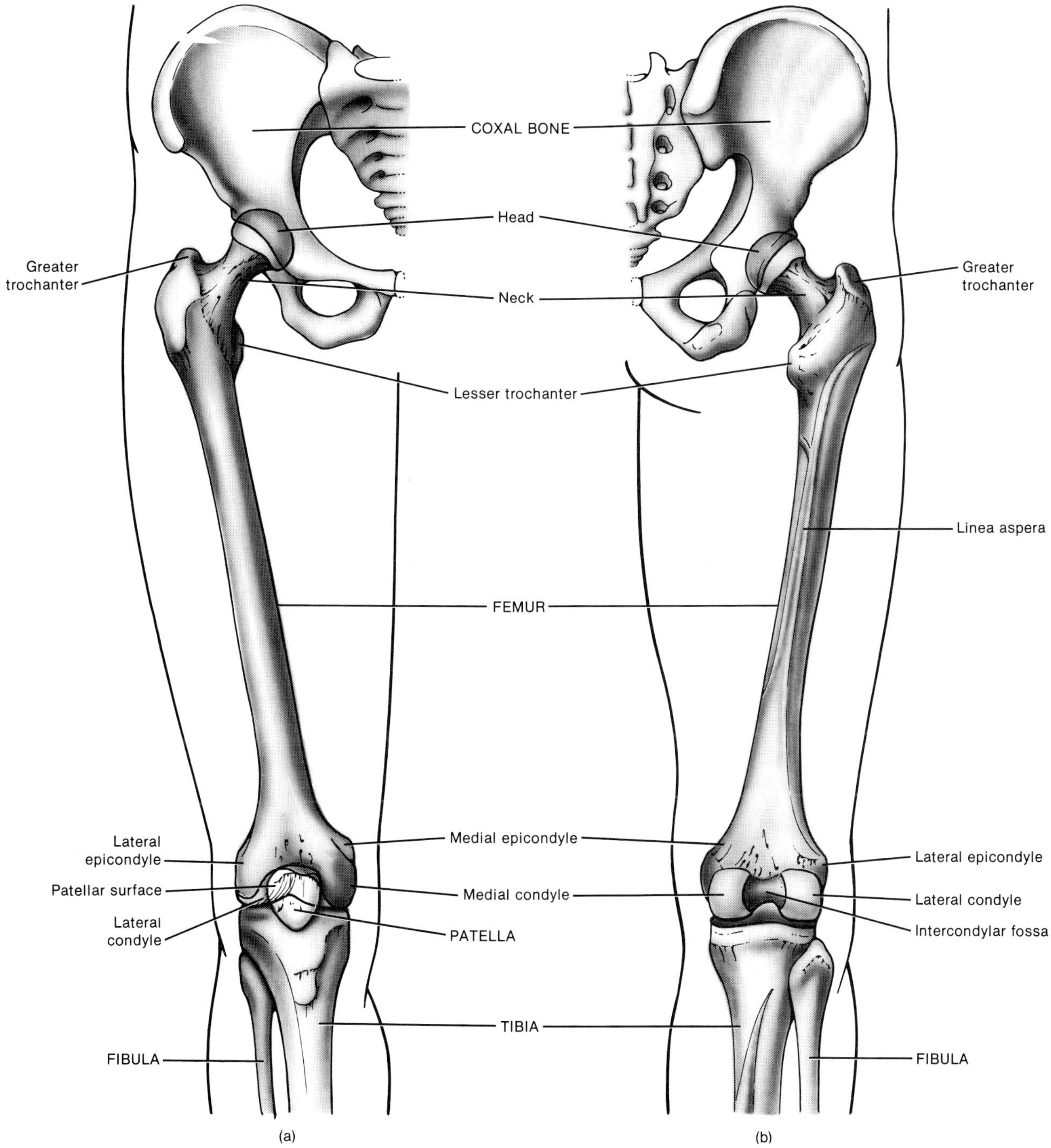

Figure 14-9 Right femur. (a) Anterior view. (b) Posterior view.

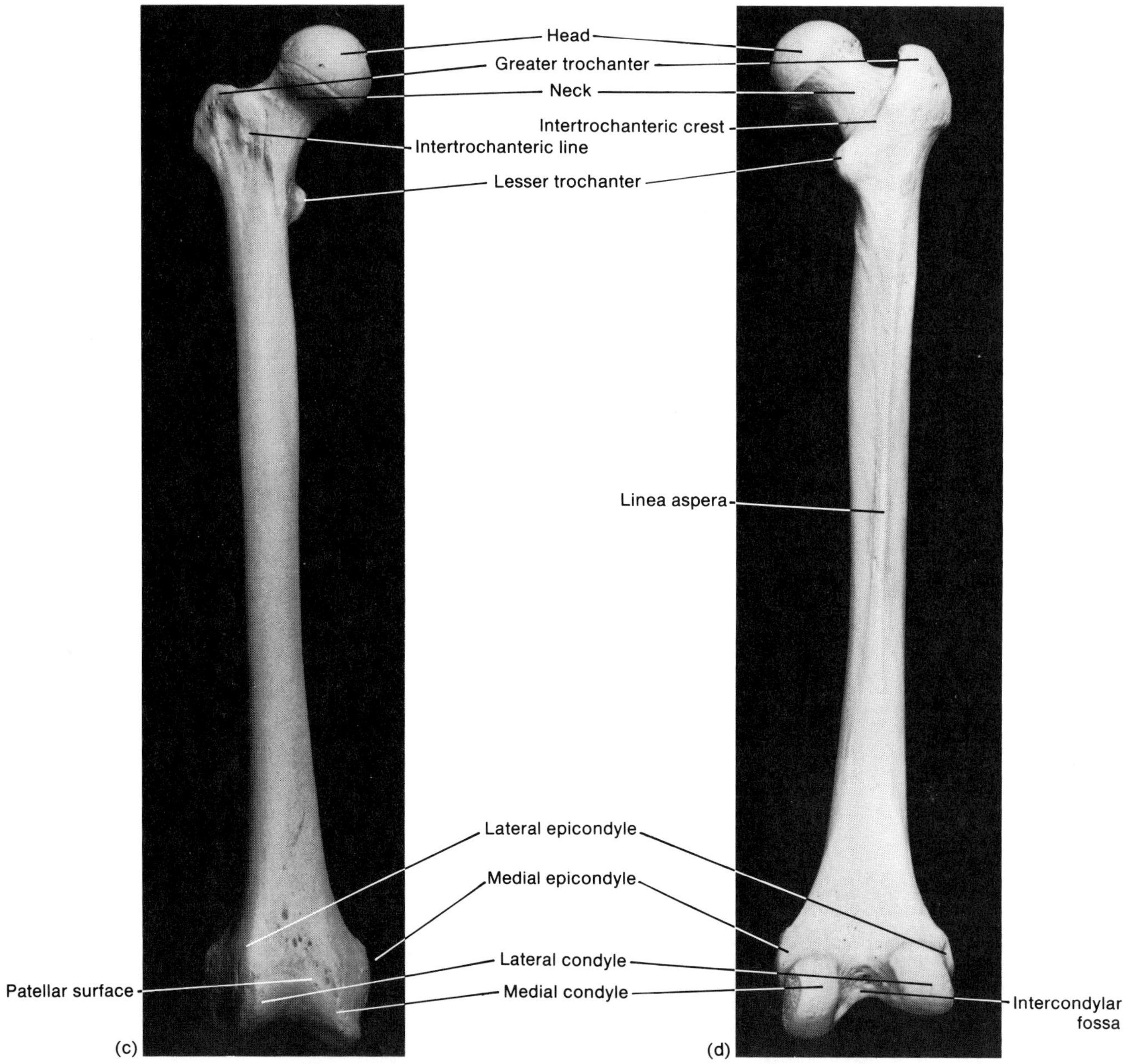

Figure 14-9 *(Continued)* (c) Anterior view. (d) Posterior view. (Photographs courtesy of Lenny Patti.)

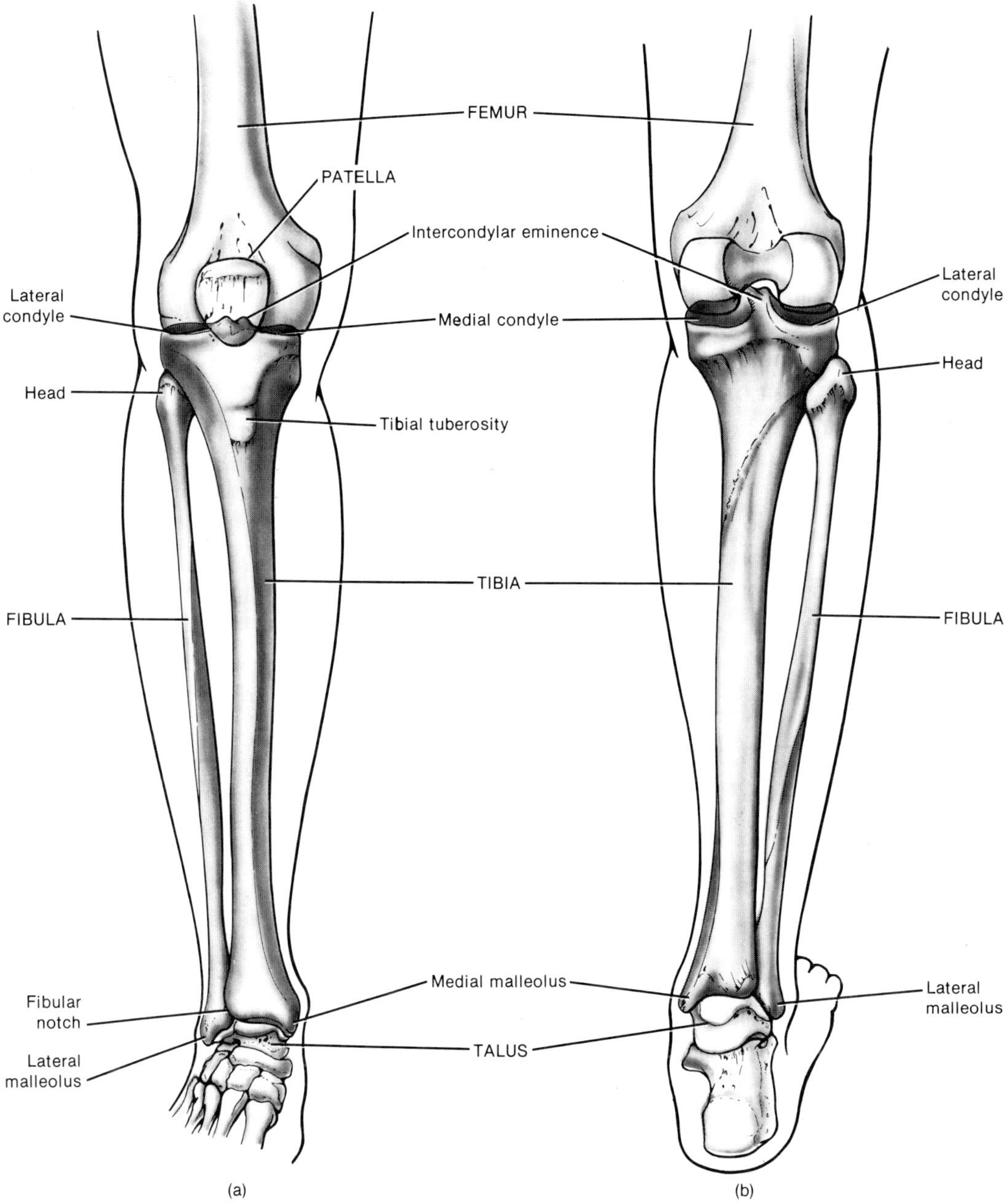

Figure 14-10 Right tibia and fibula. (a) Anterior view. (b) Posterior view.

TARSALS, METATARSALS, AND PHALANGES

The *tarsus,* or ankle, consists of seven bones. Using Figure 14-11 and the laboratory skeleton, locate the following tarsal bones.

1. *Calcaneus* (kal-KĀ-nē-us)—largest of the tarsals, forming the heel
2. *Cuboid* (KYOO-boyd)
3. *Navicular*
4. *First* (medial) *cuneiform* (kyoo-NĒ-i-form)
5. *Second (intermediate) cuneiform*
6. *Third* (lateral) *cuneiform*
7. *Talus*

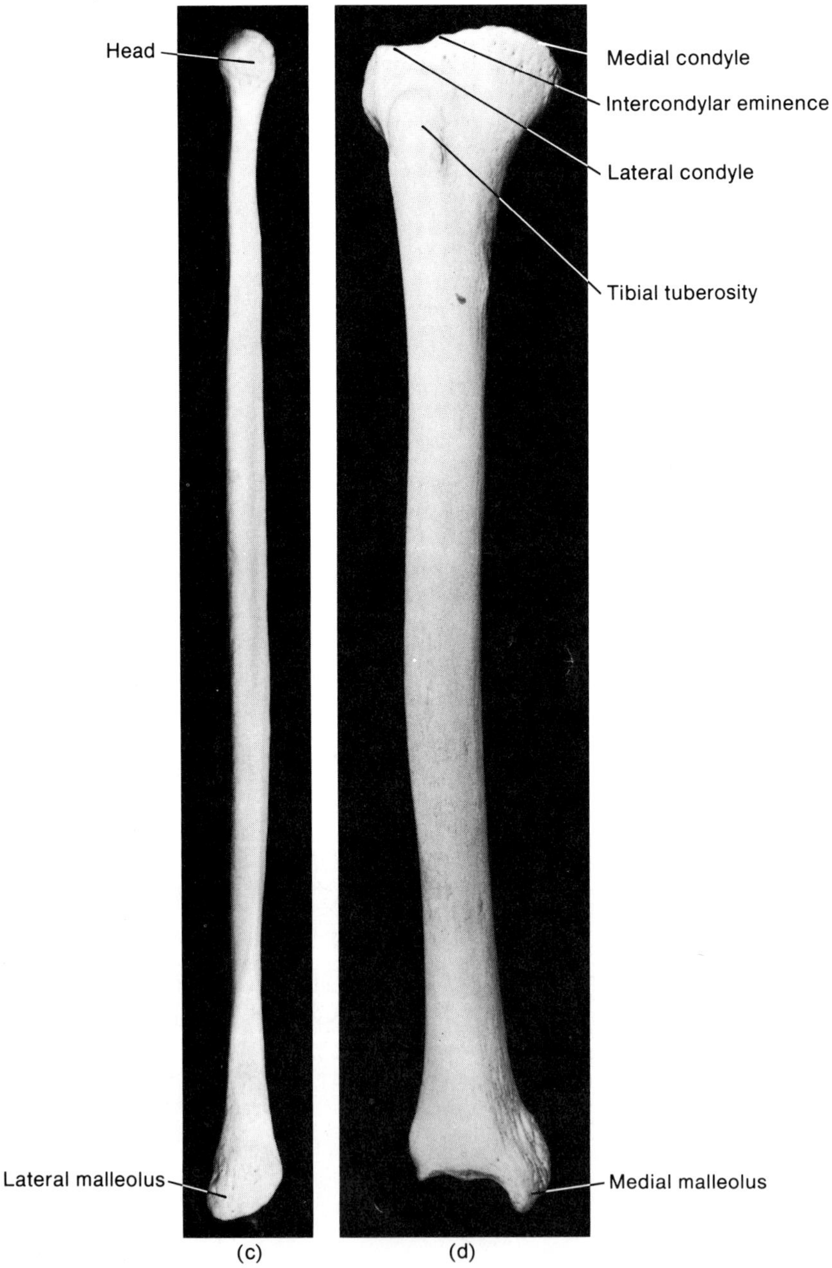

Figure 14-10 *(Continued)* (c) Right fibula, anterior view. (d) Right tibia, anterior view. (Photographs courtesy of Lenny Patti.)

The *metatarsals,* the bones of the foot, are numbered from I to V (Figure 14-11). Number 1 articulates with the "big toe."

The *phalanges* of the foot have the same number and arrangement as those of the hand. There are 14 toe bones, 2 in the big toe and 3 each in the other digits (Figure 14-11).

E. BONES OF THE PELVIC GIRDLE

The pelvic girdle serves as the point of attachment of the leg to the main axis of the body. It is made up of a total of six fused bones, three on each side. The three bones on the right and left sides form the *coxal bones* or *os coxae* (formerly called the *innominate bones*). It is not possible to see the lines of fusion of these bones, but the approximate suture lines are shown in Figure 14–12. Try to locate the approximate lines in your laboratory specimen.

The pelvic girdles of the two sexes are shaped differently, with adaptations related to childbearing (Figure 14-13). The female pelvis is broader and shallower than the male, and the bones are lighter in weight. If you have both male and female skeletons in the lab, compare them.

Study the bones that make up one coxal bone and their markings. The *ilium* is the largest of the three bones and the most superior in position. Locate the following parts (see Figures 14-12 and 14-13).

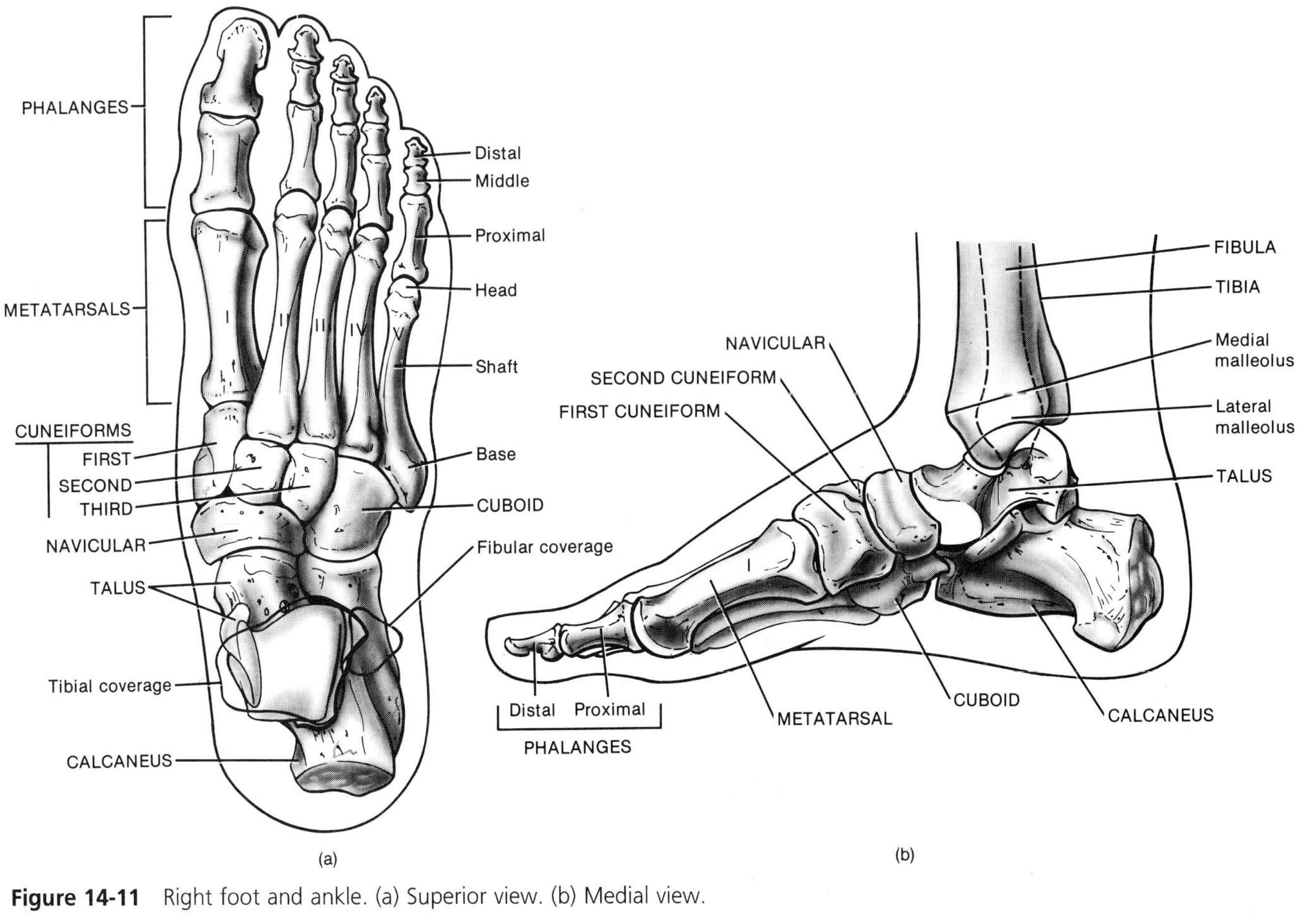

Figure 14-11 Right foot and ankle. (a) Superior view. (b) Medial view.

1. **Iliac crest**—convex superior border
2. **Anterior superior iliac** *spine*
3. **Anterior inferior iliac** *spine*
4. **Posterior inferior iliac** *spine*
5. **Greater sciatic notch** (sī-AT-ik)—concavity below the posterior inferior iliac spine
6. **Iliac fossa**—medial concavity
7. **Iliac tuberosity**—roughened area medial to the iliac fossa, point of attachment of the sacroiliac ligament
8. **Auricular surface** (o-RIK-yoo-lar)—point of articulation with the sacrum

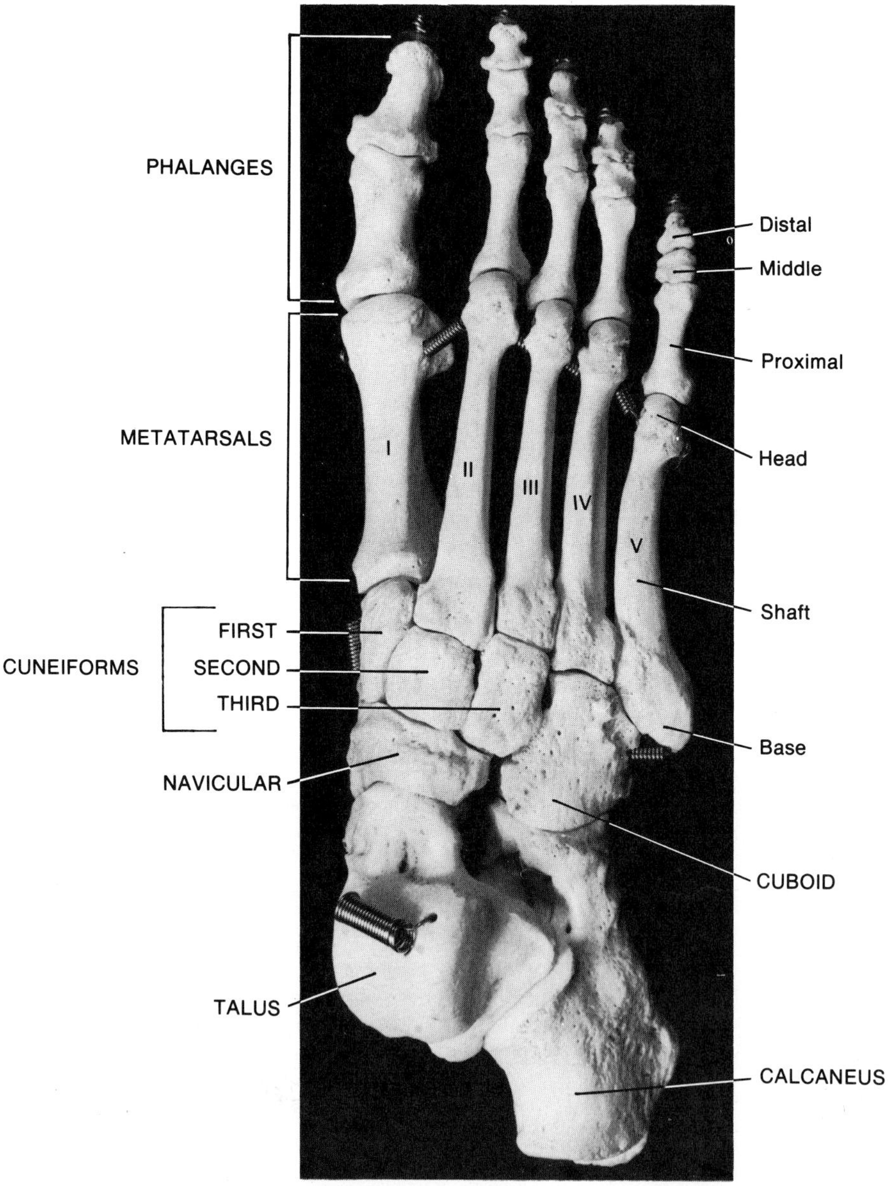

(c)

Figure 14-11 *(Continued)* (c) Superior view. (Courtesy of Lenny Patti.)

The *ischium* is the inferior and posterior part of the coxal bone. It is the part of the pelvic girdle you sit on. Locate the following parts.

1. **Ischial tuberosity**—roughened area at the most posterior and inferior point
2. **Ischial spine**—sharp projection above the ischial tuberosity
3. **Lesser sciatic notch**—concavity below the ischial spine

The *pubis* is the most anterior part of the coxal bone. At the midline, where right and left pubic bones meet, a joint is formed which is called the **pubic symphysis.** It will be discussed in the exercise on articulations.

Other markings on the pelvic girdle are:

1. **Acetabulum** (as′-e-TAB-yoo-lum)—fossa formed by the three bones which articulates with the proximal head of the femur, commonly called the hip socket
2. **Obturator foramen** (OB-tyoo-rā′-ter)—large oval hole surrounded by parts of the ischium and the pubis

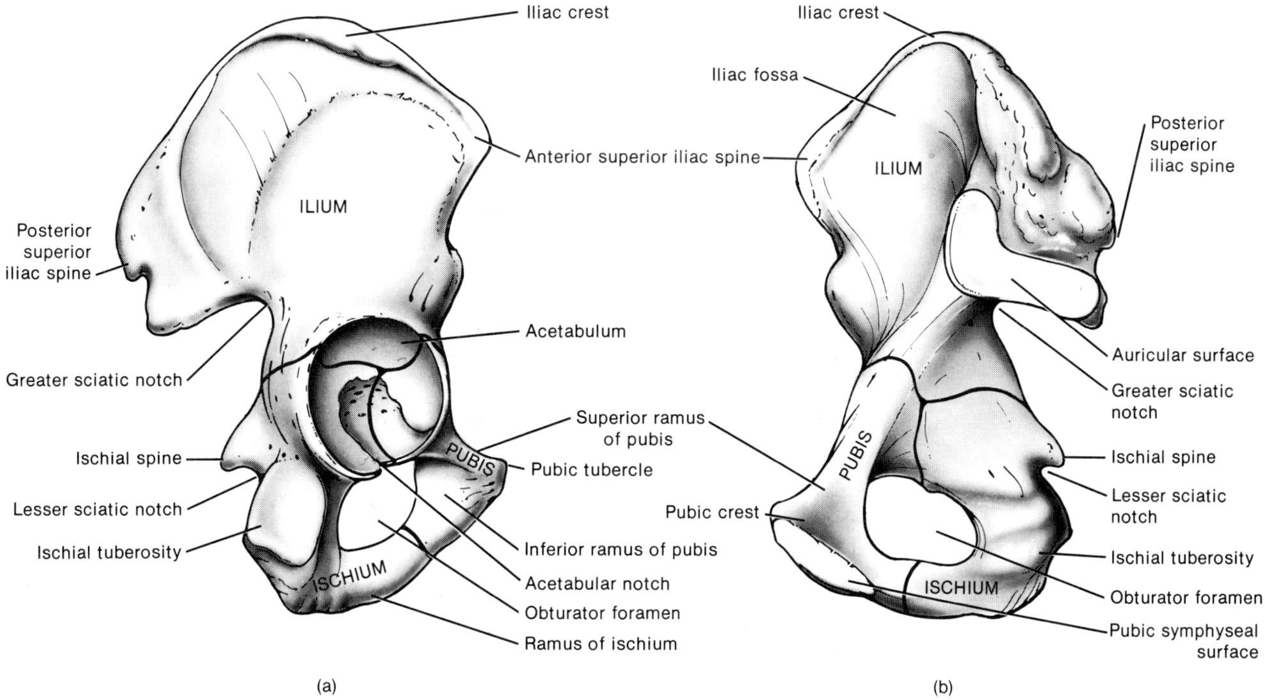

Figure 14-12 Right coxal bone. (a) Lateral view. Fusion lines are shown, but they are probably not seen on your specimen. (b) Medial view.

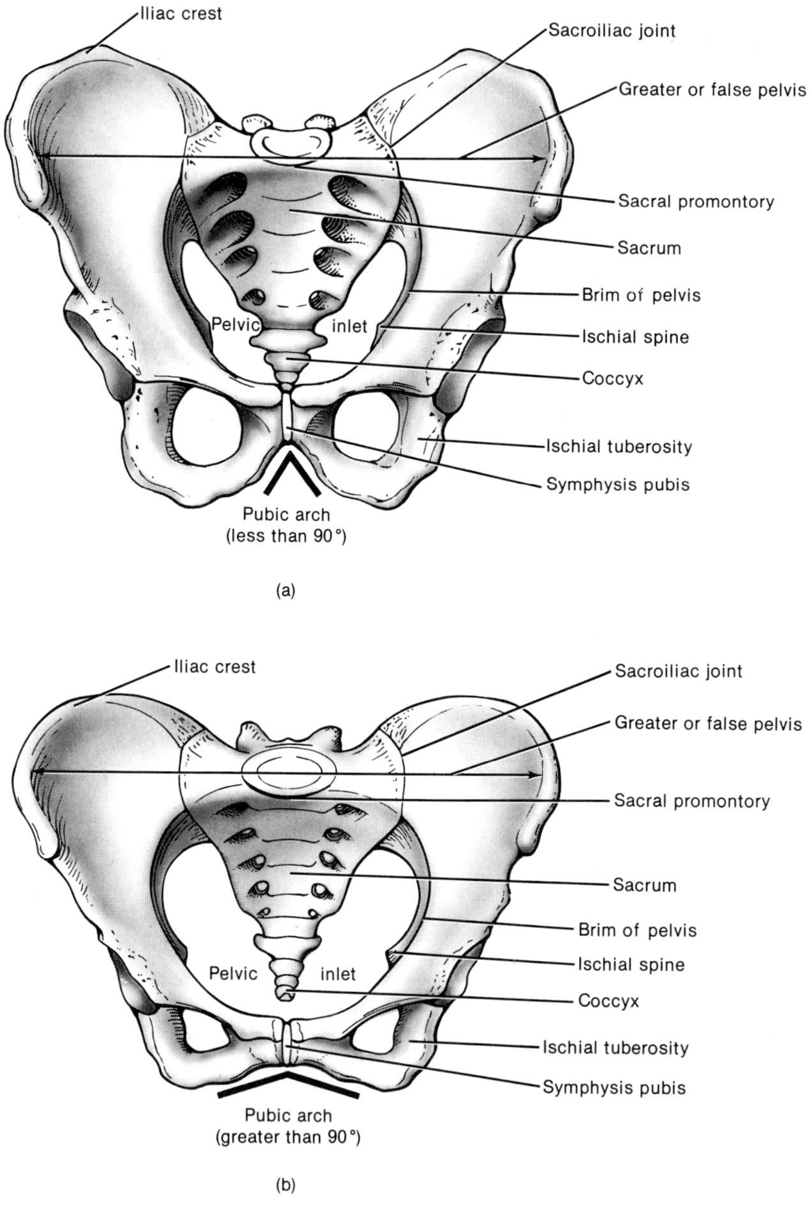

Figure 14-13 Pelvis. (a) Male pelvis in anterior view. (b) Female pelvis in anterior view.

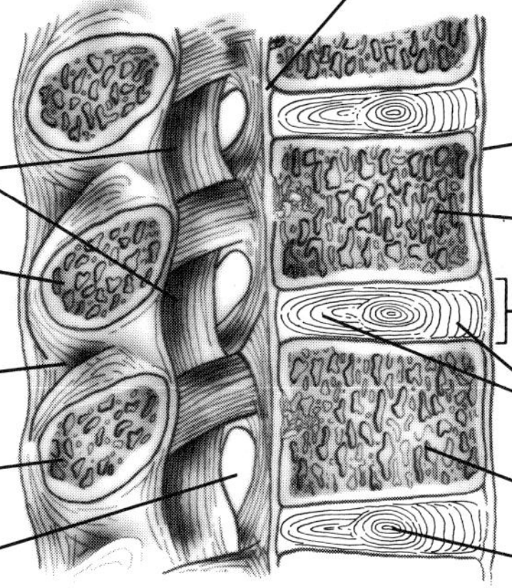

Figure 15-3 A slightly movable joint (amphiarthrosis).

way that movement in only one plane (monaxial) is allowed
Example:

a. Elbow between the distal end of the humerus and the proximal end of the ulna
b. *Knee* between the distal end of the femur and the proximal end of the tibia
c. *Interphalangeal joints* between adjacent phalanges of fingers or toes

2. **Condyloid** (KON-di-loyd) **joint** or **ellipsoidal** (ē-lip-SOYD-al) **joint**—spherical head of one bone fits into an elliptical cavity, allowing movement in two planes (biaxial)
Example:
    a. Between the distal end of the metatarsals 2 to 5 and the proximal end of the first phalanx
    b. Between the atlas of the vertebral column and the occipital condyles of the skull
    c. Between radius and ulna and adjacent carpals (wrist joint)
3. **Ball and socket joint**—rounded head of a bone fits into a capsule of another, allowing movement in three planes (triaxial)
Example:
    a. *Hip* between the proximal end of the femur and the acetabulum
    b. *Shoulder* between the proximal end of the humerus and the glenoid fossa (the most freely movable joint in the body)
4. **Pivotal** or **rotary joint**—one bone turns on another, allowing monaxial movement
Example:
    a. Between the atlas and the axis
    b. Between the proximal and distal ends of the radius and the ulna
5. **Plane** or **gliding joint**—flat articulating surfaces move side to side or forward and back on one another, allowing highly restricted biaxial movement
Example:
    a. Between adjacent carpal bones of the wrist
    b. Between adjacent tarsal bones of the ankle
    c. Between superior and inferior articular surfaces of the vertebrae
    d. Between costal demifacets of thoracic vertebrae and ribs
6. **Saddle joint**—concave surface of one bone fits over a convex surface of another, allowing biaxial movement (much like the ellipsoidal joint)
Example: Between the first metacarpal (thumb) and the greater multangular of the wrist

STRUCTURAL CLASSIFICATION OF JOINTS

On the basis of structure, joints may be classified according to the presence or absence of a joint cavity and the kind of connective tissue between the bones. There are three categories. Compare them with the functional classification system.

A. **Fibrous joint**—bones held together by fibrous connective tissue; no joint cavity; may be synarthrotic or amphiarthrotic
Example:
1. Immovable sutural joints of the skull
2. Slightly movable syndesmoses
3. Immovable gomphoses

B. **Cartilaginous joint**—bones held together by cartilage; no joint cavity; may be synarthrotic or amphiarthrotic
Example:
1. Immovable synchondroses (Figure 15-2)
2. Slightly movable symphyses

C. **Synovial joint**—bones not bound together; separated by a synovial cavity; diarthrotic
Example: Freely movable diarthroses: hinge, condyloid (ellipsoidal), ball and socket pivotal (rotary), plane (gliding), and saddle (Figure 15-4)

JOINT MOVEMENT

The types of movement allowed at freely movable joints are many and varied. There are always factors that prevent dislocation or restrict movement, such as large muscle masses that limit bending, tight tendons and ligaments, and muscle tension. However, most movements can be defined clearly and the student should try to demonstrate them on his or her own body. Try to make the following movements using Figure 15-5 as a guide.

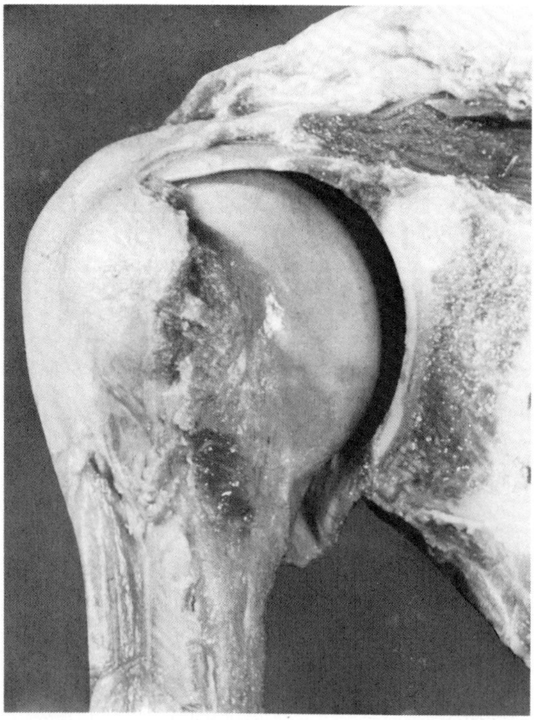

Figure 15-4 A photograph of the shoulder joint partially dissected to expose the glenoid fossa between the humerus and the scapula.

1. **Plane or Gliding**—a simple sliding movement from side to side between adjacent carpals of the wrist. Slide one vertebra on another by bending your back.
2. **Flexion**—decrease the angle between two bones. Fold the elbow.
3. **Extension**—increase the angle between two bones. Straighten the elbow (arm).
4. **Dorsiflexion**—a bending of the foot at the ankle region. With your foot on the ground, bring your toes upward.

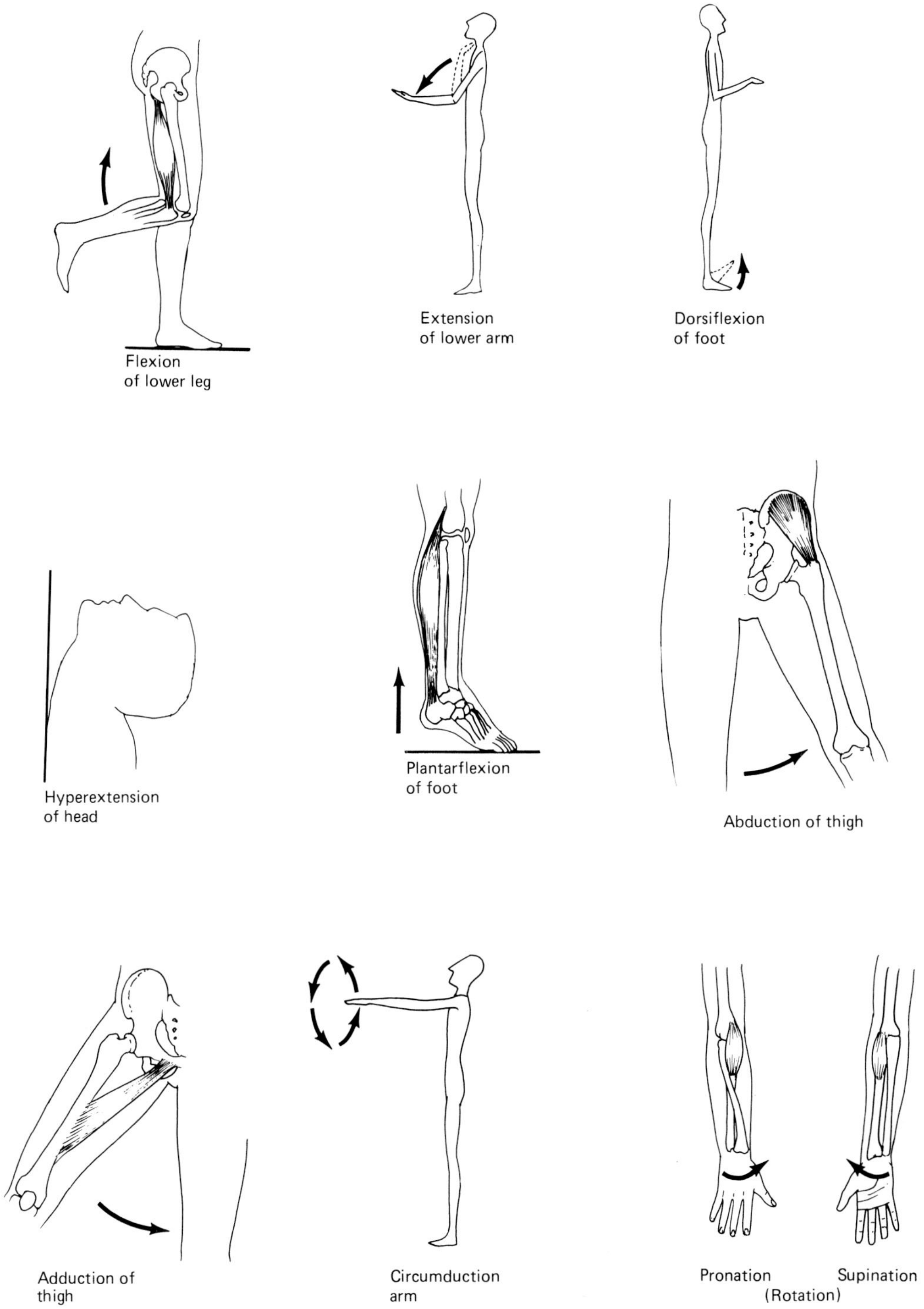

Figure 15-5 Joint movements.

5. **Hyperextension**—straighten one part on another beyond its normal position. Bend the head backward, or bend the knees backward if you can.
6. **Plantar flexion**—extend the foot on the ankle as far as possible. Stand on your toes.
7. **Abduction**—movement of an appendage away from the midline. Move your arm or leg out and away from the body. Spread your fingers.
8. **Adduction**—movement of an appendage toward the midline. Bring your arm or leg back to the midline from an abducted position.
9. **Circumduction**—while the proximal end of the appendage remains stationary, the distal end effects a circle or arc. Extend the arm and make the arc with your hand.
10. **Rotation**—one bone turns on another. Turn the atlas on the axis (turn the head). Rotate the radius on the ulna to turn the palm of the hand upward (supination) or downward (pronation). Turn the leg inward or outward by rotating at the hip.

Further, joints may be classified as:

1. *Monaxial*—movement in one plane only
2. *Biaxial*—movement in two planes
3. *Triaxial*—a wide range of movement

Questions for Review

EXERCISE **15**

ARTICULATIONS

Name Kathy Martin

Laboratory Section Mon 8:00 AM Date 4-1-91

1. Match the functional classifications of joints on the left with the structural classifications to which they belong.

a. 1 Suture
b. 2 Synchondroses
c. 1 Synostoses
d. 2 Symphyses
e. 1 Syndesmoses
f. 3 Diarthroses

1. Fibrous
2. Cartilaginous
3. Synovial

2. Give an example of a specific joint for each functional classification listed in question 1.

a. sutural joints of skull
b. epiphyseal plate
c. shoulder joint
d. intervertebral discs
e. Distal ends of the tibia + fibula
f. Elbow between distal end of humerous + proximal end of tibia

3. List the types of diarthrotic joint and give an example of each.

a. Hinge joint - monaxial - Elbow between distal end of the humerous + proximal of ulna
b. Condyloid - biaxial, between radius and ulna + adjacent carpal
c. Ball + Socket joint - triaxial - shoulder most freely movable joint in body
d. Pivotal or rotary joint - between proximal + distal ends of radius + ulna
e. Plane + gliding joint - between carpal + tarsal bones
f. saddle joint - between first metacarpal + greater multangular of the wrist

4. Match the joints on the left with the term describing its action.

a. 3 Shoulder
b. 2 Between atlas and axis
c. 4 Pubic symphysis

1. Monaxial
2. Biaxial
3. Triaxial
4. Partially movable
5. Immovable

d. ____5____ Between adjace.. skull bones

e. ____1____ Between radius and ulna

f. ____2____ Between articular surfaces of vertebrae

g. ____1____ Knee

h. ____3____ Hip

i. ____4____ Child's epiphyseal plate region in the long bone

j. ____1____ Elbow

SECTION VI

THE MUSCLE SYSTEM

The human body has over 600 skeletal muscles that account for approximately 40–45 percent of an individual's body weight. These muscles are used to move and manipulate body parts.

This section contains exercises on the dissection of cat muscles and their comparison with human muscles. The muscles are divided into groups so that a logical progression can be made to dissect the superficial muscles and the more important deep muscles. No attempt is made to cover all the cat muscles. Comparison of the cat muscles should be made with human muscles and reference should be made to the text drawings and to charts and models.

With each dissection sequence there are the following study aides: (1) a black and white photo of the specific muscle group, (2) a line drawing corresponding to the black and white photo, (3) a table summarizing the origins, insertions, and actions of cat muscles, (4) a table summarizing the origins, insertions, and actions of corresponding human muscles, and (5) color plates of cat muscle dissections.

Because of their importance in body movements, skeletal muscles have long been the subject of studies concerned with their chemical composition, ultrastructure, and mechanism of contraction. Exercises in this section also present several aspects of muscle physiology including the microscopic and submicroscopic properties of skeletal muscle fibers, the chemistry of muscular contraction, the demonstration and measurement of the degree of muscular contraction using a laboratory animal, and the equipment used to study muscle contraction, isotonic and isometric muscle contractions, muscle fatigue, and electromyography.

GENERAL REFERENCES

Berger, R. A. *Applied Exercise Physiology*. Philadelphia: Lea & Febiger, 1982.

Eisenberg, E., and L. E. Greene. "The Relation of Muscle Biochemistry to Muscle Physiology," *Annual Review of Physiology, 42:293, 1980.*

Kapit, W., R. I. Macey, and E. Meisami. *The Physiology Coloring Book*. New York: Harper & Row, 1987.

Tortora, G. J., and N. P. Anagnostakos. *Principles of Anatomy and Physiology,* 6th ed. New York: Harper & Row, 1990.

Witherspoon, J. D. *Human Physiology*. New York: Harper & Row, 1984.

SPECIFIC REFERENCES

Basmajian, J. V. "Electromyography—Dynamic Gross Anatomy: A Review," *American Journal of Anatomy,* 159:245–260, 1980.

Edwards, R. H. T. "Human Muscle Function and Fatigue." In *Human Muscle Fatigue:*

Physiological Mechanisms. Ciba Foundation Symposium, Vol. 82, pp. 1–18. London: Pittman Medical, 1981.

Grinnell, A. D., and M. A. B. Brazier (Eds.). *Regulation of Muscle Contraction: Excitation–Contraction Coupling.* New York: Academic Press, 1981.

Murray, J. M., and A. Weber. "The Cooperative Action of Muscle Proteins," *Scientific American,* 230:58–71, 1974.

Nadel, E. R. "Physiological Adaptations to Aerobic Training," *American Scientist,* 73:334–343, 1985.

Shepard, R. J. *Physiology and Biochemistry of Exercise.* New York: Praeger Publishers, 1982.

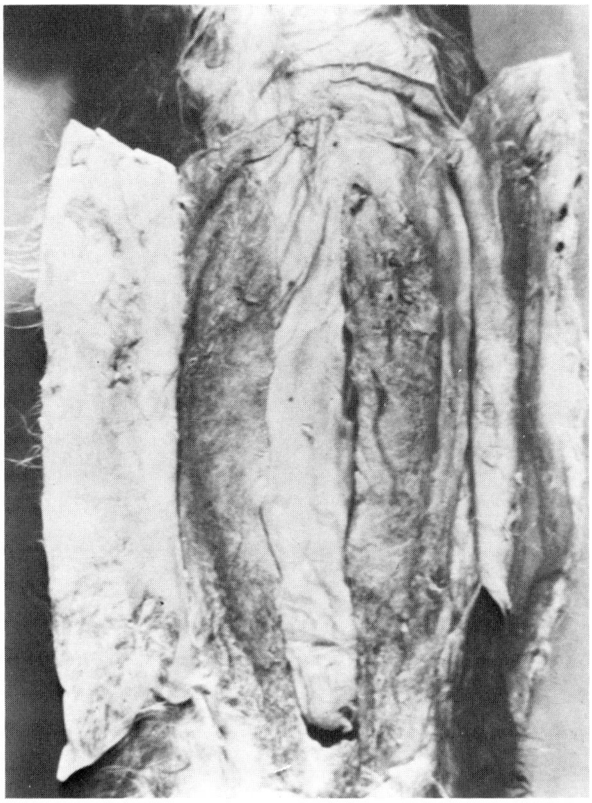

Figure 16-3 Fully developed mammary glands in female cat.

the cat with one hand and pulling the skin with the other. If the skin does not come off easily, you will have to free it by cutting (Figure 16-2c).

The last stage in the skin removal should be done very carefully if you have a male cat. In Figure 16-2d note the two large ducts leading up from the testes between the legs (a probe lies under the left duct). These ducts are an important part of the reproductive system and must *not* be cut off.

Continue pulling the skin down until you reach the external genitalia. Cut a circle around the genitalia and remove the remaining skin on the legs. The skin around the external genitalia is not removed until you study the reproductive system.

If you have a female cat, the above precautions need not be taken. Cut a circle around the anus and simply pull off the remaining skin. Occasionally a lactating female is encountered. The mammary glands may be left attached to the skin as it is being removed, or they may be removed later. Mammary glands are well developed in cats that have just given birth (Figure 16-3).

Now you have removed the skin in one piece. It may be placed under the cat while dissecting or you may keep it in the plastic bag during the laboratory period. In any event, the skin is to be retained and used to wrap the cat at the end of the lab period before storing it in its plastic bag. Wrapping the body in its skin helps to keep the cat moist between laboratory periods. Tie the skin to the body with string.

One must be careful not to let the cat dry out since it is impossible to do a good muscle dissection on a dry animal. If during the course of the semester you find that the cat is losing too much moisture, add a solution of 3 percent phenol, or a preservative purchased from a biological supply house. This solution may be placed directly on the musculature, used to remoisten the skin, or placed in small quantities in the plastic bag before it is closed with string or a rubber band at the end of the lab period.

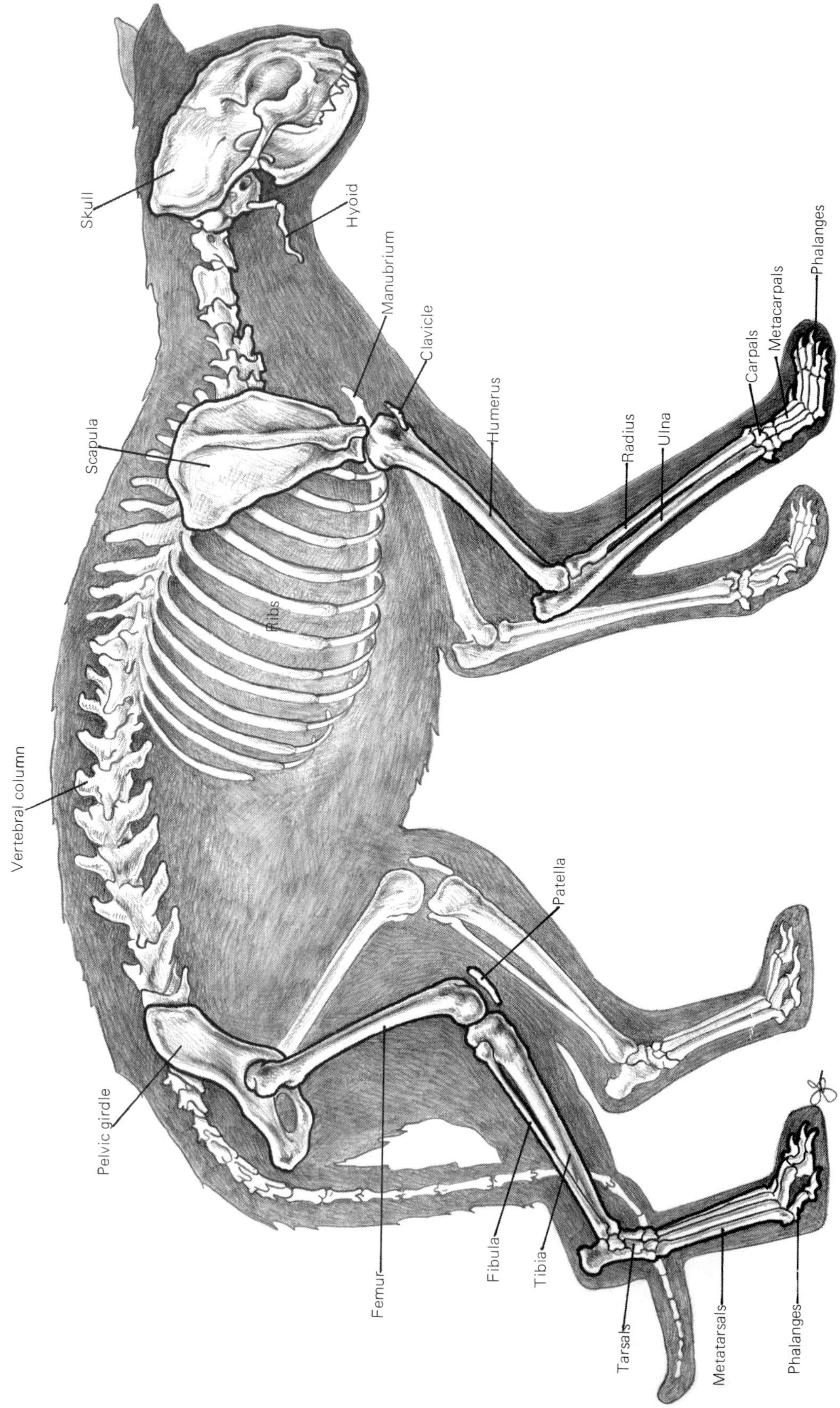

Figure 16-4 Cat skeleton.

The last step before beginning the dissection of the muscles is the removal of all extraneous material from the surface, such as bits of fat, hair, loose fascia, and remnants of cutaneous maximus muscle and platysma. If you have any doubts about the removal of any type of tissue, ask the instructor for advice so that you do not cut off something essential to your future dissection. You can always remove an unwanted substance from the surface at a later time, but you can't reattach something you have cut off and thrown away. Always proceed slowly after giving each move careful thought. When the surface is properly cleaned off, you should be able to make out some muscle outlines and see the directions of the muscle fibers. Make certain that all **cutaneous maximus** and **platysma** have been removed. Have your instructor check your cat before going on.

OVERVIEW OF MUSCLES AND MUSCLE TISSUE

Although muscular tissue can be divided into three categories on the basis of its location, the cells in muscle tissue all have one thing in common. The fibrils within them shorten and thicken and thus effect movement. Review the histology of the following muscles in Exercise 7.

1. *Visceral muscle*—found in visceral organs, largely involuntary action
2. *Cardiac muscle*—found only in the heart, involuntary
3. *Skeletal muscle*—associated with skeletal parts, voluntary
   skeletal muscle is the subject for the study of muscles in this unit.

THE STRUCTURE OF A MUSCLE

Each muscle has three main parts.

1. *Origin*—end attached to the part of the skeleton that does not move when the muscle contracts
2. *Insertion*—end attached to the part of the skeleton that moves when the muscle contracts
3. *Body* or *belly*—center of the muscle between the origin and the insertion

Most muscles are subdivided into bundles or *fasciculae.* This is an important concept to keep in mind since there is often a tendency for the student to try to subdivide a muscle into several parts simply because it seems possible to do so. The entire extent, size, and shape of a muscle must be determined *before* beginning to dissect it, and its origin and insertion should be located on the skeleton. If you have a cat skeleton available, identify the bones. You will find that they are similar to the human. Locate each origin and insertion as you go along. This manual provides photographs, line drawings, and tables indicating the origin, insertion, and action of each muscle, so that careful study of each muscle can be made *before the dissection is begun.*

Table 16-1

Muscle Action	Description
Flexion	*Flexor muscle* bends a joint or reduces the angle between two bones
Extension	*Extensor muscle* straightens a joint or increases the angle between two bones
Abduction	*Abductor muscle* carries an appendage or a part away from the midline
Adduction	*Adductor muscle* brings an appendage or a part back to the midline
Elevation	*Levator muscle* raises a part
Depression	*Depressor muscle* lowers a part
Rotation	*Rotator muscle* turns one part on another

MUSCLE ACTIONS

Use your text and your own body movements for demonstrations of muscle actions. Some of the most frequently encountered are listed in Table 16-1. A diagram of the cat skeleton (Figure 16-4) should be used to locate points of muscle origins and insertions.

DISSECTION

To dissect a muscle does not mean to cut into it, but merely to separate it cleanly between its origin and insertion from the adjacent muscles. If deeper muscles are to be viewed and dissected, it will be necessary to *transect.* To transect a muscle, find the midpoint between its origin and insertion, and cut it in half (or as close to half as possible in the case of odd-shaped muscles), perpendicular to the direction of its fibers. Fold each half back to origin or insertion. You will then be able to observe the underlying muscles. Never transect a muscle unless you are specifically instructed to do so in this manual or by your instructor.

Each muscle is surrounded by connective tissue called *fascia.* Extensions of fascia around and into the muscle are differentiated into *epimysium, perimysium,* and *endomysium,* but since we will not be separating muscles into their bundles, the term fascia is used exclusively.

Most muscles do not attach directly to the bone, but rather by extensions of fascia. Connective tissue that binds a muscle to a bone is called a **tendon.** Connective tissue that connects one bone with another is called a **ligament.** The name *aponeurosis* (ap′-ō-nyoo-RŌ-sis) is often applied to a broad, flat tendon.

The connective tissue that binds muscle together may be very tough and require cutting with a scissors or a scalpel, or it may be so soft and fragile that it can be broken by running a finger between the muscles. Do whatever you can with your fingers, resorting to scissors or scalpel only when necessary. The older the animal is, the tougher the fascia. If you have an old cat to dissect, you will find that you have to work harder to separate muscles than your neighbor who has a younger specimen.

To save time, instructions for dissection are given for one side of the cat. Corresponding photographs and line drawings in each exercise show the *left side.*

Since this is probably the first time you have made a dissection of this kind, it would be wise to take your time, think carefully before dissecting or transecting, and remember that once a wrong cut is made, it cannot be corrected (except with needle and thread). One's laboratory skill often enters into the determination of the course grade. This is the only cat you will get, so take precautions, and do not make mistakes that could be avoided simply by reading the instructions.

Immediately following this exercise there are diagrams of human musculature. Refer to these (Figures 16-5 and 16-6) and other diagrams throughout all your work on human muscles, and compare the cat muscles you have dissected with their human counterparts.

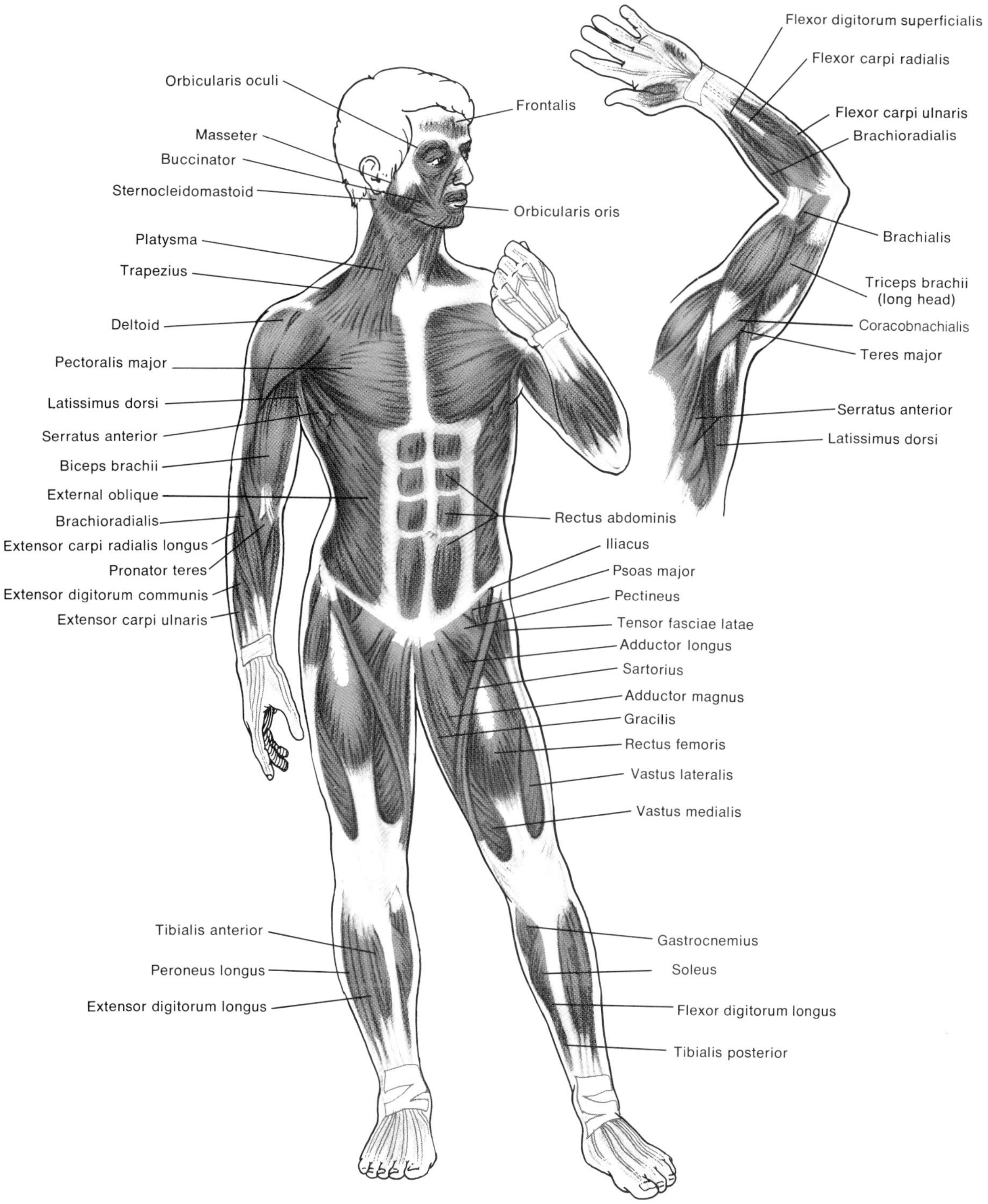

Figure 16-5 Principal superficial muscles. Anterior view.

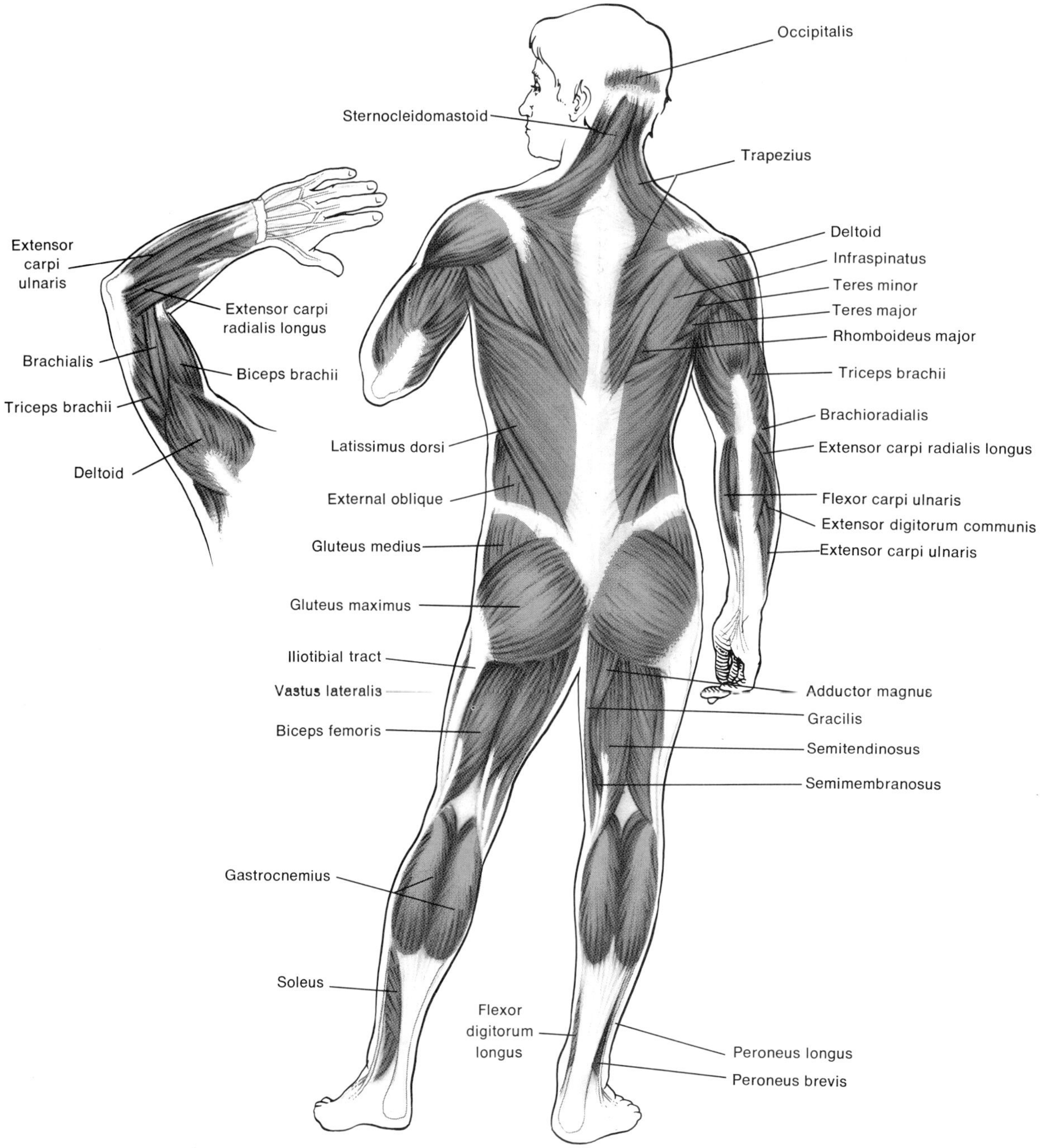

Figure 16-6 Principal superficial muscles. Posterior view.

Questions for Review

EXERCISE **16**

REMOVAL OF THE SKIN AND MUSCLE REVIEW

Name ______________________________

Laboratory Section ______________________ Date __________

1. For each of the following terms state what it is, where it is located, and what function, if any, it has.

   a. Cutaneous maximus ______________________________

   ____

   ____

   b. Platysma ______________________________

   ____

   ____

   c. Fascia ______________________________

   ____

   ____

2. How is the origin of a muscle different from the insertion? ______________

   ____

   ____

3. What is the difference between dissecting and transecting? ______________

   ____

   ____

   ____

4. Under what circumstances would it be necessary to transect a muscle? ______

   ____

5. Define the following.

   a. Tendon ______________________________

   ____

   b. Ligament ______________________________

   ____

   c. Aponeurosis ______________________________

   ____

EXERCISE 17

MUSCLES OF THE BACK

Materials

The following materials should be provided for class use:

1. Embalmed cats
2. Dissecting tools
3. Dissecting trays
4. String
5. 3 percent phenol solution

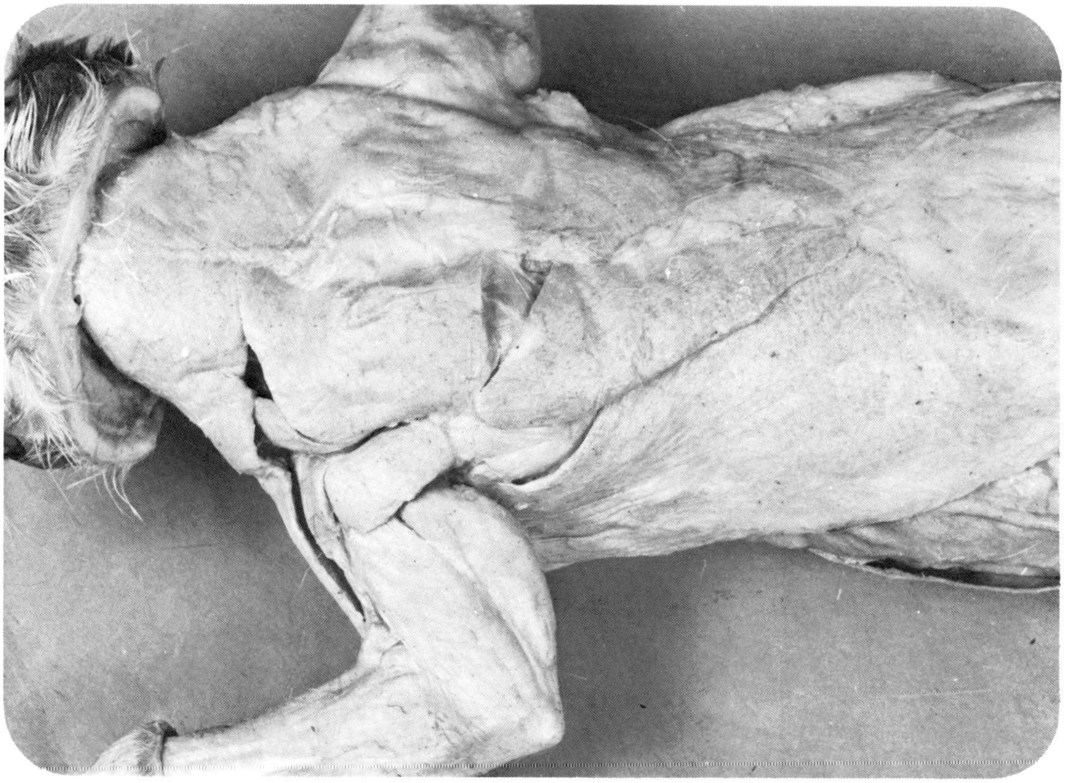

Figure 17-1a Superficial back muscles.

SUPERFICIAL BACK MUSCLES

INSTRUCTIONS FOR DISSECTION

Begin your dissection in the upper back region between the two arms. Be sure that the skin has been removed from the neck as far back as the ears (see Figure 17-1a). Determine the midline *raphe* (RĀ-fē) and locate the *trapezius group* of muscles, which have their origins on this raphe. Humans have only one trapezius muscle, but the cat has three, as shown in Figure 17-1b.

Start with the most posterior of the group, the **spinotrapezius.** This is a triangular muscle that lies over the latissimus dorsi at its posterior end and joins the acromiotrapezius anteriorly. Dissect it completely, being careful not to loosen either its origin or insertion.

Next dissect the **acromiotrapezius** and finally the **clavotrapezius,** both four-sided muscles. The clavotrapezius inserts on the very small clavicle bone embedded in the muscle between clavotrapezius and clavodeltoid. Try to feel the clavicle with your fingertips and locate it on the cat skeleton.

Underneath the acromiotrapezius and running craniad is the **levator scapulae ventralis.** Free it as much as possible. It will be seen in greater length in a later dissection.

Posterior to the spinotrapezius is the very large **latissimus dorsi.** Free it between origin and insertion. It originates by a broad, tough aponeurosis along the midline of the back. Loosen it all the way to its origin.

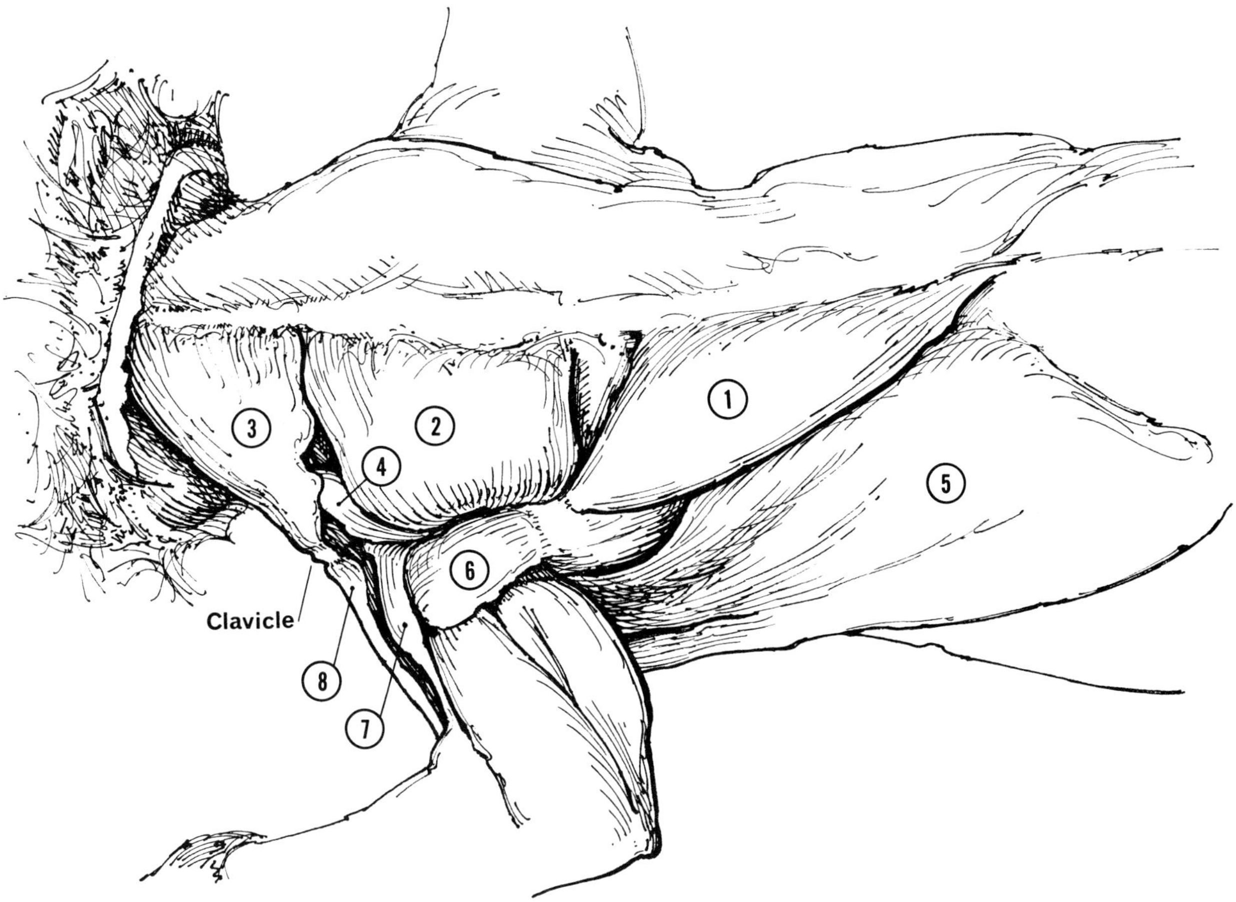

Figure 17-1b Superficial back muscles.
1. Spinotrapezius (spī-nō-tra-PĒ-zē-us)
2. Acromiotrapezius (a-KRŌ-mē-ō-)
3. Clavotrapezius (KLAV-ō-)
4. Levator scapulae ventralis (le-VĀ-tor SKAP-yoo-lē ven-TRAL-is)
5. Latissimus dorsi (la-TISS-i-mus DOR-sī)
6. Spinodeltoid (spī-nō-DEL-Toyd)
7. Acromiodeltoid
8. Clavodeltoid or clavobrachialis (klav-ō-brāk′-ēAL-is)

The deltoid group of muscles (shoulder muscles) is represented in the cat by three muscles, while in the human there is only one. The **clavodeltoid** extends down the arm from the clavicle bone and is sometimes called the **clavobrachialis.** Lying posterior to the clavodeltoid is the triangular **acromiodeltoid.** Located just over the shoulder tip, it is difficult to loosen and so should just be outlined. The most posterior of the group is the **spinodeltoid.** This is a rectangular muscle and, like the acromiodeltoid, should be outlined rather than dissected away from the bone.

Take note of the fact that the muscles in these two groups have identical prefixes, such as *acromio*trapezius and *acromio*deltoid. Making yourself aware of things like this will facilitate memorization of names.

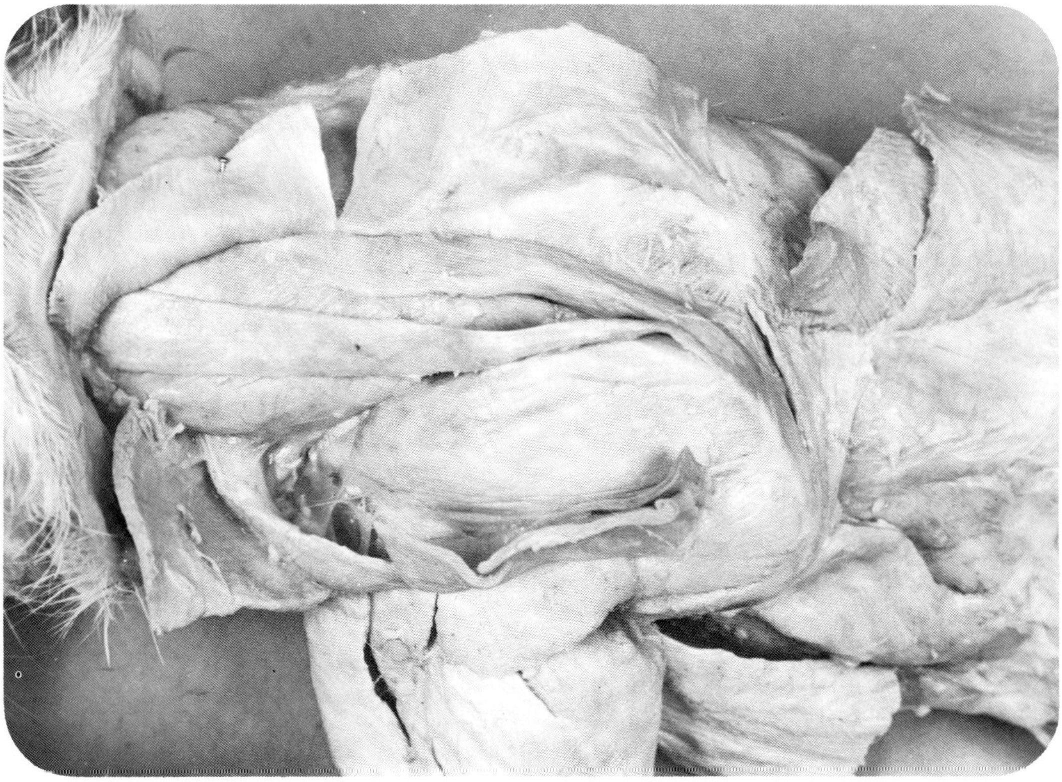

Figure 17-2a Muscles associated with scapula.

MUSCLES ASSOCIATED WITH THE SCAPULA

INSTRUCTIONS FOR DISSECTION

At this time it will be necessary to transect the trapezius group of muscles and the latissimus dorsi. Begin with the **spinotrapezius.** It is triangular in shape so that the transection line will have to be angled to divide the muscle into two separate parts. The **acromiotrapezius** and **clavotrapezius** should be transected midway. Pull back cut ends to origins and insertions. The **latissimus dorsi** is irregular in shape, and it will not be possible to get two equal halves. Angle your transection line. To pull the muscle to its origin on the midline, you will have to use scissors to complete the cut in the fascia near the posterior end of the muscle. Once this has been done, it can be pulled to the midline quite easily and reveals beneath it a thick silvery aponeurosis which is shown in Fig 17-3a and color plate 17-3.

Pull the scapula away from the body, remove excess fascia, and observe the **rhomboideus major,** the **rhomboideus minor,** and the **rhomboideus capitis** muscles (Figure 17-2a). The rhomboideus minor is larger than the rhomboideus major and lies anterior to it (Figure 17-2b). Separate them. On the lateral surface of the scapula locate the **supraspinatus, infraspinatus,** and **teres major.** The supraspinatus and infraspinatus muscles need not be dissected from the bony scapula, but be certain that the spine of the scapula is visible, showing the separation of the two muscles. Beneath the **teres major** and paralleling it is the small **teres minor.** Observe it.

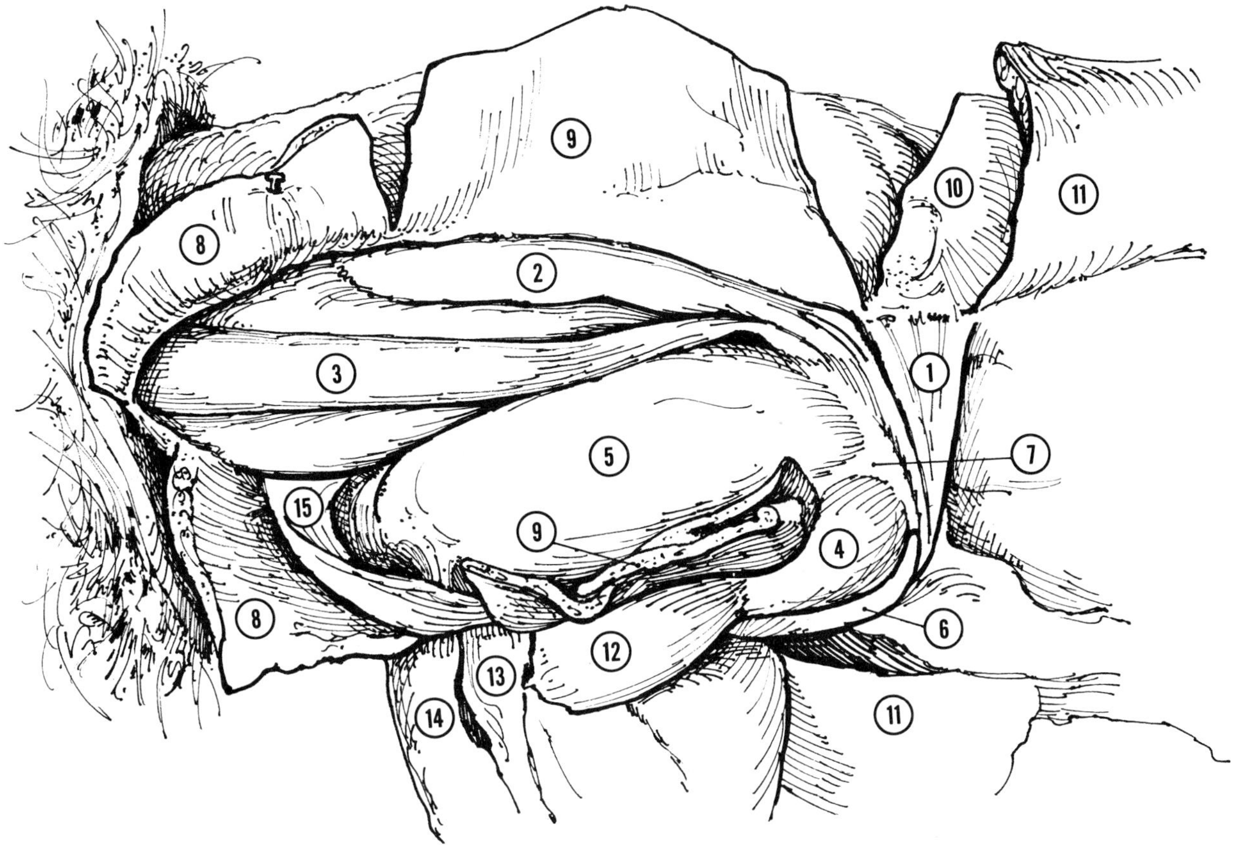

Figure 17-2b Muscles associated with scapula.

1. Rhomboideus major (rom-BOY-dē-us)
2. Rhomboideus minor
3. Rhomboideus capitis or occipitoscapularis (KA-pi-tis; ok-sip′-i-tō-skap′-yoo-LAR-is)
4. Infraspinatus (in′-fra-spī-NA-tus)
5. Supraspinatus (soo′-pra-)
6. Teres major (TER-ēz)
7. Scapular spine (bone)
8. Clavotrapezius (reflected)
9. Acromiotrapezius (reflected)
10. Spinotrapezius (reflected)
11. Latissimus dorsi (reflected)
12. Spinodeltoid
13. Acromiodeltoid
14. Clavodeltoid or clavobrachialis
15. Levator scapulae ventralis

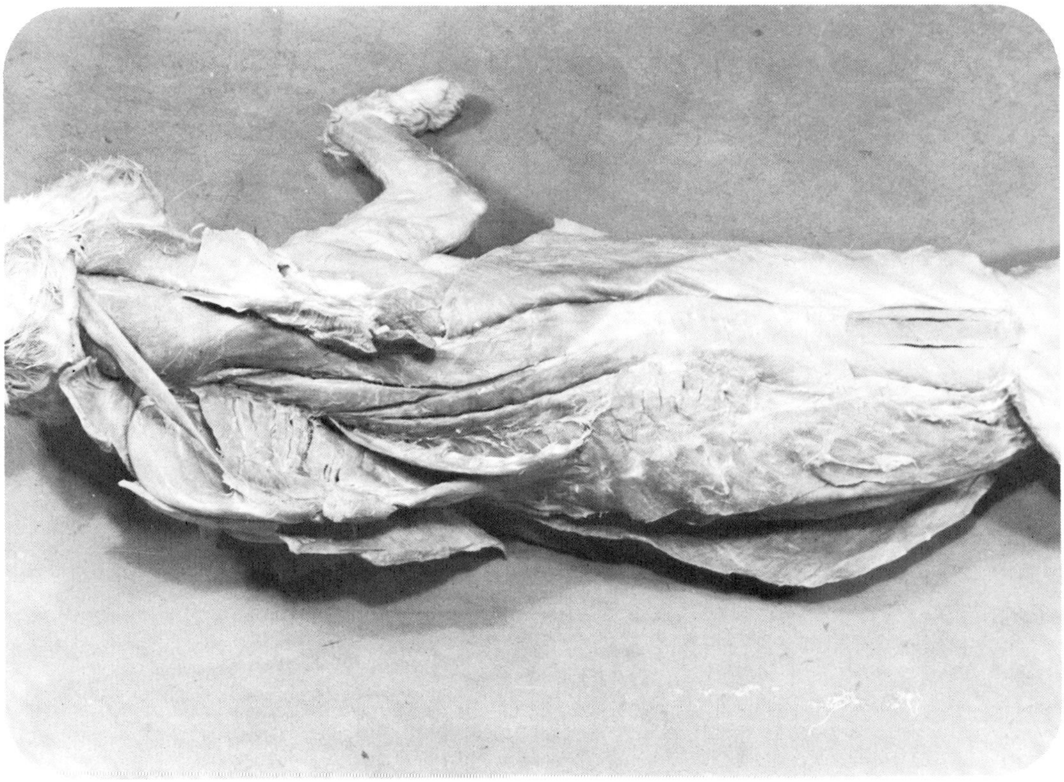

Figure 17-3a Deep back muscles.

DEEP BACK MUSCLES

INSTRUCTIONS FOR DISSECTION

If you have not already drawn the latissimus dorsi back to its origin, do so now. To observe the deeper muscles, it will be necessary to transect the **rhomboideus major** and **minor** muscles at this time. The rhomboideus muscles hold the dorsal part of the scapula to the body, and when they are transected, the scapula can be pulled away from the body to the left side of the cat (Figure 17-3a). Do not transect the rhomboideus capitis.

Anterior to the rhomboids, near the head, is a broad, flat muscle, the **splenius.** Do not dissect it, but observe its extent. Posterior to the splenius and medial in position is the **spinalis dorsi** (Figure 17-3b).

To the left of the spinalis dorsi is a very long muscle with three distinct parts: (1) the **longissimus dorsi,** at the level of the spinalis dorsi, (2) the **longissimus cervicis,** just anterior, and (3) the **longissimus capitis,** extending forward into the head region. You will not be able to see all of the last muscle, but separate as much of it as possible from the splenius. The longissimus dorsi reaches to the posterior end of the back but seems to disappear under a tough aponeurosis that is difficult to separate from the muscle at this point. To see the posterior end of the longissimus dorsi, which divides into three bundles, cut off the tough fascia as far forward as possible in a rectangular block (see photo and line drawing). This will expose four muscle bundles. The one closest to the midline is the **multifidus spinae** (not found in humans) and the other three are the longissimus dorsi.

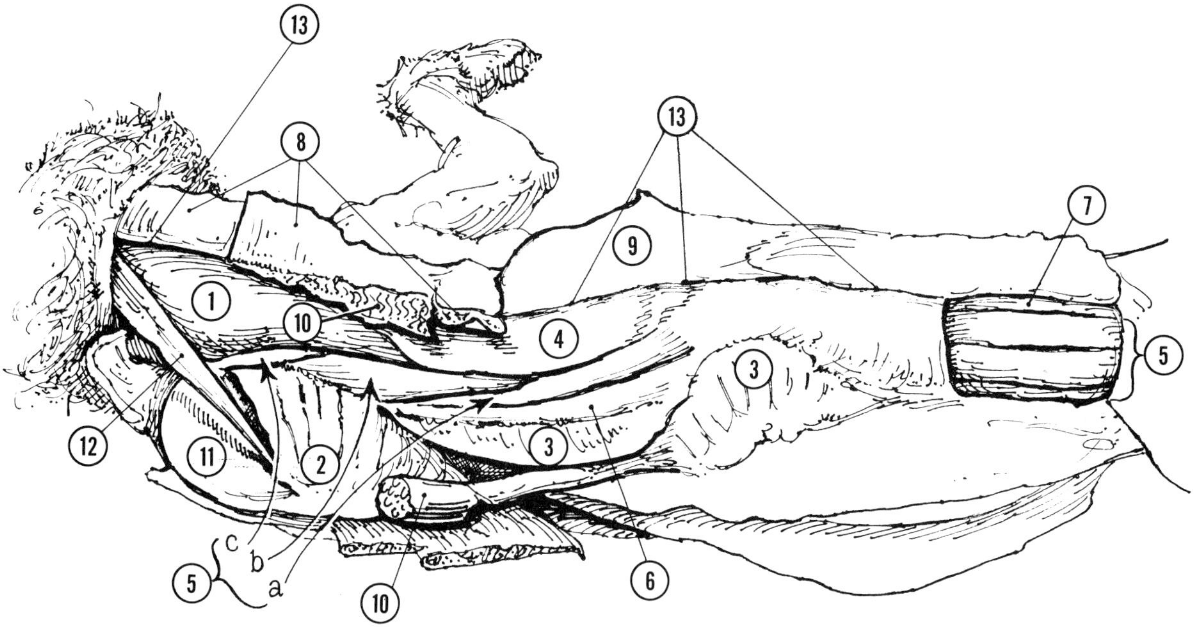

Figure 17-3b Deep back muscles.
1. Splenius (SPLĒ-nē-us)
2. Serratus ventralis (ser-RĀ-tus ven-TRAL-is)
3. Serratus dorsalis (partially reflected)
4. Spinalis dorsi (spī-NAL-is)
5. Longissimus dorsi (posterior part) (lon-JISS-i-mus)
   a. Longissimus thoracis (thō-RA-sis)
   b. Longissimus cervicis (Ser-vi-sis)
   c. Longissimus capitis (KAP-i-tis)
6. Iliocostalis (ill′-ē-ō-kos-TAL-is)
7. Multifidus spinae (mul-TIF-i-dus)
8. Trapezius group (reflected)
9. Latissimus dorsi (reflected)
10. Rhomboideus (cut end)
11. Subscapularis
12. Rhomboideus capitis
13. Midline

Return to the thoracic region. Lateral to the longissimus dorsi is a serrated muscle that originates by a thin, long aponeurosis. This is the **serratus dorsalis.** Although this aponeurosis is an origin, it must be cut in order to pull the serratus dorsalis down laterally and expose the **iliocostalis** beneath it.

Between the scapula and the ventral body region is another serrated muscle, the **serratus ventralis.** The muscle occupying the subscapular fossa is the **subscapularis.** Refer to Tables 17-1 and 17-3 for the origin, insertion, and action of the cat's back muscles.

Table 17-1 Superficial Back Muscles and Muscles Associated with the Scapula of the Cat

Muscle	Origin	Insertion	Action
Spinotrapezius	Spines of thoracic vertebrae	Spine of scapula	Pulls scapula dorsocaudad
Acromiotrapezius	Spines of cervical and thoracic vertebrae	Spine and acromion process of scapula	Holds scapula in place
Clavotrapezius	Lambdoidal crest and axis	Clavicle	Draws scapula craniad and dorsad
Levator scapulae ventralis	Occipital bone and atlas	Spine of scapula	Pulls scapula craniad
Latissimus dorsi	Spines of thoracic and lumbar vertebrae	Intertubercular groove of humerus	Pulls arm dorsad and caudad
Spinodeltoid	Spine of scapula	Deltoid ridge of humerus	Flexes humerus and rotates it outward
Acromiodeltoid	Acromion process of scapula	Deltoid ridge of humerus	Flexes humerus and rotates it outward
Clavodeltoid	Clavotrapezius muscle and clavicle	Humerus	Flexes forearm
Rhomboideus major	Spines of thoracic vertebrae	Vertebral border of scapula	Pulls scapula dorsally and medially
Rhomboideus minor	Spines of thoracic vertebrae	Vertebral border of scapula	Pulls scapula dorsally and medially
Rhomboideus capitis	Lambdoidal ridge	Vertebral border of scapula	Pulls scapula craniad
Supraspinatus	Supraspinous fossa of scapula	Greater tubercle of humerus	Extends humerus
Infraspinatus	Infraspinous fossa of scapula	Greater tubercle of humerus	Rotates humerus outward
Teres major	Axillary border of scapula	Humerus	Flexes humerus and rotates it inward

Table 17-2 Superficial Back Muscles of the Human

Muscle	Origin	Insertion	Action
Trapezius	Occipital bone and cervical and thoracic vertebrae	Spine and acromion process of scapula	Raises clavicle; adducts, elevates, and depresses scapula
Deltoid	Clavicle and spine and acromion process of scapula	Humerus	Abducts arm
Levator scapulae	Cervical vertebrae	Vertebral border of scapula	Raises scapula
Latissimus dorsi	Spines of thoracic and lumbar vertebrae, crests of sacrum and ilium, last four ribs	Intertubercular groove of humerus	Extend, adducts, and rotates arm medially; pulls shoulder down and back
Rhomboideus major	Spines of second through fifth thoracic vertebrae	Vertebral border of scapula	Moves scapula back and up and rotates it downward
Rhomboideus minor	Spines of seventh cervical and first thoracic vertebrae	Superior angle of scapula	Adducts scapula
Supraspinatus	Supraspinous fossa of scapula	Greater tubercle of humerus	Adducts arm
Infraspinatus	Infraspinous fossa of scapula	Greater tubercle of humerus	Rotates humerus laterally
Teres major	Inferior angle of scapula	Proximal end of humerus	Extends humerus; adducts and rotates arm

Table 17-3 Deep Back Muscles of the Cat

Muscle	Origin	Insertion	Action
Splenius	Spines of thoracic vertebrae	Lambdoidal ridge	Turns or raises head
Serratus ventralis	First ten ribs	Medial surface of scapula near the vertebral border	Pulls scapula forward and depresses it
Serratus dorsalis	Mid-dorsal cervical, thoracic, and lumbar raphe	On each rib	Pulls ribs craniad
Spinalis dorsi	Last four thoracic vertebrae	Thoracic and cervical vertebrae	Extends vertebral column
Longissimus dorsi, capitis, and cervicis	Sacral and caudal vertebrae	Cervical vertebrae	Extends vertebral column
Iliocostalis	Angles of posterior thoracic ribs by separate bundles	Three to four ribs anterior to origin of each bundle	Draws ribs together
Multifidus spinae	Transverse processes of lumbar vertebrae	Spines of vertebrae anterior to origin of each slip	Extends back; turns vertebral column sideways

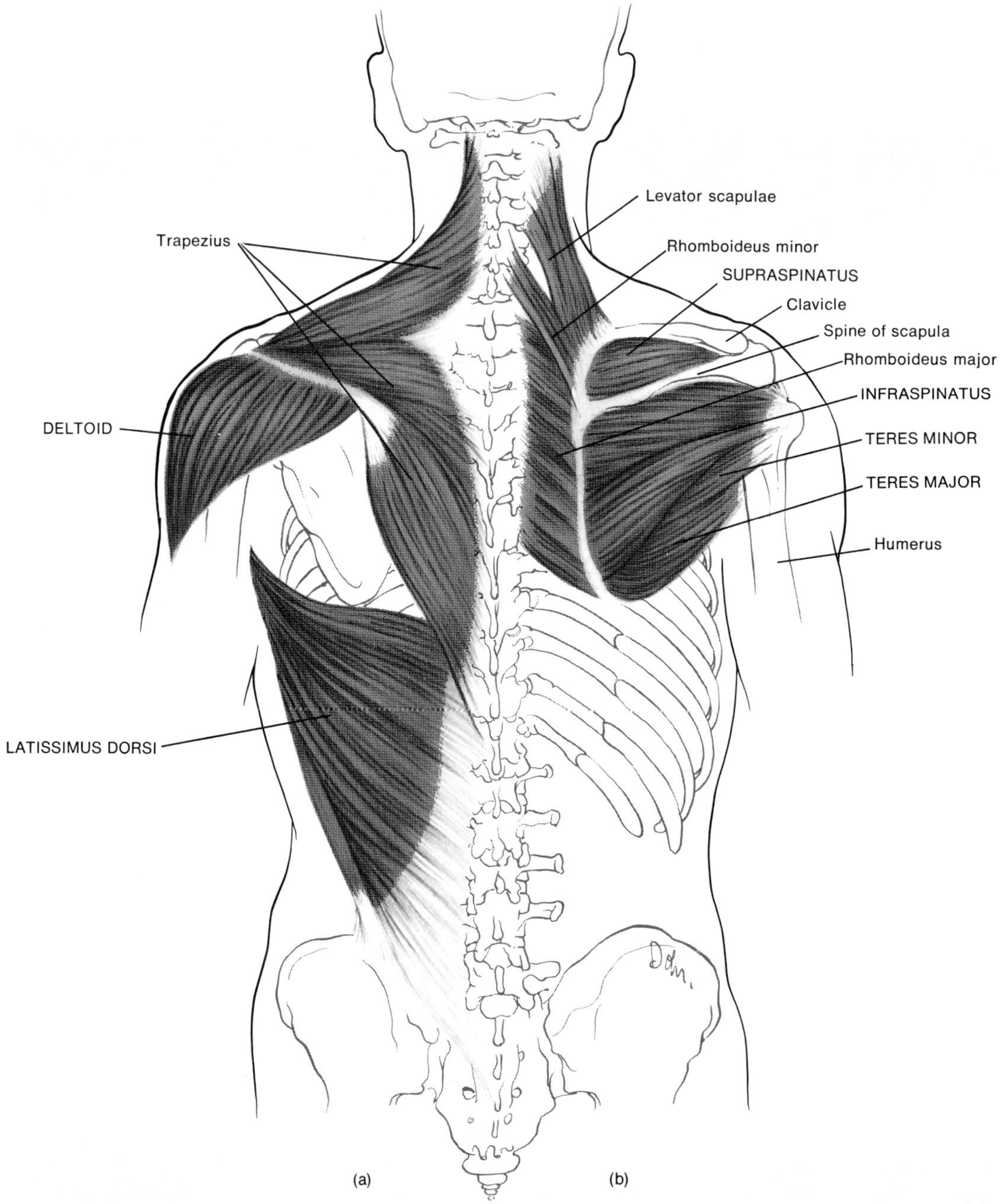

Figure 17-4 Muscles that move the arm. (a) Posterior superficial view. (b) Posterior deep view.

COMPARABLE HUMAN MUSCLES

The muscles associated with the scapula in the human comparable to those you dissected in the cat are the **trapezius,** the **rhomboids (major, minor,** and **capitis),** and the **levator scapulae.** In addition, there are the **serratus anterior** and the **pectoralis minor,** which will be dissected in a later exercise. Locate as many of these muscles on your own body as possible and make movements that demonstrate their actions (see Figures 17-4 and 17-5).

The human **supraspinatus, infraspinatus, subscapularis,** and the **teres minor** form what is called a *musculotendinous cuff* or rotator cuff over the head of the humerus. They reinforce the joint and at the same time allow a great deal of movement. Make as many shoulder joint

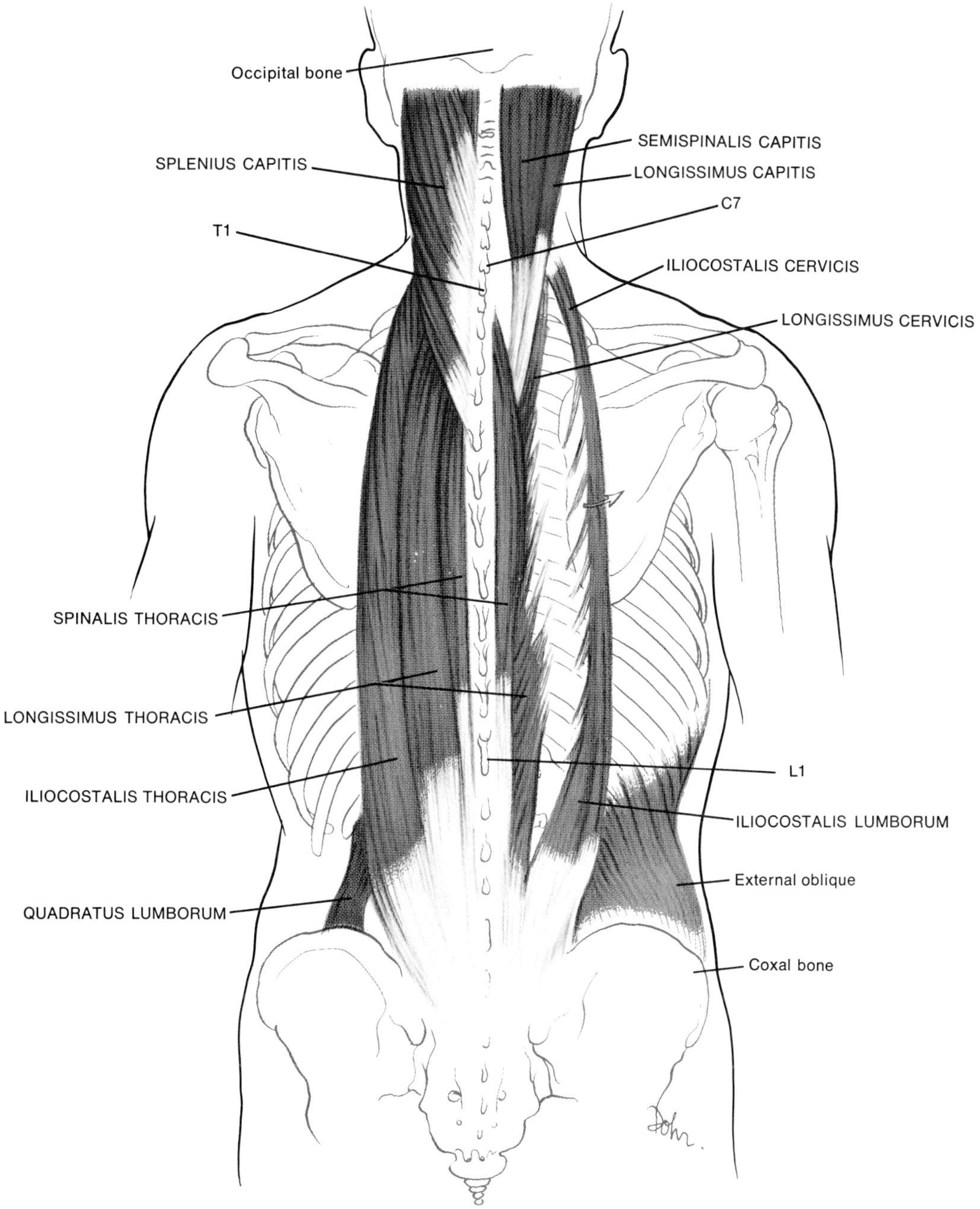

Figure 17-5 Muscles that move the vertebral column. Deep back muscles.

movements as you can, and you will see why it is called the most freely movable joint in the body.

The deep muscles of the human back stabilize the bones of the vertebral column. Many of them are short and span the distance between two vertebrae. Others are relatively long. They are postural muscles and not only maintain the upright posture but also influence the curvatures of the spinal column. They extend, flex, and rotate parts of the vertebral column.

The location of these postural muscles is between the angle of the ribs and the vertebral spines. Refer to Tables 17-2 and 17-4 for the origin, insertion, and action of the human's back muscles.

Table 17-4 Deep Back Muscles of the Human

Muscle	Origin	Insertion	Action
Splenius capitis	Transverse processes of last four cervical vertebrae	Occipital bone and mastoid process	Flexes head; rotates face
Semispinalis capitis	Articular process of cervical vertebrae 4, 5, and 6 and transverse processes of the seventh cervical vertebra and the first six thoracic vertebrae	Occipital bone	Extends head; rotates face
Spinalis thoracis	Spines of upper lumbar and lower thoracic vertebrae	Spines of upper thoracic vertebrae	Extends vertebral column
Longissimus Capitis	Transverse processes of last four cervical vertebrae	Mastoid process	Extends head, rotates face
Cervicis	Transverse processes of fourth and fifth thoracic vertebrae	Transverse processes of second through sixth cervical vertebrae	Extends cervical region of vertebral column
Thoracis	Transverse processes of lumbar vertebrae	Transverse processes of thoracic and upper lumbar vertebrae and ninth and tenth ribs	Extends thoracic region of vertebral column
Iliocostalis cervicis and iliocostalis lumborum	Iliac crest and all ribs	All ribs and transverse processes of fourth through sixth cervical vertebrae	Extends back

Questions for Review

EXERCISE **17**
MUSCLES OF THE BACK

Name ____________________

Laboratory Section ____________________ Date __________

1. Identify the muscles in Figure Q17-1 by matching the letters in the figures with the names listed below. You may color the diagram if you wish. Use your text as a guide.

_______ Spinalis thoracis

_______ Iliocostalis thoracis

_______ Semispinalis capitis

_______ Iliocostalis cervicis

_______ Splenius capitis

_______ Longissimus thoracis

_______ Longissimus cervicis

_______ Longissimus capitis

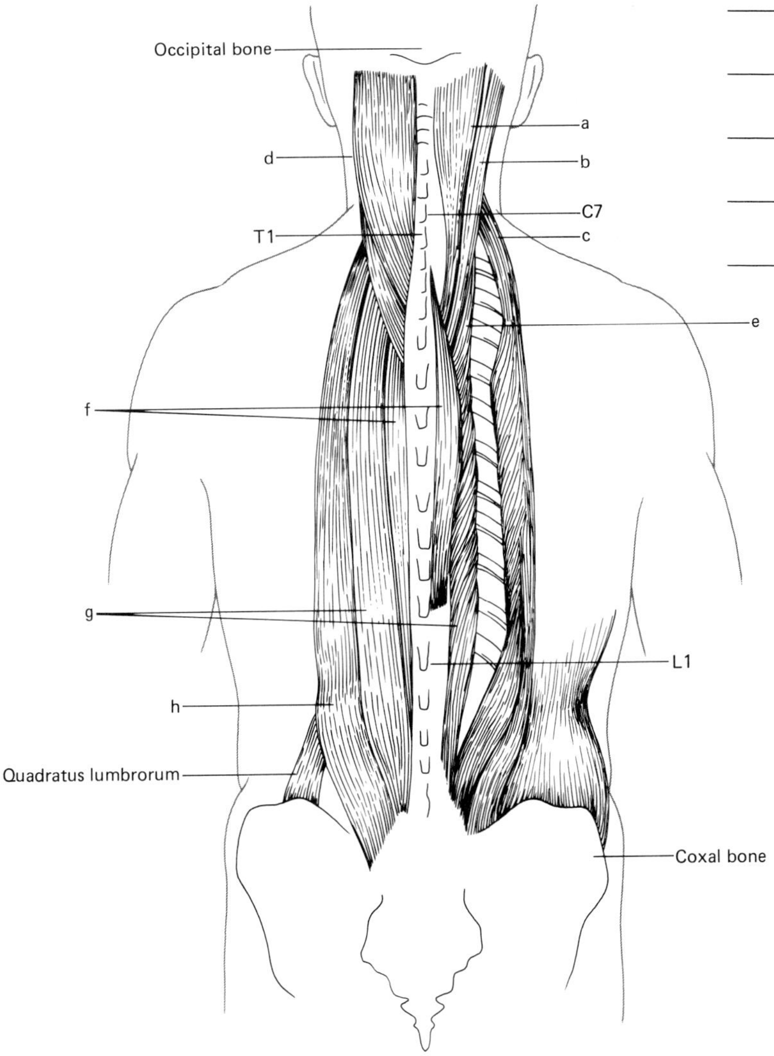

Figure Q17-1

2. Name the muscles indicated in Figure Q17-2. Color the diagram if you wish.

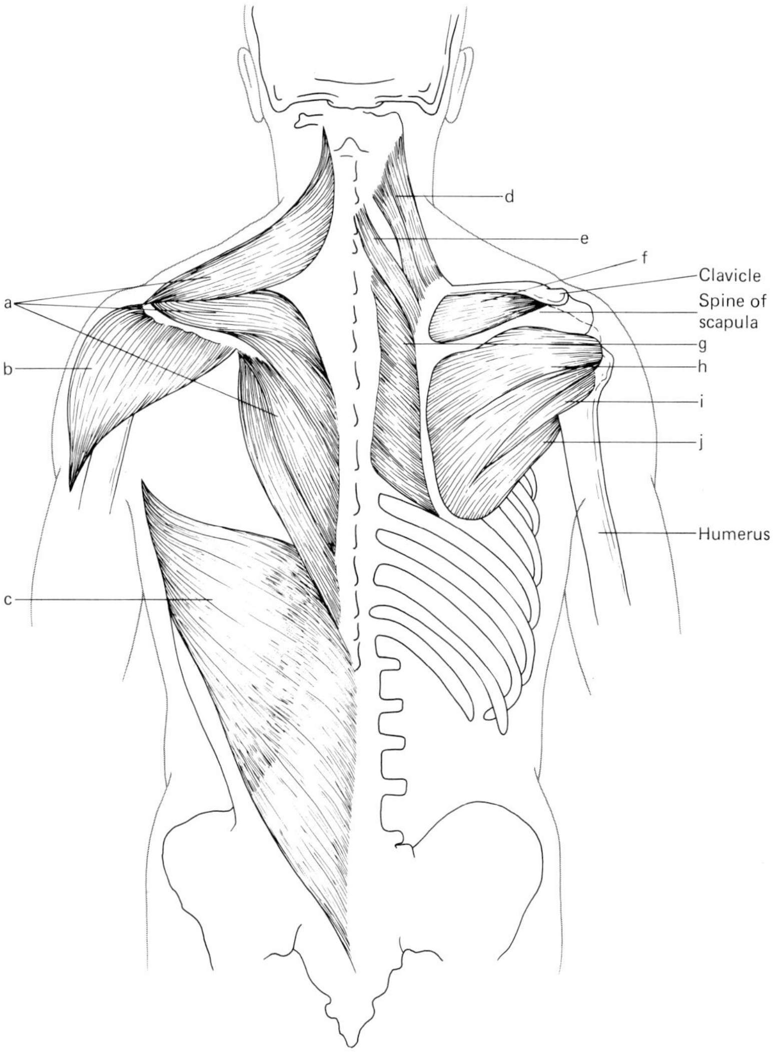

Figure Q17-2

a. ______________________________

b. ______________________________

c. ______________________________

d. ______________________________

e. ______________________________

f. ______________________________

g. ______________________________

h. ______________________________

i. ______________________________

j. ______________________________

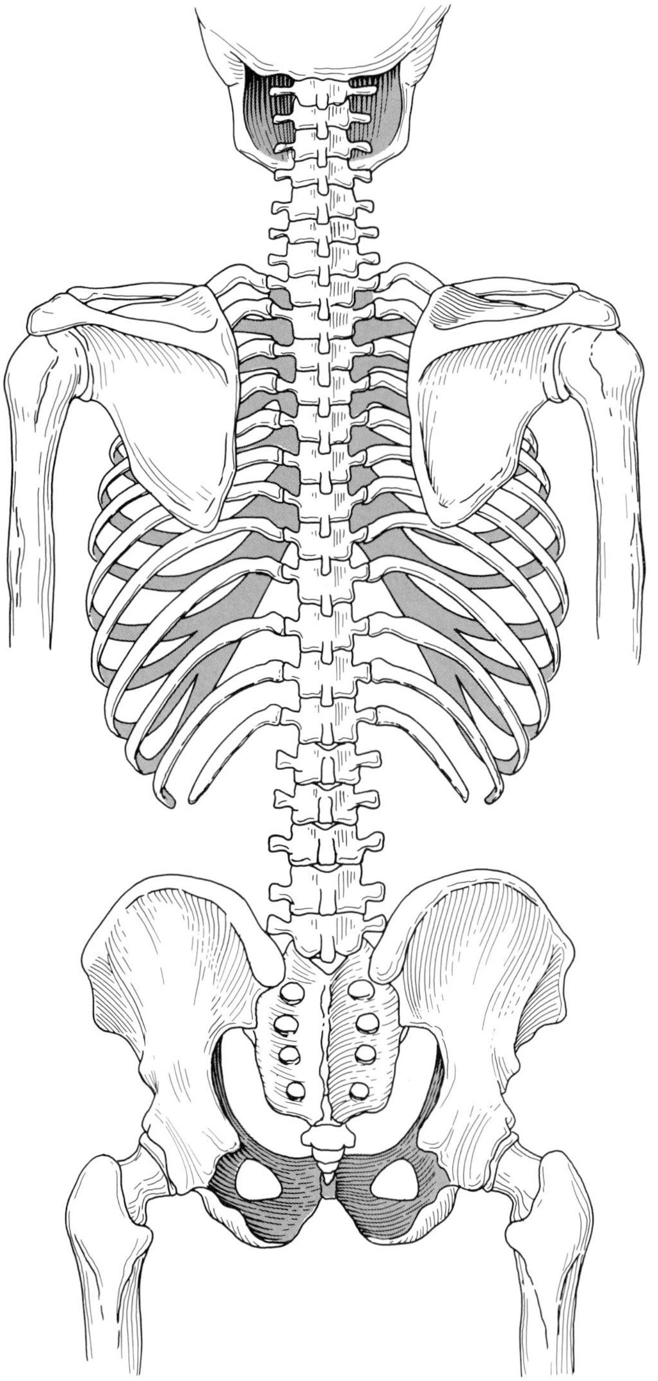

Figure Q17-3

3. Figure Q17-3 is a posterior view of the human skeleton. Draw in and label the following muscles. Review origins and insertions in Table 17-2.

a. Trapezius
b. Deltoid
c. Latissimus dorsi
d. Levator scapulae
e. Rhomboideus major
f. Rhomboideus minor
g. Supraspinatus
h. Infraspinatus

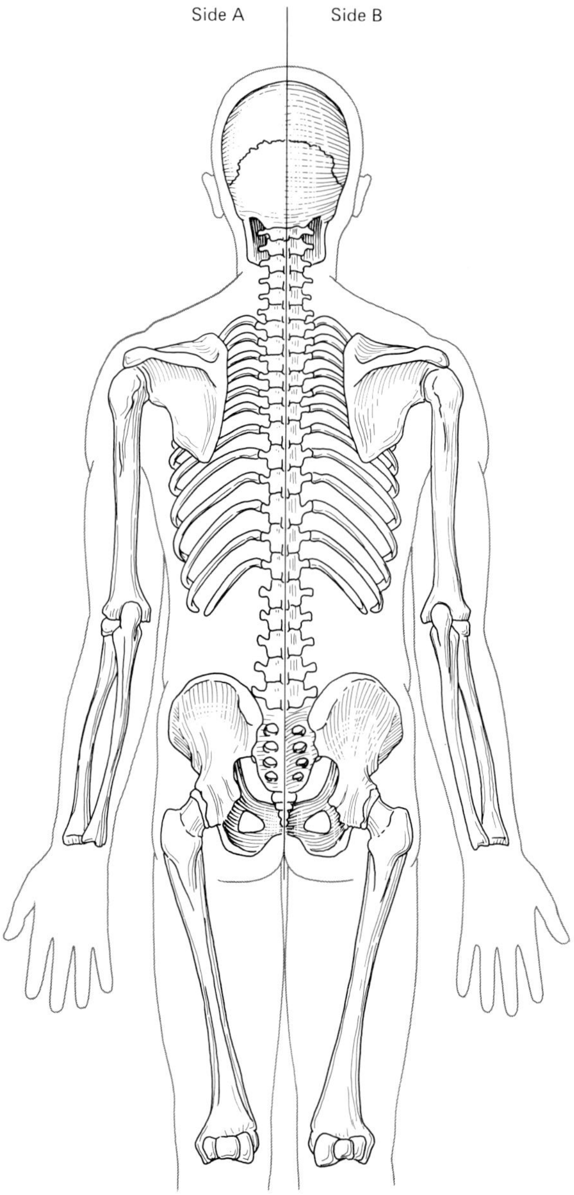

Figure Q17-4

4. Figure Q17-4 is a posterior view of the human skeleton. Draw in and label the following muscles on side A.
   1. Semispinalis capitis
   2. Longissimus capitis
   3. Iliocostalis cervicis
   4. Splenius capitis

Draw in and label the following muscles on side B.

   5. Longissimus cervicis
   6. Spinalis thoracis
   7. Longissimus thoracis
   8. Iliocostalis thoracis

EXERCISE 18

MUSCLES OF THE CHEST AND ABDOMEN

Materials

The following materials should be provided for class use.

1. Embalmed cats
2. Dissecting tools
3. Dissecting trays
4. String
5. 3 percent phenol solution

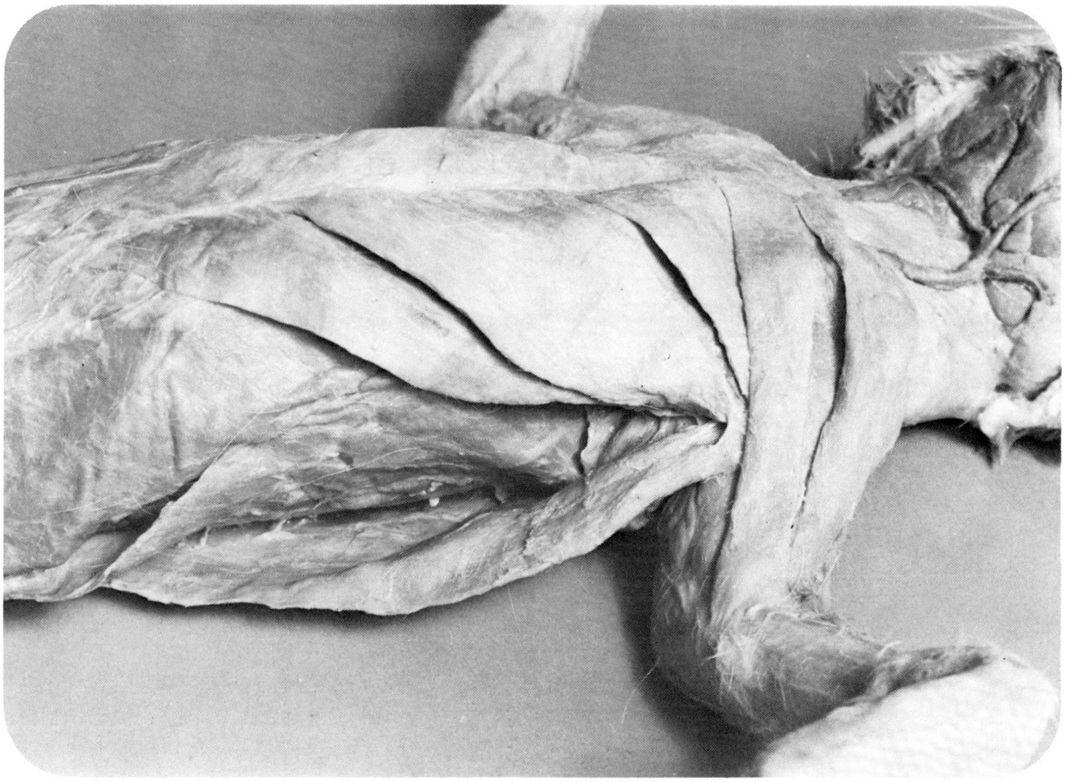

Figure 18-1a Pectoral muscles.

PECTORAL MUSCLES

INSTRUCTIONS FOR DISSECTION

Turn the cat on its back, as shown in Figure 18-1a, and find the general outline of the pectoral group of four muscles. Begin with the **pectoantebrachialis,** a long, thin, ribbonlike muscle. In most cats this is a difficult muscle to dissect because the fascia is tough. Be certain that you see its outlines before you begin, and do not dissect too deeply or you will be including part of another muscle. Separate between origin and insertion.

Next find the posterior border of the **pectoralis major** and loosen at that point. This is also a thin muscle but much broader than the pectoantebrachialis, which crosses it (see Figure 18-1b).

With your finger or a probe loosen the pectoralis major from the underlying muscle as far anteriorly as possible.

Now find the posterior border of the **pectoralis minor.** The pectoralis minor is larger than the pectoralis major and runs underneath it and the pectoantebrachialis. The pectoralis minor is the thickest muscle of this group as well as the broadest.

The last muscle of the group is the **xiphihumeralis.** Its origin on the midventral line is secure, but the insertion on the humerus is by a fragile tendon. Be careful not to tear it. There are neither pectoantebrachialis nor xiphihumeralis muscles in the human.

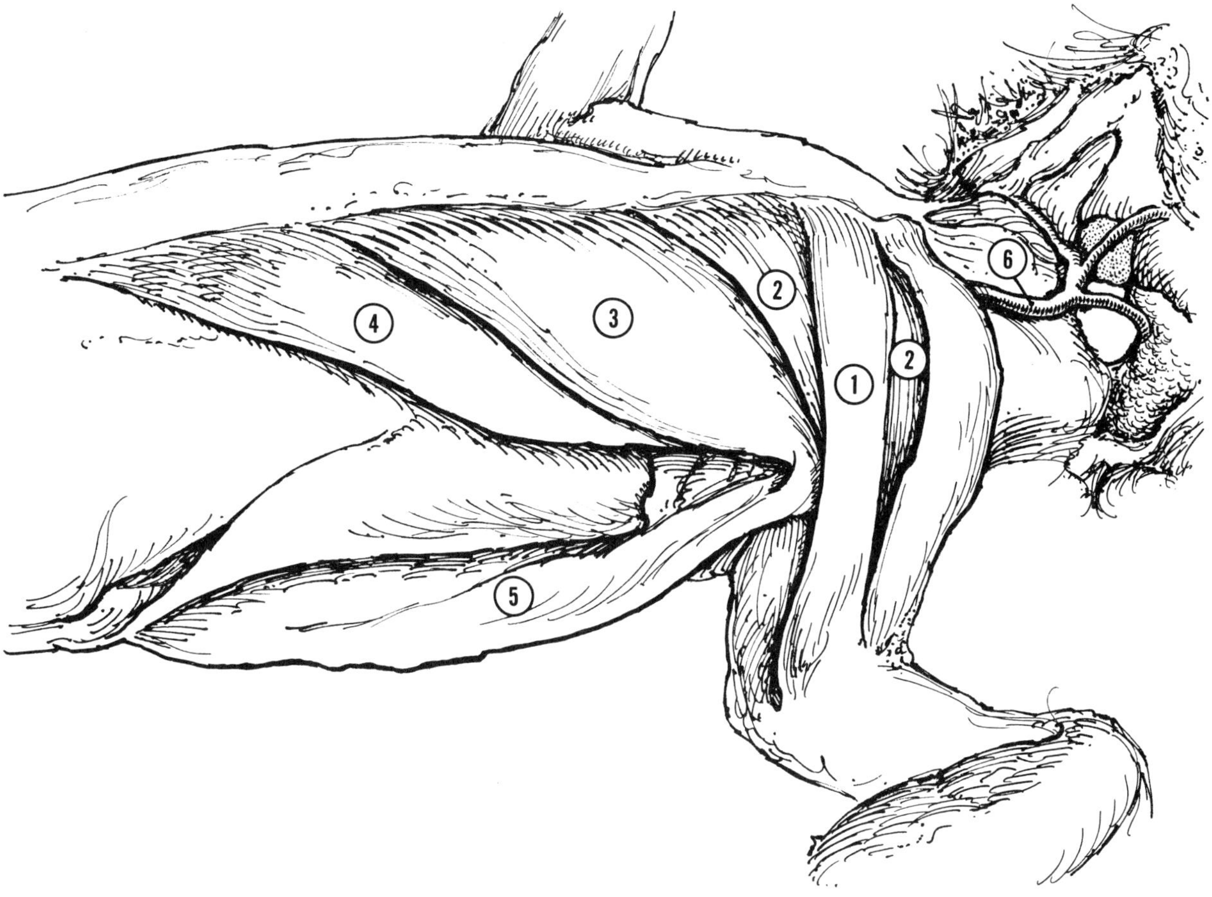

Figure 18-1b Pectoral muscles.

1. Pectoantebrachialis (pek′-tō-an′-ti-brāk-ē-AL-is)
2. Pectoralis major (pek′-tō-RAL-is)
3. Pectoralis minor
4. Xiphihumeralis (zif′-ē-hyoo′-mer-AL-is)
5. Latissimus dorsi
6. Jugular vein and tributaries

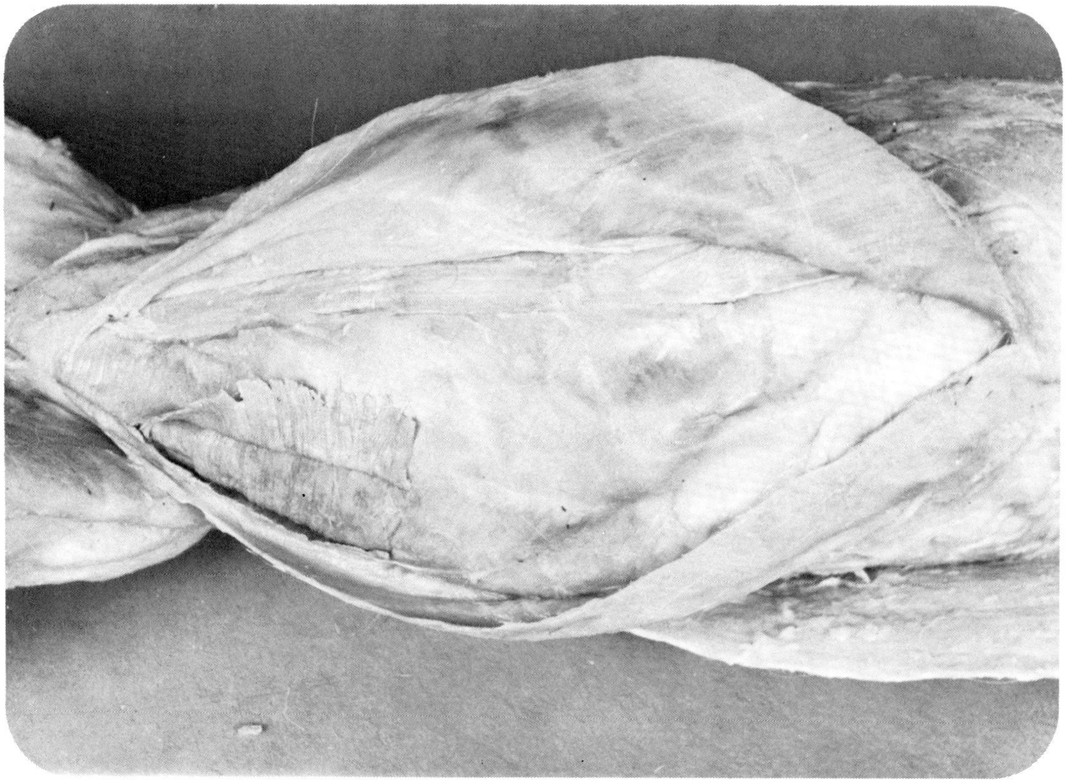

Figure 18-2a Abdominal muscles.

ABDOMINAL MUSCLES

INSTRUCTIONS FOR DISSECTION

There are three thin layers of muscle covering the abdominal cavity (Figure 18-2a). You must be careful in this dissection not to cut too deeply. Determine origin and insertion of the **external oblique,** the first layer, and mark off the midpoint between origin and insertion with a pencil or other tool. This muscle must be transected to see the one below it, and the transection line will be slightly curved due to the shape of the muscle. Begin the cut with the tip of the scissors and *gradually* extend it until you can get a probe between the external oblique and the **internal oblique,** the second layer (see Figure 18-2b). When you are sure that you have not gone too deep, extend the transection line and separate as you go along with your fingers. Bring each half back to origin and insertion.

Near the insertion on the *linea alba* (midline) you will see a long, ribbonlike muscle, the **rectus abdominis.** Expose as much of it as you can. You cannot see the anterior part of the muscle at this time, but it will be seen in a later dissection.

To separate the internal oblique from the **transversus abdominis,** the third or innermost layer, use extreme caution because both muscles are very thin. Sometimes the transversus abdominis is so thin that its fibers separate and you can see the *parietal peritoneum* (lining of the body cavity) between them. Separate as much of the internal oblique as you can and reflect the cut ends. Do not make your transection line too close to the broad aponeurosis by which the muscle inserts. Note that the direction of the fibers of the external oblique and internal oblique are almost at right angles to one another. Refer to Table 18-1 for the origin, insertion, and action of the cat's abdominal muscles.

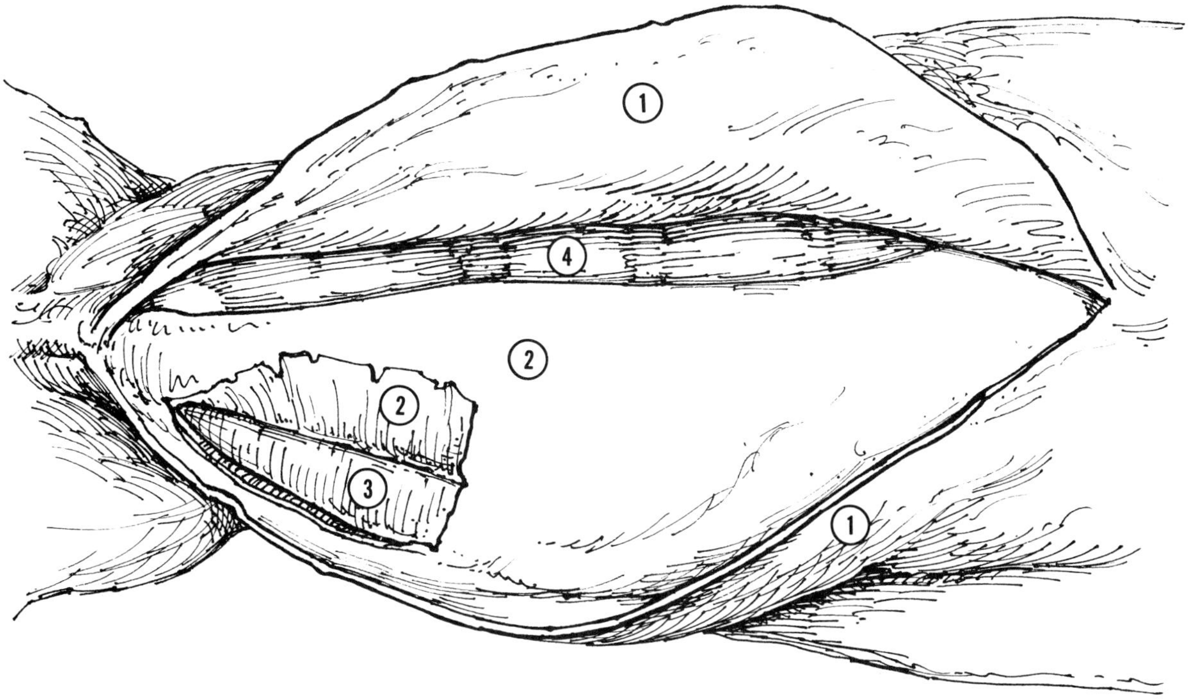

Figure 18-2b Abdominal muscles.
1. External oblique (reflected) (o-BLĒK)
2. Internal oblique (reflected)
3. Transversus abdominis
4. Rectus abdominis (REK-tus ab-DOM-i-nis)

Table 18-1 Pectoral and Abdominal Muscles of the Cat

Muscle	Origin	Insertion	Action
Pectoantebrachialis	Manubrium	Proximal end of forearm by fascia	Adducts arm
Pectoralis major	Anterior third of sternum	Proximal two-thirds of humerus	Adducts arm
Pectoralis minor	Almost entire length of sternum	Proximal half of humerus	Adducts arm
Xiphihumeralis	Xiphoid process	Proximal end of humerus	Adducts arm
External oblique	Last nine ribs	Linea alba from sternum to pubis	Compresses abdomen
Internal oblique	Lumbodorsal fascia	Linea alba	Compresses abdomen
Transversus abdominis	Costal cartilage of false ribs, transverse processes of lumbar vertebrae, and ilium	Linea alba	Compresses abdomen
Rectus abdominis	Pubic bone	Costal cartilage of first through fourth ribs	Compresses abdomen

COMPARABLE HUMAN MUSCLES

The human pectoral group consists of a **pectoralis major** and a **pectoralis minor** (see Figures 18-3 and 18-4). There is no pectoantebrachialis nor xiphihumeralis. The function of the pectoral muscles in the human is the same as in the cat, that is, to adduct the arm and the other actions listed in Table 18-2.

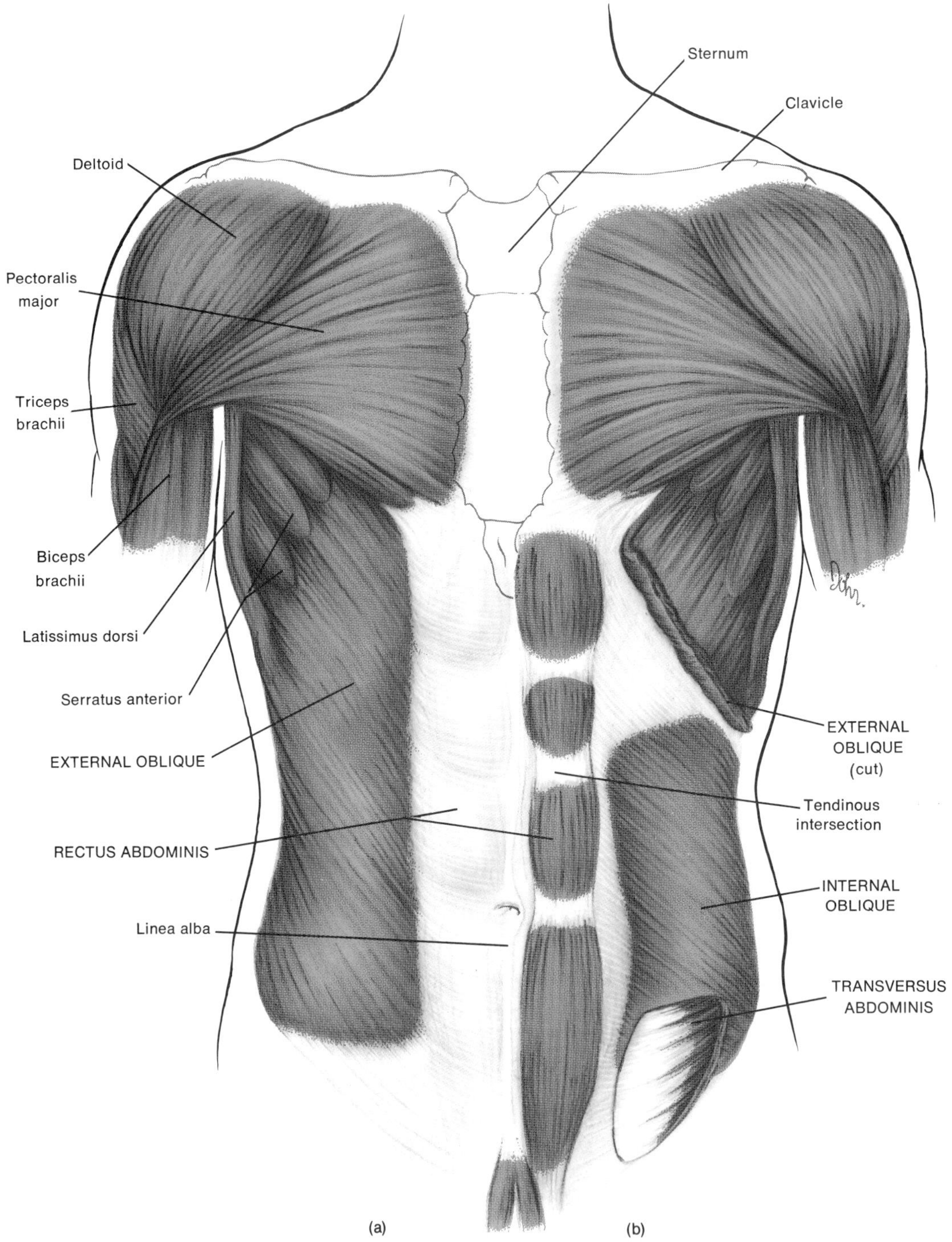

Figure 18-3

The abdominal group of human muscles is also similar to those in the cat. The **transversus abdominis** lies over the parietal peritoneum. The middle layer is the **internal oblique** and the outer layer is the **external oblique.** The **rectus abdominis** arises in segments connected by flat tendons and is enclosed in a sheath formed by the aponeuroses of the other three abdominal muscles. The segmentation of the rectus abdominis is seldom seen in the cat. The abdominal muscles compress the contents of the abdominal cavity and also contribute to flexion and rotation of the torso.

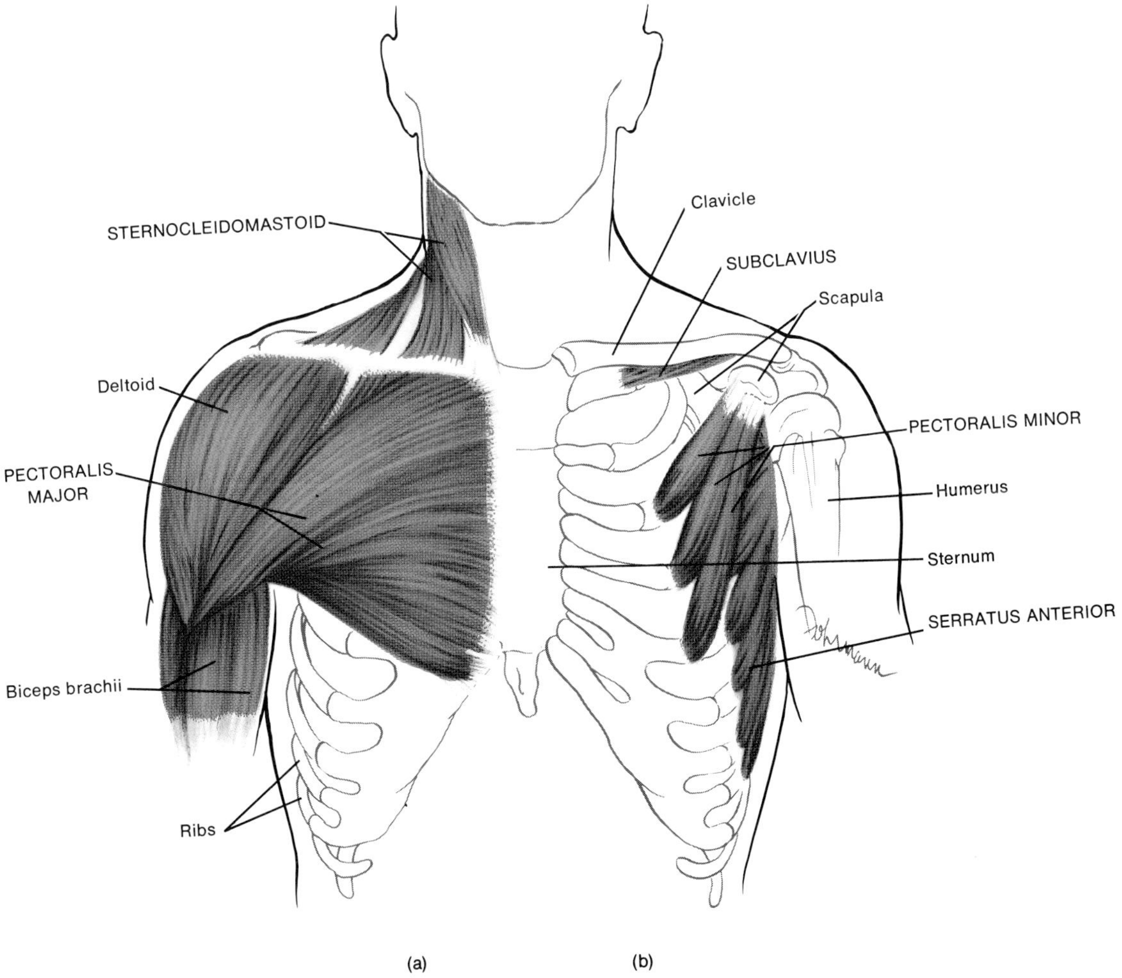

Figure 18-4

Table 18-2 Pectoral and Abdominal Muscles of the Human

Muscle	Origin	Insertion	Action
Pectoralis major	Clavicle, sternum, and cartilage of second through sixth ribs	Greater tubercle of humerus	Flexes, adducts, and rotates arm medially
Pectoralis minor	Third through fifth ribs	Coracoid process of scapula	Depresses scapula; rotates shoulder upward
External oblique	Lower eight ribs	Iliac crest and linea alba	Compresses abdomen
Internal oblique	Iliac crest, inguinal ligament, and thoracolumbar fascia	Costal cartilage of last three or four ribs	Compresses abdomen
Transversus abdominis	Iliac crest, inguinal ligament, lumbar fascia, and cartilage of last six ribs	Xiphoid process, linea alba, and pubis	Compresses abdomen
Rectus abdominis	Pubic crest and symphysis pubis	Costal cartilage of fifth to seventh ribs and xiphoid process	Flexes vertebral column

Questions for Review

EXERCISE **18**
MUSCLES OF THE CHEST AND ABDOMEN

Name ____

Laboratory Section ______________________________ Date ____________

1. Identify the muscles in Figure Q18-1 by matching the letters in the figure with the names listed below.

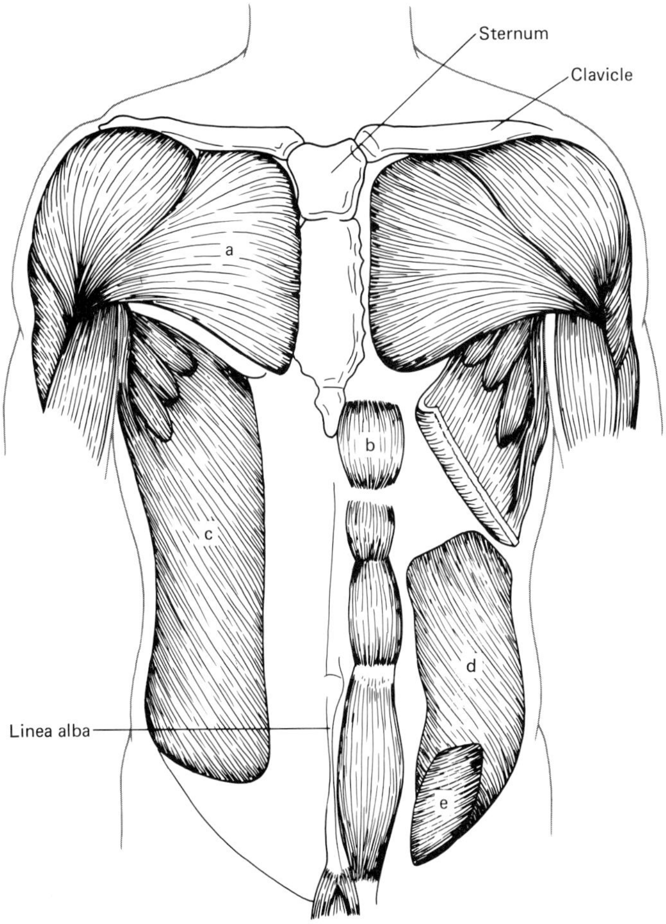

Figure Q18-1

____________ Internal oblique

____________ Transversus abdominis

____________ Pectoralis major

____________ External oblique

____________ Rectus abdominis

2. Figure Q18-2 is an anterior view of the human skeleton. Refer to Table 18-2. Draw in and label the following muscles on side A.

   a. Pectoralis major
   b. Rectus abdominis
   c. External oblique

Draw in and label the following muscle on side B.

   b. Rectus abdominis
   c. External oblique (cut)
   d. Internal oblique
   e. Transversus abdominis

Pay special attention to the accuracy of origins and insertion.

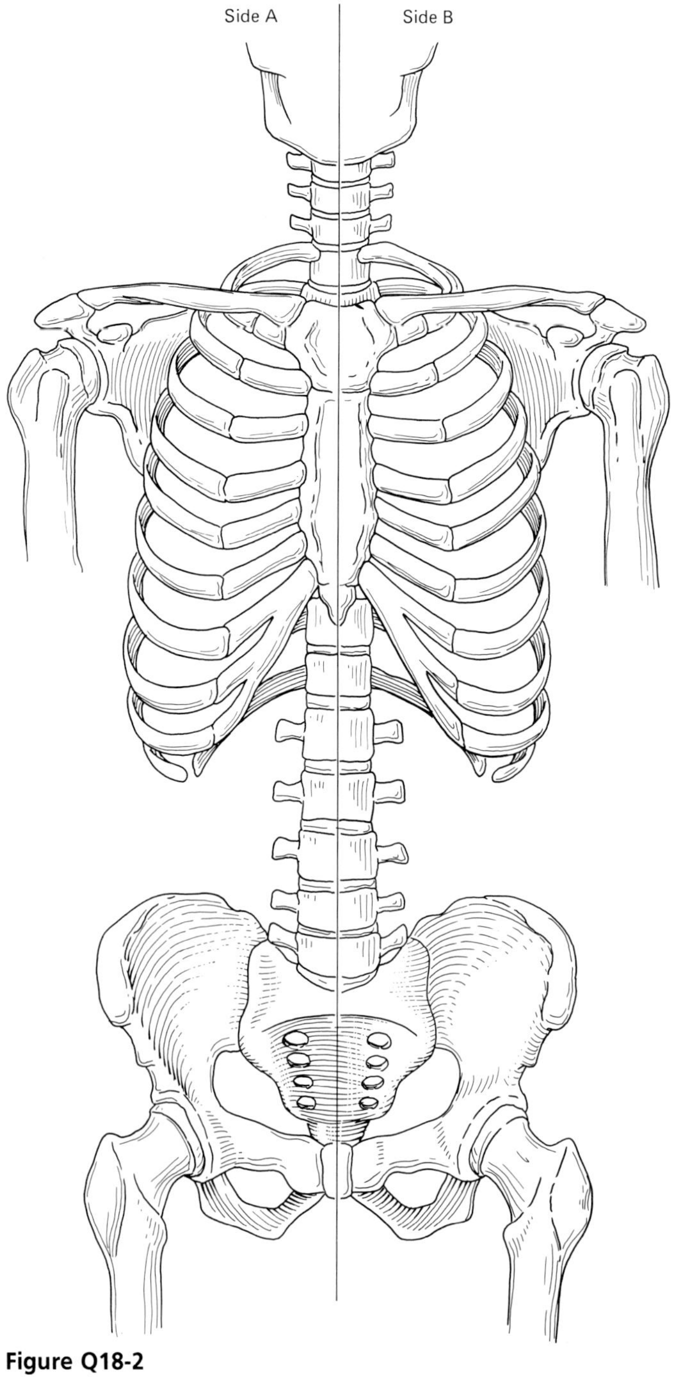

Figure Q18-2

EXERCISE 19

MUSCLES OF THE NECK, JAW, AND FACE

Materials

The following materials should be provided for class use.

1. Embalmed cats
2. Dissecting tools
3. Dissecting trays
4. String
5. 3 percent phenol solution

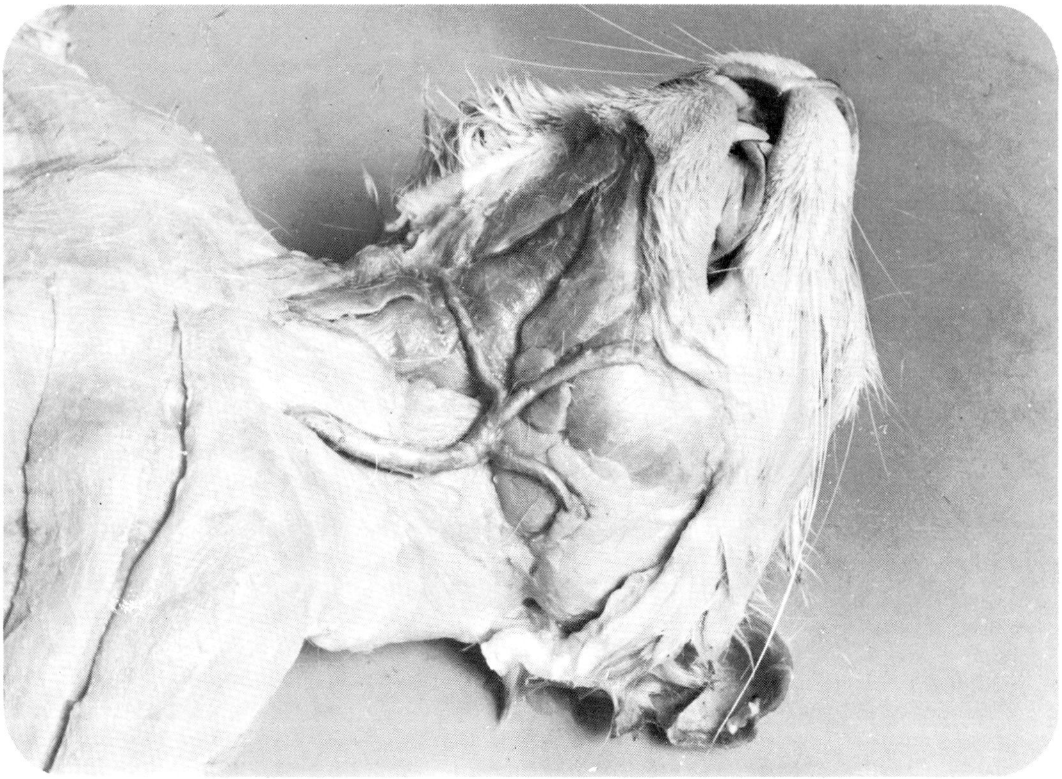

Figure 19-1a Superficial neck and jaw muscles.

SUPERFICIAL NECK AND JAW MUSCLES

INSTRUCTIONS FOR DISSECTION

The skin must be removed from the entire neck region and the left side of the face, as shown in Figure 19-1a. Remove the skin carefully. Do not remove the *jugular vein* and its tributaries and the *parotid duct.* You may also see *cranial nerve VII* passing over the cheek just below and parallel to the parotid duct. Prominent *lymph nodes* will be seen beneath the veins. Observe them. The large *submaxillary gland* lies below the *parotid gland.* Locate all these structures before beginning the identification of the muscles. Study Figure 19-1a carefully.

The muscle of the cheek is called the **masseter.** It is large and convex and is partially covered by the lymph nodes and salivary glands.

The large V-shaped muscle between the sternum and the head is the **sternomastoid.** (In the human this muscle is united with another, the cleidomastoid, to form the sternocleidomastoid. The cleidomastoid of the cat will be observed in the next dissection.)

In the middle of the neck, between the two sternomastoids, are the paired **sternohyoids.** Above the sternohyoids, with fibers running transversely, are the **mylohyoids.**

Attached to the mandible on both sides are the **diagastric** muscles.

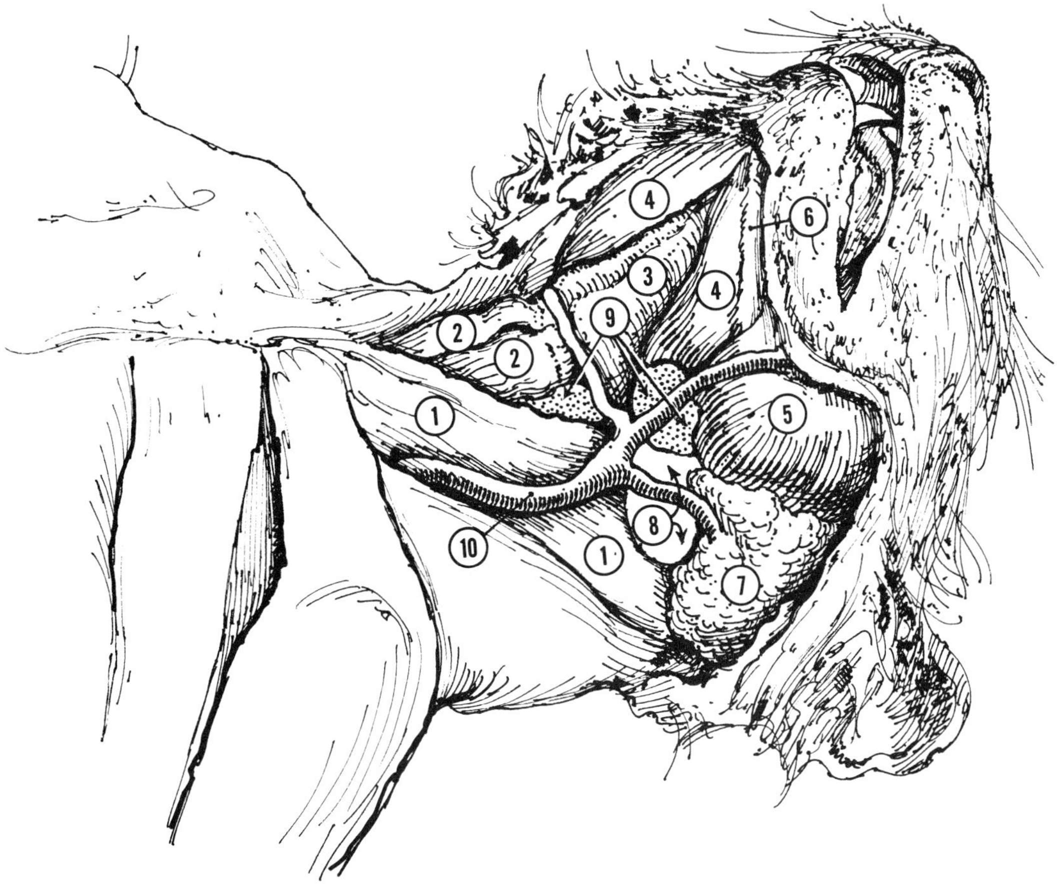

Figure 19-1b Superficial neck and jaw muscles.
1. Sternomastoid (ster′-nō-MAS-toyd)
2. Sternohyoid (ster′-nō-HĪ-oyd)
3. Mylohyoid (mī′-lō-HĪ-oyd)
4. Digastric (dī-GAS-trik)
5. Masseter (MASS-e-ter)
6. Mandible (bone)
7. Parotid gland (pa-ROT-id)
8. Submaxillary gland
9. Lymph nodes
10. Jugular vein

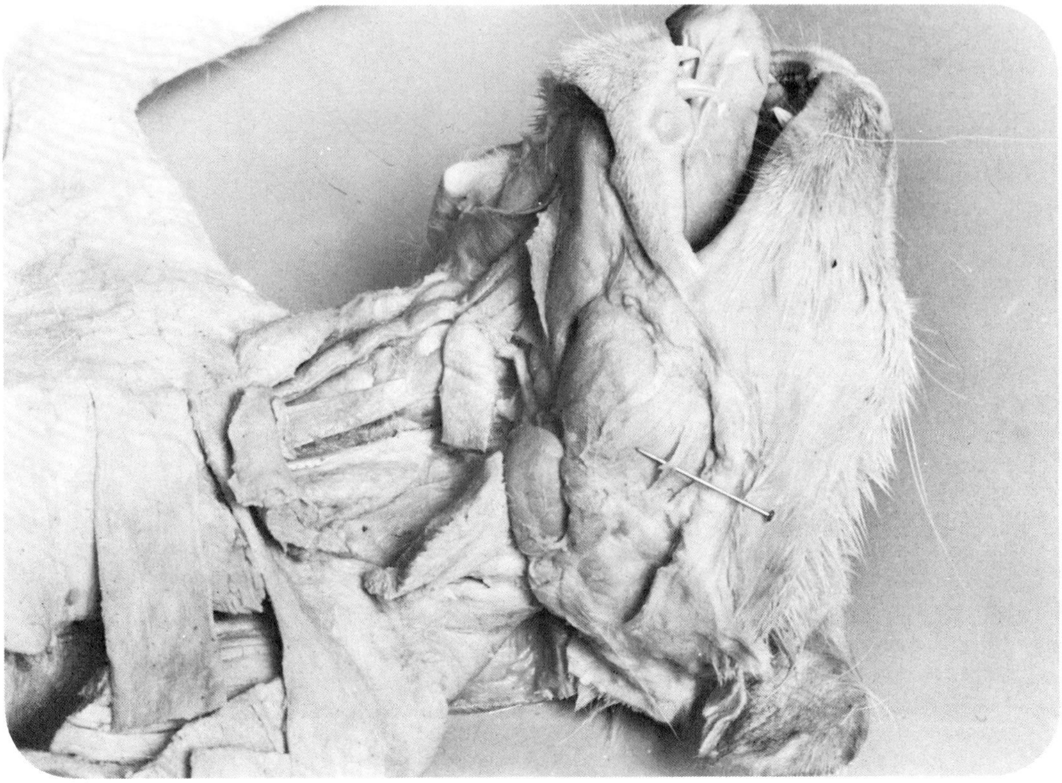

Figure 19-2a Deep neck muscles and salivary glands.

DEEP NECK MUSCLES AND SALIVARY GLANDS

INSTRUCTIONS FOR DISSECTION

Carefully dissect the veins and lymph nodes on the left side of the neck (see Figure 19-2a). If you want to preserve them (they are damaged on the right side as a result of the injection procedure), free the entire mass in one piece, clip the veins at their anterior ends and pull the whole mass posteriorly so that it may be put back in place once the dissection is completed. Identify again the muscles previously named.

Lying beneath the jugular vein at the junction between sternohyoid and mylohyoid is a small muscle, the **stylohyoid.** Its fibers run transversely (see Figure 19-2b).

Transect the sternomastoid and reflect the cut ends as far as possible to see the underlying structures. Also transect the left sternohyoid and reflect its ends. Laterally under the sternomastoid is the **cleidomastoid,** extending from the clavicle to the mastoid process.

Under the cut ends of the sternohyoid you can see two parallel muscles. Posterior in position is the **sternothyroid,** which extends upward only as far as the thyroid cartilage of the larynx. Just above the sternothyroid is the **thyrohyoid,** which spans the short distance between the thyroid cartilage of the larynx posteriorly and the hyoid cartilage of the larynx anteriorly.

Lying directly on the larynx is the **cricothyroid** muscle, which is not to be dissected but merely observed. Cut the **mylohyoid** up the midline as far as possible and reflect the cut ends laterally. This reveals paired **geniohyoid** muscles. Refer to Table 19-1 for the origin, insertion, and action of the cat's neck muscles.

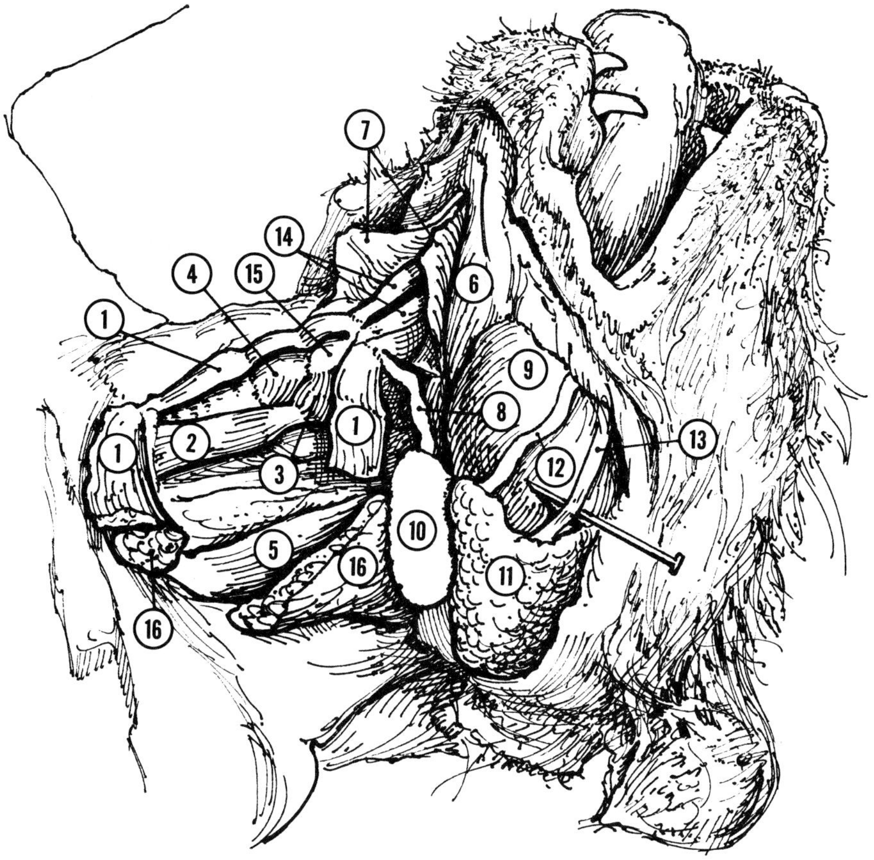

Figure 19-2b Deep neck muscles and salivary glands.
1. Sternohyoid (reflected on left side only)
2. Sternothyroid (ster′-nō-THĪ-royd)
3. Thyrohyoid (thī-rō-HĪ-oyd)
4. Cricothyroid (krī-kō-THĪ-royd)
5. Cleidomastoid (klī′-dō-MAS-toyd)
6. Digastric
7. Mylohyoid (reflected)
8. Stylohyoid (stī′-lō-HĪ-oyd
9. Masseter
10. Submaxillary gland
11. Parotid gland
12. Cranial nerve VII (facial)
13. Parotid duct
14. Geniohyoid
15. Larynx
16. Sternomastoid (reflected)

The salivary glands consist of three paired glands. You have already seen the parotid gland and its duct and the submaxillary gland. To see the sublingual gland, loosen the submaxillary gland and pull it to the cat's left side. This will reveal a much smaller cone-shaped gland, the sublingual gland, and also the submaxillary and sublingual ducts.

Table 19-1 Neck Muscles of the Cat

Muscle	Origin	Insertion	Action
Sternomastoid	Manubrium	Mastoid process	Turns and depresses head
Sternohyoid	Costal cartilage of first rib	Hyoid bone	Pulls hyoid caudally; raises ribs and sternum

Table 19-1 *(Continued)*

Muscle	Origin	Insertion	Action
Mylohyoid	Cricoid cartilage of larynx	Hyoid bone	Raises floor of mouth; pulls hyoid forward
Digastric	Occipital bone and mastoid process	Ventral border of mandible	Depresses lower jaw
Masseter	Zygomatic bone	Lateral margin near the ventral border of the mandible at the coronoid fossa	Elevates lower jaw
Sternothyroid	Costal cartilage of first rib	Thyroid cartilage of larynx	Pulls larynx caudad
Thyrohyoid	Thyroid cartilage of larynx	Hyoid bone	Raises larynx
Cricothyroid	Cricoid cartilage of larynx	Thyroid cartilage of larynx	Tenses vocal cords
Cleidomastoid	Clavicle	Mastoid process	Pulls head laterally; pulls clavicle craniad
Stylohyoid	Stylohyal bone of hyoid	Body of hyoid bone	Raises hyoid
Geniohyoid	Medial surface of mandible	Body of hyoid bone	Pulls hyoid craniad

HUMAN MUSCLES OF THE HEAD AND NECK

The largest muscle of the neck in the human is the **sternocleidomastoid.** In the cat this muscle is divided into two distinct muscles, the sternomastoid and the cleidomastoid. Its position between the clavicle, sternum, and mastoid process of the skull gives it complex capability of movement. If only the right or the left sternocleidomastoid contracts, it turns the head to the opposite side. If both right and left sternocleidomastoids contract, the head is flexed on the chest.

Many of the deep muscles of the human neck are considered in the chest and deep back of the cat. They are the **scalenus** muscles, which turn the vertebral column to one side when they contract, and the **semispinalis, longissimus,** and **splenius** muscles, which extend the head and neck. They are found in the *posterior triangle* of the neck.

The muscles associated with the hyoid bone are grouped into the **suprahyoid** muscles and the **infrahyoid** muscles, depending on whether they are found above or below the hyoid bone. The suprahyoid muscles lie in the floor of the mouth and the infrahyoid muscles are found lower in the neck around the larynx and trachea. Figures 19-3 through 19-6 and Tables 19-2 through 19-5 include the human facial muscles of the lower jaw, of the tongue and larynx, and muscles that move the head.

The neck is a complicated area. The muscles in Figure 19-7 are neck and jaw muscles shown in side view. The large *sternocleidomastoid* forms the *anterior triangle* of the neck. Using the hyoid bone as a point of reference, the muscles of the anterior triangle are divided into two groups, the *suprahyoid muscles* and the *infrahyoid muscles.* The muscles of the posterior triangle are those located posterior to the sternocleidomastoid. Muscles included in these groups are listed below the drawing. Color the muscles and compare their functions and locations with their counterparts in the cat. (Note that the muscles of the posterior triangle of the cat are discussed with the cat's deep back and chest muscles.) An anterior view of the muscles of the neck is shown in Figure 19-8. Color the muscles. Note their relationship to the larynx and the trachea.

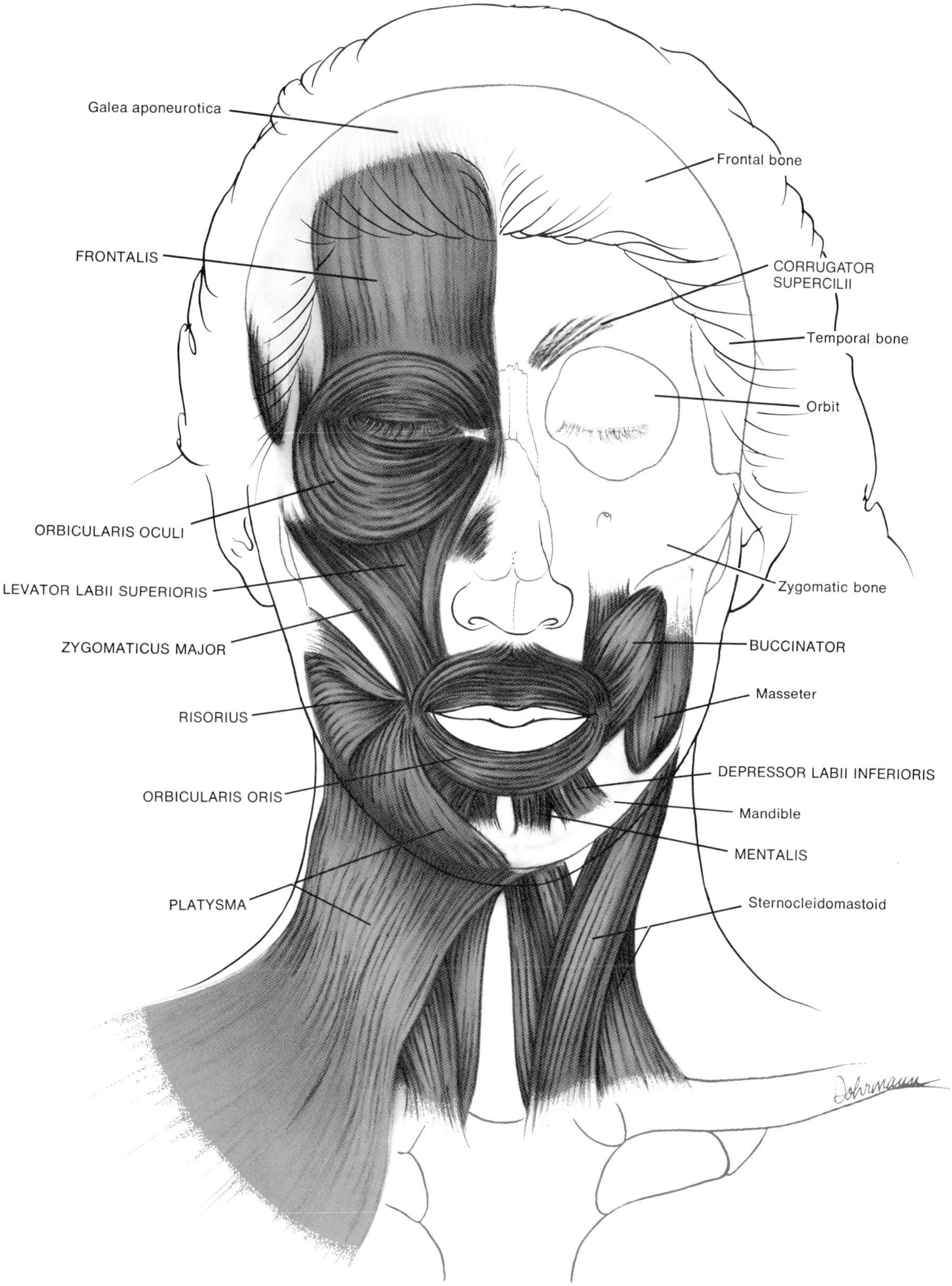

Figure 19-3 Muscles of facial expression. (a) Anterior superficial view. (b) Anterior deep view.

Table 19-2 Muscles of Facial Expression (Figure 19-3)

Muscle	Origin	Insertion	Action	Innervation
Epicranius (*epi* = over; *crani* = skull)	This muscle is divisible into two portions: the frontalis over the frontal bone and the occipitalis over the occipital bone. The two muscles are united by a strong aponeurosis, the galea aponeurotica, which covers the superior and lateral surfaces of the skull.			
Frontalis (*front* = forehead)	Galea aponeurotica	Skin superior to supraorbital line	Draws scalp forward, raises eyebrows, and wrinkles forehead horizontally	Facial nerve (VII)
Occipitalis (*occipito* = base of skull)	Occipital bone and mastoid process of temporal bone	Galea aponeurotica	Draws scalp backward	Facial nerve (VII)
Orbicularis oris (*orb* = circular; *or* = mouth)	Muscle fibers surrounding opening of mouth	Skin at corner of mouth	Closes lips, compresses lips against teeth, protrudes lips, and shapes lips during speech	Facial nerve (VII)
Zygomaticus major (*zygomatic* = cheek bone; *major* = greater)	Zygomatic bone	Skin at angle of mouth and orbicularis oris	Draws angle of mouth upward and outward as in smiling or laughing	Facial nerve (VII)
Levator labii superioris (*levator* = elevates; *labii* = lip; *superioris* = upper)	Superior to infraorbital foramen of maxilla	Skin at angle of mouth and orbicularis oris	Elevates upper lip	Facial nerve (VII)
Depressor labii inferioris (*depressor* = depresses; inferioris = *lower*)	Mandible	Skin of lower lip	Depresses lower lip	Facial nerve (VII)

Table 19-2 *(Continued)*

Muscle	Origin	Insertion	Action	Innervation
Buccinator (*bucc* = cheek)	Alveolar processes of maxilla and mandible and pterygomandibular raphe (fibrous band extending from the pterygoid hamulus to the mandible)	Orbicularis oris	Major cheek muscle; compresses cheek as in blowing air out of mouth and causes the cheeks to cave in, producing the action of sucking	Facia nerve (VIII)
Mentalis (*mentum* = chin)	Mandible	Skin of chin	Elevates and protrudes lower lip and pulls skin of chin up as in pouting	Facial nerve (VII)
Platysma (*platy* = flat, broad)	Fascia over deltoid and pectoralis major muscles	Mandible, muscles around angle of mouth, and skin of lower face	Draws outer part of lower lip downward and backward as in pouting; depresses mandible	Facial nerve (VII)
Risorius (*risor* = laughter)	Fascia over parotid (salivary) gland	Skin at angle of mouth	Draws angle of mouth laterally as in tenseness	Facial nerve (VII)
Orbicularis oculi (*ocul* = eye)	Medial wall of orbit	Circular path around orbit	Closes eye	Facial nerve (VII)
Corrugator supercilii (*corrugo* = to wrinkle; *supercilium* = eyebrow)	Medial end of superciliary arch of frontal bone	Skin of the eyebrow	Draws eyebrow downward as in frowning	Facial nerve (VII)
Levator palpebrae superioris (*palpebrae* = eyelids	Roof of orbit (Lesser wing of sphenoid bone)	Skin of the upper eyelid	Elevates the upper eyelid	Oculomotor nerve (III)

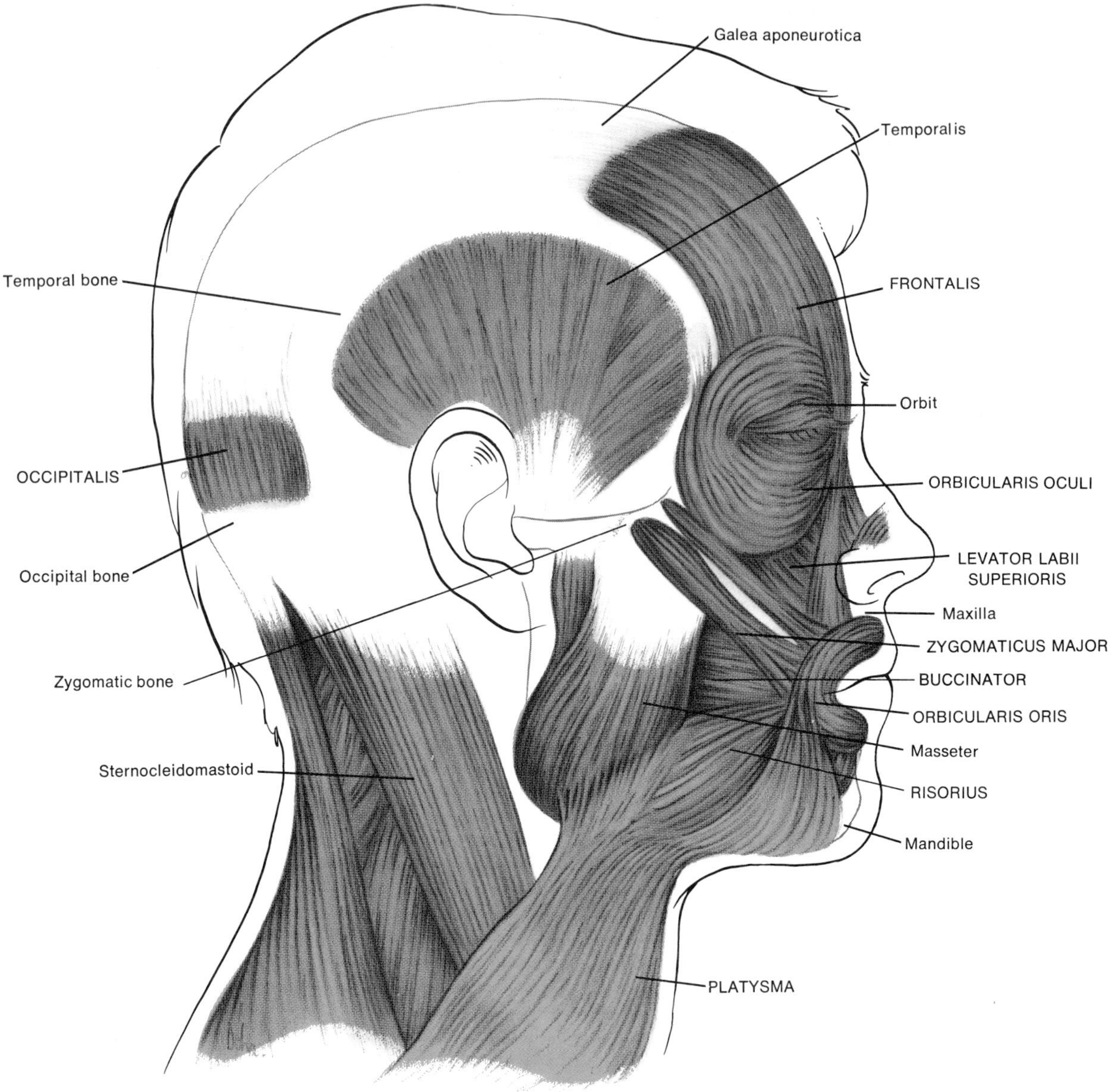

Figure 19-4 Muscles of the face and neck.

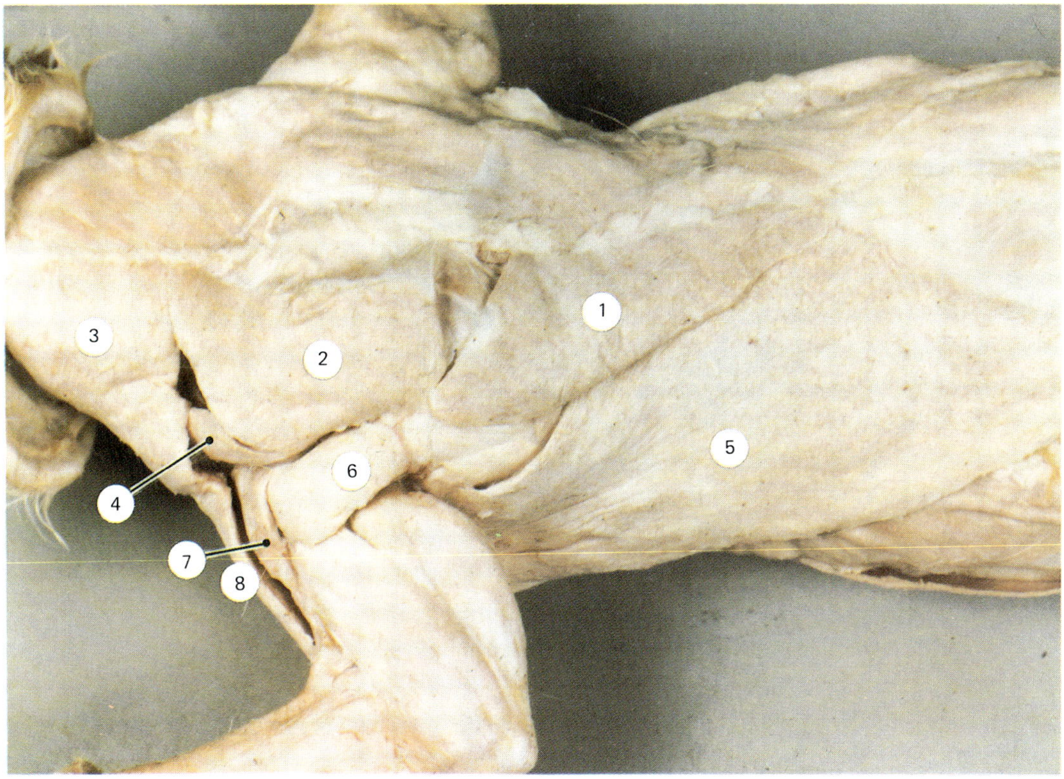

Plate 17-1 Superficial back muscles.

1. Spinotrapezius
2. Acromiotrapezius
3. Clavotrapezius
4. Levator scapulae ventralis
5. Latissimus dorsi
6. Spinodeltoid
7. Acromiodeltoid
8. Clavodeltoid or clavobrachialis

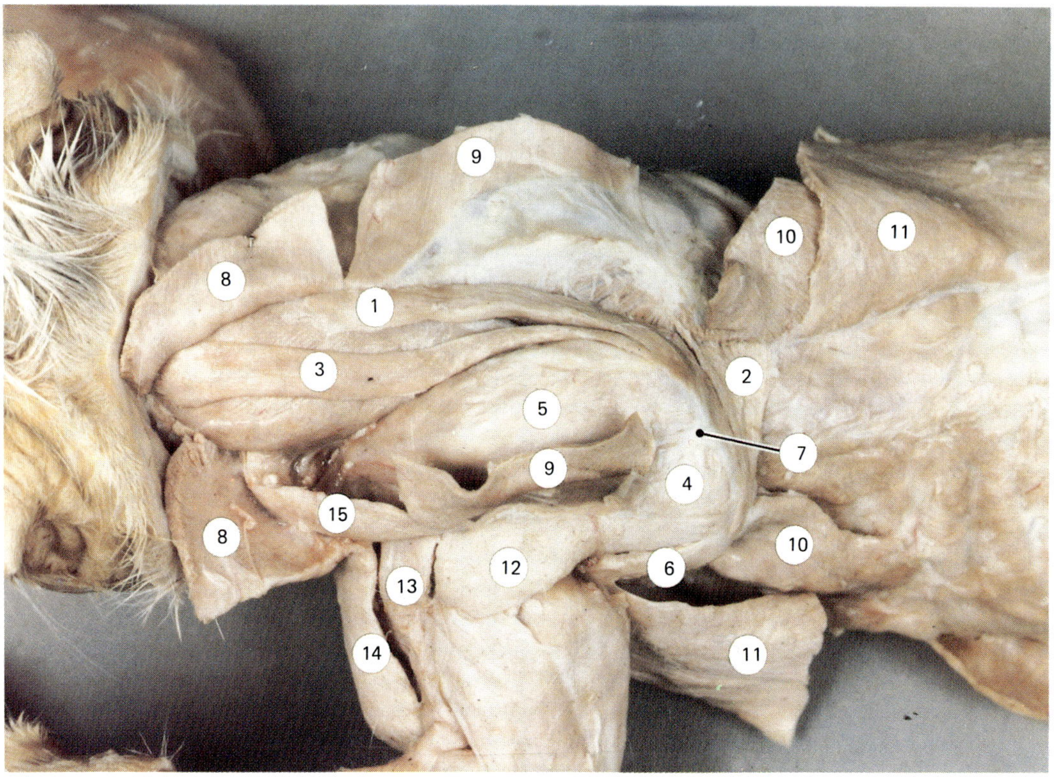

Plate 17-2 Muscles associated with scapula.

1. Rhomboideus minor
2. Rhomboideus major
3. Rhomboideus capitis or occipitoscapularis
4. Infraspinatus
5. Supraspinatus
6. Teres major
7. Scapular spine (bone)
8. Clavotrapezius (reflected)
9. Acromiotrapezius (reflected)
10. Spinotrapezius (reflected)
11. Latissimus dorsi
12. Spinodeltoid
13. Acromiodeltoid
14. Clavodeltoid or clavobrachialis
15. Levator scapulae ventralis

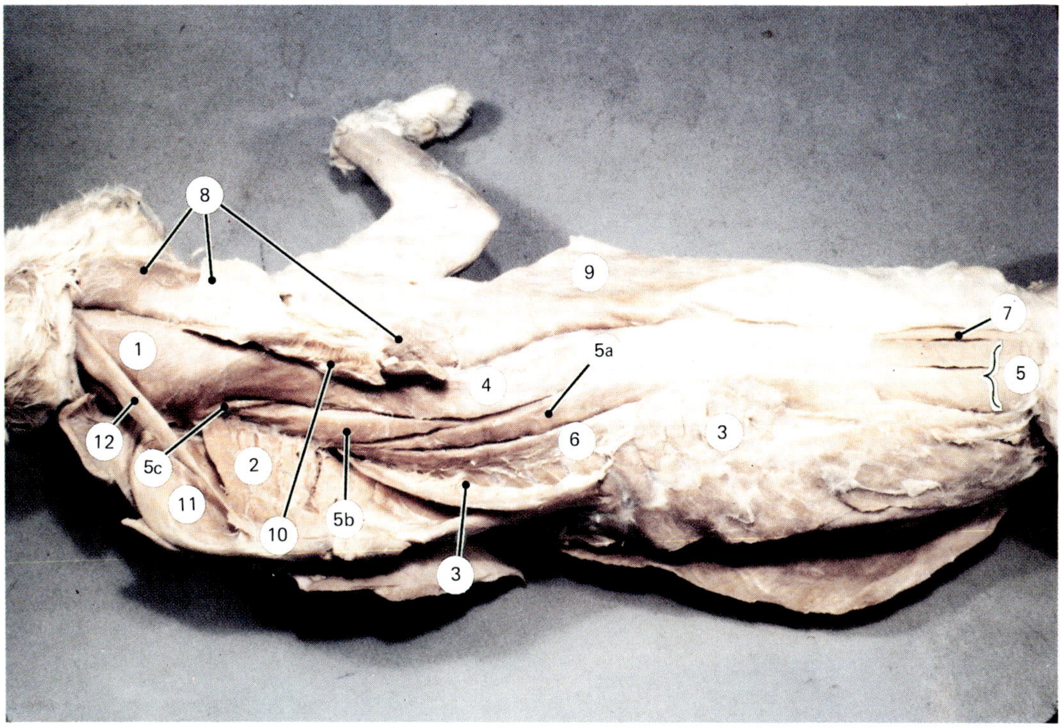

Plate 17-3 Deep back muscles.

1. Splenius
2. Serratus ventralis
3. Serratus dorsalis (partially reflected)
4. Spinalis dorsi
5. Longissimus dorsi (posterior part)
   a. Longissimus thoracis
   b. Longissimus cervicis
   c. Longissimus capitis
6. Iliocostalis
7. Multifidus spinae
8. Trapezius group (reflected)
9. Latissimus dorsi (reflected)
10. Rhomboideus (cut end)
11. Subscapularis
12. Rhomboideus capitis

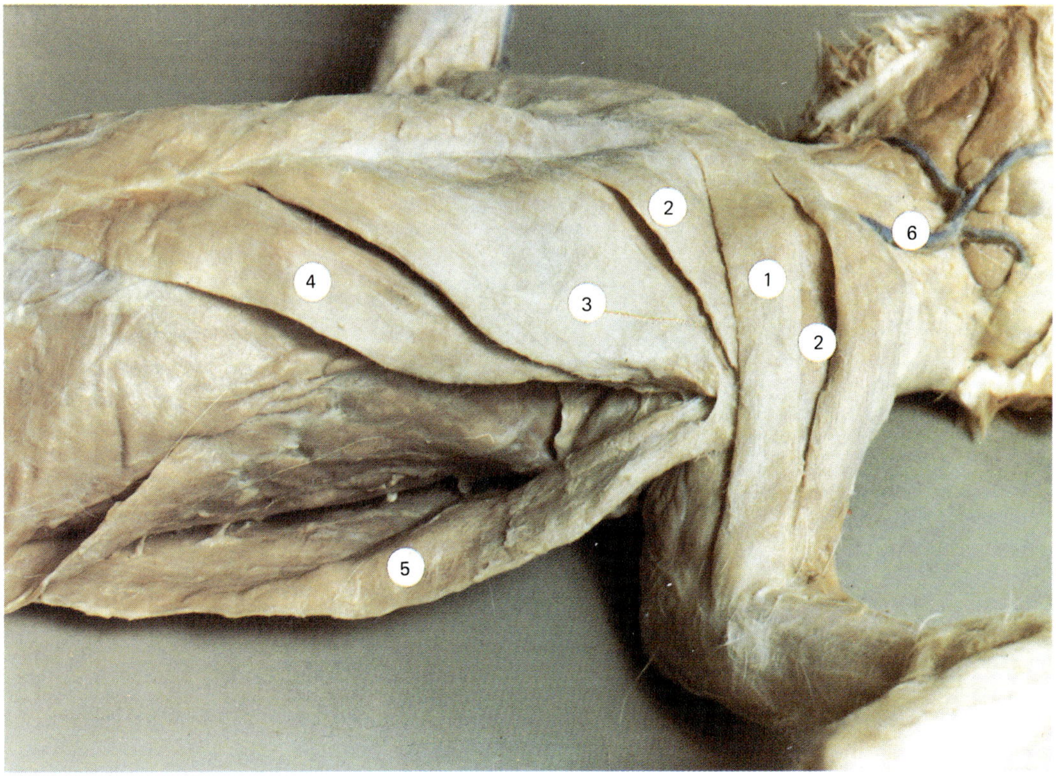

Plate 18-1 Pectoral muscles.

1. Pectoantebrachialis
2. Pectoralis major
3. Pectoralis minor
4. Xiphihumeralis
5. Latissimus dorsi
6. Jugular vein and tributaries

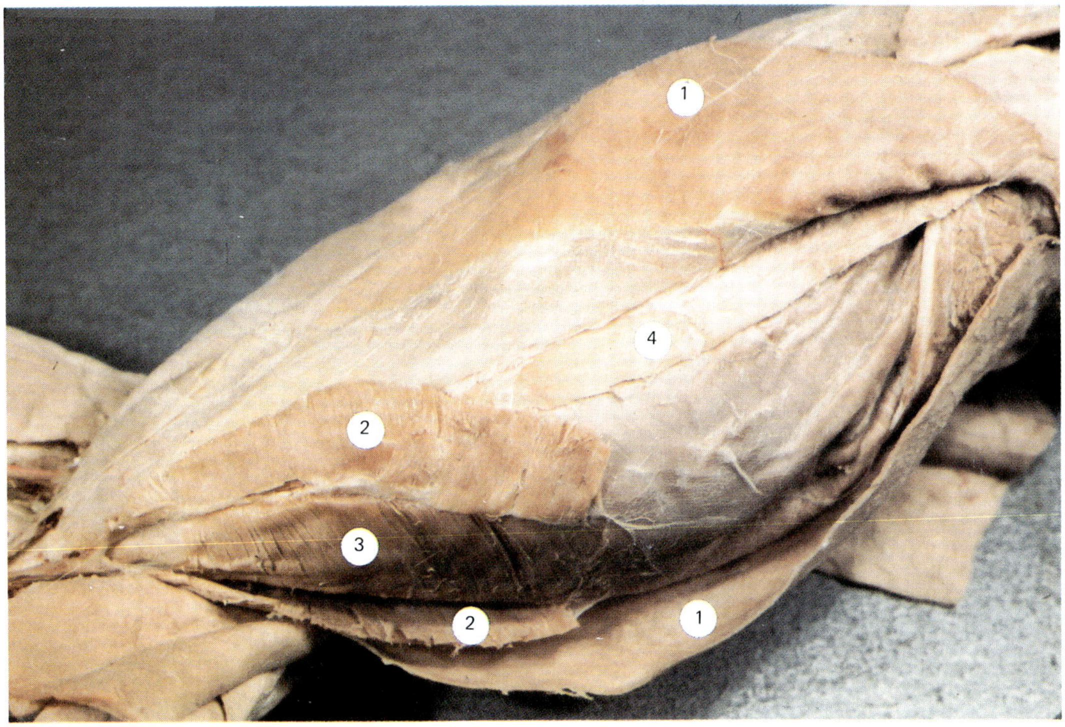

Plate 18-2 Abdominal muscles.
1. External oblique (reflected)
2. Internal oblique (reflected)
3. Transversus abdominis
4. Rectus abdominis

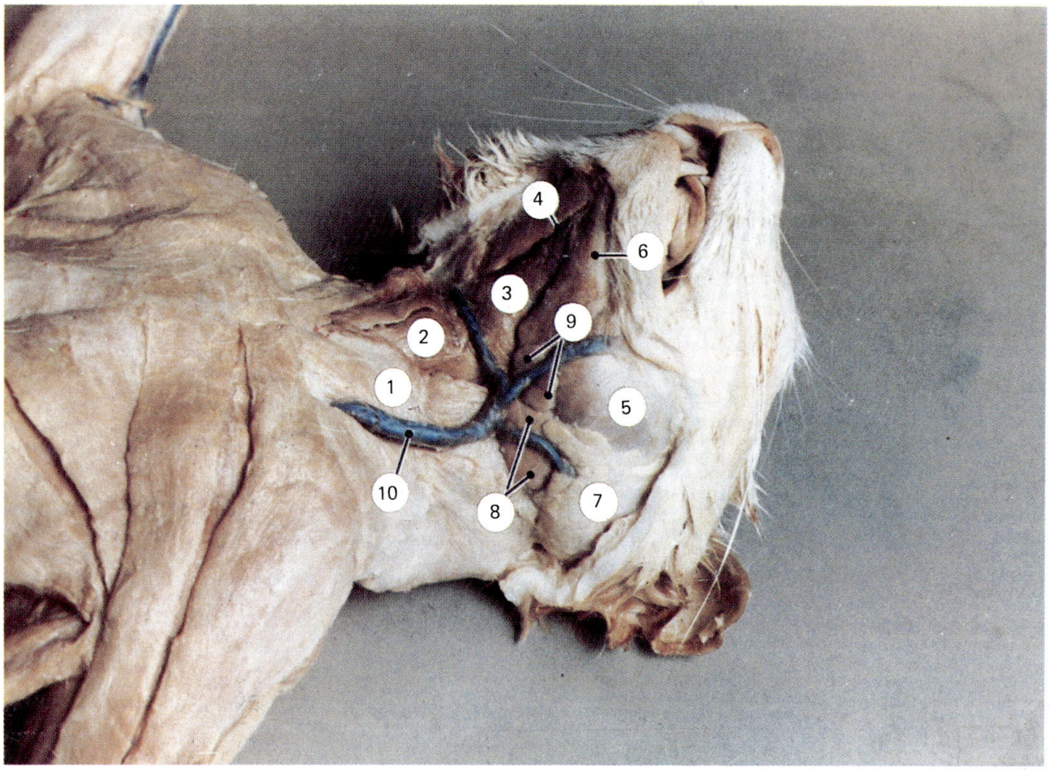

Plate 19-1 Superficial neck and jaw muscles.

1. Sternomastoid
2. Sternohyoid
3. Mylohyoid
4. Digastric
5. Masseter
6. Mandible (bone)
7. Parotid gland
8. Submandibular gland
9. Lymph nodes
10. External jugular vein and tributaries

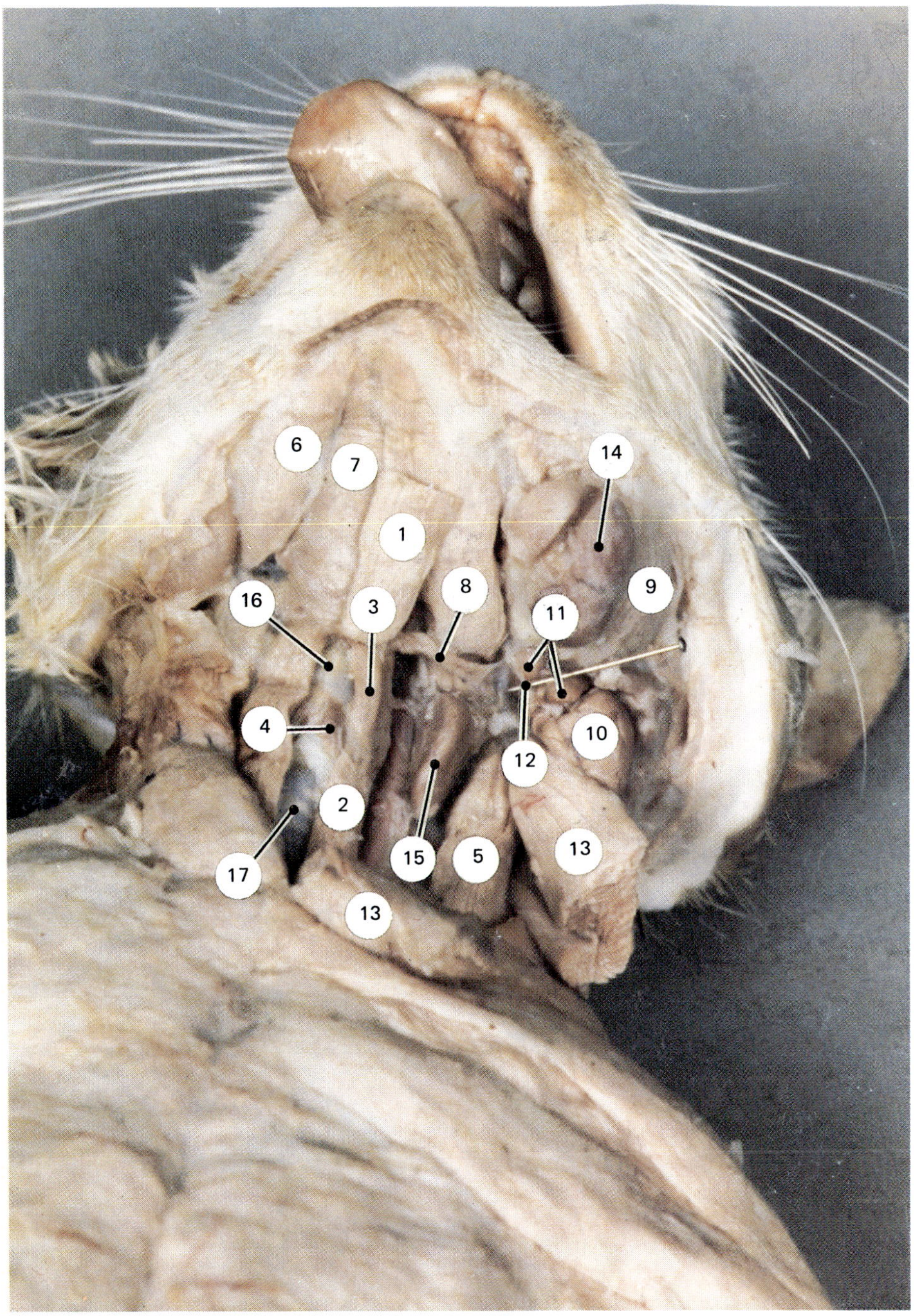

Plate 19-2 Deep muscles and salivary glands.

1. Sternohyoid (reflected)
2. Sternothyroid
3. Thyrohyoid
4. Cricothyroid
5. Cleidomastoid
6. Digastric
7. Mylohyoid
8. Stylohyoid
9. Masseter
10. Submaxillary gland
11. Sublingual gland
12. Salivary duct
13. Sternomastoid (reflected)
14. Lymph nodes (reflected)
15. Lymph node
16. Larynx
17. Trachea

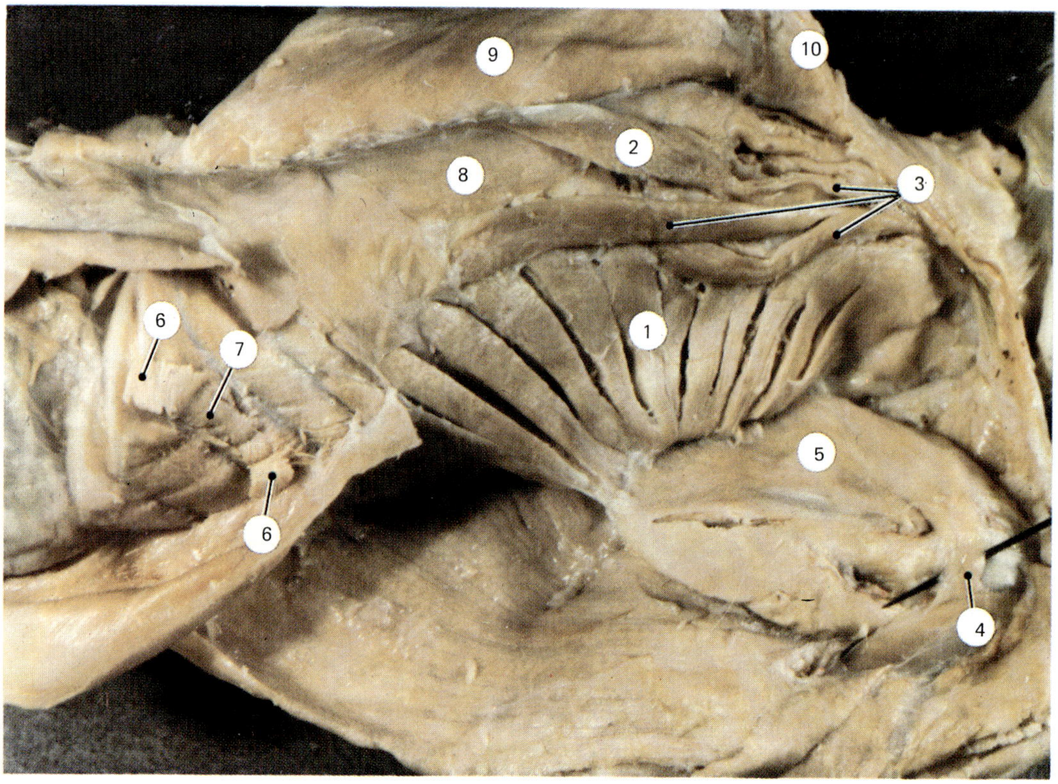

Plate 20-1 Deep chest muscles.

1. Serratus ventralis
2. Transversus costarum
3. Scalenus
4. Coracobrachialis
5. Subscapularis
6. External intercostal (reflected)
7. Internal intercostal
8. Rectus abdominis
9. Pectoralis minor (reflected)
10. Pectoralis major (reflected)

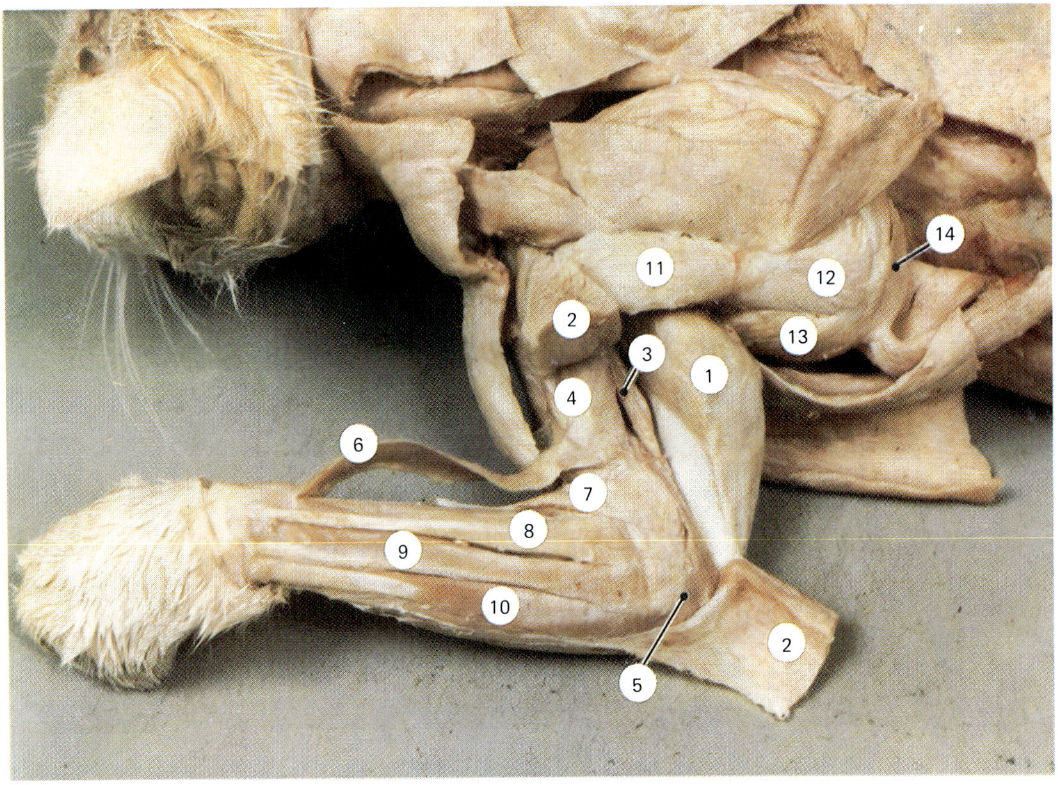

Plate 20-2 Superficial arm muscles, lateral surface.

1. Triceps brachii, long head
2. Triceps brachii, lateral head (reflected)
3. Triceps brachii, medial head
4. Brachialis
5. Anconeus
6. Brachioradialis
7. Extensor carpi radialis
8. Extensor digitorum communis
9. Extensor digitorum lateralis
10. Extensor carpi ulnaris
11. Spinodeltoid
12. Infraspinatus
13. Teres major
14. Rhomboideus major

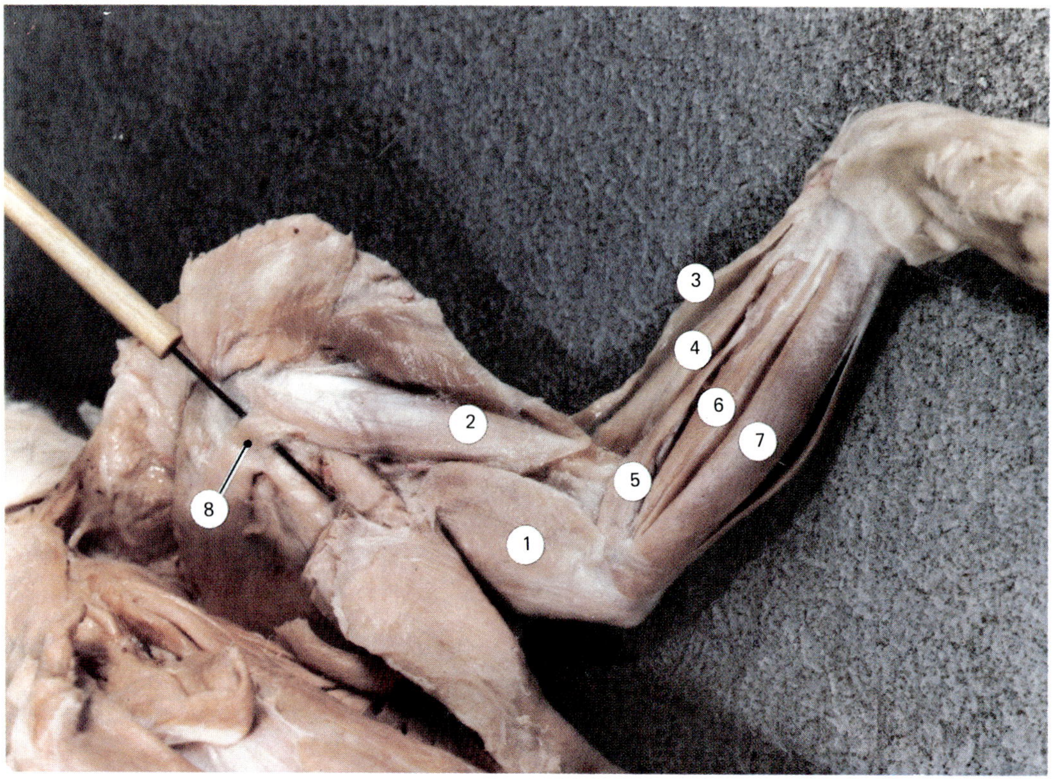

Plate 20-3 Superficial arm muscles, medial view.

1. Epitrochlearis
2. Biceps brachii
3. Brachioradialis
4. Extensor carpi radialis
5. Pronator teres
6. Flexor carpi radialis
7. Palmaris longus
8. Coracobrachialis

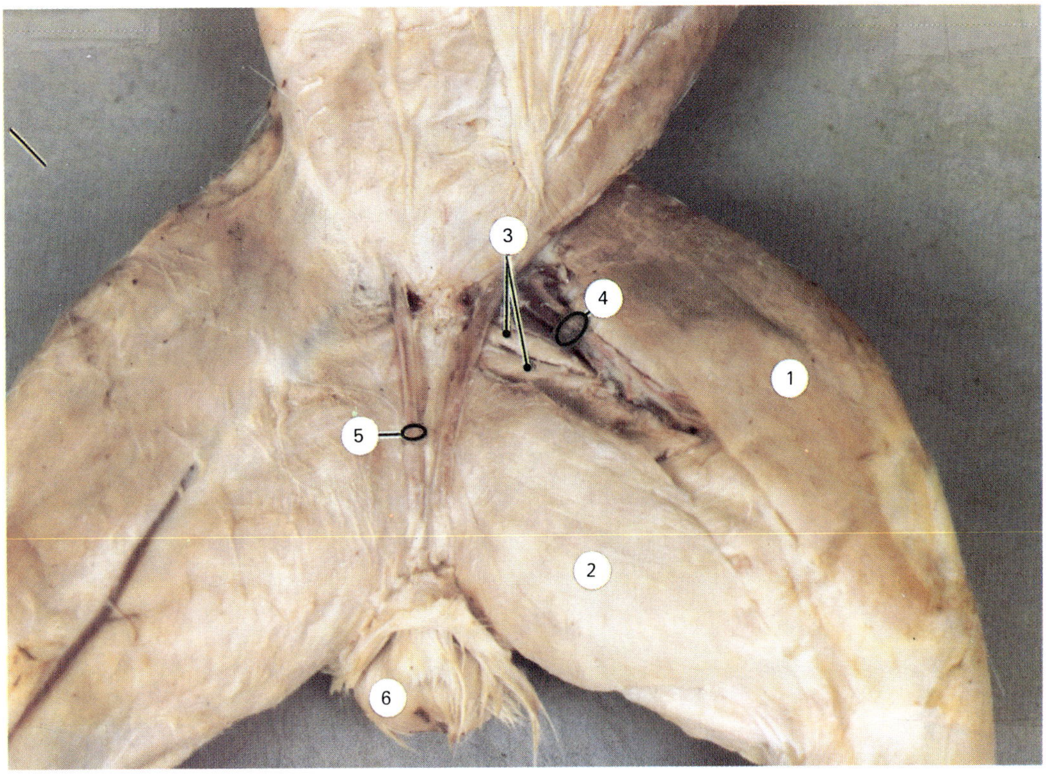

Plate 21-1 Superficial muscles of thigh, medial view.

1. Sartorius
2. Gracilis
3. Adductor muscles
4. Blood Vessels
5. Spermatic cord
6. Scrotum

Femoral triangle (3, 4)

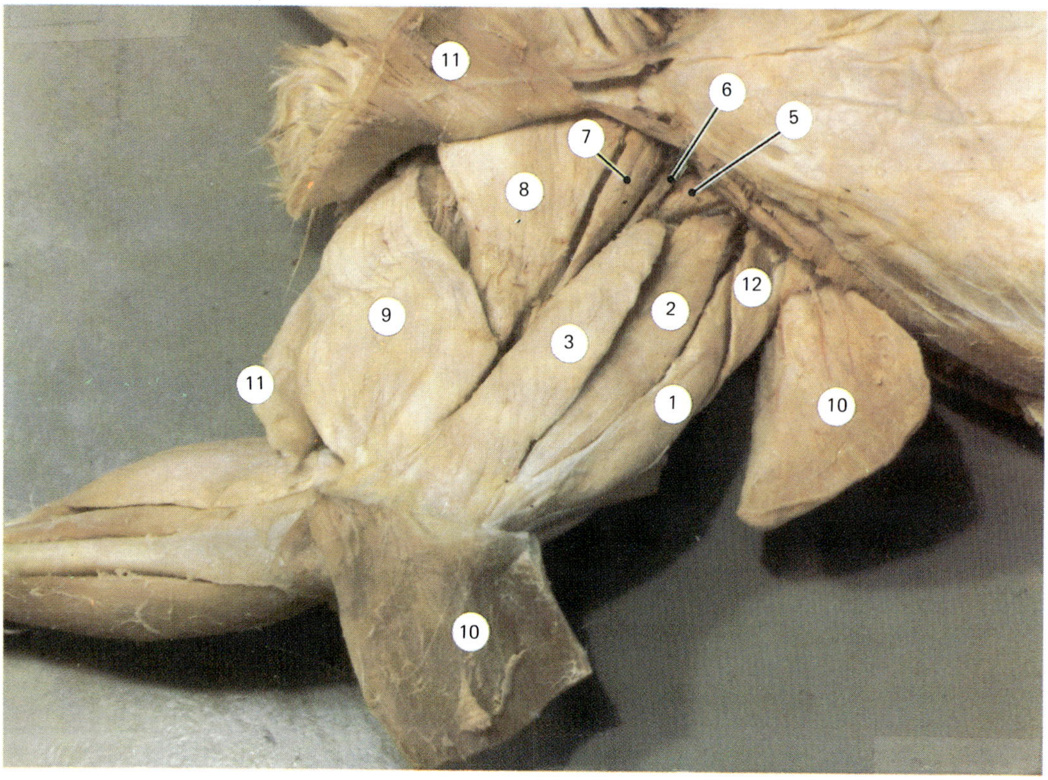

Plate 21-2 Deep muscles of medial thigh.

1. Vastus lateralis
2. Rectus femoris
3. Vastus medialis
4. Vastus intermedius (not shown)
5. Iliopsoas
6. Pectineus
7. Adductor longus
8. Adductor femoris
9. Semimembranosus
10. Sartorius
11. Gracilis
12. Tensor fasciae latae

(1–4: Quadriceps femoris)

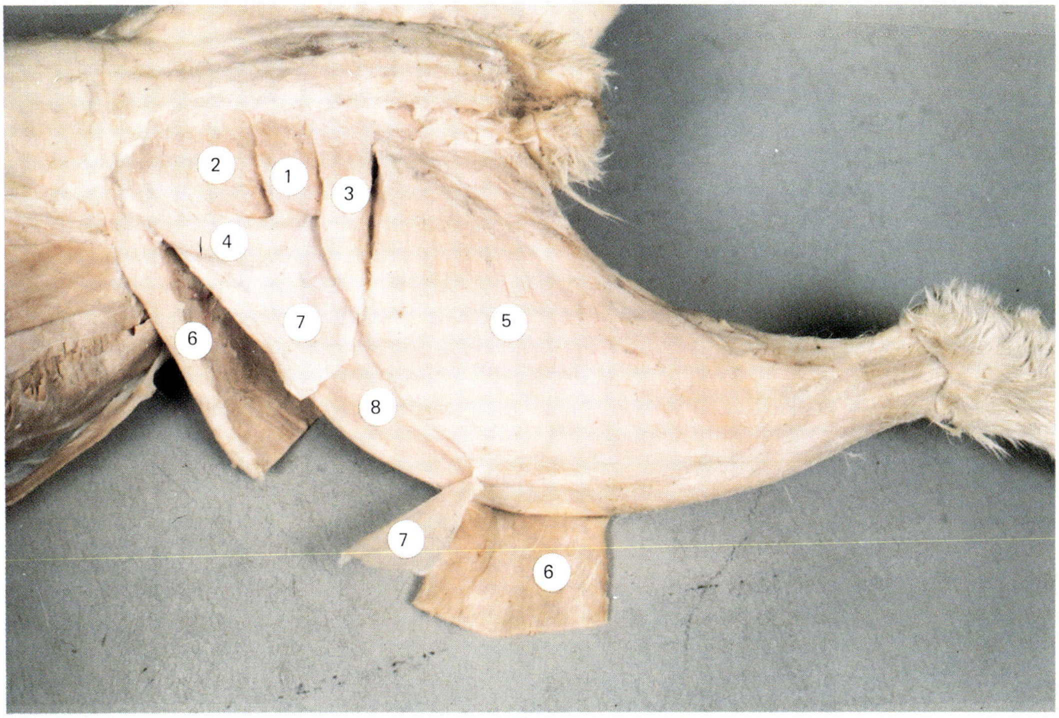

Plate 21-3 Muscles of hip and thigh.
1. Gluteus maximus
2. Gluteus medius
3. Caudofemoralis
4. Tensor fasciae latae
5. Biceps femoris
6. Sartorius
7. Fascia lata (transected)
8. Vastus lateralis

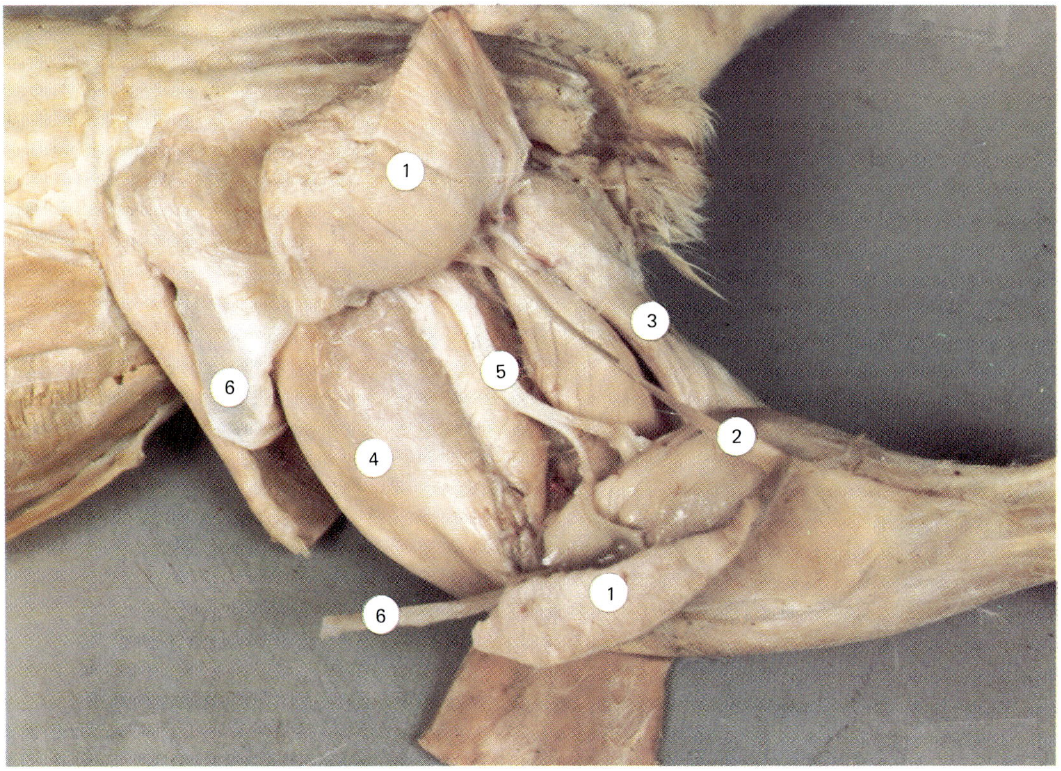

Plate 21-4 Deep muscles of thigh, lateral view.

1. Biceps femoris (reflected)
2. Tenuissimus
3. Semitendinosus
4. Vastus lateralis
5. Sciatic nerve
6. Tensor fasciae latae (fascia transected)

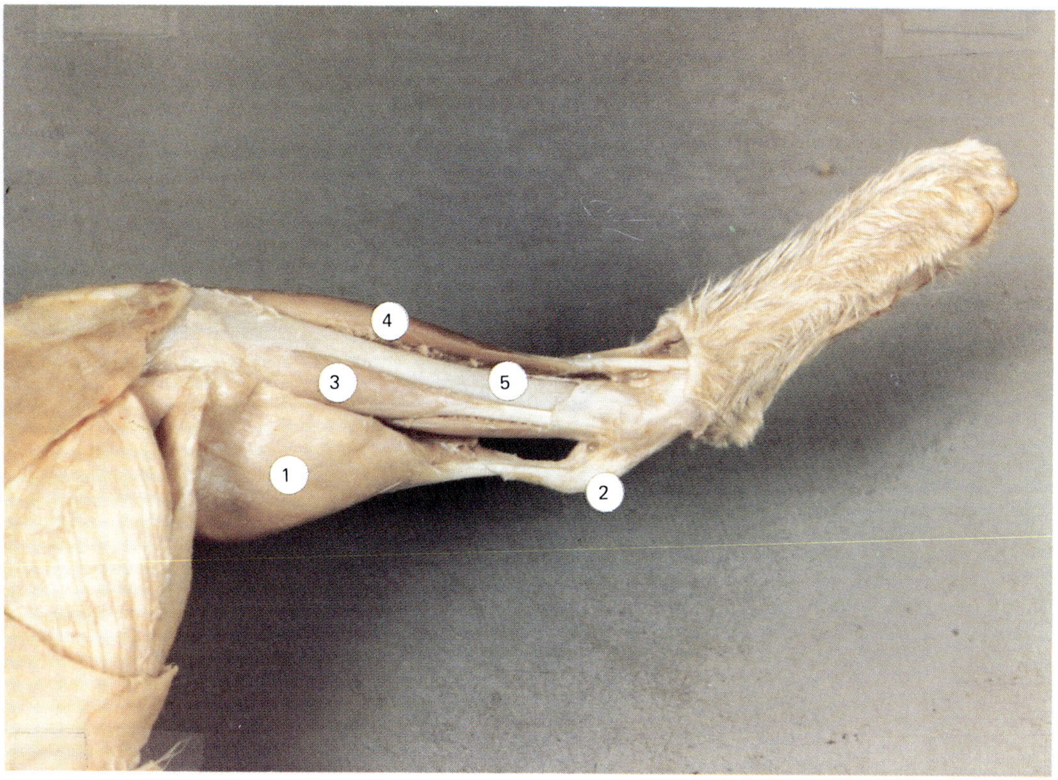

Plate 22-1 Superficial muscles of lower leg, medial view.

1. Gastrocnemius
2. Tendon of Achilles
3. Flexor digitorum longus
4. Tibialis anterior
5. Tibia (bone)

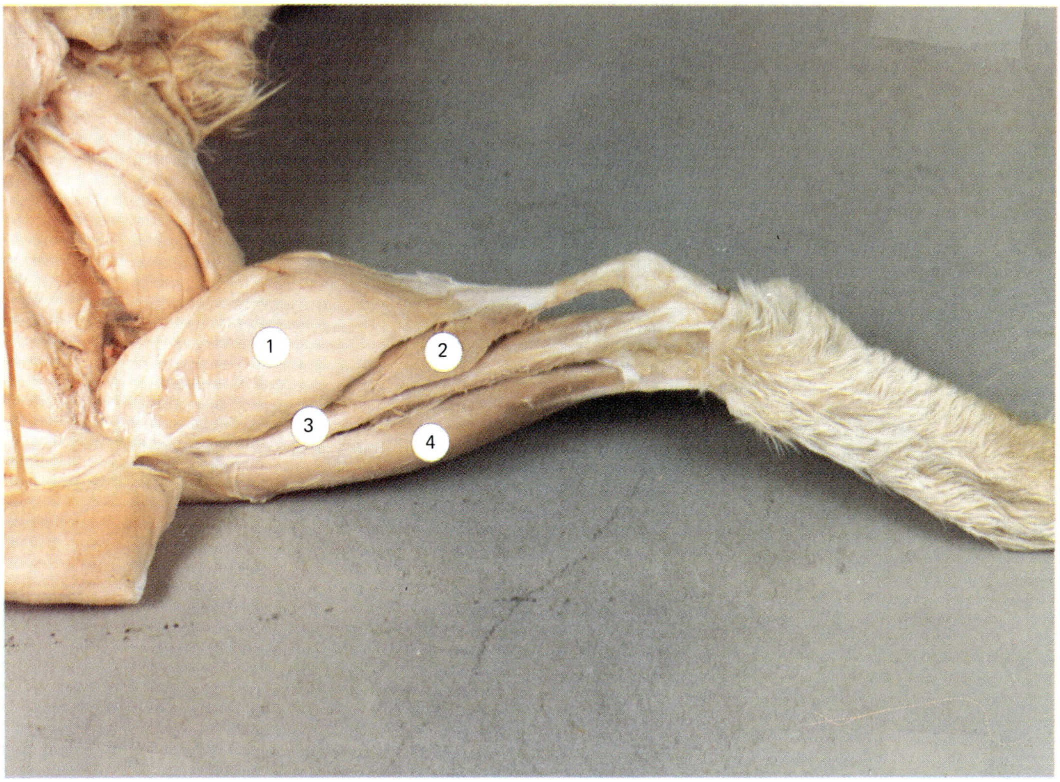

Plate 22-2 Superficial muscles of lower leg, lateral view.

1. Gastrocnemius
2. Soleus
3. Peroneus
4. Extensor digitorum longus

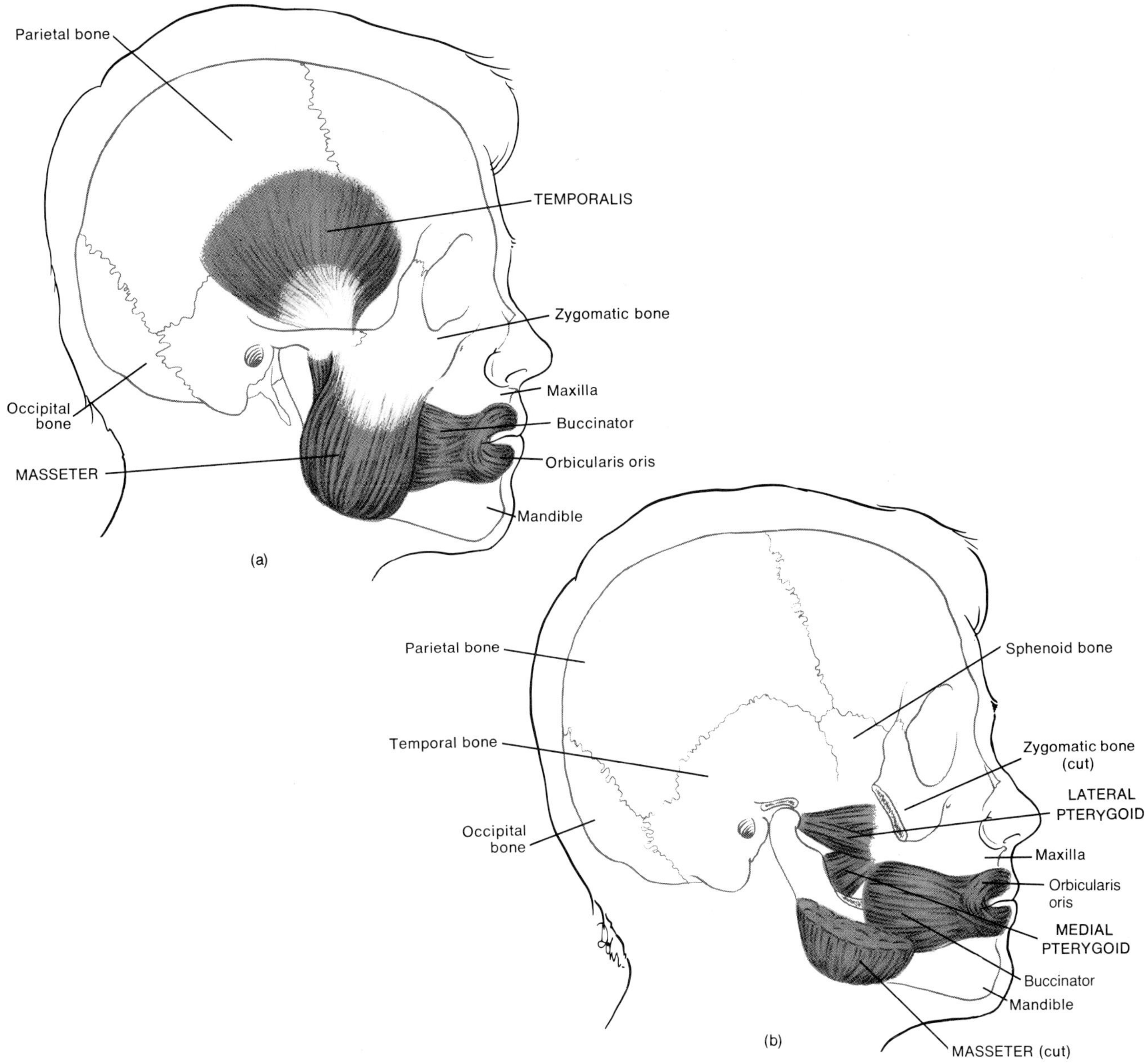

Figure 19-5 Muscles that move the lower jaw. (a) Superficial right lateral view. (b) Deep right lateral view.

Table 19-3 Muscles That Move the Lower Jaw (Figure 19-5)

Muscle	Origin	Insertion	Action	Innervation
Masseter (*maseter* = chewer)	Maxilla and zygomatic arch	Angle and ramus of mandible	Elevates mandible as in closing the mouth and protracts (protrudes) mandible	Mandibular branch of trigeminal nerve (V)
Temporalis (*tempora* = temples)	Temporal bone	Coronoid process of mandible	Elevates and retracts mandible	Temporal nerve from mandibular division of trigeminal nerve (V)
Medial pterygoid (*medial* = closer to midline; *pterygoid* = like a wing; pterygoid plate of sphenoid)	Medial surface of lateral pterygoid plate of sphenoid; maxilla	Angle and ramus of mandible	Elevates and protracts mandible and moves mandible from side to side	Mandibular branch of trigeminal nerve (V)
Lateral pterygoid (*lateral* = farther from midline)	Greater wing and lateral surface of lateral pterygoid plate of sphenoid	Condyle of mandible; temporomandibular articulation	Protracts mandible, opens mouth, and moves mandible from side to side	Mandibular branch of trigeminal nerve (V)

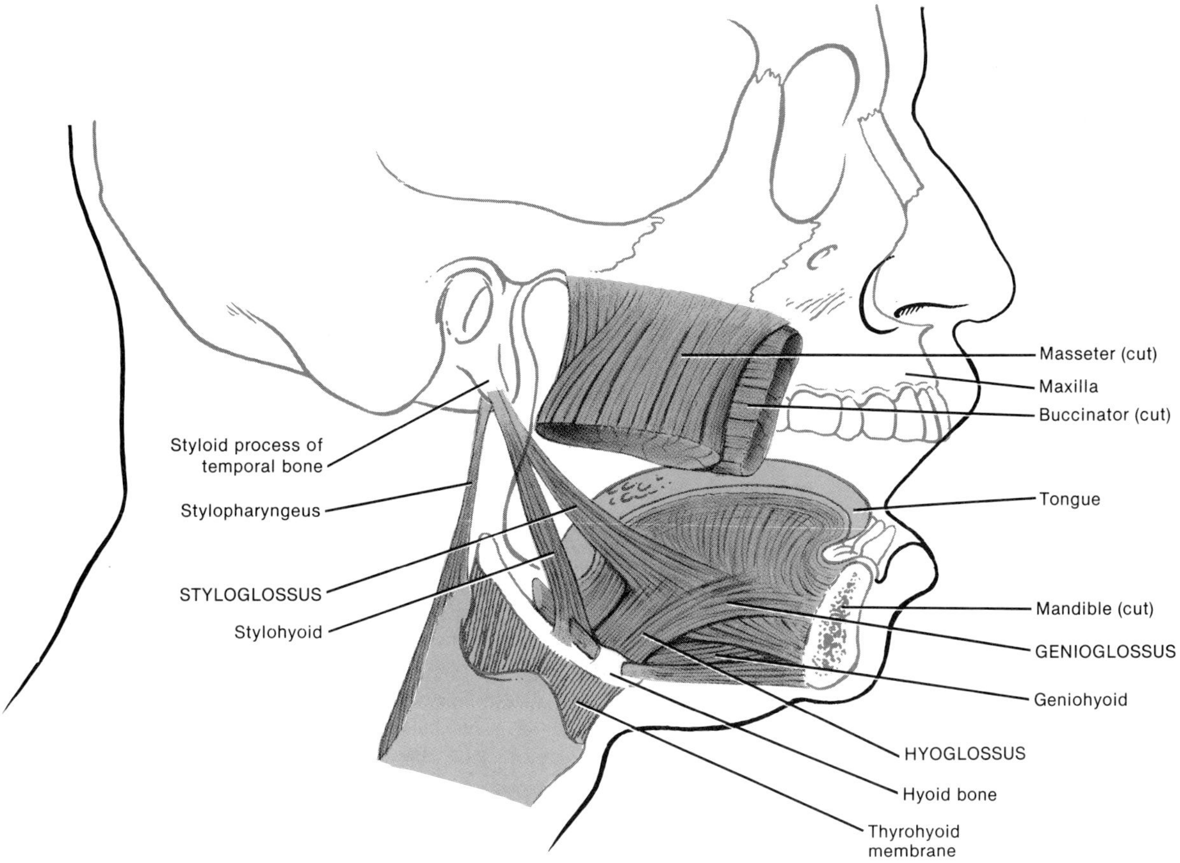

Figure 19-6 Muscles that move the tongue viewed from the right side.

Table 19-4 Muscles of the Larynx (Figure 19-9)

Muscle	Origin	Insertion	Action	Innervation
EXTRINSIC				
Omohyoid (*omo* = relationship to the shoulder; *hyoedies* = U-shaped, pertaining to the hyoid bone)	Superior border of scapula and superior transverse ligament	Body of hyoid bone	Depresses hyoid bone	Branches of ansa cervicalis nerve (C1–C3)
Sternohyoid (*sterno* = sternum)	Medial end of clavicle and manubrium of sternum	Body of hyoid bone	Depresses hyoid bone	Branches of ansa cervicalis nerve (C1–C3)
Sternothyroid (*thyro* = thyroid gland)	Manubrium of sternum	Thyroid cartilage of larynx	Depresses thyroid cartilage	Branches of ansa cervicalis nerve (C1–C3)
Thyrohyoid	Thyroid cartilage of larynx	Greater cornu of hyoid bone	Elevates thyroid cartilage and depresses hyoid bone	Cervical nerves C1–C2 and descending hypoglossal nerve (XII)
Stylopharyngeus				
Palatopharyngeus				
Inferior constrictor				
Middle constrictor				
INTRINSIC				
Cricothyroid (*crico* = cricoid cartilage of larynx)	Anterior and lateral portion of cricoid cartilage of larynx	Anterior border of inferior cornu of thyroid cartilage of larynx and posterior part of inferior border of lamina of thyroid cartilage	Produces tension and elongation of vocal folds	External laryngeal branch of vagus nerve (X)

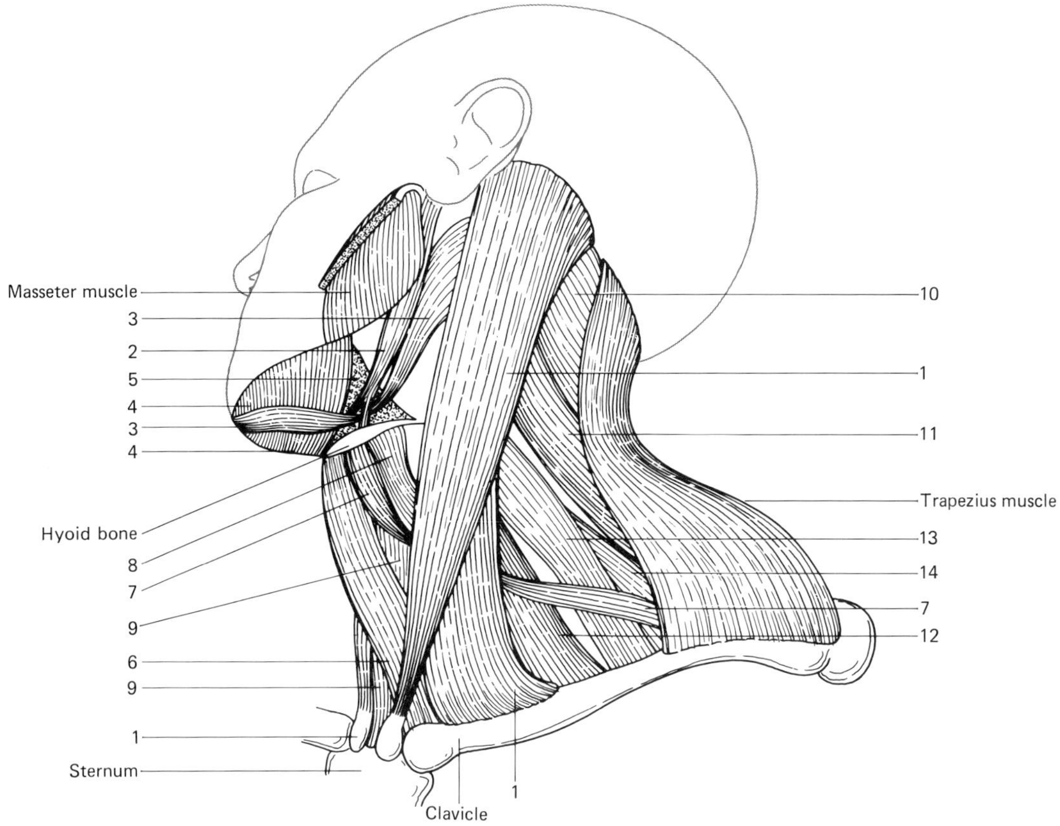

Figure 19-7 Anterior triangle muscle
1. Sternocleidomastoid

Suprahyoid muscles
2. Stylohyoid
3. Digastric
4. Mylohyoid
5. Hyoglossus

Infrahyoid muscles
6. Sternohyoid
7. Omohyoid
8. Thyrohyoid
9. Sternothyroid

Posterior triangle muscles
10. Splenius capitis
11. Levator scapulae
12. Scalenus anterior
13. Scalenus medius
14. Scalenus posterior

Table 19-5 Muscles That Move the Tongue (Figure 19-8)

Muscle	Origin	Insertion	Action	Innervation
Genioglossus (*geneion* = chin; *glossus* = tongue)	Mandible	Undersurface of tongue and hyoid bone	Depresses and thrusts tongue forward (protraction)	Hypoglossal nerve (XII)
Styloglossus (*stylo* = stake or pole; styloid process of temporal)	Styloid process of temporal bone	Side and undersurface of tongue	Elevates tongue and draws it backward (retraction)	Hypoglossal nerve (XII)
Stylohyoid (*hyoeides* = U-shaped; pertaining to hyoid bone)	Styloid process of temporal bone	Body of hyoid bone	Elevates and retracts tongue	Facial nerve (VII)
Hyoglossus	Body of hyoid bone	Side of tongue	Depresses tongue and draws down its sides	Hypoglossal nerve (XII)

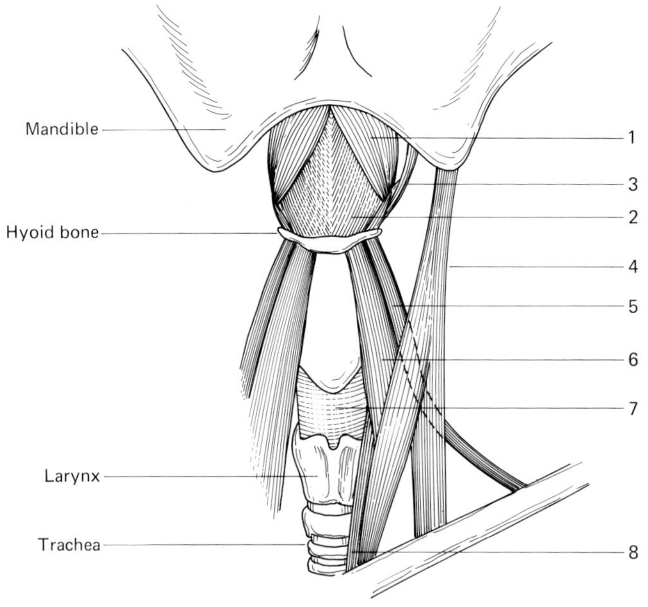

Figure 19-8 Muscles that move the tongue.
1. Digastric
2. Mylohyoid
3. Stylohyoid
4. Sternocleidomastoid
5. Omohyoid
6. Sternohyoid
7. Cricothyroid
8. Sternothyroid

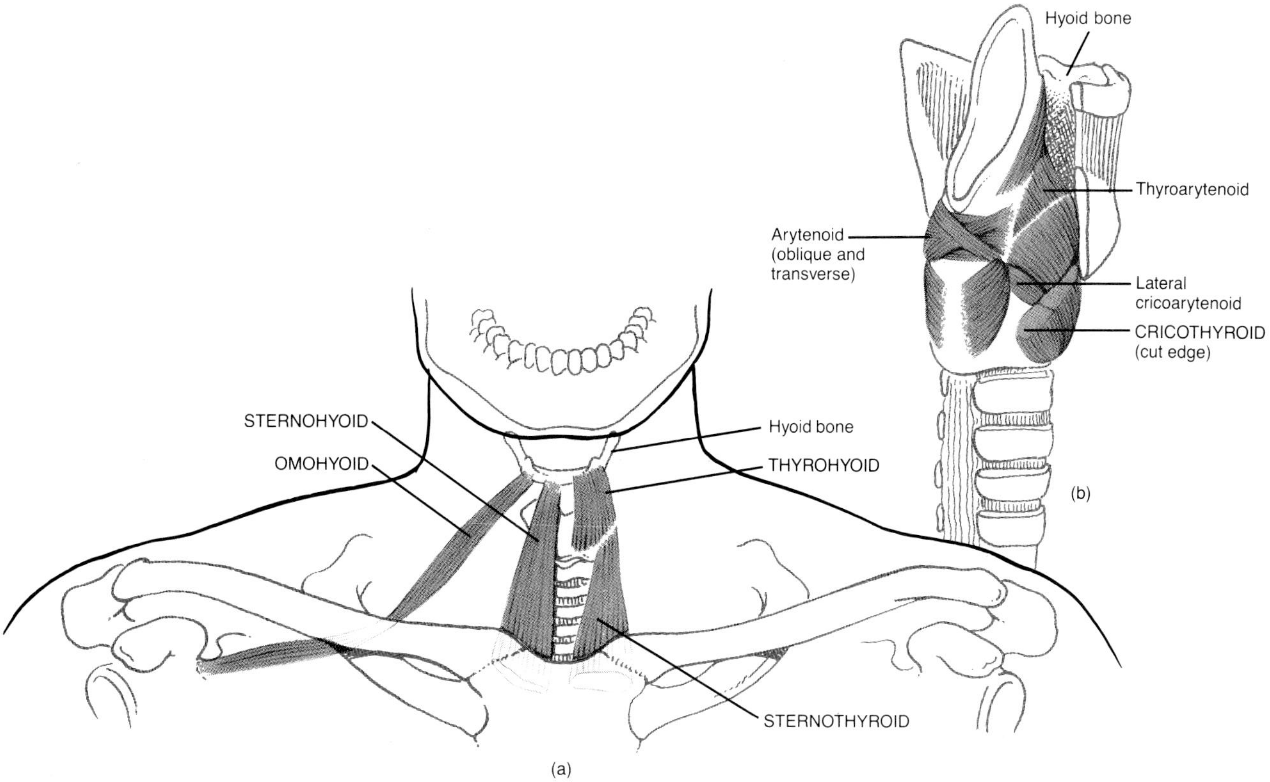

Figure 19-9 Muscles of the larynx. (a) Anterior view. (b) Right lateral view.

Questions for Review

EXERCISE **19**

MUSCLES OF THE NECK, JAW, AND FACE

Name ____

Laboratory Section ______________________________ Date ____________

1. Which human muscles perform the following functions?

   a. ______________________ Compress lips against teeth

   b. ______________________ Depress lower lip

   c. ______________________ Close the eye

   d. ______________________ Raise the eyebrows

   e. ______________________ Elevate upper lip

   f. ______________________ Elevate mandible to close the mouth

   g. ______________________ Move mandible from side to side

   h. ______________________ Draw tongue backward

   i. ______________________ Draw soft palate down on the tongue

   j. ______________________ Depress thyroid cartilage of larynx

   k. ______________________ Elongate the vocal folds

   l. ______________________ Thrust tongue forward

EXERCISE 20

MUSCLES OF THE DEEP CHEST AND ARM

Materials

The following materials should be provided for class use.

1. Embalmed cats
2. Dissecting tools
3. Dissecting trays
4. String
5. 3 percent phenol solution

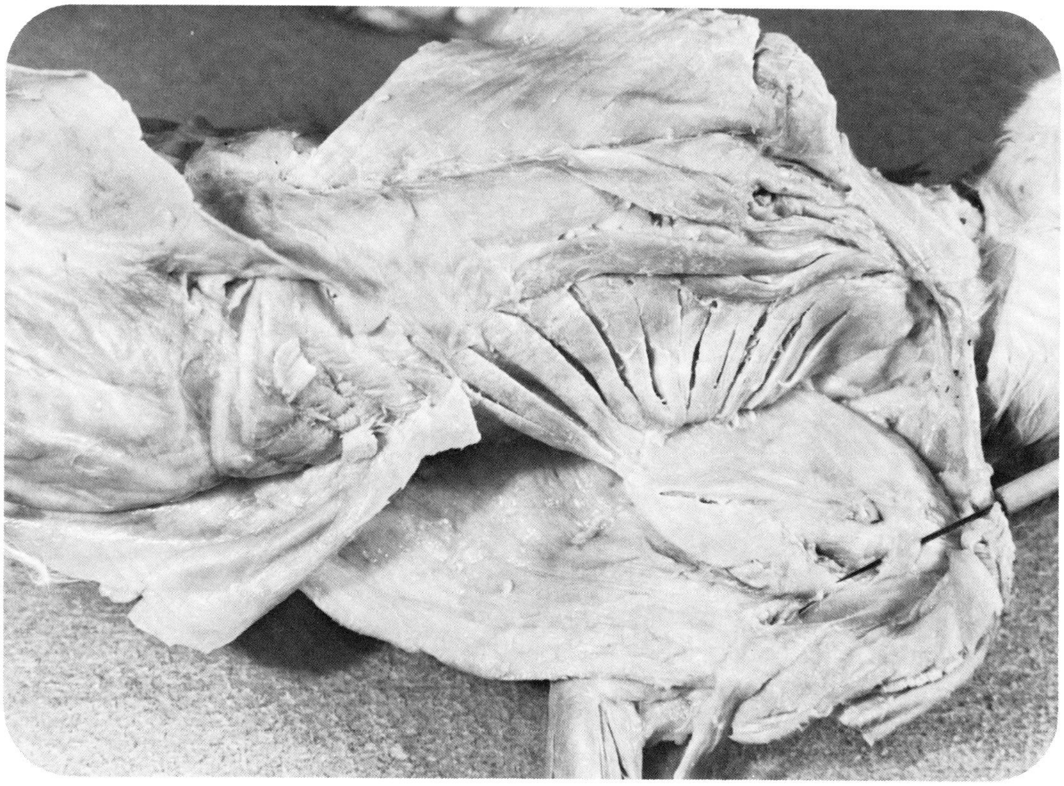

Figure 20-1a Deep chest muscles.

DEEP CHEST MUSCLES

INSTRUCTIONS FOR DISSECTION

With the cat on its back (Figure 20-1), transect the pectoral muscles—the pectoantebrachialis, the pectoralis major (which must be cut all the way to the humerus to reflect it), and finally the pectoralis minor. You may transect the xiphihumeralis, but it is not necessary.

Observe the **serratus ventralis,** the **subscapularis,** and the **teres major,** as shown in Figure 20-1. At the shoulder is a very small muscle, the **coracobrachialis.** It is no more than $\frac{1}{3}$ inch long and tightly binds the scapula to the arm. (In the human it is much larger.)

Near the midline find a small oblique muscle, the **transversus costarum,** and lateral and somewhat above it, is the **scalenus.** The most prominent part of the three-part scalenus is the **scalenus medius,** which extends furthest posteriorly. On either side of the scalenus medius are shorter slips, the **scalenus anterior** and the **scalenus posterior,** but all three unite anteriorly. (The scalenus in the human is located higher in the neck region). Separate each slip. Note also in this area the anterior end of the rectus abdominis.

Clear a space between the ribs under the external oblique muscle of all fat and fascia. Between two adjacent ribs observe the **external intercostal** muscle whose fibers run obliquely from the posterior border of one rib to the anterior border of the next rib. Under the external intercostal lies the **internal intercostal.** Separate these two muscles, keeping in mind that they are very thin. The internal intercostal can be recognized because its fibers, also oblique, run at a different angle from those of the external intercostal. Refer to Table 20-1 for the origin, insertion, and action of the cat's deep chest muscles.

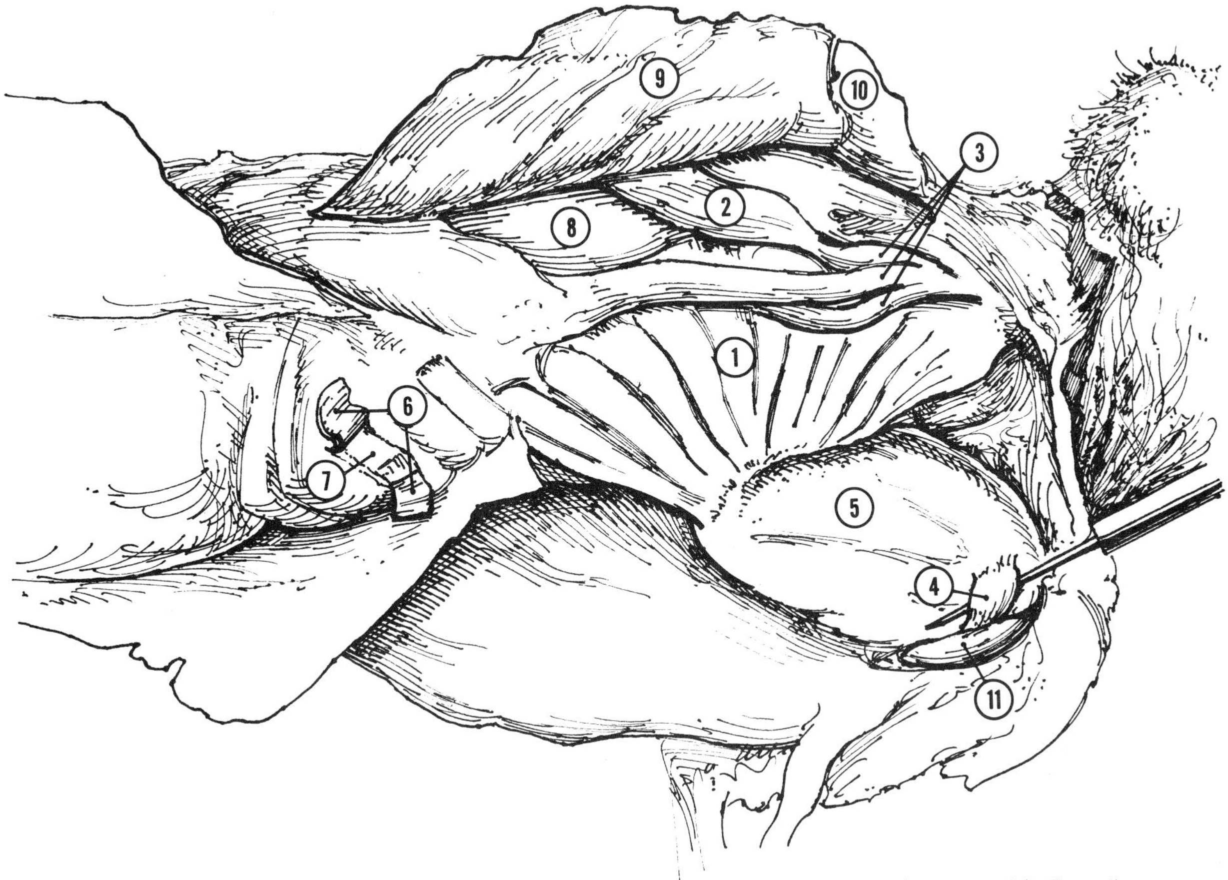

Figure 20-1b Deep chest muscles.
1. Serratus ventralis
2. Transversus costarum (kos-TAR-um)
3. Scalenus (skā-LĒ-nus)
4. Coracobrachialis (kor′-a-kō-brāk′-ē-AL-is)
5. Subscapularis (sub-skap′-yoo-LAR-is)
6. External intercostal (reflected)
7. Internal intercostal
8. Rectus abdominis
9. Pectoralis minor (reflected)
10. Pectoralis major (reflected)

Table 20-1 Deep Chest Muscles of the Cat

Muscle	Origin	Insertion	Action
Serratus ventralis	First ten ribs	Medial surface of scapula	Pulls scapula forward and depresses it
Transversus costarum	Lateral border of sternum	First rib	Pulls sternum craniad
Scalenus	Transverse processes of cervical vertebrae	Ribs	Raises ribs; flexes neck
Coracobrachialis	Coracoid process of scapula	Proximal end of humerus	Adducts humerus
Subscapularis	Subscapular fossa	Lesser tubercle of humerus	Pulls humerus medially
External intercostals	Caudal border of one rib	Cranial border of next rib caudad	Raise ribs at inhalation and thereby increase size of thoracic cavity
Internal intercostals	Cranial border of one rib	Caudal border of the next rib craniad	Draw adjacent ribs together and lower ribs

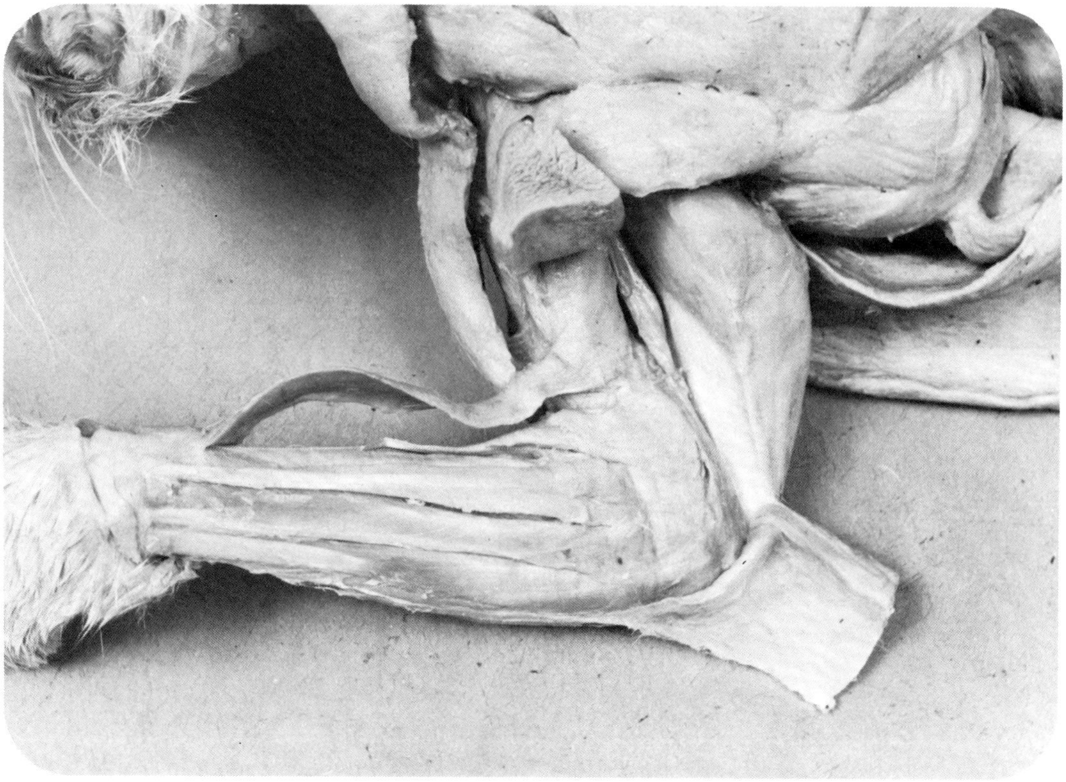

Figure 20-2a Superficial arm muscles, lateral surface.

SUPERFICIAL ARM MUSCLES, LATERAL SURFACE

INSTRUCTIONS FOR DISSECTION

Begin with the upper arm (Figure 20-2). You have already dissected the clavodeltoid which, although not an arm muscle, does cover the anterior surface. The largest superficial muscle in the upper arm is the **triceps brachii.** It is made up of three distinct muscles with three heads of origin, but it has only one point of insertion. The large mass over the posterior surface is the **long head** of the triceps. Lying next to it on the lateral surface is the **lateral head** of the triceps. Free each of these muscles between origin and insertion and then transect the lateral head of the triceps. Pull back the cut ends to reveal the smaller medial head of the triceps and next to it the **brachialis,** as seen in Figure 20-2b.

Distal to the triceps medial head is an L-shaped muscle surrounding the elbow, the **anconeus.** The anconeus need not be dissected, but merely observed. It is usually darker in color than the surrounding muscles, which helps in locating it. It is not found in humans.

In the lower arm dissect only the superficial muscles. An extension from the brachialis of the upper arm is the **brachioradialis.** The muscles next to the brachioradialis in order are the **extensor carpi radialis,** the **extensor digitorum communis,** the **extensor digitorum lateralis,** and the **extensor carpi ulnaris.**

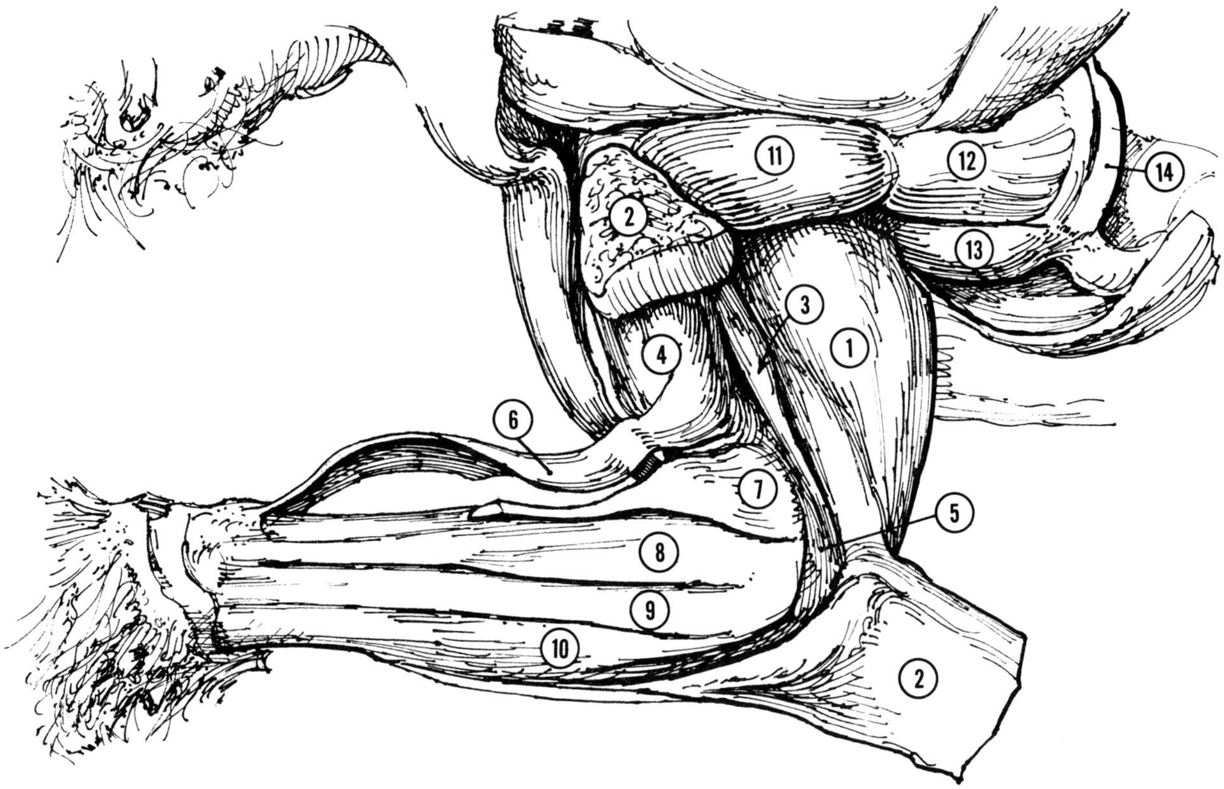

Figure 20-2b Superficial arm muscles, lateral surface.

1. Triceps brachii long head (BRĀK-ē-ē)
2. Triceps brachii lateral head (reflected)
3. Triceps brachii medial head
4. Brachialis (brāk-ē-AL-is)
5. Anconeus (an-CŌN-ē-us)
6. Brachioradialis (brāk-ē-ō-rā′-dē-AL-is)
7. Extensor carpi radialis (KAR-pē)
8. Extensor digitorum communis (dij-i-TOR-um kom-MYOO-nis)
9. Extensor digitorum lateralis
10. Extensor carpi ulnaris (ul-NAR-is)
11. Spinodeltoid
12. Infraspinatus
13. Teres major
14. Rhomboideus major

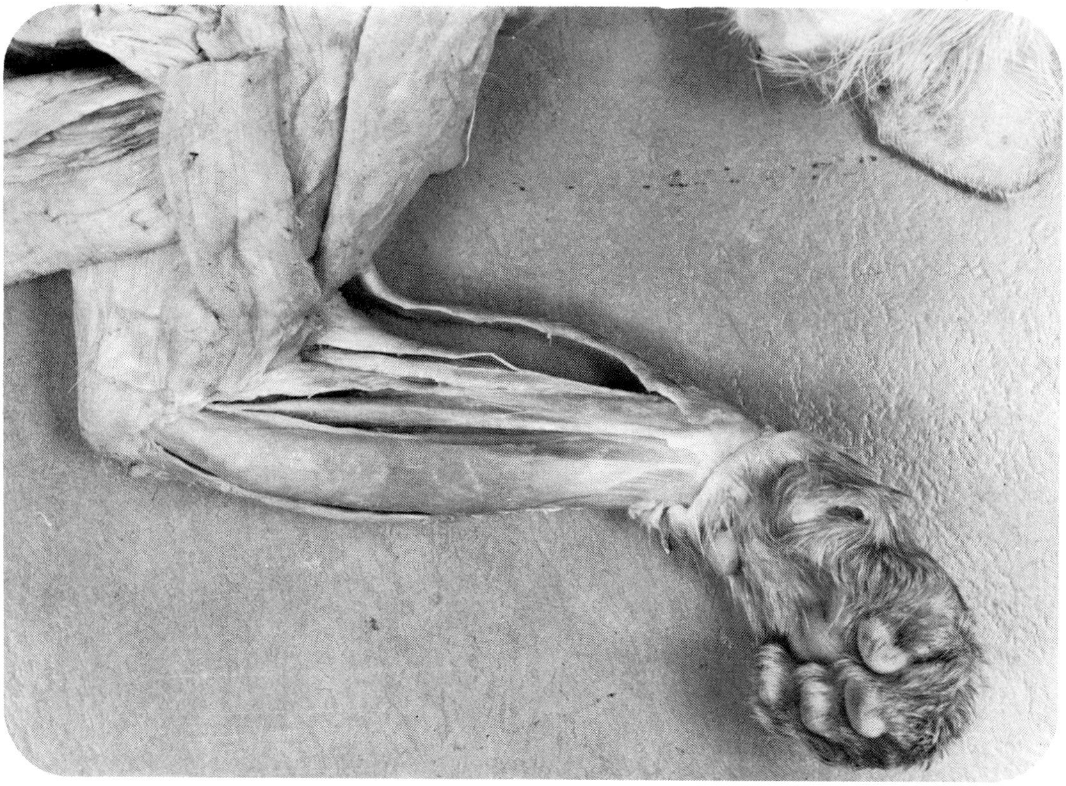

Figure 20-3a Superficial arm muscles, medial surface.

SUPERFICIAL ARM MUSCLES, MEDIAL SURFACE

INSTRUCTIONS FOR DISSECTION

The medial surface of the upper arm is almost entirely covered by a broad, flat sheet of muscle, the **epitrochlearis** (absent in humans). Dissect the epitrochlearis and note the somewhat convex **biceps brachii** muscle beneath the pectoral muscle (Figure 20-3a).

In the lower arm begin at the anterior border with the brachioradialis previously observd (see Figure 20-3b). Lying next to it is the extensor carpi radialis, which has two parts: a shorter, somewhat triangular **extensor carpi radialis brevis** and beneath it the **extensor carpi radialis longus.**

The next superficial muscles to separate are the narrow **pronator teres** and next to it the **flexor carpi radialis.** The last muscle on the medial surface is the widest of the group and is called the **palmaris longus.** Refer to Table 20-2 for the origin, insertion, and action of the cat's arm muscles.

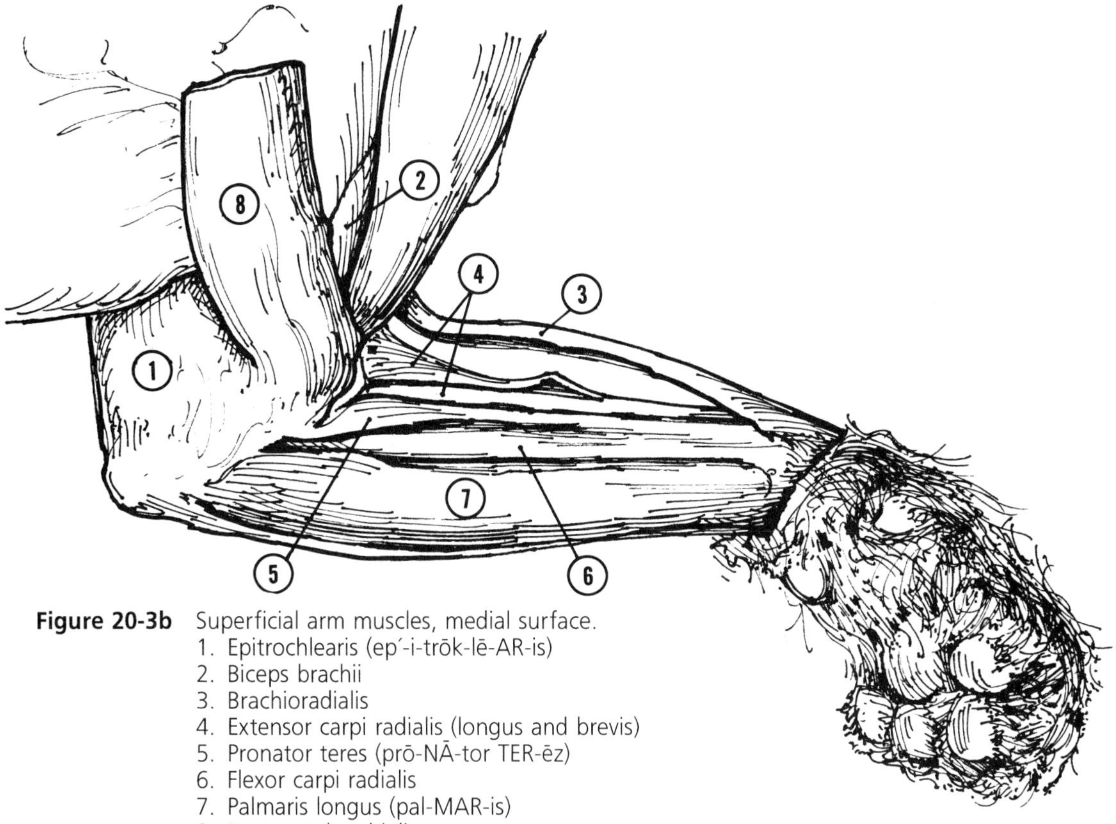

Figure 20-3b Superficial arm muscles, medial surface.
1. Epitrochlearis (ep′-i-trōk-lē-AR-is)
2. Biceps brachii
3. Brachioradialis
4. Extensor carpi radialis (longus and brevis)
5. Pronator teres (prō-NĀ-tor TER-ēz)
6. Flexor carpi radialis
7. Palmaris longus (pal-MAR-is)
8. Pectoantebrachialis

COMPARABLE HUMAN MUSCLES

As noted in Exercise 19, the muscles of the deep chest of the cat are not related easily to those of the human body. The **scalenus** is primarily a neck muscle in the human, and the **coracobrachialis** is much larger. There is no **transversus costarum** in the human.

In contrast to the deep chest, there is a great deal of similarity between cats and humans in the upper and lower arm regions. One exception is the **epitrochlearis,** a conspicuous cat muscle not found in the human. The **biceps brachii** (covered by the epitrochlearis in the cat) is more prominent in the human. The drawings of human muscles in Figures 20-4–20-6 show both the differences and similarities. Study as many as time permits, learning their origins, insertions, and actions from Tables 20-3 and 20-4.

Table 20-2 Arm Muscles of the Cat

Muscle	Origin	Insertion	Action
Triceps brachii			
Long head	Lateral border of scapula	Olecranon process of ulna	Extends forearm
Lateral head	Deltoid ridge of humerus	Olecranon process of ulna	Extends forearm
Medial head	Shaft of humerus	Olecranon process of ulna	Extends forearm
Brachialis	Lateral side of humerus	Proximal end of ulna	Flexes forearm
Anconeus	Near lateral epicondyle at distal end of humerus	Proximal end of ulna	Tenses elbow; rotates ulna
Brachioradialis	Mid-dorsal border of humerus	Distal end of radius	Supinates hand
Extensor carpi radialis longus	Supracondylar ridge of humerus	Base of second metacarpal	Extends hand
Extensor digitorum communis	Lateral surface of humerus	Tendons on four digits	Extends digits
Extensor digitorum lateralis	Lateral surface of humerus	Digits by joining tendons of extensor digitorum communis	Extends digits
Extensor carpi ulnaris	Distal end of humerus and lateral border of ulna	Base of fifth metacarpal	Extends ulnar side of wrist and fifth digit
Extensor carpi radialis brevis	Lateral epicondyle of humerus	Base of third metacarpal	Extends hand
Pronator teres	Distal end of humerus	Medial border of radius	Rotates radius for pronation
Flexor carpi radialis	Distal end of humerus	Second and third metacarpals	Flexes wrist
Palmaris longus	Distal end of humerus	Tendons on all digits	Flexes digits
Flexor carpi ulnaris	Distal end of humerus	Tendons on all digits	Flexes digits
Biceps brachii	Tuberosity near glenoid fossa and coracoid process of scapula	Radial tuberosity	Flexes and supinates forearm
Epitrochlearis	Ventral border of latissimus dorsi near insertion	Fascia on medial surface of elbow	Extends forearm and rotates ulna to supinate hand

Table 20-3 Deep Chest Muscles of the Human

Muscle	Origin	Insertion	Action
Serratus anterior	First eight or nine ribs	Vertebral border and inferior angle of scapula	Pulls scapula forward and rotates it upward
Coracobrachialis	Coracoid process of scapula	Humerus	Adducts, flexes, and rotates arm
External intercostals	Inferior border of rib above	Superior border of rib below	Raise rib at inhalation and thereby increase size of thoracic cavity
Internal intercostals	Superior border of rib below	Inferior border of rib above	Draw adjacent ribs together and lower ribs at exhalation

Table 20-4 Arm Muscles of the Human

Muscles	Origin	Insertion	Action
Triceps brachii			
Long head	Infraglenoid tuberosity	Olecranon process of ulna	Extends forearm
Lateral head	Above radial groove of humerus	Olecranon process of ulna	Extends forearm
Medial head	Below radial groove of humerus	Olecranon process of ulna	Extends forearm
Brachialis	Anterior surface of humerus	Tuberosity and coronoid process of ulna	Flexes forearm
Biceps brachii			
Long head	Near glenoid fossa of scapula	Radial tuberosity	Flexes and supinates forearm
Short head	Coracoid process of scapula	Radial tuberosity	Flexes and supinates forearm
Brachioradialis	Supracondyloid ridge of humerus	Above styloid process of radius	Flexes forearm
Flexor carpi radialis	Medial epicondyle of humerus	Second and third metacarpals	Flexes and abducts wrist
Flexor carpi ulnaris	Medial epicondyle of humerus and upper dorsal border of ulna	Pisiform, hamate, and fifth metacarpal	Flexes and adducts wrist
Extensor carpi radialis longus	Lateral epicondyle of humerus	Second metacarpal	Extends and abducts wrist
Extensor carpi ulnaris	Lateral epicondyle of humerus and dorsal border of ulna	Fifth metacarpal	Extends and adducts wrist
Flexor digitorum profundus	Body of ulna	Bases of distal phalanges	Flexes all distal phalanges
Flexor digitorum superficialis	Medial epicondyle of humerus, coronoid process of ulna, and oblique line of radius	Middle phalanges	Flexes middle phalanges of each finger
Extensor digitorum	Lateral epicondyle of humerus	Middle and distal phalanges	Extends phalanges
Extensor indicis	Dorsal surface of ulna	Tendon of extensor digitorum of index finger	Extends index finger
Supinator	Lateral epicondyle of humerus and ridge on ulna	Oblique line of radius	Supinates forearm
Pronator teres	Medial epicondyle of humerus and coronoid process of ulna	Lateral surface of radius	Pronates forearm
Palmaris longus	Medial epicondyle of humerus	Radial tuberosity	Flexes and supinates forearm
Coracobrachialis (see Figure 20-5)	Coracoid process of scapula	Middle of medial border of humerus	Flexes and adducts arm

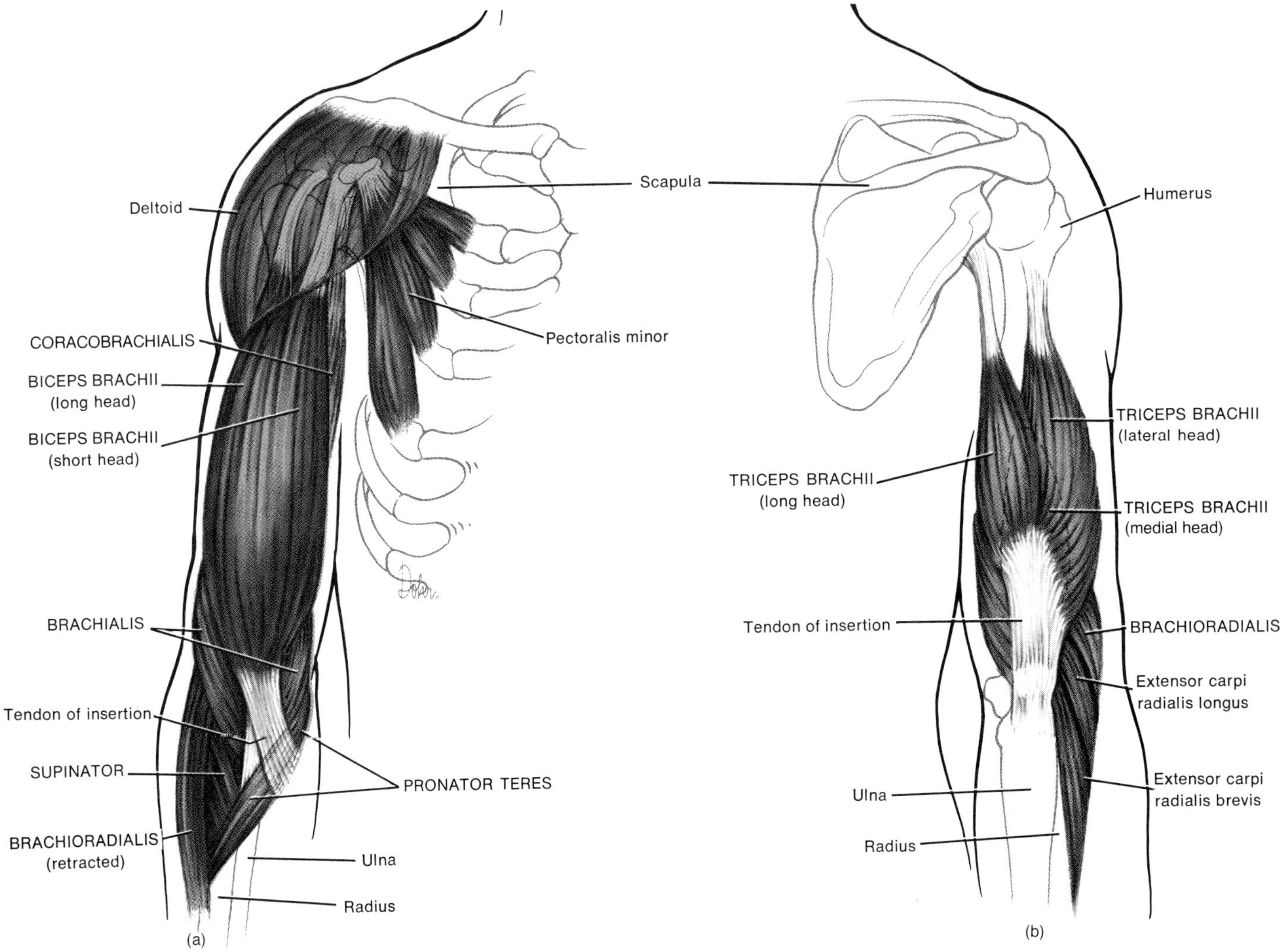

Figure 20-4 Muscles of the shoulder that move the forearm. (a) Anterior view. (b) Posteroir view.

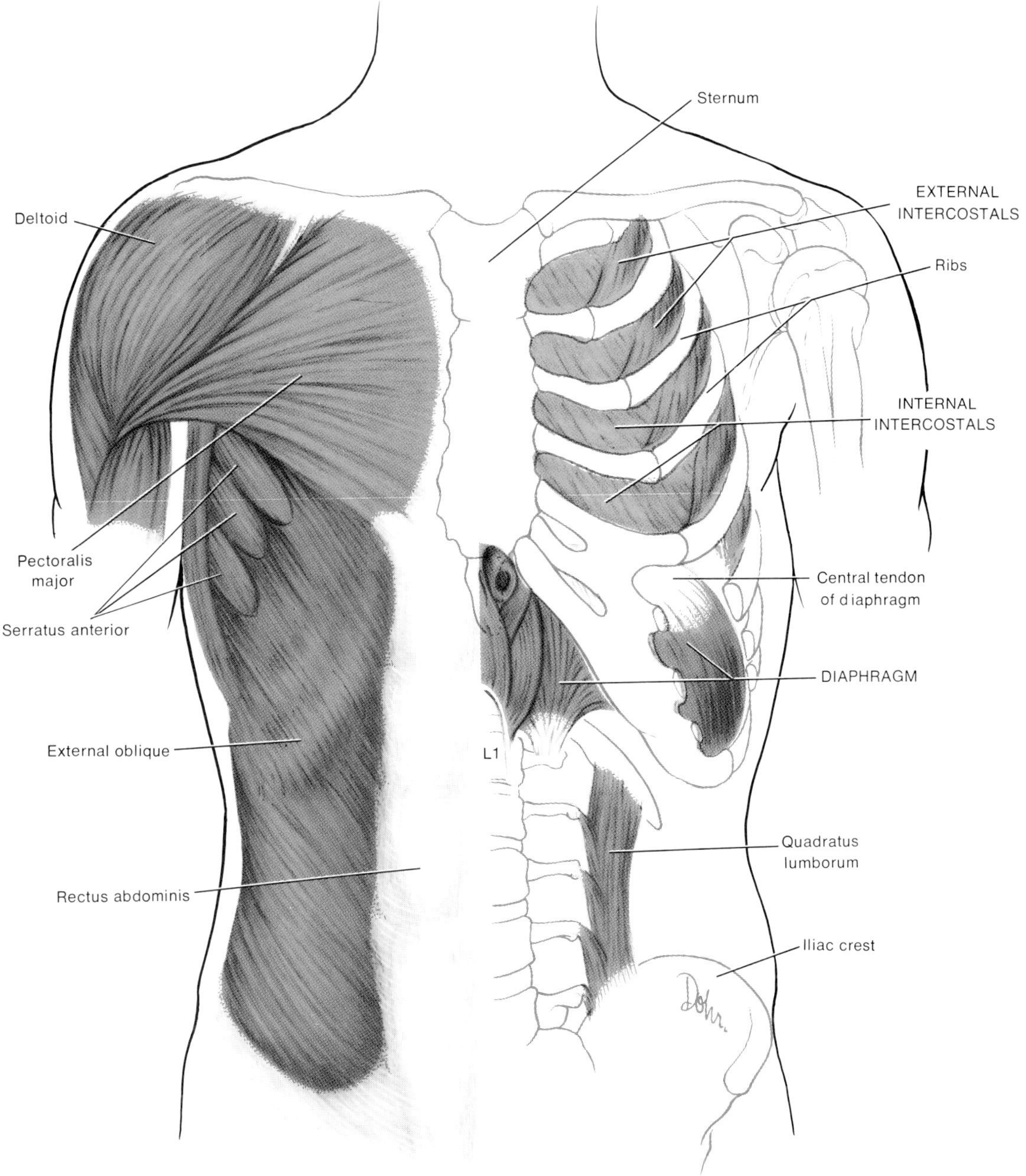

Figure 20-5 Muscles used in breathing. (a) Diagram of superficial view. (b) Diagram of deep view.

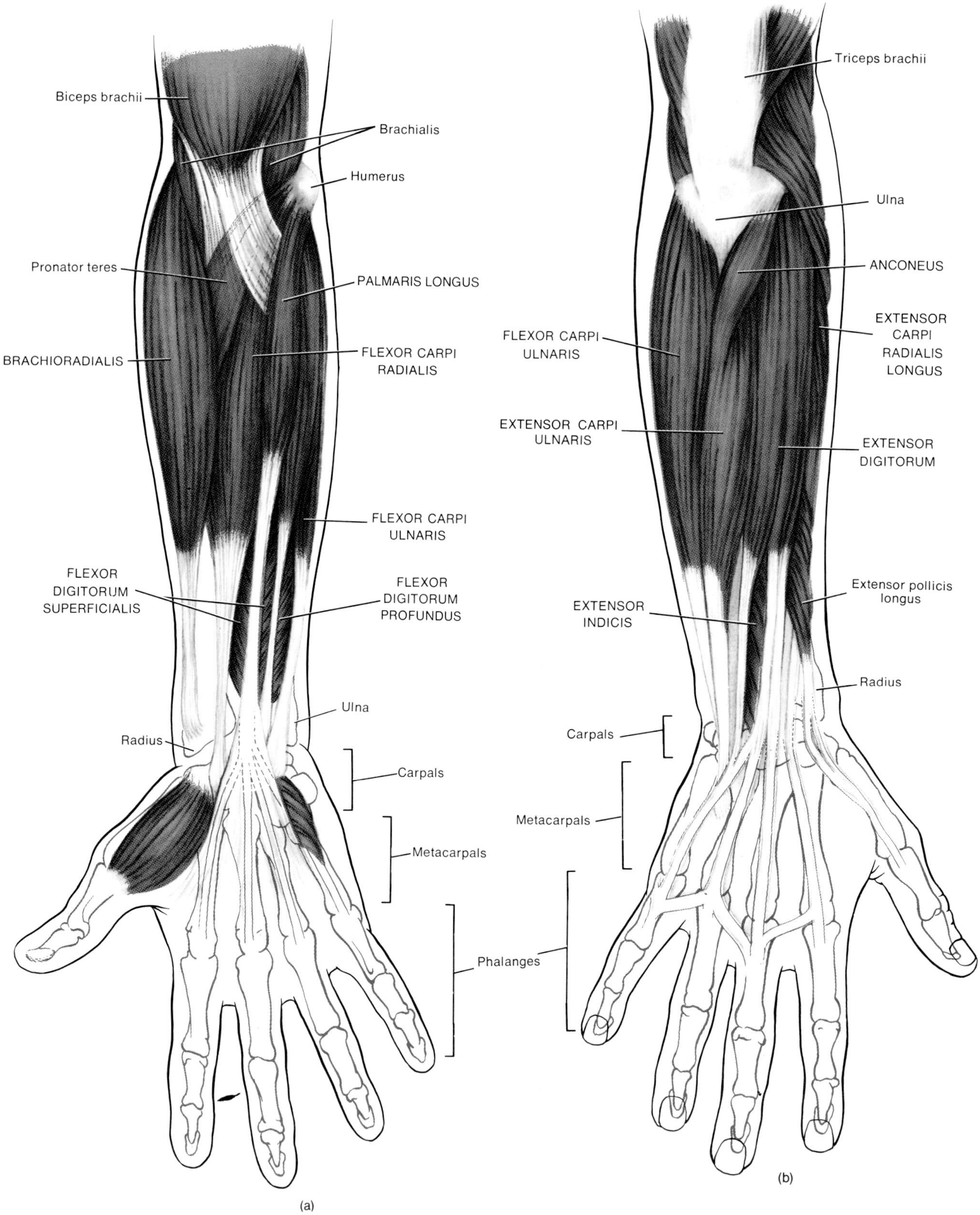

Figure 20-6 Muscles that move the wrist and fingers. (a) Anteror view. (b) Posterior view.

Questions for Review

EXERCISE **20**
MUSCLES OF THE DEEP CHEST AND ARM

Name ______________________________

Laboratory Section ______________________ Date __________

1. Complete Table Q20-1. Note how differences in location affect the action of the muscles. If the actions are similar, note that too.

Table Q20-1

Muscle	Location	Action
Coracobrachialis (cat)	a.	b.
Coracobrachialis (human)	c.	d.
Serratus ventralis (cat)	e.	f.
Serratus anterior (human)	g.	h.
Intercostal muscles (cat)	i.	j.
Intercostal muscles (human)	k.	l.

2. Study Figure Q20-1 showing the deep chest and neck muscles of the human. Compare them as to size and location with the muscles just dissected in the cat. Color this diagram if you wish.

3. Fill in the blanks.

a. ______________________ Name one muscle that flexes the forearm.

b. ______________________ Name one muscle that extends and adducts the wrist.

c. ______________________ Name one muscle that flexes and abducts the wrist.

d. ______________________ Name one muscle that extends the forearm.

e. ______________________ Name one muscle that extends a phalanx.

f. ______________________ Name another muscle that extends a phalanx.

g. ______________________ Name a muscle that pronates the forearm.

h. ______________________ Name a muscle that supinates the forearm.

i. ______________________ What is the position of the hand in supination?

j. ______________________ What is the position of the hand in pronation?

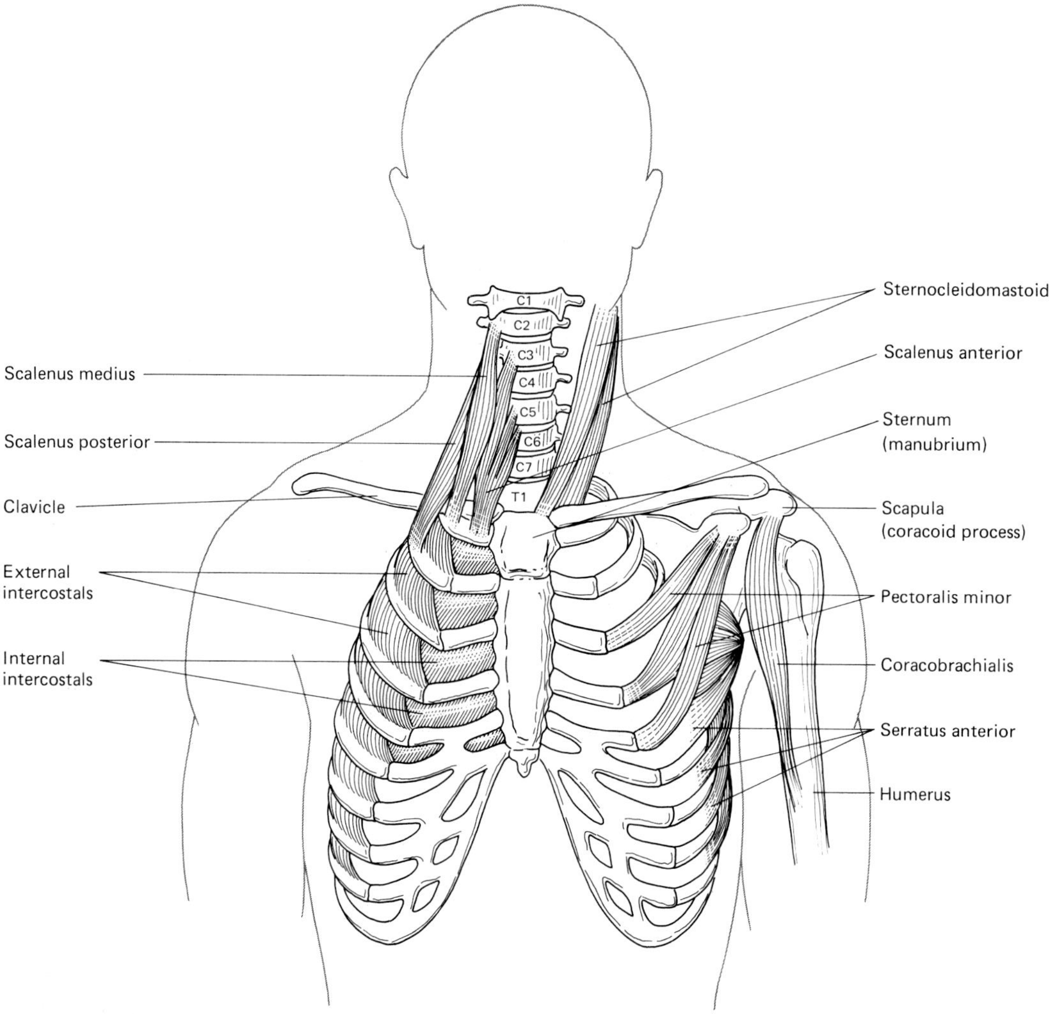

Figure Q20-1

4. Study Figure Q20-2 showing the human arm muscles. Compare each muscle to one dissected in the cat. Color the drawing for clarity.

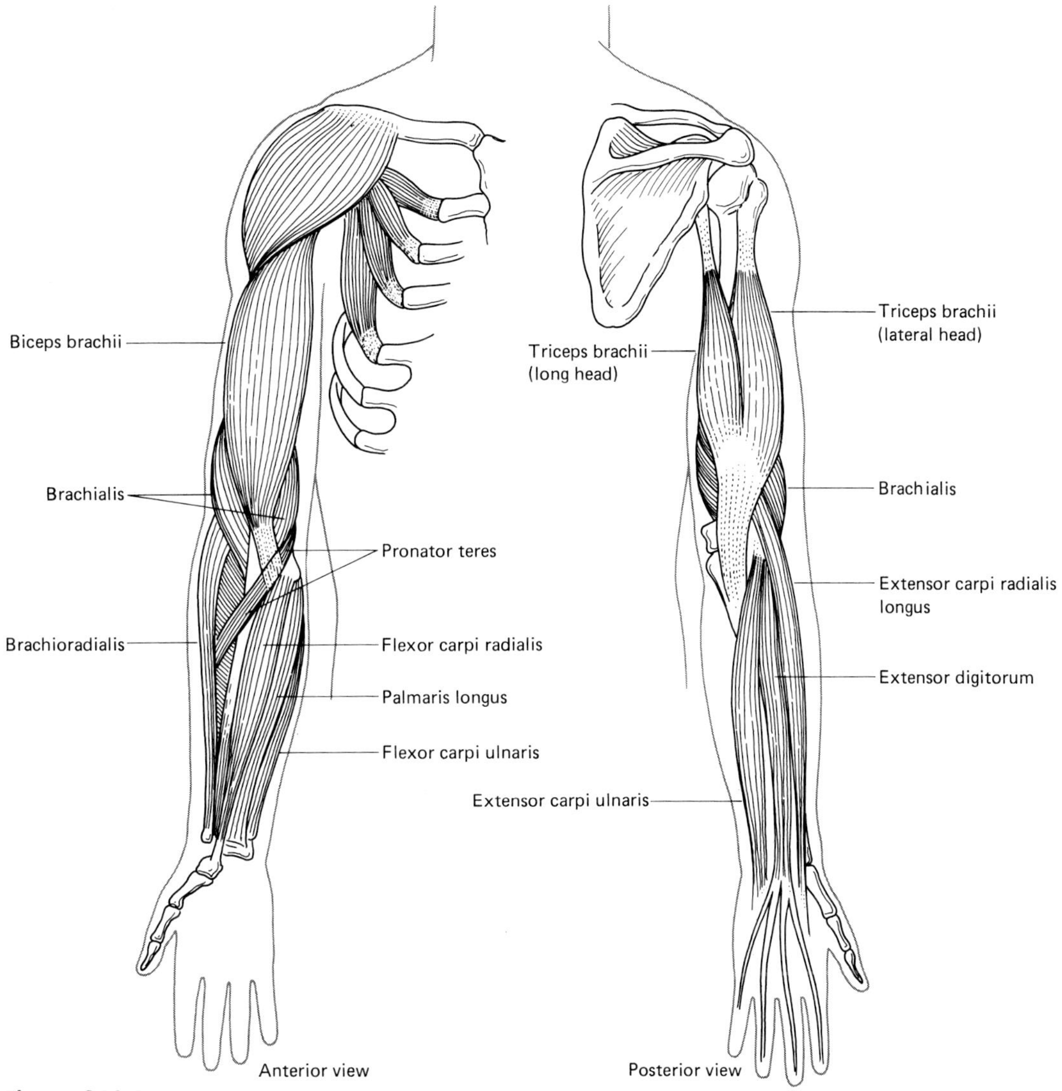

Figure Q20-2

EXERCISE 21

MUSCLES OF THE HIP AND THIGH

Materials

The following materials should be provided for class use.

1. Embalmed cats
2. Dissecting tools
3. Dissecting trays
4. String
5. 3 percent phenol solution

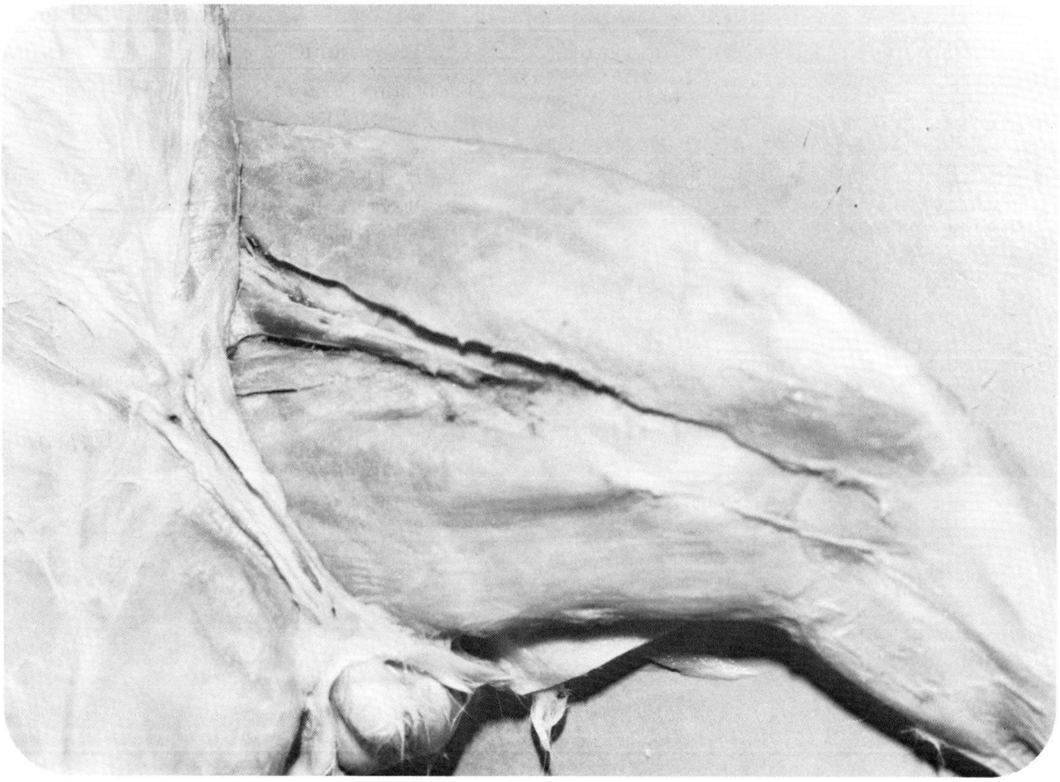

Figure 21-1a Superficial thigh muscles, medial surface.

SUPERFICIAL THIGH MUSCLES, MEDIAL SURFACE

INSTRUCTIONS FOR DISSECTION

Almost the entire medial surface of the thigh is covered by two ribbonlike muscles separated proximally by a triangular area (the *femoral triangle)* in which you see blood vessels and a nerve, as displayed in Figure 21-1a. Remove any fat and fascia that obscure these vessels and the underlying muscles. The blood vessels (Figure 21-1b) are the *femoral artery* (injected red) and the *femoral vein* (injected blue).The large nerve running parallel to the blood vessels is the *femoral nerve.*

The thin, wide muscle located laterally is the **sartorius.** Loosen it between origin and insertion. The thin, wide muscle located medially is the **gracilis.** The muscles lying in the triangular area will be considered in the next section.

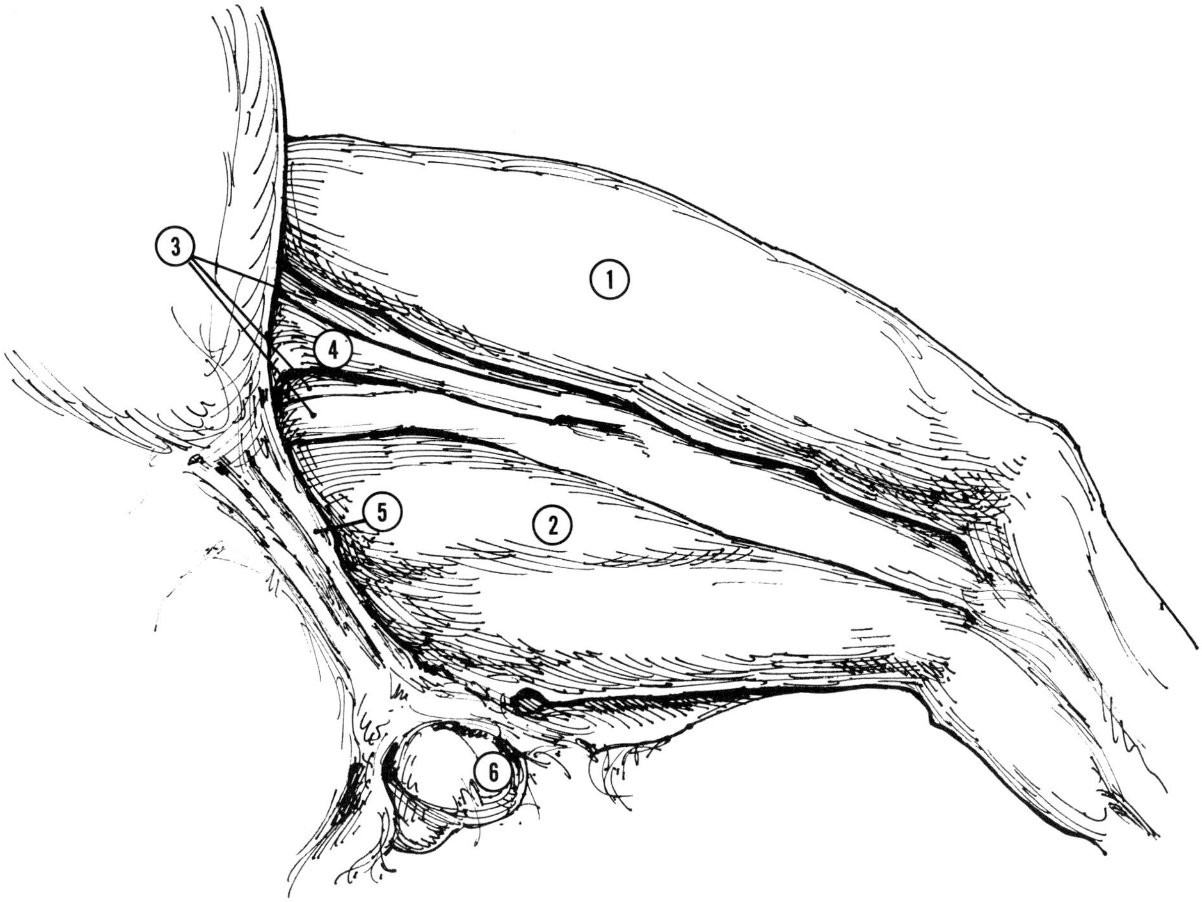

Figure 21-1b Superficial thigh muscles, medial surface.
1. Sartorius (sar-TOR-ē-us)
2. Gracilis (gra-SIL-is)
3. Adductor muscles (femoral triangle)
4. Blood vessels (femoral triangle)
5. Spermatic cord
6. Scrotum

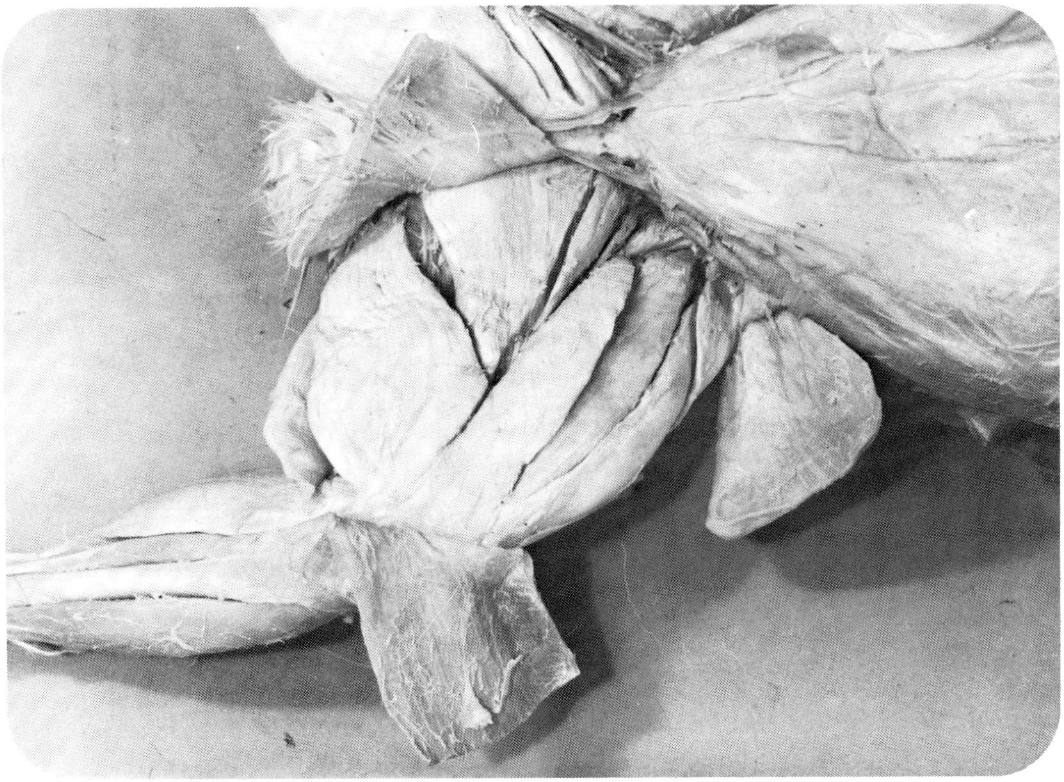

Figure 21-2a Deep thigh muscles, medial view.

DEEP THIGH MUSCLES, MEDIAL VIEW

INSTRUCTIONS FOR DISSECTION

To see this important group of muscles it will be necessary to transect both the sartorius and the gracilis muscles. Lay their ends back as far as possible and begin your observations at the lateral border.

Covering about one-half of the area of the thigh is a large muscle named the **quadriceps femoris.** The four muscles that make it up are considered separately, as they are in the human. The quadriceps separates readily into three distinct bundles: (1) a large lateral muscle, the **vastus lateralis,** (2) a middle muscle, the **rectus femoris,** and (3) a medial muscle, the **vastus medialis.** The fourth muscle of the group, the **vastus intermedius,** may be seen either by separating the rectus femoris from adjacent muscle or by transecting the rectus femoris and laying back its cut ends. (In Figure 21-2a the rectus femoris is not transected so the vastus intermedius is not shown.)

Dissect away the *femoral artery, femoral vein,* and *femoral nerve* so that you are better able to see the group of muscles beneath them. Most lateral of the group is the **iliopsoas** (composed of two muscles, the **iliacus** and the **psoas major,** but seen as one muscle in this dissection). Medial to the iliopsoas is the **pectineus,** and medial to the pectinous is the **adductor longus.**

The remaining two medial muscles are considerably larger. The **adductor femoris** lies medial to the adductor longus. The **semimembranosus,** makes up most of the medial half of the thigh. (Fig. 21–2)

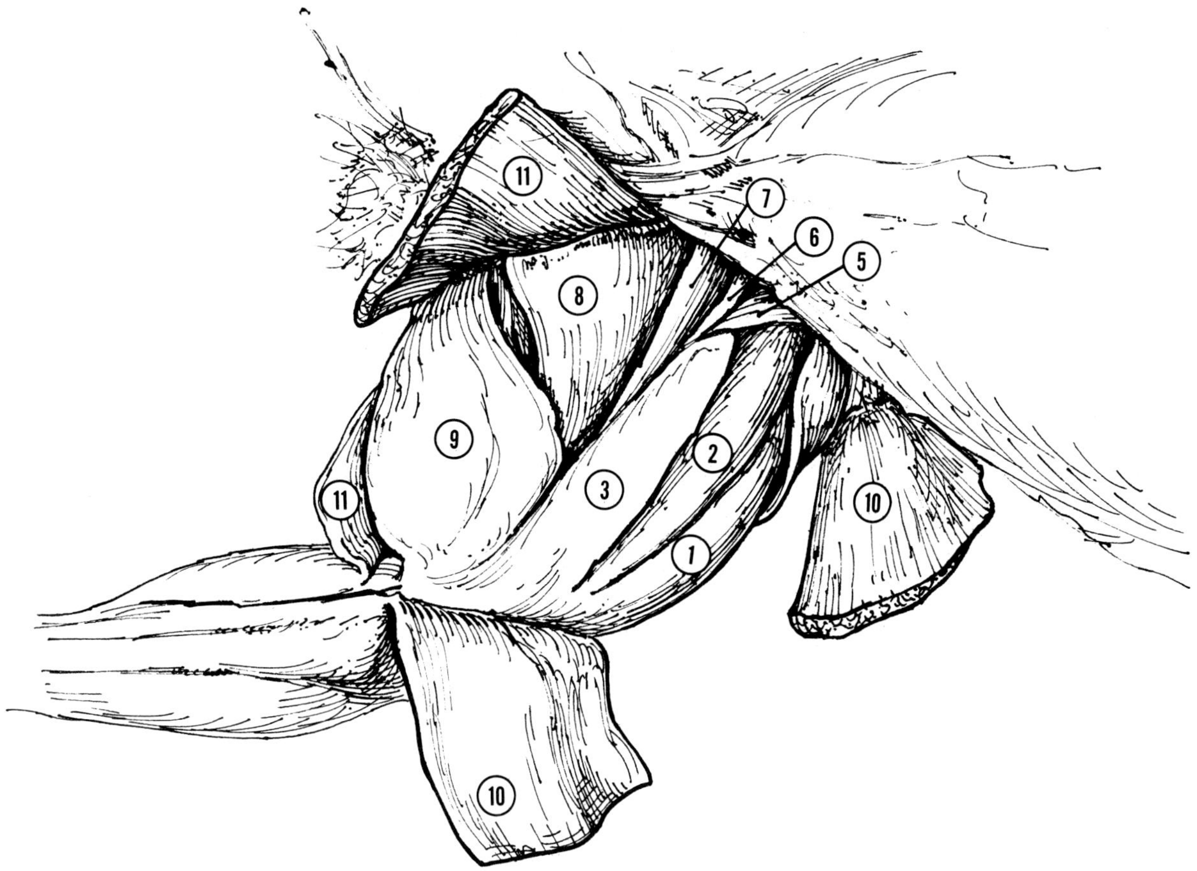

Figure 21-2b Deep thigh muscles, medial view.

1. Vastus lateralis (VAS-tus-la´-ter-AL-is)
2. Rectus femoris (REK-tus FEM-or-is)
3. Vastus medialis
4. Vastus intermedius (not shown)
5. Iliopsoas (il´-ē-ō-SŌ-as)
6. Pectineus (pek-TIN-ē-us)
7. Adductor longus
8. Adductor femoris
9. Semimembranosus (sem´-ē-mem-bre-NŌ-sus)
10. Sartorius (reflected)
11. Gracilis (reflected)

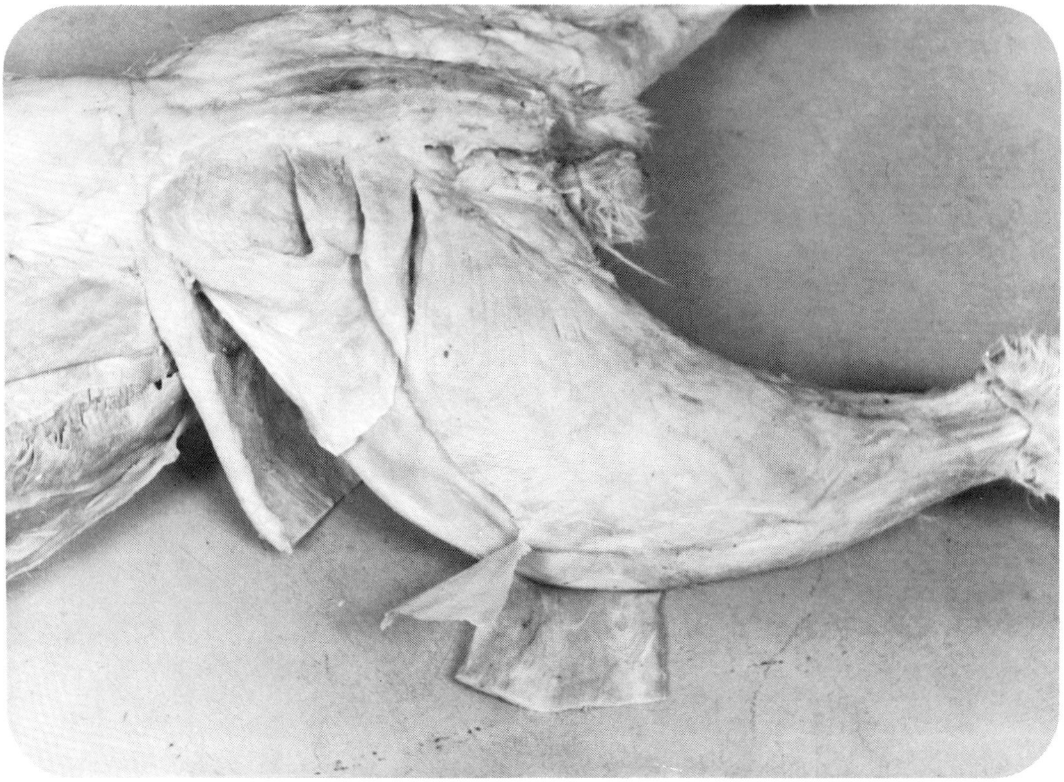

Figure 21-3a Superficial thigh muscles and hip muscles, lateral view.

SUPERFICIAL THIGH AND HIP MUSCLES, LATERAL VIEW

INSTRUCTIONS FOR DISSECTION

Turn the cat over so that the lateral surface is facing you, as in Figure 21-3a. The hip region is covered with heavy fascia, beneath which are the gluteal muscles. Remove this heavy fascia by cutting as much of it off as is necessary to expose the **gluteus medius** and posterior to it the **gluteus maximus.** Note that the gluteus maximus is the smaller of the two, contrary to the condition in the human.

Lying within the same mass of fascia that you have just removed and below the gluteus medius is the **tensor fasciae latae** (see Figure 21-3b). The tensor is an irregularly shaped muscle that not only originates in a heavy fascia but also inserts by a long broad fascia (fascia lata), almost completely covering the **vastus lateralis.** Loosen the fascia lata and transect it so that you can see the entire vastus lateralis.

Posterior to the fascia lata and the gluteus maximus is the **caudofemoralis** (a muscle not found in the human), which tapers to its insertion by a very fine tendon. Separate it from the large **biceps femoris** whose mass covers most of the lateral surface of the thigh.

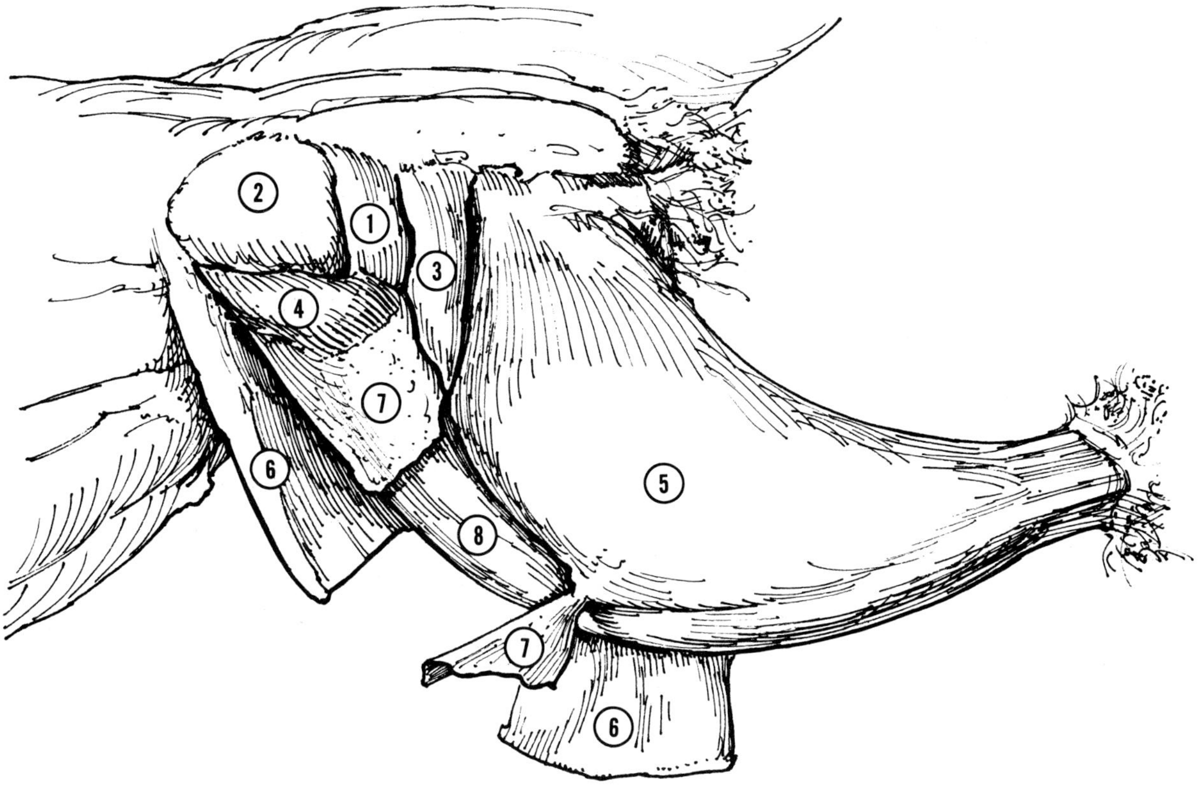

Figure 21-3b Superficial thigh muscles and hip muscles, lateral view.

1. Gluteus maximus (GLOO-tē-us-MĒ-dē-us)
2. Gluteus medius (GLOO-tē-us MAX-i-mus)
3. Caudofemoralis (kaw′-dō-fem-or-AL-is)
4. Tensor fasciae latae
5. Biceps femoris (BĪ-seps FEM-or-is)
6. Sartorius
7. Fascia lata (transected to expose vastus lateralis)
8. Vastus lateralis

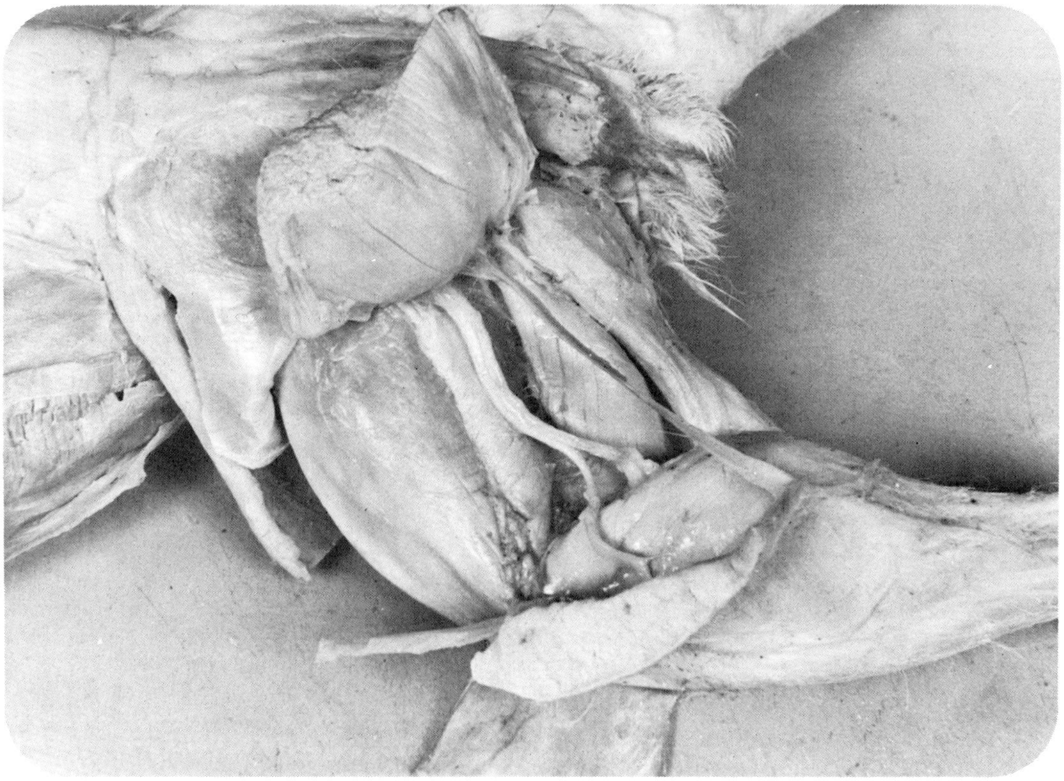

Figure 21-4a Deep thigh muscles, lateral view.

DEEP THIGH MUSCLES, LATERAL VIEW

INSTRUCTIONS FOR DISSECTION

Separate the biceps femoris from the tissue below it with your finger and a probe. Before making a transection, locate and dissect a very small muscle, the **tenuissimus,** which runs the length of the biceps femoris but is only about $\frac{1}{4}$ inch wide (not found in the human leg). Also locate the *sciatic nerve,* lying in the same general area (see Figure 21-4b). When you have done this, carefully transect the biceps femoris and draw each end to its origin and insertion.

In addition to the tenuissimus and the sciatic nerve, observe the semimembranosus already seen in the ventral view. Medial to it is the **semitendinosus.** The vastus lateralis and the adductor femoris are also prominent in this view. Refer to Table 21-1 for the origin, insertions, and actions of the cat's hip and thigh muscles.

Questions for Review

EXERCISE **21**

MUSCLES OF THE HIP AND THIGH

Name ____

Laboratory Section ______________________________ Date ____________

1. The gluteus muscles and the tensor fasciae latae are shown diagrammatically in Figure Q21-1. Color them and review their origins and insertions.

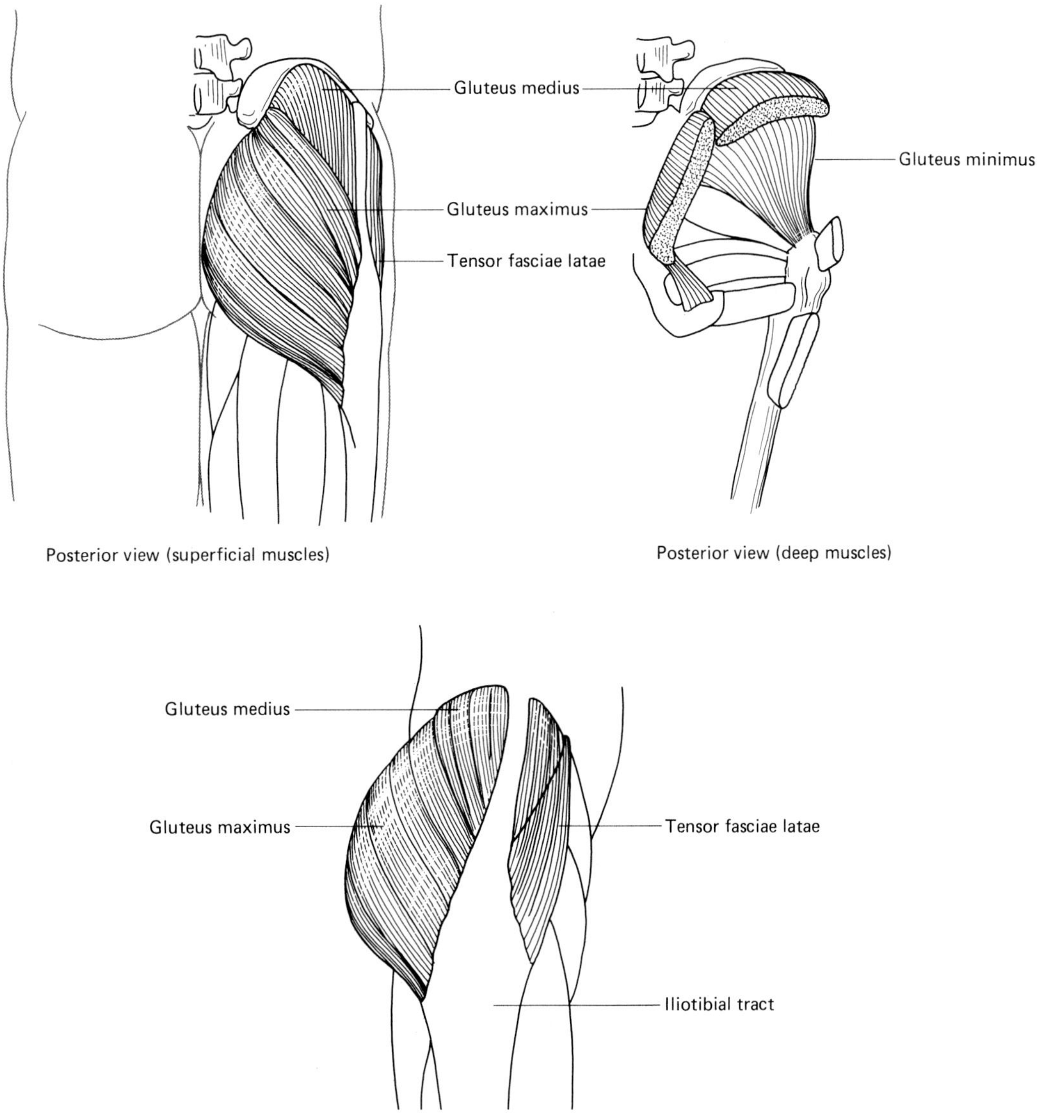

Posterior view (superficial muscles)

Posterior view (deep muscles)

Lateral view (superficial muscles)

Figure Q21-1

2. Since the human thigh muscles are so much like those of the cat, Figure Q21-2 is not labeled. Before referring to Figure Q21-6 try to identify these muscles.

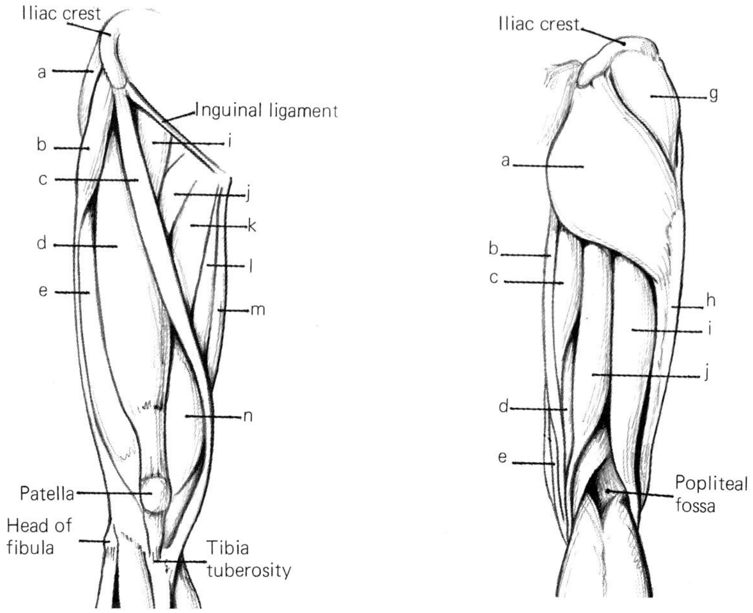

Figure Q21-2

a. ______________________

b. ______________________

c. ______________________

d. ______________________

e. ______________________

f. ______________________

g. ______________________

h. ______________________

i. ______________________

j. ______________________

k. ______________________

l. ______________________

m. ______________________

n. ______________________

o. ______________________

p. ______________________

q. ______________________

r. ______________________

s. ______________________

t. ______________________

3. What are the actions of the human gluteal muscles?

   a. Gluteus maximus ______________________________

   ____

   b. Gluteus medius ______________________________

   ____

   c. Gluteus minimus ______________________________

   ____

4. Name the muscles that can be seen in the femoral triangle of the cat. ______________

   ____

5. List the three muscles often designated as the "hamstrings."

   a. ______________

   b. ______________

   c. ______________

EXERCISE 22

MUSCLES OF THE LOWER LEG

Materials

The following materials should be provided for class use.

1. Embalmed cats
2. Dissecting tools
3. Dissecting trays
4. String
5. 3 percent phenol solution

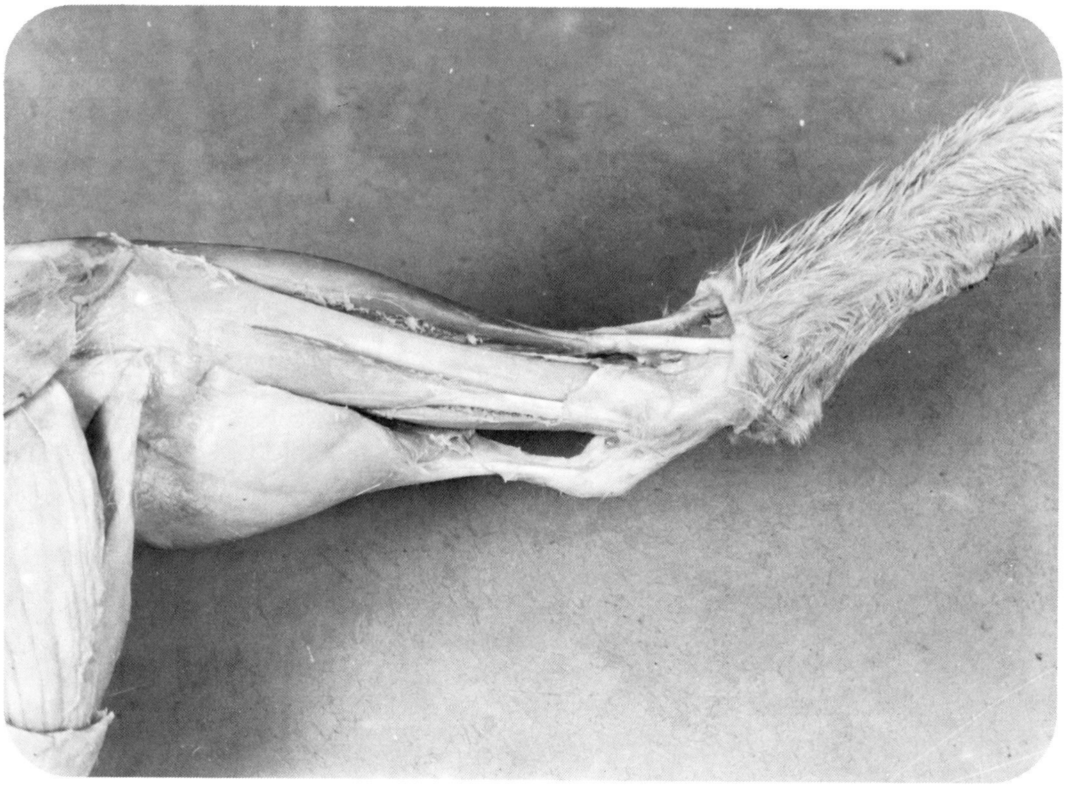

Figure 22-1a Superior lower leg muscles, medial surface.

SUPERFICIAL LOWER LEG MUSCLES, MEDIAL SURFACE

INSTRUCTIONS FOR DISSECTION

In the lower leg there is a large, prominent muscle, the **gastrocnemius** (see Figure 22-1). It can be seen on both the medial and lateral surfaces. It has two distinct heads of origin, but the two muscle masses unite to form the single very tough **Achilles tendon** or **calcaneal tendon,** which inserts on the calcaneus bone. Lying between the gastrocnemius and the tibial bone is the **flexor digitorum longus.** This muscle has two heads of origin. Locate them. On the anterior surface of the tibial bone is the **tibialis anterior** muscle, as shown in Figure 22-1.

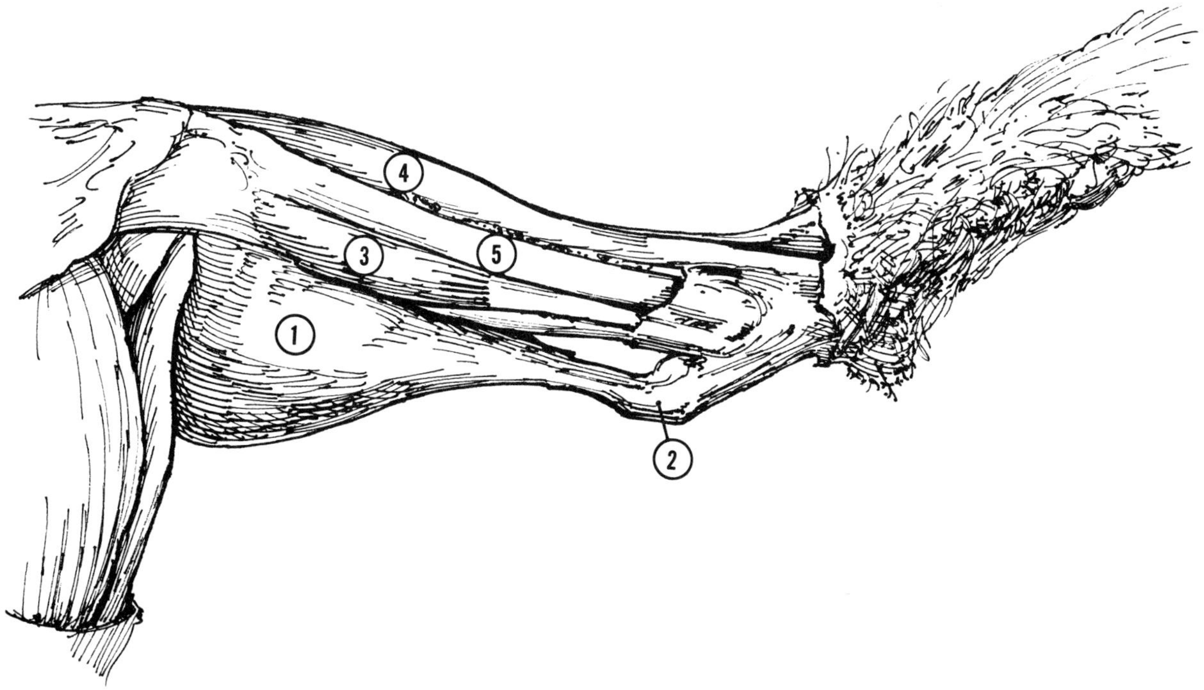

Figure 22-1b Superficial lower leg muscles medial surface.

1. Gastrocnemius (gas´-trok-NE-mē-us)
2. Achilles tendon (a-KIL-ēz)
3. Flexor digitorum longus (dij-i-TŌ-rum)
4. Tibialis anterior (tib´-ē-AL-is)
5. Tibia (bone)

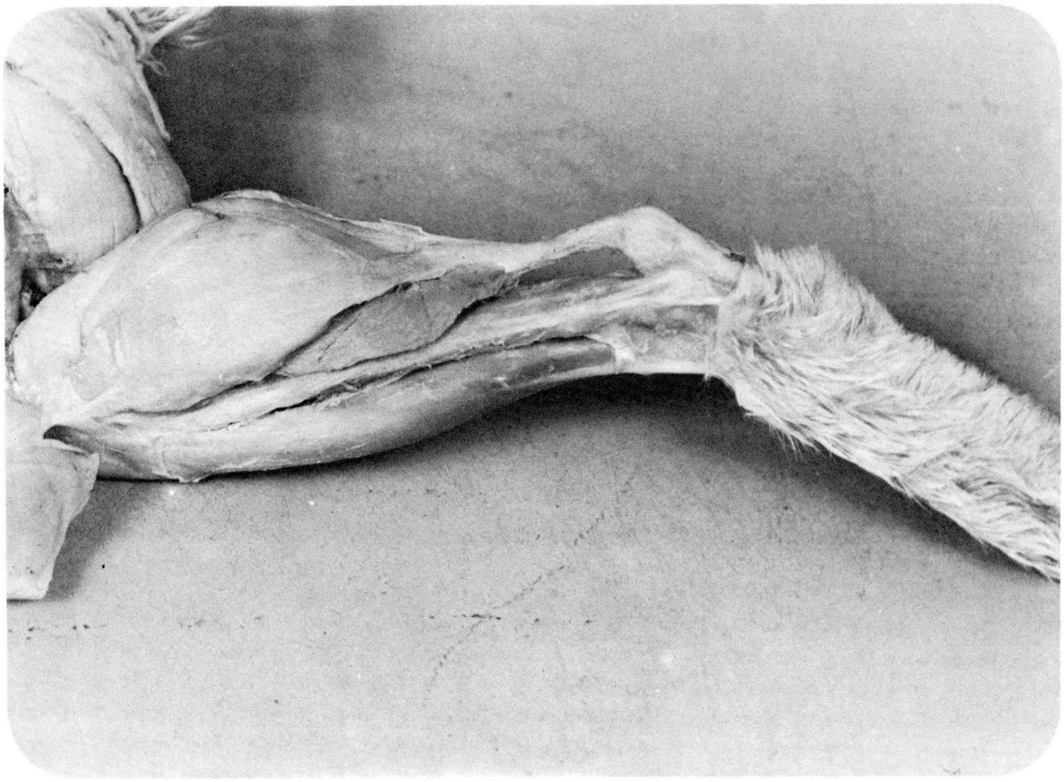

Figure 22-2a Superficial lower leg muscles, lateral surface.

SUPERFICIAL LOWER LEG MUSCLES, LATERAL SURFACE

INSTRUCTIONS FOR DISSECTION

On the lateral surface of the leg, note once again the gastrocnemius muscle and the Achilles tendon (see Figure 22-2).

Lying next to the gastrocnemius and prominent along its distal end is the small **soleus** muscle. Next to the soleus is a long, thin muscle, the **peroneus,** shown in Figure 22-2.

On the anterior border locate the **extensor digitorum longus,** which inserts by long tendons on each of the five digits. Refer to Table 22-1 for the origin, insertion, and action of their lower leg muscles.

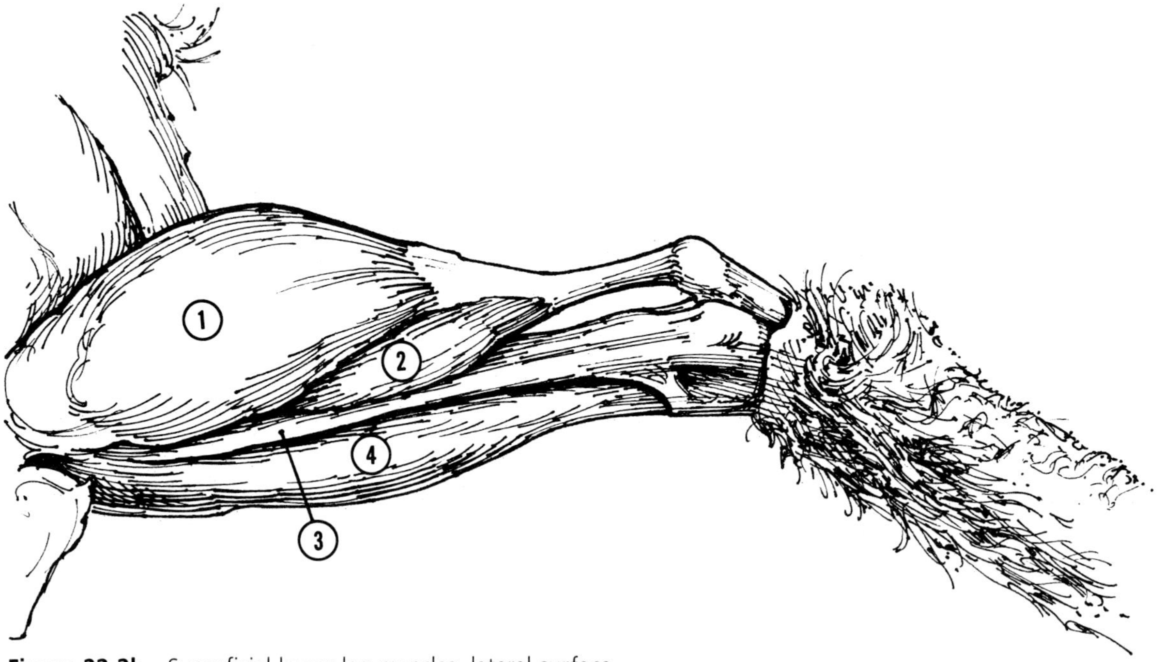

Figure 22-2b Superficial lower leg muscles, lateral surface.
1. Gastrocnemius
2. Soleus (SŌ-lē-us)
3. Peroneus (per´-ō-NĒ-us)
4. Extensor digitorum longus

Table 22-1 Superficial Lower Leg Muscles of the Cat

Muscle	Origin	Insertion	Action
Gastrocnemius	Knee fascia and distal end of femur	Calcaneus by Achilles tendon	Extends foot
Flexor digitorum longus	Distal end of tibia and head and shaft of fibula	Tendons on all digits	Flexes digits
Tibialis anterior	Proximal ends of tibia and fibula	First metatarsal	Flexes Foot
Soleus	Proximal end of fibula	Bases of metatarsals	Extends foot
Peroneus	Proximal end of fibula	Bases of metatarsals	Flexes foot
Extensor digitorum longus	Distal end of femur	All phalanges	Extends toes

COMPARABLE HUMAN MUSCLES

The muscles of the anterior lower leg (Figure 22-3) are dorsiflexors of the ankle and extensors of the toes. They are the **tibialis anterior, extensor digitorum longus,** and **peroneus tertius.**

On the lateral surface of the leg are the peroneal muscles, which evert the foot. They may also protect against excessive inversion. Specifically they are the **peroneus longus** and the **peroneus brevis.**

The **gastrocnemius** muscle covers the posterior surface of the lower leg and has two heads. It ends in the calcaneal tendon (Achilles tendon) and inserts on the calcaneus bone. It plantar flexes (extends) the foot. Beneath it is the **soleus,** which inserts on the calcaneus and also plantar flexes the foot. Stand on your toes and feel these muscles contract.

Muscles of this region are similar in cat and human. Learn as many human muscles as you can and check all origins, insertions, and actions with Table 22-2. Demonstrate the action of as many muscles of the leg as you can. Try to see or feel the contraction as you make a specific movement.

Table 22-2 Superficial Lower Leg Muscles of the Human

Muscle	Origin	Insertion	Action
Gastrocnemius	Lateral and medial condyles of femur and capsule of knee	Calcaneus by Achilles tendon	Plantar flexes (extends) foot
Soleus	Head of fibula and medial border of tibia	Calcaneus by Achilles tendon	Plantar flexes foot
Peroneus longus	Head and body of fibula and lateral condyle of tibia	First metatarsal and first cuneiform	Plantar flexes and everts foot
Peroneus brevis	Body of fibula	Fifth metatarsal	Plantar flexes and everts foot
Tibialis anterior	Lateral condyle and body of tibia	First metatarsal and first cuneiform	Flexes foot dorsally and inverts it
Tibialis posterior	Membrane between tibia and fibula	Second, third, and fourth metatarsals; navicular; third cuneiform; cuboid	Plantar flexes and inverts foot
Flexor digitorum longus	Tibia	Distal phalanges of four outer toes	Flexes toes; plantar flexes and inverts foot
Extensor digitorum longus	Lateral condyle of tibia and anterior surface of fibula	Middle and distal phalanges of four outer toes	Extends toes, flexes foot dorsally and everts it

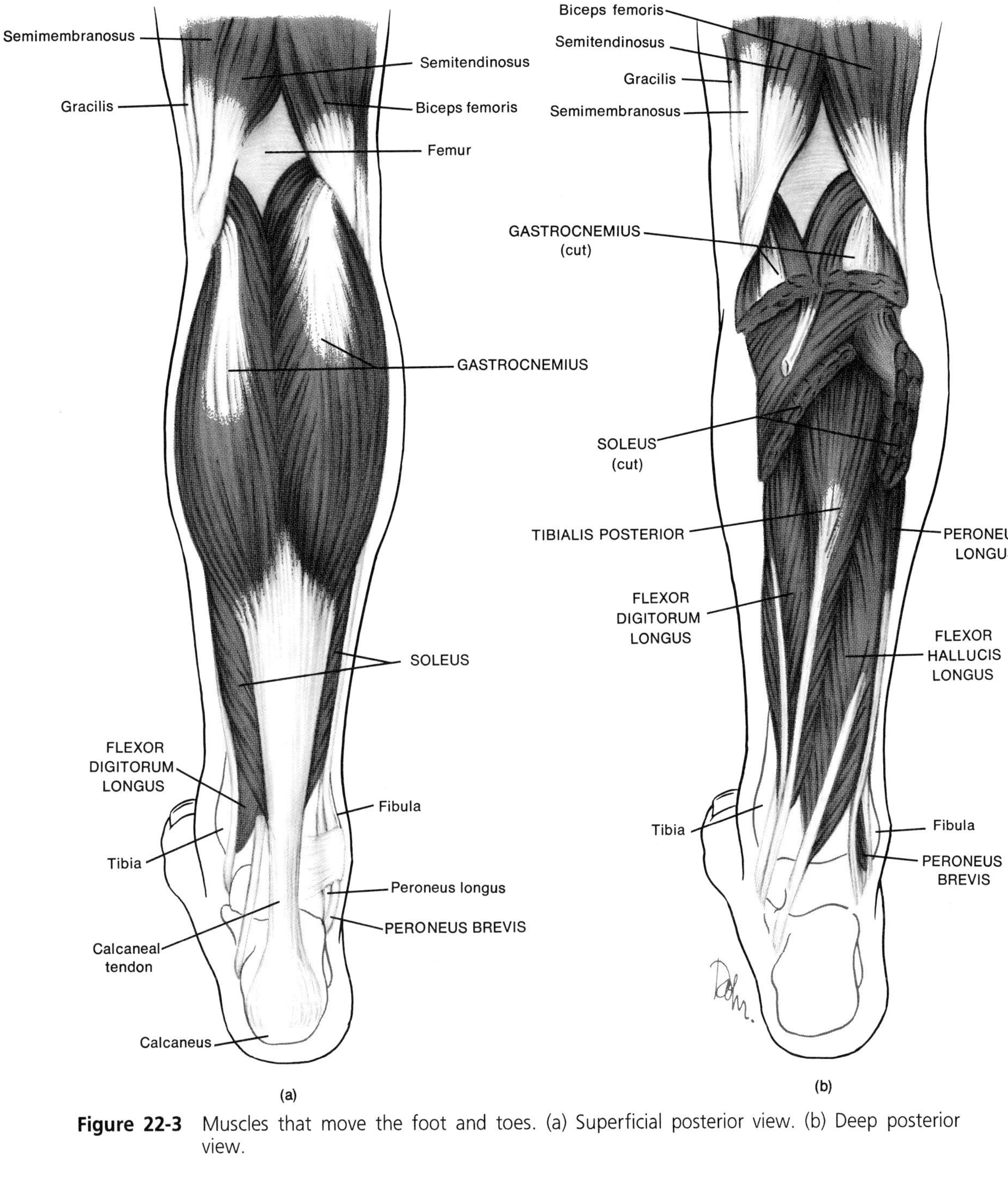

Figure 22-3 Muscles that move the foot and toes. (a) Superficial posterior view. (b) Deep posterior view.

Questions for Review

EXERCISE **22**

MUSCLES OF THE LOWER LEG

Name ____________________

Laboratory Section ____________________ **Date** ____________

1. Color the muscles of the lower leg in Figure Q22-1. Review the origin, insertion, and action of each muscle. Demonstrate as many of them as possible on your own leg.

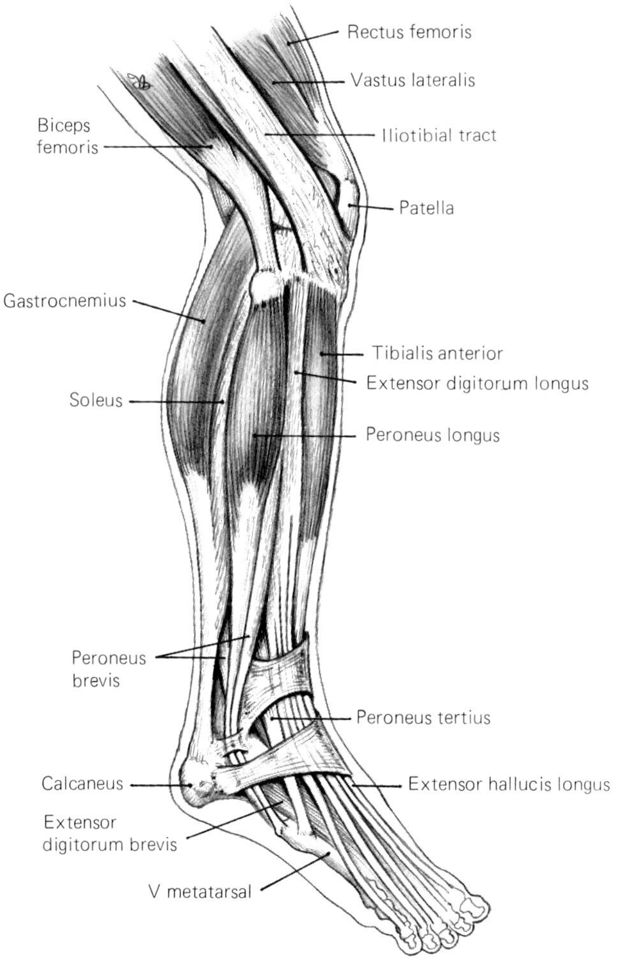

Figure Q22-1

2. Figure Q22-2 diagrams the muscles of the hip, thigh, and lower leg. Identify all the muscles and color the drawings.

Anterior view

a. ______________________ k. ______________________

b. ______________________ l. ______________________

c. ______________________ m. ______________________

d. ______________________ n. ______________________

e. ______________________ o. ______________________

f. ______________________ p. ______________________

g. ______________________ q. ______________________

h. ______________________ r. ______________________

i. ______________________ s. ______________________

j. ______________________

Lateral view

a. ______________________ i. ______________________

b. ______________________ j. ______________________

c. ______________________ k. ______________________

d. ______________________ l. ______________________

e. ______________________ m. ______________________

f. ______________________ n. ______________________

g. ______________________ o. ______________________

h. ______________________ p. ______________________

Posterior view

a. ______________________ g. ______________________

b. ______________________ h. ______________________

c. ______________________ i. ______________________

d. ______________________ j. ______________________

e. ______________________ k. ______________________

f. ______________________

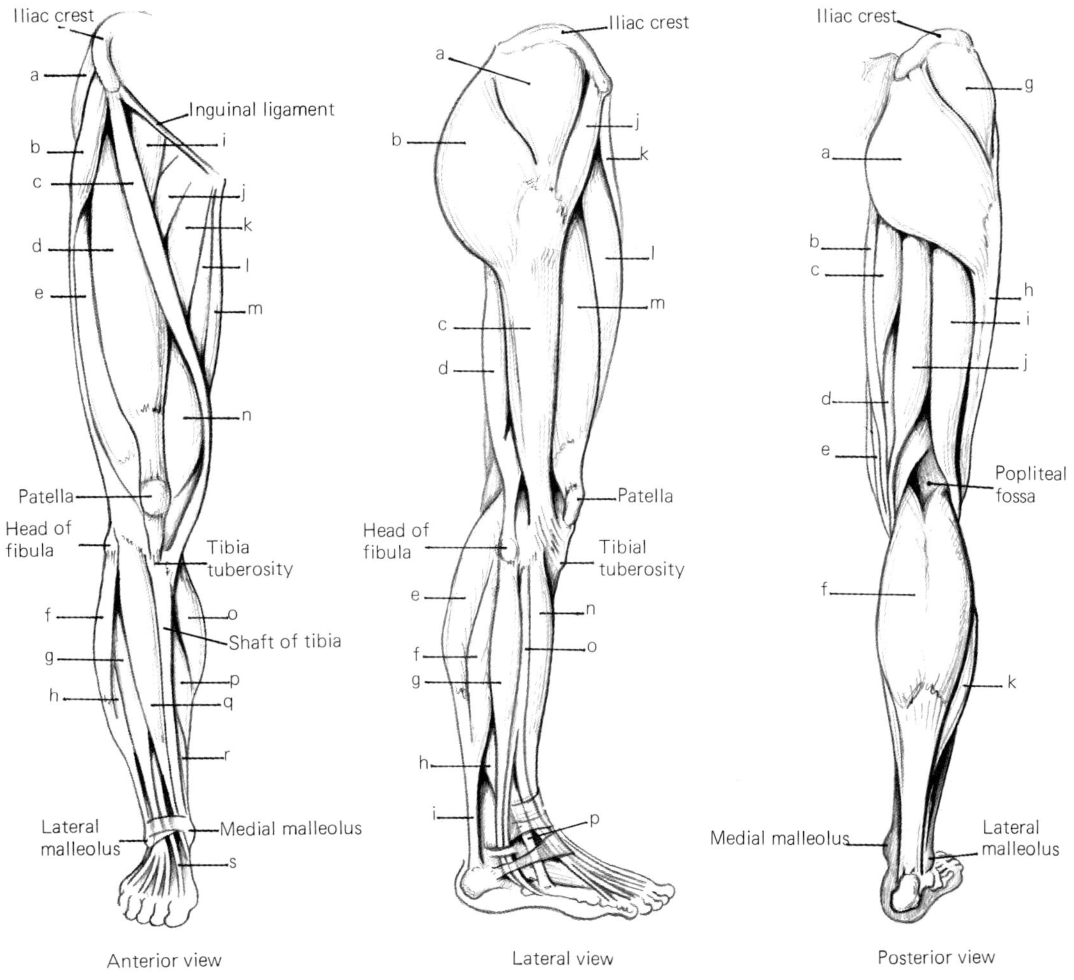

Figure Q22-2

EXERCISE 23

MUSCLE ULTRASTRUCTURE AND CHEMISTRY

Muscle tissue consists of many elongated cells called *muscle fibers,* which contain threadlike structures called *myofibrils.* Myofibrils consist of two types of smaller structures—*thin myofilaments* and *thick myofilaments.* Alternating thick and thin myofilaments are stacked parallel to one another in units called *sarcomeres.*

The sarcomere, which measures 2–3 μm in length, is bounded at both ends by a narrow band of dense material called a *Z line.* Along the length of a muscle cell there are other conspicuous bands or striations that are repeated regularly. These include the relatively clear, broad *I band,* which is bissected by the *Z line,* and the large, dense *A band* in the center of the sarcomere (Figure 23-1).

The thin myofilaments contain three proteins. They are composed mostly of *actin* and also contain *tropomyosin* and *troponin.* Tropomyosins, which are long, thin molecules, attach end to end, forming a thin filament on the surface of an actin strand. Troponin rests on the tropomyosin. The thick myofilaments are composed mostly of *myosin.*

In intact muscle one end of the thin filaments is attached to the Z line, a flat protein structure. The thin filaments extend perpendicularly from both of its surfaces. Transmission electron micrographs show that sarcomeres consist of a set of thick filaments arranged in the spaces between the thin filaments projecting from the Z lines. The two types of filament are arranged in parallel groups and overlap for part of their length. The spacing of the filaments is such that the two sets can slide past one another without restriction. This sliding motion is responsible for the shortening reaction during muscle contraction.

The energy for muscle contraction is provided by the hydrolysis of adenosine triphosphate (ATP). In addition to the four proteins, an ATP muscle contraction requires the presence of calcium ions. Calcium provides a means of regulation.

This exercise will consider the microscopic properties of muscle and the particular effects selected chemical and physical factors have on muscle contraction.

A. MICROSCOPIC FEATURES OF SKELETAL MUSCLE

Materials

1. The following prepared slides should be provided for class use:
   a. Teased skeletal muscle fibers
   b. Longitudinal section of skeletal muscle
2. Prepared slides stained to differentiate motor end plates should be provided for a class demonstration.

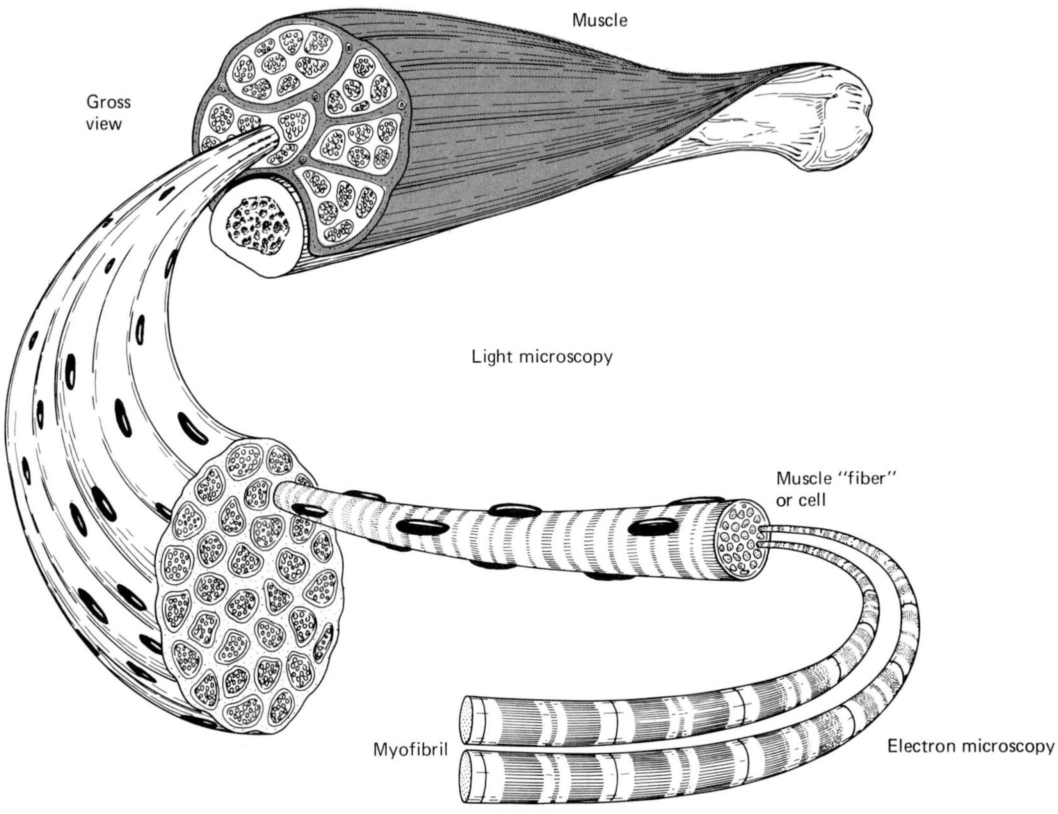

(a)

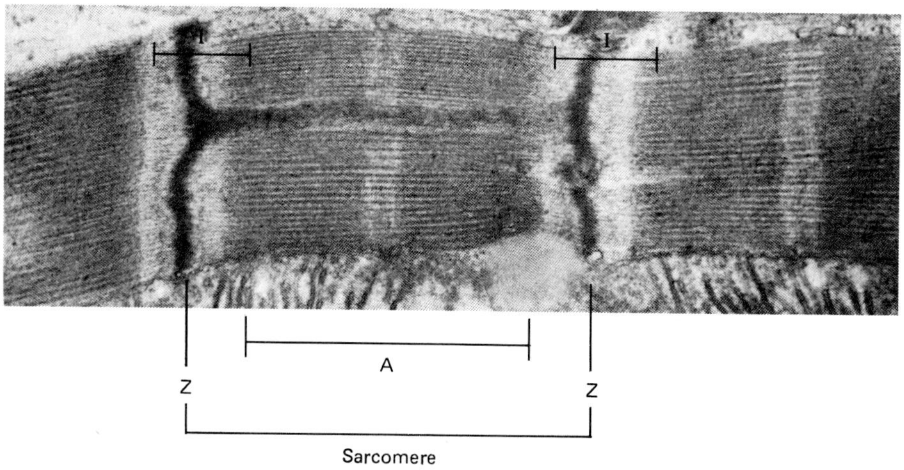

(b)

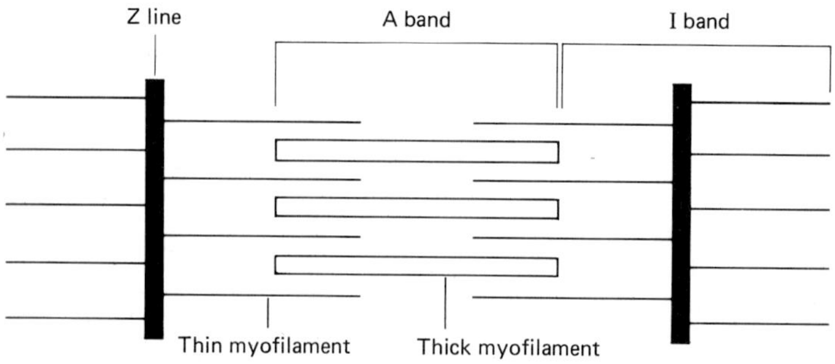

(c)

Figure 23-1 Structural and functional components of striated skeletal muscle. (a) This composite diagram shows a general relationship of skeletal muscle to bone, the components of a muscle fiber, and sarcomere. (After D. Shepro, F. Belamarich, and C. Levy. *Human Anatomy and Physiology.* New York: Holt, Rinehart and Winston, 1974.) (b) Electron micrograph of a cylindrical sarcomere. (c) The arrangement and relationships of muscle fiber components.

Procedure

This procedure is to be performed by students individually.

1. Examine a prepared slide of muscle under both low and high power objectives (Refer to Color Plates 6 and 17).
2. Select a single fiber and identify the following components: *muscle cell nuclei, sarcolemma* (a thin membrane that covers each fiber), *A bands* (repeating dark zones perpendicular to the long axis of the muscle fiber), and *I bands* (repeating light zones perpendicular to the long axis of the muscle fiber).
3. Sketch a representative field showing the components listed in step 2 in the results and observations section.
4. Examine a prepared slide of a longitudinal section of skeletal muscle. Identify the following components: *muscle cell nuclei, endomysium* (connective tissue layer surrounding each muscle fiber and external to the sarcolemma), and *local areas of contraction* (darker areas along the length of a muscle fiber).
5. Sketch a representative field showing the components listed in step 4 in the results and observations section.
6. Examine the demonstration slide showing motor end plates. Locate the following: a *motor end plate, nerve fiber,* and *nerve fiber branches.* Note that each muscle fiber has only one end plate.
7. Sketch and label a representative field in the results and observations section.
8. Answer the questions pertaining to this procedure in the results and observations section.

B. THE CHEMISTRY OF MUSCLE CONTRACTION

Materials

The following items should be provided for class use.

1. Rabbit psoas muscle strips in 50 percent glycerol (obtainable from Carolina Biological Supply Company)
2. The following prepared solutions in containers with droppers:
   a. Adenosine triphosphate (ATP)
   b. Magnesium chloride ($MgCl_2$)
   c. Potassium chloride (KCl)
   d. Three separate glycerol preparations maintained at 4°C, 25°C (room temperature), and 37°C
   e. Three separate Ringer's preparations with pH of 5, 7, and 8 maintained at 37°C
3. Microscope slides
4. Forceps
5. Millimeter rulers
6. Microscopes
7. Water bath maintained at 37°C
8. Ice bath or refrigerator to maintain appropriate solutions at 4°C
9. Eye droppers
10. Stopwatch or other timing device

Procedure 1: Effects of Temperature, Adenosine Triphosphate (ATP), Potassium Chloride (KCl), and Magnesium Chloride ($MgCl_2$) on Muscle Contraction

This procedure should be performed by students in groups of four.

1. On a glass slide place two rabbit psoas muscle fibers in a small amount of the glycerol preparation kept at 25°C (room temperature). Examine them under low power. Are striations and nuclei evident?
2. Measure the length of each fiber in millimeters. Enter your findings in the results and observations section.
3. Remove the glycerol and replace it with the glycerol preparation kept at 37°C. Wait 30 seconds. Measure the length of the fibers and enter your findings. If the length is different from the room temperature measurement at 25°C, calculate the percentage contraction by dividing the contraction distance by the fiber length at room temperature. Enter your findings in the results and observations section.
4. Next, replace the glycerol with the preparation kept at 4°C. Wait 30 seconds. Measure the length of the fibers and enter your findings. If this length is different from the length at 25°C calculate the percentage contraction as described in step 3, and enter your findings.
5. Replace the glycerol with the preparation kept at 25°C. Wait 30 seconds, and repeat steps 3 and 4 using the 4°C and 37°C glycerol solutions for 60 seconds.
6. Now replace the glycerol with the preparation kept at 25°C. Wait 30 seconds and measure the length of the fibers again. This measurement will be used as a baseline value for calculations of the effects of chemicals added to muscle preparation in the following steps.
7. Replace the glycerol with the 4°C preparation and after 30 seconds cover the fibers with the KCl solution. Wait 30 seconds and measure the length of the fibers. If there is a difference, calculate the percentage contraction as described in step 3. Enter your findings in the results and observations section.
8. Repeat step 7 using the following preparations separately, first with the 4°C and then with the 37°C glycerol preparations. Enter your findings in Table 23-1.
   a. $MgCl_2$
   b. ATP
   c. A mixture of KCl, $MgCl_2$, and ATP
9. Repeat steps 7 and 8 exposing the psoas muscle to the different preparations for 60 seconds.

Procedure 2: Effects of pH on Muscle Contraction

This procedure should be performed by students in groups of four.

1. Place two rabbit psoas muscle fibers in a small amount of glycerol (37°C) on a glass slide.
2. Measure the length of each fiber in millimeters. Enter your findings in the results and observations section.
3. Replace the glycerol with the pH 7 Ringer's solution, wait 30 seconds, measure the length of the fibers, and enter your findings in Table 23-2.
4. Add to the fibers a mixture of KCl, $MgCl_2$, and ATP. Wait 30 seconds, measure the length of the fibers, and enter your findings.
5. Repeat steps 3 and 4 using the Ringer's solution with pH 5. If there is a difference in contraction distance from that obtained with the pH 7 solution, calculate the percentage of contraction by dividing the contraction distance obtained at pH 5 by the fiber length at pH 7.
6. Repeat step 5 with the Ringer's solution with pH 8.

RESULTS AND OBSERVATIONS

MICROSCOPIC FEATURES OF MUSCLE

1. Sketch and label representative microscopic fields of the muscle specimens provided.

Teased skeletal muscle fibers   **Longitudinal section**   **Motor end plate**

2. List three areas in the human body where smooth muscle characteristically is found.

   a. ______________________________

   b. ______________________________

   c. ______________________________

3. Is striated muscle multinucleated?______________________________

   ____

EFFECTS OF TEMPERATURE, ADENOSINE TRIPHOSPHATE (ATP), POTASSIUM CHLORIDE (KCl), MAGNESIUM CHLORIDE ($MgCl_2$), AND pH ON MUSCLE CONTRACTION

1. Enter your findings in the appropriate column of Table 23-1. Note that this table combines the results obtained from procedure 1 in this part of the exercise.

Table 23-1

Muscle Fiber Condition	Length at 30 Seconds (in mm)	Percent Contraction (if applicable)	Length at 60 Seconds (in mm)	Percent Contraction (if applicable)
Glycerol (25°C)				
Glycerol (4°C)				
ATP and glycerol (4°C)				
KCl and glycerol (4°C)				

Table 23-1 *(Continued)*

Muscle Fiber Condition	Length at 30 Seconds (in mm)	Percent Contraction (if applicable)	Length at 60 Seconds (in mm)	Percent Contraction (if applicable)
$MgCl_2$ and glycerol (4°C)				
ATP, KCl, $MgCl_2$, and glycerol (4°C)				
Glycerol (37°C)				
ATP and glycerol (37°C)				
ATP and glycerol (4°C)				
KCl and glycerol (37°C)				
KCl and glycerol (4°C)				
$MgCl_2$ and glycerol (37°C)				
$MgCl_2$ and glycerol (4°C)				
ATP, KCl, $MgCl_2$, and glycerol (37°C)				
ATP, KCl, $MgCl_2$, and glycerol (4°C)				

2. Enter your findings obtained from procedure 2 in Table 23-2.

Table 23-2

Muscle Fiber Condition	Length at 30 Seconds (in mm)	Percent Contraction (if appli-cable)
Glycerol (37°C)		
Ringer's solution (pH 7)		
Ringer's solution (pH 7) and ATP, KCl, $MgCl_2$		
Ringer's solution (pH 5)		
Ringer's solution (pH 5) and ATP, KCl, $MgCl_2$		
Ringer's solution (pH 8)		
Ringer's solution (pH 8) and ATP, KCl, $MgCl_2$		

3. Was muscular contraction affected by temperature? In what way? ____________________

____

4. Was muscular contraction affected by pH? In what way? ____________________

____

5. Was muscular contraction affected by the absence of ATP? In what way? ____________

____

6. What is the role of calcium in muscle contraction? ______________________

____

Questions for Review

EXERCISE **23**

MUSCLE ULTRASTRUCTURE AND CHEMISTRY

Name ____

Laboratory Section ______________________ Date __________

1. Draw a simple diagram to show the relationships among the components of a skeletal muscle, including muscle fibers, myofibril, and sarcomere.

2. Draw a simple diagram to show the A bands, I bands, H zones, and Z lines within a sarcomere. (Include the overlapping of myosin and actin filaments.)

3. List four major proteins involved in muscle contraction.

   a. ____________________ c. ____________________

   b. ____________________ d. ____________________

4. Briefly describe the events taking place in a muscle contraction. ______________________

____

____

5. Define or explain:

   a. Sarcomere ____

   ____

   b. Myofibril ____

   ____

   c. Motor end plate ____

   ____

   d. Skeletal muscle ____

   ____

   e. Actin ____

   f. Myosin ____

EXERCISE 24

DEMONSTRATION OF PHYSIOLOGICAL RECORDING EQUIPMENT

A variety of devices are currently in use to demonstrate and record physiological activities and data accurately. Among these instruments are multichannel recording devices, known as physiographs, and the classic old style kymograph.

The physiograph is a modern device designed to be an expandable recording system. The basic components of a physiograph system generally include (1) the input transducer (converts the energy of the physiological event into a proportional electrical signal), (2) the amplifier (provides the required increase in signal strength to permit visualization), (3) the coupler (renders the input energy from the transducer or the signal from the subject compatible with the channel amplifier), and (4) output transducer (provides visual or audio representation of information). In studying physiological events such as heart action, muscle contraction, blood pressure, or cerebral activity, a detection device first is used to pick up the electrical signal from the tissue preparation or experimental subject. Examples of such devices include an electrode for electrical signals as in cases of neural activity and an input transducer for nonelectrical signals as in situations of temperature measurement.

A transducer converts the physiologic event into an electrical signal. Since physiological events often produce weak signals, an amplifier is used to enlarge the magnitude of the event. The amplified signals are converted by an output transducer into some form of visual or audio display that can be observed and measured. The display can be in the form of an electrical meter, oscilloscope, or chart recorder.

Various arrangements of physiograph components are possible. Figure 24-1 shows an arrangement example. The components of a physiograph are contained in a main frame assembly, which provides power, devices for the control and monitoring of circuitry and instrument calibration, and locations for mounting accessory plug-in units (Figure 24-2). Table 24-1 lists and describes both the basic components and accessories (Figure 24-3) used with physiographs.

The kymograph is a moving-drum recording instrument. It is the classical device used for the measurement of movements produced by muscle contractions. The kymograph is not considered here.

In this exercise, the features of physiological recording devices will be demonstrated. This exercise will serve as a general frame of reference for procedures dealing with muscle measurements of laboratory animals.

A. IDENTIFICATION OF PHYSIOGRAPH COMPONENTS

Although physiographs from different manufacturers vary in appearance or design, they generally consist of the same basic components. The parts of a physiograph and their functions are described in Table 24-1. Use this information as you examine and locate the parts of the physiograph.

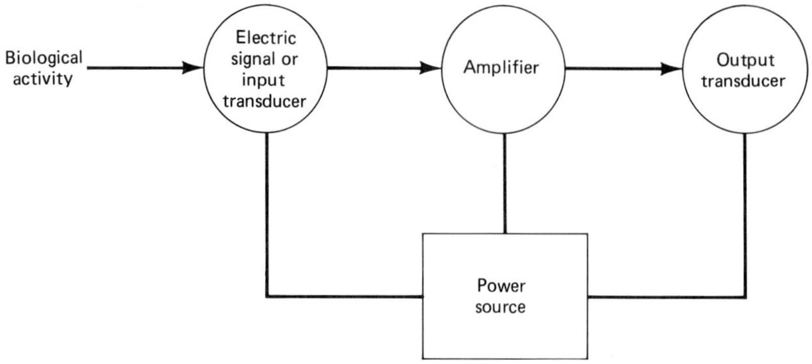

Figure 24-1 The general arrangement of physiograph instrumentation components.

Materials

The following materials should be provided for class examination and demonstration.

1. Physiographs
2. Accessory devices and attachments to include those used for ECG, pulse detection, respiratory measurements, and electromyograms
3. Reference diagrams or instruction manuals

Procedure

This procedure is to be performed by students individually.

1. Examine the physiograph and accessories provided. With the aid of Figures 24-1 and 24-2 locate and learn the functions of each component listed in Table 24-1, and others indicated by your instructor.
2. Your instructor will indicate the following control and other appropriate switches associated with the physiological functions.
   a. Turning on the instrument
   b. Changing paper speed
   c. Increasing intensity of training or recording
   d. Recording time
   e. Recording event
   f. Balancing the transducer
   g. Increasing amplifier sensitivity
   h. Stopping or starting chart movement
3. Answer items pertaining to this portion of the exercise in the questions for review section.

Table 24-1 Physiograph Components and Their Functions

Physiograph Component	Functions
Amplifier	Increases the strength of the electrical signal while keeping the proportional relationship of the electrical signal obtained from the transduced physiological activity. Amplifiers contain power supply units to convert alternating current (ac) to direct current (dc) and to either increase or decrease voltage requirements.
Amplifier controls	The types of control are determined by the function of an amplifier. Controls include: 1. *Gain control:* adjusts sensitivity of amplifier to strength of the incoming signal 2. *Mode control:* enables instrument to accept either ac or dc input signals 3. *Centering (zero) control:* moves the zero line on a recording system in either postive or negative position 4. *Calibration control:* used to check or establish the sensitivity of the amplifier.
Bellows pneumograph	Photoelectric transducer used for recording respiration.
Blood pressure transducer	Photoelectric transducer for the measurement of direct blood pressure.
Coupler	Renders the input energy from the transducer (or the signal) from the subject compatible with the channel amplifier.
Electrode	Device used to carry an electrical current into or away from a biological specimen.
Myograph	Transducer used primarily for the quantitative measurement of muscular contractions.
Oscilloscope (cathode ray tube)	Output transducer that provides a visual display of voltage signals on a fluorescent screen.
Power supply (for amplifier)	Converts alternating current (ac) to direct current (dc); either increases or decreases the voltage needed by the amplifier.
Recording device	Converts electrical energy into rotary motion; receives electrical signals from the amplifier and drives the pen across a moving chart, thereby providing a record of the event being measured.
Stimulator	Device used to produce electrical shocks of desired voltages in experimental situations.
Stylus	Writing instrument used to indicate time, beginning and ending of an event, etc.
Thermistor	Transducer that monitors temperature changes.
Transducer	Converts a physiological event into a proportional electrical signal; the most sensitive component in a transducer is a light bulb and photoelectric cell.

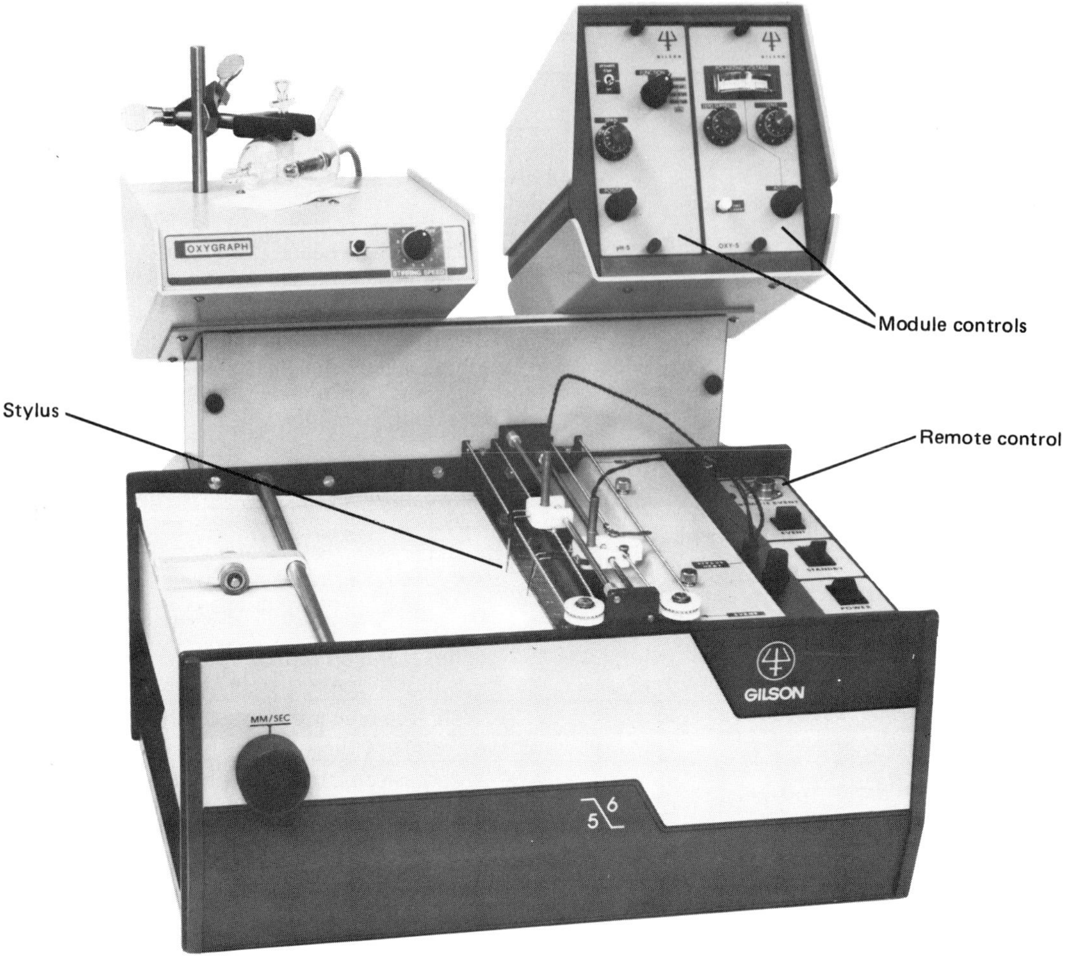

(a)

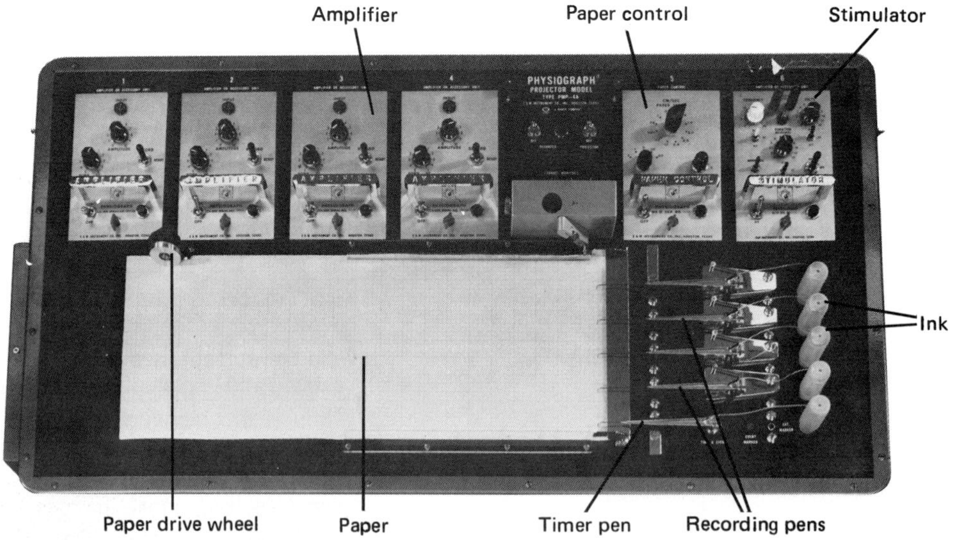

(b)

Figure 24-2 Examples of Physiographs. (a) A Gilson Biophysical Recorder and its components. (Courtesy Gilson Medical Electronics, Inc.) (b) Diagrammatic representation of Physiograph. (Courtesy Narco Bio-Systems.)

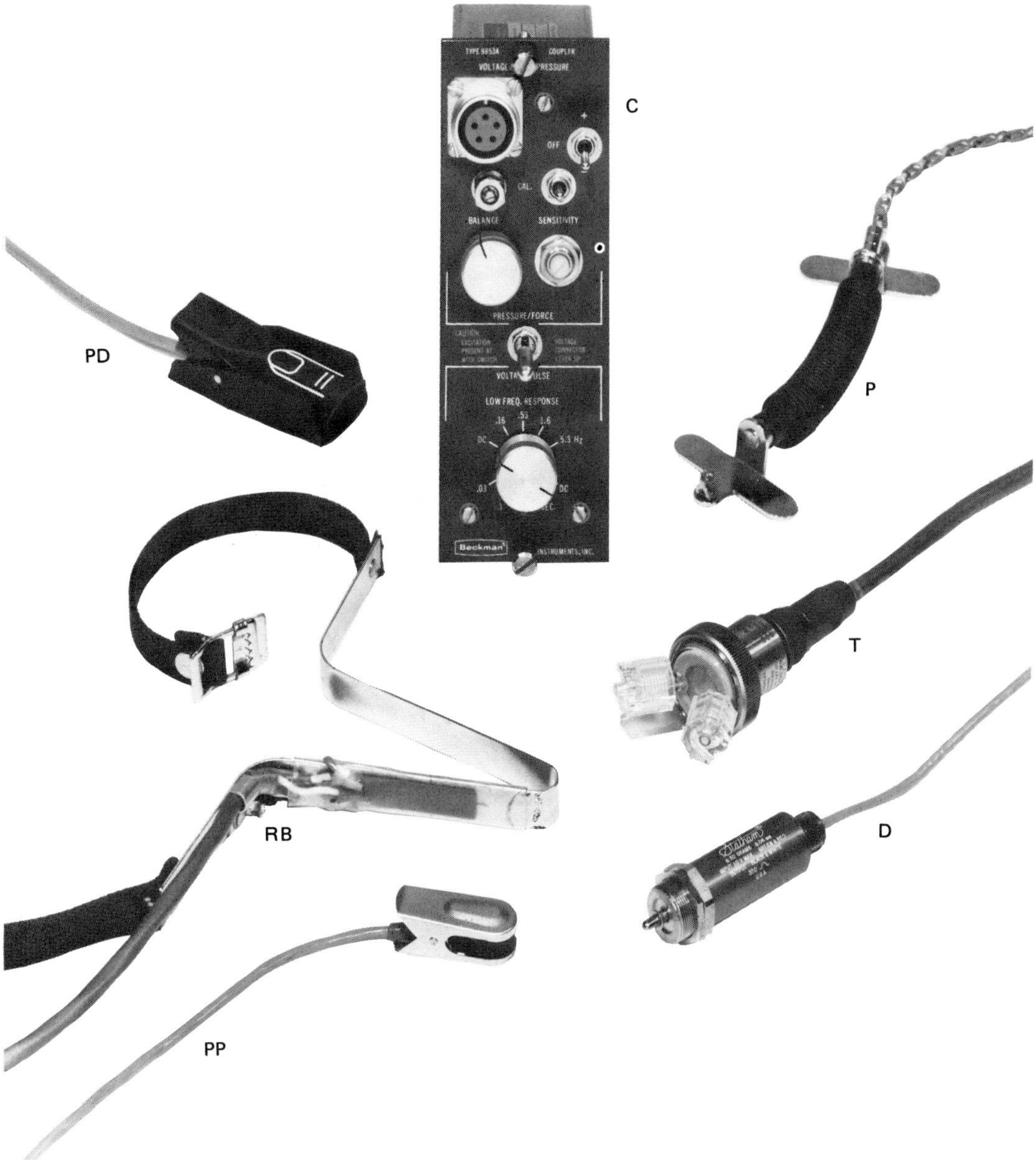

Figure 24-3 Various accessories used in physiologic measurement. The display includes: a coupler component (C), bellows pneumograph (P), general purpose transducer (T), ECG accessory pulse rate detector (D), photoelectric pulse pickup (PP), ear/nose transducer for pulse pickup (PP), respiration belt using a strain gauge bridge (RB), and photoelectric pulse detector (PD). (Courtesy of Sensor Medics Corporation.)

Questions for Review

EXERCISE **24**

DEMONSTRATION OF PHYSIOLOGICAL RECORDING EQUIPMENT

Name ____

Laboratory Section ______________________________ Date ____________

1. Distinguish between the following:

   a. Physiograph and kymograph ______________________________

   b. Input transducer and output transducer ______________________________

   ____

2. Complete Table Q24-1 by providing either the term or explanation of the physiograph component listed.

Table Q24-1

Physiograph Part	Functions
a.	Photoelectric transducer used to measure or record respiration
b.	Used to carry an electrical current into or away from a biological specimen
c.	Output transducer that displays voltage signals on a fluorescent screen
Stimulator	d.
e.	Transducer used for quantitative measurements of muscular contractions
Thermistor	f.
g.	Converts a physiological event into a proportional electrical signal
Amplifier	h.

EXERCISE 25

INTRODUCTION TO SKELETAL MUSCLE CONTRACTION USING A LABORATORY ANIMAL

Muscle is the most common tissue in the body of most vertebrates. As described in Exercise 23, it consists of *elongated cells* or *fibers* specialized for contraction. The contractile components within the muscle fiber are the *myofibrils*. These fibrils run lengthwise through the unspecialized cytoplasm of muscles or the *sarcoplasm*. Muscle bundles, called *fascicles*, are formed through the joining of fibers to other parallel muscle cells by intertwining connective tissue. The fascicles, in turn, are organized into larger units referred to as *muscle organs*.

The ability of muscle to respond to stimulus is called *irritability*. The response takes the form of a contraction. The stimulus is normally an impulse transmitted to the muscle cell membrane by the axon, or nerve fiber, of a neuron. Each fiber supplies from a few up to more than a hundred individual muscle cells. The portion of the membrane directly under the end of the axon is called a *myoneural* or *motor end plate*. The impulse causes changes within the myofibril and shortening of the sarcomere and the entire muscle cell.

In response to a single stimulus, individual muscle fibers either contract completely or do not contract at all (the all-or-none law). Muscular contraction depends on how frequently nerve impulses are received by the muscle fibers and how many fibers are stimulated. The number of individual fibers contracting also determines the strength of a muscle contraction.

Skeletal muscle contracts only when it has been stimulated adequately. It does not contract spontaneously. In this exercise isolated muscles from a frog are used to demonstrate how a whole muscle contracts and how changes in a stimulus can modify muscular contractions. Muscle used for such study is isolated and attached to appropriate equipment in such a way as to record contractions. An electrical stimulus to either the muscle itself or to an associated nerve produces a recorded response.

A single stimulus produces a single brief contraction known as a *muscle twitch*. An examination of a recorded response (Figure 25-1a) reveals three principal events: (1) the *latent period*, which is the period between the stimulus and the beginning of contraction; (2) the *contraction period;* and (3) the *relaxation period*. The extent of the twitch depends on the stimulus applied. The stimulus must reach some *threshold level* before the muscle responds at all. Increased stimulus increases to the event of contraction until some maximum is obtained. As the stimulus is increased more fibers are activated until all of them are contracting in response to the stimulus. Once a muscle is stimulated enough to contract, it briefly loses its ability to respond to additional stimuli. This is known as the *refractory period*.

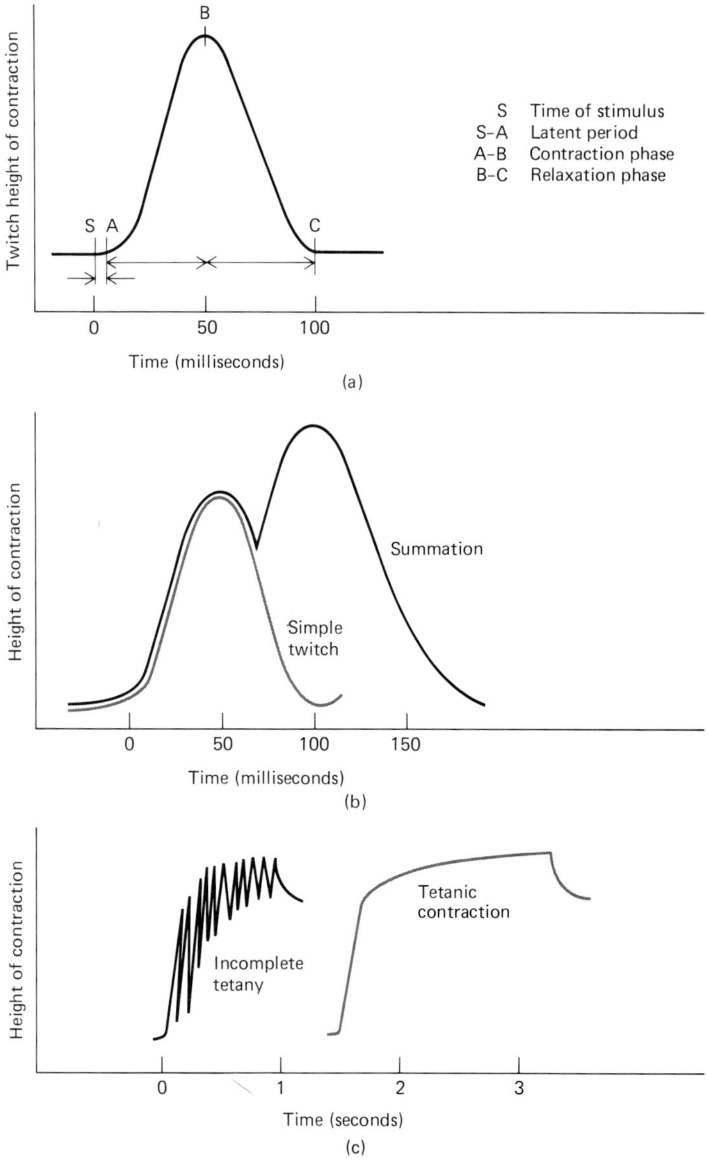

Figure 25-1 Muscle contraction responses. (a) Muscle twitch. (b) Summation of contraction. (c) Tetanic contraction.

If a second stimulus is applied after the refractory period, but before the contraction is over, the muscle will respond to both and the additive effect of both stimuli will produce a stronger contraction. This is known as a *summation of twitches* or *wave summation* (Figure 25-1b).

The application of two successive stimuli of maximum intensity causes two separate maximal contractions. If successive stimuli are applied so quickly that the muscle can relax only partially between stimuli, it remains in a state of contraction. This summation and fusion of individual twitches is known as *incomplete tetanus* (Figure 25-1c). *Complete tetanus* is the condition in which stimuli are applied so quickly that the muscle does not relax at all. Tetany is an important action because muscle twitches do not occur physiologically. Skeletal muscle contractions are all tetanic in nature, even the blinking of an eye.

A. SPECIMEN PREPARATION

A laboratory animal is used in this exercise to demonstrate certain fundamental physiological principles. To ensure the success of the experiments care must be exercised to avoid unnecessary injury, irritation, or pain to the laboratory animal. Any of these conditions will render the animal less capable of "normal" responses.

In this exercise the frog will be prepared by a procedure known as *pithing.* To pith the animal, a dissecting needle is inserted into the cranium (skull) in a downward and forward direction. The process results in the destruction of the brain and is called *single pithing* (Figure 25-2a). *Double pithing,* another process, is used to destroy the spinal cord. It is performed by inserting a dissecting needle between the first vertebra and the skull to sever the spinal cord from the brain (Figure 25-2b). *Note:* The pithed frog prepared here will be used for the continuation of the exercise.

Materials

The following materials should be provided per two students or for instructor's demonstration.

1. One normal, healthy frog
2. Dissecting tools (needle, scalpel, probe, scissors)
3. Three sterile gauze pads (2×2 inch)
4. Frog Ringer's solution
5. Thread

Procedure 1: Pithing

This procedure is to be performed by students in pairs.

1. Hold the frog in the left hand with its hind limbs extended and its dorsal surface facing toward you (Figure 25-2a).
2. Place the forefinger over the snout (nose) and bend the frog's head downward until it forms a sharp angle with the trunk of its body. Use the thumb and remaining fingers to hold the frog's limbs securely.
3. Take the dissecting needle in the right hand and run the tip down the midline of the frog's skull to locate a depression or groove between the skull and the first vertebra. This depression is about 2–3 mm behind the posterior border of the animal's eardrums (Figure 25-2a). It marks the location of the *foramen magnum,* which is a large opening of the skull through which the spinal cord emerges.
4. Push the needle in a forward and downward direction with a sharp movement through the skin, the depression, the underlying tissues, and the foramen magnum into the brain. Twist and turn the needle to destroy the brain. Remove the needle and stop any bleeding by pressing a gauze pad over the depression. This completes the *single-pithing process.*
5. To double-pith the frog, insert the needle again between the front vertebra and the skull, but in a downward or caudal direction into the spinal cord. Twist and turn the needle to destroy the spinal cord.

The legs of the frog should extend during this process indicating entry of the needle into the spinal cord. Successful double pithing results in the animal's limbs becoming limp soon thereafter. The frog is now ready for the following procedures.

Procedure 2: Preparation of Muscle

This procedure should be performed by students in pairs.

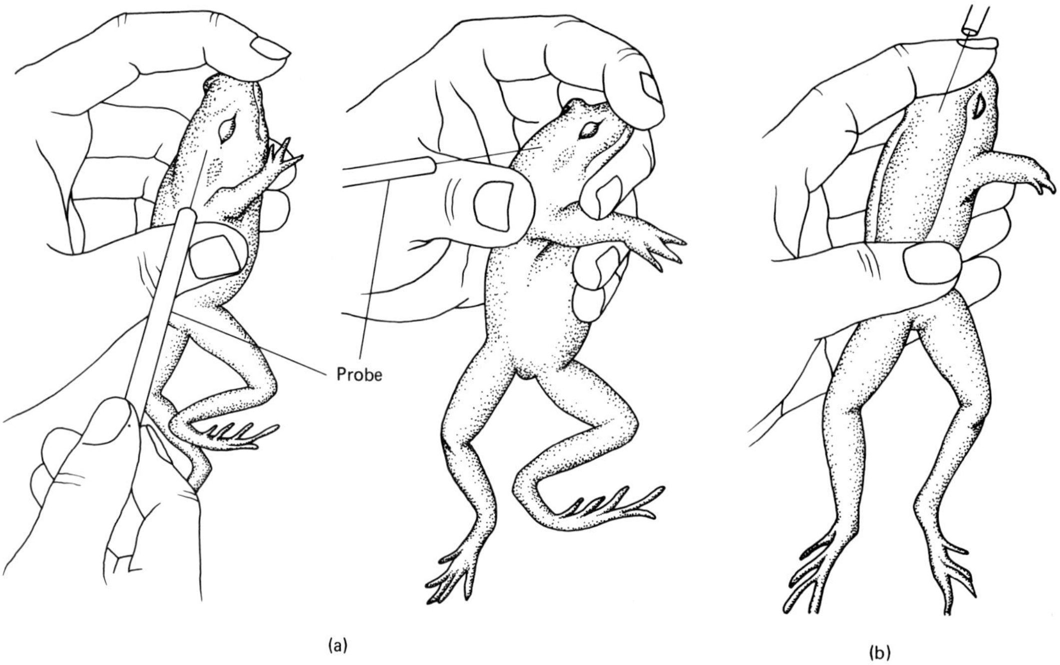

Figure 25-2 Pithing (a) The single-pithing procedure. The holding position for the process, and the location of the foramen magnum of the skull are shown. Note the bent appearance of the legs. (b) The double-pithing procedure. The insertion of the dissection needle into the spinal cord causes the frog's legs to extend and stiffen.

1. Remove the skin from one leg of the prepared frog by first cutting the skin as high as possible around the thigh. Roll the loosened skin back a short distance. Grasp it with one hand and the frog's thigh with the other. Peel the skin completely off the leg by a quick pull.
2. Use Figure 25-3a to locate the *gastrocnemius muscle.* Keep the frog's tissues moist with Ringer's solution during the remainder of the exercise.
3. Carefully insert a probe to separate the thigh muscles and to expose the sciatic nerve (Figure 25-3b).
4. Tie a piece of thread about 10 cm long (4 inches) around the sciatic nerve and cut the nerve above the tie.
5. Gently ease the nerve from its location with the thread and lay it on the surface of the gastrocnemius muscle.
6. Cut the thigh muscles first near their origins on the pelvic girdle and then just above the knee (Figure 25-3b).
7. Use the probe again to separate and free the gastrocnemius muscle from the other muscles of the frog's leg (Figure 25-3b).
8. Locate the Achilles tendon (Figure 25-3c). Tie a piece of thread about 15 cm long (6 inches) around its lowest end. Cut the tendon at the point between the thread and the animal's heel bone.
9. Lift and move the gastrocnemius muscle to one side. Cut the other muscles of the frog's lower leg and the bone (tibiofibula) just below the knee.
10. Cut the femur bone about 2.5 cm (1 inch) (Figure 25-3c). Your instructor will demonstrate the procedure to attach the nerve–muscle preparation to the recording system. Be certain to keep this material and the other leg of the frog moist with Ringer's solution.

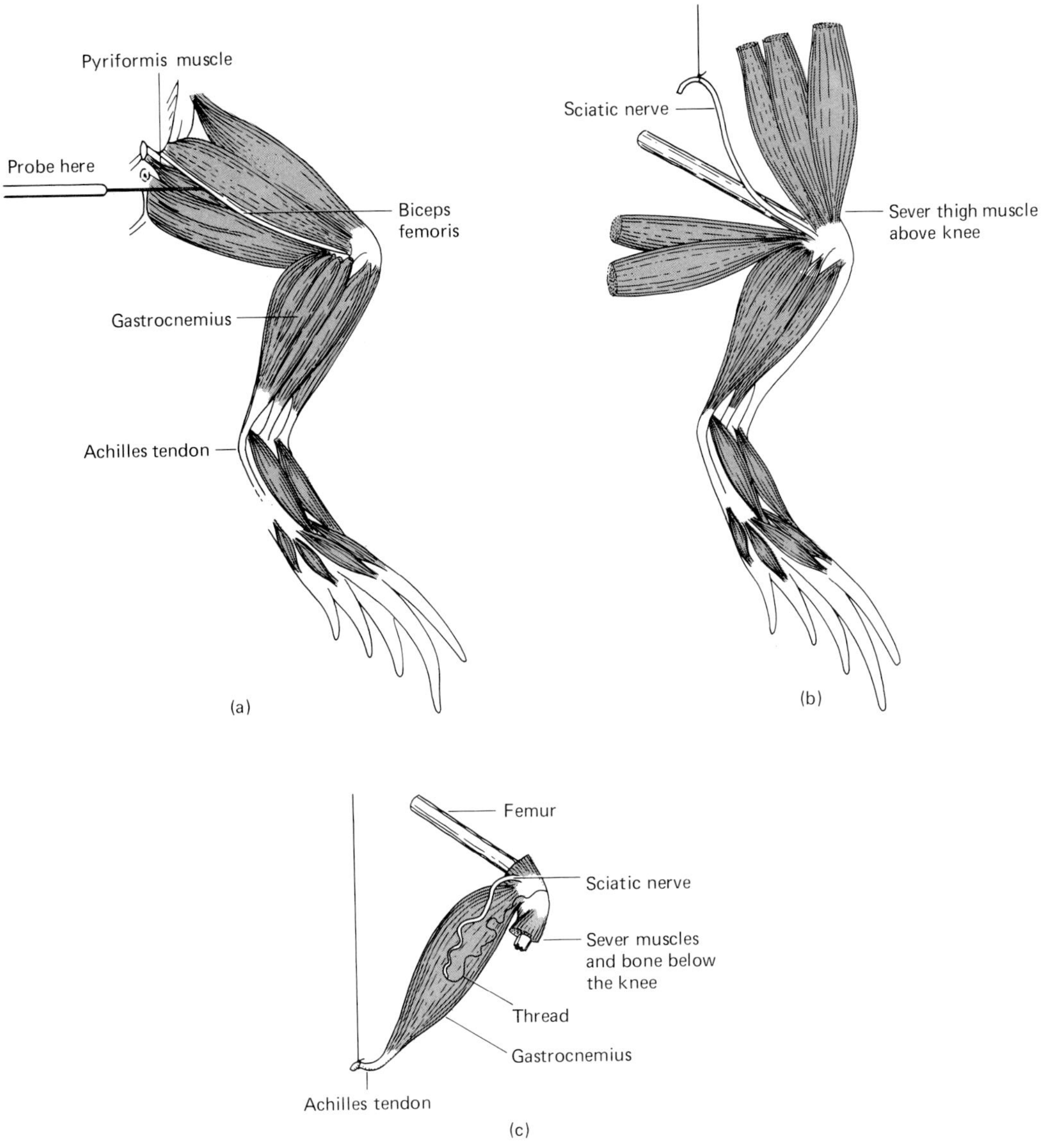

Figure 25-3 Diagrammatic representation of the musculature of the dorsal surface of a frog's leg. (a) Musculature of dorsal surface. (b) Relationship of sciatic nerve and gastrocnemius muscle. (c) Appearance of muscle preparation.

B. MUSCLE ACTIVITY MEASUREMENTS

Materials

The following materials should be provided for class use.

1. Stimulating apparatus
2. Physiograph or related recording system
3. Weight pans
4. Weights (5 and 10 g units)
5. Millimeter ruler
6. Litmus paper
7. Examples of muscle contraction tracings or records
8. Frog Ringer's solution
9. Sterile gauze pads (2 × 2 inch)

The procedures in this part of the exercise should be performed by students in groups of four. One individual should operate the equipment used. A sufficient number of tracings should be made for all members of the group.

Procedure 1: The Single Muscle Twitch

To cause a muscle twitch it is necessary to produce a *threshold stimulus,* that is, the lowest voltage of electrical stimulation that produces a contraction. As the voltage is increased, more and more motor units are activated. The lowest voltage that activates all such units and produces a maximum contraction is called the *maximal stimulus.*

A stimulator is used to regulate the strength of the stimulus (voltage) and the frequency of stimuli (number of pulses or stimuli per second). The duration of each stimulus is measured in microseconds (μsec). The muscle stimulated to contract raises the writing device so that a trace of the contraction is made.

1. Set up the stimulating apparatus and recording system as directed by your instructor. Examine the stimulator and note the position of its components.
2. Set the voltage dial of the stimulator at its lowest setting and the duration at 10 μsec.
3. Start the recording system at a slow speed.
4. Observe the recording for the presence of contractions.
5. Slowly increase the voltage by moving the voltage dial until contractions are observed.
6. Mark the response with the voltage needed to produce it. This is the threshold stimulus. Make four recordings of this response so that each member of the group can have one. Enter this value and attach a representative tracing to the results and observations section. Refer to Figure 25-1a as a guide for labeling purposes.
7. Turn off the system. Moisten the muscle preparation with Ringer's solution.
8. Set the voltage dial at the threshold value. Turn on the recording system and the stimulating system.
9. Increase the voltage by one increment (step value) and stimulate the muscle.
10. Continue to stimulate the muscle and increase the voltage by one voltage increment at a time until the muscle contractions do not increase further. The voltage at this point is the *maximal stimulus,* and the effect produced is known as the *maximal response.*
11. Mark the record and enter the value for the maximal stimulus in the results and observations section.

Procedure 2: Summation, Incomplete Tetanus, and Complete Tetanus

1. Use the same muscle preparation, stimulating apparatus, and recording system.
2. Set the voltage dial at the maximal stimulus value.
3. Stimulate the muscle at a frequency of 1 pulse per second (pps).
4. Record the effect for 15 seconds.
5. Increase the frequency to 2 pps for 15 seconds and then to 4 pps for 15 seconds.
6. Allow the muscle preparation to rest for 2 minutes. During this time moisten the preparation again.
7. Continue to double the frequency, with a rest break following, until the muscle preparation goes into complete tetany. Make sufficient records for all members of the group.
8. Label the record and attach it to the results and observations section. Be certain to indicate the following events on the record (refer to Figure 25-1).
   a. Point of maximal stimulus
   b. Wave summation

c. Incomplete tetanus
d. Complete tetanus

Procedure 3: Determination of Work Performed

1. Use the same muscle preparation, stimulating apparatus, and recording system.
2. Attach a weight pan or comparable device to the muscle preparation. Follow the directions of your instructor and make the necessary adjustments.
3. Determine the threshold stimulus. This is the value for unloaded contraction. Enter your findings in the results and observations section.
4. Add a 5 g load to the weight pan and determine the threshold stimulus. Enter your findings in the results and observations section.
5. Increase the voltage substantially above the threshold stimulus. Obtain a single muscle twitch. Mark the starting point of the stimulus application. Make enough recordings for each member of the group. Attach a labeled record in the results and observations section.
6. Continue to add 5 g loads until the muscle preparation can no longer contract. Mark the record with the total weight lifted. Measure the height or distance the weight was lifted. Enter your findings and attach a label record in the results and observations section.
7. Calculate the work performed by the muscle preparation, using the formula: work performed = weight of load × distance lifted. Work can be recorded in units of gram-millimeter.
8. Turn off the stimulating apparatus and the recording system, and moisten the muscle preparation with Ringer's solution.

Procedure 4: Muscle Fatigue

Muscle fatigue represents a loss of irritability (ability to respond to stimuli) by muscle. It is caused by the accumulation of metabolic products such as acid phosphates, carbon dioxide, and lactic acid.

1. Use the same muscle preparation, stimulating apparatus, recording system, and weight pan.
2. Add a 10 g load to the weight pan and determine the threshold stimulus. Record your finding in the results and observations section.
3. Increase the stimulus to 20 volts above the threshold value. Stimulate the muscle preparation about 15 times at a rate of 2 pps. Repeat this step until the recording line begins to slope downward toward the baseline. Mark this point. It represents the onset of fatigue. Continue to stimulate the muscle at the same rate. Determine the time it takes for the muscle preparation to reach the bottom of the record. This event is complete fatigue.
4. Turn off the stimulating apparatus and the recording system.
5. Remove the muscle tissue preparation and cut it into two parts. Press one cut end on a piece of red litmus paper and the other cut end on a piece of blue litmus paper. What happened? Enter and explain your findings in the results and observations section.

RESULTS AND OBSERVATIONS

THE SINGLE MUSCLE TWITCH

1. Threshold voltage: ______________

2. Maximal stimulus: ______________

3. Attach a labeled representation of the single muscle twitch.

4. Distinguish between maximal stimulus and maximal response. ____________________

____

____

SUMMATION, INCOMPLETE TETANUS, AND TETANUS

1. Attach a labeled representation of summation, incomplete tetanus, and tetanus.

2. Do the heights of individual contractions differ in summation? ______________

DETERMINATION OF WORK PERFORMED (GRAM-MILLIMETER)

1. Threshold stimulus in unloaded contraction: ______________

2. Threshold stimulus with first 5 g load: ______________

3. Attach a labeled representation of a single twitch with first 5 g load and intense voltage.

4. Attach a labeled representation of maximum work performed.

5. Total weight (load) with which no further contraction of the muscle preparation occurred:

____

6. Work performed by the muscle preparation:

$$\text{Work performed} = \text{load} \times \text{distance}$$

$$\_\_\_\_ = \text{g} \times \_\_\_\_ \text{mm}$$

MUSCLE FATIGUE

1. Threshold stimulus with 10 g load: ____________

2. Threshold stimulus with increased voltage: ____________

3. Time for complete fatigue to occur: ____________

4. What were the results of litmus test? Explain what happened. ____________

____

____

Questions for Review

EXERCISE **25**

INTRODUCTION TO SKELETAL MUSCLE CONTRACTION USING A LABORATORY ANIMAL

Name ______________________________

Laboratory Section ______________________ Date ____________

1. Draw and label a representation (in the form of a graph) of the events of a single muscle twitch.

2. Explain or define:

   a. Tetanus (tetany) ______________________________

   ____

   b. Threshold stimulus ______________________________

   ____

   c. Summation ______________________________

   ____

   d. Foramen magnum ______________________________

   ____

   e. Maximal stimulus ______________________________

   ____

3. Distinguish between single pithing and double pithing. ______________________________

____

4. What substances are responsible for muscle fatigue? ______________________________

____

EXERCISE 26

HUMAN MUSCLE CONTRACTION

The only action a muscle is capable of is contraction. Some other force, including gravity, must be exerted to return the muscle to its uncontracted or relaxed state. For example, the biceps (a flexor) bends the arm, and the triceps (an extensor) must contract to pull the forearm back into place and also to stretch the biceps muscle (Figure 26-1). The force involved in this case is generated by the sliding filaments (Exercise 23) and is exerted parallel to the muscle fibers.

When an individual uses the biceps muscle, for example, to lift a heavy object, the force exerted by the contracting muscle on the object is referred to as *muscle tension*. The force exerted on the muscle by the object is called the *load*. In order for a heavy object or load to be moved, muscle tension must be greater than the load, bcause these two factors are opposing forces. With *isotonic* (equal tension) contractions the muscle fibers shorten, but the tension remains the same. A contraction of this type takes place when an individual uses the biceps to lift heavy weights or other objects. With *isometric* (equal length) contraction, another type of muscle activity, a muscle's length remains essentially the same (does not shorten), but the tension changes. Contractions of this type occur when a muscle supports a load in a fixed position, for example, contraction of the quadriceps muscles of the leg when an individual is in a standing position.

Muscle tone (tonus) is a reflex stretch response to motor nerve impulses originating in the spinal cord. These impulses stimulate small numbers of different muscle fibers to contract at different times, resulting in a partial but sustained contraction of the entire muscle. Several muscles involved with body posture are associated with this muscle property. When muscles are not being stretched, as in the case of postural muscles during sleep or unconsciousness, the body collapses. The loss of muscle tone is due to an interference with the conduction of nerve impulses. It may result from certain disease states, such as poliomyelitis, or accidents

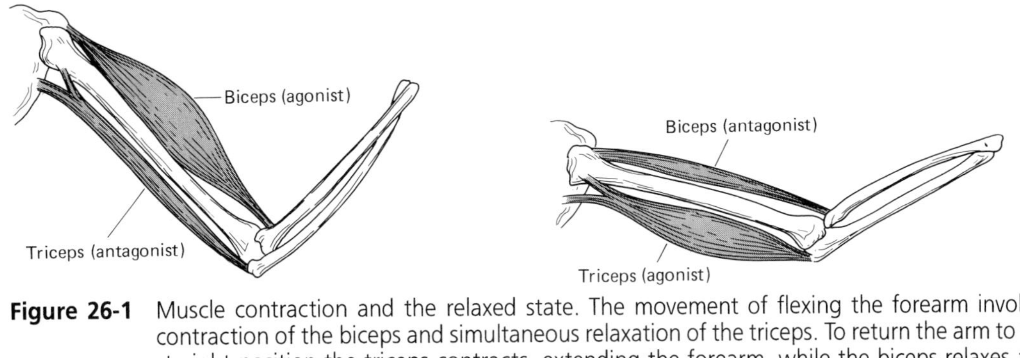

Figure 26-1 Muscle contraction and the relaxed state. The movement of flexing the forearm involves contraction of the biceps and simultaneous relaxation of the triceps. To return the arm to the straight position the triceps contracts, extending the forearm, while the biceps relaxes and yields to the action of the triceps. The muscle that performs the action is called the *agonist;* the muscle that is in opposition is the *antagonist.*

involving destruction to motor never fibers. Muscles with less than normal tone are referred to as *flaccid.*

Some interesting aspects of muscle contractions can be revealed by *electromyography,* the procedure used to record the action potentials (electric discharges) from muscles. The records are called *electromyograms.* When a nerve impulse arrives at the muscle fibers of a motor unit (a neuron and the group of muscle cells it stimulates), a reversal of the electric charge on the muscle fiber membranes occurs. Such a reversal is the action potential that spreads throughout the muscle fibers to cause contraction. Electromyograms are recorded with the aid of a needle or wire electrodes inserted directly into muscles or surface electrodes placed on the skin. The action potentials from individual motor units are picked up by the needle or wire electrodes while surface electrodes defect integrated action potentials from whole muscles. In some situations, electromyograms can be used to demonstrate which muscles are responsible for or control certain movements. For example, by placing surface electrodes over the muscles of the shin and calf of a test subject and then having the individual respond to a command such as "lean backward, sit, or stand," an electromyogram would show which muscles contract during specific movements and to what extent (Figure 26-2).

This exercise will deal with various aspects of muscle activities including isometrics.

Materials

The following materials should be provided per four students to perform procedures 1–4.

1. One 5- or 10-lb dumbbell
2. One cloth tape measure (1 meter length)

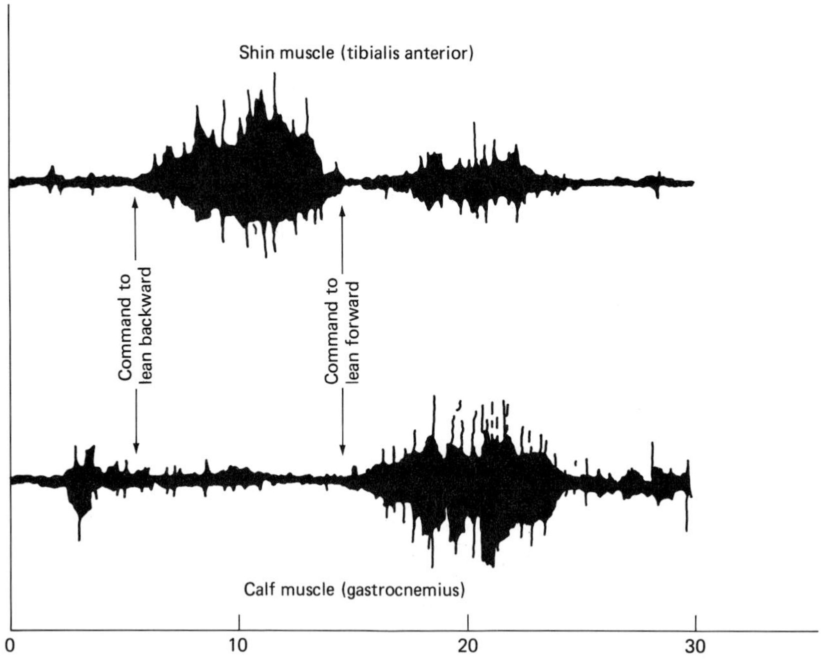

Figure 26-2 A diagrammatic representation and comparison of electromyograms obtained from human muscles of the shin and calf during the execution of specific verbal commands. The test subject first was told to lean backward, and then to lean forward. The involvement of a specific muscle and the extent of muscle contractions following the command can be seen. (Adapted from Electromyograms from J. D. Witherspoon, R. Flowers, M. Isaacson, and S. Keslensky. *Muscle Contraction* [audiovisual tape], Narco Bio-Systems, Inc., Houston, 1973.)

3. One relatively soft tennis or rubber ball
4. Stopwatch or other timing device
5. An ergograph or similarly constructed device to measure work performed
6. Weight pan or similar container to hold gram weights
7. Gram weights
   a. 500 g (2)
   b. 250 g (2)
   c. 100 g (5)
   d. 50 g (2)
   e. 10 g (10)
8. One meter stick

The following materials should be provided for class use to produce electromyograms (procedure 5).

1. Skin electrodes (8 mm or 12.5 mm diameters)
2. Electrode washers (8 mm or 12.5 mm diameters) or micropore tape for attaching electrodes to the skin
3. Skin cleanser to prepare the skin
4. Electrode gel
5. Kimwipes or paper towels
6. Sterile, disposable alcohol prep-pads (Med Pak Corp.)
7. Scissors
8. Physiograph

Procedure 1: Muscle Tone (Biceps, Triceps, Gastrocnemius, and Masseter Muscles)

This procedure should be performed by students in pairs.

1. Take turns in carrying out the following tasks. Note the degree of contraction of the biceps and triceps in each case by feeling them for tightening and/or hardening. Enter your findings and answer the questions in the results and observations section.
   a. Sit in a comfortable position with the forearms supported on a level surface and with palms down.
   b. Sit in the same position, but with palms up, and note any differences.
   c. Sit in a comfortable position. Grasp the barbell and lift it with palms up. This is known as a weight curl. (A book may be substituted for the weight.)
   d. Stand erect and allow both arms to hang freely at the sides of the body.
   e. Stand erect and hold the barbell with one hand at the side of the body.
2. Take turns in carrying out the following tasks. Note the degree of contraction of the gastrocnemius muscle. Enter your findings and answer the questions in the results and observations section.
   a. Lie down on your abdomen (prone position).
   b. Sit on the edge of a chair so that both feet touch the floor.
   c. Sit on the edge of a table so that both feet are dangling and not touching the floor.
   d. Stand erect and lock your knees for 5 seconds.
   e. Stand erect with legs relaxed.
   f. Stand on your toes with your arms extended overhead and reach toward the ceiling.
3. Locate the masseter muscle in your partner's face. Place your hands on it. Have your partner first relax and then tighten this muscle. Describe what happens in each case in the results and observations section.

Procedure 2: Distinguishing Between Isotonic and Isometric Muscle Contractions

This procedure should be performed by students in pairs.

1. With a tape measure measure the distance around the midline (circumference) of the biceps and triceps muscles of both arms under the condition of each of the following tasks. Enter your findings and answer the questions in the results and observations section.
   a. Extend both arms directly to the front of the body.
   b. Flex both arms with palms facing upward, much like a weight lifting exercise, 10 times in succession.
   c. Extend the right arm while grasping a barbell or weight substitute.
   d. Repeat step c with the left arm.
   e. Flex the right arm while grasping a barbell, 10 times in sucession.
   f. Repeat step e with the left arm.
2. With a tape measure determine the circumference of the midline calf region of both legs under the condition of each of the following tasks. Enter your findings and answer the questions in the results and observations section.
   a. Lie down on your abdomen (prone position), with gastrocnemius relaxed.
   b. Repeat step a, but contract the gastrocnemius muscles.
   c. Stand erect.
   d. Stand on toes and extend arms upward.
   e. Stand on toes with arms at the side.
   f. Sit on a table so that legs are dangling and do not touch the floor.
   g. Repeat step 2f, but try to move your feet forward and upward with someone pushing against them (to prevent any movement backward or downward).
3. Change places with your partner and repeat the steps and tasks of this procedure.

Procedure 3: Determining Muscle Fatigue

This procedure should be performed by students in groups of four.

1. Hold a tennis ball (or soft rubber ball) in your right hand. As you squeeze the ball as rapidly as possible, have your laboratory partners count the number of complete contractions during each 10-second interval for a total period of 3 minutes nonstop. Enter the count in the results and observations section.
2. Continue to time while you squeeze the ball until fatigue is complete. Record the time. This is the end of trial 1.
3. Wait 1 minute and repeat steps 1 and 2. This is the end of trial 2.
4. Wait 5 minutes and repeat steps 1 and 2. This is the end of trial 3.
5. Repeat all steps with your left hand.
6. Now have each of your laboratory partners repeat steps 1–4. Enter your findings and answer the questions in the results and observations section.

Procedure 4: Determination of Work Performed

This procedure should be performed by students in groups of four.

1. Assemble an ergograph similar to the one shown in Figure 26-3. Tie one end of the cord to a weight pan or comparable container. Have one partner hold the other end of the cord until the experiment is ready to proceed.

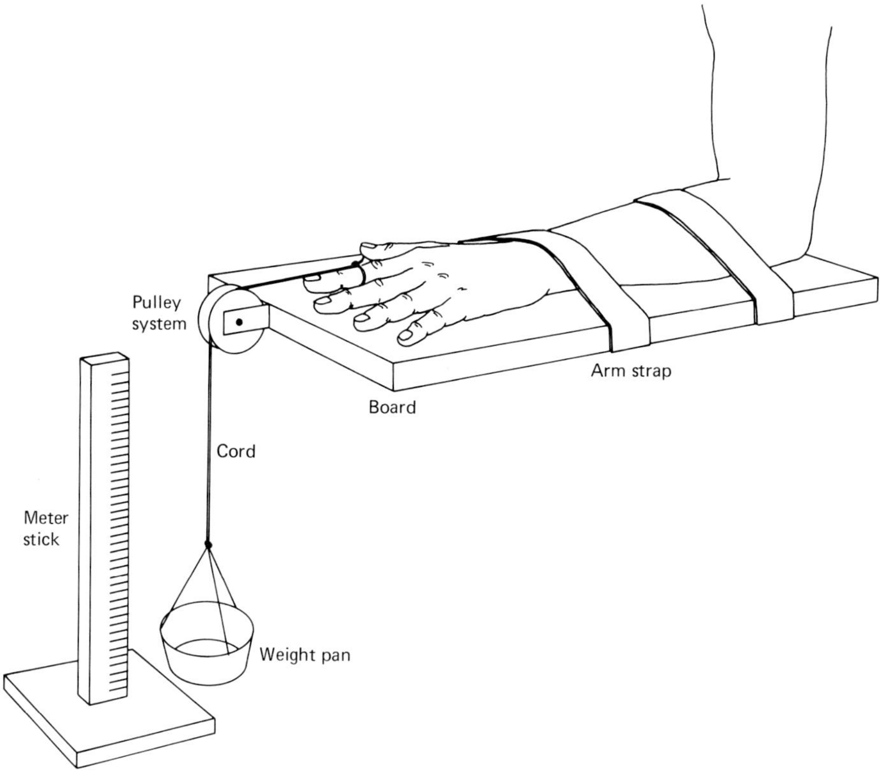

Figure 26-3 An ergograph. This device can be used to determine work performed by skeletal muscle. It consists of a board, pulley, cord, weight pan, and restraining straps.

2. Put the ergograph on a level surface and sit next to it so that you can comfortably place your left forearm upon its surface. Your fingers should be pointing toward the pulley system. Strap the forearm to the board.
3. Tie the free end of the cord to the middle phalanx of the index finger of the hand.
4. Have a laboratory partner hold a meter stick next to the weight pan and note the level at which the pan is located.
5. Have a laboratory partner add weights in 10-g increments to the pan. Lift the weight pan by raising your index finger. Continue to add weights until the weight pan cannot be lifted more than 5 mm. In carrying out this task your hand must not be lifted from the ergograph's surface nor should the wrist be flexed. Enter the maximum number of grams lifted and the distance (in cm) the weight was lifted in the results and observations section.
6. Repeat steps 3–5 with the middle finger and again with four fingers together, not including the thumb.
7. Repeat this entire procedure with your laboratory partners.
8. Calculate the work performed (weight lifted in grams × distance lifted in centimeters). Enter your calculations and answer the questions in the results and observations section.

Procedure 5: Electromyograms from Leg Muscles

This procedure should be performed by students in pairs. Your instructor will have a physiograph or similar equipment ready for class use. Refer to Exercise 24 for a general review of the component parts of a physiograph.

1. Obtain four skin electrodes and, using the individual alcohol prep-pads, wipe them and allow them to dry on a clean surface.

2. Turn on the power switch of the physiograph and allow the instrument to warm up for at least 10 minutes.
3. Using a paper towel and the skin cleaner provided, wipe the skin over the tibialis anterior (shin) and gastrocnemius (calf) muscles. See Figure 26-4
4. Apply electrode gel onto the sterilized electrode surfaces. Use a Kimwipe to smooth the gel evenly so that it fills the well (depression) between the metal surfaces and surrounding plastic or similar type of ring of the electrode.
5. Cut four 1-inch micropore tape strips or remove the paper coverings over the adhesive surfaces of the electrode washers.
6. Use the adhesive strips to apply two skin electrodes to the tibialis anterior muscle and two skin electrodes to the gastrocnemius muscle in the pattern shown in Figure 26-4. Make certain that the electrodes are placed parallel to one another and that one of the recording electrodes is placed over the proximal portion of the muscle and the second electrode is positioned over the distal portion.
7. Check to see that the electrodes are adhering securely to the skin. Press on them.
8. Make the following settings on two modules to be used for the electromyographs.
   a. Time constant setting to 0.03
   b. Sensitivity setting to between 20 and 100
   c. Gain knob setting to X100
   d. Chart speed to 1.0 cm/sec
9. Plug the electrode into the appropriate modules and turn on the physiograph to record.
10. Establish a baseline recording by having your partner relax in the standing position.
11. Next, issue the following commands to your partner to execute in the order given. Mark the electromyogram with each command. Refer to Figure 26-2.
   a. Stand straight with both legs together
   b. Sit (on the chair provided)
   c. Stand again
   d. Lean forward from the standing position
   e. Bend the knees slightly from the standing position

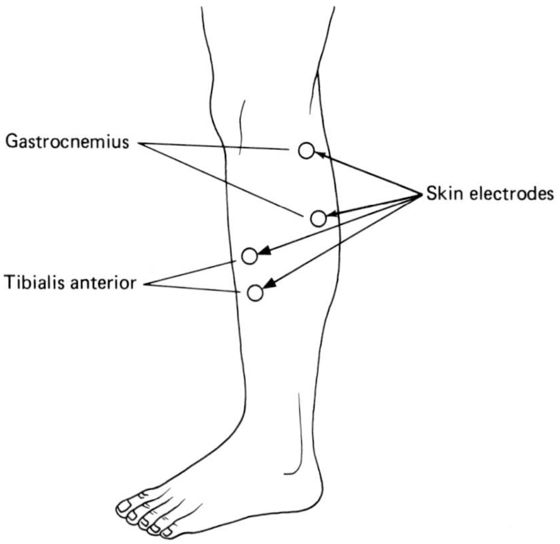

Figure 26-4 Placement of biopotential skin electrodes for electromyograms (recordings) from the tibialis anterior muscle of the shin and the gastrocnemius muscle of the calf. The two muscles may contract simultaneously, as when standing. At other times one muscle contracts while the other relaxes. This occurs when the toe end of the foot is raised.

Note the degree of contraction for each muscle by examining the electromyogram. Enter your findings and answer the questions in the results and observations section.

12. Change the chart speed to 0.5 cm/sec and repeat steps 10 and 11.
13. Change places with your laboratory partner and repeat steps 1–12.
14. Remove your electromyogram and attach it to the space provided in the results and observations section.

RESULTS AND OBSERVATIONS

MUSCLE TONE

1. Enter your findings on the biceps and triceps and those of your laboratory partner in Table 26-1.

Table 26-1

	Degree of Contraction* of Biceps and Triceps	
Task	Your Results	Partner's Results
Forearms supported with palms down		
Forearms supported with palms up		
Weight curl		
Standing with arms at side		
Standing with one hand grasping weight		

*Determine by increases in size or firmness.

2. Enter your findings on the gastrocnemius and those of your laboratory partner in Table 26-2.

Table 26-2

Task	Degree of Contraction* of Gastrocnemius	
	Your Results	Partner's Results
Body in prone position		
Sitting with feet touching the floor		
Sitting with legs dangling over an edge		
Standing with knees locked		
Standing in a relaxed manner		
Standing erect on tiptoes		

*Determine by increases in size or firmness.

3. Describe the effects observed with masseter muscle. ______________________

4. In which of the tasks performed was muscle tonus evident? ______________________

DISTINGUISHING BETWEEN ISOTONIC AND ISOMETRIC MUSCLE CONTRACTIONS

1. Enter your findings on the biceps and those of your laboratory partner in Table 26-3.

Table 26-3

	Measurements			
	Your Results		Partner's Results	
Task	Right Arm	Left Arm	Right Arm	Left Arm
Body arms extended				
Both arms flexed with palms up				
Right arm extended with weight				
Left arm extended with weight				
Right arm flexed with weight				
Left arm flexed with weight				

2. Enter your findings on the gastrocenemius and those of your laboratory partner in Table 26-4.

Table 26-4

	Measurements			
	Your Results		Partner's Results	
Task	Right Leg	Left Leg	Right Leg	Left Leg
Body in prone position with gastrocnemius relaxed				
Body in prone position with gastrocnemius contracted				
Standing erect				
Standing on toes with arms extended				
Standing on toes with arms at side				
Sitting with legs dangling				
Sitting with force applied to dangling legs				

3. Were any differences found between circumference measurements of the right and left arm? If so, explain ______________________________

____

____

4. Were any differences found between circumference measurements of the right and left gastrocnemius muscle? If so, explain. ______________________________

____

5. What variables influence the reading of measurements? ______________________________

____

DETERMINING MUSCLE FATIGUE

1. Enter your findings in Table 26-5.

Table 26-5

Student	Trial	Number of Contractions During 10-sec Intervals															
		10		20		30		40		50		60		120		180	
		R*	L	R	L	R	L	R	L	R	L	R	L	R	L	R	L
	1																
	2																
	3																
	1																
	2																
	3																
	1																
	2																
	3																
	1																
	2																
	3																

*R = right hand; L = left hand.

2. Indicate the time required for complete fatigue to occur.

Student 1: ____________

Student 2: ____________

Student 3: ____________

Student 4: ____________

3. Were all subjects able to squeeze the ball for 3 min? If not, why not? ____________

____

4. Were any differences noted in the results obtained with the right hand or left hand?

Explain. ____

____

____

____

DETERMINATION OF WORK PERFORMED

1. Enter your measurements on the ergograph and those of your laborabory partners in Table 26-6.

Table 26-6

Student	Weight Lifted (in g)			Distance Lifted (in cm)			Work Performed*		
	IF†	MF	FF	IF	MF	FF	IF	MF	FF

*Work performed = weight lifted × distance lifted.
†IF = index finger; MF = middle finger; FF = four fingers.

2. Which of the fingers performed the most work?

____

3. Did your results differ from those of your partners?

____

4. How can you account for any differences in the work performed? Explain your answer by considering the muscles involved. _______________

____

____

ELECTROMYOGRAMS FROM LEG MUSCLES

1. Place the appropriate portion of the electromyogram or sketch the results in the space provided here.

2. Enter your findings in Table 26-7. Indicate which muscle(s) contracted during the execution of the commands listed.

Table 26-7

Command	Muscles Involved: Tibialis Anterior (shin)	Muscles Involved: Gastrocnemius (calf)
Sit		
Stand		
Lean forward		
Knee bend		
Extend toe		

Questions for Review

EXERCISE **26**

HUMAN MUSCLE CONTRACTION

Name ______________________________

Laboratory Section ______________________ Date __________

1. Distinguish between isotonic and isometric contractions.

____

____

2. What is muscle tension? ______________________________

____

3. Lifting a weight exerts a force on a muscle. What is this force called? __________

____

4. Define the following:

   a. Flaccid ______________________________

   b. Muscle fatigue ______________________________

   ____

   c. Muscle tone ______________________________

   ____

   d. Electromyogram ______________________________

   ____

5. Distinguish between flexor and tensor muscles. ______________________________

____

SECTION VII

NERVOUS SYSTEM COORDINATION AND CONTROL

The nervous and endocrine systems regulate the activities and functions of the body. These two systems, however, operate at different rates. The nervous system acts rapidly, while the endocrine system operates more slowly. This section will consider the nervous system.

The nervous system is divided into the central nervous system, the peripheral nervous system, and the autonomic nervous system.

The central nervous system (CNS) includes the brain and the spinal cord. It contains the centers where connections are made between nerve fibers bringing information from receptor (sense) organs and nerve fibers carrying information to the effector organs (muscles and glands).

The peripheral nervous system consists of the nerves connecting the peripheral parts of the body with the central nervous system. Twelve pairs of nerves arise from the brain and are called the cranial nerves; 31 pairs of nerves arise from the spinal cord and are called the spinal nerves.

The autonomic nervous system is composed of nerves, nerve trunks with ganglia, and special ganglia that govern the operation of smooth muscle, cardiac muscle, and glands which for the most part are not under voluntary control.

The exercises in this section consider the general structure and organization of the nervous system as well as selected activities and functions associated with the components of the system.

GENERAL REFERENCES

Geschwind, N. "Specializations of the Human Brain," *Scientific American*, 241:180–199, 1979.

Iversen, L. L. "The Chemistry of the Brain," *Scientific American*, 241:134–149, 1979.

Netter, F. H. *The CIBA Collection of Medical Illustrations*. Vol. 1, *The Nervous System*. Summit, NJ: CIBA Pharmaceutical Products, 1953.

Shepherd, G. M. *Neurobiology*. New York: Oxford University Press, 1983.

Tortora, G. J., and N. P. Anagnostakos. *Principles of Anatomy and Physiology*, 5th ed. New York: Harper & Row, 1987.

Witherspoon, J. D. *Human Physiology*. New York: Harper & Row, 1984.

SPECIFIC REFERENCES

Dunant, Y., and M. Israël. "The Release of Acetylcholine," *Scientific American*, 252: 58–66, 1985.

Houston, J. P. *Fundamentals of Learning and Memory*, 2nd ed. New York: Academic Press, 1981.

Jones, D. G. "Ultrastructural Approaches to the Organization of Central Synapses," *American Scientist*, 69:200–210, 1981.

Stevens, C. F. "The Neuron," *Scientific American*, 241:54–65, 1979.

Williams, M. *Brain Damage, Behavior and the Mind*. New York: Wiley, 1979.

Winter, P. M., and J. N. Miller. "Anesthesiology," *Scientific American*, 252:124–131, 1985.

EXERCISE 27

ANATOMY OF THE NERVOUS SYSTEM

Materials

The following materials should be provided for class use.

1. Sheep brains (preserved)
2. Bone saw
3. Bone forceps
4. Model of human nervous system
5. Charts of human nervous system
6. Human brain (preserved)
7. Autonomic trunk and ganglia of a dissected cat specimen
8. Cat dissection specimen—Autonomic Nerve

SHEEP BRAIN

For detailed study it is best to use a sheep's brain (either fresh or preserved). It is large enough to see the parts clearly and is easily obtained. If a sheep's brain is not available and you wish to use the cat's brain, there are instructions for dissection at the end of this exercise.

MENINGES

Covering the brain are three membranes or **meninges.** Usually only the outermost and toughest membrane, the **dura mater,** is visible. The middle membrane is the **arachnoid,** and the innermost pigmented membrane is the **pia mater.**

DORSAL SURFACE

In a dorsal view (Figure 27-1) the brain reveals a large anterior portion, the **cerebrum** or **cerebral hemispheres.** Posterior to the cerebrum is the **cerebellum.** The **medulla oblongata** is located anterior to the spinal cord. If the cerebellum is gently drawn back, some of the midsection of the dorsal brain is revealed. One can see four distinct spherical bodies, the **corpora quadrigemina,** which make up the midbrain and which will be seen more clearly when the brain is cut in half. Much of the dorsal surface of the brain that is visible in lower animals is obscured by the drastic folding of both cerebral and cerebellar tissue in mammals. Refer to Figures 27-1, 27-2, and 27-3a and locate the following:

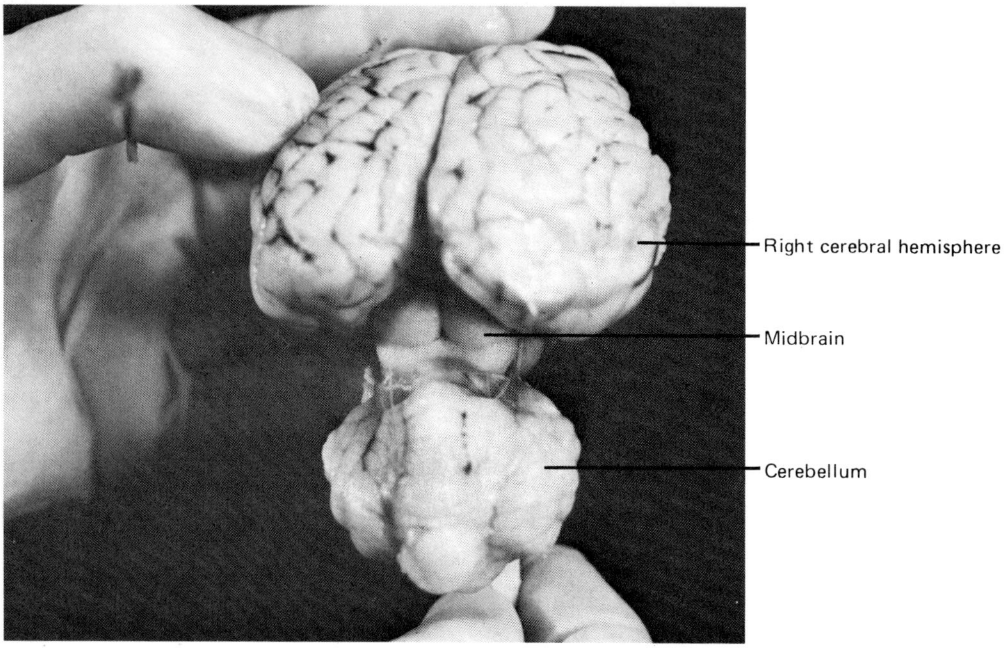

Figure 27-1 Sheep brain, dorsal view. Separate the cerebrum and cerebellum to see the midbrain.

1. Cerebral hemispheres
2. Infundibulum and pituitary gland
3. Mammillary bodies
4. Midbrain
5. Pons varolii
6. Medulla oblongata
7. Cerebellum

Turn the brain over and study the ventral surface.

VENTRAL SURFACE

Using Figures 27-2, 27-3a, and your own specimen, compare the sheep brain with the human brain (Figure 27-4).

Once again locate the **cerebral hemispheres, cerebellum,** and **medulla oblongata.** In addition, be able to identify the following:

1. **Olfactory nerve tracts,** which cross the ventral surface of the hemispheres ending in two enlarged areas, the *olfactory bulbs.*
2. **Optic nerves** cross over to form **optic tracts.** The crossover point is called the **optic chiasma.**
3. **Pituitary gland** is an endocrine gland that is located behind the optic chiasma. If it is present, it is about the size of a pea. If it has been removed it leaves a small tissue mass called the **infundibulum,** which connects the pituitary to the brain. If the infundibulum also has been removed, there is a hollow tubelike depression.
4. **Mammillary bodies** are small U-shaped regions in the midline posterior to the pituitary gland.
5. **Cerebral peduncles** are the masses surrounding the mammillary bodies and extending back about one inch. This area gives rise to the third pair of cranial nerves.

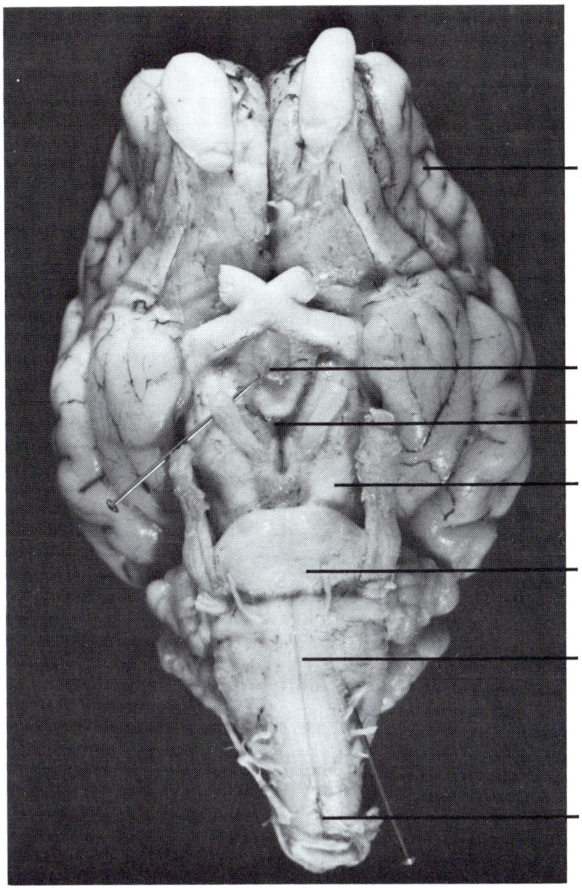

Figure 27-2 Sheep brain, ventral view (with pituitary removed). The pins underline right cranial nerve III and left cranial nerves IX, X, and XI.

6. **Pons varolii** is a bridgelike area in the cerebellar region.
7. **Medulla oblongata** has already been identified on the dorsal surface and can also be seen clearly on the ventral surface. Posterior to it is the spinal cord.
8. **Trapezium** is the right or left side of the medulla of the sheep brain.

BLOOD VESSELS

Also note the large vascular network on the ventral surface. Blood flows toward the brain in the **basilar artery,** which divides in the region of the cerebral peduncles to form circular vessels around the pituitary and the optic chiasma. This circuit, which has other contributing sources, is called the **circle of Willis.** It will be discussed in more detail in the exercise on blood vessels.

CRANIAL NERVES

The cranial nerves exit the brain on the ventral surface. They are given Roman numerals and can be identified with the help of Figures 27-2 and 27-3a. Compare the cranial nerves with their counterparts in the human brain (Figure 27-4). For specific functions of all cranial nerves, consult your textbook.

I. **Olfactory nerves** contain nerve fibers from the nasal membranes and cannot be seen in this preparation of the sheep brain. The nerve fibers synapse in the olfactory bulb. The

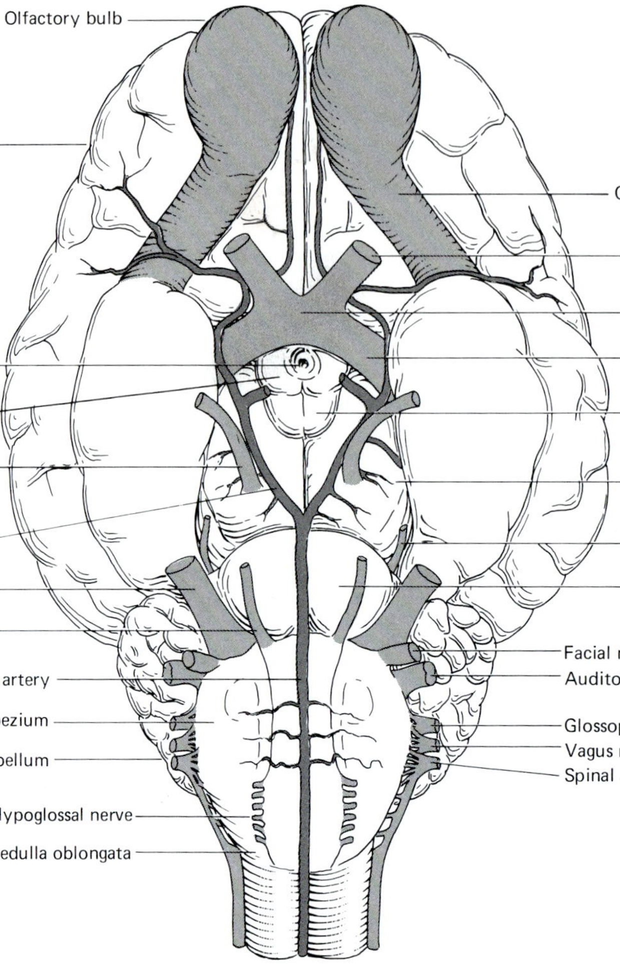

Figure 27-3a Sheep brain, ventral view. (a) Diagram shows pituitary removed and cranial nerves identified by name. (b) Photograph shows pituitary in place and cranial nerves identified by number.

second set of neurons continues in the *olfactory tract* and terminates in the olfactory areas of the cerebral hemispheres.

II. **Optic nerves** are those that arise in the retina of the eye and cross over anterior to the pituitary gland to form the *optic chiasma*. They enter the brain as the *optic tracts*. Some fibers within the nerve cross over at the chiasma, others do not. The optic nerves are the only cranial nerves that cross over.

III. **Oculomotor nerves** are medium-sized nerves posterior to the mammillary bodies. They supply four of the six pairs of extrinsic eyeball muscles with nerve fibers.

IV. **Trochlear nerves** are small and arise from the midbrain. They project from the cleft between the posterior end of the cerebrum and the pons varolii. They supply one pair of the extrinsic eyeball muscles with nerve fibers.

V. **Trigeminal nerves** are large and arise laterally from the pons. Together with cranial nerve VII they supply much of the head and face.

VI. **Abducens nerves** are small and arise medial to the trigeminal nerves from the anterior end of the medulla oblongata at the border of the trapezium. They supply one pair of extrinsic eyeball muscles.

VII. **Facial nerves** are medium sized and arise directly behind the trigeminal nerves from the trapezium area. They supply the head and face.

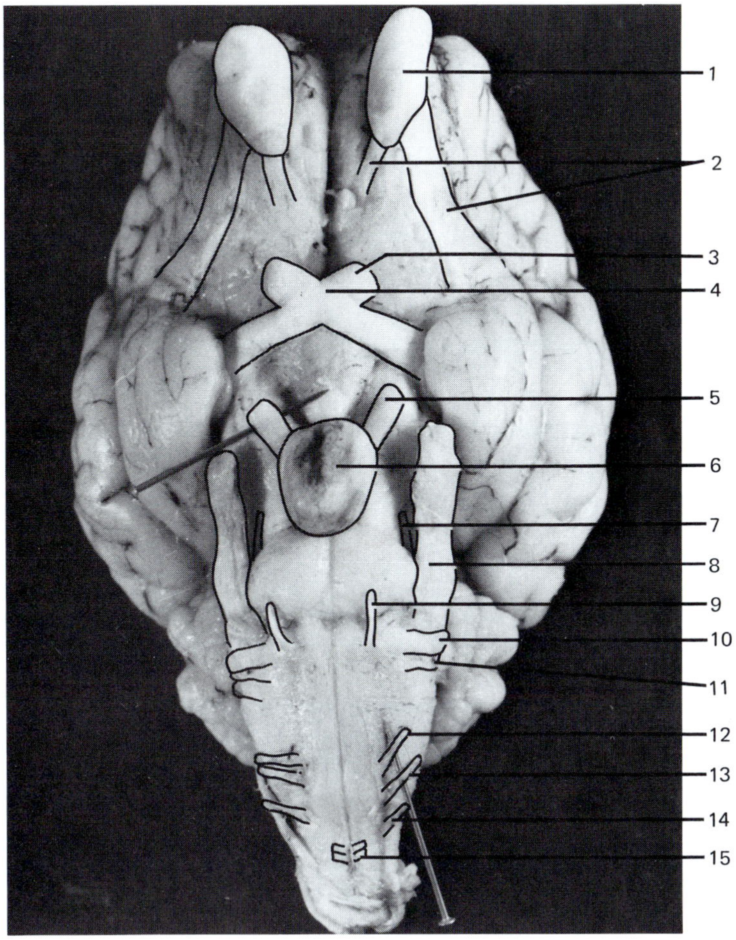

(b)

Figure 27-3b

1. Olfactory bulb
2. Olfactory tracts
3. Optic nerve (II)
4. Optic chiasma
5. Oculomotor nerve (III)
6. Pituitary gland
7. Trochlear nerve (IV)
8. Trigeminal nerve (V)
9. Abducens nerve (VI)
10. Facial nerve (VII)
11. Acoustic nerve (VIII)
12. Glossopharyngeal nerve (IX)
13. Vagus nerve (X)
14. Spinal accessory nerve (XI)
15. Hypoglossal nerve (XII)

VIII. **Acoustic (auditory, statoacoustic, or vestibulocochlear) nerves** are large, short nerves that arise from the lateral trapezium area posterior to cranial nerve VII. They are associated with the cochlea (hearing) and the semicircular canals (equilibrium).

IX. **Glossopharyngeal nerves** are small nerves that arise laterally from the medulla oblongata. They are associated primarily with the tongue and pharynx regions.

X. **Vagus nerves** are small and arise from the lateral medulla posterior to the glossopharyngeal nerves. They supply the pharynx and larynx and extend posteriorly into the thoracic and abdominal cavities to supply visceral organs with parasympathetic fibers.

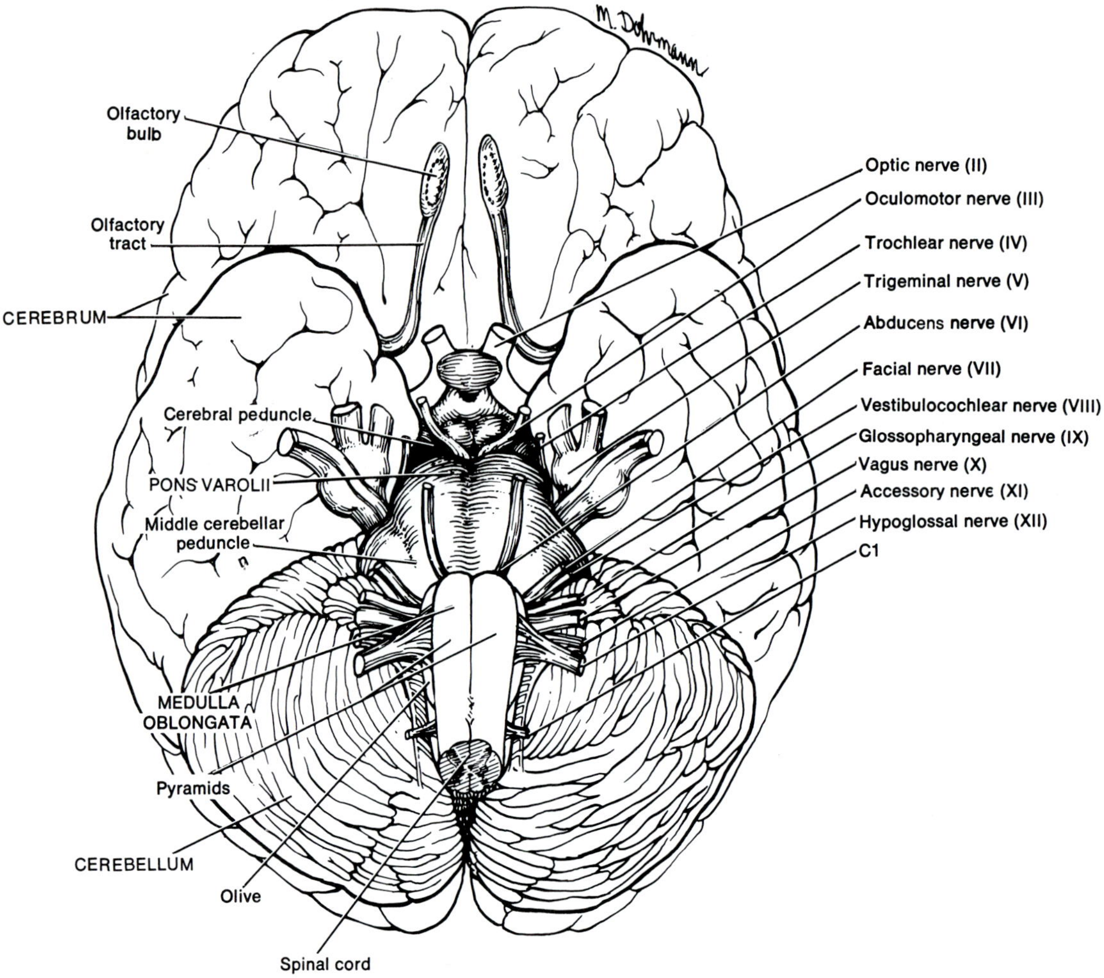

Figure 27-4 Human brain, inferior view.

XI. **Accessory nerves** have several roots of origin in the most posterior part of the lateral medulla. They innervate certain neck and shoulder muscles.

XII. **Hypoglossal nerves** also arise from several roots in the medulla medial to cranial nerves XI (in the human, slightly anterior to cranial nerves XI). They are associated with the tongue muscles.

SHEEP BRAIN IN MIDSAGITTAL SECTION

BRAIN VENTRICLES OR CAVITIES

Cut the brain in half using a sharp, long-bladed knife. Examine the cut surface and identify as many parts as possible referring to Figures 27-5 and 27-6. Compare the midsagittal section of the sheep brain with the human brain (Figures 27-7 and 27-8).

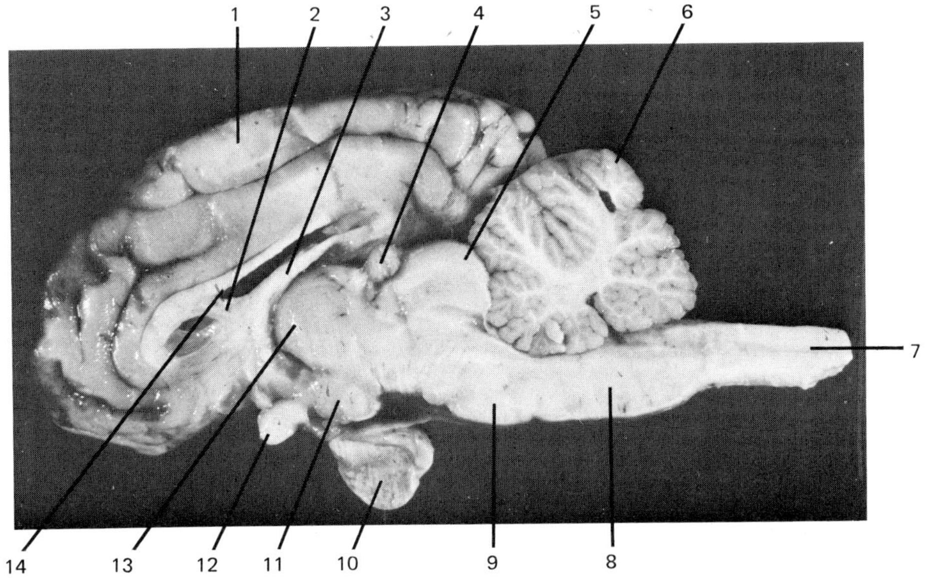

Figure 27-5 Sheep brain, midsagittal section.
1. Cerebrum
2. Septum pellucidum
3. Fornix
4. Pineal gland
5. Midbrain
6. Cerebellum
7. Spinal cord
8. Medulla oblongata
9. Pons
10. Pituitary
11. Mammillary body
12. Optic chiasma
13. Massa intermedia
14. Corpus callosum

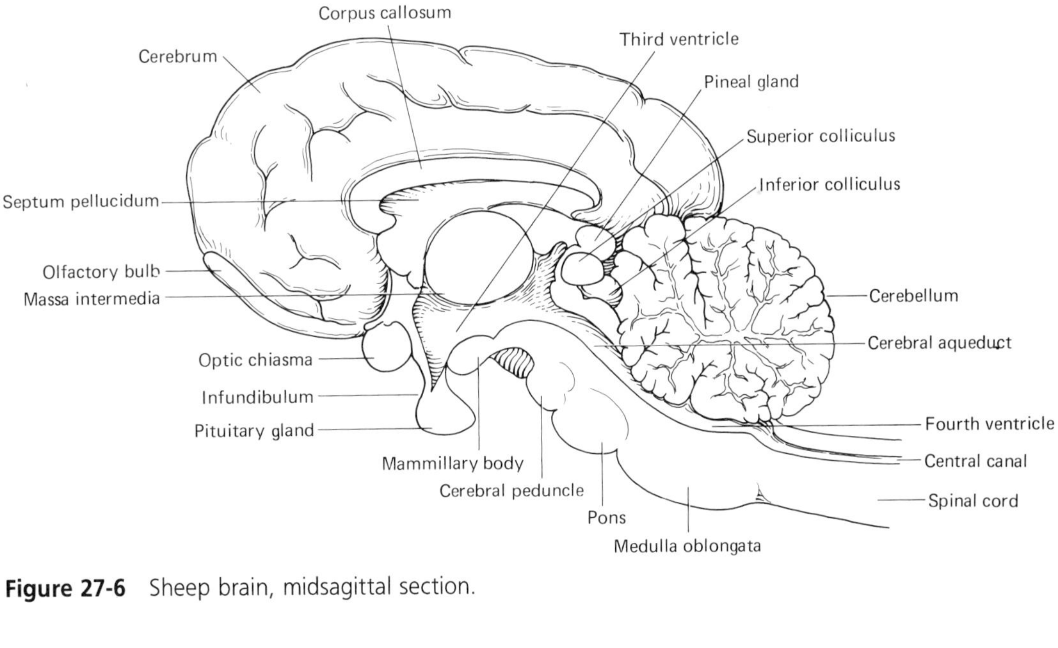

Figure 27-6 Sheep brain, midsagittal section.

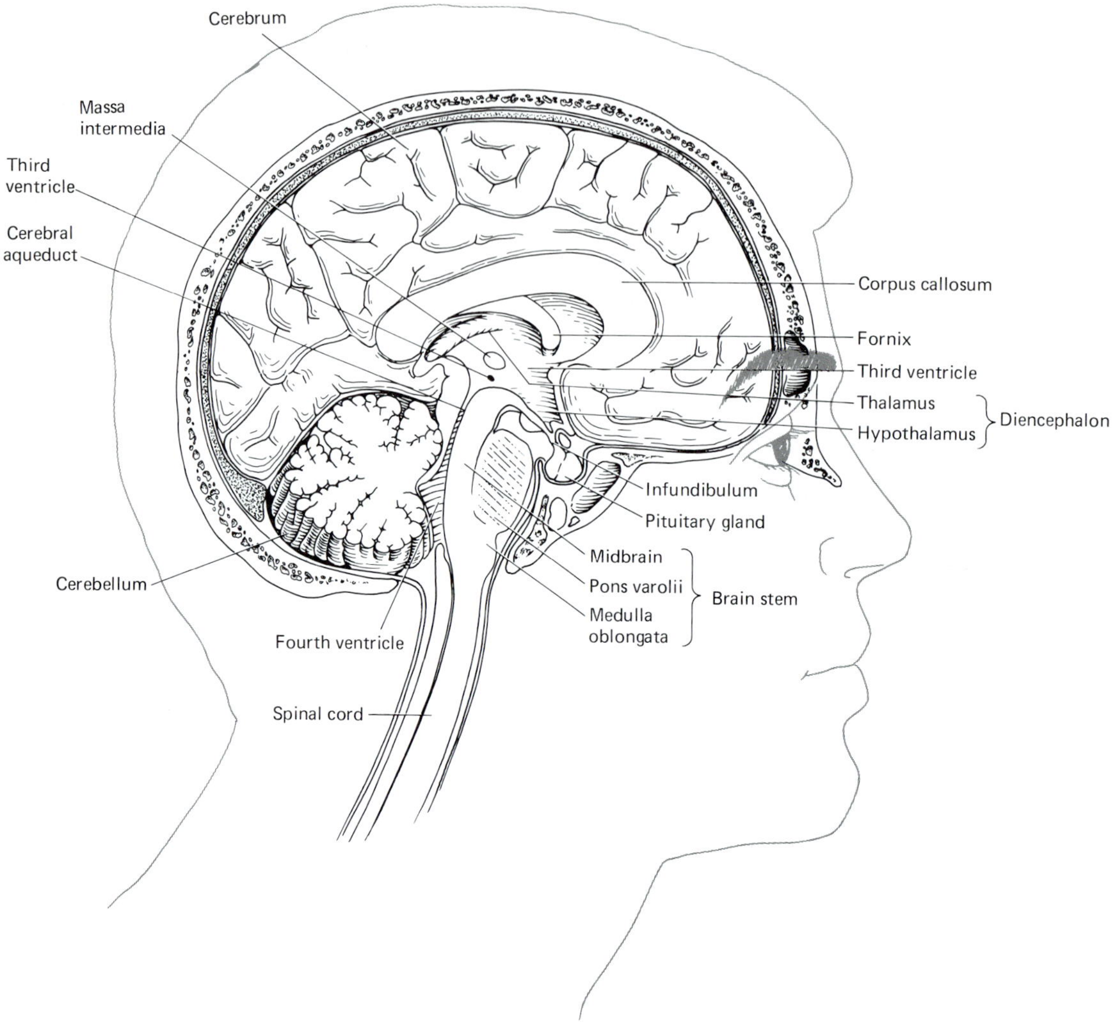

Figure 27-7 Human brain, sagittal section.

Within the brain are cavities filled with cerebrospinal fluid. They are called *brain ventricles* and are assigned Roman numerals:

I and II (lateral ventricles)—located in the cerebral hemispheres
III (third ventricle)—located in the diencephalon region
IV (fourth ventricle)—located in the medulla oblongata

The lateral ventricles connect with the third ventricle by channels called the **interventricular foramina** or **foramina of Munro.** Ventricle III is connected to ventricle IV by the **cerebral aqueduct** or **aqueduct of Sylvius.** See Figure 27-8, which shows the location of each ventricle in the human brain. The locations and shapes of the ventricles in the sheep and cat brains are similar.

BRAIN PARTS

CEREBRUM

Note the extent of folding of the **cerebral hemispheres** back over the brain stem. The large mass of cerebral tissue growing back over other parts of the brain is an evolutionary phenomenon. It has reached its greatest extent in the human brain. Near the floor of the

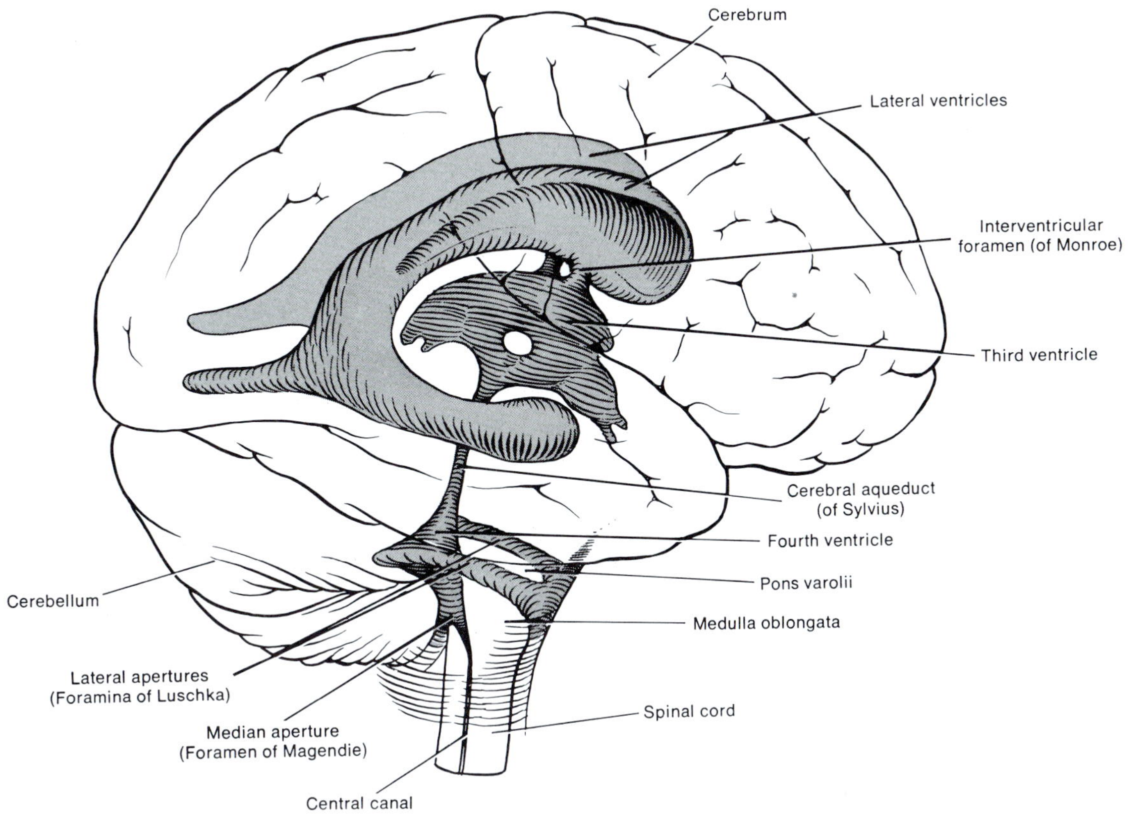

Figure 27-8 Human brain, ventricles.

cerebrum is a tough white area called the **corpus callosum** whose fibers connect the right and left hemispheres.

DIENCEPHALON

A less distinct region than the cerebrum. It consists of several structures:

1. *Pineal gland*—a small, dorsal projection of endocrine nature.
2. *Infundibulum*—a ventral downgrowth attached to the hypophysis of the pituitary gland.
3. *Anterior choroid plexus*—a network of blood vessels roofing the third brain ventricle. Together with other choroid plexuses, it contributes to the formation of cerebrospinal fluid.
4. *Fornix*—part of the olfactory pathways of the brain.
5. *Massa intermedia or intermediate mass*—gray matter that connects the right and left thalamus regions. It is centrally located.
6. *Thalamus*—represents the lateral walls of the diencephalon, which are difficult to see in a midsagittal section.
7. *Hypothalamus*—the floor of the diencephalon, which gives rise to the infundibulum.

CEREBELLUM

The second largest part of the brain. It is highly convoluted, and its gray **cortex** and inner **white matter** are distinct. The branching patterns of the white matter within the cerebellum are called the **arbor vitae** because they resemble the branches of a tree. The middle region of the cerebellum is called the **vermis.**

Pons varolii is a bridgelike region of the same embryonic area which consists primarily of fiber tracts.

Trapezium is a fiber tract area, which is considered a part of the pons. In the cat and sheep, it is a part of the medulla oblongata (Figure 27-4).

MEDULLA OBLONGATA

This is the most posterior part of the brain. It merges with the spinal cord. Within it is the fourth ventricle. Roofing ventricle IV is another blood vessel plexus, the **posterior choroid plexus.**

Note that the brain ventricles connect with the central canal of the spinal cord. If the first and second ventricles are to be located, make a cross section of the cerebral hemispheres. Refer to Figure 27-8.

THE HUMAN BRAIN

SURFACE ANATOMY OF THE CEREBRUM

Study the external anatomy of the cerebral hemispheres using Figure 27-9. If a human brain is not available, use the sheep brain. Each hemisphere is divided into four basic areas called lobes. Lobes are separated by:

Sulci (sing., *sulcus*)—shallow grooves
Fissures—deep grooves

The lobes are named as follows:

1. **Frontal lobes**—These are parts that constitute the areas beneath the frontal bone. They are located in front of the **central sulcus.**
2. **Parietal lobes**—paired regions located behind the **central sulcus** and the frontal lobes. Pay particular attention to the folds that are created by the grooves in this area. The grooves are called *gyri* (sing., *gyrus*). Two prominent gyri are:
   Precentral gyrus—located anterior to the central sulcus, it represents the location of an important motor area of the cortex.
   Postcentral gyrus—located posterior to the central sulcus, it represents the location of important sensory areas of the cortex.
3. **Temporal lobes**—located laterally and separated from the frontal lobes by the **lateral cerebral sulcus.**
4. **Occipital lobes**—occupy the area beneath the occipital bone. They are separated from the parietal lobe by the **parietooccipital sulcus.**

Each hemisphere is separated from the other by the **longitudinal fissure.** The cerebrum is separated from the cerebellum by the **transverse fissure.**

HUMAN SPINAL CORD AND SPINAL NERVES

The human spinal cord and its 31 pairs of spinal nerves should be studied from models and charts. The cord begins posterior to the medulla oblongata and continues down the middle of the back for about 16–18 inches. There are two enlargements, one in the neck region (**cervical enlargement**) and the other in the midback region (**lumbar enlargement**). During embryonic development the spinal cord does not keep pace with body growth in length and terminates at the level of the second lumbar vertebra. The last seven or eight pairs of nerves run posteriad beyond the spinal cord until they reach their designated exits, creating what looks like a bundle of hairs, appropriately named the **cauda equina**, or horse's tail. The most

(a)

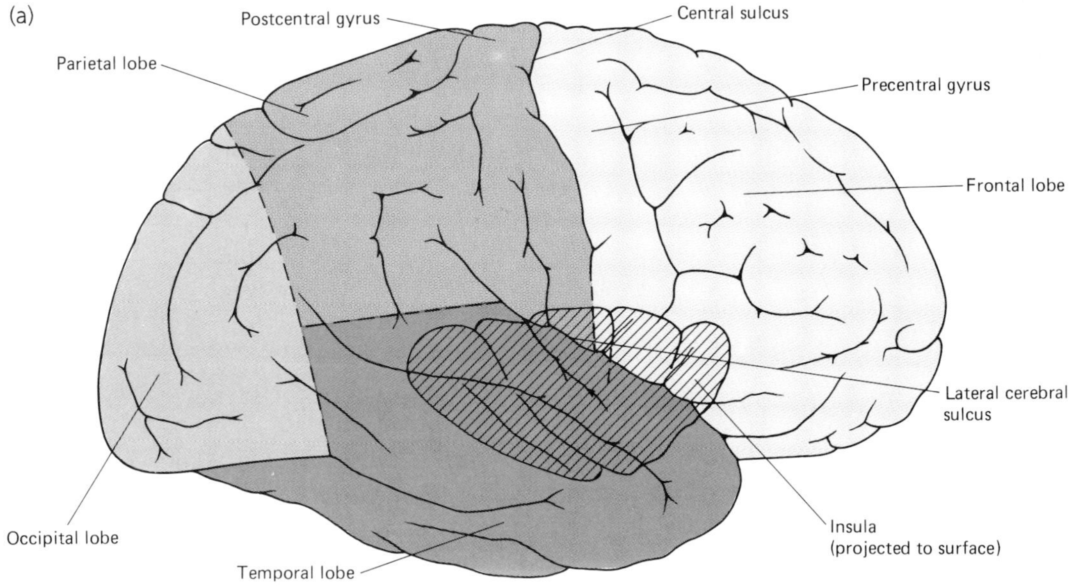

(b)

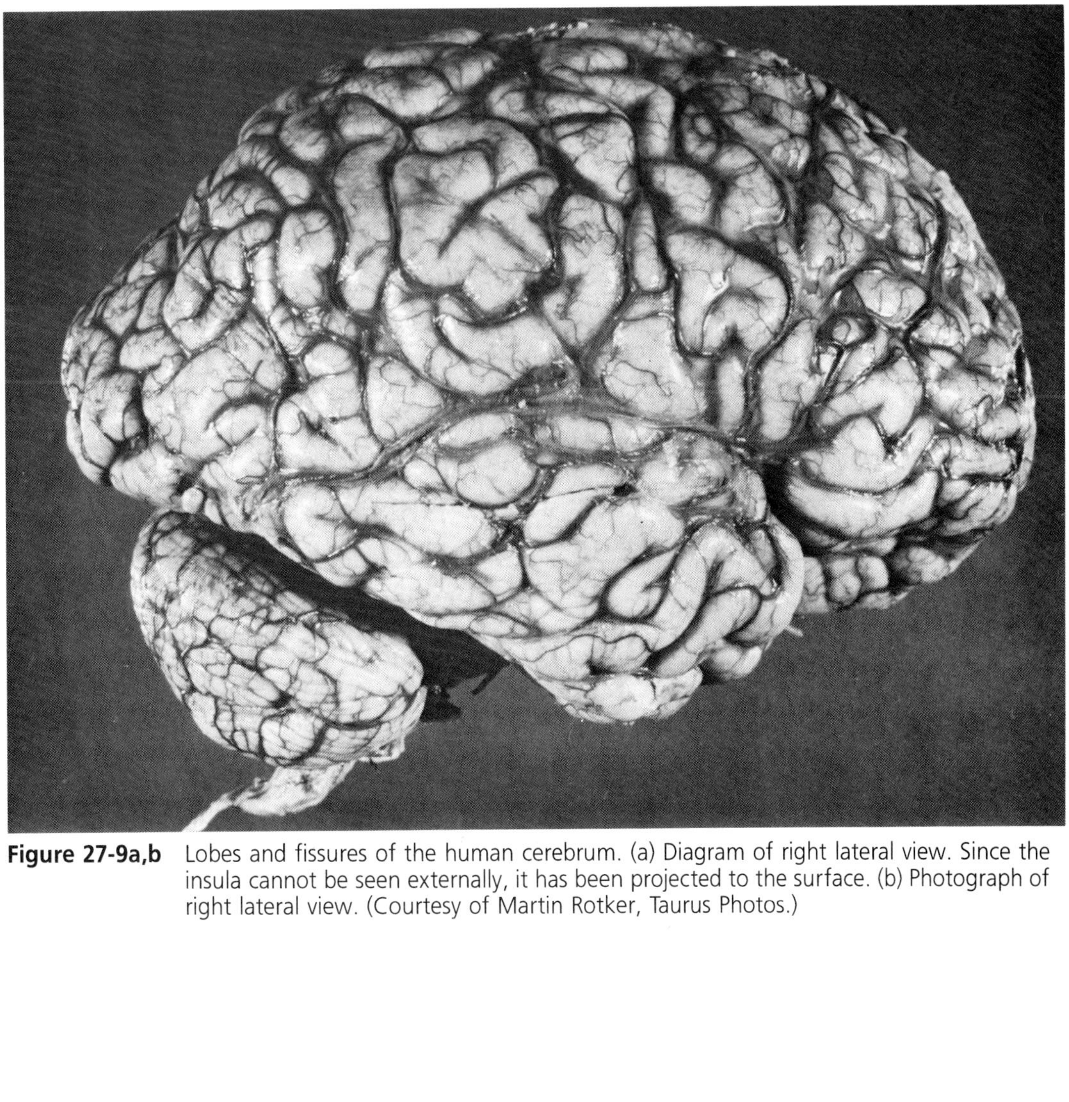

Figure 27-9a,b Lobes and fissures of the human cerebrum. (a) Diagram of right lateral view. Since the insula cannot be seen externally, it has been projected to the surface. (b) Photograph of right lateral view. (Courtesy of Martin Rotker, Taurus Photos.)

(c)

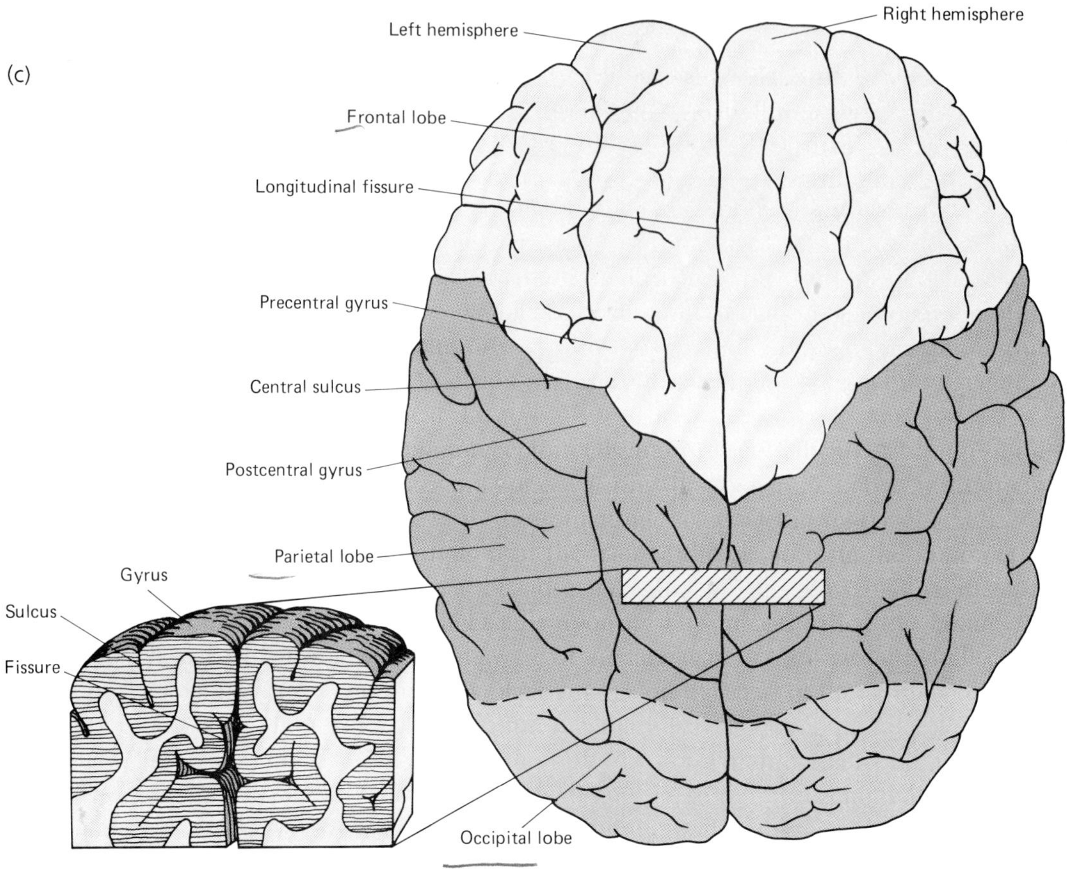

(d)

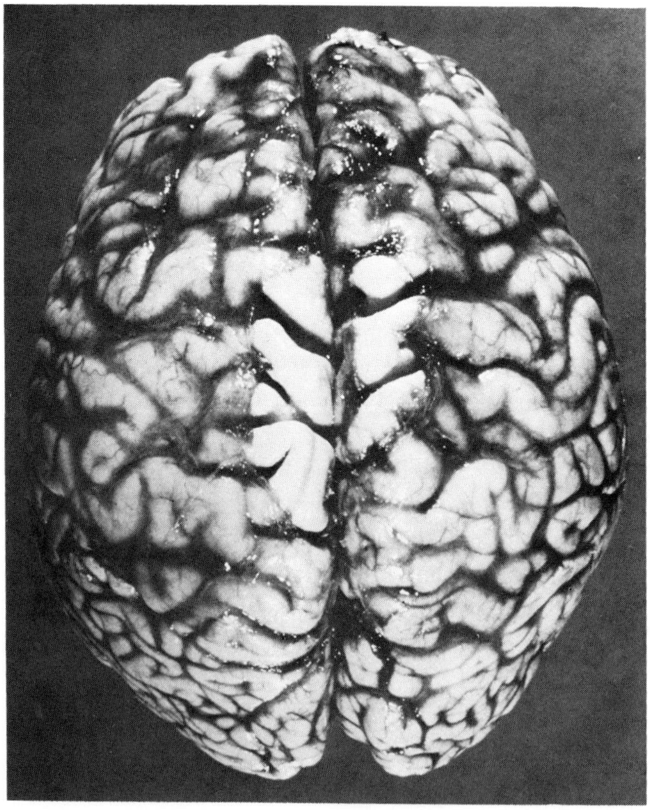

Figure 27-9c,d Lobes and fissures of the cerebrum (c) Diagram of superior view. The insert to the left indicates the relative differences among a gyrus, sulcus, and fissure. (d) Photograph of superior view. (Courtesy of Martin Rotker, Taurus Photos.)

inferior portion of the cord, in the region of the second lumbar vertebra, is called the **conus medullaris.** Extending from the conus medullaris to the last sacral vertebra is a strand of connective tissue named the **filum terminale.**

The names and numbers of each spinal nerve correspond to the names and numbers of the vertebra at their point of exit (Figure 27-10). They are classified as follows:

1. **Cervical nerves** (8 pairs)—The first cervical nerves (C_1) exit between the skull and the first cervical vertebra. The eighth cervical nerves (C_8) exit from the intervertebral foramina between the seventh cervical vertebra and the first thoracic vertebra.
2. **Thoracic nerves** (12 pairs)—The first thoracic nerves (T_1) exit from the intervertebral foramina between the first and second thoracic vertebrae. The twelfth thoracic (T_{12}) nerves exit between the twelfth thoracic vertebra and the first lumbar vertebra.
3. **Lumbar nerves** (5 pairs)—The first lumbar nerves (L_1) exit from the intervertebral foramina between the first and second lumbar vertebrae. The fifth lumbar nerves (L_5) exit between the fifth lumbar vertebra and the sacrum.
4. **Sacral nerves** (5 pairs)—The first sacral nerves (S_1) exit between fused first and second sacral vertebrae. The fifth sacral nerves (S_5) exit inferior to the sacrum.
5. **Coccygeal nerves** (1 pair)—The coccygeal nerves (Co_1) exit at the coccyx.

NERVE PLEXUSES

A nerve *plexus* is a place where two or more nerves join and exchange nerve fibers. Figure 27-10 is an example of the type of network formed by adjacent spinal nerves as they leave the spinal column. Such networks are formed on both sides of the body. Figure 27-11 shows the lumbar plexus in detail.

There are several important nerve plexuses. Using your text and charts, identify the following.

1. *Cervical plexus*—involves nerves C_1–C_5
2. *Brachial plexus*—involves nerves C_4–C_8 and T_1 and T_2
3. *Lumbar plexus*—involves nerves L_1–L_4
4. *Sacral plexus*—involves nerves L_4, L_5, and S_1–S_4

DISSECTION OF THE AUTONOMIC NERVOUS SYSTEM OF THE CAT

Using the diagram of the autonomic nervous sytem (Figure 27-12), locate all its parts on a demonstration cat. You can expose all these nerves and ganglia on your own cat more easily at the end of the course when it will be possible to remove most abdominal and thoracic organs.

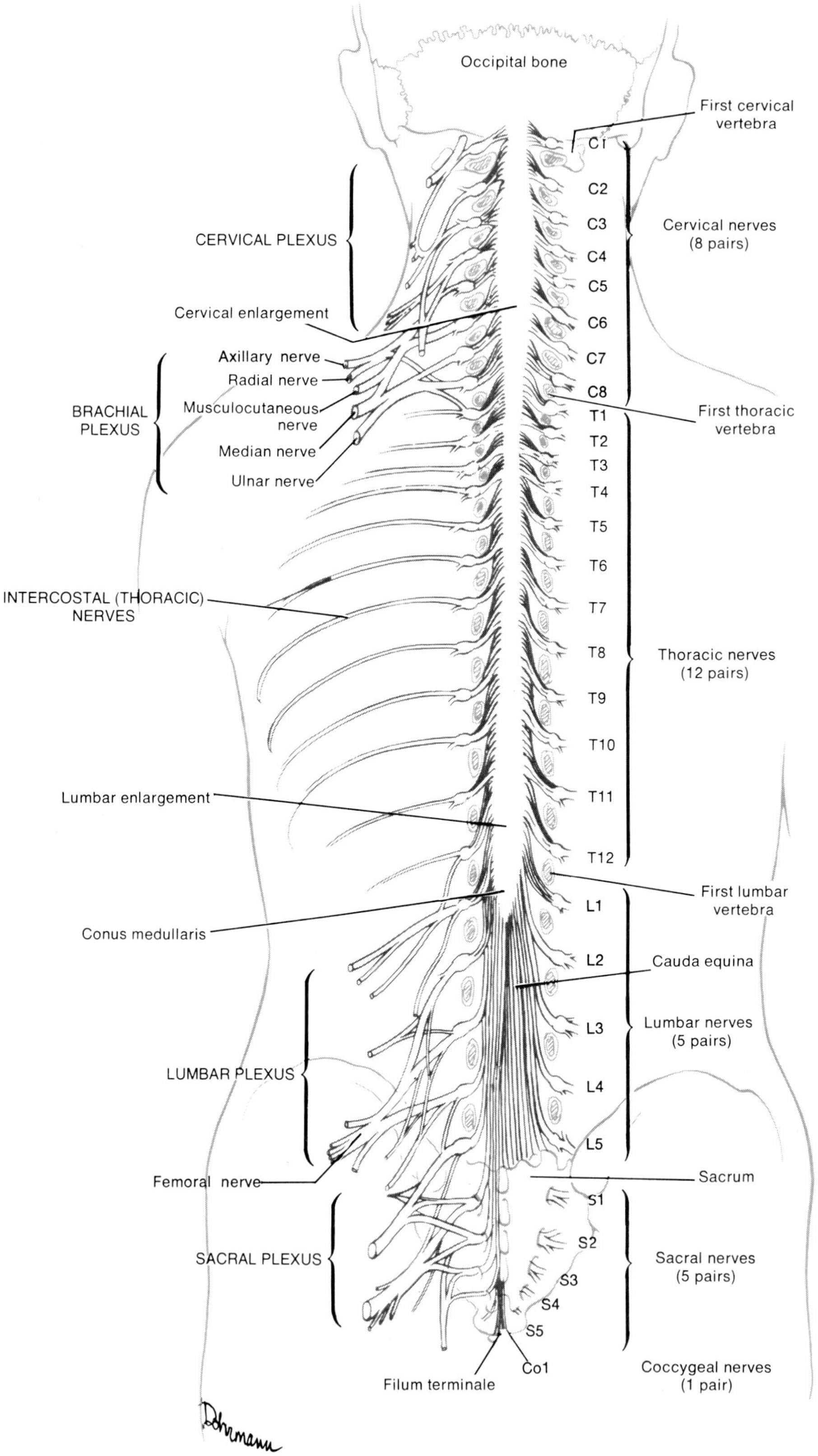

Figure 27-10 Human spinal cord and spinal nerves.

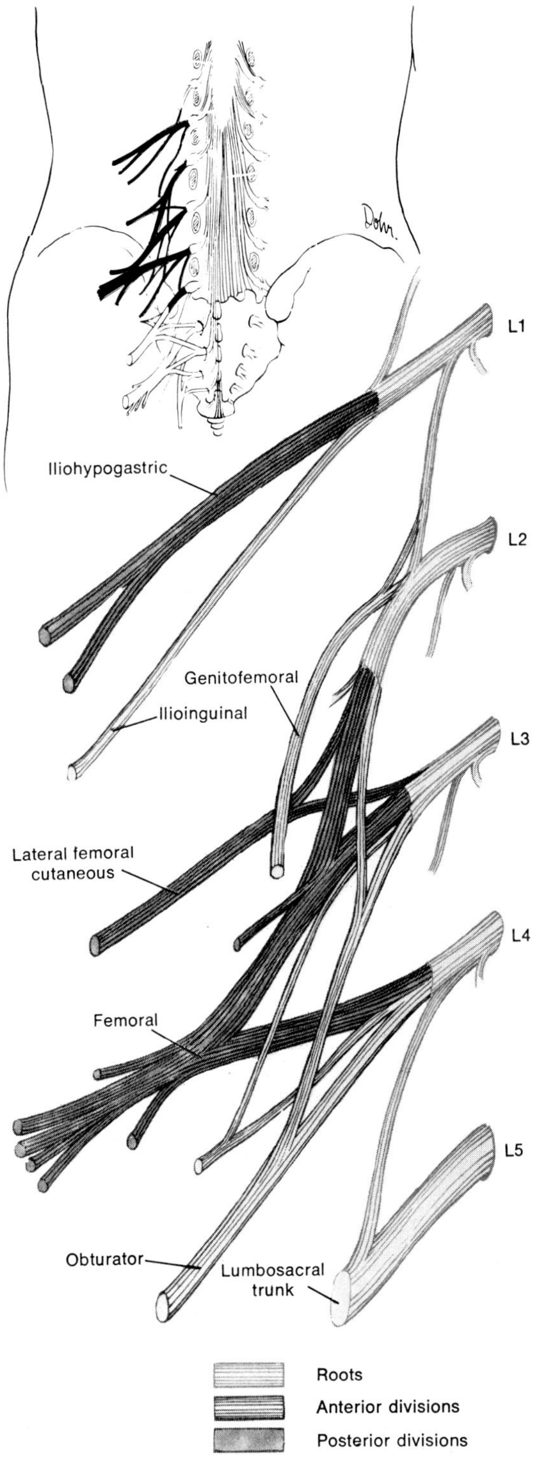

Figure 27-11 Human lumbar plexus.

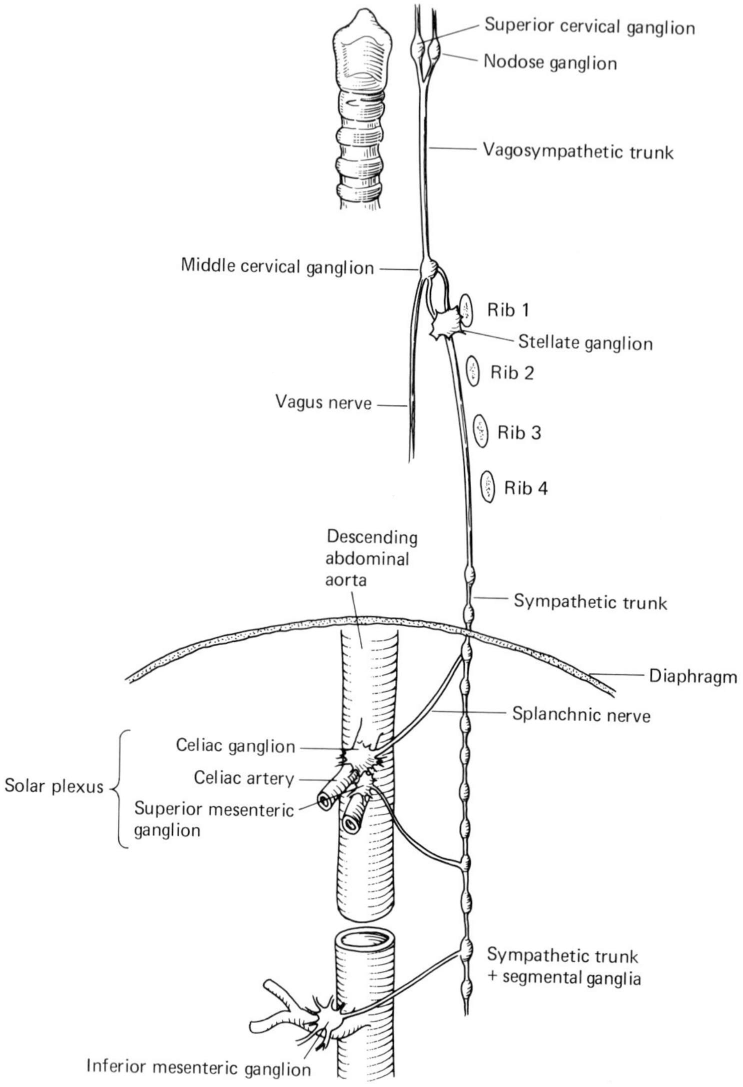

Figure 27-12 Autonomic nervous system of the cat.

REMOVAL OF THE CAT BRAIN (OPTIONAL)

This dissection may be done even though the sheep's brain has already been studied. Use extreme care in this procedure. The bone is hard and the nervous tissue is soft. When removing bone tissue, be careful not to damage the soft brain beneath it. Also guard against cutting your fingers with sharp dissecting tools. Do not use the razor-blade type scalpel because it is not durable enough for this exercise.

Begin by removing all the skin and muscle from the dorsal surface of the head (Figure 27-13a). The area to be skinned extends from the tip of the nose to the second cervical vertebra, between the ears.

When the skull is cleanly exposed, begin the removal of the bone. The ideal tool for this purpose is an autopsy saw. If this is not available, take a hammer and crack the bone as much as possible over the entire exposed surface, being careful not to injure the soft brain below. The thickest bone is in the posterior part of the skull. Remove this first with large bone forceps. Expose about $\frac{1}{2}$ inch of the spinal cord. Remove the top of the skull until the entire dorsal surface of the brain is exposed (Figure 27-13b).

Cut through the spinal cord and lift the brain out slowly by cutting the cranial nerves that hold it to the base of the skull. Leave as much of the nerves as possible so they can be identified.

Study the dorsal and ventral surfaces and the midsagittal section, following instructions for the sheep brain and using Figures 27-14 and 27-15 to identify parts of the cat brain.

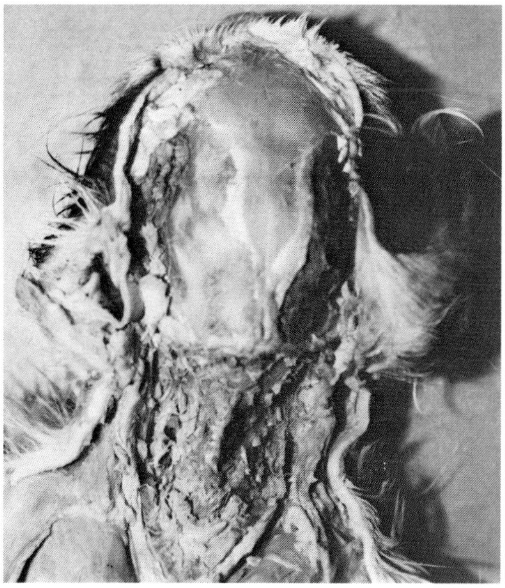

Figure 27-13a Cat skull, skin and muscle removed.

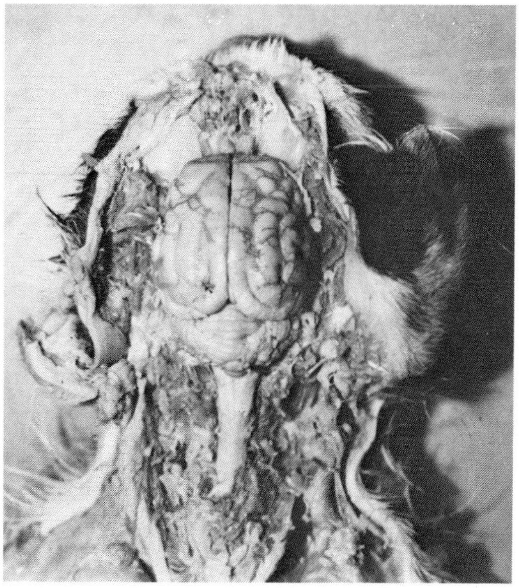

Figure 27-13b Cat skull, bone removed.

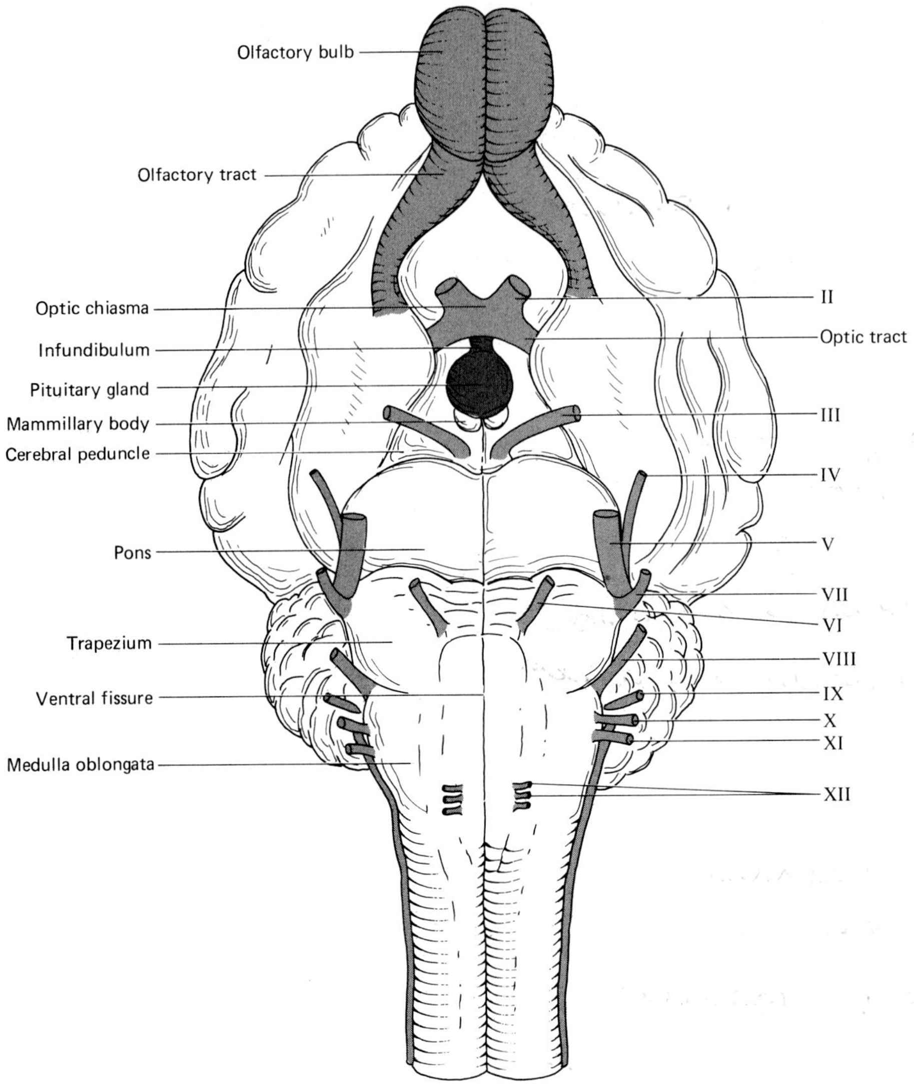

Figure 27-14 Cat brain, ventral view.

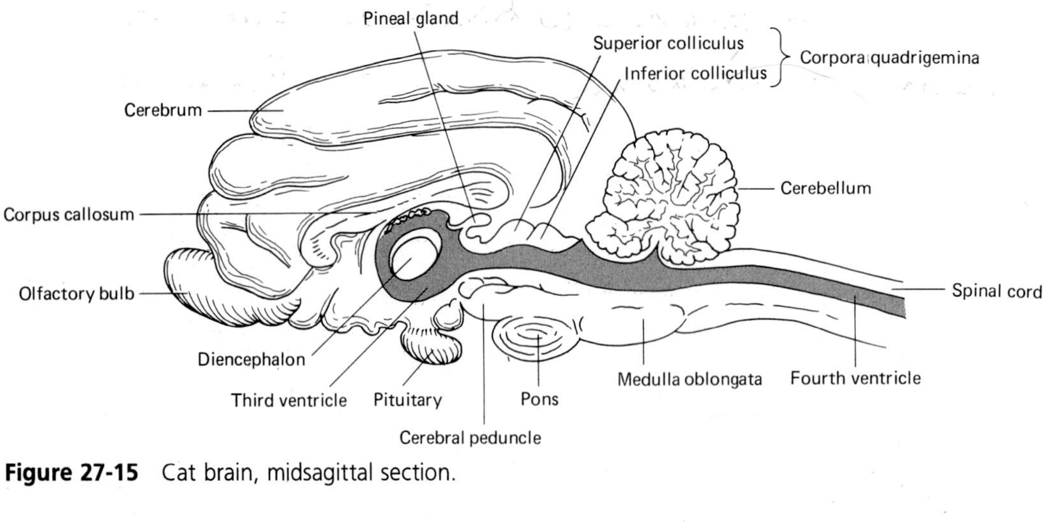

Figure 27-15 Cat brain, midsagittal section.

Questions for Review

EXERCISE **27**

ANATOMY OF THE NERVOUS SYSTEM

Name Kathy Martin

Laboratory Section Mon AM Date 4-20-90

1. Fill in the names of the cranial nerves.

I. Olfactory nerves

II. optic nerves

III. oculomotor nerves

IV. trochlear nerves

V. trigeminal nerves

VI. abducens nerves

VII. facial nerves

VIII. acoustic nerves

IX. glossopharyngeal nerves

X. vagus nerves

XI. accessory nerves

XII. hypoglossal nerves

2. Define the following.

a. Nerve plexus a place where two or more nerves join + exchange nerve fibers.

b. Ganglion a nerve that govern the operation of smoth muscle

c. Nerve a bundle of nerve fibers outside the central nervous system

d. Neuron The structional + functional unit of the nervous, composed of cell body, dendrites, and an axon. also called a nerve cell

e. Nerve fiber ______________________

f. Brain ventricle cavities filled with cerebrospinal fluid.

g. Meninx ______________________

3. Name the lobes of the cerebral hemispheres.

a. Frontal lobes c. temporal lobes

b. Parietal lobes d. Occipital lobes

4. What is a nerve plexus? a place where two or more of nerves join & exchange nerve fibers.

5. Name four nerve plexuses.

a. cervical plexus c. Lumbar plexus

b. brachial plexus d. sacral plexus

6. What is a choroid plexus? Roofing ventricle 4 a blood vessel plexus.

7. What is the difference between a gyrus and a sulcus?

gyrus is folded grooves & sulcus is shallow grooves

8. What is the difference between a sulcus and a fissure?

sulcus is shallow grooves + fissure is deep grooves

10

b. Which of the reflexes observed involved only the spinal nerve pathway?

____

____

c. Which of the reflexes observed involved both cranial and spinal nerve pathways?

____

____

Questions for Review

EXERCISE **28**

CRANIAL AND SPINAL NERVOUS RESPONSES OF THE FROG

Name ____

Laboratory Section ______________________________ Date ____________

1. Briefly describe a simple reflex arc such as the one associated with the knee-jerk reflex.

____

____

____

2. Define a reflex arc. ______________________________

____

3. Define or explain the following.

   a. Neuron ______________________________

   b. Axon ______________________________

   c. Dendrite ______________________________

   d. Synapse ______________________________

EXERCISE 29

DEMONSTRATION OF SELECTED CRANIAL NERVE AND CEREBELLAR FUNCTIONS IN THE HUMAN

The cerebellum lies at the back of the skull behind the brain stem and under the cerebral hemispheres. It is much smaller than the cerebrum but shares certain morphological features with it. As in the cerebrum, the highest functions in the cerebellum are confined to the thin layer of gray matter that makes up the cortex, and this layer is elaborately folded and wrinkled to increase its area. The folds are in fact much deeper and more closely spaced than those of the cerebral cortex.

The cerebellum plays an important role in balance, posture, and coordination. It receives a steady flow of impulses from receptors embedded in joints and muscles and from the eyes and portions of the ear. It registers the body's position, as well as its rate of movement and the coordination of muscular activities. This exercise includes a number of tests to demonstrate these functions of the cerebellum. Tests of cranial nerve function also are considered in this exercise.

The brain gives rise to 12 pairs of cranial nerves that supply motor and sensory fibers to the head, neck, and some visceral organs. These nerves form part of the peripheral nervous system that connects the brain with receptors, muscles, and glands. Table 29-1 lists the cranial nerves and their composition and function. This exercise includes tests to show the functions of selected cranial nerves.

Table 29-1 A General Summary of the Cranial Nerves

Nerve Designation	Composition	Associated Functions
I. Olfactory	Sensory	Smell
II. Optic	Sensory	Sight
III. Oculomotor	Motor and sensory	Eye movement and focusing
IV. Trochlear	Motor and sensory	Nerve supply to eye muscle
V. Trigeminal	Motor and sensory	Chewing, sensation from the head region between eyes and teeth
VI. Abducens	Motor and sensory	Eye movement
VII. Facial	Motor and sensory	Facial expression and taste
VIII. Statoacoustic	Sensory	Posture and hearing
IX. Glossopharyngeal	Motor and sensory	Swallowing, taste, and general sensation
X. Vagus	Motor and sensory	Visceral muscle movement and sensation
XI. Accessory	Motor and sensory	Swallowing and head movement
XII. Hypoglossal	Motor and sensory	Speech and swallowing

Materials

The following materials should be provided for class use.

1. Containers of
   a. Garlic
   b. Ground coffee
2. Tuning forks
3. Sterile cotton swabs and tongue depressors
4. Timers or stopwatches
5. Flashlights or penlights

Procedure 1: Cranial Nerve Function Tests

This procedure should be performed by students in pairs.

1. Take turns in performing the tasks listed for each cranial nerve. Describe your responses and those of your laboratory partner in the results and observations section.
2. Read each of the tasks first before performing them.
   a. *Olfactory nerve.* Close your eyes while your partner passes containers of garlic and ground coffee by your nose. Can you distinguish the aromas?
   b. *Optic nerve.* Determine how many words on one page of a textbook or other publication you can read in 1 minute.
   c. *Oculomotor nerve.* Keep your head still while your partner slowly moves a pencil up and down in front of you. Can you follow the movement with your eyes?
   d. *Trochlear and abducens nerves.* Look straight ahead while your partner holds a pencil about 1 meter from the side of your head and moves it forward and backward. Indicate at what point you cannot see the pencil. Have your partner determine if the pupil opens or closes more during this task.
   e. *Trigeminal nerve.* Close your mouth and clench your teeth. Have your partner hold one hand firmly under your chin. Try to open your mouth. (This is a test for a motor response.)
   f. *Facial nerve.* Carry out the following facial movements: smile and show your teeth, inflate your cheeks, wrinkle your forehead, and raise either one or both eyebrows. Can you raise each eyebrow separately?
   g. *Vestibulocochlear (acoustic) nerve.* Close your eyes. Have your partner hold a stopwatch about 1 meter away from the left or right side of your head. Can you determine the direction from which the sound is coming? Is this a test of the cochlear or vestibular portion of the nerve?
   h. *Glossopharyngeal and vagus nerves.*
      i. Open your mouth wide. Have your partner carefully touch the back of your throat. What is the response?
      ii. Have your partner hold down your tongue with a sterile tongue depressor and direct a penlight into your mouth, while you say "ah." What happens to the uvula (the fleshy mass near the back of the mouth)?
   i. *Spinal accessory nerve.*
      i. Sit down and have your partner push firmly on your shoulders. Try to raise your shoulders. (Note this does not mean stand up.) What muscles are involved?
      ii. Have your partner place his or her hands firmly on the sides of your face in front of the ears. Try to turn your head. What muscles are involved?
   j. *Hypoglossal nerve.* Stick your tongue out and try to keep it as straight as you can.

Procedure 2: Demonstration of Cerebellar Function

This procedure should be performed by students in pairs.

1. Take turns in performing the following tasks. Describe your responses and those of your laboratory partner in the results and observations section.
2. Have your laboratory partner stand approximately 2 meters in front of you. You will be the examiner. Read each of the tasks out loud and determine the ease with which they are performed.
   a. Stretch both arms to the side and move fingers rapidly.
   b. Stretch both arms to the side and snap the fingers of both hands.
   c. Turn to the side and, looking straight ahead, walk a straight line placing the heel of one foot directly in front of the toe of the other foot.
   d. Close your eyes and stand erect for 1 minute.
   e. Close your eyes and touch your nose with the index finger of your left hand.
   f. Close your eyes and touch your nose with the index finger of your right hand.
   g. Touch the index finger on my right hand. (Examiner should move the right hand accordingly.)
   h. Stand erect and move the heel of your right foot first up and then down the shin of your left leg.

RESULTS AND OBSERVATIONS

CRANIAL NERVE FUNCTION TESTS

1. Enter your responses and those of your partner in Table 29-2.

Table 29-2

Cranial Nerve Tested	Your Response	Partner's Response
Olfactory	can destinguish aroma	yes
Optic	183	152
Oculomotor	yes only to extend or eye contact only without head	same
Trochlear and abducens	pupil got bigger	same
Trigeminal	opened mouth no difficulty	same
Facial	cannot raise eyebrows seperately	same
Vestibulocochlear (acoustic)	Could hear on both sides	same
Glossopharyngeal and vagus	gag ulna moves up	same
Spinal accessory	trapezius	same
Hypoglossal	can't there is no movement!	same

2. How would you test for the sensory responses of the trigeminal nerve?

putting pressure on chin then open mouth.

3. How could you distinguish between the sensory responses of the vestibular and cochlear portions of the vestibulocochlear nerve?

Because it activates the nerve when sound is produced.

DEMONSTRATION OF CEREBELLAR FUNCTION

1. Enter your responses and those of your partner in Table 29-3. Indicate whether or not the task was performed and whether or not any difficulties were observed.

Table 29-3

Task	Your Response	Partner's Response
Move fingers quickly	no difficulty	same
Snap fingers	" "	" "
Walk a straight line in tandem without loss of balance	" "	" "
Stand erect with feet together and eyes closed without loss of balance	" "	" "
Touch index finger of each hand to the nose while eyes closed	" "	" "
Touch examiner's index finger on command	" "	" "
Move heel of one foot up and down shin of opposite leg	" "	" "

2. Suggest two other possible tasks to test cerebellar function.

head tap + stomach rub.

Questions for Review

EXERCISE **29**

DEMONSTRATION OF SELECTED CRANIAL NERVE AND CEREBELLAR FUNCTIONS IN THE HUMAN

Name Kathy Martin

Laboratory Section Mon 8:00 Date 4-21-91

1. Where is the cerebellum located in the human? Lies behind the brainstem back of the skull and under the cerebal hemispheres

2. List two functions of the cerebellum.

   a. balance b. posture

3. What are the cranial nerves? nerves that supply motor + sensory fibers to the head, neck + some visceral organs

4. How many cranial nerves are there in the human? 12 pairs

5. Give the functions of the cranial nerves listed.

   a. Trochlear eye movement + focusing

   b. Facial facial Expression + taste

   c. Glossopharyngeal swallowing taste + general sensation

   d. Vagus visceral muscle movement + sensation

   e. Hypoglossal speech + swallowing

6. Indicate which specific cranial nerve in the human could be tested by the items listed.

   a. Odor of fresh brewed coffee Olfactory nerve

   b. Reading letters on an eye chart Oculomotor nerve

   c. Flashing a penlight into an eye Trochlear

   d. Moving a stopwatch on either side of an individual whose eyes are closed

   vestibulochlear

   e. Touching the individual's uvula glossopharyngeal

EXERCISE 30

DEMONSTRATION OF REFLEXES IN THE HUMAN

As indicated in Exercise 28, discrete nervous pathways to the spinal cord or brain are involved in the performance of a reflex action. The knee-jerk response is a well-known example. Another situation involving a reflex action familiar to most is the response to touching a hot object. A stimulus received by the heat- or pain-sensitive endings in the finger is transmitted as an impulse to the spinal cord, where it relays the signal to a motor neuron. The motor neuron completes the arc by stimulating muscle contraction in the finger, moving it away from the unpleasant stimulus.

Because normal reflexes depend on normal neuron functions, reflexes often are utilized to obtain information concerning the condition of the nervous system. An anesthesiologist, for example, may use a reflex in a patient being anesthetized to determine how a particular drug is affecting the nervous system. Certain reflexes also are of importance in diagnosing nervous system disorders and in locating injured tissue. Lesions of the central nervous system may occur at all levels and result from various causes including physical injuries, infections and biochemical changes. In some situations exaggerated spinal reflexes result, while in others underactive reflexes occur. Examples of reflexes of diagnostic significance include the abdominal reflex, Achilles reflex (ankle jerk), Babinski sign, and the patellar reflex (knee jerk). Table 30-1 briefly describes both the normal and abnormal reactions for these reflexes and some of the conditions responsible for abnormal reactions. This exercise considers several types of reflex actions. These include *deep* and *superficial* reflexes. Deep reflexes are also called *jerk, myotatic* or *stretch* reflexes, and are triggered by a sharp tap on an appropriate muscle or tendon. The associated receptors for these reflexes are situated in the muscle and not in the tendon. Tapping the tendon causes the muscle to be stretched; this stretching, in turn, activates the muscle spindle, thus triggering the reflex response. Superficial reflexes, also referred to as *cutaneous reflexes,* are caused by irritating or tactile stimulation. Stroking or scratching the skin surface starts these reflexes. Figure 30–1 shows the features of certain clinically important reflexes.

Table 30-1 Normal and Abnormal Reflex Actions

Reflex	Normal Response	Abnormal Response
Abdominal	Abdominal muscles underlying the skin contract in response to stroking the side of the abdomen. Movement of the umbilicus toward the region stimulated also occurs.	Absence of reflex. Conditions causing abnormal reactions include lesions of peripheral nerves, lesions of reflex centers in chest region of spinal cord, and multiple sclerosis.

Table 30-1 *(continued)*

Achilles	Extension of the foot in response to tapping the Achilles tendon.	Absence of reflex or exaggerated reflex conditions causing first of abnormal reactions include damage to nerves supplying posterior leg muscles, damage to nerve cell in spinal cord's lumbosacral region, chronic diabetes mellitus, neurosyphilis, alcoholism, and subarachnoid bleeding. Exaggerated responses may be due to injuries of cord's sacral region.
Babinski sign (positive)	Extension of the great (big) toe in response to a light stimulation of the outer margin of the sole of the foot. Other toes may or may not respond. Children under 1½ years of age exhibit the sign because of an incomplete development of the nervous system (corticospinal tract) at this time.	Presence of reflex after 1½ years of age. Conditions causing abnormal reaction include interruption of corticospinal tract development and lesion in the upper corticospinal tract.
Patellar	Extension of the leg in response to tapping the patellar ligament.	Absence of reflex or exaggerated response conditions causing abnormal reactions include injury to nerves in the lumbar segments of the spinal cord, chronic diabetes, and neurosyphilis. Exaggerated responses may result from injury or disease in the corticospinal tract.

Materials

The following materials should be provided for class use.

1. Rubber mallets (percussion hammers)
2. Sterile cotton swabs
3. Plastic disposable tubes (15 ml capacity)
4. Lemon juice or a safe substitute
5. 5 percent glucose solution
6. Small pieces of ice
7. Rotating stools or swivel chairs without casters
8. Marking pencils or pens

Procedure: Demonstrating Common Reflex Actions

The following tests should be performed by students in pairs (except tests 4 and 6, which require three students).

1. Patellar tendon reflex (knee jerk)
   a. Have your laboratory partner sit on a table or desk so that both feet are off the floor and hang freely. (Refer to Figure 30–1a.)
   b. Locate the patellar tendon (the depression just below the kneecap) of the right leg and tap the tendon with the rubber mallet or the edge of your hand. Note the response in the results and observations section.
   c. Have your partner count backwards from 99 to 32. Tap the patellar tendon of the right leg again. Was the response greater or less than the first time?
   d. Repeat steps 2–4 with the left leg. Note the extent of the response.
   e. Change places with your partner and repeat steps a–d.
2. Achilles tendon reflex (ankle jerk)
   a. Have your laboratory partner kneel on a chair or table so that both feet hang over the edge.
   b. Locate the Achilles tendon of the right foot and tap it with a rubber hammer or the edge of your hand. Describe the response in the results and observations section.
   c. Change places with your partner and repeat steps a and b. (Refer to Figure 30–1b.)
3. Babinski sign
   a. Have your partner remove the shoe and sock from the right foot.
   b. Take the tip of a rubber mallet or the blunt end of a pen and run it on the sole of the foot firmly in one upward movement from the heel to the transverse arch. Record your findings in the results and observations section. (Refer to Figure 30–1c.)
   c. Change places with your partner and repeat steps a and b.
4. Static equilibrium response
   a. Have one laboratory partner stand erect and still with both feet as close together as possible. This position should be maintained for 2 minutes. Note the kinds and amount of movement necessary to keep the body in this position. Record your findings in the results and observations section.
   b. Have your laboratory partner repeat step a, but with eyes closed. (*Note:* Two students should stand by to prevent a fall.) Enter your findings in the results and observations section.
   c. Change places with your laboratory partners and repeat steps a and b.
5. Spatial orientation
   a. Have your laboratory partner stand erect with both arms extended to the sides and then bring the tips of the index fingers together.
   b. Repeat step a with eyes closed until the index fingers are brought together successfuly at least three times.
   c. Have your partner sit in a swivel chair with eyes closed. Twirl your partner around rapidly 10 times. Repeat step b. Record your findings in the results and observations section.
   d. Change places with your partner and repeat steps a–c.
6. Ciliospinal reflex
   a. Have one laboratory partner sit comfortably in a chair. Observe the partner's pupils.
   b. Have another laboratory member unexpectedly touch a small piece of ice to the back of the seated individual. Note the response of the pupils. Record your findings in the results and observations section.
   c. Change places with your partners and repeat steps a and b.
7. Salivary reflex
   a. Have your partner rinse his or her mouth with water.
   b. Collect all saliva secreted during a 3-minute period in a plastic tube. Mark the top level

of fluid in the tube. Empty the tube as indicated by your instructor.

c. Dip a cotton swab in the glucose solution and wet the underside of your partner's tongue with the solution. Collect the saliva secreted during a 3-minute period and mark the fluid level.

d. Dip another swab in lemon juice and repeat step c. In which case was more saliva produced? Enter your findings in the results and observations section.

e. Change places with your partner and repeat steps a–d.

8. Convergence reflex

a. Have your laboratory partner sit comfortably. Place a pencil or pen about 1 meter in front of his or her face. Note the position of the eyes.

b. Slowly move the pencil toward the nose just short of touching it. Note the position and appearance of the eyes in the results and observations section.

c. Change places with your partner and repeat steps a and b.

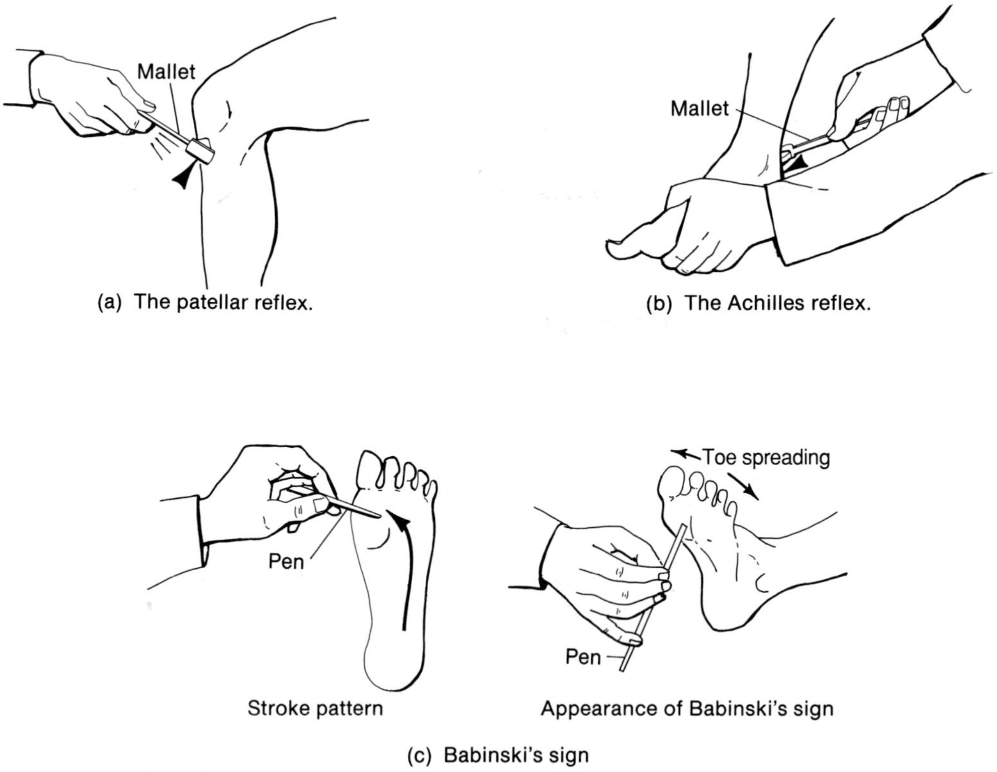

Figure 30–1 Selected clinically important reflexes.

RESULTS AND OBSERVATIONS

1. Enter in Table 30-2 brief descriptions of your reactions and those of your partner in the reflex testing procedures.

Table 30-2

Reflex Procedure	Your Response	Laboratory Partner's Response
Patellar tendon reflex Right leg	leg shot foward	same
Right leg after counting	same	" "
Left leg	" "	" "
Left leg after counting	" "	" "
Achilles tendon reflex	twitches	" "
Babinski sign	toes spread	" "
Static equilibrium Eyes open	some wobbling	" "
Eyes closed	more wobbeling	" "
Spatial orientation Eyes open	could contact	" "
Eyes closed	same	" "
After twirling	could touch but not finger	" "

Table 30-2 (*continued*)

Reflex Procedure	Your Response	Laboratory Partner's Response
Ciliospinal reflex	larger pupils	same
Salivary reflex Normal	7½ m.m.	" "
With glucose solution	normal	" "
With lemon juice	produced (8 mm) more saliva	" "
Convergence reflex	same	" " eyes closed

2. Were differences in patellar tendon responses observed? Explain. 2 no
same reactions

3. What is the medical significance of the Babinski response? it can help detect
an incomplete developement of the nervous system (corticospinal tra
an interuption of this tract & lesions in the upper tract

4. What type of nerve is stimulated in the ciliospinal reflex? optic (eye, pupil)

5. Can the salivary reflex by controlled? Explain.
yes, threw stimulation.

6. What structures in the body contribute to an individual's static equilibrium?
depends on neuron functions,

7. Distinguish between strabismus and myopia.
strabismus - both eyes don't focus
myopia - causes eyes to focus

Questions for Review

EXERCISE **30**

DEMONSTRATION OF REFLEXES IN THE HUMAN

Name ______________________________

Laboratory Section ______________________ Date __________

1. What is a reflex? a response automaticly done, like the knee reflex

2. Diagram a reflex arc.

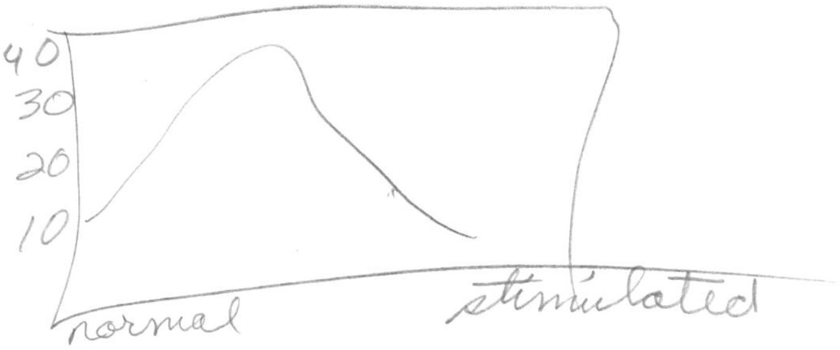

3. In the following tests for reflex action what body structure is stimulated?

   a. Knee jerk pateller tendon Reflex

   b. Babinski sign toe spreading a sharp object on bottom foot

   c. Ankle jerk Achilles tendon reflex

4. With the aid of your text and other references define or explain the following.

   a. Tonus - state of partial substaned contraction which keeps muscles sleghtly taut.

   b. Flaccid - soft limp flabby.

   c. Ipsilateral - one the same side as opposed to contralateral (knee jerks)

   d. Reciprocal innervation contraction of muscles which are stimulated by sympathetic nerves.

5. What is a flexion or withdrawal response? a movement that decreases the ankle between parts of a joint,

6. Distinguish between deep and superficial reflexes. Give one example of each type.

Deep reflexes are jerk, myotatic, stretch reflexes sharp tap on muscle & tendon, Superficial reflexes - cutaneous reflexes → irritating or tactile stimulation

SECTION VIII

THE ENDOCRINE SYSTEM

The endocrine glands provide a communication and regulatory system in addition to the nervous system. While the two supplement one another or perform separately, the interplay between endocrine secretions and nerve impulses functions to control total human behavior.

The endocrine secretions, or *hormones,* do not belong to any one category of chemical compounds. They have varying chemical structures. Some are proteins, while others are amines (amino acid derivatives) or steriods (types of lipids).

The concentration of a hormone in the blood is controlled by a negative feedback mechanism that regulates production, storage, and elimination processes. Sensors read the status of the body's internal environment and direct appropriate responses. The presence of a greater than normal concentration of a hormone will initiate a response to inhibit the gland's production. Glands may be under direct neural control or under the control of other hormones or both.

The exercises in this section point out the location of the human endocrine glands and demonstrate selected properties of the human endocrine system.

GENERAL REFERENCES

Gorbman, A., W. W. Dickhoff, S. R. Vigna, N. B. Clark, and C. L. Ralph. *Comparative Endocrinology.* New York: Wiley, 1983.

Greenspan, F. S., and P. H. Forsham (Eds.). *Basic and Clinical Physiology.* Los Altos, CA: Lange Medical Publications, 1983.

Hardy, R. N. *Endocrine Physiology.* Baltimore: University Park Press, 1982.

Tortora, G. J., and N. P. Anagnostakos. *Principles of Anatomy and Physiology,* 6th ed. New York: Harper & Row, 1990.

Williams, R. H. (Ed.). *Textbook of Endocrinology,* 6th ed. Philadelphia: Saunders, 1980.

Witherspoon, J. D. *Human Physiology,* New York: Harper & Row, 1984.

SPECIFIC REFERENCES

Carmichael, S. W., and H. Winkler. "The Adrenal Chromaffin Cell," *Scientific American,* 253:40–49, 1985.

Dupre, J. "Insulin Therapy: Progress and Prospects," *Hospital Practice,* 18:171–179, 1983.

Golde, D. W., and J. C. Gasson. "Hormones That Stimulate the Growth of Blood Cells," *Scientific American,* 259:62–70, 1988.

Reichlin, S. "Somatostatin," *The New England Journal of Medicine,* 309:1495–1501, 1983.

Reiter, R. J. (Ed.). *The Pineal Gland.* New York: Raven Press, 1984.

Tepperman, J. *Metabolic and Endocrine Physiology,* 4th ed. Chicago: Year Book Medical Publishers, 1980.

Volpe, R. "Autoimmune Thyroid Disease," *Hospital Practice,* 19:141–158, 1984.

EXERCISE 31

THE LOCATION AND HISTOLOGY OF HUMAN ENDOCRINE GLANDS

Many endocrine glands are *organs* that secrete chemical regulators, or *hormones*, into the blood. The approximate locations of endocrine glands are shown in Figure 31-1. Hormones are carried by the vascular system to all the organs of the body. Some hormones, however, only affect the metabolism of specific organs; these organs are called the *target organs* of the hormone (Figure 31-1). Other hormones influence every body cell. Hormones are not enzymes; they affect the metabolism and activities of their target organs by regulating the activity of specific enzymes in these and related structures. In this way, the endocrine system plays an important role in the regulation of (1) growth and development, (2) metabolic homeostasis, and (3) reproduction. Since hormones are powerful chemicals, their circulating levels must be carefully controlled.

When the level of a hormone is adequate to meet existing body needs, its further release is stopped. This form of control is called *negative feedback*.

Overactivity or underactivity of an endocrine gland are examples of malfunction. Conditions of this type cause endocrine diseases such as those listed in Figure 31-2. When a gland secretes an excessive amount of a specific hormone, it is *hyperactive*. This type of activity may be caused by various conditions including an enlarged gland or a tumor. A gland failing to secrete its hormones or simply secreting inadequate quantities is considered to be *hypoactive*. Conditions that may cause such limited activity include infection, tumors, or the unfavorable effects of surgical or radioactive treatments. Glands that are decreased in size and consequently secrete inadequate amounts of hormones are referred to as being atrophied.

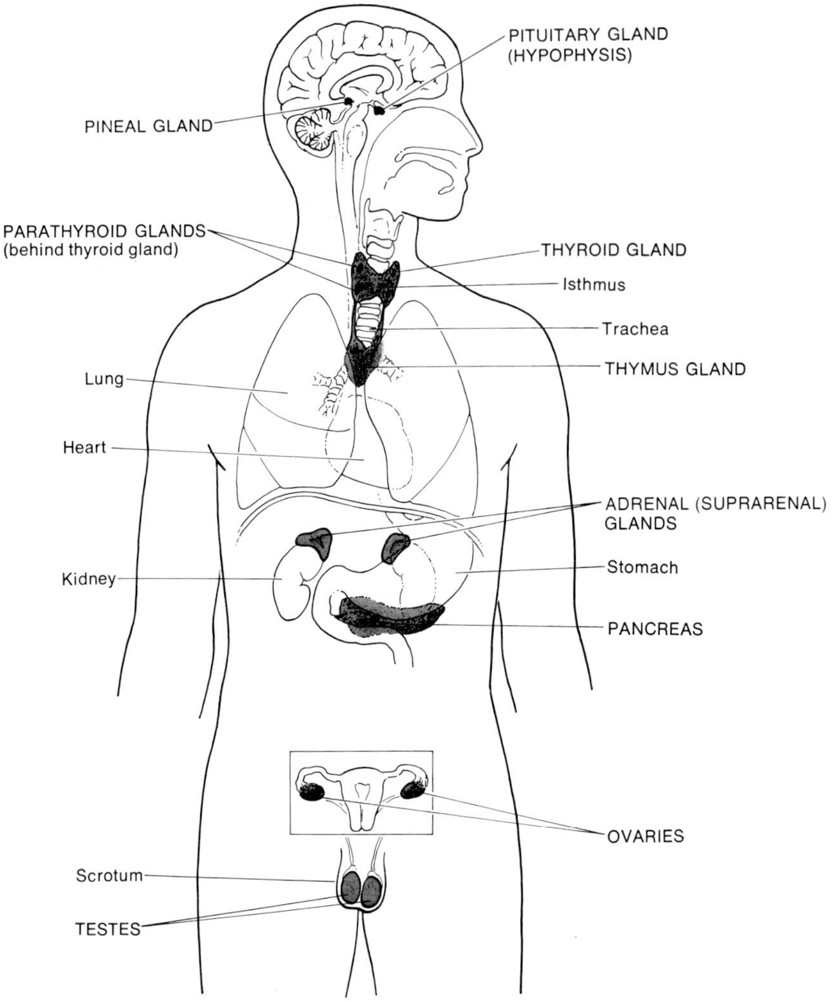

Figure 31-1 Location of the human endocrine glands.

This exercise will present the location, microscopic features of normal endocrine glands, and some pathological conditions commonly found in humans.

A. ENDOCRINE GLANDS, LOCATION AND BRIEF DISCUSSION

Materials

The following materials should be provided for class use.

1. Human torso
2. Head models
3. Human skull

Endocrine glands secrete into the blood and do not have ducts. They therefore are called the *ductless glands.* All other large secretory masses produce substances that are carried to their specific destinations by one or more ducts. These are called *duct glands* or *exocrine glands.* The products of endocrine glands are called *hormones,* while products of exocrine glands are given such names as saliva, bile, and pancreatic juice. The advantage that an

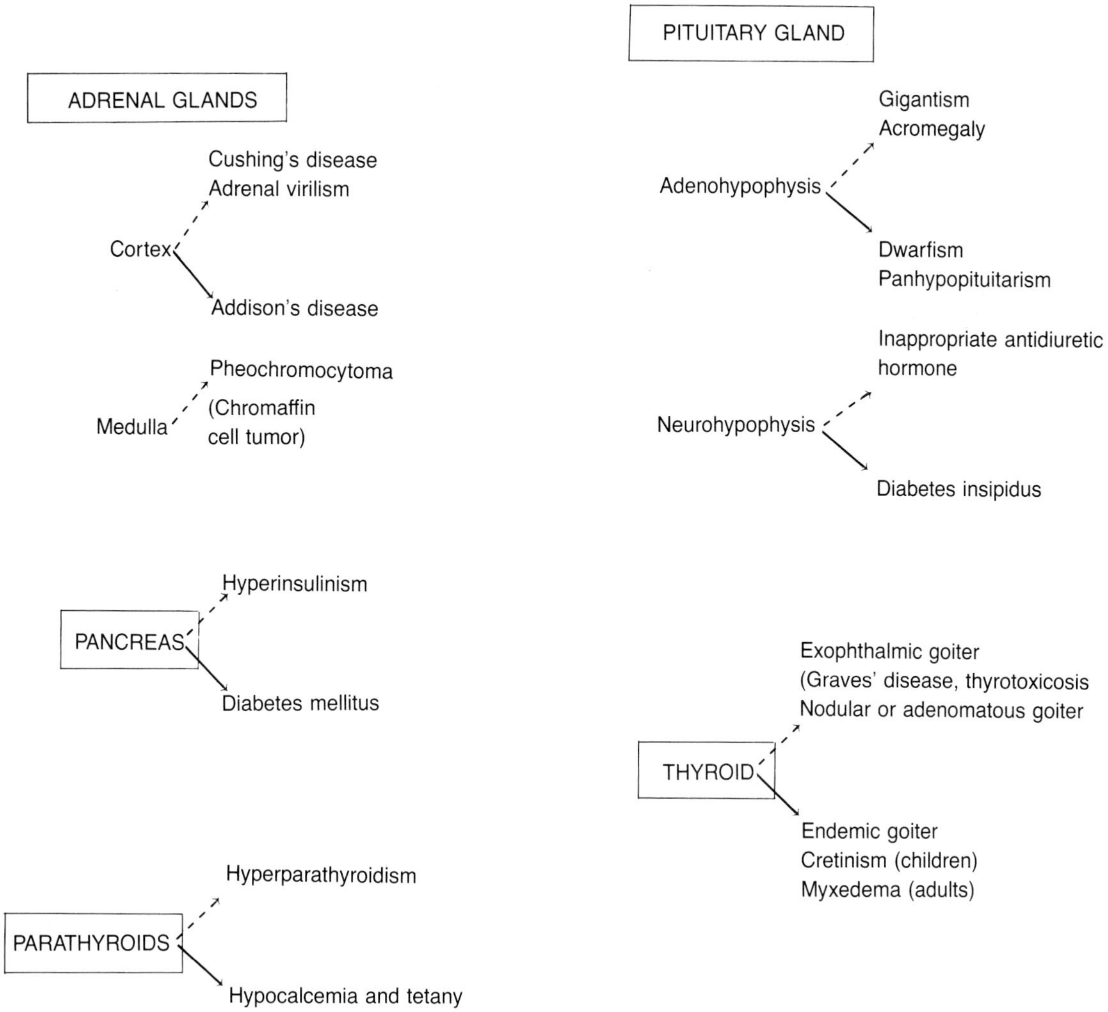

Figure 31-2 Endocrine diseases. The hyperactivity of a gland is indicated by a dotted line. The hypoactivity is indicated by a solid line.

endocrine gland has is obvious. Once in the blood, the secretions may be carried to all cells of the body.

The major endocrine glands are shown in Figure 31-1.

PITUITARY GLAND

The pituitary gland (Figure 31-3) is located at the base of the brain within the hypophyseal fossa of the sella turcica of the sphenoid bone. It is suspended by a stalk called the *infundibulum* and is composed of two major parts:

1. *Anterior Lobe (Adenohypophysis).* This part of the pituitary arises from embryonic mouth tissue and produces the following hormones:
   a. *Somatotropin or Human Growth Hormone (HGH).* This hormone regulates growth of body tissue but specifically acts on the epiphyseal cartilages of the long bones (refer to Exercise 15) to effect growth in height. Hypersecretion of this hormone in childhood may result in *gigantism* and a hyposecretion in a *midget* or *pituitary dwarf.* A hypersecretion in the adult may cause *acromegaly,* a disfiguring condition in which some growth areas in the skeleton still respond and grow absurdly out of shape, and others, such as those that affect height, do not continue to grow.

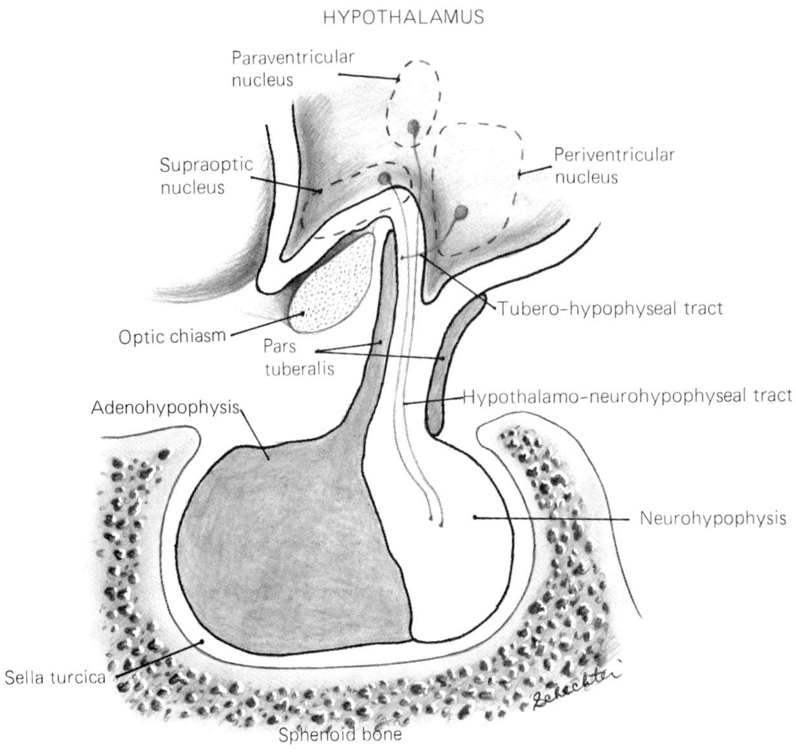

Figure 31-3 Human pituitary gland.

b. *Prolactin.* This hormone acts on the mammary glands causing them to produce milk before and after childbirth. The following tropic hormones are those that act as stimuli on other endocrine tissue:
c. *Thyrotropic Hormone (TH).* This hormone regulates the growth, development, and activity of the *thyroid gland.*
d. *Adrenocorticotropic Hormone (ACTH).* This hormone regulates the growth and development of the *adrenal cortex.*
e. *Follicle Stimulating Hormone (FSH).* This is one of two *gonadotropic hormones.* It stimulates and regulates the early developmental stages of the ovarian (Graafian) follicle in the female and the spermatozoa in the seminiferous tubules of the male testis. It has other functions as well.
f. *Luteinizing Hormone (LH).* This hormone finishes what FSH has begun in the ovary and the testis. The follicle containing the ovum or egg completes its development and ovulates. The spermatozoa are anatomically completed and released from the seminiferous tubules. In the male, the luteinizing hormone often is called the *interstitial cell stimulating hormone (ICSH).*

A small cleft in the anterior lobe of the pituitary gives rise to the *intermediate lobe.*

2. *Posterior Lobe (Neurohypophysis).* The hormones found in this part of the gland are produced in the brain and are stored in the posterior lobe.
   a. *Oxytocin.* This is a hormone that is found only in the female and stimulates uterine contractions. It is one of the major factors in initiating labor at the time of birth.
   b. *Antidiuretic Hormone (ADH)* or *Vasopressin.* This is a hormone that causes reabsorption of water from the distal convoluted tubules in the kidney, thereby preventing severe water loss. The hormone also causes a rise in blood pressure. Hyposecretion leads to *diabetes insipidus.*

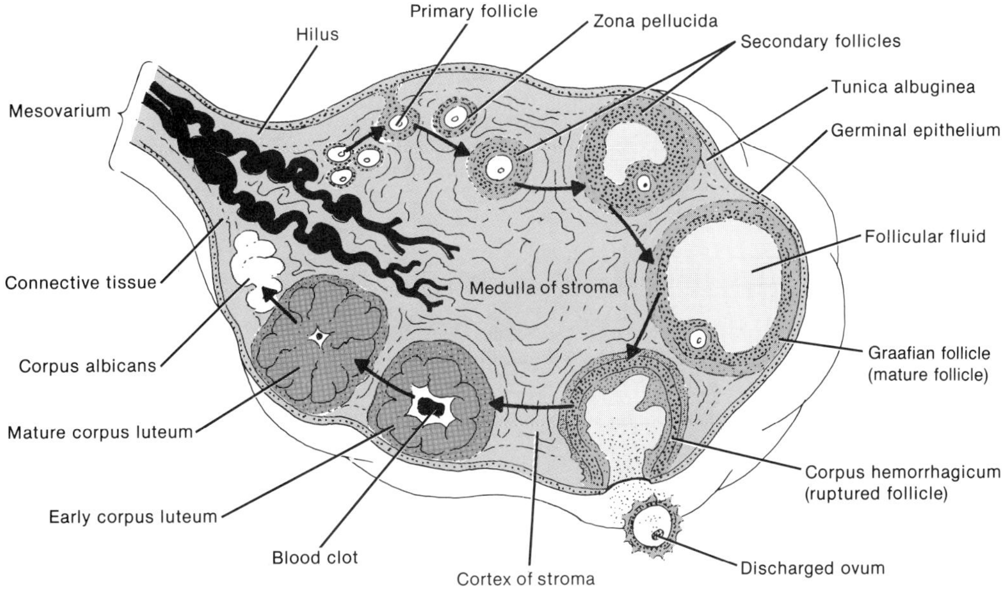

Figure 31-4 Diagram of the parts of an ovary seen in sectional view. The arrows indicate the sequence of development stages that occur as part of the ovarian cycle.

OVARY

In addition to producing ova (eggs), the ovary (Figure 31-4) contains two hormone-producing tissues:

1. The *follicle cells* which make up the Graafian follicle produce *estrogen,* the female sex hormone responsible for the development and maintenance of the female secondary sexual characteristics. Estrogen has many other functions.
2. The *corpus luteum* is the ovarian structure that remains after the egg has ovulated. Its cells secrete the hormone *progesterone.*

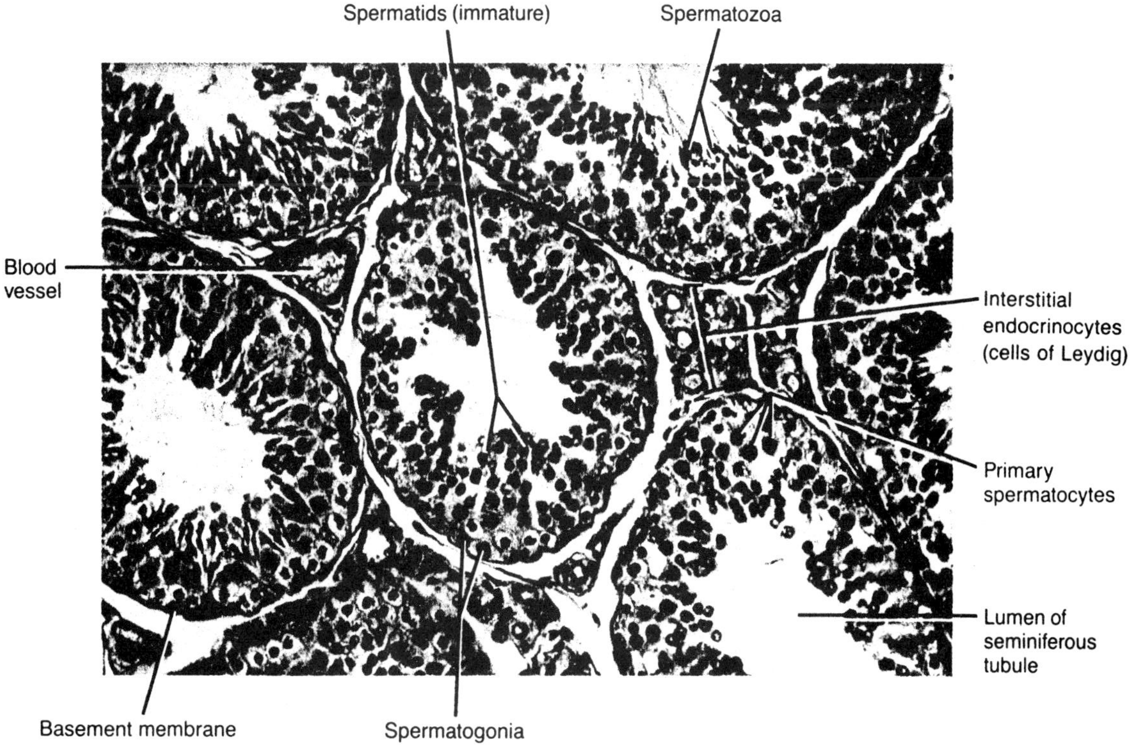

Figure 31-5 Histology of the testes. Several seminiferous tubules show stages of spermatogenesis.

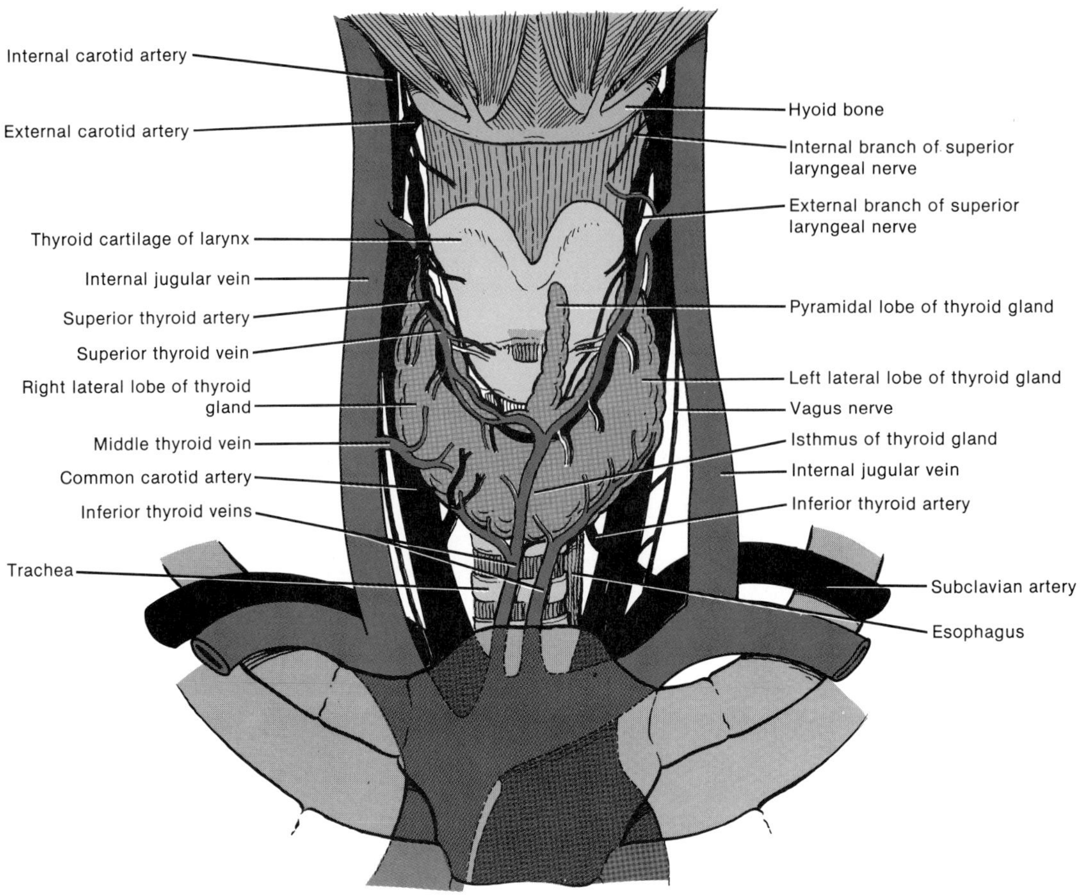

Figure 31-6 Human thyroid gland, anterior view.

TESTIS

In addition to producing and releasing spermatozoa, the testis (Figure 31-5) contains interstitial cells that produce the male sex hormone *testosterone.* This hormone controls the growth and development of the male secondary sexual characteristics.

THYROID GLAND

The thyroid gland (Figures 31-6 and 31-7) is a large mass of tissue on the anterior surface of the trachea just inferior to the larynx. It is composed of right and left lateral lobes connected by a slim *isthmus* on the anterior surface of the trachea. A large quantity of iodine is found in this gland. It produces three hormones:

1. Thyroxin (T4)
2. Triiodothyronine (T3)
3. Thyrocalcitonin

Thyroxin and triiodothyronine control the metabolic rate of all cells in the body, while thyrocalcitonin regulates blood calcium.

Hypothyroidism results in a slow-moving, sleepy, obese individual, while *hyperthyroidism* produces the opposite. A severe lack of the thyroid hormones in the fetus or in the child results in *cretinism.* In the adult a similar physiological need produces an individual with *myxedema.* Lack of iodine in the diet also produces an abnormal gland that increases in size

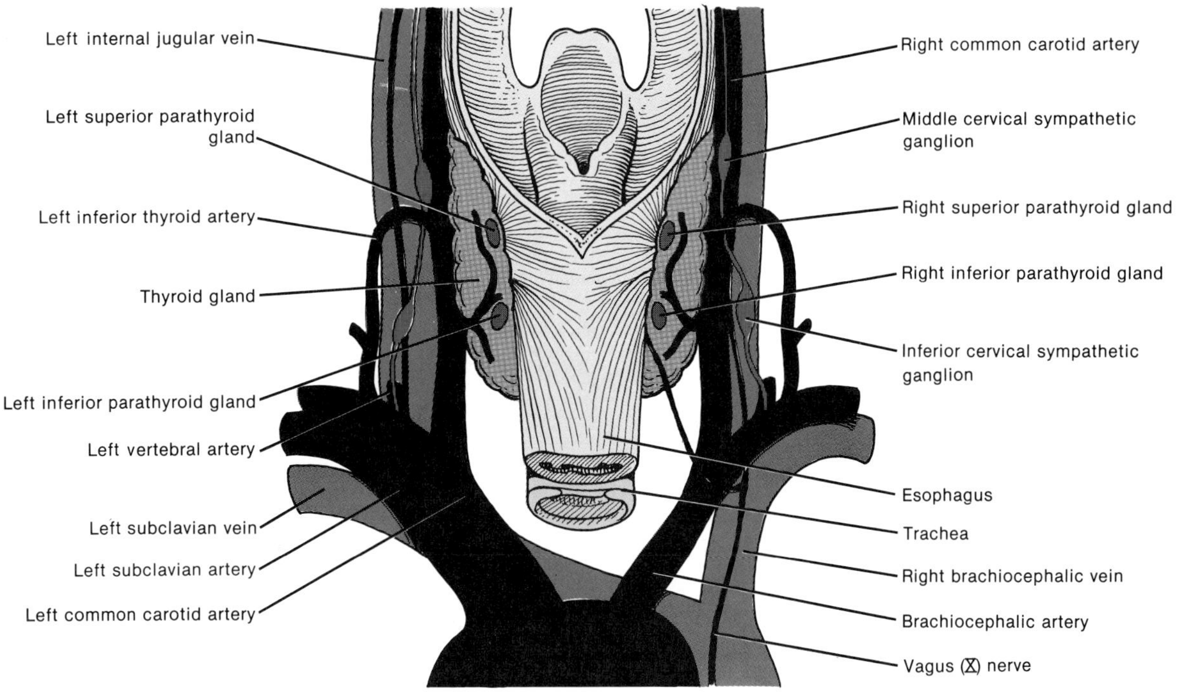

Figure 31-7 Human thyroid gland, posterior view, showing parathyroid glands.

due to *compensatory hypertrophy,* creating a *goiter,* which is an enlarged thyroid gland. A severe hyperthyroid condition causes another type of goiter that sometimes is called *exophthalmic goiter.* The two types of goiter are not the same since one is *endemic* and caused by iodine lack in the area, while the other is a failure of the thyroid gland to function properly even in the presence of iodine-rich soil and water.

PARATHYROID GLANDS

The parathyroid glands (Figure 31-7) are usually four in number (two pairs embedded in the posterior medial surface of each thyroid lobe). The number is not constant, however. Some people have only one pair and others have three pairs of the glandular masses. They arise from embryonic visceral arch tissue and produce the hormone *parathormone (PTH),* which controls calcium and phosphate metabolism in the body. *Hypoparathyroidism* results in brittle bones and low blood calcium and *hyperparathyroidism* results in soft bones, loss of teeth, and high blood calcium.

ISLETS OF LANGERHANS OF THE PANCREAS

The pancreas is a large exocrine gland that produces powerful digestive enzymes that are carried to the small intestine by ducts. Between the enzyme-producing tubules there are islands of cells called *alpha cells* and *beta cells* (the islets of Langerhans; see Figure 31-8). The beta cells produce *insulin,* which enables the body cells to obtain blood sugar needed for physiological oxidation. A hyposecretion of this hormone causes *diabetes mellitus.* The alpha cells secrete the hormone *glucagon,* which, by converting stored glycogen to glucose, increases the blood sugar.

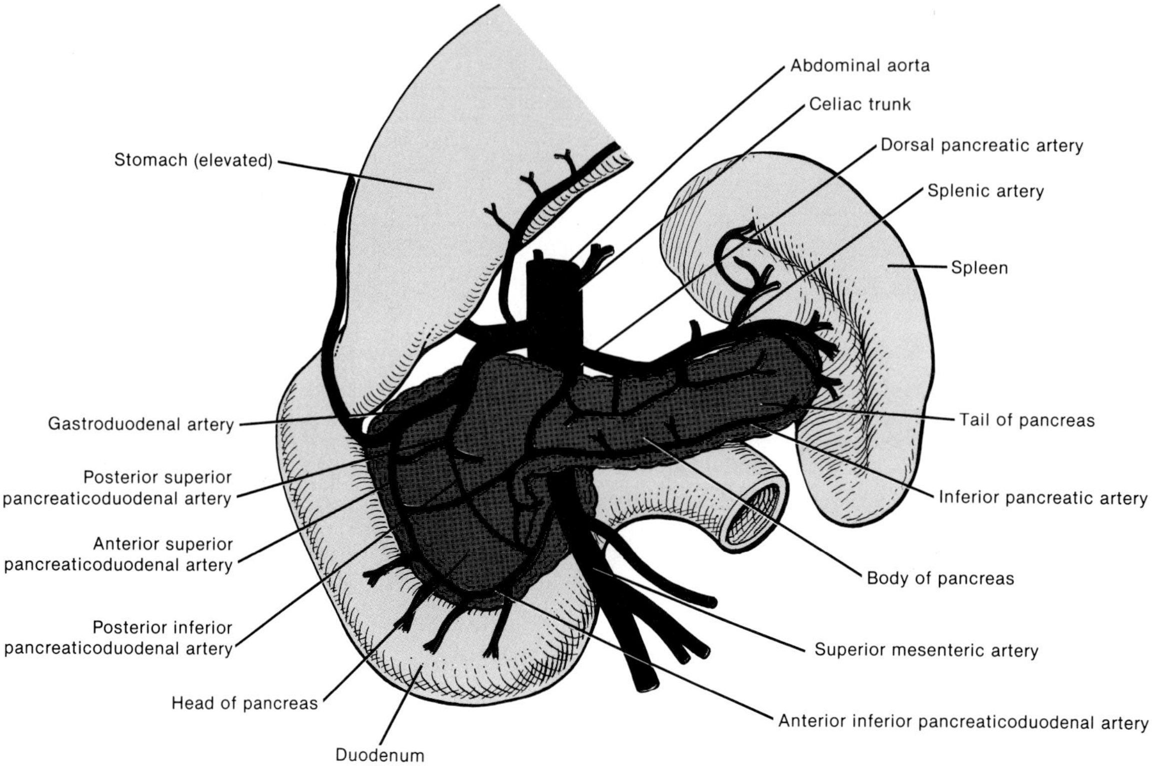

Figure 31-8 Human pancreas.

SUPRARENAL OR ADRENAL GLANDS

The suprarenal glands (Figure 31-9) are located above the kidneys in the human. In most other animals they are found lying beside the kidneys—hence the names suprarenal and adrenal. The gland has two distinct regions that have different embryonic origins:

1. The *suprarenal medulla* is the core of the gland and it originates from embryonic neural crest cells. The hormones produced in this region are called *epinephrine* (adrenaline) and *norepinephrine*. They have a function similar to that of the sympathetic division of the autonomic nervous system: they prepare the body for an emergency when secreted in large amounts.
2. The *suprarenal cortex* secretes many hormones, which are divided into categories:
   a. *Glucocorticoids. Cortisone* is an example. This group of hormones act as anti-inflammatory agents, function in protein digestion and storage, and provide resistance to stress.
   b. *Mineralocorticoids. Aldosterone* is an example. These hormones regulate sodium and potassium metabolism.
   c. *Adrenal Sex Hormones* or Gonadocorticoids. These estrogens and androgens are secreted in minute quantities and have little effect in the normal body. However, they can cause sexual anomalies when secreted in large amounts as the result of a tumor. A hyposecretion of the adrenocorticoids may result in a debilitating and progressive condition called Addison's disease. Hypersecretion leads to Cushing's syndrome.

PINEAL GLAND

The pineal gland (see Figure 31-1) is a very small gland known to secrete quantities of the

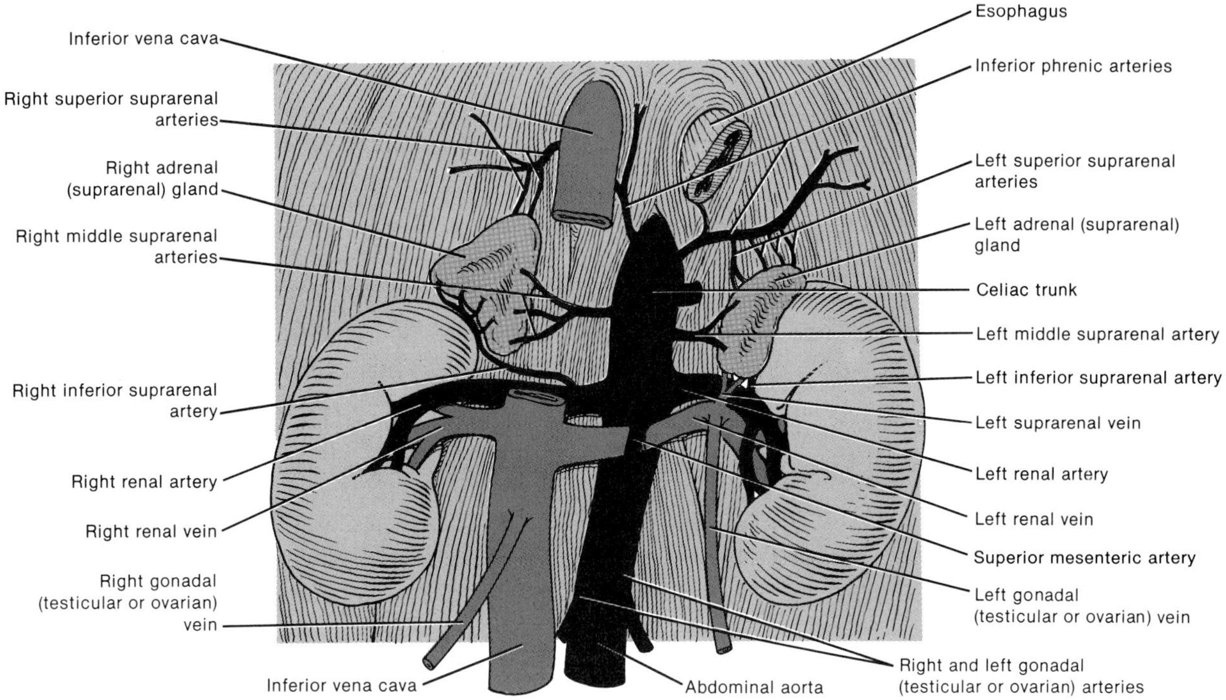

Figure 31-9 Human suprarenal glands.

hormones *melatonin* and *adrenoglomerulotropin*. The gland is a dorsal outgrowth of the epithalamus of the brain. Its hormones are not well understood.

It should be pointed out here that these same endocrine organs are found in the cat. When you study the internal organs in your dissection of the cat, try to locate each one of them.

B. HISTOLOGY OF SELECTED ENDOCRINE GLANDS

Materials

The following prepared slides should be provided for class use.

1. Adrenal gland
2. Pancreas
3. Thymus gland (prepuberty)
4. Thymus gland (adult)
5. Thyroid gland

Procedure: Microscopic Examination of Selected Endocrine Glands

This procedure is to be performed by students individually.

ADRENAL GLANDS

1. Examine a prepared slide of an adrenal gland with the low power or scanning objective. Refer to your textbook (Color Plates 18–21) and identify the following parts.
   a. *Capsule*—thin, outer covering
   b. *Cortex*—outer section
   c. *Medulla*—central region

2. Examine the cortex more closely and locate its three zones.
   a. *Zona glomerulosa*—region next to the capsule containing thin layer of cells in clumps or even arcs
   b. *Zona fasciculata*—thicker central zone containing cells arranged in long cords
   c. *Zona reticulosa*—zone next to the medulla containing cords of cells arranged in a netlike fashion.
3. Examine the medulla. Its cells are arranged in irregular groups with several blood vessels separating them.
4. Sketch and label a general representation of the different parts of the adrenal gland in the results and observations section.

PANCREAS

1. Examine the prepared slide of the pancreas with the low power objective. (Refer to Color Plate 22.)
2. Identify the following portions of the gland.
   a. *Acini*—small tubules that consist of clusters of cells and a lumen (the exocrine part of the pancreas)
   b. *Islets of Langerhans*—clusters of alpha and beta cells surrounded by capillaries and exocrine structures, small granules in alpha cells darker than those found in beta cells
3. Sketch and label a general representation of the different parts of the pancreas in the results and observations section.

THYMUS GLAND

1. Examine the two forms of thymus gland preparation with the low power and high power objectives.
2. Note the cell types in the preparation. (Refer to Color Plate 23.)
3. Answer the questions in the results and observations section.

THYROID GLAND

1. Examine the thyroid gland slide with the low power objective.
2. Identify the circular, single-celled layers called *follicles* within the specimen. Such follicles usually enclose solid appearing materials known as *colloids,* which contain several proteins including thyroxin. Some follicles will be empty. (Refer to Color Plate 24.)
3. Sketch and label a representative portion of the prepared slide in the results and observations section.

THYROID GLAND AND GOITER

1. Examine the thyroid gland preparation from a goiter case, with the low power objective. (Refer to Color Plate 25.)
2. Compare this specimen with the normal thyroid gland preparation. Are there any obvious differences? If so, sketch a representative portion in the results and observations section.

PARATHYROID

1. Examine the prepared slide of the parathyroid gland with the low power objective first and then with the high power objective.
2. Identify the principal and oxyntic cells of the gland. (Refer to Color Plates 26 and 27.)
3. Sketch and label a representative portion of the gland in the results and observations section.

RESULTS AND OBSERVATIONS

1. Adrenal gland

Cortex

Medulla

2. Pancreas

3. Thymus

Adult

Puberty

a. Which cell types are more numerous in the adult thymus gland? ______________

b. Which cell types are more numerous in the prepuberty thymus gland? ______________

c. Why are there such differences between the two preparations? ______________

____

____

d. What is a T lymphocyte? ______________________________________

____

e. What is a B lymphocyte? ______________________________________

____

4. Thyroid gland

Normal

Goiter

a. Were there any differences between the normal and goiter thyroid glands?

____

b. What is a goiter? ______________________________

____

5. Parathyroid gland

Questions for Review

EXERCISE **31**

THE LOCATION AND HISTOLOGY OF HUMAN ENDOCRINE GLANDS

Name ______________________________

Laboratory Section ______________________ Date ____________

1. Complete Table Q31-1 by indicating at least one specific effect that results from the hyposecretion or hypersecretion of the hormones indicated.

Table Q31-1

Hormone	Hormonal Level	Effects
Glucocorticoids	Hyposecretion	a.
	Hypersecretion	b.
Parathyroid hormone	Hyposecretion	c.
	Hypersecretion	d.
Thyroxin	Hyposecretion	e.
	Hypersecretion (adult)	f.
Antidiuretic hormone	Hyposecretion	g.
Human growth hormone	Hypersecretion (childhood)	h.
	Hypersecretion (adult)	i.

2. Complete Table Q31-2.

Table Q31-2

Endocrine Gland	Hormone	Associated Activity
a.	b.	Regulates growth is controlled by human growth hormone releasing factor
c.	Follicle-stimulating hormone	d.
e.	f.	Stimulates mammary glands to secrete milk after birth
g.	h.	Stimulates water reabsorbtion by the kidneys
i.	Parathyroid hormone	j.
Thymus (Adult)	k.	l.
m.	Prolactin	n.
o.	Insulin	p.
Alpha cells of the pancreas	q.	r.
s.	Thyroxin	t.

3. Distinguish between an exocrine and an endocrine gland. ______________________

____

4. Identify the parts of the thyroid gland indicated in Figure Q31-1.

a. ______________________ c. ______________________

b. ______________________ d. ______________________

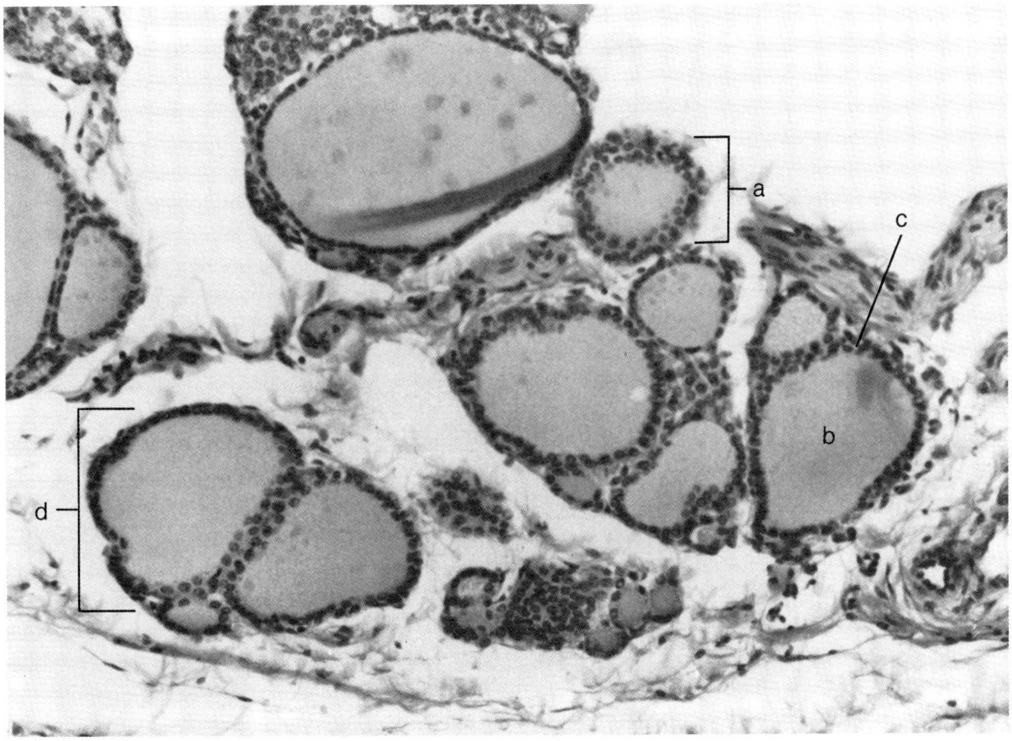

Figure Q31-1

SECTION IX

THE SENSES

Specialized sensory receptors pick up environmental stimuli and transmit them to the central nervous system. Mechanical, chemical, and visual receptors each receive different types of stimuli. The nature of the sensation provoked by such stimuli is determined by the brain. Table IX-1 describes several sensory receptors.

The exercises in this section consider the general structure of the eye and several procedures to test normal functioning, the structure and functioning of the ear, the sense of smell and its relationship to taste, and various sensory properties of the skin. Attention also is given to several types of optical illusions including those involving the interpretation of spatial relationships.

Table IX-1 The Sensory Receptors

Sensory Receptor	Location	Type of Receptor and Type of Stimulus
Rods and cones	Retina of each eye	Visual—light
Hair cells		
For converting sound waves to nerve impulses	Organ of corti in cochlear duct of inner ear	Mechanical—vibrations or movements of fluid in inner ear
For positioning of head and body	Utricle of inner ear	Mechanical—movements of CaCo$_3$ (calcium carbonate) particles in direction of gravity
For changes in movement of body	Ampulla of semicircular ducts of inner ear	Mechanical—rapid movement of $CaCO_3$ particles
Taste buds	Papillae of tongue	Chemical—specific chemicals in solution
Olfactory cells	Olfactory epithelium in nasal cavity	Chemical—dissolved gaseous substances
Touch and pressure	Dermis of skin, smooth muscles of visceral organs	Mechanical—touch
Temperature (heat and cold)	Dermis of skin, smooth muscles of visceral organs	Mechanical—hot and cold
Pain	Dermis of skin, smooth muscles of visceral organs	Mechanical—all intense stimuli
Proprioceptors	Muscles, tendons, joints	Mechanical—movement

GENERAL REFERENCES

Barlow, H. B., and J. D. Mollon (Eds.). *The Senses.* New York: Cambridge University Press, 1982.

Pickles, J. O. *An Introduction to the Physiology of Hearing.* New York: Academic Press, 1982.

Rodieck, R. W. *The Vertebrate Retina, Principles of Structure and Function.* San Francisco: Freeman, 1976.

Tortora, C. J., and N. P. Anagnostakos. *Principals of Anatomy and Physiology,* 6th ed. New York: Harper & Row, 1990.

Witherspoon, J. D. *Human Physiology.* New York: Harper & Row, 1984.

SPECIFIC REFERENCES

Cytowic, R. E. "Tasting Colors, Smelling Sounds," *The Sciences,* September/October, 32–37, 1988.

Doty, R. L. "Olfactory Communication in Humans," *Chemistry of the Senses,* 6: 351–376, 1981.

Loeb, G. E. "The Functional Replacement of the Ear," *Scientific American,* 252: 104–111, 1985.

Koretz, J. F., and G. H. Handelman. "How the Human Eye Focuses," *Scientific American,* 259:92–99, 1988.

Wertenbaker, L. *The Eye.* Washington, DC: U.S. News Books, 1981.

Wolfe, J. M. "Hidden Visual Processes," *Scientific American,* 248:94–103, 1983.

EXERCISE 32

THE ANATOMY OF THE EYE AND THE EAR

A. ANATOMY OF THE EYE

Materials

The following materials should be provided for class use.

1. Fresh eyes of sheep or cow
2. Preserved eyes of sheep or cow
3. Model of human eye
4. Chart of human eye
5. Model of human ear
6. Demonstration—mammalian ear ossicles
7. Human skull

DISSECTION OF SHEEP OR COW EYE:

Outer surface of eyeball and extrinsic eyeball muscles

Examine the surface of the eyeball. It is almost round with a clear surface in front and a muscular area surrounded by fat in the rear. Examine the posterior surface carefully before removing the fat. Locate the *optic nerve* and isolate it so that it will not be removed accidentally with the fat.

Dissection of the extrinsic eyeball muscles in the sheep. Mammalian eyes have six extrinsic muscles per eye plus a set of *retractor muscles* which are not present in the human. Using Figure 32-1, locate the following muscles:

Superior rectus	Superior oblique
Inferior rectus	Inferior oblique
Lateral rectus	Retractor bulbi
Medial rectus	

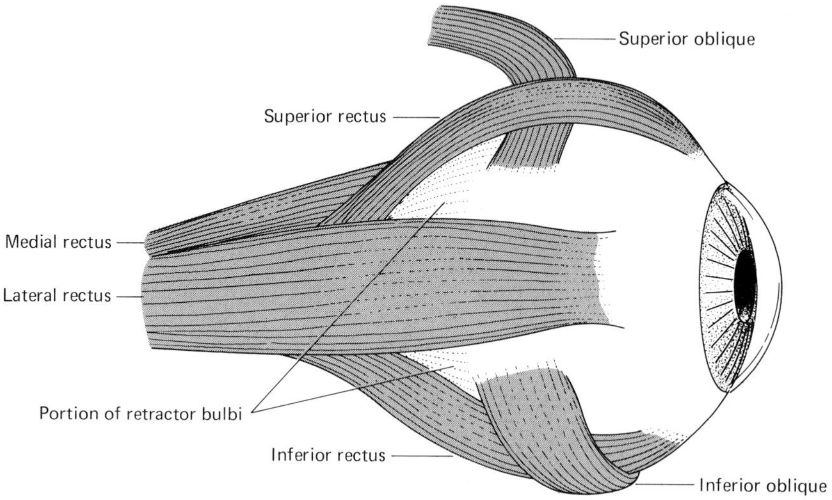

Figure 32-1 Extrinsic eyeball muscles of a sheep eye.

OPENING THE EYEBALL

Hold the sheep or cow eye between thumb and forefinger with the thumb on the posterior surface and the forefinger on the cornea (Figure 32-2). Make an incision about halfway between poles, dividing the eye into anterior and posterior halves. Continue this incision around the equator with your scissors. Separate the two halves on a dissecting tray or in a pan of water. Compare the parts of the eye of your dissected specimen with those shown in Figure 32-3 and answer the questions in the results and observation section.

The anterior portion of the eyeball contains the thick cornea and the circular ciliary body and iris. The latter two are pigmented and appear black in contrast to the translucency of the cornea. The aqueous fluid, or aqueous humor, in this part of the eye has probably drained away and will not be seen. The aqueous fluid in the living specimen is found in two chambers of the eye: the *anterior chamber* between the cornea and the iris and the *posterior chamber* between the iris and the lens. It is secreted by the cornea and drained into the Canal of Schlemm, a venous sinus. Refer to your textbook for a more complete discussion.

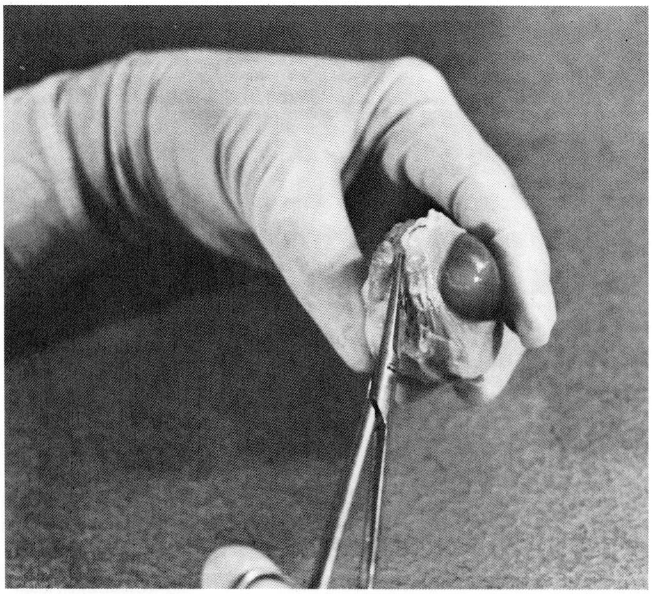

Figure 32-2 Opening the eye around its circumference with scissors.

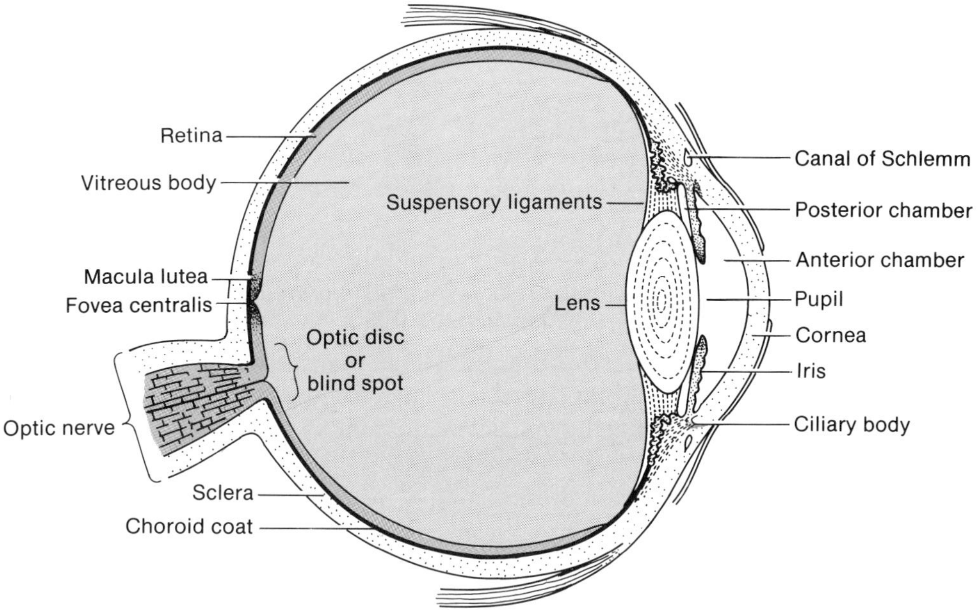

Figure 32-3 Mammalian eye in sagittal section.

The cavity within the eyeball between the lens and the retina contains a jellylike substance called the *vitreous body*. This viscous fluid creates pressure within the eyeball and keeps it from collapsing. Embedded in it is the large, transparent lens. The lens usually remains attached to the vitreous jelly in dissection specimens by black strands, the *suspensory ligaments.*

Remove the lens and vitreous body and find the *retina,* a thin, transparent membrane forming the inner coat of the eyeball. It contains the sensory receptors (rods and cones). In a sheep or cow eye it has a shiny, greenish-blue layer called the *tapetum,* which improves night vision. The tapetum is not found in the human eye.

The region where the retinal fibers become the optic nerve is called the *optic disc* or *blind spot.* In the next exercise you will demonstrate the blind spot in your own eyes. Close to the optic disc on the retina is the yellowish (in the living animal) *macula lutea.* In the center of the macula lutea is the *fovea centralis,* the region of sharpest vision with a high concentration of cone cells.

OBSERVATIONS OF THE CAT EYE AND HUMAN EYE

The layer outside the choroid coat is the *sclerotic coat* or *sclera.* It is tough and gives much support to the eyeball. The extrinsic eyeball muscles insert on the sclerotic coat and have their origin on the bony orbit of the skull. When they contract, they move the eyeball.

Examine the cat's eye without removing it from the body. Observe the eyelids. There are upper eyelids, lower eyelids, a *nictitating membrane* which represents a third eyelid, and a *medial caruncle,* located in the nasal angle of the eyelid folds. The nictitating membrane is a fold of tissue that can be released to cover the entire corneal surface. Only a remnant of the nictitating membrane is found in the human eye and it is called the *plica semilunaris.* Examine your eye with a mirror, and examine your lab partner's eye. Locate the plica semilunaris which can be seen lateral to the *medial caruncle.* It is vestigial. The conjunctiva, Fig. 32–3, contains blood vessels and covers most of the cornea. Over the center of the eyeball, the conjunctiva merges with the stratified epithelium of the cornea.

The portion of the eyeball from the lens to the retina contains the jellylike *vitreous body,* which creates pressure within the eyeball to keep it from collapsing.

B. ANATOMY OF THE HUMAN EAR

Structurally the human ear consists of three regions: the *external* or *outer ear,* the *middle ear,* and the *internal* or *inner ear* (Figure 32-4a).

The *outer ear* is made up of the *auricle,* or *pinna,* and the *external auditory canal.* The auricle is the large, soft flap on the side of the head which is supported by cartilage. Within it is the opening to the external auditory canal, a tube about 2.5 cm (1 inch) long which terminates at the tympanic membrane, or eardrum. The human auricle is vestigial, but the external auditory canal functions as a funnel that traps high-frequency vibrations and directs them to the middle ear. The skin lining this canal contains hair follicles and *cerumen* or *wax glands.*

The *middle ear* is an air-filled cavity in the temporal bone. It extends from the eardrum to the cochlea. The *tympanic membrane* arises from all three embryonic germ layers but generally is considered to be part of the middle ear. Attached to the tympanic membrane are three ear ossicles which have developed from the embryonic visceral arches and are therefore a part of the visceral skeleton. The ossicles, named malleus (hammer), incus (anvil), and stapes (stirrup), bridge the span of the middle ear and terminate at the *cochlea.* Also located in the middle ear is the opening to the *eustachian tube* which leads to the pharynx.

The lower region of the *inner ear* is known as the *cochlea.* It contains the sensory apparatus for hearing and is shaped like a snail shell, coiled 2¾ times. The entire structure is enclosed in a bony labyrinth within the temporal bone.

The stapes of the middle ear lies up against the *oval window* of the cochlea. Sound waves passing over the ear ossicles cause the membrane to vibrate, setting into motion the fluid in the cochlear channels, called the *scala vestibuli, scala tympani,* and *scala media.* The membranes separating these fluid-filled canals are the *vestibular membrane* and the *basilar membrane.* The basilar membrane supports the *organ of Corti* in the *scala media,* which is the sensory part of the ear in contact with the cranial nerve VIII. At the end of the scala tympani is another membraneous window, the *round window,* which serves as an outlet for sound wave vibrations in the cochlea. Just above the cochlea of the inner ear are three small semicircular canals. These function together as an organ of balance.

The parts of the ear are too small to make dissection feasible and are almost inaccessible, lying deep within the petrous region of the temporal bone. The location of the parts of the ear will be studied using models and diagrams. Trace the path of a sound wave by following the arrows in Figure 32-4b.

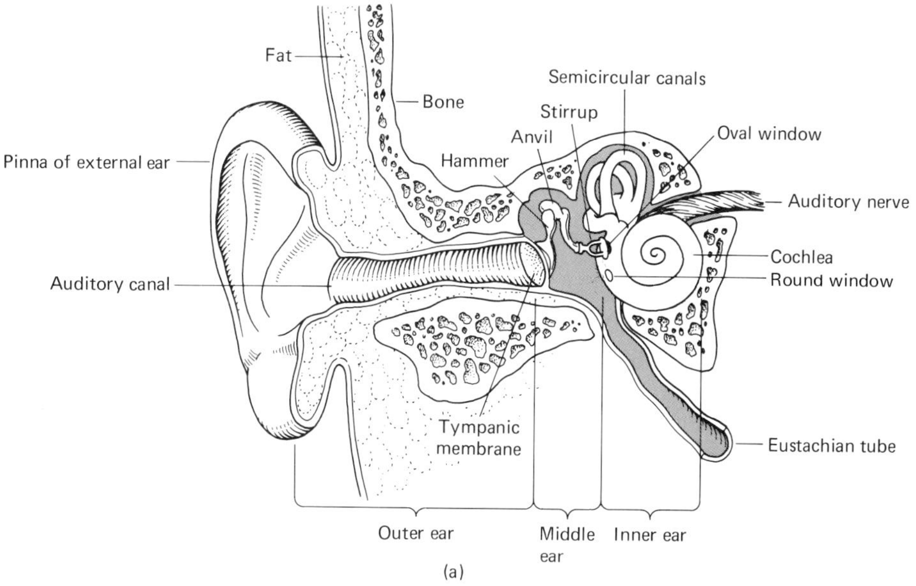

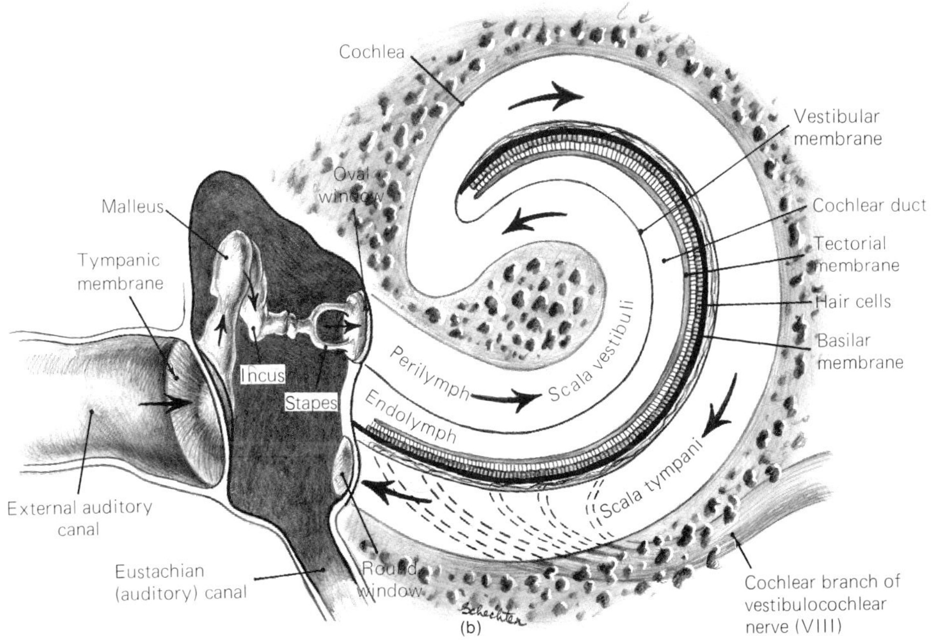

Figure 32–4 Human ear (a) The outer, middle, and inner ears and their parts. (b) Enlarged view of the cochlea, showing the pathway of sound vibrations.

With the aid of an otoscope (Figure 32-5a) make an examination of the external auditory canal and the tympanic membrane. Your instructor will explain the use and cleaning of the instrument.

EXAMINING THE EAR

If an otoscope (Figure 32-5) is available, make an examination of the external auditory canal and the tympanic membrane. Your instructor will explain the use and cleaning of the instrument. Carefully wipe the tip of the scope with a cotton square containing disinfectant.

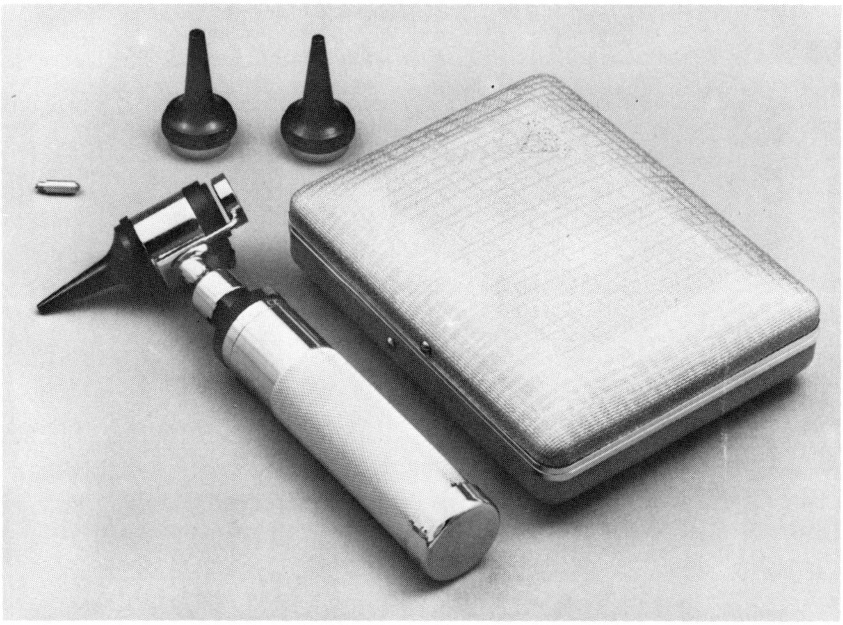

(a)

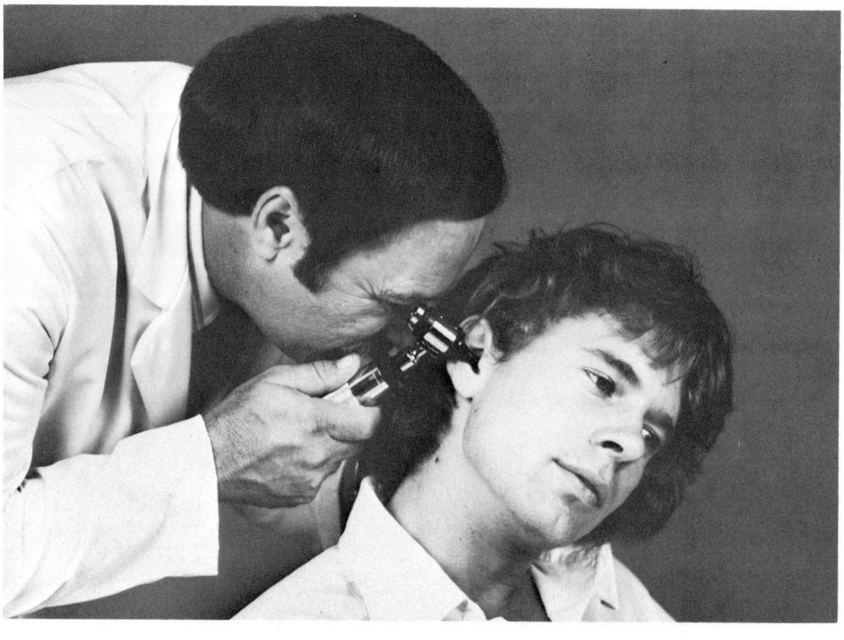

(b)

Figure 32–5 The otoscope.

C. EXAMINING THE EAR

Materials

The following materials should be provided for class use.

1. Model of the ear
2. Chart of the anatomy of the ear
3. Otoscope
4. Disinfectant and cotton

RESULTS AND OBSERVATIONS

DISSECTION OF THE EYE

1. Were you able to find all the parts of the eye shown in Figure 32-3? List the parts that you were able to observe. ~~yes~~, black lines, vitreous body, suspensory ligaments, retina, all of them, tapetum, optic disc, blind spot, lens

EXAMINING THE EAR

1. Description of view through otoscope. little hairs,

Quiz Test 32 & 33

Questions for Review

EXERCISE **32**

THE ANATOMY OF THE EYE AND THE EAR

Name Kathy Martin

Laboratory Section 4-29-91 Date 4-29-91

1. Label the parts of the mammalian eye in Figure Q32-1. Use the key below.

a. Sclerotic coat
b. Choroid coat
c. Retina
d. Vitreous chamber
e. Posterior chamber
f. Anterior chamber
g. Fovea
h. Optic disc (blind spot)
i. Cornea
j. Lens
k. Iris
l. Ciliary body
m. Suspensory ligaments
n. Pupil
o. Eyelid
p. Eyelash
q. Optic nerve

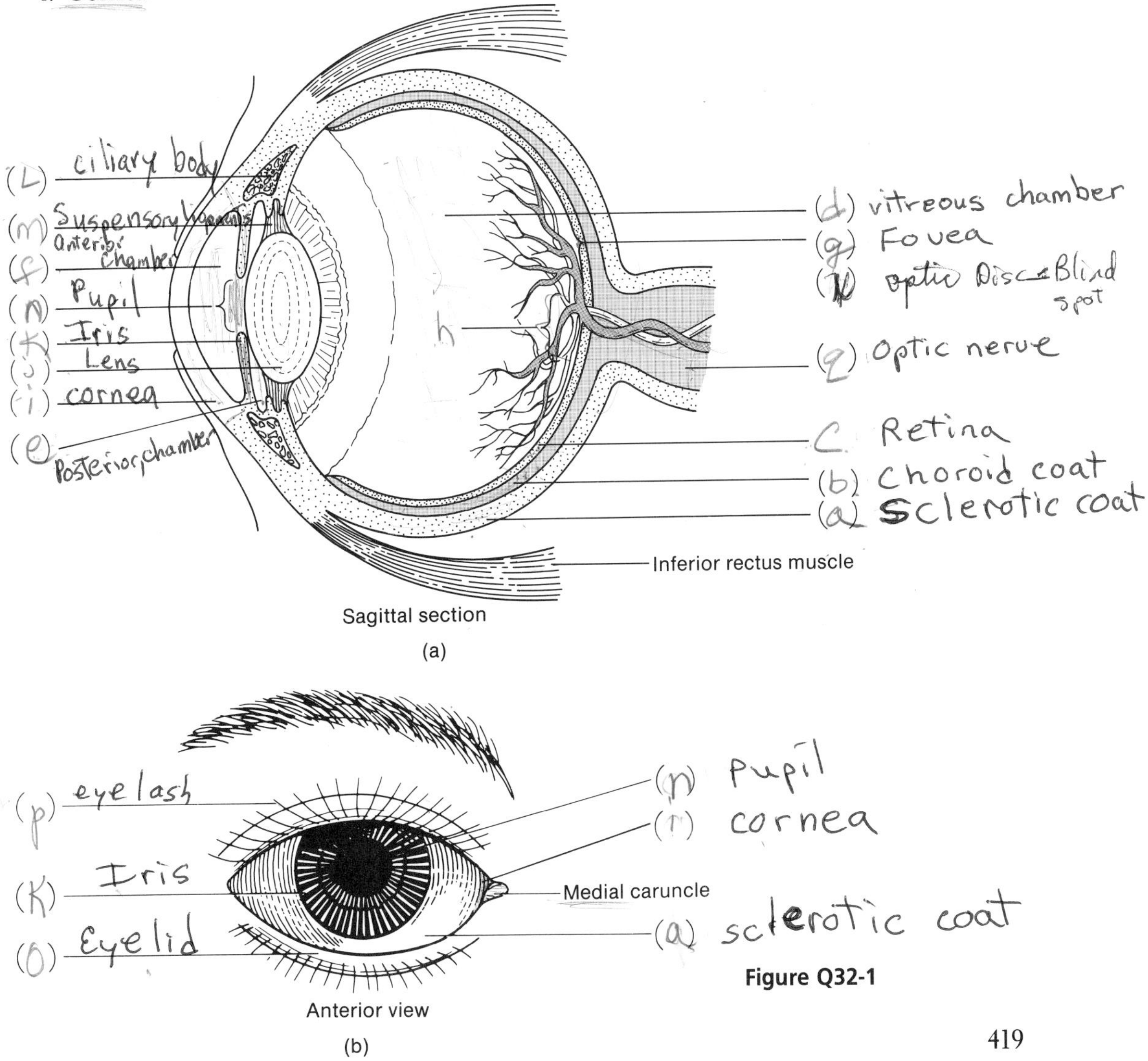

Figure Q32-1

2. Trace the pathway of light from the outer surface of the eye to the brain. Some of the steps are provided. You fill in the rest.

a. Conjunctiva
b. cornea
c. anterior chamber
d. pupil
e. posterior chamber
f. Lens
g. vitreous chamber
h. retina
i. optic nerve
j. Brain

3. Label the parts of the ear in Figure Q32-2. Use the key below.

a. External auditory canal
b. Tympanic membrane
c. Cochlea
d. Eustachian tube opening
e. Oval window
f. Round window
g. Scala media
h. Basilar membrane
i. Organ of Corti
j. Scala tympani
k. Scala vestibuli
l. Vestibular membrane
m. Malleus
n. Incus
o. Stapes
p. Tectorial membrane

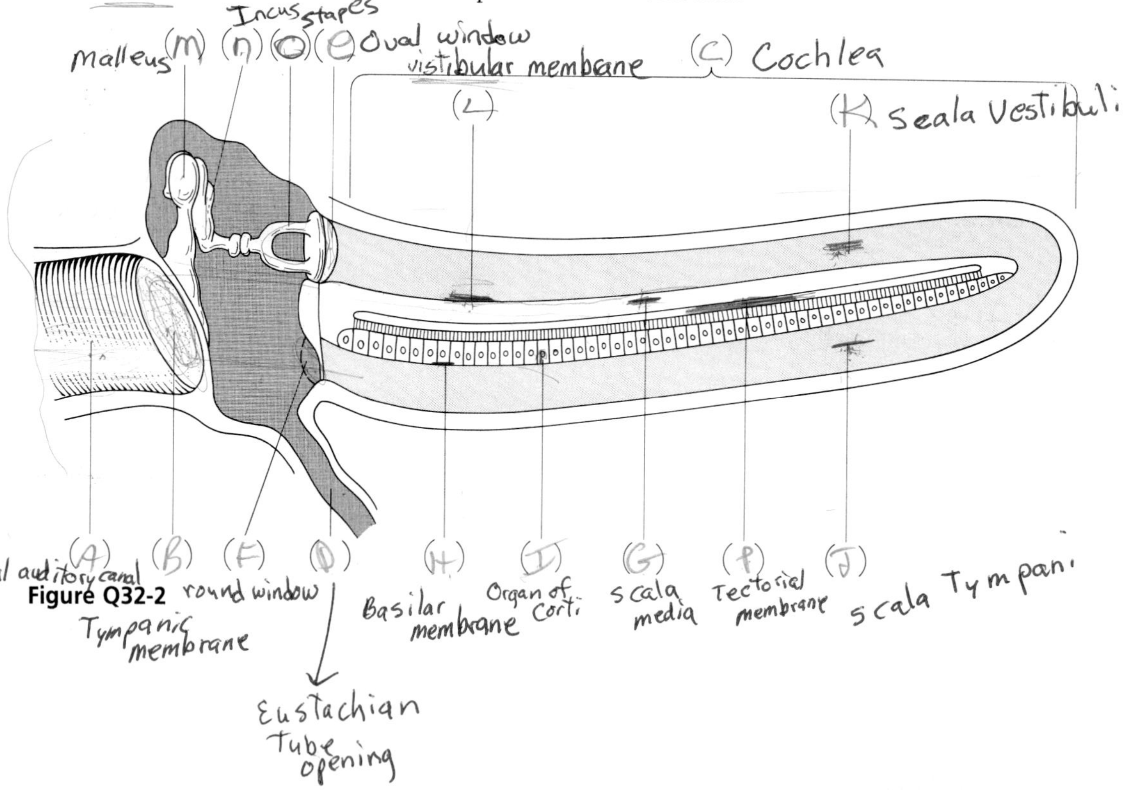

Figure Q32-2

EXERCISE 33

THE PHYSIOLOGY OF VISION

Light rays reflected from objects pass into the eye through the cornea and on through the anterior chamber and the lens. As light passes through these structures, it is focused on a light sensitive region, the retina (Figure 33-1). The iris regulates the amount of light entering the eye, and the position and shape of the lens can be shifted slightly to allow the eye to focus on objects at varying distances. This exercise will include a physical examination of the eye with the aid of an ophthalmoscope (Figure 33–2).

Tiny inverted images are focused on the retina, and the varying intensities and wavelengths of the light energy stimulate combinations of the photoreceptors, the cones and rods, to initiate a series of nerve impulses. These impulses are coordinated by other neurons in the retina and are transmitted by the optic nerves to the optic centers of the brain for final interpretation and integration. Procedures commonly used to study functioning related to this series of events are included in this exercise. They are a test of *visual acuity* (the degree of detail the eye can distinguish), a test for *astigmatism* (the inability of the eye to focus lines in all directions), and determination of the location of the blind spot.

The pupil is the opening of the center of the iris which controls light transmission. In the darkness the diameter of the pupil is wide, while in bright light it decreases. The dimensions of the pupil are controlled by the muscles of the iris which are under reflex control. Procedures to test pupillary reflexes are included in this exercise.

Sensation of any color can be obtained by the appropriate mixture of three colors of light: red, green, and blue. In color blindness, a sex-linked genetic disorder, the number of colors that may be distinguished by an individual is greatly reduced. With humans the most common type is red-green color blindness. A less common type is an inability to distinguish between blues and yellows. In the most severe forms of the genetic disorder all objects appear to be some shade of gray. Approximately 8.0 percent of males and 0.5 percent of females exhibit some degree of color blindness. A test for color blindness is included in this exercise.

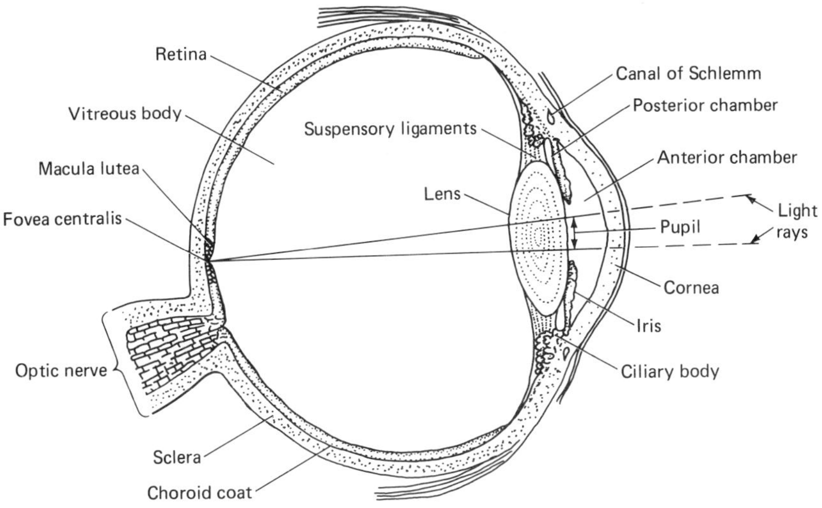

Figure 33-1 The components of the human eye shown in saggital (lengthwise) section. The focusing of light rays onto the *fovea centralis,* the area of sharpest vision also is shown.

Various drugs cause changes in the diameter of the pupil. For example, homatropine, which is a drug used in examinations, makes the pupil larger than normal. Drugs such as cocaine and epinephrine also cause a similar reaction. Morphine brings about a constriction of the pupil size.

Individuals constantly make judgments related to their perception of the size and distance of the objects they see. Such estimations depend on several factors including the relative positions of an individual's eyes, the size of the retinal images, the amount of detail seen, and the color and clarity of the outlines of objects. A false interpretation of a visual sensation results in an optical illusion. Examples of such illusions include objects appearing farther away or larger than they really are. A portion of this exercise will consider depth perception measurement and demonstrate optical illusions.

The retina is the location of the photoreceptor cells and is essentially an extension of the brain. Before light can reach the rods and cones, it must first penetrate between the many nerve cell bodies and processes that form the remaining layers of the retina. These nerve fibers all come together in a thick bundle near the back of the eye to form the *optic nerve,* which completely interrupts the retina. Because there are no receptors in this area, no sensation of vision is produced when light falls upon it. This area is called the *blind spot.*

Most individuals are normally quite unaware of the large gap in their visual field because the movements of their eyes continuously place images on that part of the retina where the sharpest vision occurs, the central fovea in the macula lutea. The macula lutea lies to the side of the blind spot, and it forms a disk, the outer rim of which contains both rods and cones (Figure 33-1). Toward the center the rods disappear, so that the central area or fovea contains only the cones. This is the area transmitting the sharpest image to the brain. The existence of the blind spot will be demonstrated in this exercise.

The photoreceptors, rods and cones, contain light-sensitive pigments that decompose when they absorb light energy. Rods contain a purple pigment known as *rhodopsin.* In the presence of light, rhodopsin breaks down into the yellowish substance *retinene,* which is made from vitamin A, and a colorless protein known as *scotopsin.* This photochemical dissociation of rhodopsin is called the bleaching reaction. In bright light the sensitivity of retinal rods is greatly reduced.

Individual cones are composed of retinene combied with one of three different types of visual proteins. There are three types of cones, namely, red, blue, and green. The particular color experienced by an individual depends on the particular set or combination of cones stimulated by a light source. If all three types of cones are stimulated, a person will see white; if none of the cones is stimulated, a person will see black. Cones are less sensitive than rods to light. They provide color vision and greater visual acuity. In situations when an individual becomes adapted to a bright light in the form of a light bulb and then quickly views a blank surface such as a wall, the image of the light bulb will appear. This effect is called a positive afterimage and results from the continued activity (firing) of the photoreceptors. Usually after a short period of time, the negative afterimage, a dark image of the bulb, will appear against a lighter background. This effect is caused by the bleaching of the visual pigment of the involved receptors. This exercise also will consider visual pigment activity.

A. EXAMINING THE EYE

Materials

The following materials should be provided for class use.

1. Model of the eye
2. Chart of the anatomy of the eye
3. Ophthalmoscope
4. A darkened room
5. Timers
6. Disinfectant and cotton

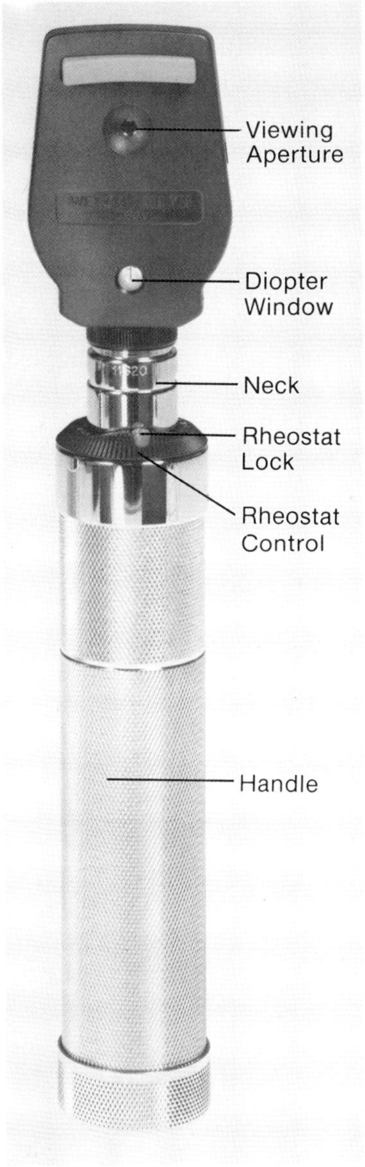

(a)

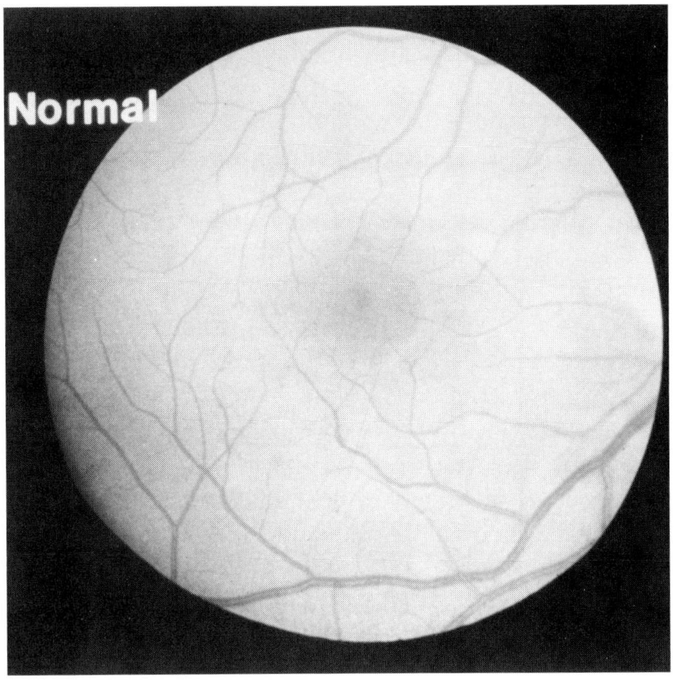

(b)

Figure 33-2 (a) The ophthalmoscope and its application. (b) The appearance of a normal retina as seen through an ophthalmoscope.

PROCEDURE

This procedure is to be performed by students in pairs.

1. Review the locations of the components of the eye (Figure 33–1) and of the retina (Figure 33–2b).
2. Examine the parts of the ophthalmoscope. Locate all of the parts indicated in Figure 33–2a. Pay particular attention to the lens selection disk, rheostat control, and the viewing aperture.
3. Take the ophthalmoscope in one hand and turn it on by depressing the rheostat lock button and rotating the rheostat control. Hold the instrument about 5 cm (2 inches) above the desk and rotate the rheostat control over its entire range of settings. Note how the intensity of the light changes from very bright to very dim with the movement of the control.

4. Set the lens selection disk on "O" and the light source on the green spot of the instrument. The ophthalmoscope is now ready for the eye examination procedure.
5. Next using Figures 33–3a through 33–3d practice holding the ophthalmoscope and the proper positions for examining both the right and left eyes. (Note that the ophthalmoscope is held in the right hand for the examination of the right eye and in the left hand for the left eye.)
6. In a darkened room have your partner seated on your right and facing you.
7. Grasp the ophthalmoscope, resting your index finger on the lens selection disk (see Figure 33–3c), and check to see that the "O" setting is in the diopter window and that the light is on.
8. Instruct your partner to set the timer for 1 minute and to look straight ahead at a fixed object or spot at his or her eye level. (*A safe time of light exposure is one minute.*)
9. Begin viewing your partner's right eye at a distance of about 12 inches as shown in Figure 33–3a. Shine the beam of light onto the pupil and examine the lens and vitreous body of the eye. A red reflex (red light reflection) will appear as you look through the pupil. If the image seen is not a sharp one, adjust the focus with the lens selection disk of the ophthalmoscope.
10. Move to within 4 cm (2 inches) of your partner, while still keeping the pupil in focus (Figure 33–3b). Try to direct the ophthalmoscope's beam of light toward the pupil's edge rather than its center. (The red reflection should be more obvious.)
11. Look for the *optic disk* of the retina and adjust the focus to obtain a sharp image. The blood vessels radiating from the optic disk should be clearly in view. Follow the path of at least two blood vessels to the edge of the retina.
12. Locate and identify the *macula* by having your partner look directly into the light beam for *1 second only*. DO NOT INCREASE THIS EXPOSURE TIME. Can you see any blood vessels in the macula?
13. Examine the entire surface of the retina. Pay particular attention to surface depressions, protrusions (bumps) or any other irregularity. Bring any unusual condition to the attention of your instructor.
14. Keeping in mind the 1-minute beam exposure time, examine the outer limits (periphery) of the retina. Ask your partner to look up so that you can see the superior portion of the retina, and to look down so that the inferior portion of the retina can be observed.
15. Shut off the ophthalmoscope, and answer the questions in the results and observations section.
16. Change places with your partner and repeat steps 1 through 15.

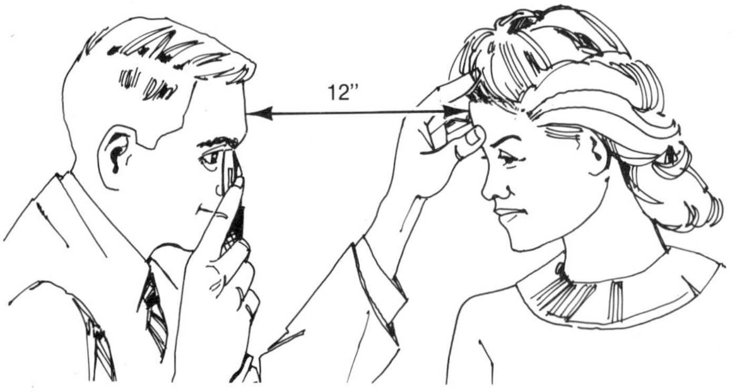

(a) The position for viewing the lens and vitreous body of the right eye.

(b) The position for examining the retinal surfaces of the right eye

Figure 33-3 Proper positions during eye examinations.

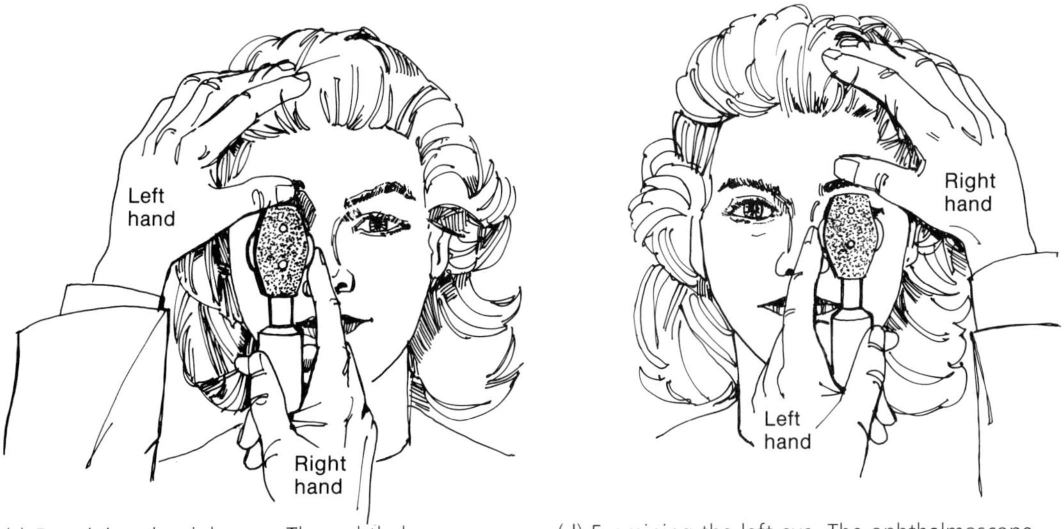

(c) Examining the right eye. The ophthalmoscope is held with the right hand.

(d) Examining the left eye. The ophthalmoscope is held with the left hand.

Figure 33-3 (*Continued*)

B. VISUAL ACUITY AND ASTIGMATISM TESTING

Materials

The following materials should be provided for class use.

1. Snellen chart
2. Astigmatism test chart
3. Blank 3 × 5 inch cards
4. Masking tape
5. Yardstick

Procedure 1: Snellen Visual Acuity Test

This procedure should be performed by students in pairs.

1. Place the Snellen chart in a well-lighted area of the laboratory (Figure 33-4).
2. Use the yardstick to measure a distance of 20 feet on the floor from the Snellen chart. Put a short piece of masking tape on the 20 foot site.
3. If you are wearing corrective lenses (eyeglasses or contact lenses), remove them. Face the Snellen chart and cover your left eye with a blank 3 × 5 inch card.
4. Read the letters of the fourth line from the top of the chart. Next, select and read the smallest line of letters you can see clearly. As you call out the letters, have your laboratory partner write them in the space provided in the results and observations section.
5. Repeat steps 3 and 4 with the right eye.
6. If you normally wear corrective lenses, repeat steps 3–5 with them on.
7. Determine the visual acuity for each eye without and with glasses. The measure of visual acuity states the distance at which the chart is read (20 feet) and the distance indicated on the chart for the smallest line the individual can read. For example, an individual standing 20 feet from the chart who can read line 2, but nothing smaller, is said to have a visual acuity of 20/100.
8. Compare the letters read with the actual letters on the Snellen chart. Enter the Snellen letter in the spaces provided in the results and observations section.

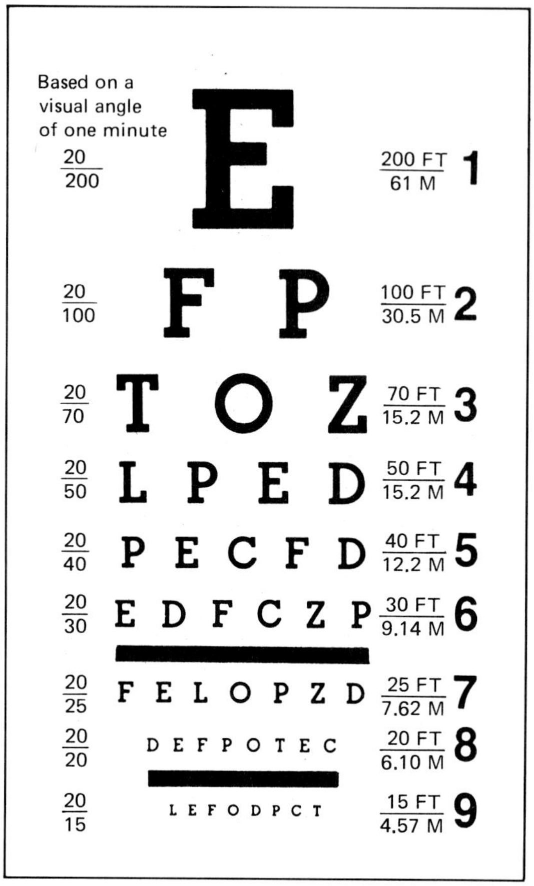

Figure 33-4 The Snellen chart. This chart is used to determine if an individual has normal or abnormal vision. The standard of normal vision (20/20) is the ability to stand 20 feet from the chart and be able to read line 8.

Procedure 2: Astigmatism Testing

This procedure should be performed by students in pairs.

1. Place the astigmatism test chart in a well-lighted area of the laboratory.
2. Use the yard stick to measure a distance of 10 feet from the chart. Put a short piece of masking tape on the 10-foot site.
3. If you are wearing corrective lenses, remove them. Test each eye separately. Look at the chart and determine which, if any, of the line groups appear exceptionally thicker and darker than others. Indicate such line groups by number in the space provided in the results and observations section.
4. If you wear corrective lenses, repeat step 3 with them on.

C. DEMONSTRATION OF BLIND SPOT

Materials

The following materials should be provided per two students.

1. Five coins
2. Four blank 3 × 5 inch cards
3. One millimeter ruler

Procedure

This procedure should be performed by students in pairs.

1. Place the five coins in a straight line about 8 millimeters apart.
2. Look at the coins and concentrate your view on the middle coin.
3. Close your right eye and hold a blank card over it. How many coins can you see? Have your partner record your findings in the results and observations section.
4. Repeat step 3 with your left eye closed.
5. Assist your laboratory partner with this procedure and record the findings in the results and observations section.

D. PUPILLARY REFLEXES

Materials

The following materials should be provided per two students.

1. One small flashlight (or penlight)
2. One millimeter ruler

Procedure 1: Light Reflex

This procedure should be performed by students in pairs.

1. Hold the millimeter ruler under one eye of your laboratory partner and determine the diameter of the pupil.
2. With the ruler in the same position flash a light into the eye. Quickly measure the diameter of the pupil.
3. Remove the light source and record your findings in the results and observations section.
4. Change places with your partner and repeat steps 1–3.

Procedure 2: Accommodation Reflex

This procedure should be performed by students in pairs.

1. Hold the millimeter ruler under one eye of your laboratory partner and determine the diameter of the pupil.
2. With the ruler in the same position have your partner look first at an object located at the far end of the room and second at your shoulder. Measure the diameter of the pupil in each situation.
3. Record your findings in the results and observations section.
4. Change places with your partner and repeat steps 1–3.

E. DEPTH PERCEPTION TESTING

Materials

The following materials should be provided for class use.

1. Depth Perception Testing System with the following items in place.
   a. Calibrated base with adjustment string attached
   b. Two slider bases equipped with arrow holders and arrows
2. Sunglasses
3. Commercially prepared disposable alcohol pads
4. Container for used alcohol pads

Procedure

This procedure should be performed by students in pairs.

1. Examine the Depth Perception Testing System provided. Make certain that all appropriate parts are operational. Note that one of the slider bases is attached to the adjustment string. Refer to Figure 33-5.
2. Place the testing system flat on a table or desk in a well-lighted area.

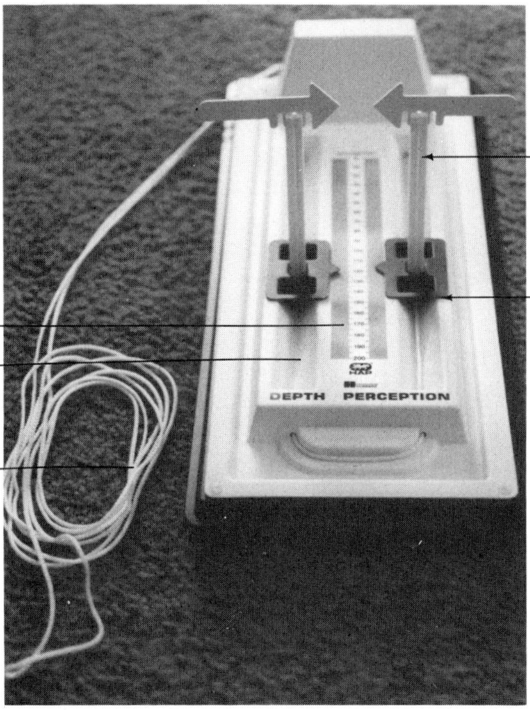

Figure 33-5 A depth perception tester. The components shown include arrow holders (AH), troughs (T), slider bases (SB), scale (S) and adjustment string (AS).

3. Have your laboratory partner sit comfortably on a chair situated 8 feet away from the depth Perception Testing System. The zero end of the scale should be closest to the chair. If your partner wears glasses or contact lenses have them removed.
4. Ask your partner to close his or her eyes while the unattached slider base is moved to some position along the testing system's millimeter scale (Figure 33-5). Make certain the arrows are facing one another and that they are at the eye level of your partner.
5. Give the adjustment string to your partner to align the two arrows by moving the slider base attached to the string.
6. When your partner is satisfied with the alignment, note the position of the string-attached (movable) slider base relative to the unattached one. Record any difference shown on the Depth Perception Testing System scale in millimeters in Table 33-6 in the results and observation section.
7. Repeat steps 4–6 using another position for the attached slider base on the testing system's scale.
8. Have your partner repeat steps 4–7, first with the right eye closed. Record all findings in Table 33-6, in the results and observations section.
9. Repeat step 4–8 with your partner wearing a pair of sunglasses. Record all findings in Table 33-7.
10. Change places with your partner and perform steps 1–9.
11. Answer questions listed in the results and observations section.

F. OPTICAL ILLUSIONS

Materials

The following materials should be provided for class use.

1. Millimeter rulers
2. Magnifying glasses (optional)

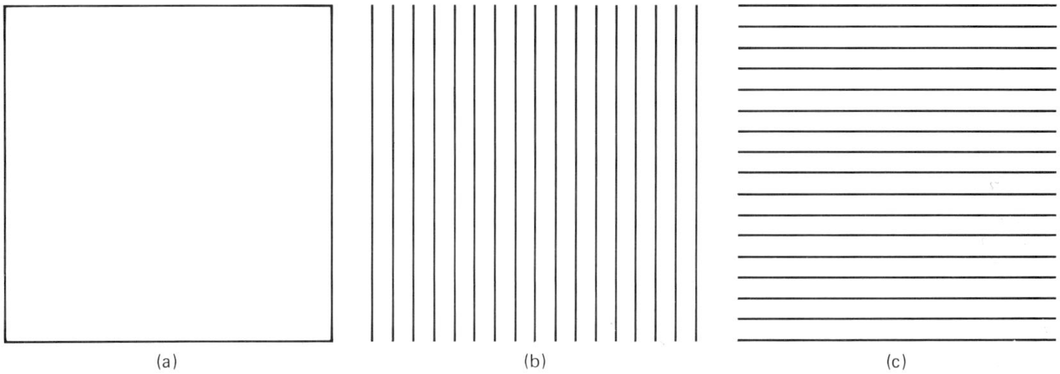

Figure 33-6 Optical illusion. Which square appears to be smallest?

Figure 33-7 Optical illusion. Which horizontal line is longer?

Procedure

This procedure should be performed by students individually.

1. Examine the three squares in Figure 33-6.
2. Determine visually which of the squares is smaller and which of the squares is taller. Record your findings in the results and observations section.
3. Measure each square with a ruler. Record your findings in the results and observations section.
4. Examine the diagrams in Figure 33-6 and 33-7.
5. Determine visually if the horizontal lines are of equal length. Record your findings in the results and observations section.
6. Measure the lines with a ruler. Record your findings in the results and observations section.
7. In the space provided in the results and observations section construct an optical illusion diagram of your own making. Test it with your laboratory partner.

F. TEST FOR COLOR BLINDNESS

Materials

The following materials should be provided for students in groups of four.

1. One Holmgren Color Vision Test
2. One Ichikawa or Ishihara Color Blindness Test
3. Correct explanation keys for color blindness tests

Procedure

This procedure should be performed by students in groups of four.

1. Each member of the group should take the Holmgren Color Vision Test.
   a. Select and match 12 different wool strands with those mounted in the binder. List responses.
   b. Consult key and list key colors. Circle incorrect responses in the results and observations section.
2. Each member of the group should take the Ichikawa or Ishihara Color Blindness Test.
   a. Examine each of the plates in the test booklet.

b. Enter your impressions as to the symbols on these plates and the key explanation for your responses in the results and observations section.

3. Compare the findings obtained with both color blindness tests and answer the questions in the results and observations section.
4. The instructor should compile test results of the whole class. Determine the number of individuals who are color-blind. Answer the questions in the results and observations section.

G. VISUAL PIGMENT ACTIVITY

Materials

The following materials should be provided for class use.

1. Laboratory lamps (40 watt bulbs) without covers or flashlights
2. Squares of bright, glossy red, yellow, green, and blue paper (4 × 4 inch) mounted in the center of the individual solid, black poster boards or papers (8½ × 60 inch)
3. Squares of bright, glossy, red, yellow, green, and blue paper (4 × 4 inch), with center-cut holes, mounted in the center of individual white poster boards or papers (8½ × 11 inch)
4. White poster board or paper sheets (8½ × 11 inch)

Procedure

This procedure should be performed by students in pairs.

1. Have your laboratory partner sit comfortably approximately 6 feet in front of a blank wall, and a table with a lamp without a shade or cover.
2. Turn on the lamp and direct your partner to stare at the bright light for 20 seconds and then immediately to look away toward the blank wall.
3. Ask what afterimages appeared. Record the findings in the results and observations section.
4. Stare at the bright light once again for 20 seconds. Now have your partner immediately hold in front of you a white poster board with a red paper square. Record a description of the types of afterimages seen in the results and observation section.
5. Did you see the colored square?
6. Next, have your partner first hold a black poster board with a red paper square in front of you. Stare at the square for 60 seconds. Quickly have your partner hold a white poster board in front of you. Note the color of the afterimages. Did you see a colored square? Answer the questions in the results and observations section.
7. Repeat step 6 with the yellow, green, and blue paper squares.
8. Change places with your partner and repeat steps 1–8.

RESULTS AND OBSERVATIONS

X EXAMINING THE EYE

1. Description of the right eye as seen through the ophthalmoscope. Red, you can see blood supply

2. Description of the left eye as seen through the ophthalmoscope. same

3. Were you able to find all the parts of the eye shown in Figure 33–2b? List the parts that you were able to observe. optic disc,

4. Were any unusual structures observed? No

VISUAL ACUITY AND ASTIGMATISM TESTING

1. Enter the results of the Snellen test in Table 33-1.

Table 33-1

Eye Condition	Letters Read	Actual Sequence on Chart	Visual Acuity
Left eye (without glasses)	could all	top to bottom	2/15
Left eye (with glasses, if applicable)	DFMPTH	DFMPTH	20/40
Right eye (without glasses)	e all	top to Bottom	20/15
Right eye (with glasses, if applicable)	4ZDTF	DFMPTH	20/30

2. Enter the results of the astigmatism test in Table 33-2.

Table 33-2

Eye Condition	Thicker and Darker Line Groups
Without glasses	none darker) same
With glasses	7+1

DEMONSTRATION OF BLIND SPOT

1. Enter your findings and those of your laboratory partner in Table 33-3.

Table 33-3

Student	Number of Coins Seen	
	Right Eye Closed	Left Eye Closed
Kathy	5	5
todd	5	5
Becky	5	5
	5	5

2. Would you expect to see more or fewer coins if the distance between coins were increased? less

3. What is the blind spot? ______________________________

____

PUPILLARY REFLEXES

1. Enter your findings on the light reflex in Table 33-4.

Table 33-4

Student	Pupil Diameter	
	Without Light	With Light
Kathy	5 mm	4 mm
Becky	5 mm	4 mm

2. Enter your findings on the accommodation reflex in Table 33-5.

Table 33-5

Student	Pupil Diameter		
	At Start	Distant Object	Near Object
M.M.	5 mm	5 mm	7 mm
Todd	4 mm	4 mm	5 mm

DEPTH PERCEPTION TESTING

1. Enter your findings and those of your laboratory partner obtained with conditions of normal lighting in Table 33-6.

Table 33-6

Student	Depth Perception Measurement Difference in Millimeters					
	Both Eyes Open		Right Eye Closed		Left Eye Closed	
	1	2	1	2	1	2
____________(without glasses)						
____________(with glasses, if applicable)						
____________(without glasses)						
____________(with glasses, if applicable)						

2. Was alignment achieved in one or both attempts with having both eyes open? ______

____

3. Was alignment achieved with either the right eye or left eye closed? ______________

4. a. If alignment was not achieved with both eyes open, or with either eye closed, which situation produced a greater millimeter distance away from the fixed slider base.

____

b. Offer an explanation for any differences that occurred. ______________

____

5. Enter your findings and those of your partner obtained with the use of sunglasses in Table 33-7.

Table 33-7

Student	Depth Perception Measurement Differences in Millimeters					
	Both Eyes Open		Right Eye Closed		Left Eye Closed	
	1	2	1	2	1	2
______(without glasses)						
______(with glasses, if applicable)						
______(without glasses)						
______(with glasses, if applicable)						

6. a. Was alignment more difficult with the sunglasses? ______________

____

b. Explain your findings. ______________

____

____

OPTICAL ILLUSIONS

1. Which of the squares appears to be smallest? ___b___ Which of the squares appears to be tallest? ___c___

2. Record the dimensions of the squares in Table 33-8.

Table 33-8

Student	Height (in mm)	Length (in mm)
Kathy	44 m	44 mm
todd	" "	" "
Buffy	" "	" "

3. Which of the horizontal lines appears to be longer? a

4. Record the length of the horizontal lines in Table 33-9.

Table 33-9

Line	Length
(a)	2 1/4 in 57 mm
(b)	2 1/2 in 57 mm

5. Construct your own optical illusion diagram.

COLOR BLINDNESS TESTS

1. Enter the responses to the Holmgren Color Vision Test in Table 33-10.

Table 33-10

Thread Color	Student Response 1	Student Response 2	Student Response 3	Student Response 4	Key Color

2. Enter the responses to the Ichikawa or Ishihara Color Blindness Test in Table 33-11.

Table 33-11

Plate	Student Response				Key Explanation
	1	2	3	4	
1					
2					
3					
4					
5					
6					
7					
8					
9					
10					
11					
12					
13					
14					
15					
16					
17					
18					
19					
20					
21					
22					
23					
24					

3. Were the findings obtained with the two color blindness tests similar? If not, explain the results obtained. ______

4. Summarize class results in Table 33-12.

Table 33-12

Students	Number Color-Blind	Type of Color Blindness	Percent Color-Blind
Female			
Male			
All			

5. How many color-blind individuals were unaware of their color blindness? ______

VISUAL PIGMENT ACTIVITY

1. Did you see all color squares as afterimages in the situations of the experiment? If not, which colors were not seen? No Red

2. Were the center holes in any of the colored squares observed in the afterimages? If so, which ones? Yes all but black background

3. What causes an afterimage? the image stays on retina

Questions for Review

EXERCISE **33**

THE PHYSIOLOGY OF VISION

Name ______________________________

Laboratory Section ______________________ Date __________

1. Define or explain the following.

   a. Visual acuity the degree of detail that the eye can distinguish

   b. Astigmatism the inability of the eye to focus lines in all directions.

   c. Color blindness a sex linked genetic order cannot distinguish certain colors

2. What is an optical illusion? When objects appear to be different than they really are.

3. List five factors that can affect judgment of size and distance.

   a. position of eye

   b. size of retinal images

   c. amount of detail

   d. color of object

   e. clairity of the outline

4. List two pupillary responses and the normal result for each.

   a. light reflex - pupil got small with light

   b. Accomodation reflex - distant object, light pupil got smaller - closer - larger

5. Are the following Snellen chart test results considered normal, better than normal, or worse than normal?

a. 20/20 normal
c. 20/100 Bad
b. 20/15 better than normal
d. 70/100 almost blind

EXERCISE 34

THE PHYSIOLOGY OF THE EAR

The ear has two distinct functions. The auditory component is concerned with the sensation of hearing, while the vestibular portion provides information to the brain about the movement and position of the head. The sensory elements of both components are hair cells in the inner ear (see Figure 32-4). These are excited by the bending of hairs and are referred to as mechanoreceptors.

The ear responds to mechanical disturbances produced by a vibrating body. For example, if the arms of a tuning fork are struck, they will swing back and forth, and a mechanical disturbance of molecules will be created through the surrounding air in all directions. Such vibrations are perceived by the mechanoreceptors as sound. The human ear can distinguish among the properties of sound which include intensity, pitch, and timbre. Intensity or loudness is measured in units called decibels. Pitch is determined by the frequency of vibrations, with higher frequencies producing higher pitched sounds. The range of frequencies to which the human ear is sensitive is from 20 to 20,000 hertz (vibration cycles per second). Timbre refers to the quality of sound and depends on the presence of several frequencies.

Sound is transmitted through the ear by first passing through the external auditory meatus and causing the tympanic membrane to vibrate. These vibrations are transferred to the bones of the middle ear and amplified as they reach the oval window. Vibrations of the oval window cause the fluid of the scala vestibuli to vibrate. This disturbance causes certain portions of the basilar membrane to bulge into the cochlear duct. Basilar membrane movement stimulates the hair cells of the organ of Corti which results in the conduction of impulses along the cochlear nerve. Vibrations of the fluid of the scala vestibuli are transferred to the scala tympani and spread as the round window bulges toward the middle ear (See Figure 32-4b for a visual representation of this process.) Interference or damage to this conduction system or to the nerve fibers or receptor cells results in hearing loss or deafness. Deafness is a severe form of hearing loss that makes everyday auditory communication difficult or impossible. Based on the part of the hearing mechanism that fails to function adequately, hearing loss or deafness can be classified into three main types, namely, conductive, sensory-neural, and central hearing loss.

Conductive hearing loss may be caused by simple situations such as the accumulation of wax in the ear. However, it also may be due to an inflammation and infection of the middle ear, known as otitis media, or to otosclerosis, a condition in which new bone growth may prevent the normal vibration of the stapes in the oval window.

Sensory-neural hearing loss results from injury or disease involving the organ of Corti or the cochlear nerve. Causes of this form of hearing loss include intense noise, ear damaging drugs, bacterial and viral infections of the inner ear, and aging.

Central hearing loss results from defects in the auditory tracts or centers in the brain. Several causes are associated with this condition and include traumatic injury, tumors, infections, blockage of blood vessels in the brain, and a progressive hearing loss that occurs with age known as presbycusis.

The vestibular apparatus, or nonauditory labyrinth, is the component of the inner ear that detects specific movements and positions of the head. Compensatory responses of the muscles of the neck, trunk, and limbs are initiated by the vestibular apparatus and help to keep the body in balance and in a proper orientation with respect to gravity. Motion sickness, a temporary disturbance, and Meniere's disease are known to interfere with the functioning of the organs of Corti. Motion sickness results from repetitive changes in the rate and direction of movement and from conflicting visual and vestibular signals. Meniere's disease may be caused by allergies, infections, or trauma. It results in an irritation or injury to the inner ear leading to the accumulation of an excess of endolymph.

Procedures in this exercise consider certain properties of sound and an individual's sensitivity to sound.

HEARING TESTS

Materials

The following materials should be provided for class use.

1. Watch, clock, or other small device with audible ticking sound
2. Tuning fork
3. Rubber reflex hammer
4. Individual cotton plugs
5. Yardstick
6. Masking tape

Procedure 1: Auditory Acuity

This procedure should be performed by students in pairs.

1. Have your laboratory partner sit down with eyes closed. Carefully insert a cotton plug into the left ear.
2. Hold a ticking watch or clock in line with your partner's right ear approximately 12 feet away.
3. As you move slowly toward your partner, have your partner indicate exactly when the ticking sound is heard.
4. Stop and measure the distance between the sound source and your partner's ear.
5. Repeat this procedure with the other ear. Enter your findings in the results and observations section.
6. Change places and repeat steps 1–5.
7. Answer the questions in the results and observations section.

Procedure 2: Sound Localization

This procedure is to be performed by students in pairs.

1. Have your laboratory partner sit down with eyes closed.
2. Bring a ticking watch into the hearing range of your partner from the following locations:
   a. Front of head
   b. Rear of head
   c. Left side of head
   d. Right side of head
   e. One arm's length above head on the right side
   f. One arm's length above head on the left side

3. Have your partner point to the position from which the sound is coming.
4. Indicate whether the sound localization was correctly identified in the results and observations section.
5. Change places and repeat steps 1–4.

Procedure 3: Rinne Test (Test for Middle-Ear Deafness)

This procedure is to be performed by students in pairs.

1. Have your partner sit down.
2. Locate the mastoid process of the temporal bone behind the left ear.
3. Strike a tuning fork and place the handle of the vibrating fork against the mastoid. (The sensation felt is bone conduction.)
4. Have your partner indicate when vibrations are no longer heard and immediately hold the tuning fork just outside the external ear. (In middle-ear deafness, no sound is heard by air conduction.)
5. Describe your partner's responses in the results and observations section.
6. Change places and repeat steps 1–5.

Procedure 4: Weber Test (Test for Middle-Ear Deafness and Nerve Defects)

This procedure is to be performed by students in pairs.

1. Have your partner sit down in a relatively quiet room.
2. Strike the tuning fork and place the handle of the vibrating fork against the middle of your partner's forehead.
3. Have your partner indicate the direction from which the sound appears to be coming. (With normal hearing the sound appears to be referred to the middle of the head. With middle-ear deafness the sound is detected in the defective ear. With nerve defects vibrations are received better with the normal ear.)
4. Describe your partner's responses in the results and observations section.
5. Change places and repeat steps 1–4.

RESULTS AND OBSERVATIONS

EXAMINING THE EAR

1. Description of view through otoscope. could see hairs.

HEARING TESTS

1. Enter your findings in the auditory acuity test in Table 34-1.

Table 34-1

Student	Distance at Which Sound Detected (Right Ear)	Distance at Which Sound Detected (Left Ear)
Brandi	one foot	" "
Shana	" "	6 in.

2. Was the auditory acuity of both ears the same? For you? No For your partner? Yes

3. Enter your findings in the sound localization test in Table 34-2. Indicate correct sound localization by a "+" and an incorrect identification by a "−."

Table 34-2

Student	Front	Rear	Left Side	Right Side	One Arm's Length to Left	One Arm's Length to Right
Gary	+	+	+	+	+	X
Vallery	+	+	+	+	+	+

4. a. Describe your response to the Rinne test. slight vibration behind ear, + could still hear

b. Describe your partner's response. same

5. a. Describe your response to the Weber test. feel vibration on head

b. Describe your partner's response. heard it all over

Questions for Review

EXERCISE **34**

THE PHYSIOLOGY OF THE EAR

Name ______________________________

Laboratory Section ______________________ Date ____________

1. Give the two distinct functions of the ear and the specific portions concerned with each.
Sensation of hearing + provides information to the brain.

2. What is sound? *Is different vibrations.*

3. List three specific properties of sound which the ear can distinguish.
 a. *intensity* c. *timbre*
 b. *pitch*

4. List the three main types of hearing loss and the particular portion of the hearing mechanism involved.
 a. *conductive hearing - was otitis media*
 b. *sensory neural - organ of corti or chochlear nerve*
 c. *central hearing - audatiory orach of anterian brain*

5. Briefly describe two tests for the detection of middle ear deafness.
 a. *Rinne test - hit tunning fork put end in middle forehead and should be felt on hole head*
 b. *Weber test - hit tunning fork put behind ear when vibration stops put by ear to see if sound cor still be heard*

6. What parts of the ear are associated with balance? *vestibular apparatus.*

7. List and briefly describe two disturbances involved with the functions of the organs of balance.

   a. motion sickness

   b. Menier's disease

8. Define or explain the following.

   a. Otitis media middle ear

   b. Otosclerosis a condition in which new bone growth may prevent the normal vibration of stapes in the oval window

   c. Presbycusis hearing loss that occurs with age

   d. Decibel measuring units to measure sound

EXERCISE 35

THE PHYSIOLOGY OF SMELL AND TASTE

Chemoreceptors that respond to chemical changes in the external environment are referred to as exteroceptors. These include the sense of smell, or olfactory receptors that respond to gaseous molecules in the air, and the sense of taste, or gustatory receptors that respond to chemicals dissolved in food or drink.

The receptors for the sense of smell are located in the olfactory cells in the mucous membrane lining of the upper portion of the nasal cavity. Only volatile substances in solution can stimulate the receptors. The gaseous particles enter the nasal cavity by diffusion from the circulating air in the external environment. Glands that surround olfactory cells secrete fluid in which the particles dissolve and this solution acts as a chemical stimulus to a nerve impulse.

In humans the sense of smell is less important than in other animals. However, it does play a role in starting the flow of certain digestive juices at mealtime. What have you experienced when you were quite hungry and smelled the aroma of food? The sensation of smell also can play a role in reviving a person who has fainted. Smelling salts stimulate the endings of the trigeminal cranial nerve and produce pain. This effect initiates respiratory reflexes and eventually results in reviving individuals who have fainted.

The epithelial surface of the tongue contains taste receptors which are specialized cells grouped into barrel-shaped structures called taste buds. (See Color Plate 28.) Each taste bud consists of a taste pore, which communicates with the external environment, supporting cells, and receptor cells, which have little hairs (microvilli) that project through the taste pore. Molecules dissolved in saliva at the tongue's surface interact with the microvilli causing receptor cells to release a neurotransmitting chemical. This response, in turn, triggers sensory nerve endings in the taste buds.

Despite the fact that individuals experience a variety and range of taste sensations, there are only four basic responses, namely sensations of sweet, sour, salty, and bitter. Figure 35-1 shows the general locations of the taste sensations.

Several experiments demonstrating the features of the sensation of smell and taste are considered in this exercise.

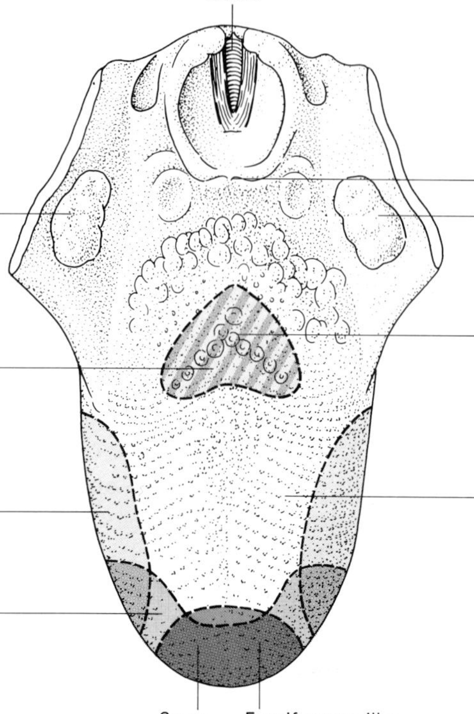

Figure 35-1 General structure of the tongue. Areas of specific taste reception on the tongue are indicated.

Materials

The following materials should be provided for class use.

1. Containers (with tops or plugs) of
   a. Freshly ground coffee
   b. Garlic powder
   c. Oil of cloves
   d. Oil of wintergreen
   e. Oregano
   f. Vanilla extract
   g. Vinegar
   h. Orange or lemon peels
   i. Perfume
   j. Distilled water
   k. Five percent NaCl solution
   l. Five percent sucrose solution
   m. 0.5 percent quinine-sulfate solution
2. Equally cut slices of apple, onion, and potato in separate and covered containers
3. Cotton swabs
4. Tongue depressors
5. Container for the disposal of swabs and tongue depressors
6. Distilled water for drinking
7. Paper towels

Procedure 1: Demonstration of Olfactory Stimulus

This procedure is to be performed by students in groups of four.

1. Dip a cotton swab into the perfume and gently squeeze out the excess fluid on the inner side of the container.
2. Apply the perfume to the forearm of one of your laboratory partners.
3. Note the time elapsed before each of your other laboratory partners first detects the aroma of the perfume after its application.
4. Note the time elapsed from application until the aroma is no longer detectable for each partner.
5. Record your findings and answer the questions in the results and observations section.

Procedure 2: Stimulus Identification

This procedure is to be performed by students in groups of four.

1. Smell the contents of each of the containers.
2. Close all containers. Wait 5 minutes.
3. Close your eyes. Have a laboratory partner open each container and pass it slowly about 3 inches from your nose. Your exposure should be no longer than 3 seconds for each container.
4. Identify each stimulus by name. Have your partner record whether each response is correct or incorrect in the results and observations section.
5. Immediately after the first round of testing, repeat steps 3 and 4 *two* more times.
6. Repeat steps 1–5 with each member of your group.
7. Answer the questions in the results and observations section.

Procedure 3: Taste Receptor Distribution

This procedure is to be performed by students in groups of four.

1. Have your partner rinse his or her mouth with distilled water and then use a paper towel to dry the surface of the tongue. Dispose of the towel in an appropriate container.
2. Moisten a clean cotton swab with the 5 percent sucrose solution and remove any excess fluid by pressing the swab against the inside surface of the container.
3. Touch several regions on your partner's tongue with the moistened swab and ask your partner to indicate the type of sensation and where it is detected. Indicate the site(s) on the sketch of the tongue in the results and observations section.
4. Have your partner rinse the mouth with distilled water again, and repeat steps 2 and 3 with the 5 percent NaCl, vinegar, and 0.5 percent quinine-sulfate solutions.
5. Change places with your partner and repeat steps 1–4.
6. Compare your responses.
7. Answer the questions in the results and observations section and those pertaining to this portion of the exercise in the questions for review section.

Lab

Procedure 4: The Relationship Between Smell and Taste

This procedure is to be performed by students in pairs.

1. Have your laboratory partner sit comfortably with eyes closed and nostrils held shut.
2. Place a small piece of apple or potato on your partner's tongue. Instruct your partner not to swallow and to identify the material placed on the tongue. Note whether each identification is correct or incorrect in the results and observations section.
3. Have your partner rinse his or her mouth with water.

4. Repeat step 2 with the other food, either apple or potato. Record your findings in the results and observations section.
5. Repeat steps 2–4 with the nostrils open.
6. Place a piece of onion on your partner's tongue. Remove it and repeat steps 1–5. Are the results obtained after the use of the onion similar to those obtained without the onion?
7. Change places with your partner and repeat steps 1–6.

RESULTS AND OBSERVATIONS

DEMONSTRATION OF OLFACTORY STIMULUS

1. Enter your findings on Table 35-1.

Table 35-1

Student	Time Elapsed Until Odor Detected (in min)	Time Elapsed Until Odor No Longer Detected (in min)
Justin	instantly	3 min.
Kim	instantly	6 min 4 sec.
Barb	instantly	13 min.
Erin	instanly	28 min.

2. Did the laboratory partner on whom the perfume was applied detect the aroma earlier than the other laboratory partners? No Did the laboratory partner on whom the perfume was applied detect the aroma after the others could no longer smell it? Explain. Because it was closer to that person

STIMULUS IDENTIFICATION

1. Enter all findings in Table 35-2. Indicate correct responses by a "+" and incorrect responses by "−."
2. Did the number of correct responses increase or decrease with each repetition of the test? No

Table 35-2

Stimulus	Student Justin			Student Barb			Student Kathy		
	Test 1	Test 2	Test 3	Test 1	Test 2	Test 3	Test 1	Test 2	Test 3
garlic	X	X	X	X	X	X	X	X	X
vanilla	X	X	X	X	X	X	X	X	X
peppermint	X	X	X	X	X	X	cloves	X	X
coffee	oregeno	X	X	X	X	X	X	X	X
cloves	X	X	X	X	X	X	X	X	X
Orengo	X	X	X	X	peppermint	X	X	X	X

TASTE RECEPTORS DISTRIBUTION

1. Circle and label the tongue outline in Figure 35-2 accordingly to indicate the location where a particular taste sensation was detected.

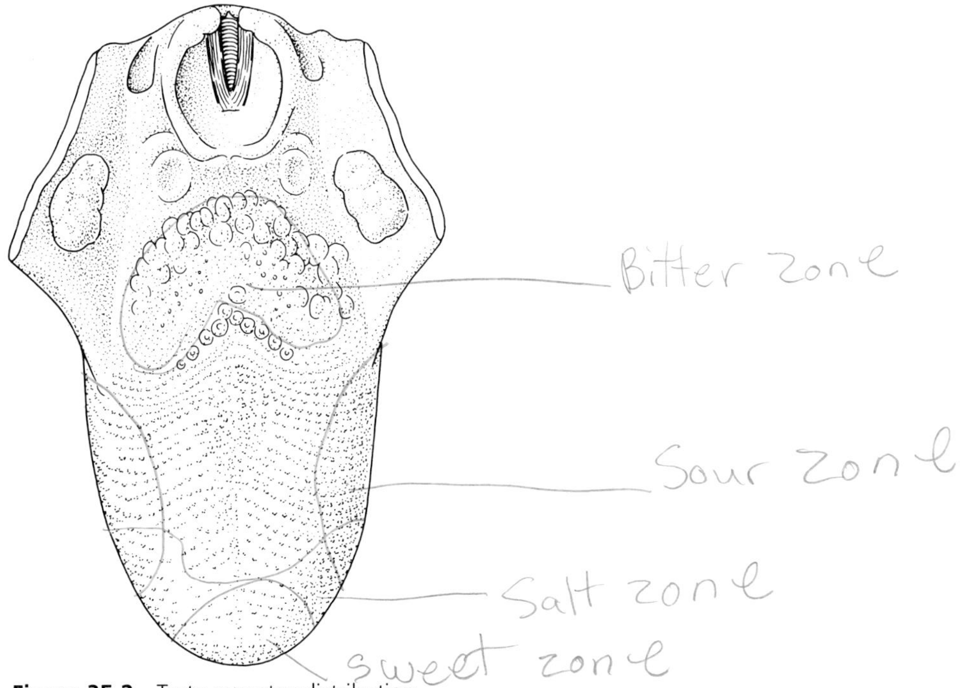

Figure 35-2 Taste receptor distribution.

2. a. Were all taste sensations present on your tongue?

yes

b. Were any taste sensations particularly more definite? If so, which one(s)? yes

sweet

3. a. Did any of the taste receptor locations overlap? If so, which one(s)? yes

Sour, Salt Sweet

b. How can this situation be explained? because there on

the tip of tongue in small area

THE RELATIONSHIP BETWEEN SMELL AND TASTE

1. Enter your findings in Table 35-3. Indicate correct identifications by a "+" and incorrect identifications by a "−."

Table 35-3

	Before Onion				After Onion			
	Nostrils Closed		Nostrils Open		Nostrils Closed		Nostrils Open	
Student	Potato	Apple	Potato	Apple	Potato	Apple	Potato	Apple
Justin	X No		No	Yes	No			Yes
Kim		X	No	Yes		Yes	No	
Barb		No	No	Yes	No	Yes		No
Erin		X	No	Yes		Yes	Yes	

Questions for Review

EXERCISE **35**

THE PHYSIOLOGY OF SMELL AND TASTE

Name Kathy Martin

Laboratory Section Mon 8:00 Am Date 4-28-91

1. Where are the receptors for the sense of smell located?

the olfactory cells in the mucous membrane lining of the upper portion of nasal cavity

2. What types of substances stimulate the receptors of smell?

volatile substances in solution

3. On what physical process does the sensation of smell depend? ______

4. Indicate the location of specific taste reception on the tongue in Figure Q35-1.

a. sweet c. Bitter

b. sour d. no specific zone

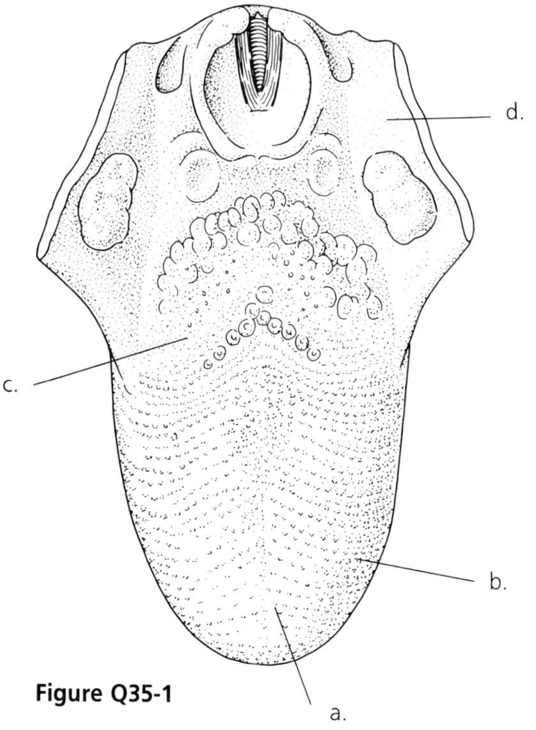

Figure Q35-1

5. Explain the relationship between the sensation of smell and the use of smelling salts (ammonium salts) to revive an individual who has fainted. smelling salts stimulate the endings of the trigeminal cranial nerve & produce pain the initiates respiratory reflexes + revives fainting peo

6. How would a cold affect an individual's sense of taste?
by pluging off the nose & block the senses

EXERCISE 36

CUTANEOUS SENSORY SYSTEMS

The sensations of touch include those of pressure, heat, cold, and pain. In humans and many other animals many of the receptors of these sensations are located in the skin (see Figure 9-1). Touch receptors enable many animals to monitor their environments.

Structure and different functions serve as the basis on which sensory organs can be categorized. From a structural standpoint, the sense organ can be the dendrites of sensory neurons which are either free, as in the case of those found in the skin mediating pain and temperature, or enclosed within non-neural structures, such as the pressure receptors in the skin. Other sense receptors, including the photoreceptors in the eye (Exercise 33) and taste buds (Exercise 35) synapse with sensory dendrites.

As indicated earlier, the structures of certain skin receptors for the different sensations are quite similar. The types of impulses they send to the brain are also very much alike. The impulse that registers heat, for example, is exactly the same as that for cold or for touch. Humans experience them as different sensations because they are transmitted to different centers in the brain. If the pathways were changed so that the cold receptors sent impulses to the heat center in the brain, one would probably feel a sensation of heat every time a piece of ice was touched.

Pain receptors are nerve endings not covered by a myelin sheath. Upon injury of the nerve ending an impulse is sent to the brain which is interpreted as pain. Even if the nerve ending itself is not injured, it can pick up signals from those that are and can transmit the same message. In a condition called *referred pain,* the sensation is not projected back to the point of stimulation. Pain receptors are more numerous than receptors for heat, cold, or touch, especially in humans.

The application of cold at times has been used to reduce or eliminate pain in certain clinical situations, such as times of tooth or mouth pain, bone marrow aspiration (removal), and postoperative recovery. The two forms of cold therapy currently in use are in ice massage (the direct application of ice with a circular or back and forth motion) and cold packs. The areas in which cold is applied include the skin site over or surrounding pain, a region between the pain and the brain, the opposite side of the body corresponding to the pain site, acupuncture points, and apparently unrelated areas. This exercise contains an experiment to test the effect of cold applications.

In this exercise, consideration will be given to sensory receptors that are (1) *thermoreceptors,* which respond to heat and cold, (2) *mechanoreceptors,* such as those for touch and pressure, which are stimulated by deformation of the sensory dendrite cell membrane; and (3) *nociceptors,* or pain receptors, which are stimulated by chemicals released from damaged cells.

A. MICROSCOPIC IDENTIFICATION OF THE SKIN AND SKIN RECEPTORS

Materials

The following materials should be provided for class use.

1. Prepared slides of human skin showing both epidermis and dermis
2. Prepared slides of human skin showing skin receptors

Procedure

This procedure is to be performed by students individually.

1. Examine the prepared slides of human skin.
2. With the aid of Figure 11-1 and your textbook identify the specific layers of the epidermis and the various glands typically found in the dermis.
3. Examine the preparation of human skin and identify the skin receptors present. Refer to Color Plates 29 and 30.
4. Answer the questions for review at the end of the exercise.

B. SKIN RECEPTORS

Materials

The following materials should be provided for class use.

1. Pens or markers with washable ink
2. Millimeter rulers
3. Metal compass or dividers with points blunted
4. Large beakers (500 ml) with water maintained at the following temperatures
   a. 4°C (cold)
   b. 25°C (room temperature)
   c. 45°C (hot)
5. Metal probes or teasing needles
6. Ice with or without plastic cups
7. Paper or cloth towels
8. Timers

Procedure 1: Heat and Cold Receptors

This procedure is to be performed by students in pairs.

1. On the underside of your laboratory partner's forearm use a fine-tip marking pen to draw a 4 × 14 mm rectangle divided into 2 mm squares (total of fourteen 2-mm squares). Allow the grid to dry.
2. Place a metal probe in each of the temperature-controlled beakers of water. Allow probes to remain in the beakers for 5 minutes before use.
3. Touch one square with a probe from one of the beakers. Touch the same square with the probes from the other beakers, one at a time. Ask your partner to identify each touch as cold, room, or hot, and record in the results and observations section whether each response is correct or incorrect.
4. To maintain the temperature of the probes, return them to the appropriate beaker frequently during the experiment.
5. Repeat step 3 on each of the remaining squares. Vary the order in which the probes for the three beakers are used.
6. Change places with your partner and repeat steps 1–5.

Questions for Review

EXERCISE **36**

CUTANEOUS SENSORY SYSTEMS

Name Kathy Martin

Laboratory Section Mon 8,00 Date 4-28-91

1. Identify the parts of the skin in Figure Q36-1.

a. hair shaft
b. sebaceous gland
c. stratum corneum
d. stratum lucidium
e. epidermis
f. meissners corpuscle
g. ~~nerve fibers~~
h. sweat gland
i. dermis
j. subcutanious tissue
k. arrector muscle
l. root of hair plexus root of plexus
m. bulb of hair
n. ~~hair follicle~~ hair follicle ~~root~~

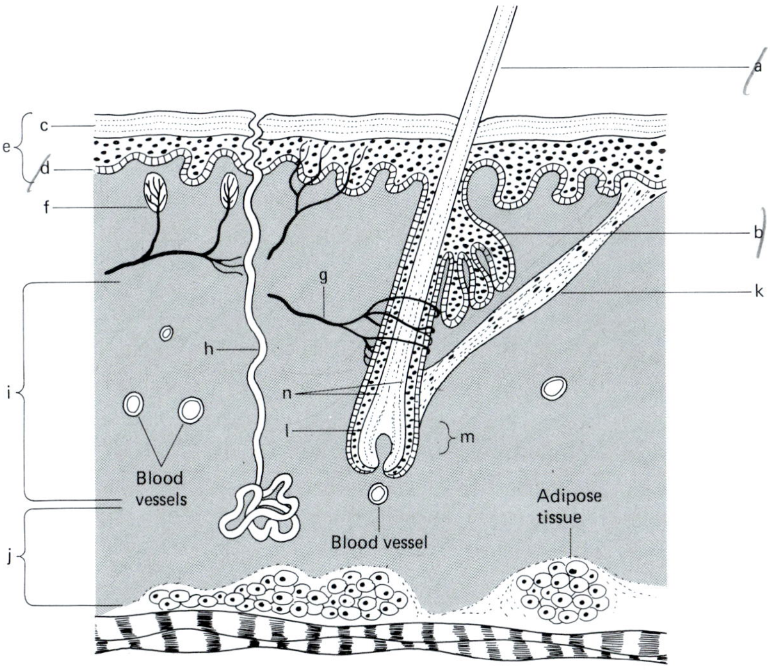

Figure Q36-1

2. What is a receptor? a sense organ or specialyed distal end of a sensory neuron that recieves stimuli from enviroment

3. List the specific cutaneous sensations.

a. pressure
b. heat
c. cold
d. pain
e. touch

4. Indicate the locations of the receptors listed.

a. Touch lower layers of epidermis
b. Pressure hypodermis, + lower layers of epidermis
c. Cold dermis
d. Heat lower layers of dermis
e. Pain lower layers of epidermis

5. What function do pain receptors serve? to send impulses to the brain and enterpet these impulses as pain

6. List two sense receptors in the skin that are free:

a. pain
b. temperature

7. List a sense receptor in the skin that is enclosed within a non-neural structure. pressure

8. Briefly describe the functions and/or activities of the following sense receptors in skin.

a. Thermoreceptor respond to heat + cold
b. Mechanoreceptor respond touch + pressure
c. Nociceptor pain receptors

SECTION X

CAT DISSECTION: AN INTRODUCTION TO THE VISCERAL ORGANS

It is not always possible to dissect human cadavers. The cat is a good substitute because its anatomy is similar to the human, and it is relatively inexpensive and easily available. The perfect situation is for each student to dissect his own specimen, but if finances do not permit this, it is possible to assign 2 or 3 students to each specimen. Throughout these exercises, references will be made to human anatomy, especially where it varies in important ways from the anatomy of the cat. Frequent comparisons with photos and line drawings of the human will make it possible to discuss human organ systems while working on the cat. Never forget that the ultimate goal in the class is to learn the human anatomy.

This exercise can be used to acquaint the student with the internal organs. Major organs can readily be identified, regions delineated and organ systems pointed out. Gently move the exposed organs about so that you can appreciate their depth. Do not cut anything out unless specifically told to do so by your instructor. Carefully separate one part from another without tearing membranes, and try to find as many of the organs mentioned in Exercise 37 as possible before going on to the anatomy and physiology of each system in detail in other exercises.

EXERCISE 37

AN EXAMINATION OF THE INTERNAL ORGANS OF THE CAT

Materials

The following materials should be provided for class use.

1. Cat dissection specimens
2. Dissecting tools
3. Bone forceps
4. Large dissecting trays
5. Charts and models of human organ systems

OPENING THE ABDOMINAL CAVITY

To open the abdominal cavity of the cat, place the animal on its back and begin a midventral incision line from a point between the legs. Continue this line anteriorly until you reach the diaphragm. The diaphragm can be seen or felt as you move forward. Do not cut through the diaphragm at this time since its anatomy is better appreciated if it is in one piece. When you reach the diaphragm level, cut laterally both to the right and to the left, following the contour of the diaphragm, which is a curved line. Also cut laterally on both sides near the beginning of your first incision line. Now you may fold the two flaps of muscle to the side and expose the entire abdominal cavity (Figure 37-1a and 37-1b). Gently move the abdominal organs so that the diaphragm can be observed.

Begin your study with the muscular diaphragm, which is divided into two parts: the *vertebral portion* and the *sternocostal portion*. The origins of the slips of the sternocostal portion are on the *xiphoid process* of the sternum and the last five ribs. The muscle fibers interdigitate with those of the *transverse abdominis* muscles. The origin of the vertebral portion is by a strong tendon from the ventral surface of the second, third, and fourth lumbar centra. The origin divides into two parts with muscle fibers that surround the openings of the aorta and the esophagus.

The muscles of the diaphragm contract in such a way as to lower the dome-shaped structure during inhalation, thereby expanding the thoracic cavity above it. During exhalation the muscles relax and the diphragm rises, decreasing the size of the thoracic cavity and forcing air out of the lungs.

Make a brief examination of the contents of the body cavity. All organs will be studied in detail in the next few exercises, but for now, try to become familiar with the major organs, membranes, and cavities.

OPENING THE THORACIC CAVITY

After the diaphragm has been studied, continue to cut forward to the neck region. Begin above the diaphragm about $\frac{1}{2}$ inch to the left of the midline. The ribs attach to the sternum at

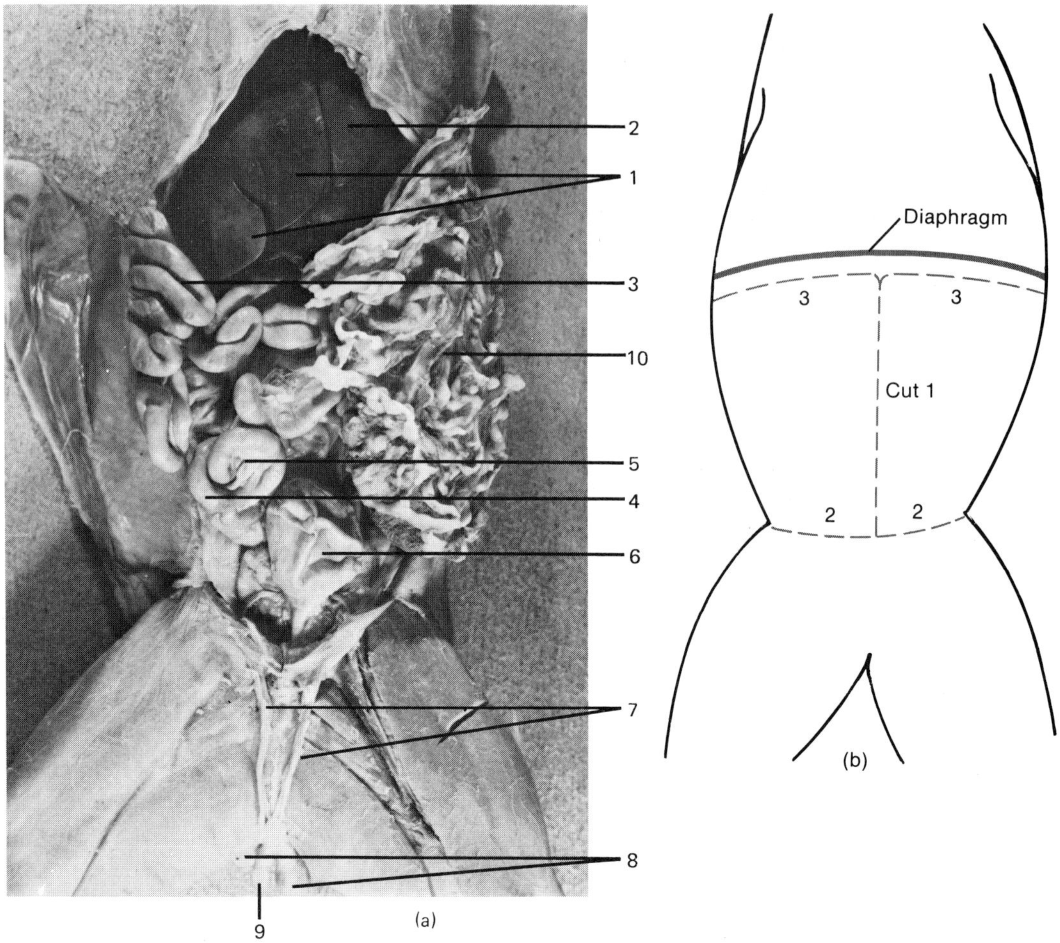

Figure 37-1 Opening the abdominal cavity.
1. Liver, right medial lobe
2. Liver, left medial lobe
3. Jejunum
4. Ileum
5. Cecum
6. Urinary bladder
7. Spermatic cord
8. Testis
9. Penis
10. Greater omentum

the midline by *costal cartilages* (observe them on the cat skeleton), and it will be possible to cut through these cartilages with your scissors. If you encounter too much resistance, use the bone forceps. When you have cut through the first costal cartilage, continue cutting forward until you reach the neck region. Cut transversely if necessary. Open the thoracic cavity on both right and left sides (Figure 37-2a and 37-2b). In the midline there are some important blood vessels that must be severed as close to the chest plate as possible.

Another method of opening the thoracic cavity is shown in Figure 37-3. The chest plate is removed as a triangular piece and finally severed at the neck region. It may be saved and used to cover the thoracic viscera between laboratory sessions.

You now have exposed both the abdominal and thoracic cavities and are ready to begin a brief examination of the organs.

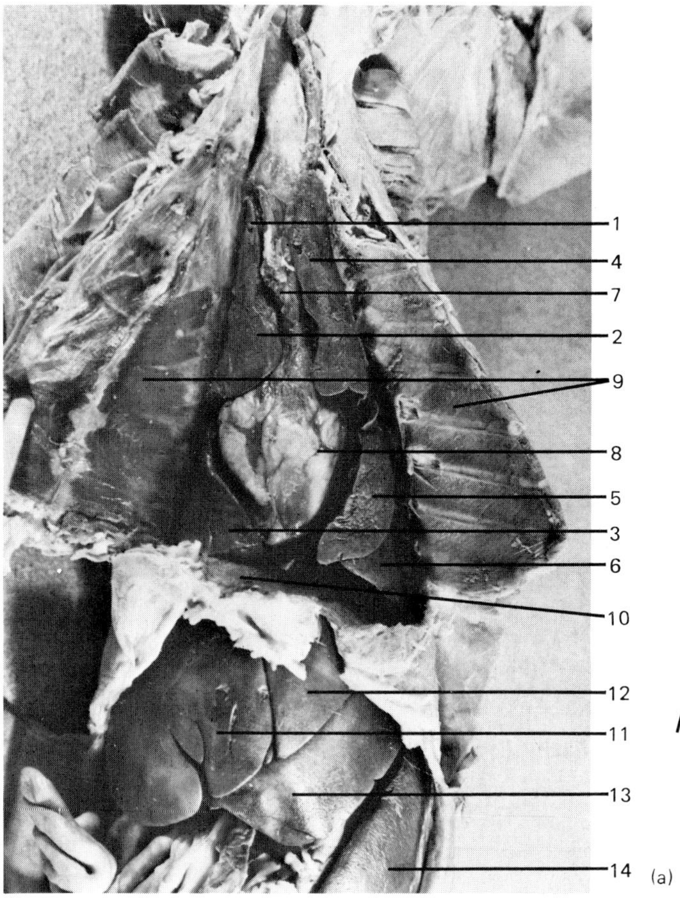

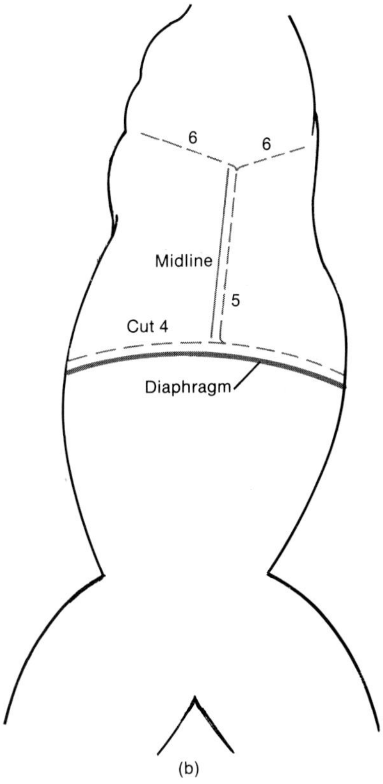

Figure 37-2 Opening the thoracic cavity.
1. Lung, right anterior lobe
2. Lung, right medial lobe
3. Lung, right posterior lobe
4. Lung, left anterior lobe
5. Lung, left medial lobe
6. Lung, left posterior lobe
7. Thymus gland
8. Heart
9. Chest plate
10. Diaphragm
11. Liver, right medial lobe
12. Liver, left medial lobe
13. Liver, left lateral lobe
14. Spleen

ORGANS IN THE ABDOMINAL CAVITY

Examine the organs in the abdominal cavity. Covering a large part of the ventral body cavity region in the cat is a broad, fatty membrane, the *greater omentum*. If you wish to preserve this mesentery, it can be loosened, freed, and folded to the left side of the animal. If, after observation, you want to cut it off, free it carefully from all organs to which it is attached. It usually is discarded when heavily infiltrated with fat. Figure 37-1 shows the superficial abdominal viscera.

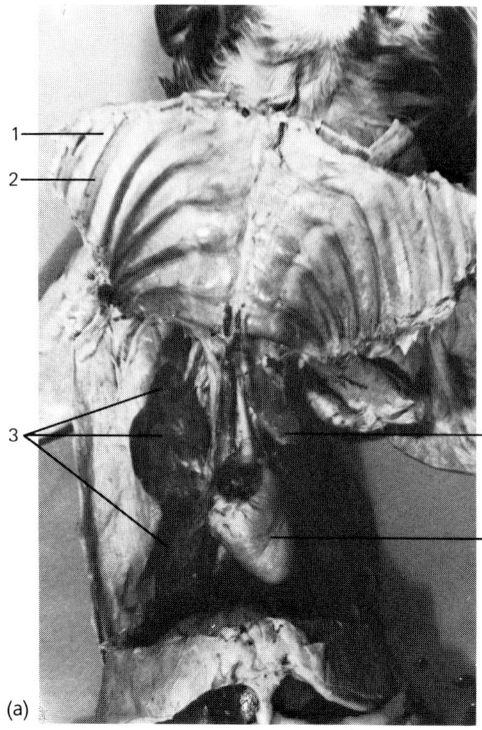

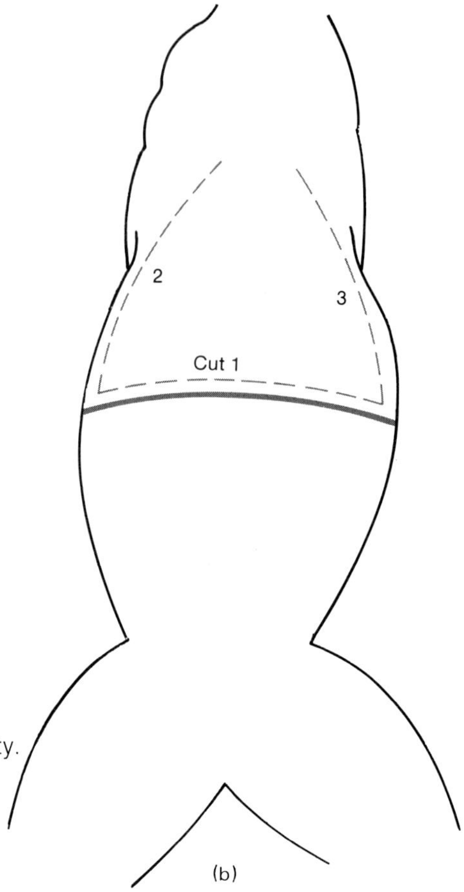

Figure 37-3 Alternate method of opening the thoracic cavity.
1. Chest plate
2. Rib
3. Lung lobes
4. Heart
5. Thymus gland (reflected)

THE LIVER

The liver is the largest organ in the abdominal cavity. It is usually dark brown in color because of the large amount of dried blood in it and becomes very hard and somewhat brittle in the preserved animal. Gently loosen its lobes to identify the following five parts.

1. *Right medial lobe*—largest of the lobes. The greenish *gallbladder* is located in a cleft within it.
2. *Right lateral lobe (quadrate lobe)*—lateral and dorsal to the right medial lobe. It also has a deep cleft.
3. *Right caudate lobe*—small, taillike lobe that lies deep and is centrally located. It is covered by a mesentery, which may be removed to expose it.
4. *Left medial lobe.*
5. *Left lateral lobe.*

THE DIGESTIVE TRACT

Under the left side of the liver is the J-shaped *stomach.* It begins at the *esophagus,* a small portion of which can be seen below the diaphragm, and it ends at the *pylorus,* which merges into the small intestine.

Follow the small intestine along its entire length. The *duodenum* is about 14–16 cm long and consists of two limbs. The *descending limb* runs 8–10 cm caudad of the pylorus. It then makes a U-shaped bend, extending cranially for 4–5 cm (the *ascending limb).* From that point the small intestine turns posteriorly and is called the *jejunum.*

The **jejunum** and the **ileum** make up the rest of the small intestine. Connecting all parts of the small intestine is a large mesentery, the **mesentery proper,** and within it is a large lymph node (not found in the human), the **pancreas of Aselli.**

The **ileum** enters the *large intestine* or *colon,* which has a much greater diameter. It consists of the **ascending colon, transverse colon, descending colon,** and **rectum.** The rectum opens to the outside through the *anus.*

At the beginning of the ascending colon, just beyond the ileum, there is a small blind sac, the **cecum.** The cat has no *appendix.*

Examine the area between the duodenum and the stomach. You will see a glandular organ, the **pancreas,** which has a longitudinal portion running alongside the duodenum called the *head of the pancreas* and a transverse portion called the **tail of the pancreas.**

On the cat's left side is a large, dark organ, the *spleen.*

UROGENITAL ORGANS

At the posterior end of the abdominal cavity is the ventrally located **urinary bladder.** The size will vary with the amount of urine it contained at the time of death, which determined its distention.

Located deep toward the dorsal surface are two large, bean-shaped organs, the right and left **kidneys.** Above and medial to each kidney is a small hard *adrenal gland.* Extending from the kidney is the long, yellowish **ureter,** which terminates at the *urinary bladder.* The kidneys and ureters lie beneath the **parietal peritoneum** (retroperitoneal).

If you have a female cat, you will find **uterine horns** in the same area as the ureter. Capping each is a small oval **ovary** surrounded by the **ostium** of the short **oviduct.** The **testis** in the male cat is located outside the body cavity in the **scrotal sac,** which was pointed out at the time the cat was skinned. One of the sperm-carrying ducts may be noted at this time. It is the **ductus deferens,** which loops over the posterior end of the ureter and enters the *urethra* posterior to the urinary bladder. The skin will be removed from the *scrotum* when the reproductive system is studied in detail, and the **ductus deferens** will be seen in its full length.

ORGANS IN THE THORACIC CAVITY

The most conspicuous organs in the thoracic cavity are the **lungs.** Lung tissue is soft, flexible, and dark brown in color. The lungs are composed of six lobes.

1. Right anterior lobe
2. Right medial lobe
3. Right posterior lobe
4. Left anterior lobe
5. Left medial lobe
6. Left posterior lobe

A seventh lobe, the **mediastinal lobe** on the right side, is recognized by some.

Lying in the midline among the lobes of the lung is the *heart.* It is surrounded by a tough sac, the **pericardium,** and will be dissected in a later exercise. Above the heart in a young cat is glandular tissue, which will vary in mass in different cats. It is the **thymus gland** (Figures 37-2 and 37-3). The younger the cat, the more thymus tissue will be found. An old cat (usually with yellow fat) may have only a trace of the gland or none at all. The thymus gland has a similar fate in humans, reaching its full size in childhood.

Along the dorsal wall is the flattened **esophagus.** Follow it to the **diaphragm** and see where it expands to form the **stomach** below the diaphragm. Lying ventral to the esophagus is the **trachea,** supported by cartilaginous rings. At the anterior end of the trachea is the **larynx,** which contains the **vocal cords.** Situated just posterior to the larynx are the lobes of the **thyroid gland,** connected by a narrow isthmus.

Major blood vessels in both the abdominal and thoracic cavities have been injected with a colored latex so that they can be traced. With few exceptions the arteries are injected red and the veins blue. The blood vessels will be dissected in more detail in a later exercise.

THE ENDOCRINE GLANDS

The following glands are found in the abdominal cavity.

1. **Islets of Langerhans.** The alpha and beta cells, which produce the hormones, are located within the *pancreas,* which can be seen between the stomach and the duodenum of the small intestine. It will be studied in Exercise 54 which discusses organs of the digestive system.
2. **Adrenal Glands.** These are small, hard masses of tissue located above and medial to the kidneys. See Figure 31-9.
3. **Ovaries.** The female reproductive organs also produce hormones. They are located in the dorsal part of the abdominal cavity lying anterior to the uterine horns.
4. **Testes.** The male reproductive organs, which produce testosterone in addition to spermatozoa, are located in the scrotum outside the body cavity. Although these organs originate in the body cavity, they migrate into the scrotal folds before birth. See Figure 37-1. The main spermatic duct, the *vas deferens (ductus deferens),* can still be seen in the body cavity looping over the ureter before entering the urethra.

The following glands are found in the thoracic cavity.

1. **Thyroid Gland.** This gland is located in the neck region on either side of the trachea just below the larynx. It appears as two oval masses which are connected by a small isthmus dorsally.
2. **Thymus Gland.** This will appear as a large mass lying over the midthoracic region above the heart. It decreases in size with the age of the cat, and in an old specimen it may be entirely absent.
3. **Parathyroid Glands.** These are comprised of four masses located in the thyroid tissue on the dorsal surface. They are encapsulated, however, and have no physiological connection with the thyroid gland.

The following glands are found in the head region.

1. **Pituitary Gland.** This gland has been studied in the exercise on the nervous system. It is located on the ventral surface of the diencephalon region of the brain.
2. **Pineal Gland.** This tiny gland also has been pointed out in the exercise on the nervous system. It is a dorsal outgrowth of the diencephalon region.

Questions for Review

EXERCISE **37**

AN EXAMINATION OF THE INTERNAL ORGANS OF THE CAT

Name ____

Laboratory Section ______________________ Date ____________

1. In the right-hand column, arrange the organs listed on the left in order according to the passage of food through the digestive tract from mouth to anus. Locate all organs in the cat.

a. Ileum	Mouth
b. Ascending colon	____________
c. Pharynx	____________
d. Descending colon (rectum)	____________
e. Esophagus	____________
f. Jejunum	____________
g. Mouth	____________
h. Transverse colon	____________
i. Stomach	____________
j. Cecum	____________
k. Duodenum	____________
l. Anus	Anus

2. Name six organs visible in the thoracic cavity of the cat.

a. ____________

b. ____________

c. ____________

d. ____________

e. ____________

f. ____________

SECTION XI

THE CARDIOVASCULAR SYSTEM

The human circulatory system performs a number of essential functions that include transporting various substances from one part of the body to another, defending the body against a wide variety of disease-causing microorganisms, assisting in the repair of damaged tissue, playing an important role in acid–base balance, and regulating body temperature. The cells of the body depend on the circulatory system for oxygen and essential nutrients and the removal of wastes. If the circulation to any region of the body is stopped, all the cells within that region soon die.

The cardiovascular system is the portion of the circulatory system that includes the heart and blood vessels. This system serves the body by moving blood between the cells of the body and the organs of the digestive, endocrine, integumentary, respiratory, and urinary systems, which communicate with the external environment. In performing this function, the heart works as a pump, forcing blood through blood vessels. These blood vessels, which form a closed system of channels, transport blood and allow for the exchange of gases, nutrients, and wastes between the blood and cells of the body. Although slight variations in heart activity occur normally, marked changes may be indications of cardiac abnormalities or disease. Specific exercises demonstrate the structure and organization of the components of the cardiovascular system. Other exercises in this section include the procedures used to demonstrate the normal and abnormal activities and processes associated with the cardiovascular system.

GENERAL REFERENCES

Cohn, P. F. *Clinical Cardiovascular Physiology*. Philadelphia: Saunders, 1984.

Netter, F. H. *The CIBA Collection of Medical Illustrations*. Vol. 5, *The Heart*. Summit, NJ: CIBA Pharmaceutical Products, 1969.

Tortora, G. J., and N. P. Anagostakos. *Principles of Anatomy and Physiology*, 6th ed. New York: Harper & Row, 1990.

Witherspoon, J. D. *Human Physiology*. New York: Harper & Row, 1984.

SPECIFIC REFERENCES

Brown, A. M. "Receptors Under Pressure: An Update on Baroreceptors." *Circulation Research*, 46:1–100, 1980.

Goldberger, A. L. *Myocardial Infarction: Electrocardiographic Differential Diagnosis*, 3rd ed. St. Louis: Mosby, 1984.

Oliver, M. F. "Diet and Coronary Heart Disease," *British Medical Bulletin*, 37:49–58, 1981.

Pickering, J. G., et al. "Blood Pressure During Normal Daily Activities, Sleep and Exercise," *Journal of the American Medical Association*, 247:992–996, 1982.

Ross, J., Jr. "The Failing Heart and the Circulation," *Hospital Practice*, 18:151–169, 1983.

EXERCISE 38

THE HEART AND ASSOCIATED BLOOD VESSELS OF THE CAT AND SHEEP

Materials

The following materials should be provided for class use.

1. Cat dissection specimens
2. Fresh sheep or beef hearts
3. Dissecting instruments
4. Model of the human heart
5. Charts of the human heart

CAT HEART

Locate the heart in the thoracic cavity. It is surrounded by a tough membrane called the *pericardium* or *pericardial sac.* Examine the sac carefully. Note that a pair of nerves attach to it laterally. These are the *phrenic nerves,* which supply the muscular diaphragm. Free the phrenic nerves (or at least the one on the left side of the cat) from the pericardial sac. Slit the pericardial sac and fold it back out of the way—or remove it entirely.

Identify all parts of the heart that can be seen without removing it from the body. Locate the *left* and *right ventricles,* the *left* and *right atria* (seen externally as the flaplike auricles), and the *coronary artery* and its branches on the surface of the heart.

By lifting the ventricles slightly and pulling the heart toward the cat's left side you can see the large veins entering the right atrium. Identify the *superior vena cava,* which returns blood to the heart from the head and shoulders, and the *inferior vena cava,* which returns blood from the organs below the diaphragm.[1] The *pulmonary artery,* which carries blood from the right ventricle to the lung, can be seen prominently in ventral view. Remove enough connective tissue from the anterior region to expose the large *aorta,* which arches to the left just anterior to the heart. The *pulmonary veins,* which empty into the left atrium, can be seen if the ventricles are lifted and pulled to the cat's right side.

[1]NOTE: The student should be aware that the blood vessels in the cat were named to correspond with those in the human, so that it is common to find the adjectives *superior* and *inferior* replacing the more correct terms *anterior* and *posterior* for the four-legged animal. To be consistent with terminology used in most textbooks on the cat and sheep, the terms *superior* and *inferior* will be used in these exercises.

SHEEP HEART

For a more complete study of the anatomy of the mammalian heart (which is remarkably similar in all species) use the freshly thawed or preserved sheep heart. As in the cat, the terms *superior* and *inferior* will be used. Orient the heart so that you are looking at its ventral side. Find the following parts using Figures 38-1 and 38-2: *left ventricle, right ventricle, left atrium, right atrium, pulmonary artery, aorta, superior vena cava, inferior vena cava,* and the *pulmonary veins.* There are four pulmonary veins, two from each lung, but they may be cut off in your specimen. Insert a probe into all major blood vessels to determine the chambers with which they connect.

To open the heart, begin at the end of the *pulmonary artery* and cut back into the right ventricle (Figure 38-3a). The valve at the exit of the pulmonary artery from the right ventricle is the *pulmonary semilunar valve.*

Open the ventricle carefully and observe the delicate membranelike *tricuspid valve,* which separates the right ventricle from the right atrium (Figure 38-3b). The valve is held to the *papillary muscle* of the ventricle by connective tissue strands called *chordae tendineae,* which prevent the valve from being forced into the atrium when the ventricle contracts.

Now cut through the tricuspid valve into the right atrium (Figure 38-3c). The muscle

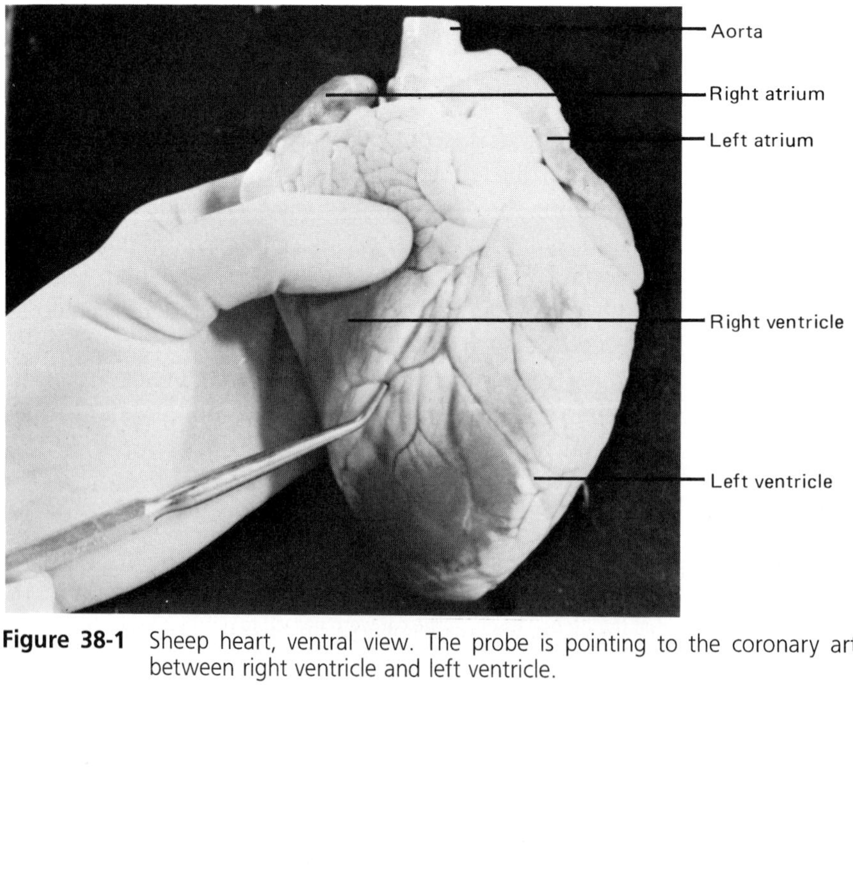

Figure 38-1 Sheep heart, ventral view. The probe is pointing to the coronary artery at the septum between right ventricle and left ventricle.

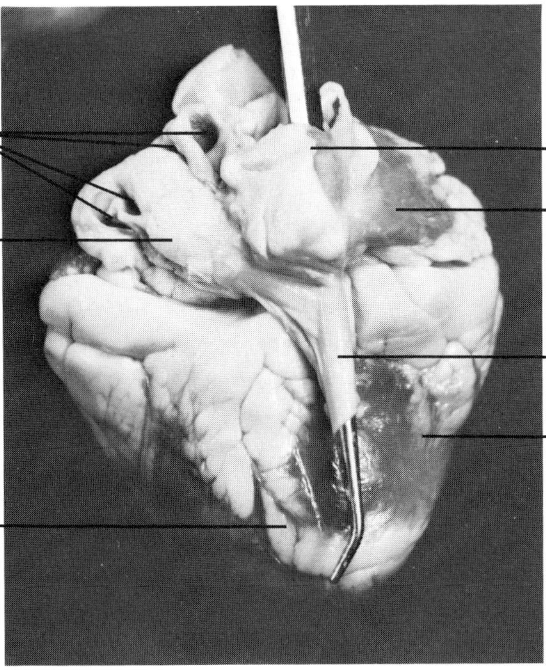

Figure 38-2 Sheep heart, dorsal view. The probe passes through right atrium from the superior vena cava to the inferior vena cava.

layer or *myocardium* of the atrium is thinner than that of the ventricle, and when not filled with blood, the atrial wall collapses. Open the atrium wide and locate the openings from the three veins that empty into it: *superior vena cava, inferior vena cava,* and *coronary sinus.* The latter, located between the *venae cavae,* is the expanded end of the *coronary vein,* which drains the heart tissue. The walls between right and left sides are called *septa.* The *interatrial septum* separates right and left atria, and the *interventricular septum* separates the two ventricles. The muscle of the atrium is called the *pectinate muscle.*

Now examine the left side of the heart. With your scalpel cut through the *left atrium* and the *left ventricle,* through the *mitral* or *bicuspid valve* which separates them, to the apex of the heart. Like the tricuspid valve, the mitral valve is held to the ventricular wall by chordae tendineae. Opening into the left atrium are the four pulmonary veins, which return oxygenated blood to the heart from the lungs. (This is an example of a vein that carries oxygenated blood. Another example is the umbilical vein of the fetus, which brings blood back from the *placenta.*)

Examine the left ventricle. Note the thickness of the muscle. The left ventricle pumps blood through the aorta to distant parts of the body and must contract with greater force than the right ventricle. Locate the *aortic semilunar valve,* which leads to the aorta. Cut into the aorta and open it to see the two large openings near its base (Figure 38-3d). These are the openings into the *right* and *left coronary arteries,* which carry oxygenated blood to the heart muscle.

Examine the area where the *arch of the aorta* passes close to the *pulmonary artery.* Find a small ligament connecting the two (the *ligamentum arteriosum*). This is a remnant of the *ductus arteriosus,* an important fetal lung bypass, which closes and becomes ligamentous in the adult heart. Consult your text for a discussion of the fetal circulation.

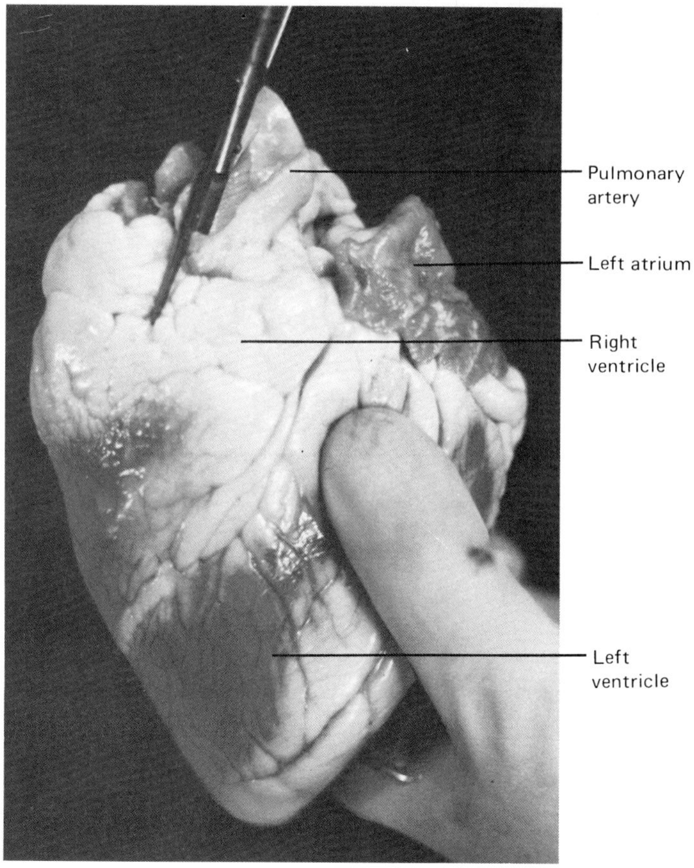

Figure 38-3a A scissors cut through the pulmonary artery into the right ventricle.

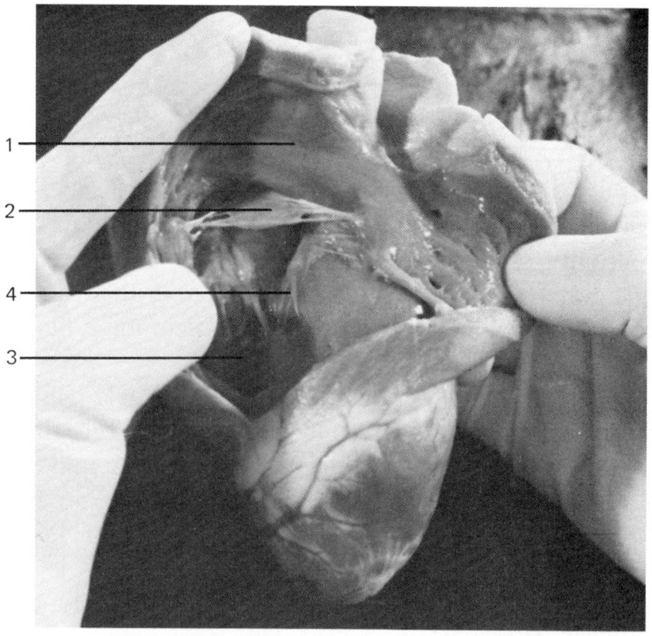

Figure 38-3b Right ventricle opened to show tricuspid valve.
1. Right atrium
2. Tricuspid valve
3. Right ventricle
4. Chordae tendineae

Figure 38-3c Right ventricle and right atrium opened to show the entire tricuspid valve and chordae tendineae. Probe runs beneath the tricuspid valve.
1. Right atrium
2. Tricuspid valve
3. Right ventricle

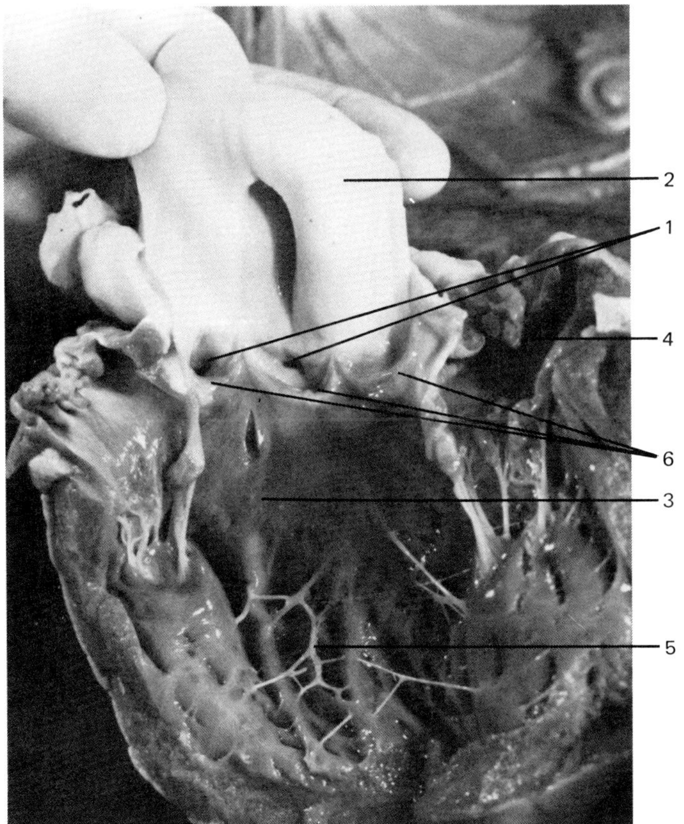

Figure 38-3d Aorta cut lengthwise and held open to show the coronary artery openings.
1. Openings to coronary artery
2. Aorta
3. Left ventricle
4. Left atrium
5. Chordae tendineae
6. Aortic semilunar valves

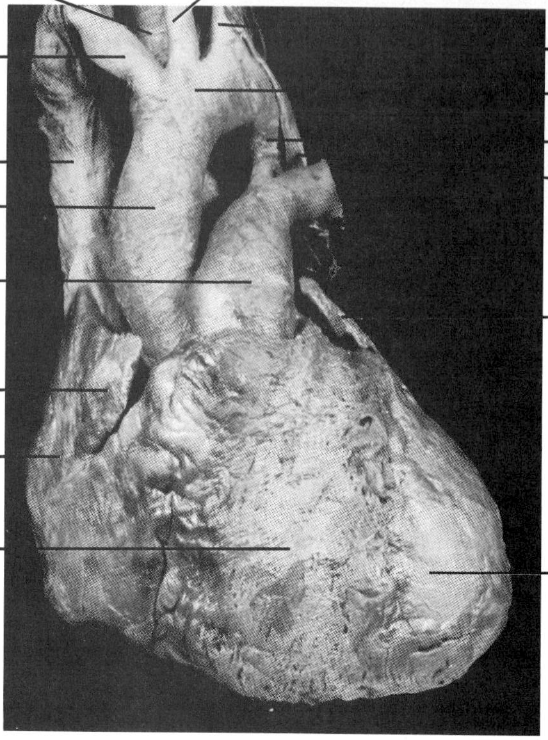

Figure 38-4 Structure of the heart. Photograph of anterior external view. (Courtesy of J.A. Gosling, P.F. Harris, et al., *Atlas of Human Anatomy,* Gower Medical Publishing Ltd., 1985.)

HUMAN HEART

A photograph of the human heart is shown in Figure 38-4. Its size and position in the thoracic cavity is much like that of the sheep. It is located in the *mediastinal cavity* and is surrounded by a tough pericardial sac, which is composed of two layers. The inner layer corresponds to the visceral peritoneum, and since it adheres tightly to the heart muscle, it is called the *epicardium.* The outer pericardial layer corresponds to the parietal peritoneum and lies very close to the inner layer. A potential cavity exists between the two layers of the pericardium. It is filled with fluid to prevent friction and is called the *pericardial space.* Figure 38-5 shows the relationship of the heart to major blood vessels associated with it.

HEART STRUCTURE AND BLOOD PATHWAY IN THE HUMAN HEART

The human heart is a four-chambered muscle. The two anterior chambers, the *atria,* are rather thin-walled, but the inferior chambers, the *ventricles,* are thick-walled. Separating the two atria is a muscular septum called the *interatrial septum.* Similarly, the ventricles are separated by a muscular *interventricular septum.* Valves between atria and ventricles (atrioventricular valves) are named for their anatomical configuration. The *tricuspid valve,* on the right side, has three flaps and the *biscuspid valve,* on the left side, has two flaps. The latter is often called the *mitral valve.* Associated with the right atrium are three main veins: the *superior vena cava,* the *inferior vena cava,* and the *coronary sinus.* Blood returning to the heart through these veins passes through the tricuspid valve into the right ventricle. When the right ventricle contracts and the tricuspid valve closes, the blood leaves the heart

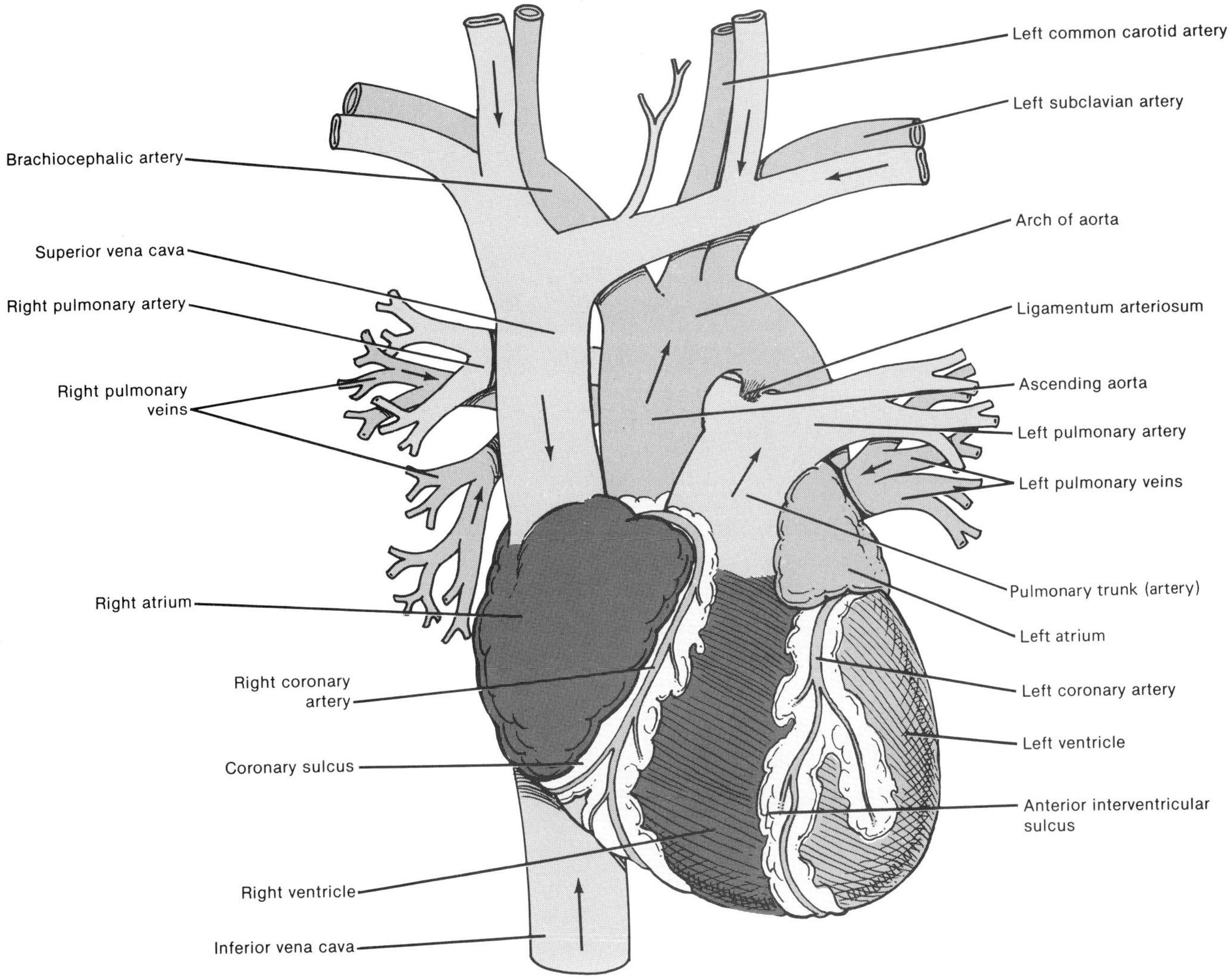

Figure 38-5 Structure of the heart. Anterior external view.

by way of the *pulmonary artery,* which terminates in pulmonary capillaries. Here the gases oxygen and carbon dioxide are exchanged, and the newly oxygenated blood returns to the left atrium in the *pulmonary vein.* When the left atrium contracts, blood flows through the mitral valves into the left ventricle. The mitral valve closes when the left ventricle contracts, and blood leaves the heart through the aorta to be distributed to all parts of the body. This tracing of a single drop of blood through the pulmonary circuit can be noted in the diagram at the end of this exercise (Figure Q38-1). Of course, it must be remembered that *both* atria and *both* ventricles contract simultaneously.

The tricuspid and mitral valves are made of fibrous tissue that is lined by the endothelial inner lining of the heart. They are not muscular. Both heart valves are held down to the ventricular muscle by connective tissue cords, the *chordae tendineae.* The large vessels that leave the ventricle (the pulmonary artery and the aorta) also possess valves at their bases to prevent backflow. These valves are called the *semilunar valves.* Individually they usually are referred to as the *pulmonary valve* and the *aortic valve.* All four valves are designed so that they permit blood flow in only one direction and do not allow for backflow unless they are damaged.

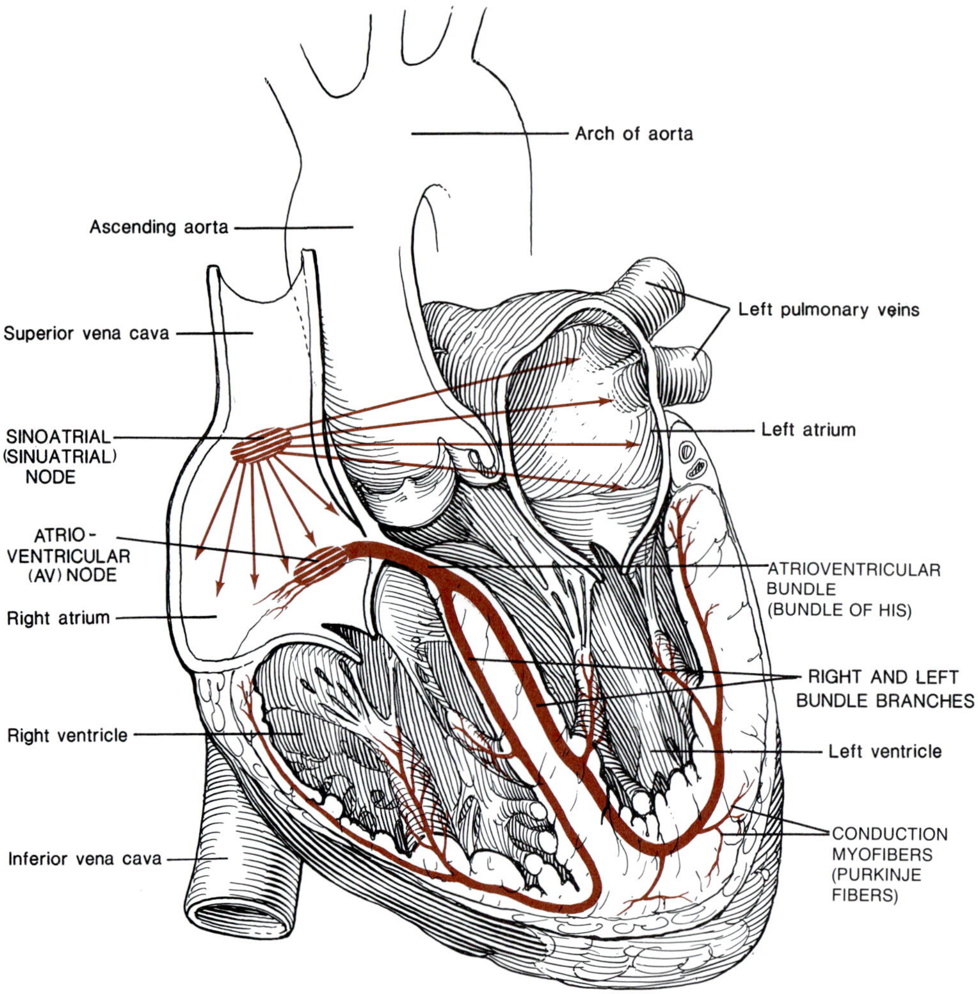

Figure 38-6 Location of intrinsic nervous system of the heart.

The microscopic nature of the myocardium of the heart was studied in Exercise 7. It is striated, branched, and has centrally located nuclei and intercalated discs. The histology of cardiac muscle should be reviewed at this time. On opening the heart chambers one finds a difference in the appearance of the muscle. In the atria, the muscles appear to be ridged like the teeth of a comb, and are called *pectinate muscles.* In the ventricles, the muscles project as small cones and are called *papillary muscles.* Irregular ridges and folds in the ventricular musculature are called *trabeculae carnae.*

CONDUCTION SYSTEM OF THE HEART

In addition to the extrinsic nerves from the autonomic nervous system, which regulate the beat of the heart, there is an *intrinsic* nervous system within the heart muscle which originates from embryonic cardiac muscle cells and which consists of the following parts (Figure 38-6):

Sinuatrial node (SA node)—the pacemaker in the right atrium
Atrioventricular node (AV node)—in right atrium near ventricles

Atrioventricular bundle (bundle of His)
Right and left bundle branches
Purkinje cells—extending into the muscle of each ventricle

A chick embryo heart can be seen to beat before any nerve fibers have formed. If the beating heart of a pithed frog is removed from the animal's body and placed in a dish of water, it will continue to beat. This indicates that nerve tissue, which initiates the beat, is incorporated into the muscular tissue of the heart.

Questions for Review

EXERCISE **38**

THE HEART AND ASSOCIATED BLOOD VESSELS OF THE CAT AND SHEEP

1. Identify the parts of the mammalian heart in Figure Q38-1.

a. ______________________ i. ______________________

b. ______________________ j. ______________________

c. ______________________ k. ______________________

d. ______________________ l. ______________________

e. ______________________ m. ______________________

f. ______________________ n. ______________________

g. ______________________ o. ______________________

h. ______________________

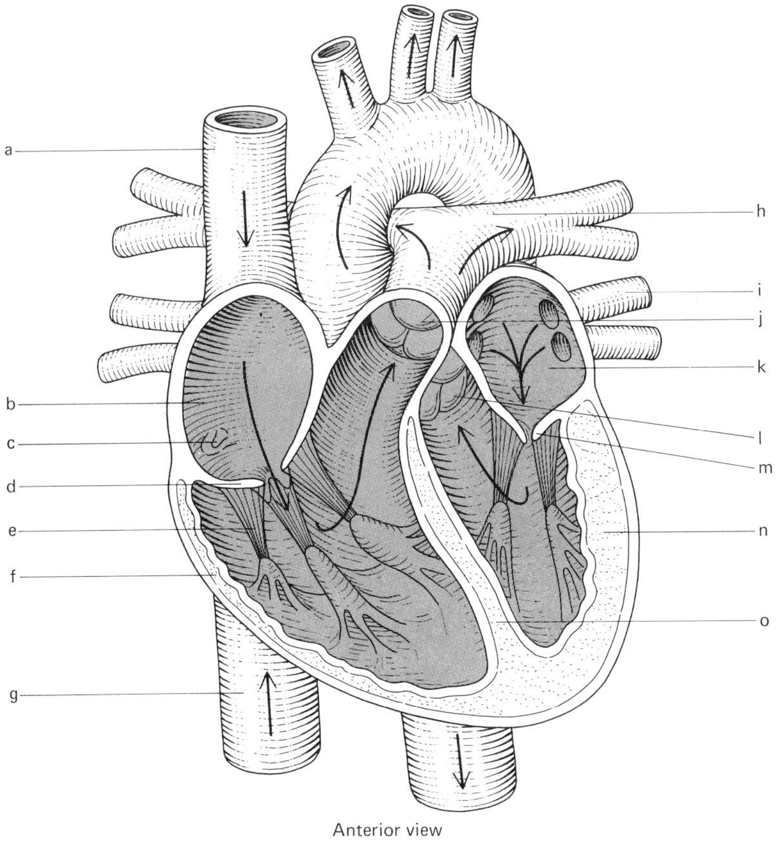

Figure Q38-1

EXERCISE 39

BLOOD VESSELS

Materials

The following materials should be provided for class use.

1. Cat dissection specimens
2. Dissecting tools
3. Charts of human blood vessels
4. Demonstration of the *portal system* in the cat

MAJOR ARTERIES OF CAT THORAX, NECK, AND SHOULDER

As you locate the arteries discussed in this exercise compare them with Figure 39-1. Dissect on the left side of the cat whenever possible.

Return to the heart. Locate again the *pulmonary artery* that leaves the right ventricle and carries blood to the lung. Leaving the left ventricle is the *aorta*. The first part of the aorta that gives rise to the *coronary arteries* is called the *ascending aorta*. The *arch of the aorta* makes a sharp turn to the left and extends posteriorly as the *descending thoracic aorta*.

Arising from the region of the aortic arch are large arteries that supply the head and shoulder regions. On the right side of the arch find a large straight artery that runs anteriorly for about an inch before branching. This is the *brachiocephalic artery*. Its branches are the *right subclavian artery* to the arm and the *right* and *left common carotid arteries*, which pass anteriorly alongside the trachea and give rise to a number of smaller branches in the neck and head region (labeled in Figure 39-1).

The subclavian artery reaches the underarm region, where it is called the *axillary artery*, and then passes into the arm, where it is known as the *brachial artery*. Some important branches of the subclavian, axillary, and brachial arteries are shown in Figure 39-1.

Examine the left side of the aortic arch. Note that there is no brachiocephalic artery here. The *left subclavian artery* arises directly from the arch of the aorta and has the same branches as the right subclavian.

MAJOR ARTERIES OF CAT ABDOMINAL CAVITY

Follow the *descending thoracic aorta* through the diaphragm into the abdominal cavity, where it is known as the *descending abdominal aorta* (Figure 39-2). Just below the diaphragm the aorta gives off a short unpaired vessel, the *celiac artery*, which has three important branches: the *hepatic artery* to the liver, the *left gastric artery* to the stomach, and the *splenic artery* to the spleen, stomach, and pancreas.

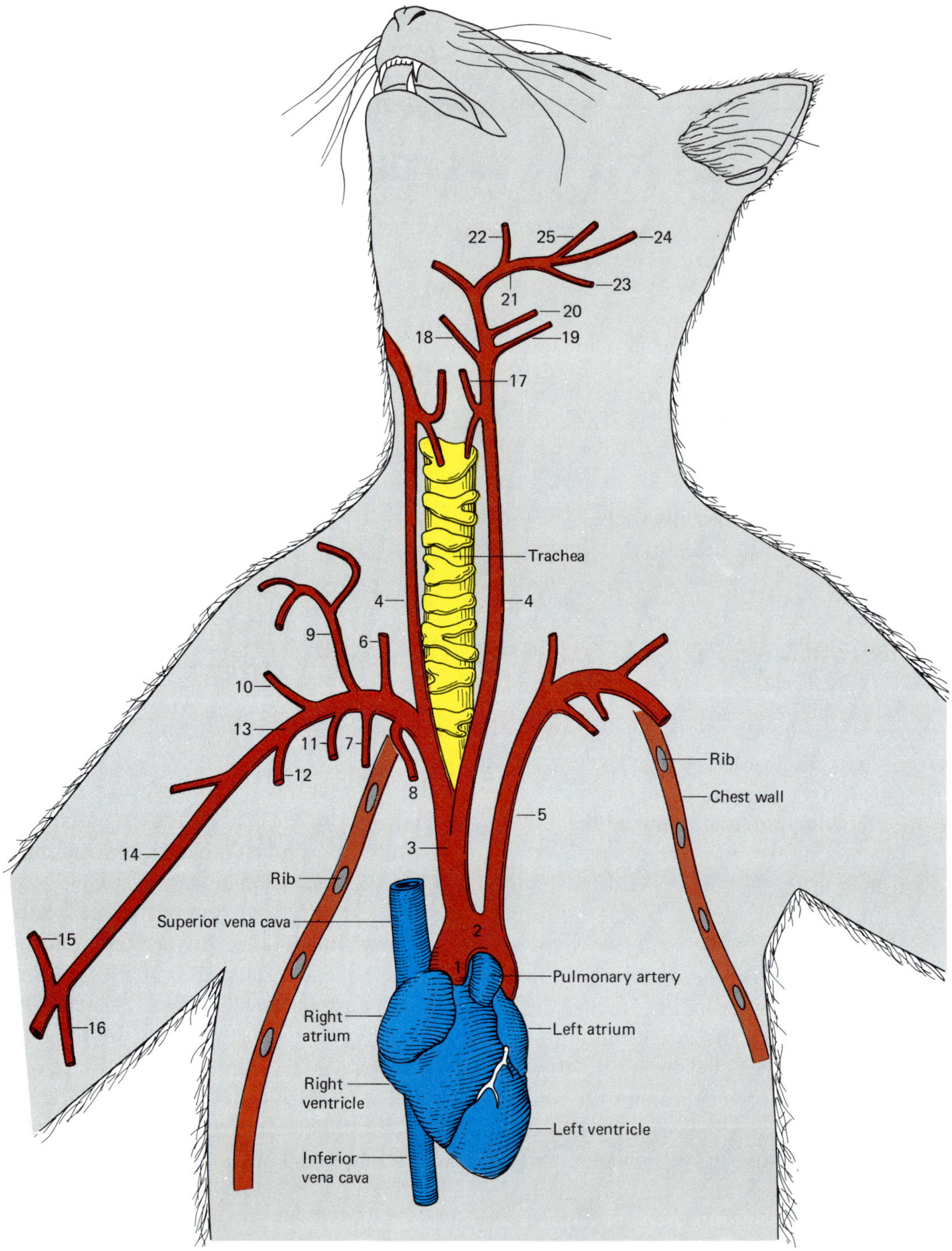

Figure 39-1 Major arteries in the thoracic cavity, neck, and shoulder in the cat.

1. Ascending aorta
2. Arch of the aorta
3. Brachiocephalic
4. Common carotid
5. Subclavian
6. Vertebral
7. Costocervical
8. Internal mammary (thoracic)
9. Thyrocervical
10. Subscapular
11. Ventral (anterior) thoracic
12. Long thoracic
13. Axillary
14. Brachial
15. Radial collateral
16. Ulnar collateral
17. Superior thyroid
18. Laryngeal
19. Occipital
20. Internal carotid
21. External carotid
22. External maxillary
23. Posterior auricular
24. Superficial temporal
25. Internal maxillary

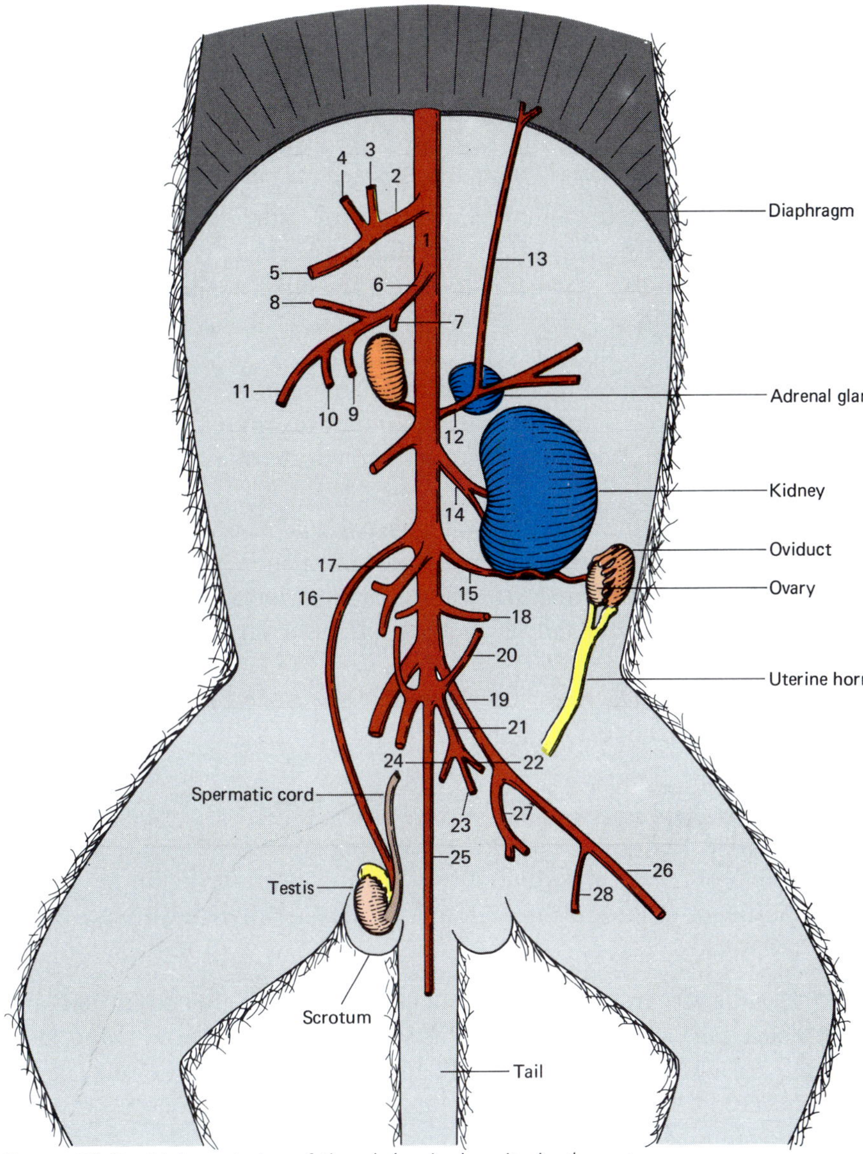

Figure 39-2 Major arteries of the abdominal cavity in the cat.

1. Descending abdominal aorta
2. Celiac
3. Hepatic
4. Left gastric
5. Splenic
6. Superior mesenteric
7. Middle colic
8. Inferior pancreatico-duodenal
9. Ileocolic
10. Right colic
11. Intestinal
12. Adrenolumbar
13. Phrenic
14. Renal
15. Ovarian
16. Spermatic (testicular)
17. Inferior mesenteric
18. Iliolumbar
19. External iliac
20. Umbilical
21. Internal iliac
22. Middle hemorrhoidal
23. Inferior gluteal
24. Superior gluteal
25. Caudal (medial sacral)
26. Femoral
27. Deep femoral
28. Muscular branch

Return to the aorta. A short distance posterior to the celiac artery is the *superior mesenteric artery,* which supplies the small intestine, pancreas, and part of the large intestine. It has several branches labeled on Figure 39-2.

The next major arteries along the aorta are the small, paired *adrenolumbar arteries,* which supply the adrenal gland and muscles of the back and the diaphragm. Posterior to the adrenolumbars are the large, paired *renal arteries* to the kidneys. Posterior to the renal arteries are the paired *genital arteries.* If you have a male cat, this artery passes caudad to the testis and is called the *spermatic* (testicular) *artery.* If you have a female cat, it is called the *ovarian artery* and supplies the ovary. The genital arteries are small in sexually immature cats, but large in older, mature cats.

Caudad the genital arteries is an unpaired vessel, the *inferior mesenteric artery,* which supplies the large intestine. Some of its branches are labeled on Figure 39-2. The next major branches of the aorta are the paired *iliolumbar arteries* to the back muscles. Seven pairs of *lumbar arteries* to the muscles of the back branch off the aorta dorsally. Lift the aorta to see them.

Posterior to the ilolumbar arteries the aorta gives off large *external iliac arteries,* which pass out of the body cavity into the legs. The aorta continues for a short distance below the external iliacs and gives rise to paired *internal iliac arteries* laterally and a single *caudal artery* (medial sacral), which continues into the tail. Branches of the internal iliacs are labeled on Figure 39-2.

The artery that penetrates the thigh from the abdominal cavity becomes known as the *femoral artery.* Find its branches labeled on Figure 39-2.

COMPARABLE HUMAN ARTERIES

Compare the arterial system of the cat with that of the human shown in Figure 39-3. The relationship between the major arteries of the cat and human is strikingly similar. Some differences in pattern are noted.

1. The left common carotid artery in the human branches directly from the aortic arch and not from the brachiocephalic artery. However, this same condition will be found from time to time in the cat. The length of the brachiocephalic artery will vary in cat specimens. Students should compare dissections in this area to determine how much variation occurs among animals of the same species.
2. The human has a *common iliac artery* but the cat does not. In the cat, the *external iliac artery* and the *internal iliac artery* are branches of the aorta.
3. The *internal carotid* at its point of origin from the *common carotid* in the cat is very small and often difficult to find. It is more prominent in the human.
4. The *left gastric* and the *hepatic arteries* of the cat sometimes have a common origin from the celiac. There are other minor variations of the celiac branches in the human.
5. The *medial sacral* of the human is very small corresponding to the absence of a functional tail.

MAJOR VEINS OF CAT THORAX, NECK, AND SHOULDER

Veins return blood to the heart. Most of them transport unoxygenated blood, which will be pumped to the lung where its carbon dioxide will be exchanged for oxygen. Two veins that carry oxygenated blood toward the heart are the *pulmonary veins* and the *umbilical vein* of the fetus. The *pulmonary veins* return oxygenated blood from the lungs directly to the left

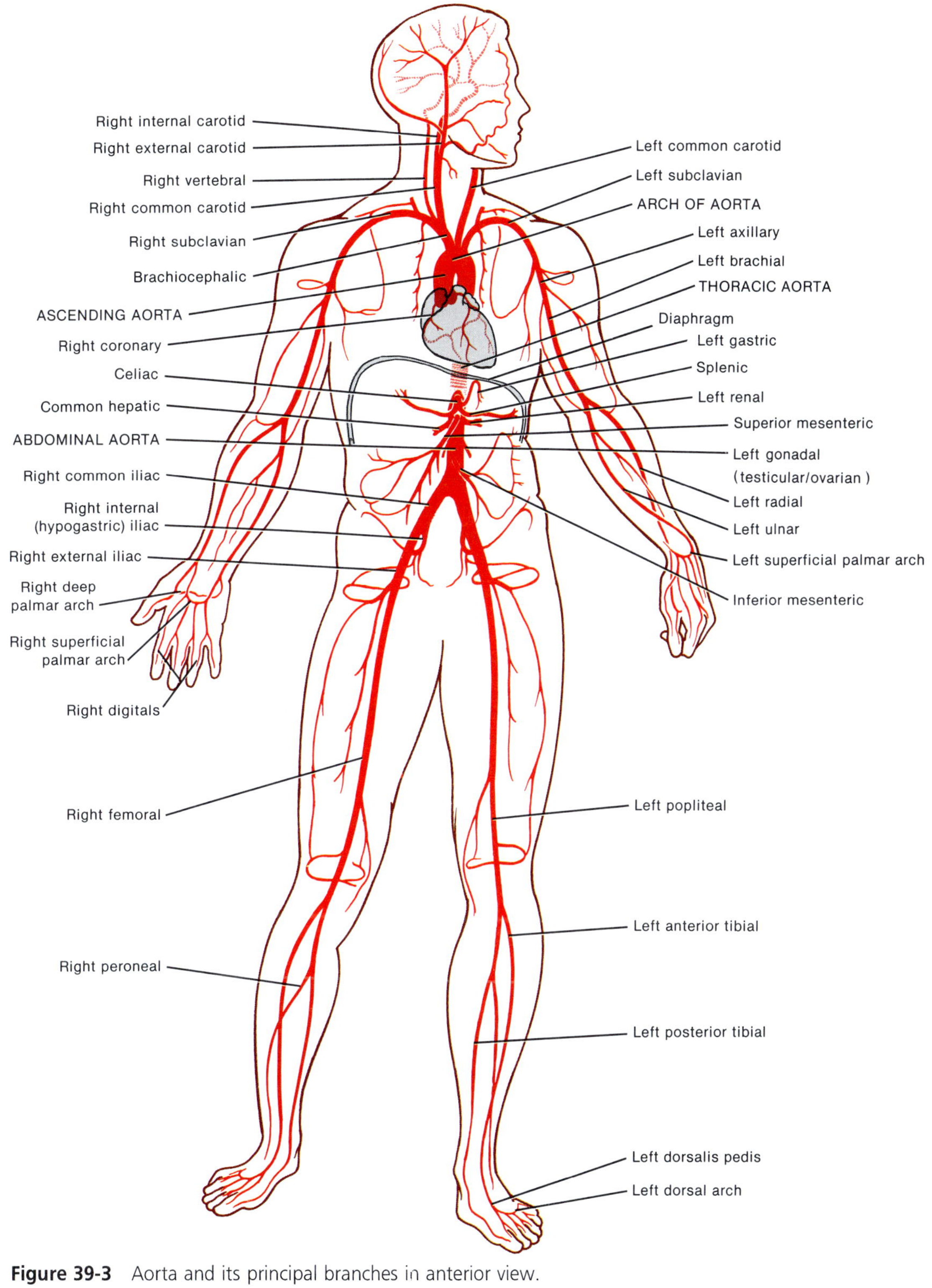

Figure 39-3 Aorta and its principal branches in anterior view.

atrium of the heart and the *umbilical vein* returns oxygenated blood from the placental blood of the fetus to the maternal venous system.

The logical place to begin the tracing of the veins is at the heart, but it must be kept in mind that as you move away from the heart you are working back to tributaries of major veins and that the blood flow is always toward the heart. Locate the *superior vena cava* again and follow it away from the heart (Figure 39-4). The small *internal mammary vein* from the chest wall empties into it on the ventral surface. This vein was cut when the thorax was opened. Lift the right side of the heart slightly and you will see a large vein emptying into the dorsal side of the *superior vena cava* from the abdominal cavity. This is the *azygos vein* from the muscles of the back and the intercostal muscles. Flowing into the superior vena cava about an inch above the heart are the paired *brachiocephalic veins.*

Working back from the left or right brachiocephalic vein find the jugular vein, which soon branches into *internal jugular vein* and *external jugular vein* from the head and neck regions. Trace the external jugulars to the neck region where they are connected by the *transverse jugular vein* observed when muscles in that region were dissected. Veins draining the head are shown in Figure 39-4.

The *subclavian vein* from the arm also joins the brachiocephalic vein in this area. The subclavian vein is known as the *axillary vein* in the underarm region and the *brachial vein* in the arm. Another prominent vein that empties into the venous channel at this point is the *subscapular vein* from the shoulder.

Locate once again the *pulmonary veins* from the lungs that empty into the left atrium.

MAJOR VEINS OF THE CAT ABDOMINAL CAVITY

Trace the *inferior vena cava* into the abdominal cavity through the diaphragm (Figure 39-5). Just below the diaphragm it receives a pair of veins from the liver, the *hepatic veins.* Follow the inferior vena cava posteriorly and locate the *adrenolumbar veins,* the *renal veins,* the *spermatic* (testicular) or *ovarian veins,* and the *iliolumbar veins.* These veins drain the same organs supplied by their arterial counterparts. Often, the left spermatic or ovarian veins terminate at the renal vein rather than the inferior vena cava. See Figure 39-5.

The venous drainage channel from the abdominal viscera is a large *portal vein,* which receives tributaries from the abdominal organs. A *portal system* of veins is one that begins in capillaries and ends in capillaries. In this case the terminal capillaries are in the liver. This has great significance for the study of the physiology of these organs. Refer to Figure 39-6 for a diagram of the portal system of the cat. (Since it is not injected with colored latex in your cats, it is difficult to study in detail unless you have a special preparation in which the veins of this system are injected with a yellow dye.) However, do locate the main portal vein, which is located beneath the common bile duct. If uninjected, it is brown.

In the posterior abdominal region the inferior vena cava is formed by the confluence of paired *common iliac veins.* Near the point of juncture find the single *caudal vein* or medial sacral vein from the tail and, somewhat more posteriorly, the *internal iliac vein* and the *external iliac vein.* Trace the external iliac vein to the leg where it is called the *femoral vein.*

See Figure 39-5 for smaller tributaries of the venous system. Find as many of these veins as you can. (Some cats are better injected than others.)

The venous system of the cat is notorious for its variability. Be alert for all deviations from the normal pattern. Compare your dissection with others in the class.

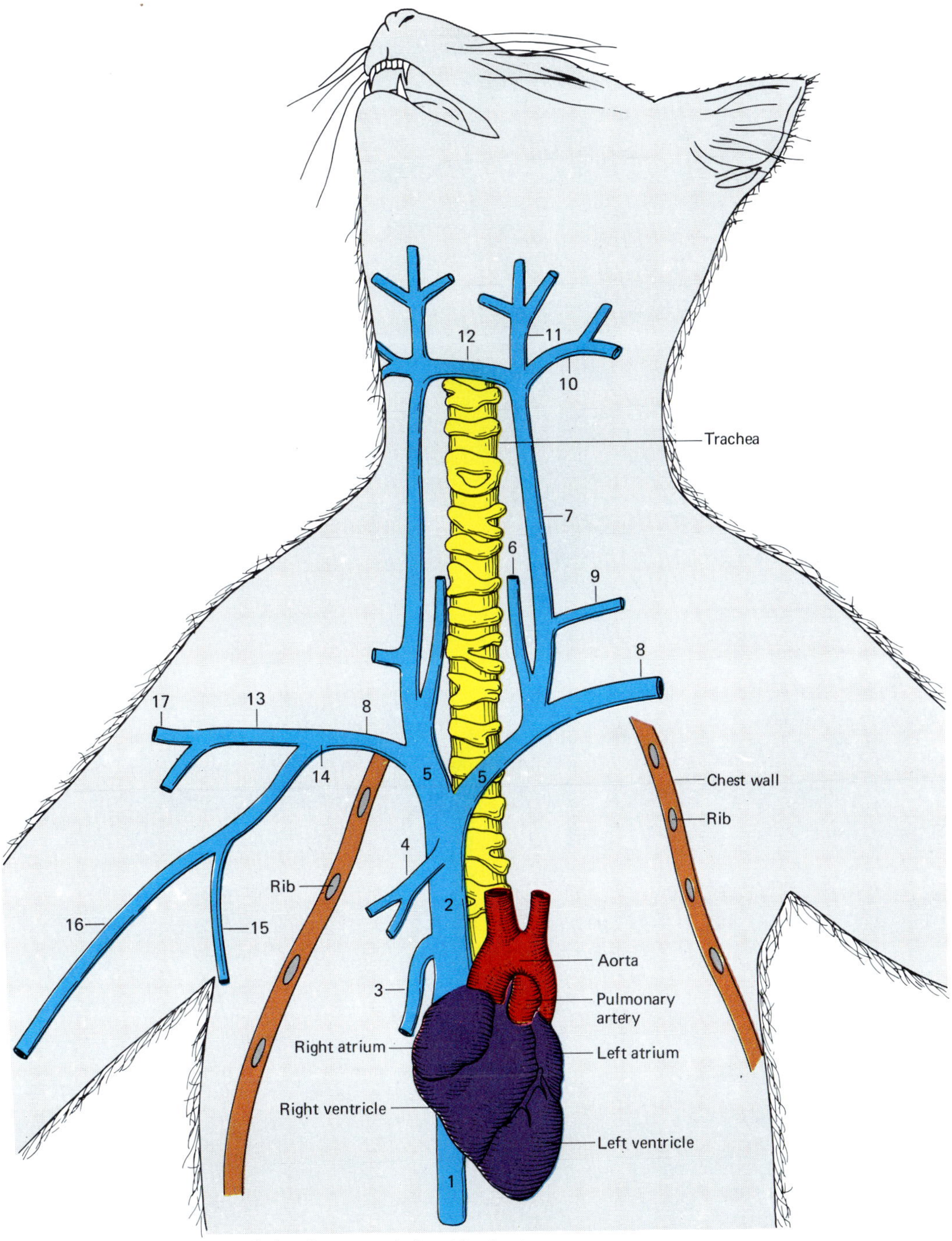

Figure 39-4 Major veins of the thorax and shoulder in the cat.

1. Inferior vena cava
2. Superior vena cava
3. Azygos
4. Internal mammary (thoracic)
5. Brachiocephalic
6. Internal jugular
7. External jugular
8. Subclavian
9. Transverse scapular
10. Posterior facial
11. Anterior facial
12. Transverse jugular
13. Subscapular
14. Axillary
15. Long thoracic
16. Brachial
17. Posterior humeral circumflex

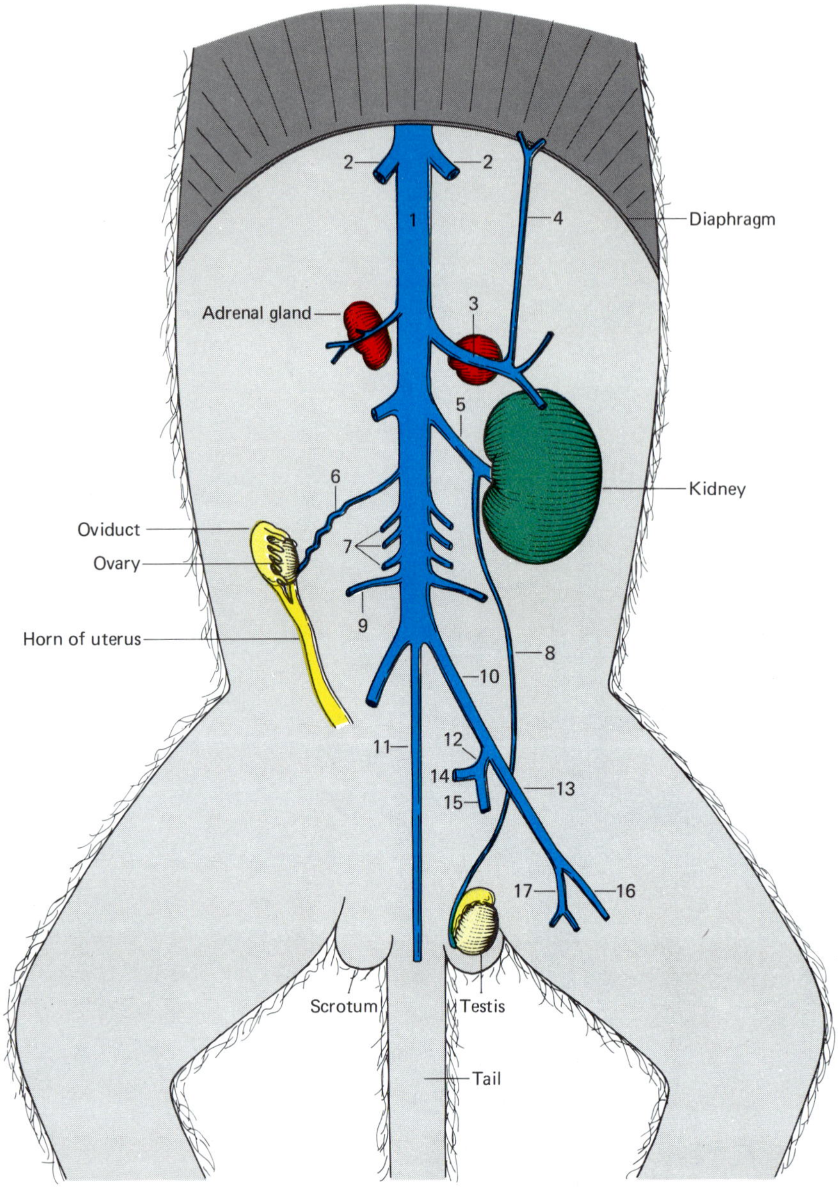

Figure 39-5 Major veins of the abdomen and leg in the cat.

1. Inferior vena cava
2. Hepatic
3. Adrenolumbar
4. Phrenic
5. Renal
6. Ovarian
7. Lumbar
8. Spermatic (testicular)
9. Iliolumbar
10. Common iliac
11. Caudal (medial sacral)
12. Internal iliac
13. External iliac
14. Inferior gluteal
15. Middle hemorrhoidal
16. Femoral
17. Deep femoral

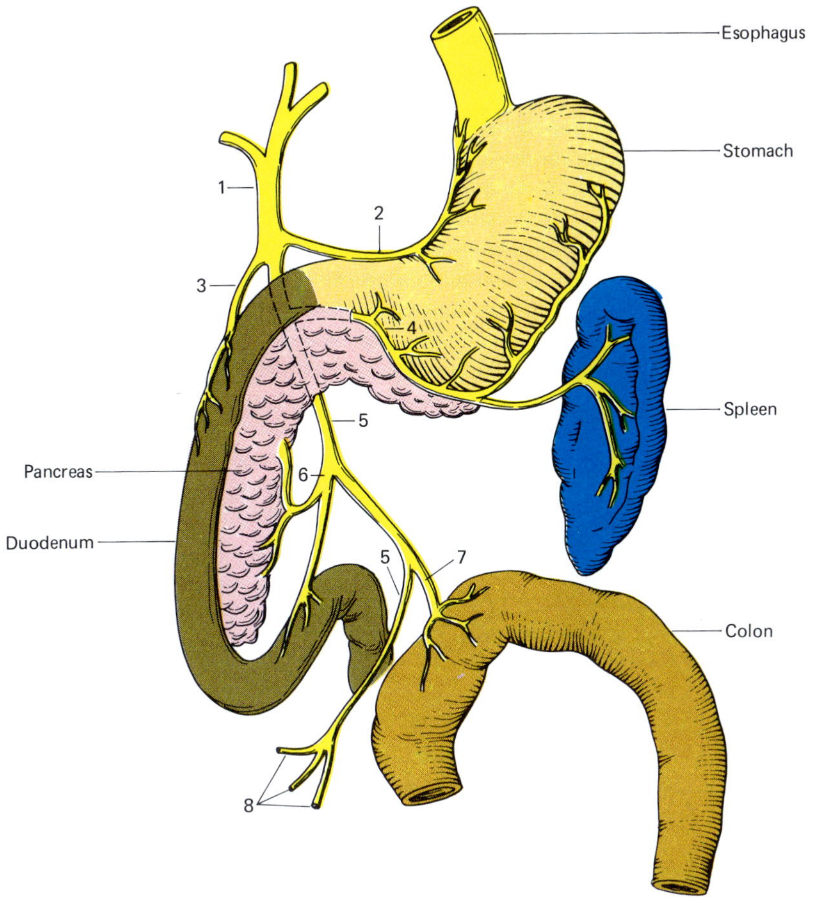

Figure 39-6 Major tributaries of the portal system in the cat.
1. Portal vein
2. Coronary
3. Superior pancreatico-duodenal
4. Gastrosplenic
5. Superior mesenteric
6. Inferior pancreatico-duodenal
7. Inferior mesenteric
8. Intestinal

COMPARABLE HUMAN VEINS

A diagram of the human venous system (Figure 39-7) is shown for comparison with that of the cat. Study the similarities and the differences.

Some of the major differences in the areas we studied are:

1. The *external* and *internal jugular* veins of the human flow separately into the *brachiocephalic,* whereas the cat usually has a short *common jugular* vein. Check your cat dissection to determine whether or not it differs from the human in this minor way.
2. In the cat the *external jugular* is larger than the *internal jugular.* This is reversed in the human.
3. There is no *transverse jugular* vein in the human.
4. The human has an *anterior jugular* vein. It branches from the *external jugular* vein and is not present in the cat.
5. There are four pulmonary veins that open into the left atrium of the human. In the cat there are three main pulmonary veins that open into the atrium by expanded veins called *dorsal, sinistral,* and *dextral sinuses.* Each of the three main pulmonary veins is formed by two veins for a total of six (one from each major lung lobe).

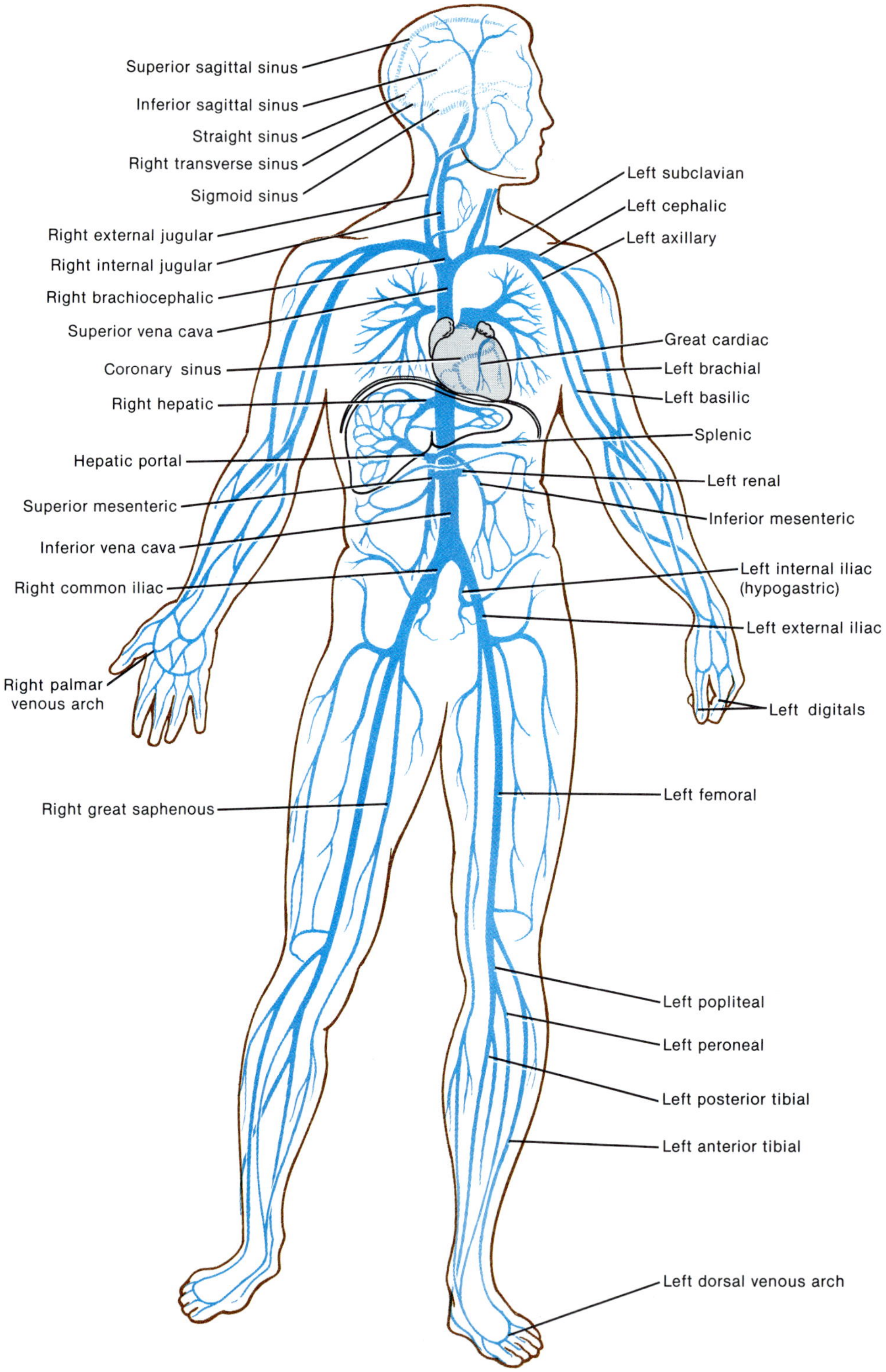

Figure 39-7 Principal veins in anterior view.

6. The cat has a small *ulnar* vein joining the *brachial* midway down the forearm. In the human both *ulnar* and *radial* veins join the *brachial* at the elbow and are about equal in size.
7. An important arm vein in the human is the *basilic*. It is not found in the cat.

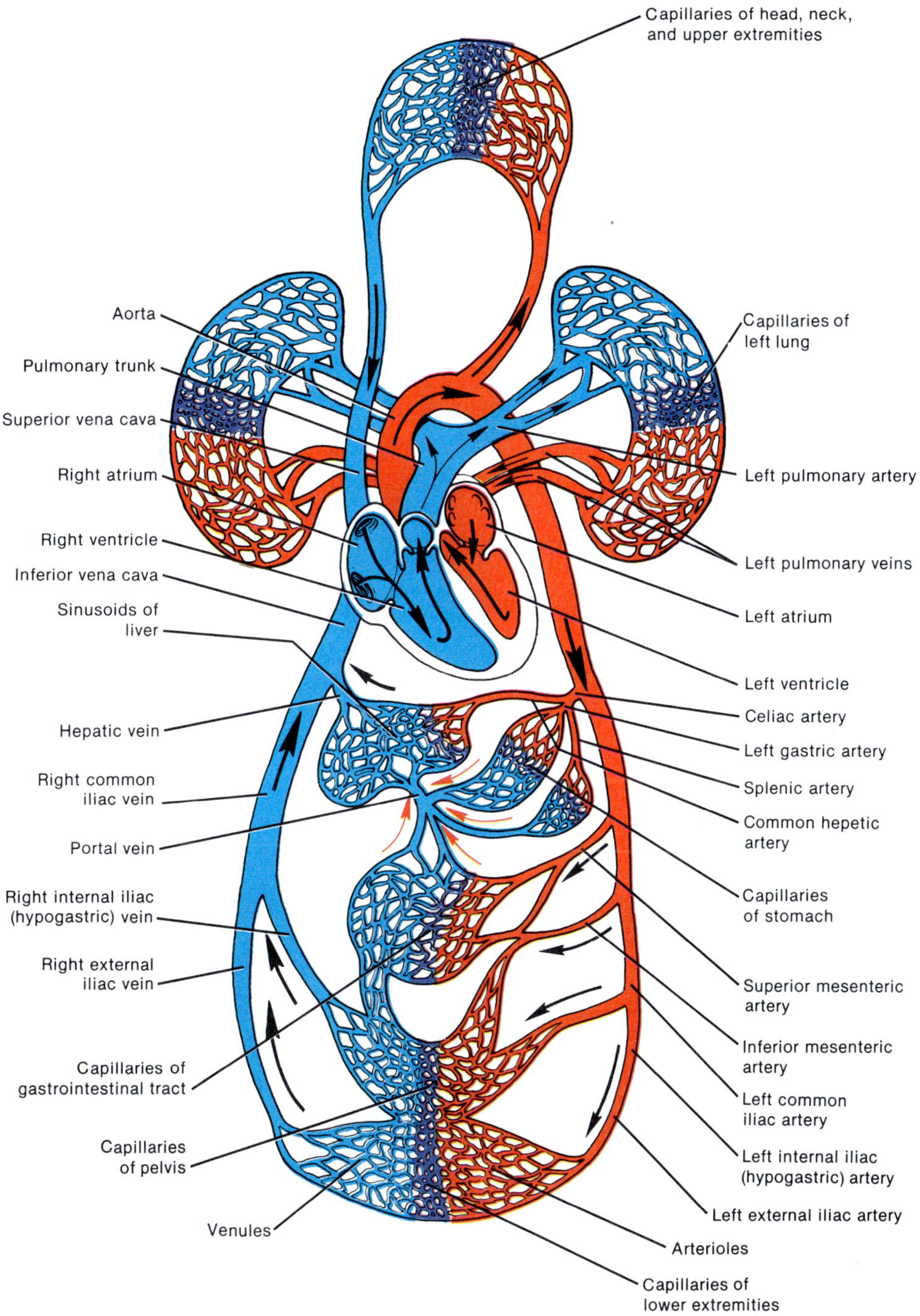

Figure 39-8 Circulatory routes. Systemic circulation is indicated by heavy black arrows, pulmonary circulation by thin black arrows, and hepatic portal circulation by thin colored arrows.

RIGHT ARM

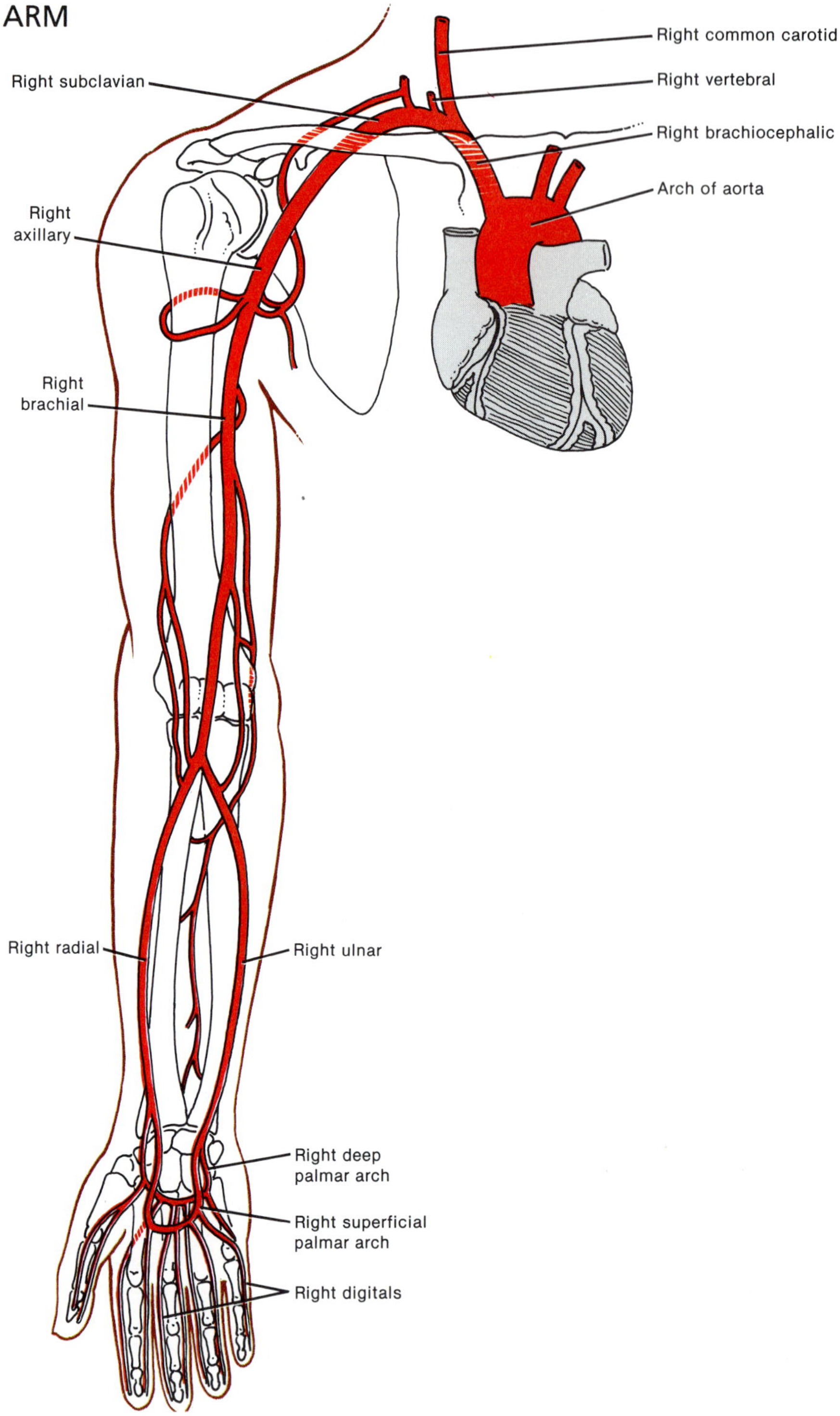

Figure 39-9a Arteries of the right upper extremity in anterior view.

OVERALL CIRCULATORY PATTERN IN THE HUMAN

Use the diagram in Figure 39-8 to trace blood from one part of the body to another. On the diagram red indicates oxygenated blood, blue unoxygenated blood, and purple a capillary

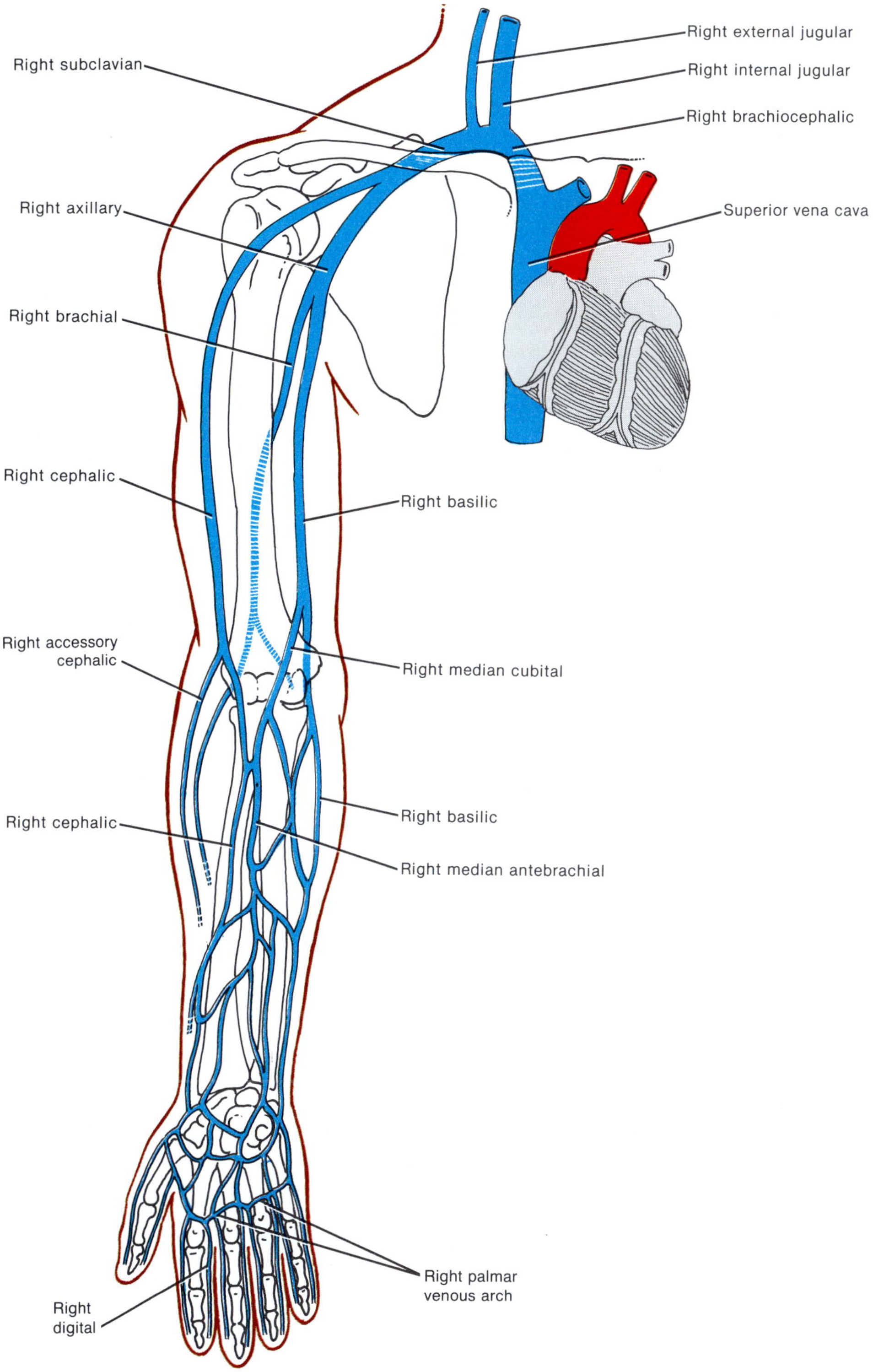

Figure 39-9b Veins of the right upper extremity in anterior view.

bed where oxygen and carbon dioxide are being exchanged at the tissue level. Arrows indicate the direction of flow. Figures 39-9, 39-10, 39-11, and 39-12 show in detail the vessels of the appendages, head, neck, and brain. Become familiar with the flow pattern in these areas.

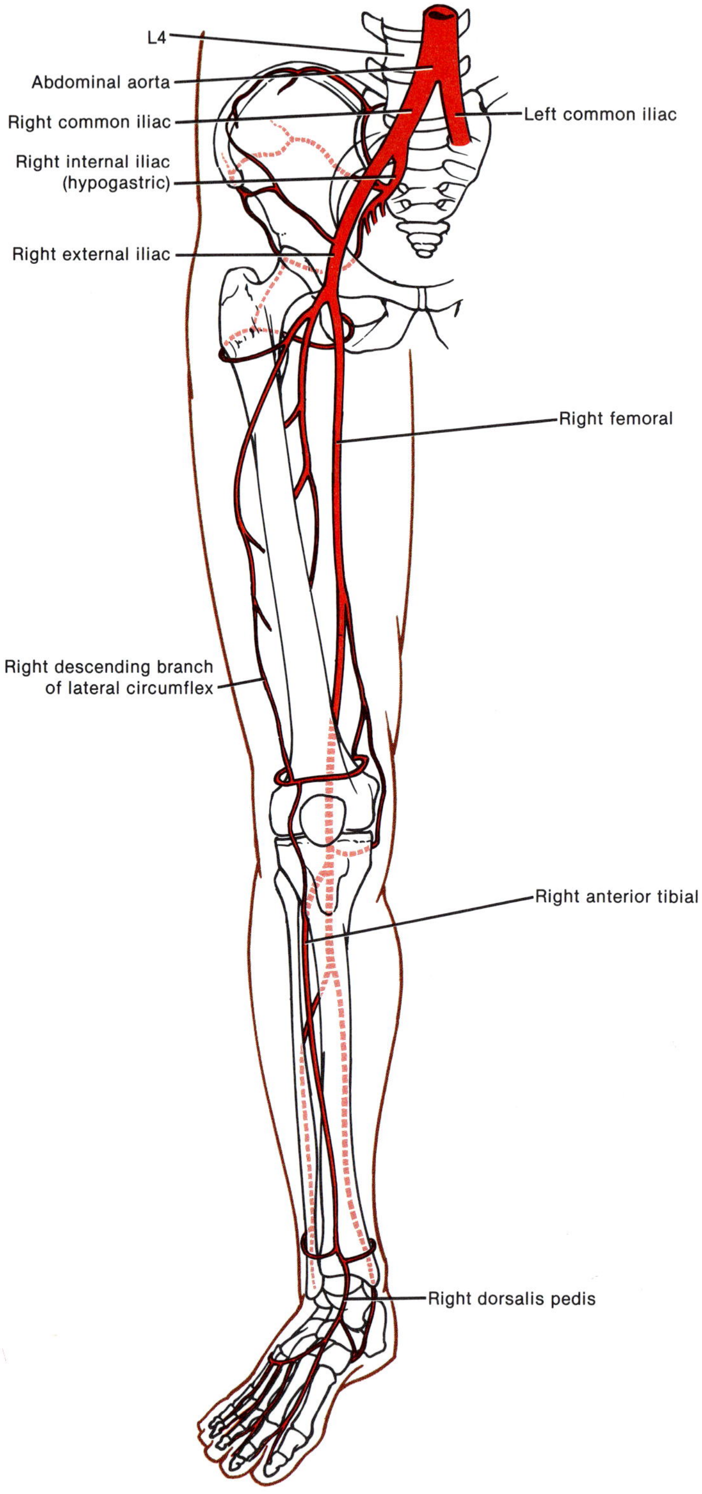

Figure 39-10a Arteries of the pelvis and right lower extremity.

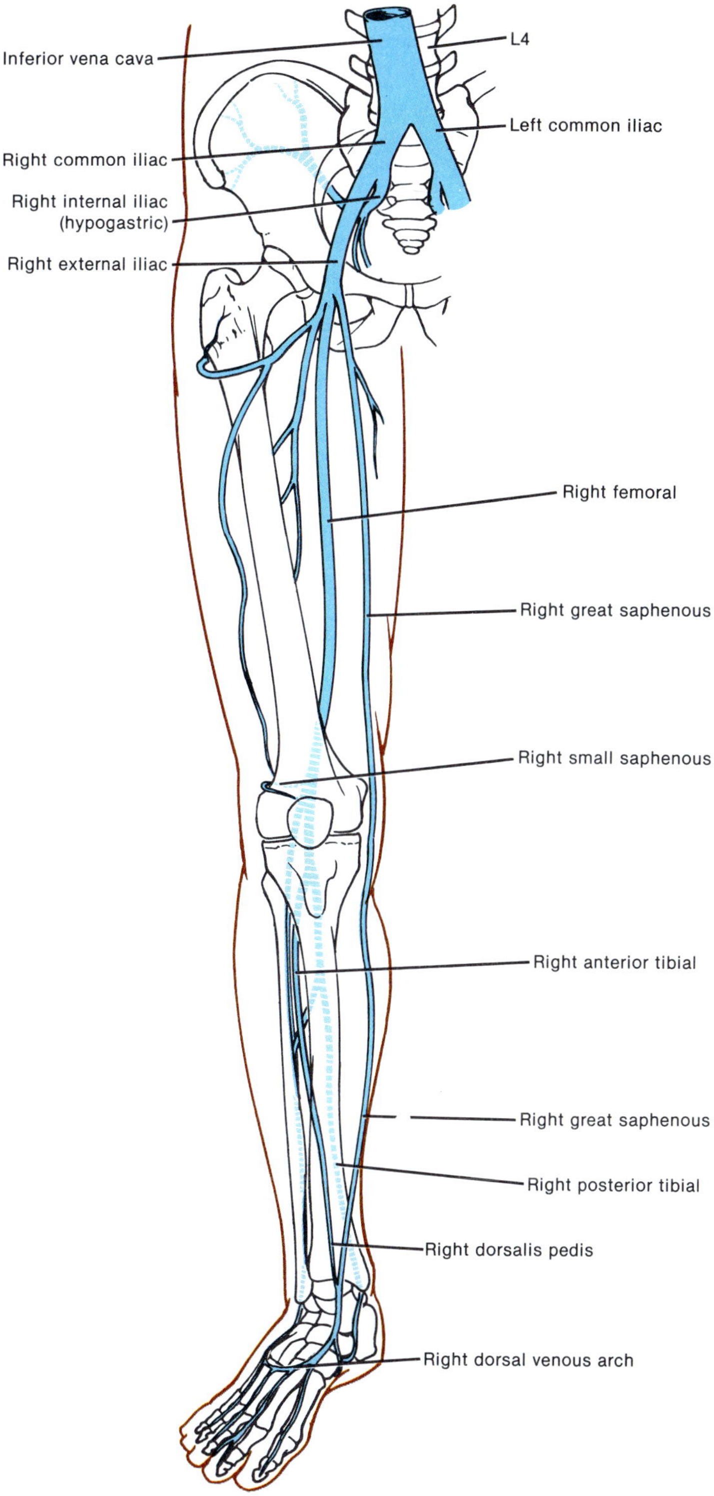

Figure 39-10b Veins of the pelvis and right lower extremity.

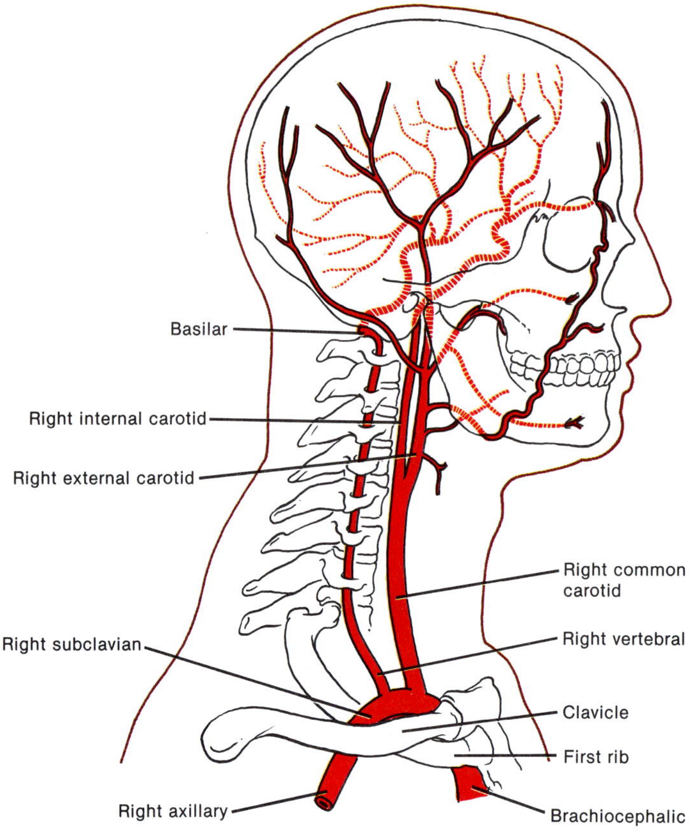

Figure 39-11a Arteries of the head and neck, right lateral view.

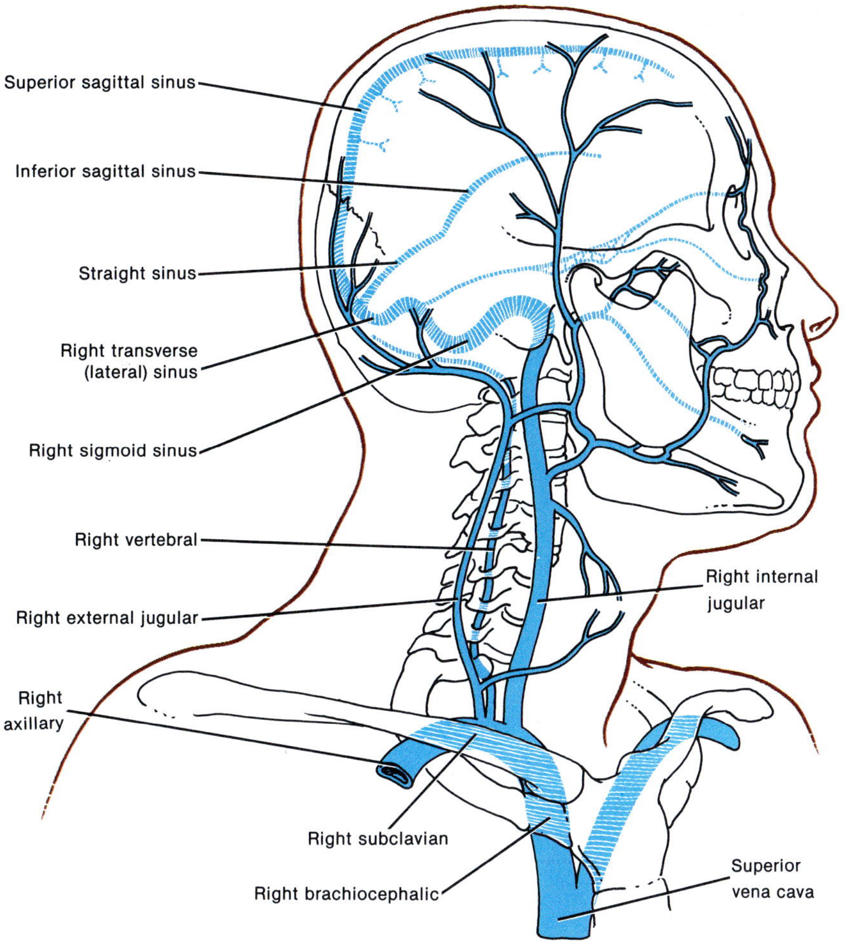

Figure 39-11b Veins of the head and neck, right lateral view.

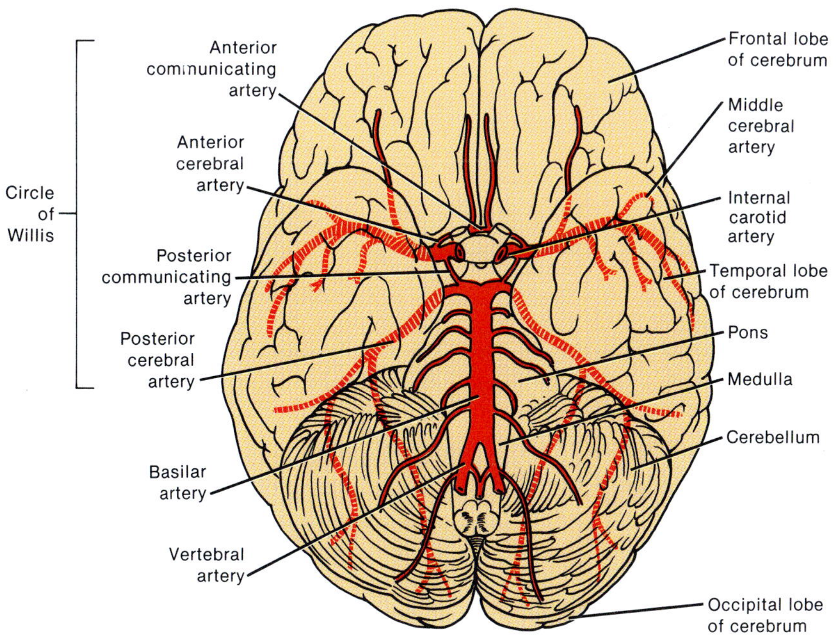

Figure 39-12 Arteries of the base of the brain. Note the arteries that comprise the cerebral arterial circle (circle of Willis).

Questions for Review

EXERCISE **39**

BLOOD VESSELS

Name ______________________________

Laboratory Section ______________________ Date ____________

1. Trace the pathway of a drop of blood from the head to the leg. Some of the steps are provided. You fill in the rest. Include the pulmonary circuit, cavities, and valves.

a. External jugular vein
b. ______________
c. ______________
d. ______________
e. ______________
f. Right ventricle
g. ______________
h. ______________
i. Lung capillaries
j. ______________
k. ______________
l. ______________
m. Left ventricle
n. ______________
o. ______________
p. Arch of aorta
q. ______________
r. ______________
s. ______________
t. ______________
u. Femoral artery

2. Trace the pathway of a drop of blood from the myocardium of the heart to the arm.

a. Coronary vein
b. ______________
c. ______________
d. ______________
e. Right ventricle
f. ______________
g. ______________
h. ______________
i. ______________
j. Left atrium
k. ______________
l. ______________
m. ______________
n. ______________
o. ______________
p. Brachiocephalic artery
q. ______________
r. ______________
s. ______________
t. Radial artery

3. Fill in the chart below:

Supplied by	Organ	Drained by
	Brain	
Hepatic artery	(1)	(2)
(3)	(4)	Great saphenous vein
(5)	Finger	(6)
(7)	Stomach	(8)
Inferior mesenteric artery	(9)	(10)
(11)	(12)	Renal vein
(13)	Spleen	(14)
Coronary artery	(15)	(16)
(17)	Adrenal gland	(18)
(19)	(20)	Axillary vein

EXERCISE 40

DEMONSTRATION OF THE NORMAL CARDIAC CYCLE

The heart consists of specialized muscle that has a fundamental physiological property of rhythmicity. This feature accounts for the uniformly paced alternating periods of contraction and relaxation. In the human the heartbeat is regulated by the *sinoatrial node,* known as the *SA node* or *pacemaker,* located in the wall of the right atrium.

We will be concerned with the electrical activity associated with the contraction of heart muscle. Records of such electrical activity, referred to as bioelectrical potentials, may be made by suitable equipment, and are of importance in the diagnosis and treatment of various disorders and disease states.

The *electrocardiograph* is the instrument that monitors electrical activity of the heart. It amplifies the impulses so that they are strong enough to activate a recording device. Electrodes attached to the surface of the body receive impulses from the heart. The electrodes usually are made of metal and are slightly concave to make good contact with the skin. The amplified impulse causes a recording needle to move up and down on the recording paper. The needle may be a pen, which marks the paper with ink, or a stylus, which inscribes on special paper that does not require ink. The paper moves under the needle at a rate determined by the scale of the graph and therefore the record appears as a series of waves (Figure 40-1). The record produced by the electrocardiograph is called an

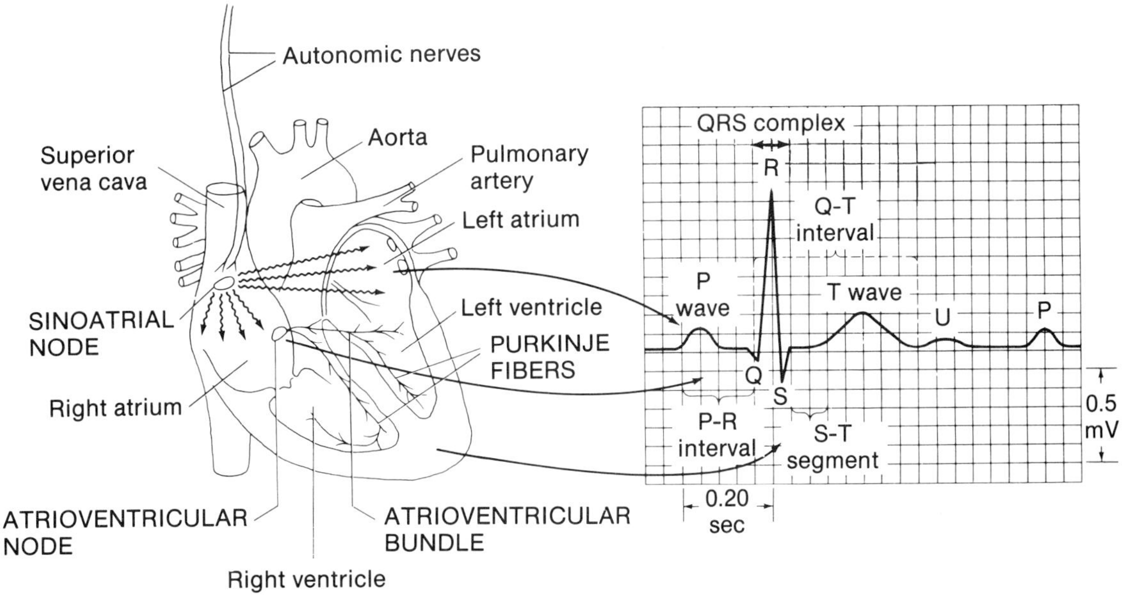

Figure 40-1 The relationship of selected heart components to the conduction of electricity in the heart as indicated by a normal electrocardiogram.

electrocardiogram abbreviated *ECG* or *EKG*. An ECG often is obtained with a *polygraph,* a recording device that operates on the principle just described but that is capable of monitoring impulses from several different physiological activities in addition to that of the heart. Depending on the instrument used, the record is superimposed on either a standardized graph or other appropriate material. The vertical lines indicate time in seconds and the horizontal lines amplitude in millivolts (mV). The space between two vertical lines represents 0.04 second; the space between two horizontal lines represents 0.1 mV.

The procedure used in taking an ECG involves the application of a jelly containing an electrolyte which forms a low-resistance surface between the skin and electrode. This material aids the conduction of electrical signals. Paper pads presaturated with an electrolyte solution also can be used in place of the jelly. The electrodes are held securely in place by means of straps. The particular arrangement of two electrodes is called a *lead.* The position of the electrodes on the body, the manner in which they are wired to the recording system, and any and all extra movements of the individual undergoing testing will influence the final record.

The component waves of a normal ECG are shown in Figure 40-1. The major components include the *P wave,* which indicates the spread of excitation from the pacemaker over the atrial musculature; the *QRS wave (complex),* which represents the depolarization of the ventricular heart muscle that immediately precedes ventricular systole; and the *T wave,* which represents the repolarization of the ventricular muscle that occurs just before ventricular diastole. A U wave is sometimes seen following the T wave and is thought to represent late repolarization of the ventricles. This exercise considers the equipment and procedure used in obtaining an ECG.

The direction of depolarization depends on the heart's orientation and on the particular instant the heart cycle is being recorded or examined. From a medical standpoint it is sometimes desirable to record from more than two sites and even in two or three dimensions. With the particular technique of *vectorcardiography,* the continuous recording of the heart's electrical activity can be achieved. This type of continuous display of the depolarization –repolarization process as it moves over the heart may reveal important abnormalities. For example, an increase in the size (hypertrophy) of one ventricle shifts the average or mean direction of depolarization during the cardiac cycle. This change occurs because it takes longer for the ventricle to depolarize and may be the result of a narrowing of the aortic semilunar valve.

In situations where there is a narrowing of the pulmonary semilunar valve or congenital defects such as the tetralogy of Fallot, a shift of the depolarization toward a hypertrophied right ventricle would occur.

Depolarization pattern abnormalities also occur with a heart block, which is a delay of the electrical. impulse of even a complete blockage of an impulse in its progress over the conduction system, or as a result of a heart attack.

This exercise also will consider the determination of the mean electrical axis (direction) of depolarization of the cardiac cycle. This determination can be made by recording the QRS complex voltages from two different dimensions using two different leads, namely, Leads I and III. Lead I provides a horizontal axis of observation (from left arm to right arm) while Lead III has an axis of about 120° (from left arm to left leg). The recordings from Leads I and III should show the normal mean electrical axis of the ventricles about 59° (Figure 40-2a).

Materials

The following materials should be provided for class use or demonstration purposes.

1. Electrocardiograph systems and appropriate cables
2. ECG electrodes (metal or disposable paper pads)

3. Tubes of electrode jelly
4. Disposable abrasive pads or brushes
5. Commercially packaged sterile alcohol prep-pads
6. Sterile gauze pads (4 x 4 inch)
7. Paper towels
8. Examples of normal and abnormal electrocardiograms

Procedure 1: The Normal ECG

The instructor will demonstrate the principles, calibration, and correct use of the electrocardiograph system. This procedure should be performed on a group basis.

1. Before connecting the electrodes and taking an electrocardiogram, check the following items.
   a. Sufficient paper and recording ink (if needed) are available.
   b. The recording pen or device is properly centered.
   c. The recording speed has been set for 25 mm/sec.
   d. The preamplifier has been set at 10 mm/mV.
2. Ask your laboratory partner to remove any materials that could interfere with taking a recording (remove wristwatch or bracelet, roll up sleeves, roll down stockings, and so forth). Have your partner lie down on a table or cot in a comfortable and relaxed position.
3. Disinfect both wrists and the area on the inside of the legs just above both ankles with sterile alcohol prep-pads.
4. Apply a small amount of electrode jelly to the skin where the electrodes will be placed and to the surface of the electrodes.
5. The following three leads will be used in this exercise.
   a. *Lead I*—electrodes attached to the left and right wrists
   b. *Lead II*—electrodes attached to the right wrist and left leg
   c. *Lead III*—electrodes attached to the left wrist and left leg
6. Obtain a record with each lead. Usually six ECG complexes (six complete cycles of P wave, QRS wave, and T wave) for each lead will be sufficient. Be sure to mark the record with your partner's name.
7. Have your partner exercise vigorously for 3–5 minutes. Jogging, running in place, or going up and down stairs are suitable forms of exercise. Again, obtain a record with each lead.
8. After the recording is complete, remove the leads and clean the electrodes. The subject should remove the electrode paste with gauze or paper towels.
9. Obtain representative segments of the ECG. Attach and label them in the results and observations section.
10. Change places with your partner and repeat steps 1–9.

Procedure 2: Mean Electrical Axis of the Ventricles

This procedure should be performed on a group basis after the instructor has demonstrated the correct use of the electrocardiograph system.

1. Repeat steps 1–4 of Procedure 1 in this exercise.
2. Leads I and III with a sensitivity setting of 1 will be used in this procedure.
3. Obtain a record with each lead. Be certain to mark the record with your partner's name.
4. With the Lead I record, obtain the following information items and enter your findings and calculations in the appropriate spaces of the results and observations section.

a. Find the QRS complex and count the number of small boxes (millimeters) that it projects above the baseline.
b. Find the Q and S waves and count the number of millimeters each projects below the top of the baseline. Add the sum of the two downward deflections.
c. Algebraically add the two values obtained for the upward and downward deflections.

5. Repeat step 4 with Lead III and enter your findings and calculations in the appropriate spaces of the results and observations section.
6. Find and mark the number that corresponds to the total value obtained with Lead I on the horizontal axis in Figure 40-2b. With the aid of a ruler, make and extend a vertical line from your mark to the top of the figure.
7. Find the mark the number that corresponds to the total value obtained with Lead III on the slanted axis located on the left-hand side of Figure 40-2b. With the aid of a ruler, make and extend a straight line down to the right-hand side of this figure.
8. With the aid of a ruler, draw an arrow line from the centrally located circle in Figure 40-2b that intersects the lines made earlier on the figure. Extend the arrow to the edge of the outer circle and record the value that represents the mean electrical axis of the ventricles. Enter your finding in the results and observations section.
9. An example of this procedure is shown in Figure 40-2a. The values for Leads I and III were 10 and 10, respectively.

RESULTS AND OBSERVATIONS

THE NORMAL ECG

1. Attach labeled segment of electrocardiograms made during relaxation period.
   a. Lead I

   b. Lead II

c. Lead III

2. Attach labeled segment of electrocardiograms made immediately after vigorous exercise.
   a. Lead I

   b. Lead II

   c. Lead III

3. Are there any noticeable differences in the records taken with the three leads? If so, how would you explain the differences? ______________________________

____

____

THE MEAN ELECTRICAL AXIS OF THE VENTRICLES

1. Enter your findings for Lead I in the spaces provided.
   a. Number of millimeters that the QRS complex projects *above* the baseline; upward deflections ______
   b. Sum in millimeters of the Q and S waves that project *below* the baseline; downward deflections ______
   c. The algebraic sum of the upward and downward deflections (if the resulting total is a negative number, enter it as such) ______
2. Enter your findings for Lead III in the spaces provided.
   a. Number of millimeters that the QRS complex projects *above* the baseline; upward deflections ______
   b. The sum in millimeters of the Q and S waves that project *below* the baseline; downward deflections ______
   c. The algebraic sum of the upward and downward deflections (if the result is a negative number, enter it as such) Total ______
3. Mean electrical axis of the ventricles was found to be ______
4. List four heart conditions that could produce depolarization pattern abnormalities.

   a. ______________ c. ______________

   b. ______________ d. ______________

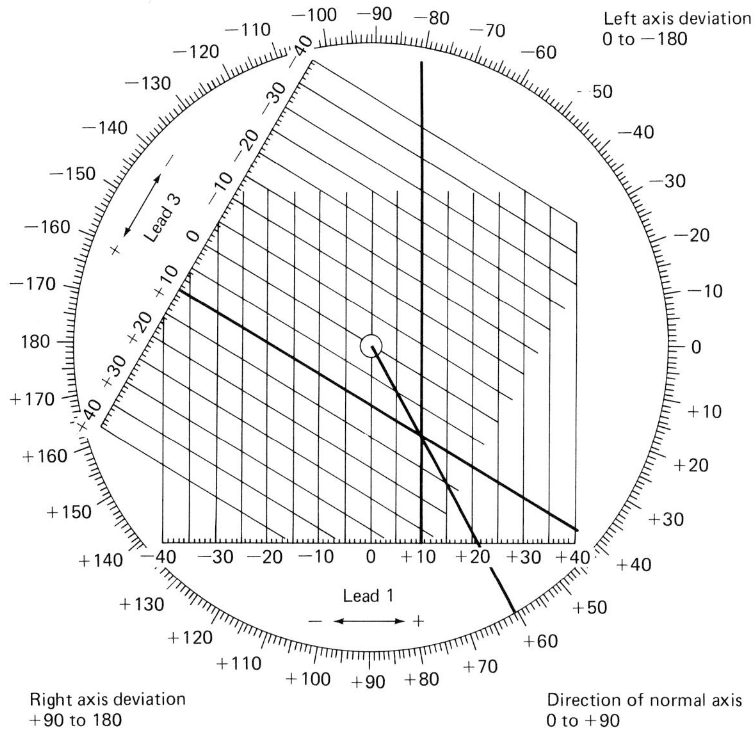

(a)

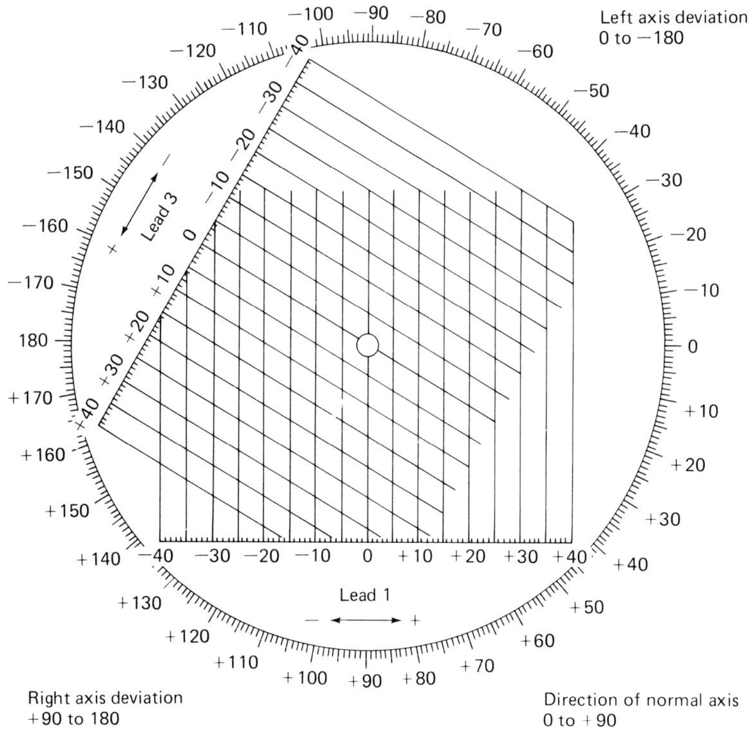

(b)

Figure 40-2 Grids used for the determination of the mean electrical axis of the ventricles. (a) Grid showing an example of how to determine the mean electrical axis of the ventricles. The results show a value of 59°. (b) A grid to be used by the student for Procedure 2.

Questions for Review

EXERCISE **40**

DEMONSTRATION OF THE NORMAL CARDIAC CYCLE

Name ______________________________

Laboratory Section ______________________ Date ____________

1. What is an electrocardiogram? ______________________________

____

2. What is a lead? ______________________________

____

3. Identify the specific components of the electrocardiogram shown in Figure Q40-1.

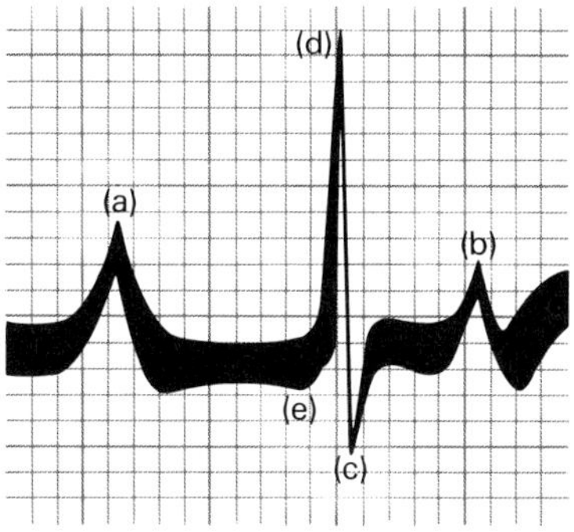

Figure Q40-1

a. ______________

b. ______________

c. ______________

d. ______________

e. ______________

4. With the aid of your textbook or other reference indicate what each of the following components of an electrocardiogram represents.

   a. P wave ______________________________

   b. QRS complex ______________________________

   c. T wave ______________________________

   d. Q-T interval ______________________________

5. What is a normal cardiac cycle? ______________________________

____

6. With the aid of your textbook or other reference determine the effect of the following chemical or physical factors on rate and strength of the heartbeat.

   a. Temperature ______________________________

b. Acetylcholine ______________________________

c. Epinephrine ______________________________

d. Digitalis ______________________________

7. Define or explain the following.

a. Fibrillation ______________________________

____

b. Heart block ______________________________

____

8. Distinguish between diastole and systole. ______________________________

____

9. a. What is a myocardial infarction? ______________________________

____

____

b. Can a myocardial infarction be detected in an ECG? If so, how it is indicated?

____

____

10. a. Would you consider the ECG in Figure Q40-2 normal?

b. If not, what features are abnormal? ______________________________

____

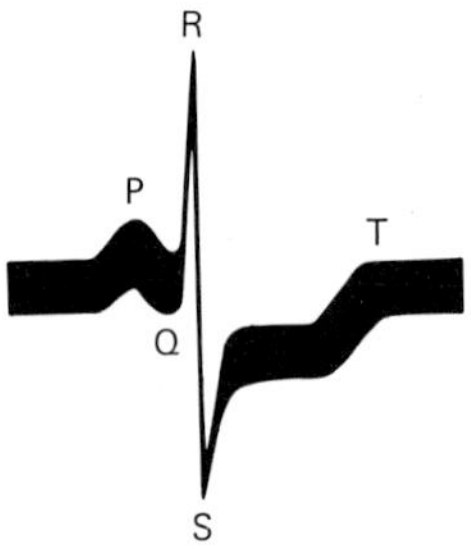

Figure Q40-2

11. Using the following recordings from Leads I and III of an electrocardiogram and the grid chart provided in Figure Q40-3, determine the electrical axis of the ventricles. Show your calculations.

Grid chart

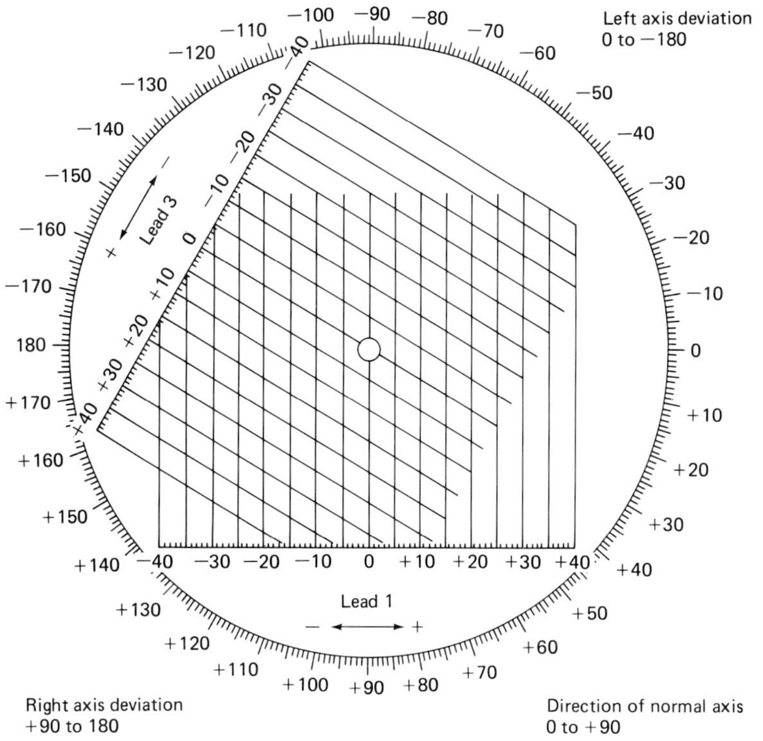

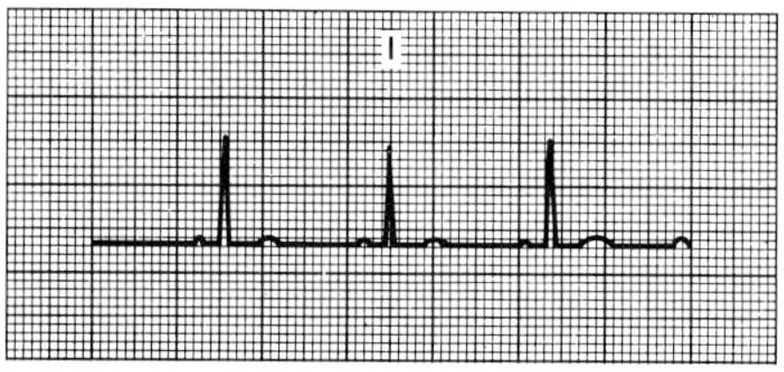

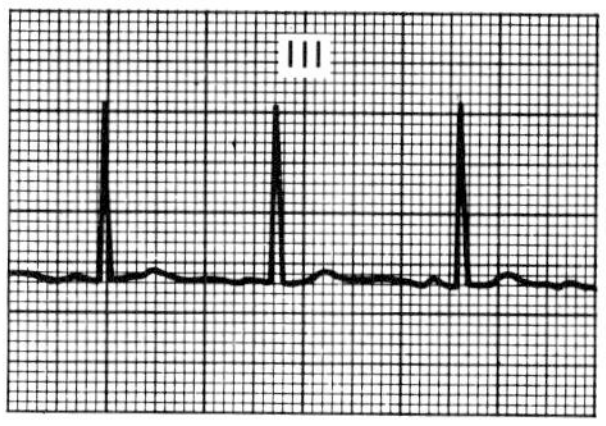

Upward deflection: ____________

Downward deflection: ____________

Lead I Result: ____________

Upward deflection: ____________

Downward deflection: ____________

Lead III Result: ____________

Figure Q40-3

12. Considering that the normal mean electrical axis of the ventricles is 59°, was the value obtained in question 11 normal? ______________________________

EXERCISE 41

BLOOD PRESSURE AND PULSE DETERMINATION

When the ventricles of the heart contract, blood is pumped under pressure into the pulmonary artery and aorta. This pressure exerts a force against the walls of the vessels through which blood flows. Within any blood vessel, artery, vein, or capillary, the pressure depends on the volume of blood present, the size of the vessel, the rate of blood flow, and any resistance to blood flow. Blood pressure is the basis for maintaining a functional and stable movement of blood from the heart, to the various regions of the body, and back into the heart. Because of its importance a blood pressure determination is a major part of any physical examination. In a medical context blood pressure refers to the pressure of the arterial system.

The routine method for measuring arterial blood pressure is indirect and usually employs a *sphygmomanometer.* This device basically consists of an inflatable cuff, an inflation pumping device, a pressure gauge, and an exhaust control valve.

A stethoscope is used to amplify the sounds within the artery used for the determination. Usually the brachial artery is chosen for convenience. In the procedure the pressure cuff is first inflated around the upper arm to stop the flow of blood temporarily (Figure 41-1). This collapses the artery. When a stethoscope is placed above the brachial artery, no sound can be heard. The pressure in the cuff is gradually reduced until the blood flows through the vessel with each systole of the heart. Detection of a thumping sound indicates that the cuff pressure is about equal to the systolic pressure in the artery. The reading on the pressure gauge when this first sound is detected corresponds to the individual's *systolic blood pressure.* The artery will remain open now during the systole of the heart. The sound continues with each heart cycle. Reducing the cuff pressure further enables the artery to remain open even during diastole. Blood flows freely now and the thumping sound disappears. The reading at which the sound disappears corresponds to the *diastolic blood pressure.*

Blood pressure usually is measured in millimeters (mm) of mercury (Hg). The arterial blood pressure is expressed as the systolic blood pressure value over the diastolic pressure value. Table 41-1 lists values for males and females. The difference between the systolic and diastolic pressures is known as the *pulse* pressure. Various factors influence arterial blood pressure, including age, emotional state, sex, weight, and strenuous exercise.

Today additional methods have been developed for arterial pressure determinations. Electronic devices incorporating both sound and light indicators are available for rapid determinations.

The rate and characteristics of the arterial pulse are extremely important since it is indicative of heart action. This pulsation of the blood through the arteries is in response to the contraction of the heart. The number of pulsations per minute is called the *pulse rate* or *heart rate.* It can be felt in any artery that lies near the surface of the body or over a bone or other firm tissue. One of the sites commonly used to measure pulse rate is the undersurface of the wrist behind the thumb, where the radial artery pulsates against the radius bone. Individuals with a rapid heart rate are said to be exhibiting *tachycardia,* while those persons with a very slow rate exhibit *bradycardia.*

This exercise considers several approaches to determination of blood pressure and pulse rate.

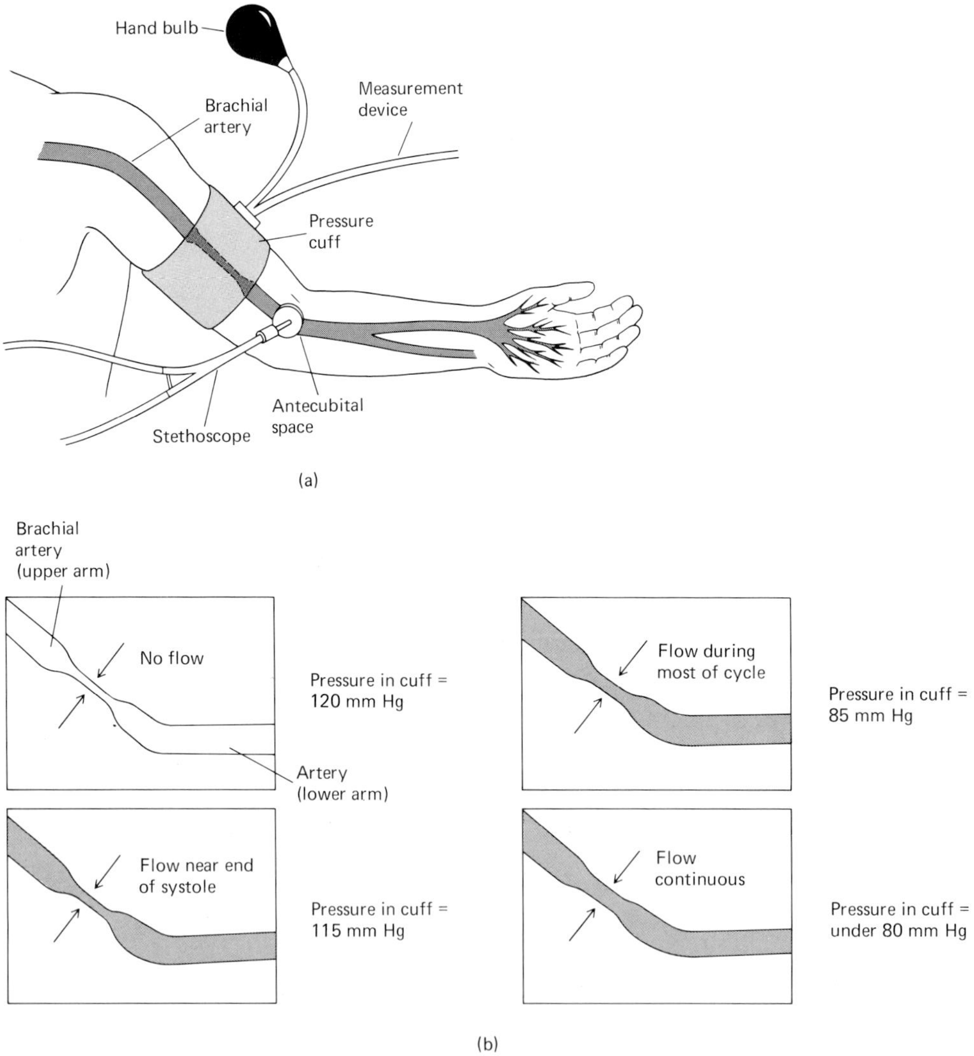

Figure 41-1 Measuring blood pressure. (a) Position of a sphygmomanometer cuff and stethoscope during measurements using the brachial artery. (b) Diagrammatic representation of the appearance of the brachial artery at different pressures in the cuff.

Table 41-1 Selected Blood Pressure Values, Normal Range

Sex	Age	Systolic	Diastolic
Females	16–24	100–130	60–85
	25–29	102–130	60–86
	30–34	102–135	60–88
	35–39	105–140	65–90
	40–44	105–150	65–92
	45–49	105–151	65–96
	50–54	110–165	70–100
	55–59	110–170	70–100
	60–64	115–175	70–100
Males	16–18	105–135	60–86
	19	105–140	60–88
	20–24	105–140	62–88
	25–29	108–140	65–90
	30–39	110–145	68–92
	40–44	110–150	70–94
	45–49	110–155	70–96
	50–54	115–160	70–98
	55–59	115–165	70–98
	60–64	115–170	70–100

A. BLOOD PRESSURE DETERMINATIONS

Materials

The following materials should be provided for class use.

1. Stethoscope
2. Standard sphygmomanometer
3. Electronic sphygmomanometer
4. Astropulse 10 blood pressure monitor
5. Digimax 5000 electronic blood pressure monitor
6. Disinfectant
7. Sterile gauze pads or cotton

Procedure 1: Examination of Blood Pressure Determination Devices

This exercise is to be performed by students individually. Knowing the parts of equipment and their normal positions is extremely important if equipment is to be used effectively.

1. Examine the various devices provided for the measurement of blood pressure and pulse determinations.
2. With the aid of Figures 41-2a–e identify the parts and note the positions of these devices.

Procedure 2: Auscultation with Standard Sphygmomanometer

This procedure is to be performed by students in pairs.

1. Ask your laboratory partner to sit comfortably and place the left forearm palm up on a table. The upper arm should be at heart level.
2. With Figures 41-2a–e identify the parts and note the positions of these devices.
3. Examine the *sphygmomanometer cuff* to see if the area to be placed against the brachial artery is marked. If the cuff is not marked, hold the inflatable portion of the cuff on the inside of the upper arm while you wrap the long end of the cuff around the arm. The cuff should fit snugly. Inflate the cuff until it does not fall. Ask the instructor for

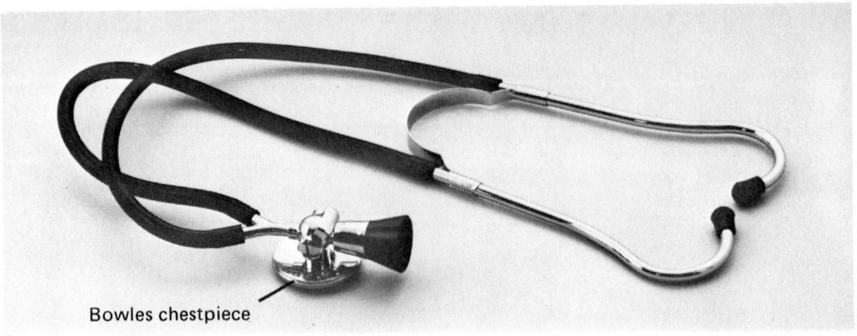

(a)

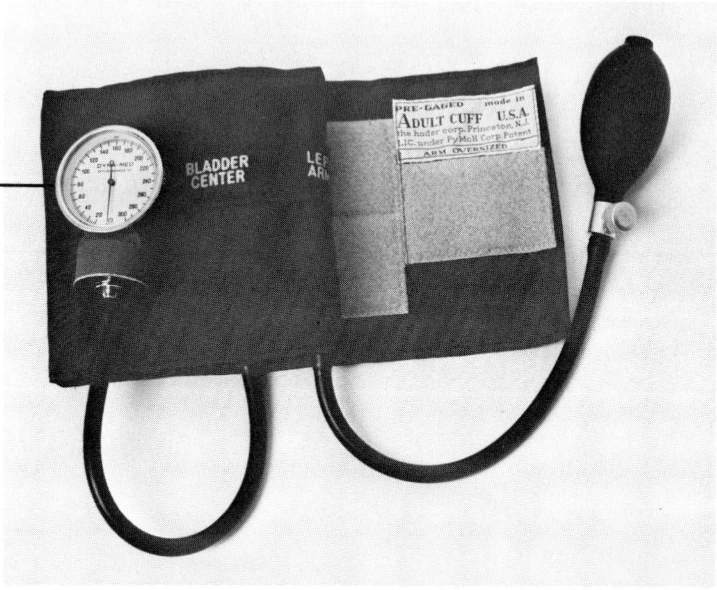

(b)

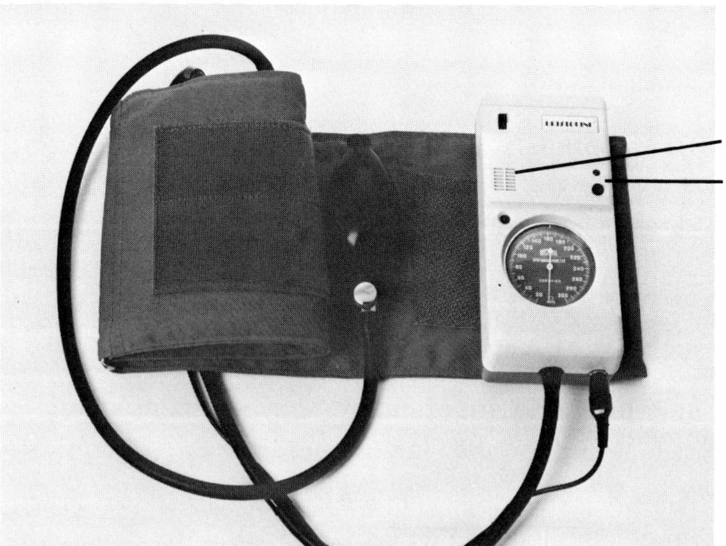

(c)

Figure 41-2 Instruments used for blood pressure measurement. (a) Combination stethoscope with a Bowles chestpiece. (b) Hare Deluxe Sphygmomanometer. This device has several convenient features including a cuff which indicates the right size for different individuals and a clearly marked bladder center which shows where to position the cuff over the artery. (c) Electronic sphygmomanometer. This solid state electronic device can be used to read systolic and diastolic blood pressure using both a flashing light and sound. A highly sensitive and functional microphone located in the cuff replaces the use of the stethoscope. (Courtesy of Dyna-Med, Inc., Carlsbad, CA.)

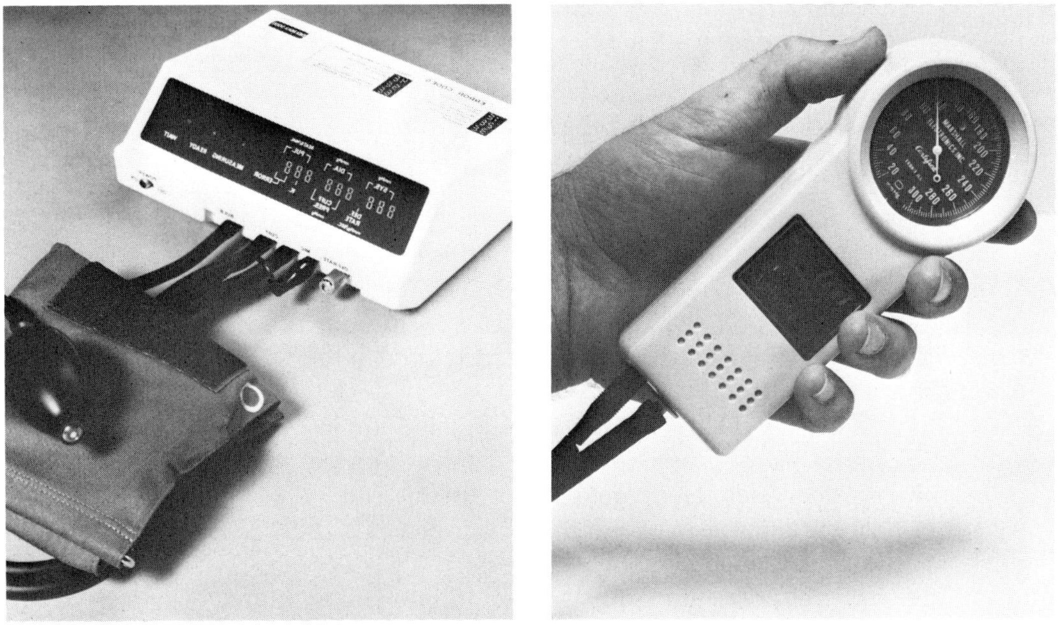

(d) (e)

Figure 41-2 *(Continued)* (d) Digimax 5000 Electronic Blood Pressure Monitor. This device can provide a simultaneous digital readout of diastolic and systolic pressure and the pulse rate. (e) Gauge of the Astropulse 10 Blood Pressure Monitor. This instrument uses electronic sound and light indicators as well as a standard needle indicator. A microphone located in the cuff eliminates the need for a stethoscope. (Courtesy of Dyna-Med, Inc., Carlsbad, CA.)

assistance if you have any problems with the placement of the cuff. Place the measurement gauge or mercury column in a position close enough to be read easily.

4. Insert the earpieces of the stethoscope in the ears and place the stethoscope bell or diaphragm above the brachial artery at the antecubital space (see Figure 41-1a). What do you hear?
5. Open the valve on the rubber bulb and inflate the cuff to 150 mm Hg. (Do not go higher without the recommendation of your instructor.) Close the valve. (*Note:* You should work quickly from this point.)
6. Open the valve on the rubber bulb slightly and listen carefully for pressure sounds. As the cuff pressure is decreased, vibrations of the arterial wall are caused by the spurting of blood through the narrowed artery under the cuff into the fully opened artery below the cuff. These vibrations are called *Korotkov's sounds.* The pressure at which these sounds are detected is the *systolic pressure.* Record this pressure in the results and observations section.
7. Slowly open the valve further while you listen for the total dissappearance of all sounds. The point at which no sound can be detected is the *diastolic pressure.* Record this value in the results and observations section.
8. Determine the systolic and diastolic pressures with your partner standing and after running up and down several flights of stairs. Record your findings in the results and observations section.
9. Calculate the pulse pressure for each situation using the following formula:

$$\text{Pulse pressure} = \text{systolic pressure} - \text{diastolic pressure}$$

Enter your findings and answer the questions in the results and observations section.

10. Repeat the entire procedure using the right arm. Record your findings in the results and observations section.
11. Change places with your laboratory partner and repeat steps 1–10.

Procedure 3: Astropulse 10 Blood Pressure Monitor

This procedure is to be performed by students in pairs.

1. Check to see if the unit is ready to use. The battery should be in place and the microphone jack plugged into the receptacle at the bottom of the unit.
2. Ask your laboratory partner to sit comfortably with the left palm facing up and the arm in line with the heart.
3. Be certain the end of the pressure cuff is inserted through the "D" ring and slip the cuff over the left arm (Figure 41-3).
4. Turn the cuff so that its blue dot is on the inside of the arm just above the elbow. Pull the end of the cuff so that it fits firmly on the arm, and wrap any extra cuff under and around the arm. Press the Velcro tape together to secure the cuff.
5. Place or hold the pressure gauge so that it can be easily read. Now you are ready to determine the systolic and diastolic pressures.
6. With the power off, close the air-flow valve on the bulb by turning it in a clockwise direction. Inflate the cuff quickly by repeatedly squeezing the bulb until the needle reaches the point marked 150 mm on the gauge. (In some cases it may be necessary to inflate the cuff beyond 150 mm. Consult with your instructor.)
7. Turn on the power switch. You will hear a beep and see a light flash on the gauge as the

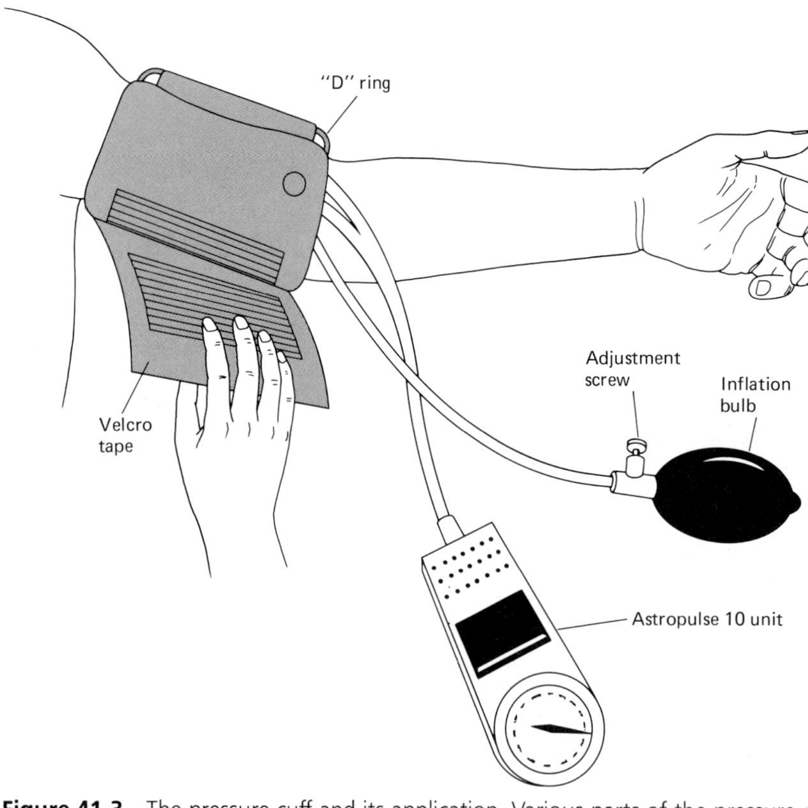

Figure 41-3 The pressure cuff and its application. Various parts of the pressure cuff, the air flow system, and the positioning of the arm are shown. The entire procedure can be performed by one individual.

power enters the unit. (The beep and light will not occur again until the pressure is reduced.)

8. Open the valve slightly in a counterclockwise direction so that the pressure falls at the rate of 2–4 mm/sec (one mark on the gauge equals 2 mm).
9. Continue to deflate the cuff at the rate of 2–4 mm/sec. When the pressure in the cuff is equal to the *systolic pressure* in the artery, the beep will sound and the light will flash. Note the reading on the gauge.
10. Allow the pressure to continue to drop at the same rate by opening the valve. The beeping and flashing will continue. Watch the gauge. At the moment the beeping and flashing stop, take a reading. This is the *diastolic pressure value.*
11. Repeat steps 1–10 until you can carry out the procedure easily.
12. Obtain readings for systolic and diastolic pressures in the sitting, standing, and after exercise situations. Record the values and answer the questions in the results and observations section.
13. Change places with your partner and repeat steps 1–12.
14. Experiment and carry out the procedure on yourself.

B. PULSE RATE DETERMINATIONS

Materials

The following materials should be provided for class use.

1. Electronic pulse monitor.
2. Paper towels or other material to wipe fingers.
3. Stopwatch

Procedure 1: Examination of Pulse Rate Determining Device

This procedure is to be performed by students individually.

1. Examine the electronic pulse monitor provided.
2. With the aid of Figure 41-4 familiarize yourself with the parts of this device.

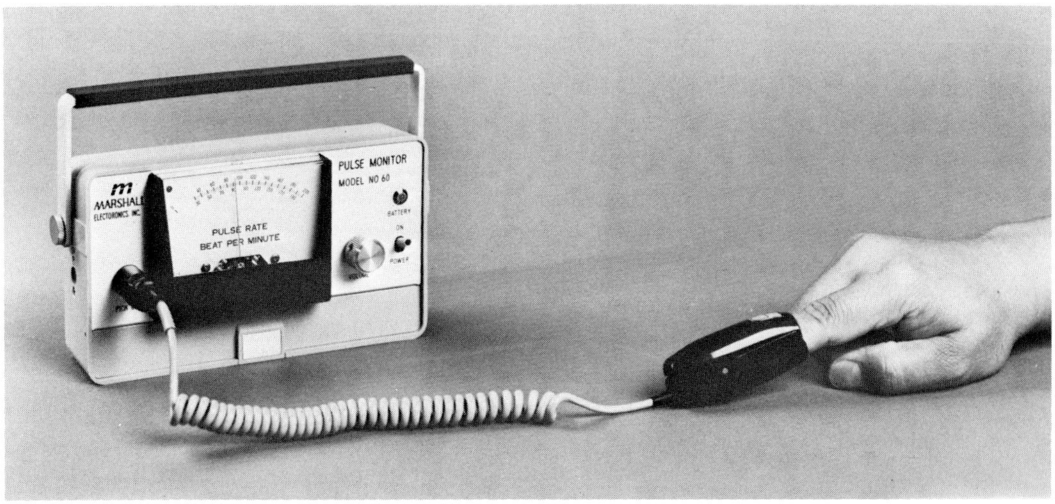

Figure 41-4 Electronic pulse monitor. The pulse rate of an individual can be determined simply by inserting the index finger into the sensitive photoelectric sensor(s). The pulse rate registers on the scale dial. (Courtesy of Dyna-Med, Inc., Carlsbad, CA.)

Procedure 2: Radial Pulse Determinations

This procedure is to be performed by students in pairs.

1. Ask your laboratory partner to sit comfortably and hold the forearm of one hand with palm up toward you.
2. Take the wrist of this hand and lightly place your index and middle fingers on the surface facing you in the area behind the thumb (Figure 41-5). Your thumb should be holding the underside of the wrist.
3. Feel for the beating sensation in this area. Do not use your thumb since it has a pulse of its own.
4. Count the number of beats that occur per 15 seconds. Multiply the beats by 4 to obtain the pulse rate for a minute.
5. Repeat this procedure three times and obtain an average value. Record your findings in the results and observations section.
6. Take the pulse rate of your partner in the standing position and after exercise. Running up and down a flight of stairs three times or running rapidly in place for 2 minutes would be sufficient exercise.
7. Enter your finding and answer the questions in the results and observations section.
8. Change places with your laboratory partner and repeat steps 1–7.

Procedure 3: Electronic Pulse Monitoring

This procedure is to be performed by students in pairs.

1. Turn on the electronic pulse device.
2. Ask your laboratory partner to sit comfortably next to the measuring device.
3. Insert your partner's index finger into the photoelectric sensor and obtain a reading.
4. Repeat this step. Was the reading the same? Enter your findings in the results and observations section.
5. Obtain pulse rates with your partner standing and after exercise. Enter your findings in the results and observations section.
6. Change places with your partner and repeat steps 1–5.
7. Turn off the electronic device after use.

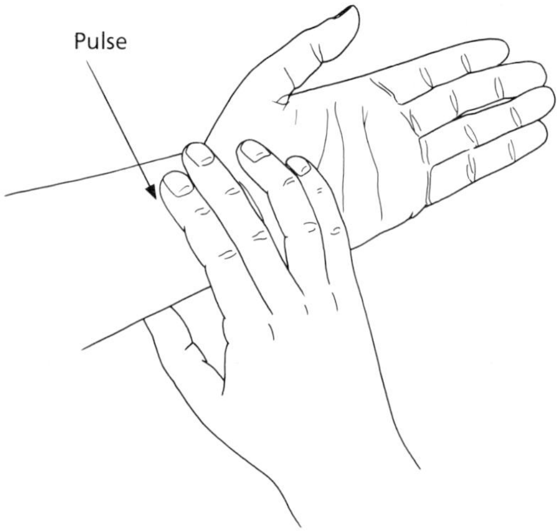

Figure 41-5 Position of the fingers for radial pulse determination.

Procedure 4: Carotid Pulse Determinations

This procedure is to be performed by students in pairs.

1. Ask your partner to run in place for 2 minutes and then sit down.
2. Place your fingers (no thumb) on either side of the partner's larynx. Gently press inward and toward the back of the neck. Continue to apply gentle pressure until a pulse is felt.
3. Once the pulse region is located, practice the procedure until only two fingers are necessary to make a pulse determination. Record your findings and answer the questions in the results and observations section.
4. Repeat step 1 and determine the radial pulse rate (refer to procedure 2). Enter your findings in the results and observations section.
5. Change places with your partner and repeat steps 1–4.

RESULTS AND OBSERVATIONS

BLOOD PRESSURE DETERMINATIONS

1. Enter the blood pressure readings determined by auscultation in Table 41-2. Record your own and your partner's systolic (S) and diastolic (D) pressure. Space is provided for measurements of other students.

Table 41-2

	Seated				Standing				After Exercise			
	Left Arm		Right Arm		Left Arm		Right Arm		Left Arm		Right Arm	
Student	S	D	S	D	S	D	S	D	S	D	S	D

2. Enter pulse pressure results in Table 41-3.

Table 41-3

	Seated		Standing		After Exercise	
Student	Left Arm	Right Arm	Left Arm	Right Arm	Left Arm	Right Arm

3. Compare your pressures with those listed for your age group in Table 41-1. Are your values similar to those listed? ____________

4. Are the values for your partner similar to those listed? ____________

5. Did this procedure cause you any problems? Explain. ______________________________

____

6. Enter the blood pressure readings determined by the Astropulse 10 Blood Pressure Monitor in Table 41-4.

Table 41-4

	Seated		Standing		After Exercise	
Student	Systolic	Diastolic	Systolic	Diastolic	Systolic	Diastolic

7. Was the electronic device easier to use than the standard sphygmomanometer? Explain.

____

____

8. Were the determinations comparable to those obtained by other methods used?

____

9. Was your blood pressure value within the normal range? ______________

10. Was your laboratory partner's value within the normal range? ______________

PULSE RATE DETERMINATIONS

1. Enter your findings obtained with the radial pulse determinations (R) and with the electronic monitor (E) in Table 41-5.

Table 41-5

	Seated		Standing		After Exercise	
Student	R	E		E	R	E

2. If an electronic monitoring device was provided, were the determinations comparable to those obtained by other methods used? ______________

3. Enter your findings comparing radial and carotid pulse determinations in Table 41-6.

Table 41-6

Student	Radial Pulse Determination	Carotid Pulse Determination

Questions for Review

EXERCISE **41**

BLOOD PRESSURE AND PULSE DETERMINATION

Name ______________________________

Laboratory Section ______________________ Date __________

1. List three factors that affect heart rate.

   a. ______________________________

   b. ______________________________

   c. ______________________________

2. a. What is pulse rate? ______________________________

   ____

   b. What is the average value for an adult? ______________________________

   ____

3. What sound is heard with a stethoscope when the brachial artery is collapsed? ______

4. List the major conditions that should be met while taking an individual's blood pressure.

   ____

   ____

5. Identify the device shown in Figure Q41-1 and identify the specific labeled parts.

   a. Device __________ c. __________

   b. __________

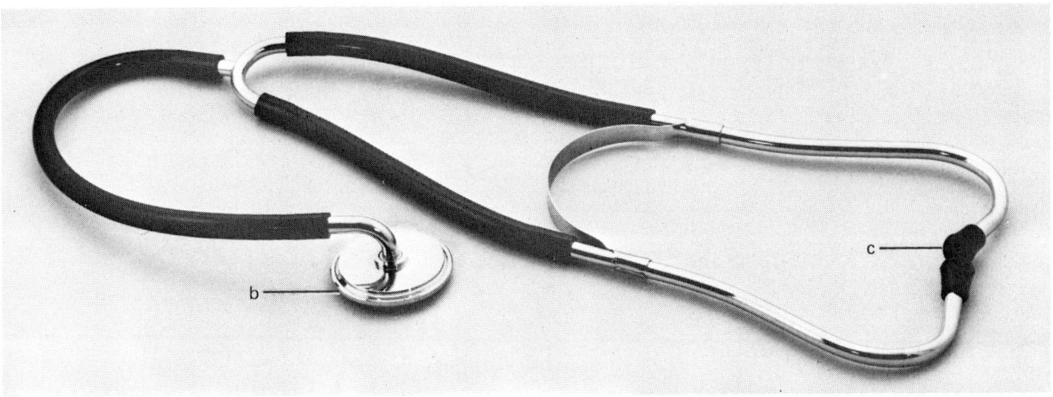

Figure Q41-1 (Courtesy, Dyna-Med, Inc., Carlsbad, CA.)

6. Which artery is usually used to take an individual's pulse rate? ______________

7. Complete the following formula.

   Pulse rate (beats/min) × stroke volume (ml) = ______________ (liters/min)

8. What are Korotkov's sounds? ______________

   ____

9. Define the following.

   a. Bradycardia ______________

   b. Tachycardia ______________

10. With the aid of your textbook and other references indicate the effects of the following conditions on arterial blood pressure.

    a. Vasoconstriction ______________

    b. Increased blood volume ______________

    ____

    c. Arteriosclerosis ______________

    ____

    d. Increased aldosterone concentration ______________

    ____

11. What is hypertension? ______________

    ____

EXERCISE 42

THE EFFECTS OF TEMPERATURE ON HEARTBEAT

There are several factors that influence the heart rate. These include age, body size, sex, certain hormones, and body temperature. This exercise uses a laboratory animal to demonstrate the effects of different temperatures on the rate and strength of the heartbeat.

In the human and other warm-blooded animals the rate and strength of the heartbeat increase with a rise in temperature up to a certain point. This point coincides with the most functional temperature (optimal) for enzymatic reactions to occur. With further substantial increases in temperature, enzymes and structural proteins are denatured. With a substantial decrease in temperature, the heart rate slows and the rate of chemical reactions decreases.

To demonstrate the effects of temperature we must obtain accurate, permanent tracings of functional heart activity. Since a variety of systems, for example, kymographs, chart recorders, and physiographs, can be used for such purposes, describing each one completely is not possible. The recording system for this exercise and for the following one includes a device to pick up information, such as a transducer that changes input signals (information) into a mechanical lever movement; one or more amplifiers to increase the strength of the transduced signal; and a readout component to display the amplified signal in some form of record or tracing.

Materials

1. The following materials should be provided for students in groups of four.
   a. Dissecting instruments (scalpel, scissors, dissecting pins, forceps)
   b. One dissecting pan
   c. One live frog
   d. One recording system (mechanical or electronic)
   e. One ring stand and clamps
   f. One small fishhook (size 10) or similar device
   g. Two glass or metal rods.
2. The following materials should be provided for class use.
   a. Ringer's solution in plastic squeeze bottles: three bottles for each group, one kept at 4°C (refrigerator temperature), one kept at 25°C (in water bath), and one kept at 37°C (in water bath)
   b. Two water baths, one set at 25°C and the other at 37°C.
   c. Thread
   d. Stopwatch
   e. Beaker of water at 4°C (ice water)

Procedure 1: Preparation of Laboratory Animal

This procedure is to be performed by students in groups of four. Review the anatomical orientation described in Exercise 8 before carrying out this procedure.

1. Double pith the frog. (Refer to Exercise 25 for the pithing procedure.)
2. Place the double-pithed frog on a dissecting pan with its ventral surface up. Pin the feet and lower jaw to the pan.
3. With a pair of sharp scissors make a midventral incision through the skin, from the pubic symphysis to the midline of the mandible to expose the underlying muscle layers. Pin the skin flaps to the dissecting pan (Figure 42-1a).
4. Remove the exposed muscle layers completely by cutting laterally on each side of the pectoral and pelvic girdles.
5. Carefully cut away the sternum to expose the frog's heart beating within a thin pericardium (Figure 42-1c).
6. Grasp the pericardium with a pair of forceps so that you can slit it open with scissors to expose the heart. (Be careful not to puncture the heart.)
7. Apply Ringer's solution kept at 25°C to keep the heart and other exposed organs moist. Continue to do this periodically during the experimentation period.
8. Identify the frog's atria, ventricle, and sinus venosus (Figure 42-1c). (The sinus venosus is the pacemaker in the frog, comparable to the human sinoatrial node.)
9. Tie a piece of fine thread, about 45 cm (18 inches) long, to a small fishhook (Figure 42-1d).
10. Either grasp or support the tip of the frog's ventricle with your fingers. Insert the fishhook held by forceps as close to the tip of the heart as possible (Figure 42-1e). Continue to apply the Ringer's solution.
11. Pass the free end of the thread through the appropriate portion of the transducer. Carefully adjust the thread so that there is no slack. Avoid pulling too tightly so that the heart is displaced from its normal location.
12. Tie the thread securely and continue to the next procedure.

Procedure 2: Normal Heart Action

This procedure is to be performed by students in groups of four.

1. Observe the rate (beats per minute) of the frog's right and left atria, the ventricle, and the sinus venosus.
2. Repeat step 1 two more times and obtain an average value. Enter your findings in the results and observations section.
3. Connect the transducer and heart lever components to the mechanical or electronic recording device provided, as directed by your instructor.
4. Turn on the recording system and trace the frog's cardiac cycle for 5 minutes. Keep the frog's heart and other exposed parts moist with the application of 25°C Ringer's solution.
5. Compare the tracings to Figure 42-2. Label the following components of the frog's cycle on your tracing: atrial systole, atrial diastole, and ventricular systole. Attach a small portion of the tracing in the results and observations section.
6. Turn the system off when you are through with this procedure.

Procedure 3: Effects of Temperature

This procedure is to be performed by students in groups of four or as a class demonstration.

1. Place several glass or metal rods in a beaker of water at 4°C.
2. Gently apply a chilled rod to the right atrium for 1 minute. Count the beats per minute of the atria, the sinus venosus, and the ventricle. Enter your findings in the results and observations section.

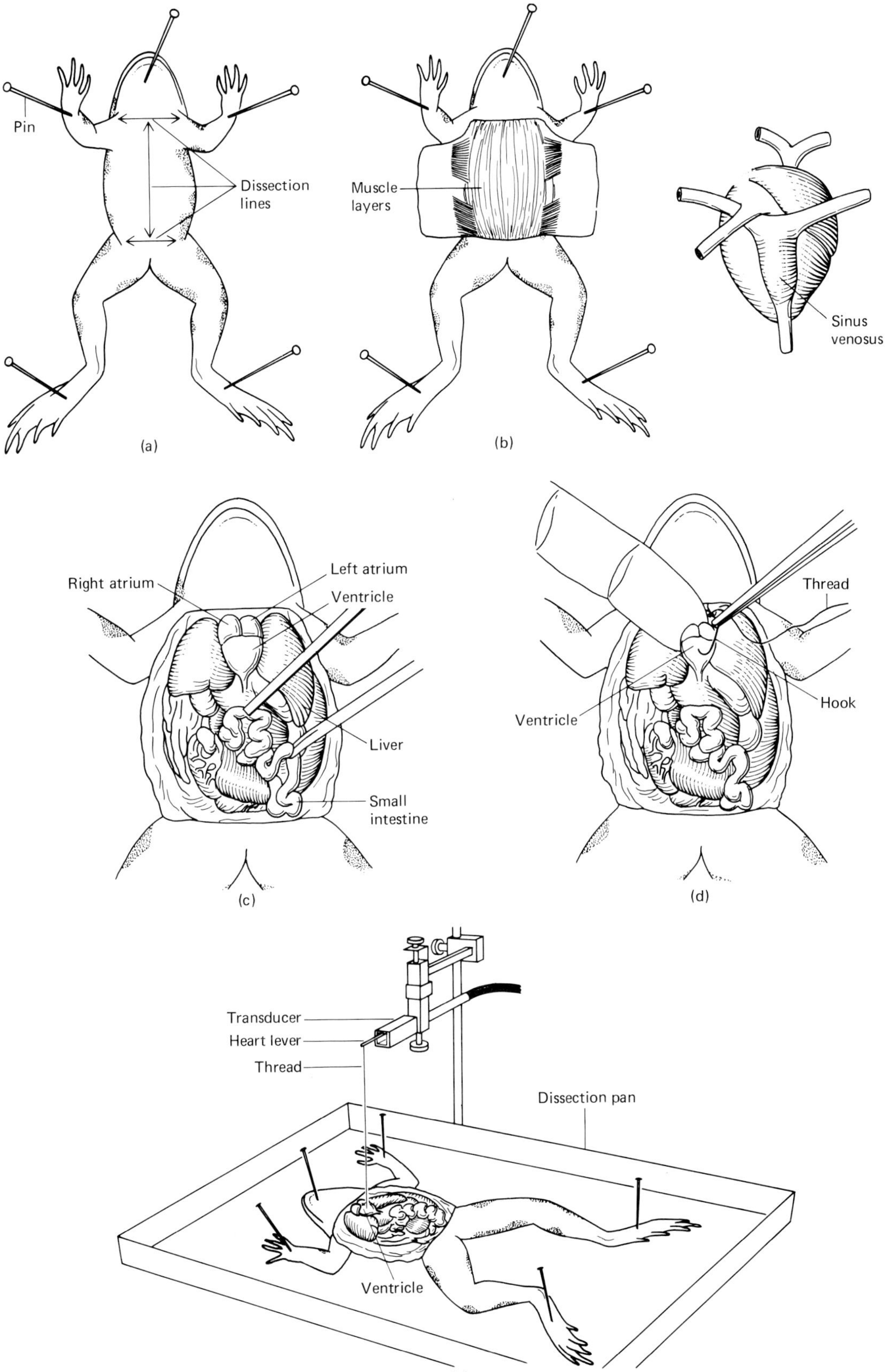

Figure 42-1 Preparation of the laboratory animal. (a) Pinning the animal to the dissection pan and dissection lines. (b) Midventral incision through skin and exposure of muscle layers. (c) Exposure of heart. Components of the heart from the ventral view are shown in the animal. The dorsal view is shown on the right. (d) Fishhook attachment. (e) Heart lever attachment. The heart lever shown here is a light-weight rod. The thread attached to the heart by the fishhook is tied to the end of the lever.

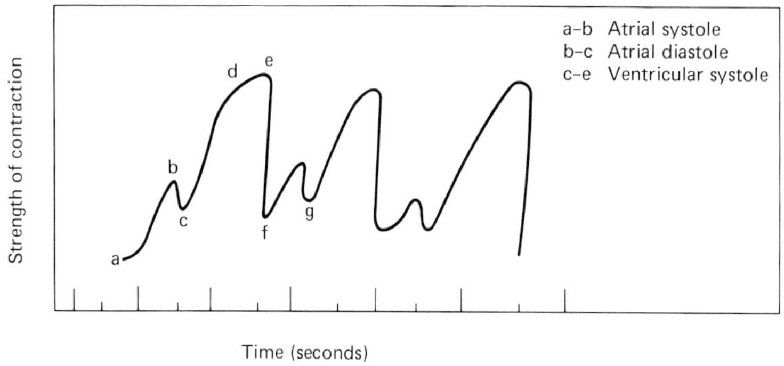

Figure 42-2 A normal frog cardiac cycle showing atrial systole, atrial diastole, and ventricular systole. The strength of contractions are plotted against time measured in seconds.

3. Restore the heart to room temperature by rinsing it with 25°C Ringer's solution.
4. Repeat steps 3 and 4 applying the rod first to the left atrium and then to the ventricle.
5. Connect the transducer and heart lever components to the mechanical or electronic recording device provided, as directed by your instructor.
6. Rinse the heart with 25°C Ringer's solution. When the heart appears to be contracting normally, make a control record. It should be compared to the normal cycle record made earlier.
7. Rinse the heart with 4°C Ringer's solution.
8. Remove the rinsing fluid as best you can with an eyedropper and add more of the 4°C solution so that the frog's heart is totally immersed in the fluid.
9. Start the recording system, as directed by your instructor, and obtain a tracing of the frog's cardiac cycle for 5 minutes. Label this tracing 4°C. Attach a section of it in the results and observations section.
10. Observe the heart rate at this temperature. Enter your findings in the results and observations section.
11. Restore the heart to room temperature with the application of 25°C Ringer's solution.
12. Repeat steps 7–11 using the Ringer's solution with a temperature of 37°C. Attach a label tracing and enter your findings on heart rate in the results and observations section.
13. Answer the questions in the results and observations section.
14. This pithed frog can be used in the next exercise.

RESULTS AND OBSERVATIONS

1. Indicate in Table 42-1 the heart rate of each chamber when heart is bathed in 25°C Ringer's solution. (The remaining columns in the table are to be filled during procedure 3.)

Table 42-1

	Number of Beats per Minute		
Heart Chamber	At 25°C	At 4°C	At 37°C
Right atrium			
Left atrium			
Sinus venosus			
Ventricle			

2. Attach labeled tracings or sketches of frog's cardiac cycle at different temperatures. (The records at 4°C and 37°C will be made during procedure 3.)
   a. 25°C

   b. 4°C

   c. 37°C

3. a. Indicate in Table 42-2 the heart rate when treated metal or glass rods are used to cool different heart chambers.

Table 42-2

	Number of Beats per Minute			
Heart Chamber Cooled	Right Atrium	Left Atrium	Ventricle	Sinus Venosus
Right atrium				
Left atrium				
Ventricle				
Sinus venosus				

   b. Does it appear that one part of the frog's heart has more influence over heart rate than other parts? ______________________________

____

____

4. From your observations in this exercise can a general statement be made as to the role of the central nervous system in controlling heart action? ______________________________

____

____

5. a. Are the effects of cold temperatures on heartbeat reversible? ____________________

____

b. What practical application can you suggest for your observations? ______________

____

Questions for Review

EXERCISE **42**

EFFECTS OF TEMPERATURE ON HEARTBEAT

Name ______________________________

Laboratory Section ______________________ Date ____________

1. What effect on heart rate would you expect in a young child with a fever of 39°C?

____

2. List six factors and their specific effects on heart rate.

   a. ______________________________

   b. ______________________________

   c. ______________________________

   d. ______________________________

   e. ______________________________

   f. ______________________________

3. Would you expect a frog or an elephant to have more heartbeats per minute? Explain.

____

____

____

4. Do ions influence the rate and strength of heart contractions? Explain. ____________

____

____

EXERCISE 43

THE EFFECTS OF VAGUS NERVE INHIBITION AND EPINEPHRINE ON THE CARDIOVASCULAR SYSTEM

Heart muscle is capable of spontaneous, rhythmic self-excitation. Even when all nerves to the heart have been cut, certain cells of this organ have a more rapid rate of firing than others. In mammals a number of such cells form a small mass of specialized conducting cells called the sinoatrial (SA) node or pacemaker. This node is embedded in the right atrial wall near the entrance of the superior vena cava. It generates the impulses that stimulate the heart muscle fibers to contract and thereby sets the basic heart rate.

The rate set by the node can be altered by impulses from the autonomic nervous system. Many fibers from the parasympathetic (vagus) and the sympathetic nervous systems end on the SA node and other regions of the heart. Parasympathetic fibers reach the heart by means of the vagus nerve. Impulses transmitted along these fibers cause the release of acetylcholine, which decreases the heart rate and contraction strength. Impulses transmitted along sympathetic components result in an increase in heart rate. Cutting the sympathetic nerves slows the heart rate.

The heart rate also may be altered by chemical factors, in addition to nervous system influences. One such chemical is the hormone epinephine (adrenaline). Epinephrine is the principal hormone of the adrenal medulla and is one of several substances that make up the group of catecholamines. These body chemicals are the emergency or stress hormones that aid animals to meet both physical and emotional crises and changes. The release of epinephrine, which is under direct neural control, produces a number of effects such as increased heart activity, elevated blood pressure, and glycolysis.

Factors such as temperature and other hormones also affect heart rate, but they are of lesser importance. This exercise will consider the effects of vagus nerve inhibition and epinephrine on the heart activity of a laboratory animal, the frog. In this animal the pacemaker is the sinus venosus.

Materials

The following materials should be provided for students in groups of four.

1. Dissecting instruments (scalpel, scissors, dissecting pins, forceps)
2. One double-pithed frog (used in Exercise 25)
3. One dissecting pan
4. One recording system (mechanical or electronic)
5. Ringer's solution
6. One milliliter hypodermic syringe and 26-gauge needle
7. One electronic stimulator and event-timer marker

8. Epinephrine and acetylcholine preparations (1:1000 dilutions)
9. Thread

Procedure 1: Normal Heart Action

This procedure is to be performed by students in groups of four.

1. Determine the heart rate of the frog's heart as described in procedure 2 of Exercise 42. Use the recording system provided to make a tracing of the normal cardiac cycle.
2. Attach a small portion of the tracing in the results and observations section. This record will serve as a basis for comparing the effects of the drugs used in the following procedures.

Procedure 2: Effect of Vagus Nerve Inhibition (Use of Acetylcholine)

This procedure is to be performed by students in groups of four.

1. Keep all parts of the frog's body moist with Ringer's solution.
2. Obtain a hypodermic syringe and needle containing acetylcholine (1:1000 dilution). Examine the units on the barrel of the syringe.
3. Turn on the recording system and observe the contractions for a period of 3 minutes.
4. Locate the frog's liver and slowly inject 0.1 cc of the acetylcholine into the organ. Mark the tracing to indicate the time of the injection.
5. Determine the effect of the drug on heart rate. Enter your findings in the results and observations section.
6. Attach a portion of the tracing showing the effects of acetylcholine in the results and observations section.
7. Turn off the recording system.

Procedure 3: Effect of Epinephrine

This procedure is to be performed by students in groups of four.

1. Keep all parts of the frog's body moist with Ringer's solution.
2. Obtain a hypodermic syringe and needle containing epinephrine (1:1000 dilution). Examine the units on the barrel of the syringe.
3. Turn on the recording system and observe the contractions for a period of 3 minutes. Is the cycle normal?
4. Locate the frog's liver and slowly inject 0.1 cc of epinephrine into the organ. Mark the tracing to indicate the time of injection.
5. Determine the effect of the drug on the heart rate. Is it faster or slower? Enter your findings in the appropriate portion of the results and observations section.
6. Attach a portion of the tracing showing the effects of epinephrine in the results and observations section.
7. Turn off the recording system.

Procedure 4: Effect of Repeated Exposures to Acetylcholine and Epinephrine

This procedure is to be performed by students in groups of four.

1. Keep all parts of the frog's body moist with Ringer's solution.
2. Turn on the recording system and observe the contractions for a period of 3 minutes or longer to obtain a normal cycle.
3. Obtain two hypodermic syringes, one containing acetylcholine (1:1000) and the other epinephrine (1:1000).
4. Inject 0.1 cc of epinephrine into the frog's liver. Mark the tracing to show the injection.

Observe the heart rate.

5. Wait 2 minutes and inject 0.1 cc of the acetylcholine in a similar manner. Again mark the tracing to show the injection. Observe the heart rate.
6. Wait 2 minutes and inject another 0.1 cc of epinephrine into the liver. Mark the tracing to show the injection. Determine how long it takes before the frog's heart cycle returns to a normal rate. Enter this finding on a portion of a tracing showing the return to a normal cycle.
7. Repeat step 5. Attach a representative tracing in the results and observations section.
8. Turn off the recording system.

Procedure 5: Effects of Large Dosages of Epinephrine and Acetylcholine

This procedure is to be performed by students in groups of four.

1. Repeat procedure 3 of this exercise but increase the doses of epinephrine (1:1000) and acetylcholine (1:1000) to 0.3 cc.
2. Attach a portion of the tracing showing the individual effects of the large dose of each chemical in the results and observations section.
3. Determine how long it takes for the heart cycle to return to normal.
4. Turn off the recording system.
5. Answer the questions in the results and observations section.

RESULTS AND OBSERVATIONS

1. Indicate in Table 43-1 the number of beats per minute made by the heart chambers under normal conditions. (The remaining columns of the table will be filled in during later procedures.)

Table 43-1

	Number of Beats per Minute				
		Acetylcholine Dosages		Epinephrine Dosages	
Heart Chamber	Normal	Small	Large	Small	Large
Right atrium					
Left atrium					
Ventricle					

2. Attach labeled tracings or sketches of frog's cardiac cycle under different conditions.
   a. Normal

b. After initial epinephrine exposure

c. After initial acetylcholine exposure

d. After second exposure to epinephrine following acetylcholine

e. After second exposure to acetylcholine following epinephrine

f. After a large epinephrine dose

g. After a large acetylcholine dose

3. By what means other than injection of acetylcholine can vagus nerve inhibition be produced? ____

____

4. Describe the observable differences between effects produced by epinephrine and those produced by acetylcholine. Use the labeled tracings. Consider the number of beats per minute and the strength of individual beats (heights in mm).

____

____

____

5. Compare the recovery times of the heart to a normal cycle after acetylcholine and epinephrine exposure.

   a. After acetylcholine ____________

   b. After epinephrine ____________

   c. After repeated exposure to acetylcholine after epinephrine ____________

   d. After large epinephrine dose ____________

Questions for Review

EXERCISE **43**

THE EFFECTS OF VAGUS NERVE INHIBITION AND EPINEPHRINE ON THE CARDIOVASCULAR SYSTEM

Name ______________________________

Laboratory Section ______________________ Date ____________

1. What is the SA node? ______________________________

____

____

2. Would the SA node cease to function if the heart were removed from the body? Explain.

____

____

____

3. List three specific factors that influence heart rate and give the effects of each.

   a. ______________________________

   b. ______________________________

   c. ______________________________

4. What are the body sources of epinephrine and acetylcholine?

____

5. How will the following influence heart rate?

   a. Cutting of the vagus nerve ______________________________

   ____

   b. Direct injection of epinephrine into the heart ______________________________

   ____

   c. Electrical stimulation of the vagus nerve ______________________________

   ____

SECTION XII

BLOOD AND THE IMMUNE SYSTEM

The circulatory system, which consists of the blood, heart, and blood vessels, provides a link between the body's internal components and its external environment. The exercises in this section will deal specifically with the blood and its components and will demonstrate their involvement in providing vital support for cellular activities and the maintenance and protection of a favorable cellular environment. In addition, attention is given to conditions that interfere with the production of red blood cells or that lead to an increased level of red blood cell destruction.

The two major components of blood are *formed elements*—including erythrocytes (red blood cells), leukocytes (white blood cells), and platelets—and *plasma*, a pale yellow fluid. Plasma contains about 90 percent water, with the remaining 10 percent consisting of carbohydrates, hormones, lipids, proteins (albumin, enzymes, fibrinogen, immunoglobulins or antibodies, and other globulins), salts, and vitamins. Several laboratory procedures are used to determine the concentrations of blood components (Figure XII-1). Because the results of such tests are of diagnostic importance, instrumentation improvements have been made so that procedures can be performed more rapidly and accurately.

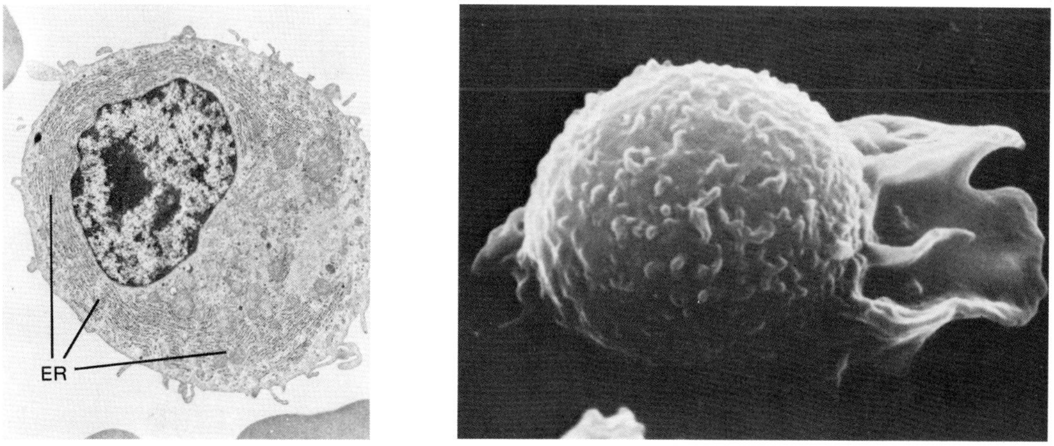

(a) (b)

Figure XII-1 Normal and abnormal blood cells. (a) An electron micrograph of an activated plasma cell. This cell type is the immunoglobulin producer of the body. Note the presence of the large quantity of endoplasmic reticulum (ER) within the cell. (From D. Douer, A. Weinberger, M. Djaldetti, J. Asherov, I. Pick and J. Pinkhas. *Acta Hematologica,* 62:81–85, 1979.) (b) A scanning micrograph showing a leukemia cell originating from lymphoid tissue.

Several exercises consider procedures used to demonstrate the presence and functions of the components of blood. In addition, attention is given to conditions that interfere with the production of red blood cells or that lead to an increased level of red blood cell destruction.

GENERAL REFERENCES

Tortora, G. J., and N. P. Anagnostakos. *Principles of Anatomy and Physiology,* 6th ed. New York: Harper & Row, 1990.

Winthrobe, M. M., et al. *Clinical Hematology,* 8th ed. Philadelphia: Lea & Febiger, 1981.

Witherspoon, J. D. *Human Physiology.* New York: Harper & Row, 1984.

SPECIFIC REFERENCES

Dvorak, A. M., and H. F. Dvorak. "The Basophil," *Archives of Pathology and Laboratory Medicine,* 103:551–557, 1979.

Evans, V. J. "Platelet Morphology and the Blood Smear," *Journal of Medical Technology,* 1:689–695, 1984.

Golde, D. W., and J. C. Grasson. "Hormones That Stimulate the Growth of Blood Cells," *Scientific American,* 259:62–70, 1988.

MacKinney, A. A., Jr. (Ed.). *Pathophysiology of Blood.* New York: Wiley, 1984.

Morse, E. E., and B. Jacobs. "Blood Donation and Its Aftereffects," *American Scientist,* 73:68–69, 1985.

Payne, B. A., R. V. Pierre, and M. A. Morris. "Use of Instruments to Obtain Red Blood Cell Profiles," *Journal of Medical Technology,* 2:379–387, 1985.

EXERCISE 44

BLOOD SMEAR PREPARATION AND DIFFERENTIAL STAINING

Blood cells must go through a development cycle in the same way a human matures during development. In healthy individuals only mature cells are seen upon examination of blood specimens. Immature or abnormal cells can be observed in various disease states. Thus the examination of a patient's blood may be extremely important in the detection of abnormalities or the diagnosis of disease.

An essential part of a hematological examination is a properly prepared blood smear or spread. Such a smear is made by obtaining a drop of blood from the patient, placing it on a slide, and then spreading it, with the aid of a second slide, into a thin layer. After the preparation dries, any one of several staining procedures, such as Wright's or Giemsa, can be applied.

The examination of blood for abnormalities and the determination of the percentage of various types of cells may be diagnostic in themselves (Figure 44-1), or they may serve as a basis for additional tests. For example, the presence of sickled cells in the blood is diagnostic for the condition of sickle cell anemia (Figure 44-1), whereas an elevated white cell count may be a sign that a disease exists but not an identification of the disease.

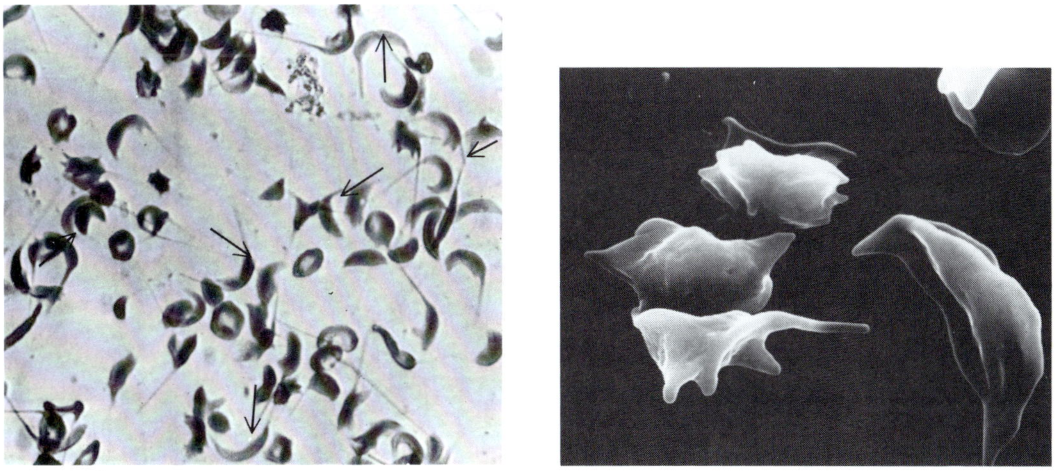

(a) (b)

Figure 44-1 The appearance of sickle cells. (a) A light microscopic view of a red blood cell preparation showing a large number of sickled cells (arrows). (From N. M. Rumen, *Blood,* 45:45–48, 1975.) (b) A scanning micrograph of sickled cells.

This exercise concentrates on formed elements of blood. The formed elements of blood include red blood cells *(erythrocytes),* white blood cells *(leukocytes),* and platelets *(thrombocytes).* Erythrocytes carry oxygen to tissues and remove carbon dioxide from them. Leukocytes function in inflammation and immune responses of the body. There are five types of leukocytes, divided into two general categories, granulocytes and agranulocytes. The *granulocytes,* cells containing cytoplasmic granules, consist of neutrophils, eosinophils, and basophils (so named for the staining characteristics of their granules; see Color Plate 4). Neutrophils serve as essential agents in an individual's defense against various bacteria and some fungi. Eosinophils have a role in defending against disease agents (pathegens) such as protozoa and worms. The exact function of basophils is not well understood. These cells are believed to be associated with allergic reactions and inflammation. The *agranulocytes,* cells without granules, consist of lymphocytes and monocytes (Color Plate 4). Monocytes are critical in the body's defense against intracellular disease-causing agents such as viruses and certain bacteria. Lymphocytes also help in recognizing and destroying several types of pathogens. The platelets together with coagulation factors in plasma control the clotting of blood. The diameters of the formed elements are: erythrocytes, approximately 7 μm; lymphocytes, 9–18 μm; monocytes, 24 μm; granulocytes, 10–14 μm; and platelets, 2–4 μm.

Determination of the proportion of each type of leukocyte in the blood is made by means of the *differential count* procedure. Out of a sample of 100 white blood cells observed on a stained blood smear, the normal number for each type of cell is:

Basophils	0–1
Eosinophils	1–4
Neutrophils	60–70
Large lymphocytes	0–3
Small lymphocytes	25–30
Monocytes	4–8
	90–116

Detection of abnormalities associated with erythrocytes, such as anemias and malignancies, can be made from examination of blood smears. *Anemia* is a below normal number of erythrocytes or a below normal hemoglobin content in the cells. In general it is caused by blood loss, by an interference with erythrocyte production in the bone marrow, by accelerated erythrocyte destruction, or by combinations of these factors. The appearance of erythrocytes is one method of classifying anemias. The cells may be described in terms of size and hemoglobin content. Combinations of the following properties are encountered in examination of blood smears.

Erythrocyte Size

Macrocytic—cells are larger than normal
Normocytic—cells are normal size
Microcytic—cells are smaller than normal

Hemoglobin Content

Hyperchromic—cells contain more than normal amount
Normochromic—cells contain normal amount
Hypochromic—cells contain less than normal amount

Anemias resulting from the decreased production of erythrocytes can be caused by various deficiencies including iron, vitamin B_{12}, destruction of bone marrow by chemical and physical agents, and malignancies of different types. Anemias caused by the increased destruction of red blood cells occur in cases of severe infections, sickle cell anemia, incorrect transfusions, and hemolytic disease of the newborn.

Various conditions also are associated with leukocytes. *Leukopenia,* or low white blood cell count, often is induced by virus infections and treatments of malignancies by radiation or chemicals. An above normal white blood cell count may be provoked by several conditions (Table 44-1).

Table 44-1 Conditions Associated with White Blood Cell Increases

Cell Type	Condition Causing Increase in Cell Number
Basophil	Granulocytic leukemia
Eosinophil	Allergies, worm infections, Hodgkin's disease, certain leukemias
Lymphocyte	Hepatitis, infectious mononucleosis, measles, mumps, tuberculosis, lymphocytic leukemia
Monocyte	Hodgkin's disease, monocytic leukemia, tuberculosis
Neutrophil	Granulocytic leukemia, pus-producing bacterial infections, poisons

The leukemias are a group of diseases primarily affecting the lymphatic, mononuclear, and phagocytic systems and the bone marrow. Overproduction of leukocytes and the presence of great numbers of immature and abnormal forms of white blood cells in the bone marrow (Color Plates 31) and blood-forming organs are characteristic features of these diseases. Many of the immature and abnormal leukemic cells gain access to the circulating bloodstream. These rapidly reproducing leukemic cells fill the bone marrow spaces, prevent immature red blood cells from maturing, and interfere with platelet formation. Leukemic cells also spill into regions of the body and accumulate in organs with no blood-forming capability.

Three types of leukemia are recognized: granulocytic (myelogenous), lymphatic, and monocytic (Color Plates 32–34). This classification is based on the particular leukocyte involved. Leukemias also are grouped according to other factors such as the duration of the disease—acute (rapid onset) or chronic (of long duration)—and the prominence of immature and abnormal cell types in the bone marrow and blood. In granulocytic (myelogenous) leukemia, an excessive number of granulocytes and the bone marrow cells that give rise to them (precursors) appear in the blood. The precursors for granulocytes are called myeloblasts and myelocytes. With lymphatic leukemia, an abnormal number of lymphocytes and their precursors (lymphoblasts and lymphoid cells) appear in the blood and bone marrow. In monocytic leukemia, an abnormal number of immature monocytic leukocytes are found in the blood (Figure 44-2a). One other type of leukemia, known as plasma cell (multiple) myeloma, is included here even though the cells involved normally do not circulate in the bloodstream. Plasma cells, as described in the section introduction, are immunoglobulin-producing cells. When abnormal plasma cells spill over from locations such as the marrow of the ribs and skull into the peripheral blood, a condition of plasma cell leukemia exists (Figure 44-2b). In this exercise slides from several types of leukemia states will be placed on demonstration to emphasize the appearance of abnormal and immature leukocytes.

This exercise considers preparation of a blood smear, a differential count procedure, and examination of normal and abnormal blood specimens. Exercise 45 contains procedures using blood cell counting equipment.

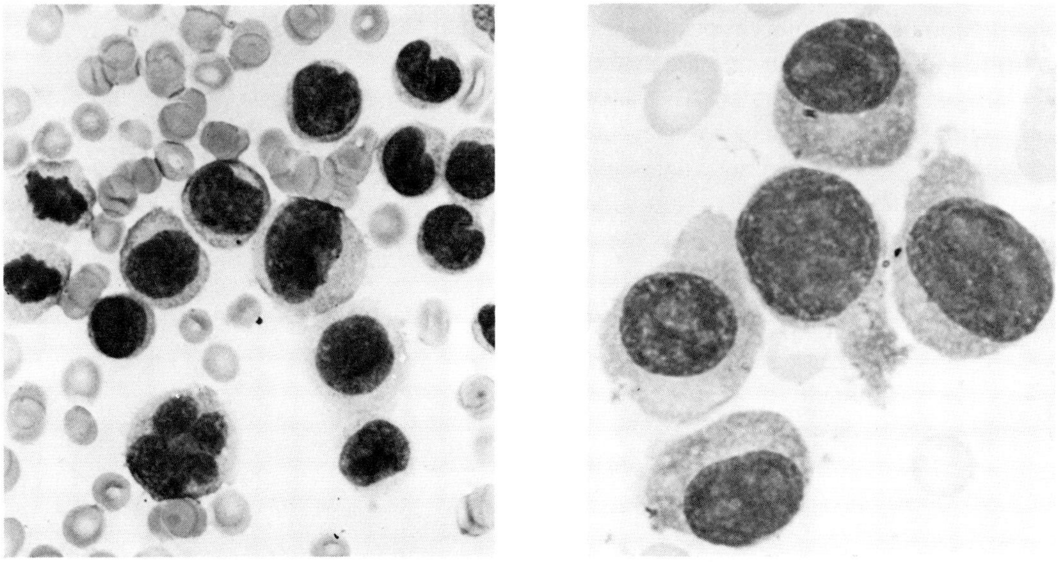

(a) (b)

Figure 44-2 Examples of the leukemias. (a) Leukemic cells with distinct monocytic morphology. (From I. Mirchandani, P. Tabaczka, and M. Palutka, *American Journal of Hematology,* 12:139–147, 1982.) (b) The appearance of plasma cell leukemia. The majority of the cells shown are plasma cells. (From D. Douer, A. Weinberger, M. Djaldetti, J. Asherov, I. Pick, and J. Pinkhaus, *Acta Hematologica,* 62:81–85, 1979.)

Materials

1. The following materials should be provided per four students.
   a. Commercially prepared sterile alcohol prep-pads
   b. Sterile, disposable blood lancets
   c. Sterile gauze pads (2 × 2 inch)
   d. Wright's stain and buffer solution for Wright's stain (in dropper bottles or other suitable containers)
   e. Staining rack
   f. Containers for the disposal of lancets, pads, and used slides
2. The following materials should be provided for class use.
   a. Clean, lint-free glass slides
   b. Immersion oil
   c. Xylene and lens paper
   d. Demonstration slides of blood smears representing normal blood and noninfectious anemic states
      i. Normal blood smears
      ii. Sickle cell anemia
      iii. Macrocytic, hyperchromic anemia
      iv. Microcytic, hypochromic anemia
      v. Hemolytic anemia
   e. Demonstration slides of blood smears from selected microbial infections
      i. Representative blood stages of malaria parasites, *Plasmodium falciparum* or *P. vivax*
      ii. African sleeping sickness (*Trypanosoma brucei* variety *gambiense*) infection
      iii. Infectious mononucleosis

f. Demonstration slides of blood smears from selected leukemic conditions
   i. Acute myelomonocytic leukemia
   ii. Chronic granulocytic leukemia
   iii. Acute lymphoblastic leukemia
g. Twelve demonstration microscopes

Procedure 1: Blood Smear Preparation

This procedure is to be performed by students individually, but it will require the assistance of another student.

1. Wipe the finger from which a sample is to be taken with an alcohol-moistened (but not saturated) gauze pad. Do not use a finger that will be involved later in holding a pencil or other such activity. (See Exercise 6 for this procedure.)
2. Unwrap the blood lancet.
3. Hold the finger in a downward position and squeeze it gently.
4. Have your laboratory partner puncture the ball of the finger with the lancet.
5. Discard the first two drops of blood by wiping them off with a fresh pad.
6. Keeping the finger in a downward position, touch one end of a clean glass slide to a newly formed blood drop. Put the slide on a flat surface, wet side up.
7. Place a fresh, sterile pad on the punctured area and hold it tightly in place until the bleeding stops.
8. Pull one edge of a second slide into the blood drop (Figure 44-3a). Allow the blood to run across its width (Figure 44-3b).
9. With the second slide at about a 30° angle, move it toward the opposite end of the first slide (Figure 44-3c). A good smear should have a feathered edge.
10. Scratch your initials in the center of the preparation and allow it to air dry.

Procedure 2: Staining and Differential Count

1. Place your blood smear on the staining rack, or as directed otherwise by your instructor.
2. Apply enough Wright's stain to cover the entire preparation.
3. Allow the stain to remain in contact with the smear for approximately 1 minute.
4. Add the buffer solution (in drops) until the buffer and stain are in equal proportion. Blow gently on the slide to mix the two solutions.
5. Allow the resulting mixture to stay in contact with the smear for 4 minutes.
6. Pour off the solution and flood the slide with buffer solution in order to wash the preparation.
7. Air dry or blot dry the smear. No cover slip is needed.

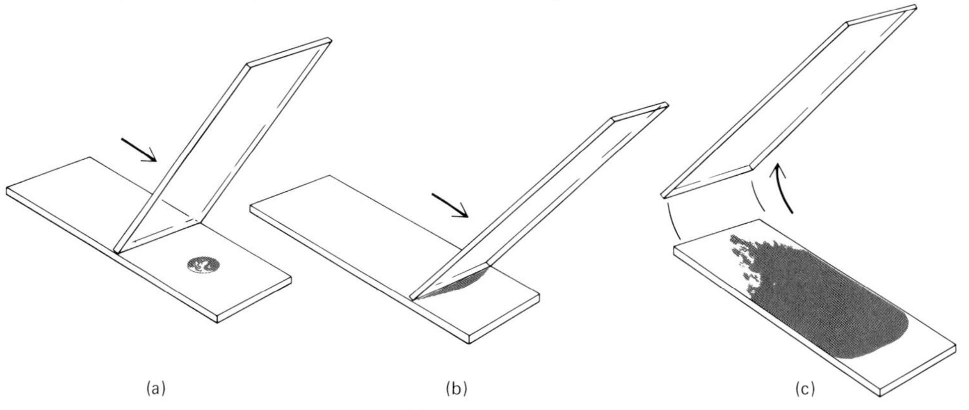

Figure 44-3 Blood smear preparation. (a) Pull the second slide into the drop of blood. (b) Allow blood to spread across its width. (c) Push the slide toward the end of the first slide.

8. Examine the demonstration slides of normal blood smears.
9. Perform a differential count, that is, count the first 100 white blood cells observed and note the type of each cell. Enter your findings in the results and observations section.
10. Examine the red blood cells in your smear and in your laboratory partner's. Answer the relevant questions in the results and observations section.

Procedure 3: Demonstration of Normal Blood and Noninfectious Anemic States

1. Your instructor will emphasize the distinguishing features of a normal blood smear as compared to those of anemic states.
2. Examine all the demonstration slides provided for this portion of the exercise. Refer also to Color Plate 4.
3. Make sketches of the distinguishing features noted in the results and observations section and answer all questions pertaining to this exercise in the questions for review section.

Procedure 4: Demonstration Slides of Blood Smears from Selected Microbial Infections

1. Your instructor will emphasize the major distinguishing features of the preparations.
2. Examine all the demonstration slides. Refer to Color Plates 35–37.
3. Note all distinguishing features. Make sketches of these in the results and observations section.
4. Note whether or not the infectious disease agents are found inside or outside erythrocytes.

Procedure 5: Demonstration Slides from Selected Leukemic Conditions

1. Your instructor will emphasize the major features of the preparations. Also refer to Color Plates 31–34.
2. Pay particular attention to the presence of immature blood cells, as well as the smaller than normal size of others.
3. Make sketches of typical leukemic cells in the results and observations section and answer questions associated with this portion of the execise.
4. Answer any items pertaining to this and previous portions of this exercise in the questions for review section.

RESULTS AND OBSERVATIONS

1. Enter your findings for the differential count in Table 44-2.

Table 44-2

Type of White Cell	Number per 100	Average Number per 100*
Lymphocyte		
Monocyte		
Eosinophil		
Basophil		
Neutrophil		

*Enter normal values here as indicated in your textbook or other reference.

2. Did you observe any unusual cells in the demonstration slide? If so, describe them.

____

____

3. Were your red blood cells and those of your partner similar to those in the demonstration slides of a normal blood smear? Describe any differences. ______________________

____

4. Sketch distinguishing features of the blood preparations representing normal blood and noninfectious anemic states.

Normal smear

Macrocytic, hyperchromic anemia

Microcytic, hypochromic anemia

Sickle cell anemia

5. Sketch distinguishing features of blood smear preparations from selected microbial infections.

African sleeping sickness infection

Infectious mononucleosis

Malaria infection with *Plasmodium falciparum*

6. In which of the demonstration slide preparations did you find disease agents actually inside erythrocytes?

____

7. Sketch distinguishing features of leukemic blood cells.

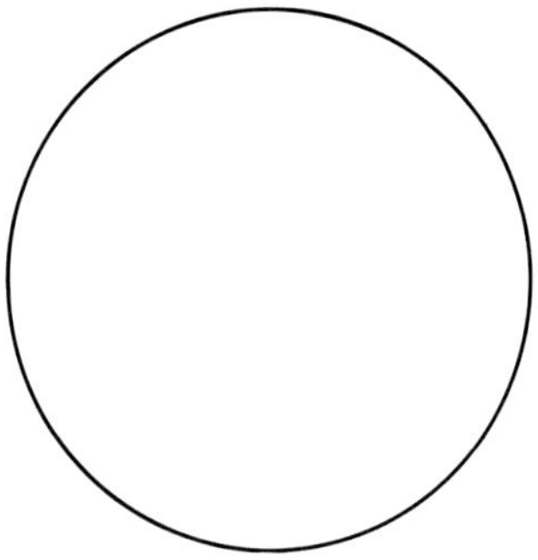

Acute myelomonocytic leukemia

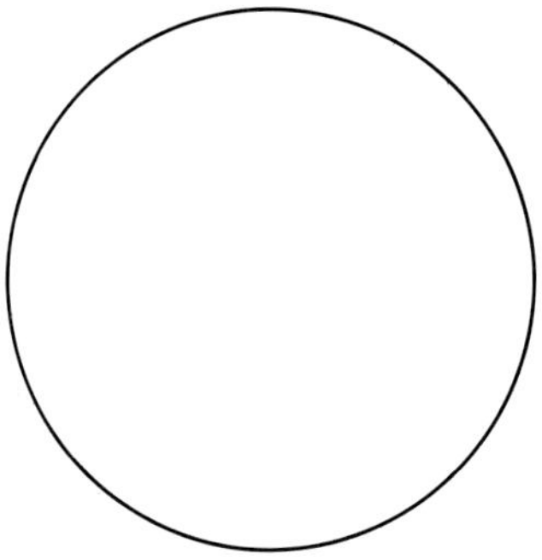

Chronic granulocytic leukemia

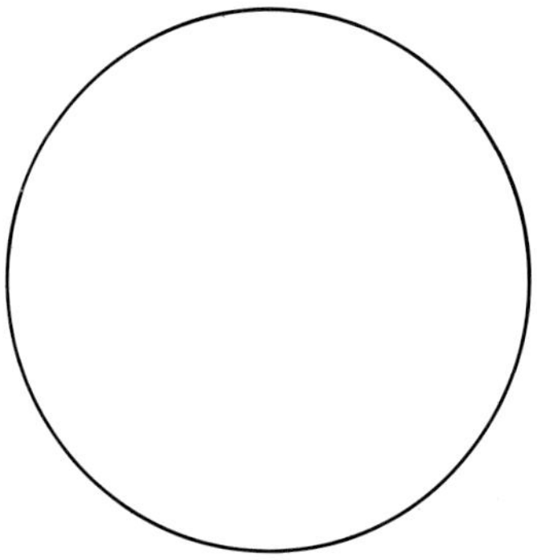

Acute lymphoblastic leukemia

Chronic lymphocytic leukemia

Questions for Review

EXERCISE **44**

BLOOD SMEAR PREPARATION AND DIFFERENTIAL STAINING

Name ____

Laboratory Section ______________________________ Date ____________

1. Identify the blood cells shown in the blood smear in Figure Q44-1.

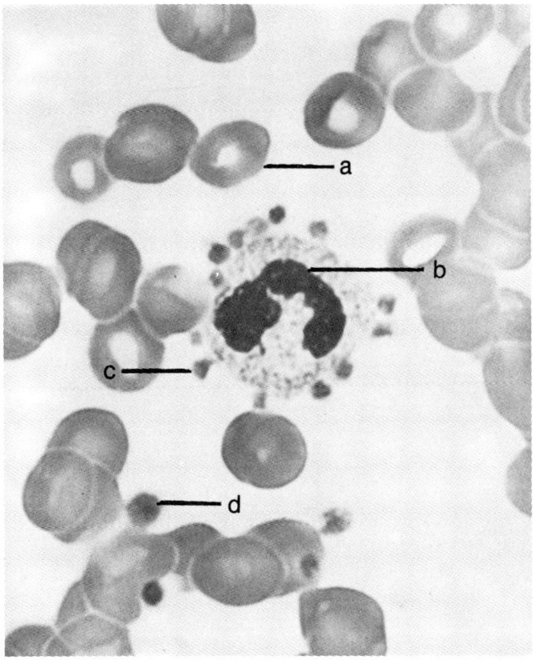

Figure Q44-1

a. ______________________ d. ______________________

b. ______________________ e. ______________________

c. ______________________ f. ______________________

2. Give the function of or explain the activity of the following.

a. Differential stain ____

____

b. Lymphocyte ____

c. Monocyte ____

d. Eosinophil ______

e. Neutrophil ______

f. Basophil ______

g. Platelet ______

3. Would you consider the micrograph in Figure Q44-2 typical of a normal blood smear? If not, explain your answer. ______

______

______

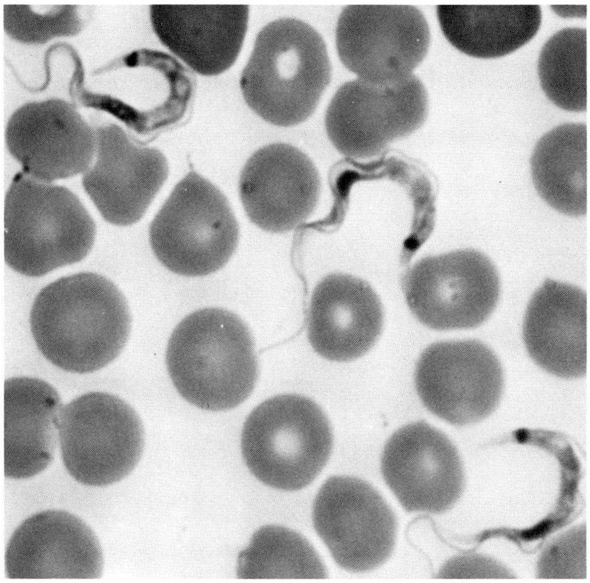

Figure Q44-2

4. List three infectious diseases of the circulatory system.

a. ______

b. ______

c. ______

5. Do hereditary diseases of the blood exist? If so, name two.

a. ______

b. ______

6. Which of the following micrographs in Figure Q44-3 contain macrocytic, normocytic, or microcytic erythrocytes? Use a micrograph of a normal smear as a guide.

   a. ______________________________

   b. ______________________________

   c. ______________________________

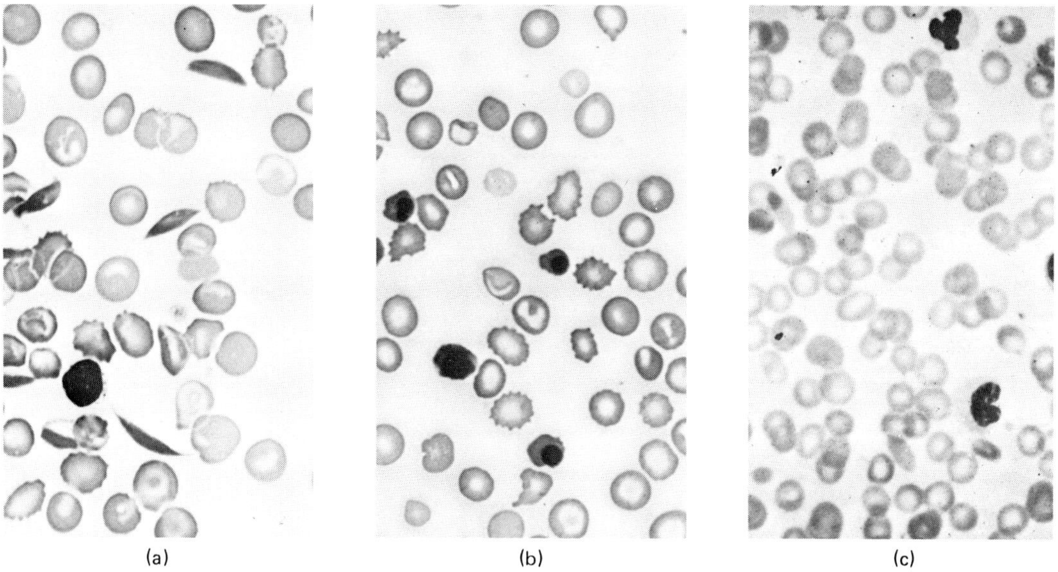

Figure Q44-3

EXERCISE 45

BLOOD CELL COUNTS AND THE HEMATOCRIT (PACKED CELL VOLUME)

The total blood volume in an adult human is about 6.1 liters or approximately 7.5 percent of body weight. About 45 percent of blood consists of formed elements: erythrocytes, leukocytes, and platelets. The remaining 55 percent, the fluid portion, is called plasma.

The number of the formed elements bears a direct relationship to an individual's state of health. This exercise emphasizes red and white blood cells. The total number of red blood cells in an adult human is approximately 25 trillion. These cells survive for 100–120 days within the circulatory system. In laboratory tests the normal average red blood cell count is usually in the range of 4.5–5.5 million cells per cubic millimeter (mm^3) of blood. Males tend to have a red blood cell count about 20 percent higher than females at the same age and level of physical activity. The red cell count increases proportionately as the level of physical activity increases over a period of time. Thus, a well-trained athlete can be expected to have a higher count than a bed-ridden hospital patient.

In the healthy adult the normal number of circulating white blood cells is about $7000/mm^3$ of blood. This value is somewhat higher in children. Newborns have counts of about $16,000/mm^3$. If the white blood cell count is quite high, over 10,000, the condition of *leukocytosis* exists. If, on the other hand, the final count is 5000 or below, a condition of *leukopenia* exists. Leukocytosis occurs in situations of acute infections. It is also evident in leukemia, but in this case the white blood cells are abnormal. Leukopenia generally is associated with viral infections such as chickenpox, measles, mumps, and poliomyelitis. The survival or life span of white blood cells in the blood is difficult to determine exactly.

One laboratory method used to determine the number of blood cells incorporates the use of an instrument called a *hemocytometer* counting chamber and diluting pipettes (Figure 45-1). The counting chamber is designed to hold exactly 1 mm^3 of a diluted blood sample. On the bottom of the chamber is a glass surface etched in a precise grid. The blood cell determination involves counting cells in several squares of the grid and obtaining an average number. That number is multiplied by a factor that takes the amount of dilution into account.

More advanced approaches to making blood cell counts incorporate an electronic counting device. As a suspension containing either red or white blood cells passes through a narrow opening at high speed, each cell is counted automatically. Excellent reproducibility and accuracy generally are obtained with electronic counting devices. The laboratory offers a number of procedures that serve as valuable aids in the diagnosis and management of various disease states. The *hematocrit,* or *packed cell volume,* is one such procedure. It is an important measurement of red blood cell concentration and is used as a test for the detection of an anemic condition.

The hematocrit is the ratio of red blood cells to whole blood by volume. When subjected

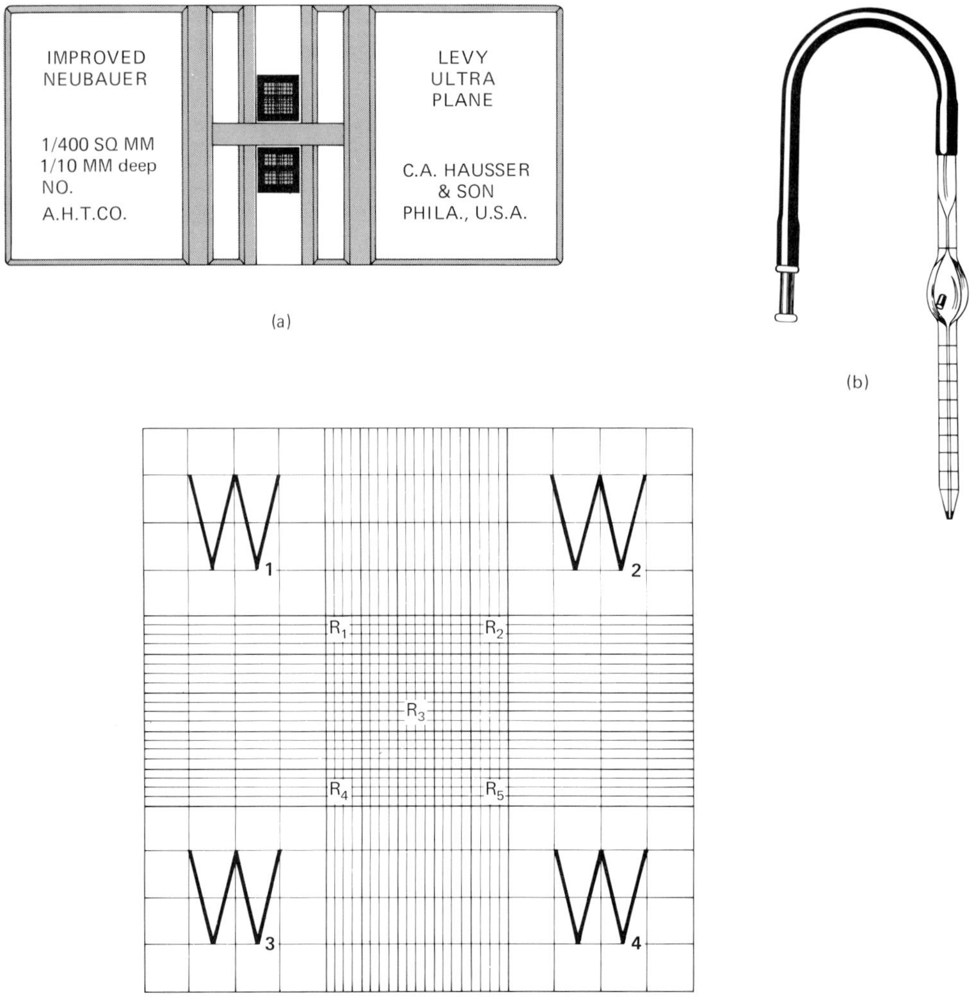

(c)

Figure 45-1 Equipment for blood cell counting. (a) Hemocytometer counting chamber. (b) Red blood cell pipette and tubing. Note the presence of a mixing bead in the bulb portion. (c) Enlargement of grid on the glass surface of the counting chamber. The areas used for red cell counts are labeled R_1 through R_5 and those for white cell counts are labeled W_1 through W_4.

to centrifugal force, red blood cells separate from plasma and the other constituents in blood; they become "packed" together. Once separated the volumes can be measured to obtain the ratio (Figure 45-2). In a normal condition such a preparation consists of approximately 45 percent packed red blood cells, 1 percent white blood cells and platelets, and 54 percent plasma. The normal average hematocrit values for adult males is 47 percent (range 40–54) and for adult females 42 percent (range 37–47). Hematocrits of 15 percent or lower are characteristic of anemia and significant red blood cell loss. Conditions such as polycythemia and malignancies involving the red bone marrow in which there is an overproduction of red blood cells produce hematocrits of 65 percent and higher. The hematocrit also will be performed in this exercise.

A. BLOOD CELL COUNTS

Materials

1. The following materials should be provided per two students.
   a. One *hemocytometer* and cover glass (a Neubauer hemocytometer is recommended)

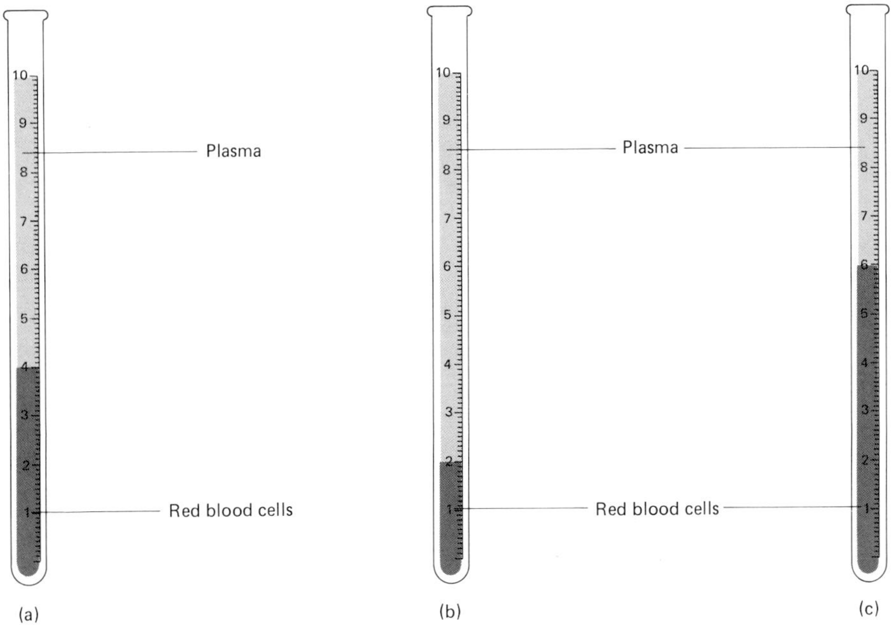

Figure 45-2 Hematocrit determination. A hematocrit is the ratio of red blood cells to whole blood. The appearance of three specimens after centrifuging in Wintrobe tubes is shown. (a) Normal blood. (b) Low hematocrit or serious anemia. (c) elevated hematocrit as in a condition of dehydration resulting from a severe burn.

   b. Two Unopette Reservoirs (Becton-Dickinson) for red blood cell determinations
   c. Two Unopette Reservoirs (Becton-Dickinson) for white blood cell determinations
   d. Four capillary pipette assembly systems (Becton-Dickinson)
   e. One miscroscope
   f. One hand counter
2. The following materials should be provided for class use.
   a. Individual, sterile alcohol prep-pads
   b. Sterile blood lancets
   c. Kim wipes or other tissue material
   d. Disposal container

Procedure 1: Red Blood Cell Count

This procedure is to be performed by students in pairs as in other procedures where partners change places so that each does the experiment (still work in pairs and materials are distributed to pairs). Blood samples for both red blood cell and white blood cell counts can be obtained at the same time. (*Note:* Different colored reservoirs are needed.)

1. Examine the Unopette Reservoir for red blood cell determinations and the capillary pipette assembly (Figure 45-3). Identify the reservoir chamber, diluent fluid, protective shield on the capillary pipette, and the pipette itself. Note the color of the reservoir's bottom surface. It should be red.
2. Hold the reservoir on a flat surface in one hand, and hold the pipette assembly in the other hand. Push the tip of the pipette shield firmly through the diaphragm in the reservoir's neck (Figure 45-4a). Pull out the assembly unit and remove the protective shield from the pipette assembly with a twist.
3. Wipe the finger of your partner from which the blood sample is to be taken with an alcohol-moistened gauze square.

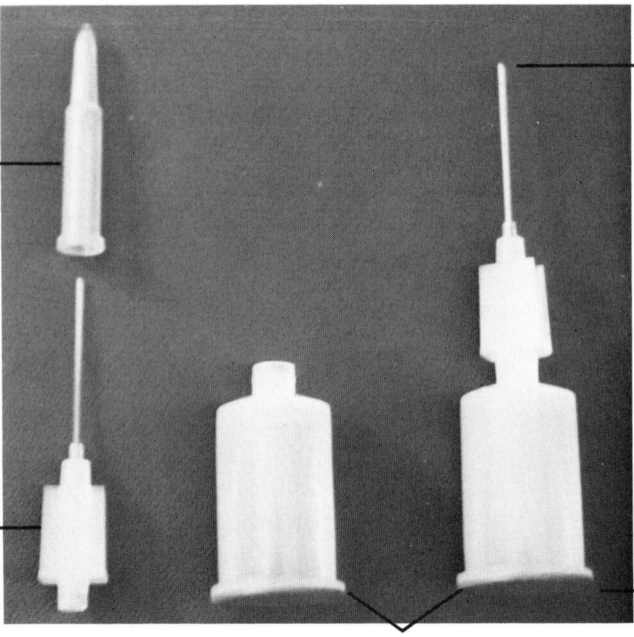

Figure 45-3 Components of a Unopette Reservoir System. The reservoirs used for red blood cell counts have a red undersurface; those used for white blood cell counts have a white undersurface.

4. Hold the finger in a downward position and prick it with a sterile lancet. Wipe away the first drop and gently squeeze the finger to provide a free flow of blood.
5. Holding the pipette almost horizontally, touch the tip of it to the forming blood drop (Figure 45-4b). The pipette will fill by capillary action. When the blood reaches the end of the capillary bore in the neck of the pipette the filling action will stop. Carefully wipe any excess blood from the pipette's surface with a Kim wipe or similar wiping tissue.
6. Squeeze the reservoir slightly to force out a small amount of air. While still maintaining pressure on the reservoir, cover the opening of the overflow chamber of the pipette with your index finger and push the pipette securely into the reservoir neck (Figure 45-4c).
7. Release the pressure on the reservoir and remove your finger from the pipette opening. Negative pressure will draw the blood into diluent fluids.
8. Mix the contents of the reservoir chamber by squeezing the reservoir gently two or three times. (Squeeze gently so the diluent is not forced out of the chamber.) Also invert the reservoir gently a few times (Figure 45-4d).
9. Remove the pipette from the chamber, reverse its position, and replace it on the reservoir (Figure 45-4e). This converts the system into a dropper assembly (Figure 45-4f).
10. Squeeze a few drops out of the system into a container or wipe with tissue to clean the capillary bore. Now you are ready to fill the blood counting chamber.
11. Wipe the glass cover of the hemocytometer, and position it as shown in Figure 45-5a.
12. Hold the end of the pipette and squeeze the sides of the reservoir so that you can deposit a small drop of the diluted specimen onto the polished surface of the counting chamber next to the edge of the cover glass. *Be careful not to allow the chamber to overfill.* If the chamber overfills, charge (fill) the other side of the chamber.
13. Carefully place the charged hemocytometer on the microscope stage, and focus with the low power objective to bring the small (R) grid areas into clear view (Figure 45-5b).
14. Then move the high power objective into place, focus, and, with a hand counter or by

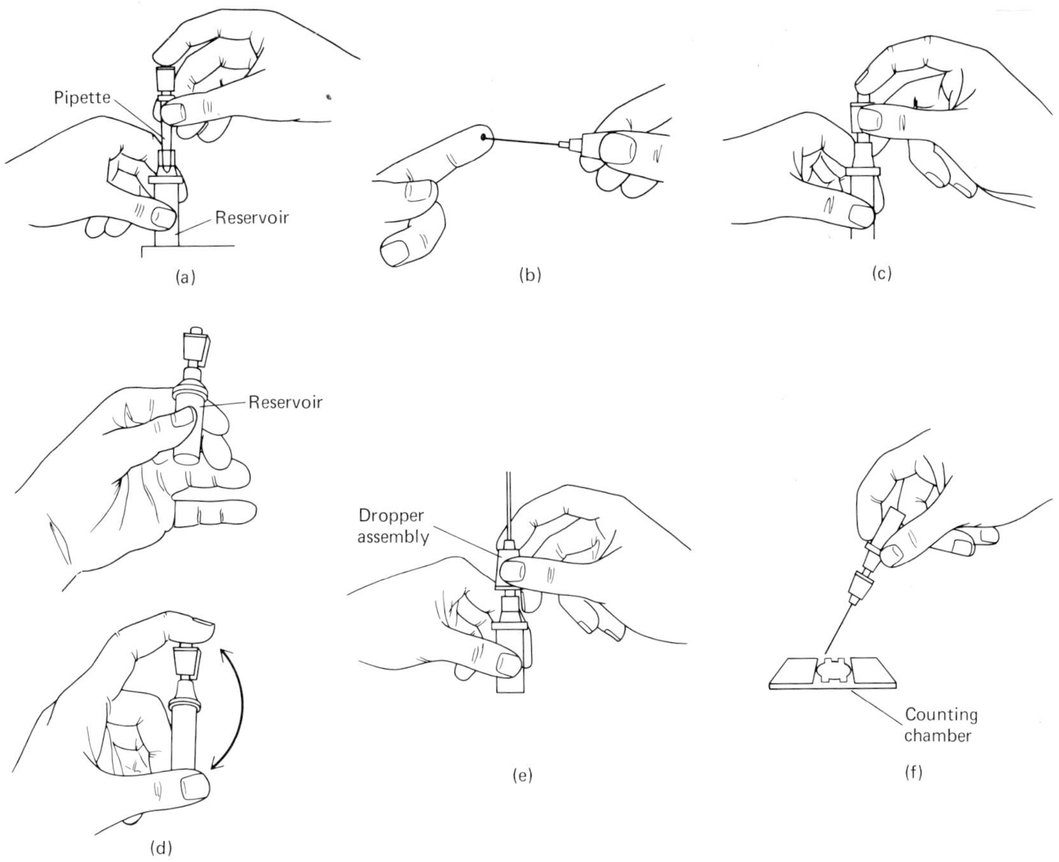

Figure 45-4 Preparation of the Unopette Reservoir System. (a) Puncturing the diaphragm. (b) Obtaining a blood sample. (c) Introducing the sample. (d) Mixing the contents of the system by squeezing and inverting. (e) Converting to a dropper assembly. (f) Filling (charging) the blood counting chamber. (Modified from *RBC Determination for Manual Methods* and *WBC Determination for Manual Methods,* Becton-Dickinson, Division of Becton, Dickinson and Company.)

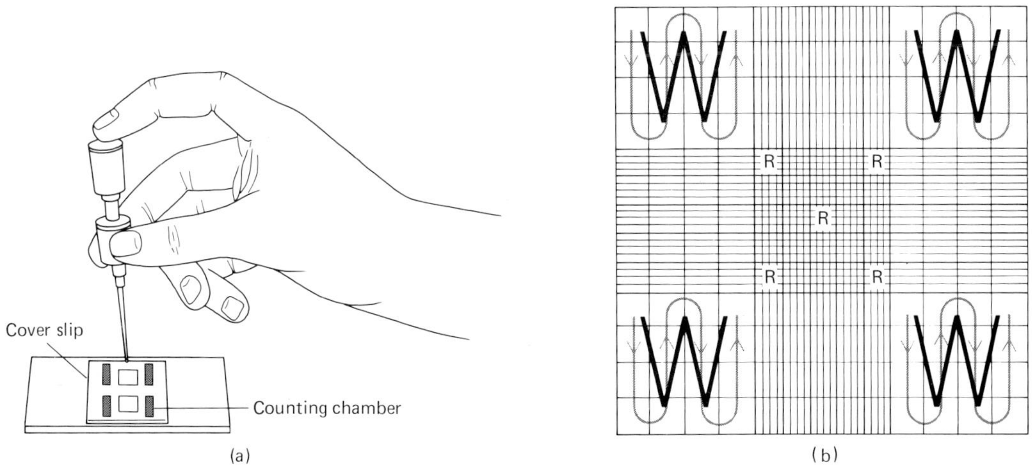

Figure 45-5 The use of the hemocytometer. (a) Positioning the Unopette and filling the chamber. It is important not to overfill the chamber. A properly filled chamber has a blood specimen only within the space between the cover glass and the counting area. (b) Direction to follow for white cell counts.

other means of counting, determine the number of cells in each of the five specified areas. On the edges of the squares count only the cells that touch the lines on the *left* and *top* sides.

15. Multiply the total number of cells counted in the five squares by 10,000 to obtain the red blood cell count for 1 mm^3. Record your findings in the results and observations section.
16. Rinse the hemocytometer, wipe it dry, and repeat steps 11–15. Was your second count similar to the first?
17. Repeat step 16.
18. Have your partner repeat the entire procedure.
19. Dispose of all materials as directed by your instructor.
20. Answer the questions in the results and observations section.

Procedure 2: White Blood Cell Count

This procedure is to be performed by students in pairs.

1. Follow steps 1–12 of procedure 1, but use the Unopette chamber for white blood cell determinations. It has a white bottom surface.
2. Place the hemocytometer on the microscope stage. With the low power objective focus on the chamber area to bring the four large (W) corner regions into view (see Figure 45-5b).
3. With a hand counter or by other means, determine the number of cells in each of the four specified areas. On the edges of the squares, count only the cells that touch the lines on the *left* and *top* sides.
4. Multiply the total number of cells by 50 to obtain the white blood cell count for 1 mm^3. Record your findings in the results and observations section.
5. Rinse the counting chamber, wipe it dry, and repeat steps 1–4. Was your second count similar to the first?
6. Repeat step 5.
7. Have your partner repeat the entire procedure.
8. Dispose of all materials as directed by your instructor.
9. Answer the questions in the results and observations section.

B. THE HEMATOCRIT (PACKED CELL VOLUME)

Materials

1. The following materials should be provided per four students.
   a. Commercially prepared individual sterile alcohol prep-pads
   b. Sterile, disposable blood lancets
   c. Sterile gauze pads (2 × 2 inch)
   d. Heparinized capillary tubes
   e. Clay-Adams Seal-Ease
2. The following materials should be provided for class use.
   a. Microhematocrit centrifuge
   b. Microhematocrit tube reader (Clay-Adams) or a substitute

Procedure

This procedure is to be performed by students in groups of four.

1. Laboratory partners should obtain blood specimens from one another. Wipe the finger

from which the sample is to be taken with an alcohol-moistened sterile pad. Hold the finger in a downward position and gently squeeze it. Puncture the ball of the finger with the lancet. Discard the first two drops of blood. (Refer to Exercise 6 for a description of the procedure.)

2. Place the marked (red) end of a heparinized capillary tube in the forming blood drop and allow the blood to flow downward to fill about two-thirds of the tube (Figure 45-6a).
3. Put a finger over the unmarked end of the tube and gently stick the red end into the Seal-Ease (Figure 45-6b). This seals the blood-containing end of the tube.
4. Place the prepared blood specimen into a slot of the microhematocrit centrifuge with the sealed end of the capillary tube next to the rubber ring guard at the outer rim of the instrument (Figure 45-6c). Do not centrifuge the specimen until there are additional specimen tubes. There must be an even number of tubes for proper balance. Use the specimens of your laboratory partners.
5. Check to be sure the centrifuge lid is properly fastened. A wrench is usually provided for this purpose.
6. Set the centrifuge time for 4 minutes and turn on the instrument.
7. After centrifugation place your capillary tube on or next to the tube reader and determine the hematocrit reading. There are usually directions on the instrument to help you make your determination.
8. Record your findings and answer the questions in the results and observations section.

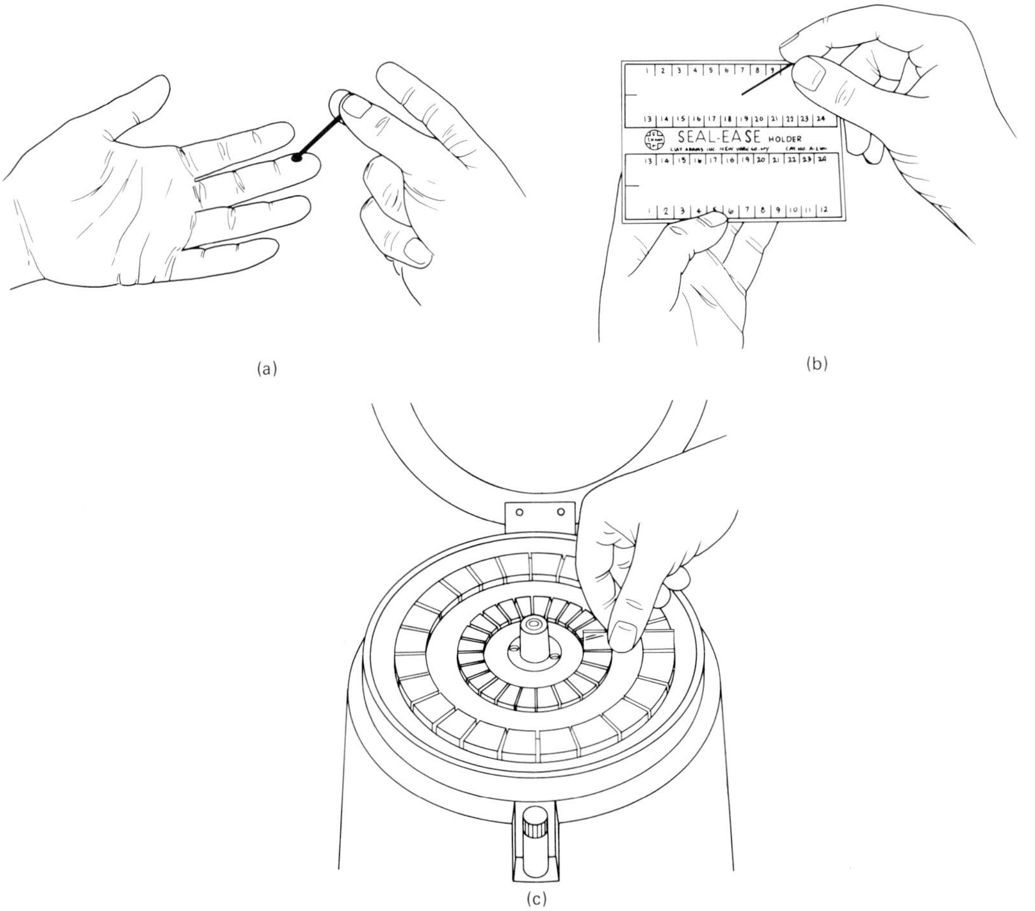

Figure 45-6 General steps of the hematocrit procedure. (a) Specimen collection. (b) Sealing capillary tube. (c) Inserting specimen in hematocrit for centrifugation.

RESULTS AND OBSERVATIONS

1. Enter the results of both red blood cell (RBC) and white blood cell (WBC) counts in Table 45-1.

Table 45-1

Student	Sex	Number of RBCs Counted	RB Count per mm^3	Number of WBCs Counted	WBC Count per mm^3

2. Indicate the calculation formula for the following.
   a. Red blood cell determinations

   ____

   b. White blood cell determinations

   ____

3. Enter the hematocrit determinations for members of your laboratory group in Table 45-2.

Table 45-2

Student	Sex	Hematocrit	Normal or Abnormal

Questions for Review

EXERCISE **45**

BLOOD CELL COUNTS AND THE HEMATOCRIT (PACKED CELL VOLUME)

Name ______________________________

Laboratory Section ______________________ Date ____________

1. a. Are human red blood cells nucleated? ____________

   b. Are all mammalian red blood cells nucleated? ____________

2. Describe human erythrocytes by filling in the following information.

   a. Function ______________________________

   b. Shape ______________________________

   c. Diameter ______________________________

   d. Average life span ______________________________

   e. Normal number per mm^3 in the female ______________________________

   f. Normal number per mm^3 in the male ______________________________

3. a. What is hemoglobin? ______________________________

   ____

   b. What heavy metal has as its chief function the synthesis of hemoglobin?

   ____

4. What is anemia? ______________________________

   ____

5. List three causes of anemias.

   a. ____________________ c. ____________________

   b. ____________________

6. Where are red blood cells formed in the human adult? ____________________

   ____

7. What is the study of blood called? ______________________________

8. What is a hemocytometer? ______________________________

____

9. In Table Q45-1 determine the final red blood cell counts and indicate whether the results are normal or abnormal.

Table Q45-1

Microscopic Count	Sex	Red Blood Cell Count (per mm^3)	Normal or Abnormal
1000	Male	a.	b.
300	Female	c.	d.
150	Female	e.	f.
195	Male	g.	h.

10. In Table Q45-2 determine the final white blood cell counts and indicate whether the results are normal or abnormal.

Table Q45-2

Microscopic Count	Sex	White Blood Cell Count	Normal or Abnormal
900	Male	a.	b.
100	Female	c.	d.
500	Female	e.	f.
62	Male	g.	h.

11. Are the dilution pipettes for red blood cell and white blood cell counts different? Explain. ____

____

12. How many and which squares are counted in a red blood cell count? ______________

____

13. How many and which squares are counted in a white blood cell count? ____________

____

14. What is a hematocrit? ____

____

15. What are the percentages of the components of a blood sample in a normal hematocrit?

____

16. The results of 12 hematocrit determinations are listed in Table Q45-3. Indicate whether they are normal or abnormal.

Table Q45-3

Adult Male Patient	Hematocrit Value	Normal or Abnormal	Adult Female Patient	Hematocrit Value	Normal or Abnormal
1	36	a.	1	32	g.
2	45	b.	2	40	h.
3	54	c.	3	61	i.
4	63	d.	4	47	j.
5	42	e.	5	22	k.
6	40	f.	6	37	l.

17. Distinguish between plasma and serum. ______________________________

____

____

18. What is the normal hematocrit value for the following?

a. Adult female ____________ b. Adult male ____________

EXERCISE 46

COAGULATION TIME AND HEMOGLOBIN DETERMINATIONS

Blood normally is maintained in liquid form as long as it remains in blood vessels. However, when blood comes out of a vessel, it is changed within a short time to a soft, jellylike mass, a *blood clot.* The liquid released by the clot is called *serum.* The physiological process whereby blood is so changed is called *coagulation.* Clot formation is a vital mechanism that prevents excessive blood loss from the body.

Examination of a magnified blood clot shows that it is composed of a network of delicate fibers in which both red and white blood cells are enmeshed. The fibers consist of *fibrin,* a substance derived from one of the plasma proteins, *fibrinogen.* The conversion of fibrinogen to the solid fibrin is a complicated chemical process involving enzyme systems.

In recent years the enzymatic reactions associated with clotting have been formalized into a "cascade" or "waterfall" scheme involving at least 13 factors (Figure 46-1). It is widely held that the different coagulation factors exist in plasma as inactive proenzymes that are converted to active enzymes during the clotting process. The reactants appear in sequence with the product of the first reaction operating as enzyme to catalyze the second reaction. The second reaction yields a product that is an enzyme to catalyze the third reaction, and so on. Once activated, the sequence must continue to completion.

The complex process of coagulation is often greatly simplified into five stages, with the first three related to the formation of the visible blood clot, the fourth to clot retraction, and the fifth to dissolution of the clot.

Stage 1. Formation or release of thromboplastin (the initiator of clotting)
Stage 2. Prothrombin converted to thrombin (by thromboplastin and other coagulation factors)
Stage 3. Fibrinogen converted to fibrin (by thrombin and other coagulation factors)
Stage 4. Clot retraction or syneresis (consolidation and contraction of fibrin threads)
Stage 5. Fibrinolysis (clot dissolution)

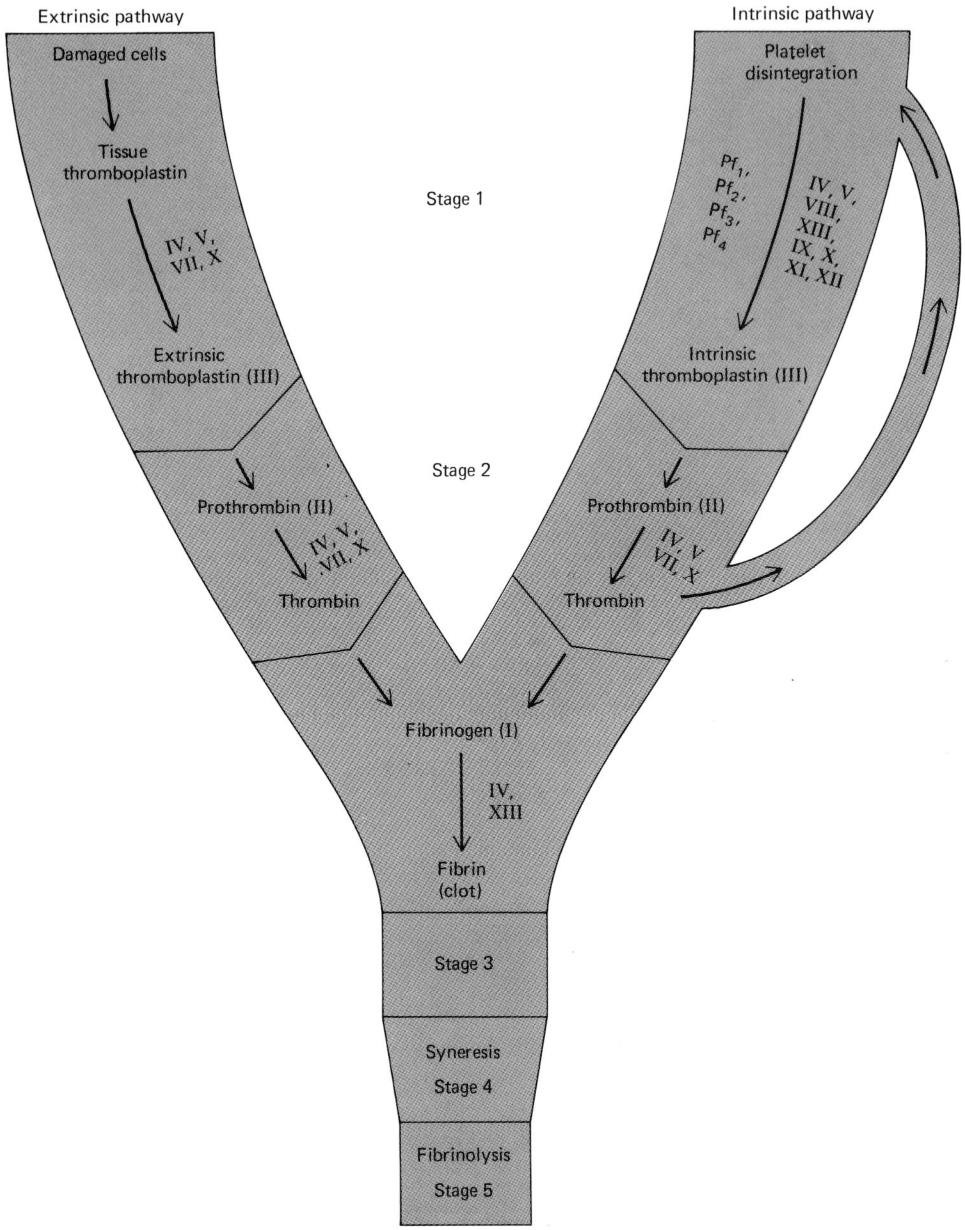

Figure 46-1 Blood clotting mechanism. Both intrinsic and extrinsic pathways are shown. Roman numerals are plasma coagulation factors (the substance which had been assigned number VI is no longer used in coagulation theory). Pf_1 to Pf_4 are platelet coagulation factors.

All five stages may be occurring continuously in minute amounts in the normal coagulation process. Table 46-1 lists the plasma clotting factors.

Table 46-1

Number	Name
Plasma clotting factors	
I	Fibrinogen
II	Prothrombin
III	Tissue thromboplastin (extrinsic or intrinsic)
IV	Calcium ions
V	Proaccelerin, or labile factor
VII	Proconvertin, serum prothombin conversion accelerator (SPCA), or stable factor
VIII	Antihemophilic factor (AHF)
IX	Plasma thromboplastin component (PTC), or Christmas factor
X	Stuart–Prower factor, or Stuart factor
XI	Plasma thromboplastin antecedent (PTA)
XII	Hageman factor
XIII	Fibrin stabilizing factor (FSF)
Platelet factors (Pf)	
Pf_1	Platelet accelerator
Pf_2	Thrombin accelerator
Pf_3	Platelet thromboplastin factor
Pf_4	None assigned formally

Depending on the manner in which thromboplastin is introduced, the clotting process may proceed along one of two routes, the *extrinsic* or the *intrinsic* pathway. The extrinsic pathway begins when a blood vessel is ruptured, and damaged tissues surrounding it or the ruptured area of the blood vessel itself release a lipoprotein called *tissue thromboplastin*. This substance reacts with coagulation factors in plasma to form *extrinsic thromboplastin (extrinsic prothrombin activator)*.

The intrinsic pathway involves the clumping of platelets, their disintegration, and the release of platelet coagulation factors into the plasma. These factors in turn react with certain other coagulation factors to form *intrinsic thromboplastin (intrinsic prothrombin activator)*.

Coagulation or bleeding disorders may be the result of a variety of factors such as platelet abnormalities, blood vessel (vascular) abnormalities, or coagulation factor deficiencies. The laboratory tests available to detect or screen for these disorders are described in Table 46-2.

Table 46-2 Coagulation Disorder Screening Tests

Test	Brief Description
Clotting or bleeding time	A check for platelet aggregation and primary clot formation. This test is a helpful indication of platelet abnormality, whether in number or function. Individuals with vascular problems will have an abnormal bleeding time.
Platelet count	Determination of the quantity of platelets.
Prothrombin time (PT)	A test to detect deficiencies in the extrinsic clotting system (Figure 46-1) and defects with Factors I, II, VII, or X in the presence of tissue thromboplastin. Individuals on anticoagulant therapy will have a prolonged PT.
Activated partial thromboplastin time (APTT)	A test to detect deficiencies in the intrinsic clotting system (Figure 46-1) and defects with Factors I, V, VIII, IX, X, XI, and XII. Individuals on anticoagulant therapy will have a prolonged APTT test result.

Three tests associated with clotting and related disorders, clotting time, prothrombin time (PT), and activated partial thromboplastin time (APTT), are considered in this exercise. The PT and APTT tests complement one another and can be used together with other laboratory tests to determine the basis of bleeding disorders associated with prolonged clotting time. Table 46-3 shows test results that can occur in case of clotting factor defects.

Table 46-3 Clotting Factor Defects: Results with Prothrombin Time (PT) and Activated Partial Thromboplastin Time (APTT)

Clotting Factor Defect	PT Result	APTT Result
V	Abnormal	Abnormal
VII	Abnormal	Normal
VIII	Normal	Abnormal
IX	Normal	Abnormal
X	Abnormal	Abnormal
XI	Normal	Abnormal
XII	Normal	Abnormal

This exercise also considers *fibrinolysis,* or clot dissolution. Fibrinolysis is the process responsible for the enzymatic degradation of fibrin thrombi which results in the formation of *fibrin degradation products* (FDP). The major fibrinolytic system, known as the *plasminogen–plasmin system,* consists of five components—plasminogin, plasmin, plasminogen activators, plasminogen activation inhibitors, and plasmin inhibitors. Normal serum contains less than 10 micrograms per milliliter (μg/ml) of FDP. Tests used to measure FDP levels are of value in cases involving recovery from abdominal operations, kidney transplant, or childbirth. Such tests also serve as an aid in the diagnosis of pulmonary embolism and renal disease. The Thrombo-Wellco test is a rapid and reliable procedure for the semiquantitative determination of FDP (Figure 46-2).

Hemoglobin is responsible for gaseous transport in the body. The hemoglobin content of the blood usually is expressed in grams per 100 milliliters (g/100 ml) or sometimes as a percentage of normal. The normal hemoglobin content of the blood is about 14.5–15 g/100 ml. Tests for the determination of hemoglobin content are included in this exercise.

Normal hemoglobin (HbA) is a conjugated protein consisting of heme, an iron-containing nonprotein, and a globin (protein) portion composed of two pairs of polypeptide

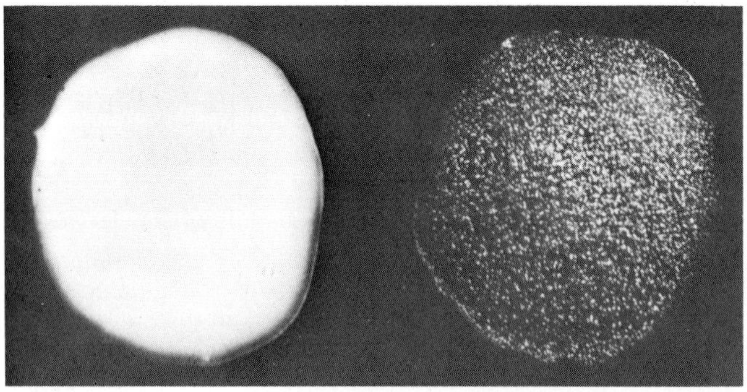

Figure 46-2 Result patterns with the Thrombo-Wellco test/Latex test. A negative result appears on the left. On the right a positive clumping or agglutination reaction can be seen. In a positive test latex particles coated with antibodies against fibrin degradation products (FDP) react with fibrin degradation products in serum from a blood sample. (Courtesy of Wellcome Reagents Division, Burroughs Wellcome Co., Research Triangle Park, NC.)

chains. The amino acids and sequence of their arrangement in each polypeptide are under genetic control. If a gene causes a substitution for an amino acid or a change in the amino acid sequence in one of the polypeptides, an abnormal molecule of hemoglobin forms. Such genetic changes (mutations) may result in an anemia.

One of the better known gene-controlled defects is sickle cell anemia. In sickle cell anemia hemoglobin S (HbS) is produced; under conditions of low oxygen tension, HbS tends to form crystals in red blood cells that result in the characteristic sickle shape distortion. In homozygous individuals (individuals in whom both genes for hemoglobin code for HbS), severe hemolytic sickle anemia occurs. In heterozygous individuals (individuals who have one gene for HbS and one for HbA), the hemolytic anemia is comparatively mild. Screening tests for HbS are available. This exercise demonstrates this type of procedure.

The presence of hemoglobin in tissues can be an important sign of various abnormal body conditions, such as gastrointestinal ulcers, hiatal hernia, postoperative problems, and colorectal cancer. This exercise includes a procedure to test for the presence of hemoglobin released from lysed cells (occult blood) in fecal specimens. This procedure, the Hemoccult test, is rapid, quite simple to perform, and relatively easy to interpret. It should be noted that the ingestion of certain oral medications or foods can cause false reactions. For example, aspirin, some hormones, and reserpine can cause occult bleeding. Large amounts of ascorbic acid (vitamin C) produce false negative results.

A. COAGULATION (CLOTTING) TIME

Materials

The following materials should be provided for class use.

1. Sterile alcohol prep-pads (Med Pak Corp.)
2. Sterile, disposable blood lancets
3. Sterile gauze pads
4. Nonheparinized capillary tubes (0.5-mm diameter)
5. Stopwatch or other timing device
6. Glass files
7. Container with disinfectant for disposal of glass

Procedure

This procedure is to be performed by students in pairs.

1. Disinfect the surface of a finger with an alcohol prep-pad and puncture it with the lancet to produce a free flow of blood (as described in Exercise 6). Wipe away the first blood drop.
2. Hold the finger in a downward position and place a nonheparinized capillary tube into and below the newly forming blood drop (Figure 46-3a). Allow the tube to fill as completely as possible by capillary action. (Prevent air bubbles from forming by keeping the open end of the tube totally in the blood drop.)
3. Have your partner start timing as soon as the tube is filled to capacity.
4. After 1 minute use the file to make a small surface scratch near one end of the tube. Break the tube on the score line and separate the small piece from the rest of the tube. Look for the presence of coagulation (Figure 46-3b). Continue to break off pieces of the tube at 1 minute intervals until coagulation is observed. Record your findings in the results and observations section.

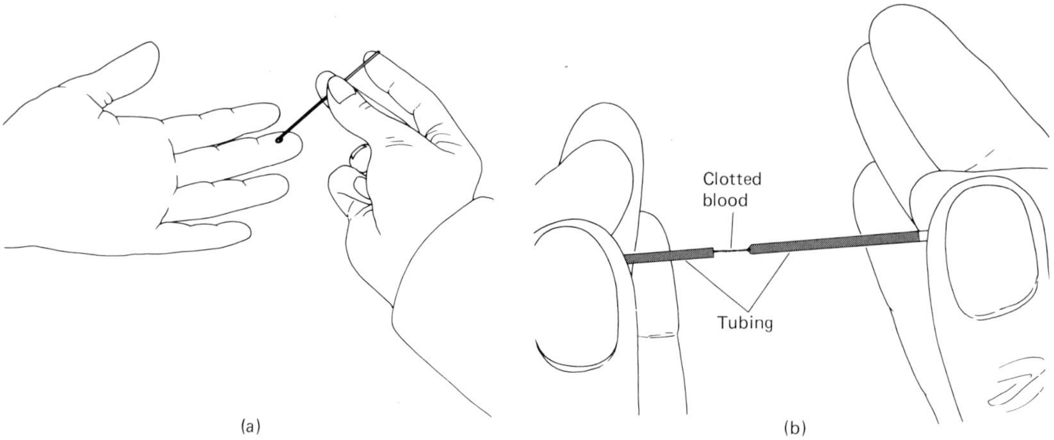

Figure 46-3 Steps in determining coagulation time. (a) Obtaining a specimen. (b) Breaking portions of capillary tube to find coagulation time.

5. Change places with your laboratory partner and repeat steps 1–4. Answer the questions in the results and observations section.
6. Dispose of the capillary tube as directed by your instructor.

B. PROTHROMBIN TIME (PT)

Materials

The following materials should be provided for class use.

1. Simplastin or other commercial source of tissue thromboplastin
2. Plasma samples kept at 37°C
3. Plastic test tubes
4. 1-ml pipettes, graduated in 0.1 ml, with rubber bulbs
5. 37°C water bath with a rack for tubes
6. Stopwatch or other timing device
7. Container with disinfectant for disposal of pipettes

Procedure

This procedure is to be performed by students in pairs or in groups of four.

1. Your instructor will review the use of a pipette.
2. Pipette 0.2 ml of the thromboplastin into a small test tube. Place the tube in the 37°C water bath.
3. Remove the tube containing the thromboplastin from the water bath and pipette 0.1 ml of warmed plasma sample into it. Shake gently to mix.
4. Have your partner start timing as soon as the two materials are mixed.
5. Place the tube in the water bath again.
6. At least every 4 seconds remove the tube, tilt it, and look for the formation of a fibrin clot.
7. Stop the procedure as soon as a clot appears. Record the time in the results and observations section.
8. Repeat steps 2–7 and try to obtain a more accurate time for formation of the fibrin clot.
9. Answer the questions in the results and observations section.

C. ACTIVATED PARTIAL THROMBOPLASTIN TIME (APTT)

Materials

The following materials should be provided for class use.

1. Activated cephaloplastin
2. Calcium chloride ($CaCl_2$) 0.02 molar solution
3. Two plasma samples
4. Plastic test tubes
5. Test tube racks
6. 1-ml pipettes, graduated in 0.1 ml, with rubber bulbs
7. 37°C water bath with a rack for tubes
8. Stopwatch or other timing device
9. Marking pens
10. Container for disposal of pipettes

Procedure

This procedure is to be performed by students in pairs or in groups of four.

1. Your instructor will review the use of pipettes.
2. Obtain three test tubes. Label two of the tubes 1 and 2 to correspond to the labels on the plasma samples; label the third tube control.
3. With a 1-ml pipette introduce 0.1 ml of activated cephaloplastin into each tube. Dispose of the pipette as indicated by the instructor.
4. With separate pipettes, introduce 0.1 ml of the plasma samples into the correspondingly labeled tubes. Dispose of the pipettes as indicated.
5. Incubate all tubes at 37°C for 3 minutes.
6. With another 1-ml pipette introduce 0.1 ml of warmed $CaCl_2$ into each tube and immediately start the timer.
7. Observe the tubes for the appearance of fibrin thread formation.
8. Place the tubes in the 37°C water bath and gently shake them continuously. Examine the tubes at 20 second intervals. Holding and tilting the tubes against a bright light helps to detect the fibrin threads.
9. Note the time required for the reaction to occur. Normal reaction time is approximately 40 seconds or less.
10. Record your findings in the results and observations section and answer the questions pertaining to this exercise in the questions for review section.

D. DETECTION OF FIBRIN DEGRADATION PRODUCTS (FDP)

Materials

The following materials should be provided for the instructor's use.

1. Latex test suspension (contains polystyrene latex particles coated with sheep antibodies against FDP)
2. Positive control serum
3. Negative control serum
4. Two unknown serum preparations, diluted 1:5 in buffer
5. Glycine saline buffer
6. Glass slides

7. Wooden applicator sticks
8. Clean Pasteur pipettes with rubber bulbs
9. Wax pencil or other marking device

Procedure

This procedure is to be demonstrated by the instructor.

1. Use the marking device to make and number four separate circles on a glass slide.
2. With separate Pasteur pipettes, place 1 drop of each specimen in the circles as follows:
   a. Circle 1, unknown specimen 1
   b. Circle 2, unknown specimen 2
   c. Circle 3, known positive serum
   d. Circle 4, known negative serum
3. Gently shake the container of latex test suspension. Add 1 drop of the suspension to each circle.
4. Stir each combination with separate applicator sticks.
5. Rock the slide back and forth for 2 minutes and observe for clumping (agglutination). Agglutination indicates the presence of more than 10 μg/ml of FDP.
6. Record your findings and answer the questions in the results and observations section.

E. HEMOGLOBIN DETERMINATIONS

Materials

The following materials should be provided for class use.

1. Commercial sterile alcohol prep-pads
2. Sterile, disposable blood lancets
3. Sterile gauze pads
4. Tallquist test paper
5. Tallquist reference scale
6. Hemo Cue hemoglobinometer
7. Hemo Cue microcuvettes
8. Kim wipes or other tissue material or lens paper
9. Isopropyl alcohol.
10. Rubber gloves
11. Container of disinfectant

Procedure 1: Tallquist Method

This procedure is to be performed by students in pairs.

1. Remove a piece of test paper from a Tallquist booklet and place it on a flat surface before you.
2. Obtain a blood sample (as described in Exercise 6). Wipe away the first drop of blood.
3. Touch the second forming blood drop to the center of the Tallquist paper.
4. Wipe the finger and apply a sterile gauze pad to stop any additional bleeding.
5. After 15 seconds compare your blood sample with the Tallquist reference scale. Record your hemoglobin estimate in grams of hemoglobin per 100 ml of blood in the results and observations section. (Refer to Color Plate 38.)
6. Repeat steps 1–5 with your laboratory partner.
7. Answer the questions in the results and observations section.

Procedure 2: Hemoglobinometer Method

This procedure is to be performed by students in pairs.

1. Examine the hemoglobinometer and its parts (Figure 46–4a). Note the display window for hemoglobin measurements.
2. Have your partner put on a pair of rubber gloves before performing any blood work.
3. Obtain a blood sample (as described in Exercise 6). Wipe away the first two drops.
4. Position the finger so that one large drop of blood forms. It should be big enough to fill the microcuvette completely.
5. Introduce the microcuvette's tip into the center of the blood drop as shown in Figure 46–4b.
6. Fill the cuvette in one continuous process. Wipe off any excess blood on the outside surface of the cuvette. (NOTE: If a second sample is to be taken from the same finger, it should be done immediately after the first has been taken. Wipe away the remains of the first drop of blood and take a second sample from a second drop.
7. Place the microcuvette in the cuvette holder and push it into the Hemo Cue to the stop point.
8. Wait 45 seconds and read the hemoglobin value shown on the display window.
9. Enter your findings in the results and observations section. Note that the value is given in grams/liter.
10. Change places with your partner and repeat steps 2 through 9.

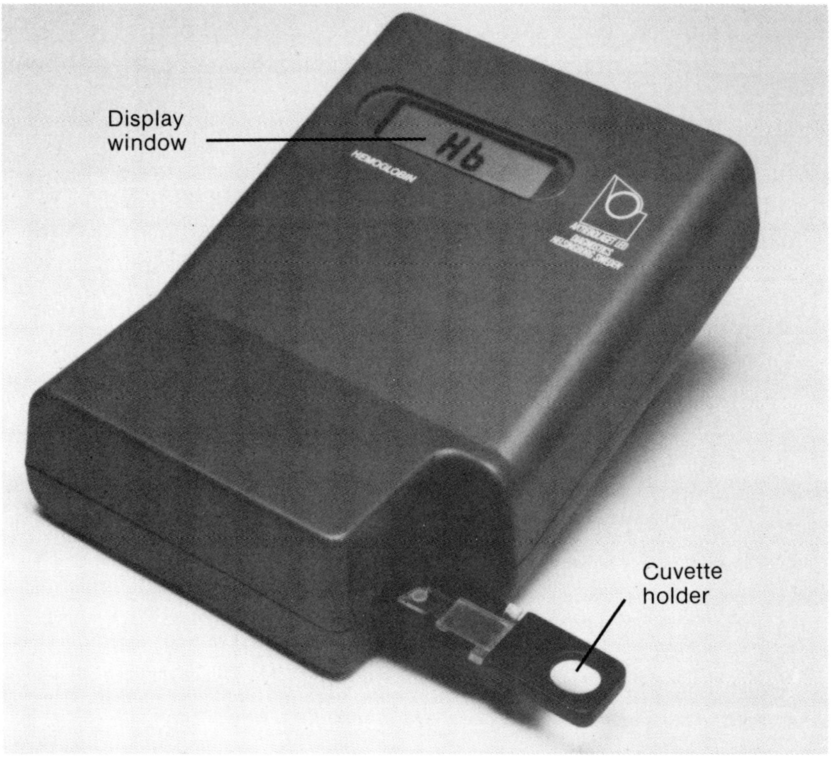

Figure 46-4 The Hemo Cue tm hemoglobinometer and specimen preparation. (a) The Hemo Cue and its parts. (Courtesy of Medical Equipment Designs, Inc. Laguna Hills, California.)

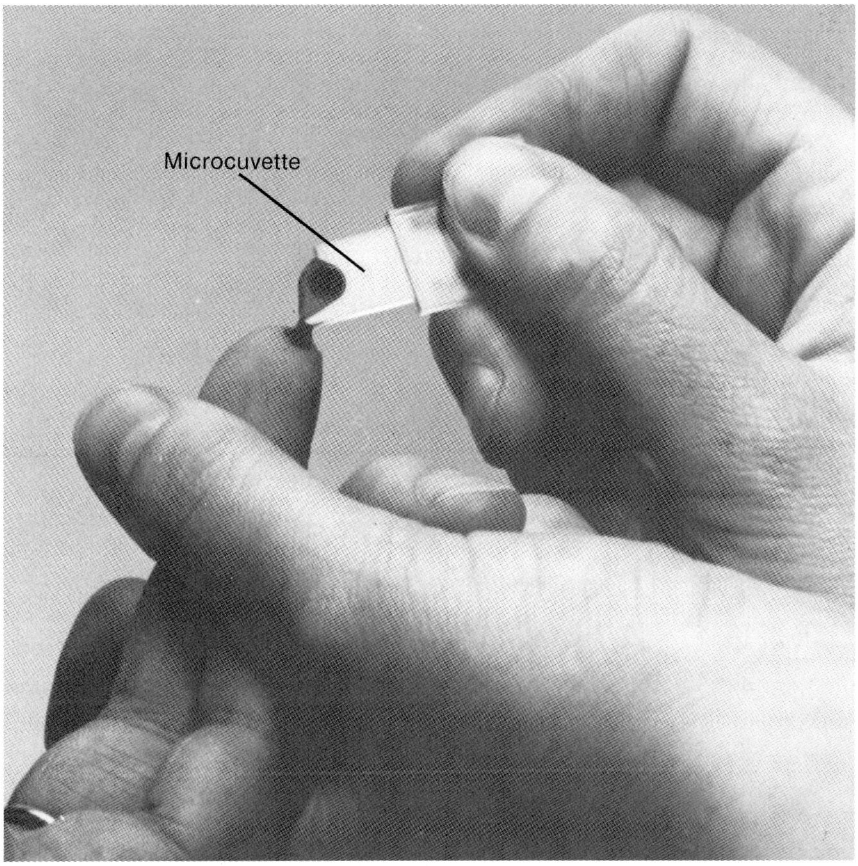

Figure 46-4 *(Continued)* (b) Filling the microcuvette. Note the positions of the fingers. (Courtesy of Medical Equipment Designs, Inc. Laguna Hills; California.)

F. SICKLE HEMOGLOBIN ASSAY DEMONSTRATION

Materials

The following materials should be provided for the instructor's use.

1. Five Isolab Sickle Hb Assay tubes
2. Five whole blood samples to include
   a. One normal specimen
   b. One containing hemoglobin AS (heterozygous for sickle cell trait)
   c. One containing hemoglobin SS (homozygous for sickle cell trait)
   d. Two unknowns
3. Five 1-ml serological pipettes, graduated in 0.1 ml, with rubber bulbs
4. Container for disposal of pipettes
5. Marking pen
6. Tabletop centrifuge and six centrifuge tubes (optional)
7. Balance (optional)
8. Distilled water bottle for balancing

Procedure

This procedure is to be demonstrated by the instructor.

1. Label the five Isolab Sickle Hb Assay tubes as follows to represent the different blood samples:
   a. N for normal specimen
   b. HbAS for heterozygous sickle cell trait
   c. HbSS for homozygous sickle cell trait
   d. 1 for unknown sample 1
   e. 2 for unknown sample 2
2. With separate 1-ml pipettes, add 0.1 ml or 2 drops of each blood sample to their correspondingly labeled tubes.
3. Tighten the caps of each tube and shake them vigorously for at least 10 seconds.
4. Allow the tubes to stand at room temperature for 5 minutes.
5. Incubate the tubes in an upright position at room temperature for 2 hours. *Alternatively*, if a tabletop centrifuge is available, spin all tubes (after balancing them) at approximately

3400 rpm for 5 minutes. If distinct layers are not formed, continue to centrifuge the tubes until layers appear.

6. Examine the tubes for the presence of 3 layers—upper, interface (middle), and lower—and interpret the results according to those described in Table 46-4.

Table 46-4 Hemoglobin Assay Results and Interpretations

	Hemoglobins		
Layers	Normal	HbAS	HbSS
Upper	Grey white to pale pink	Grey white	Grey white
Interphase	Pink	Dark red	Dark red
Lower	Red	Shades of pink	Straw yellow to orange

7. Enter your findings in the results and observations section and answer any questions pertaining to this procedure.

G. DETECTION OF FECAL OCCULT BLOOD

Materials

The following materials should be provided for the instructor's use.

1. Three Hemoccult slides containing high-quality filter paper impregnated with guaiac resin
2. One container of Hemoccult developing solution (a mixture of hydrogen peroxide and denatured alcohol)
3. One occult blood positive specimen
4. One occult blood negative specimen
5. One or more occult blood unknown specimens
6. Applicator sticks
7. Disposal container with disinfectant

Procedure

This procedure is to be demonstrated by the instructor. It is a rapid, simple test for the detection of hemoglobin released from lysed cells (occult blood) in fecal material.

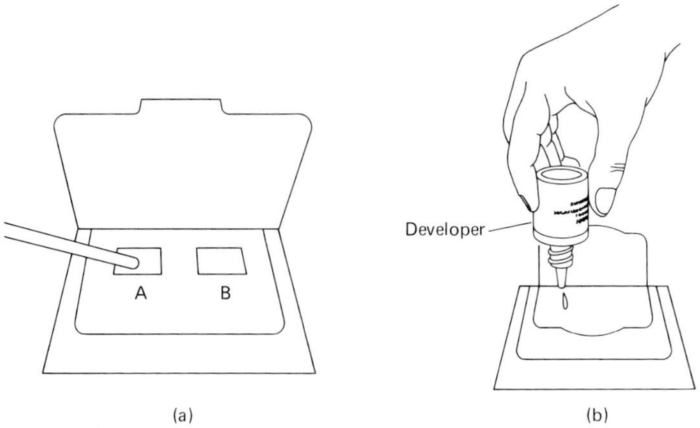

Figure 46-5 Steps for the detection of occult blood in fecal material. (a) Applying specimens. (b) Adding developer.

1. With an applicator smear a small amount of material taken from one portion of a positive specimen on the inside surface of box A of a Hemoccult slide (Figure 46-5a).
2. Using the other end of the applicator, obtain another sample from a different part of the fecal specimen and smear it on the surface of box B.
3. Close the cover of the Hemoccult slide.
4. Turn the Hemoccult slide over and open the perforated window portion.
5. Apply 2 drops of the developer solution to each smear (Figure 46-5b). Wait 30–40 seconds.
6. Determine the test results. The presence of a light blue color around the smear is a positive reaction for occult blood. No detectable blue color is considered a negative reaction. Refer to Color Plate 39. Record the findings in the results and observations section.
7. Repeat steps 1–6 with the negative and unknown specimens.
8. Answer the questions in the results and observations section.

RESULTS AND OBSERVATIONS

COAGULATION (CLOTTING) TIME

1. Enter your findings and those of your laboratory partner in Table 46-5. Indicate the presence of coagulation with a "+" and its absence by a "−".

Table 46-5

	Coagulation Time (in minutes)					
Student	1	2	3	4	5	6

2. What is a normal clotting time for blood? ____________

3. Were the coagulation times of most students within normal limits? (Check the clotting times of at least 12 students before answering this question.) What variations did you observe? ____________

____

4. List three factors that increase clotting time.

   a. ____________ c. ____________

   b. ____________

PROTHROMBIN TIME (PT)

1. How long did it take for the plasma to clot? ____________

2. How would this test be performed with a blood sample? ____________

____

____

3. List two factors that increase prothrombin time.

a. ____________________ b. ____________________

ACTIVATED PARTIAL THROMBOPLASTIN TIME (APTT)

1. Enter your findings in Table 46-6.

Table 46-6

Reaction Time	Sample 1	Sample 2	Control

2. Were the test results easy to read and interpret? __________

3. Were any abnormal reactions observed? If so, which ones? ____________________

____

4. If an individual had a defect in Factor X, would you expect to see an abnormal APT reaction? __________

DETECTION OF FIBRIN DEGRADATION PRODUCTS (FDP)

1. Enter your findings in Table 46-7. Indicate the presence of clumping by a "+" and its absence by a "−."

Table 46-7

Specimen	Presence of Agglutination
Unknown 1	
Unknown 2	
Known positive serum	
Known negative serum	

2. Which of the unknown samples produced a positive result? __________

3. What does FDP represent? ____________________

____

4. List three situations in which FDP determinations are useful.

a. ____________________ c. ____________________

b. ____________________

HEMOGLOBIN DETERMINATIONS

1. Enter your findings and those of your laboratory partner in Table 46-8.

Table 46-8

Student	Sex	Hemoglobin Value, Tallquist (in g/100 ml)	Hemoglobin Value, Hemoglobinometer (grams/liter)

2. Did the hemoglobin values obtained with both methods agree? ____________

3. If not, what factors could account for the differences? ____________

SICKLE HEMOGLOBIN ASSAY DEMONSTRATION

1. Enter your findings in Table 46-9.

Table 46-9

	Assay Tubes				
Layers	Normal	HbAS	HbSS	Unknown 1	Unknown 2
Upper					
Interphase					
Lower					

2. Which, if any, of the unknown samples exhibited positive results for HbAS or HbSS?

____

DETECTION OF FECAL OCCULT BLOOD

1. Enter the results of the Hemoccult test in Table 46-10.

Table 46-10

Specimen	Box A	Box B
Positive		
Negative		
Unknown 1		
Unknown 2		

2. Were the test results easy to read and interpret? ____________

3. What result would you expect in the case of an individual with a slowly bleeding ulcer?

Is this result reliable? ______________________________

Questions for Review

EXERCISE **46**

COAGULATION TIME AND HEMOGLOBIN DETERMINATIONS

Name ______________________________

Laboratory Section ______________________ Date ____________

1. Describe four laboratory tests associated with the clotting activity of blood.

   a. ______________________________

   b. ______________________________

   c. ______________________________

   d. ______________________________

2. Distinguish between the intrinsic and extrinsic pathways of clotting. ______________________________

   ____

   ____

   ____

3. How many factors are known to be involved with the blood coagulation process?

   ____

4. List the stages of blood coagulation and indicate which factors are associated with each.

   a. ______________________________

   b. ______________________________

   c. ______________________________

   d. ______________________________

   e. ______________________________

5. Which of the blood clotting factors require vitamin K for their formation? ____________

   ____

6. What are platelets and where are they formed? ______________________________

   ____

7. Indicate the prothrombin time (PT) and activated partial thromboplastin time (APTT) reactions that can be expected with defects of the clotting factors listed in Table Q46-1.

Table Q46-1

Defective Clotting Factor	PT Reaction	APTT Reaction
VIII		
V		
X		
XI		
XII		
IX		
VII		

8. What is the physiological function of hemoglobin? ______________________

____

9. How is the hemoglobin content of blood usually expressed? ______________________

10. Give the normal hemoglobin content ranges for the following.

a. Adult female ____________ c. 4-year-old child ____________

b. Adult male ____________ d. 1-year-old child ____________

11. What is sickle cell anemia? ______________________

____

12. Of what importance is occult blood detection in fecal specimens? ______________________

____

EXERCISE 47

IMMUNOHEMATOLOGY

In 1900 Karl Landsteiner observed that the mixing of red blood cells and sera from two different human donors, followed by incubation at body temperature, resulted in the agglutination of the red cells. This phenomenon, known as *isohemagglutination,* enabled Landsteiner and others to demonstrate the presence in every individual's erythrocytes of specific, genetically determined antigens, called *agglutinogens.* At the present time more than 50 blood factors are known to exist in the human species. Examples include A, B, O, AB, M, N, S, Rh_o, Hr_o, and rh′. Blood types are based on the presence of these factors. For example, an individual whose erythrocytes contain only agglutinogen B is said to have blood type B. The major classification of blood types is the ABO blood grouping system.

In addition to having specific blood group antigens on erythrocytes, the blood of certain humans contains antibodies that react with the erythrocytes from individuals of other blood types, causing them either to agglutinate or to lyse. The antibodies, called *agglutinins,* are found in the blood plasma and/or serum. Usually the antibodies of an individual are not directed against the factors normally present in that individual's blood. Most individuals have antibodies against factors that are absent from their blood cells. For instance, a person of blood type O has antibodies against agglutinogen A (called *anti-A, antibody a,* or *α antibody*) and against agglutinogen B (*anti-B, antibody b,* or *β antibody*). Table 47-1 lists the characteristics of the major blood types. Antibody concentrations can be increased as a result of transfusions or in the case of women who have born children possessing different blood characteristics.

Table 47-1 Selected Properties of the Major Human Blood Groups

Blood Type	Agglutinogen (Antigen) on Red Blood Cells	Agglutinin (Antibody) in Plasma
A	A	Anti-b
B	B	Anti-a
AB	A and B	Neither anti-a nor anti-b
O	Neither A nor B	Anti-a and anti-b

Because AB individuals do not possess antibodies against any of the other major blood types, they are referred to as *universal recipients.* Since type O blood cells have no agglutinogens to serve as antigens in another person's blood, members of this group are referred to as *universal donors.*

In 1937 Landsteiner and A. S. Wiener uncovered antisera that clumped (agglutinated) rhesus monkey blood as well as cells of 85 percent of Caucasians. This phenomenon was independent of the reactions obtained with the four major blood types. The factor was designated Rh for "rhesus." Several other factors subsequently were found to be involved in a complex Rh-Hr system. Two different nomenclature schemes commonly used to designate the factors in the Rh-Hr system are shown in Table 47-2. These factors are now recognized as the principle cause of erythroblastosis fetalis, or the hemolytic disease of the newborn, and as the cause of many transfusion reactions.

Table 47-2

Wiener Scheme	Fisher–Race Scheme
RH_0	D
Hr_0	d
rh′	C
hr′	c
rh″	E
hr″	e

The determination of an individual's blood group can be made by testing his or her red cells with standardized anti-a or anti-b sera. The mixtures are observed for the presence or absence of agglutination. Table 47-3 shows the patterns of reactions that are used for blood type determinations.

Determining the Rh factors of an individual's red cells is obtained by agglutination tests using the appropriate antisera—anti-rh_o, − rh′ and − rh″ (Color Plate 40). The classic and clinically most important of the factors is Rh_o(D).

Both false positive and false negative reactions may be encountered in testing. To prevent this, suitable control measures are used; these include pretesting of sera for activity, dilution of cells or sera in physiological saline if a reaction has not occurred after a sufficient length of time, and incorporation of tests with known cells and sera.

The blood group antibodies are of significant importance in cases of transfusions. The procedure known as compatibility testing or cross-matching is carried out to prevent a transfusion reaction as a consequence of blood incompatibilities between the recipient and donor. Specifically, the cross-matching technique is designed to detect any incompatibility between the recipient's serum and the donor's cells (known as the *major cross-match)* or between the recipient's cells and the donor's serum (the *minor cross-match).* After the blood specimens have been typed as to the ABO, Rh, and any other factors that appear to be indicated, different mixtures of donor's and recipient's cells and sera are made. Agglutination or lysis of the red cells from either the donor or recipient is considered to be indicative of an incompatible situation.

In certain clinical conditions, which include hemolytic anemia, incompatible transfusions, and the hemolytic disease of the newborn, unusual antibodies are present on the surface of blood cells. These globulins are not detected in major or minor cross-matches. The detection of such antibodies is accomplished by the Coombs' or antiglobulin test. The test employs the use of antihuman globulins (Coombs' reagent). This diagnostic material is prepared by successive injections of human serum into a suitable laboratory animal such as a rabbit or goat. The serum of the immunized animal will form antibodies against the globulin of human serum. This material is collected and used in the detection of human antibodies on the surface of red blood cells. The procedure known as the direct Coombs' test is performed by mixing red blood cells with the Coombs' reagent. Agglutination of the cells is a positive test demonstrating the presence of human antibodies. The direct Coombs' test will be demonstrated in this exercise.

In recent years the medico-legal aspects of blood grouping have assumed great importance. Standard blood grouping procedures provide information such as the compatibility of blood donors and recipients, identification of blood stains, detection of rare blood types, potential dangers in pregnancy, and facts that help decide cases of disputed parentage. In this exercise the procedures for blood typing and cross-matching will be considered.

Table 47-3 Reaction Patterns for Blood Type Determinations

	Agglutination When Mixed with Antiserum	
Blood Type	anti-a	anti-b
A	+	−
B	−	+
AB	+	+
O	−	−

A. BLOOD TYPING

Materials

1. The following materials should be provided per four students.
   a. Commercially prepared sterile alcohol prep-pads
   b. Sterile gauze pads (2 × 2 inch)
   c. Sterile blood lancets
   d. Eight glass slides
   e. Wooden applicator sticks
   f. Container for the disposal of the lancets and applicator sticks
2. The following materials should be provided for class use.
   a. Blood typing antisera—anti-A, anti-B, and anti-D (Rh_0)
   b. Unknown blood specimens

Procedure 1: ABO and Rh_0 Determinations

This procedure is to be performed by students in groups of four.

1. Place 1 drop of each antiserum on separate areas of the blood typing slide.
2. Obtain a blood sample from one of your laboratory partners (as described in Exercise 6).
3. Allow 1 full-sized drop of blood to fall directly into each drop of antisera.
4. Mix the components in each area, using different ends of the applicator sticks for each mixture.
5. If a reaction does not develop after 2 minutes, warm the slide over a desk lamp or similar device and gently shake the slide from side to side.
6. Refer to Table 47-3 and Color Plate 40 and interpret the reactions obtained.
7. Record the reactions obtained with your blood and your laboratory partner's in the results and observations section.

Procedure 2: Unknown Blood Sample

1. Obtain an unknown blood sample from your instructor.
2. Determine the major blood type and Rh factor or factors using procedure 1 of this exercise and Color Plate 40 as a guide.
3. Record your findings in the results and observations section. Check your results with the instructor.

B. CROSS-MATCHING

Materials

The following materials should be provided for class use. (*Note:* All plasma or serum preparations should be heated to 56°C for 30 minutes prior to use.)

1. Donor blood cell preparation, type A, in physiological saline
2. Donor blood cell preparation, type O, in physiological saline

3. Recipient plasma or serum preparation, type A
4. Recipient plasma or serum preparation, type O
5. Two recipient plasma or serum preparations, blood type unknown
6. Recipient blood cell preparation, type A, in physiological saline
7. Recipient blood cell preparation, type O, in physiological saline
8. Donor plasma or serum preparation, type A
9. Donor plasma or serum preparation, type O
10. Two donor plasma or serum preparations, blood type unknown
11. Glass slides
12. Pasteur pipettes with rubber bulbs
13. Wooden applicator sticks

Procedure

This procedure is to be performed by students in groups of four.

1. Perform the four major cross-matches using the combinations of donor cell preparations and recipient serum preparations indicated in Table 47-5 in the results and observations section.
2. Observe for the presence of agglutination. Record your findings and answer the questions in the results and observations section.
3. Perform the four minor cross-matches using the combinations of recipient cell preparations and donor serum preparations indicated in Table 47-6 in the results and observations section.
4. Observe for the presence of agglutination. Record your findings and answer the questions in the results and observations section.

C. DEMONSTRATION OF THE DIRECT COOMBS' TEST

Materials

The following materials should be provided for class demonstration.

1. One saline-washed suspension of sensitized red blood cells (coated with human gammaglobulins)
2. One saline-washed suspension of nonsensitized red blood cells
3. One saline-washed suspension of unknown red blood cells
4. Coombs' reagent
5. Glass slides
6. Pasteur pipettes with bulbs
7. Wooden applicator sticks
8. Marking pen
9. Container for the disposal of applicator sticks and pipettes

Procedure

The Instructor will demonstrate this procedure.

1. Label three separate areas of a glass slide, "+", "−", and "X" for an unknown, respectively. Make the markings near the edge of the slide.
2. Place one drop of each blood cell suspension near its labeled area.
3. Add 1 drop of the Coombs' reagent to each blood drop.
4. Mix the components in each area using a different applicator stick for each combination.
5. Examine the slide for agglutination during a 3-minute period.
6. Answer the questions in the results and observations section.

RESULTS AND OBSERVATIONS

BLOOD TYPING

1. Enter your findings in the ABO and Rh_0 determinations in Table 47-4. Indicate agglutination by a plus sign and no agglutination by a minus sign.

Table 47-4

Student	Reactions with Antisera			Blood Type
	anti-a	anti-b	anti-d (rh_O)	

2. Enter your findings on the unknown blood sample.

   a. Sample number __________

   b. ABO type __________

   c. Rh_0 type __________

CROSS-MATCHING

1. Enter your findings for the major cross-matches in Table 47-5. Indicate an incompatibility by a plus sign and compatibility by a minus sign.

Table 47-5

Donor Cells Blood Type	Recipient Blood Type			
	A	O	Unknown 1	Unknown 2
A				
O				

2. Which of the unknown combinations for the major cross-matches showed incompatibility? __________

3. Enter your findings for the minor cross-matches in Table 47-6.

Table 47-6

Recipient Cells Blood Type	Donor Blood Type			
	A	O	Unknown 1	Unknown 2
A				
O				

4. Which of the combinations for the minor cross-matches showed incompatibility?

____

5. Give at least two possible causes for the incompatibilities observed. ___________

____

DIRECT COOMBS' TEST

1. What reaction indicates a positive reaction?

____

2. Was the unknown specimen positive or negative?

____

3. What is the purpose of the Coombs' test?

____

____

Questions for Review

EXERCISE **47**

IMMUNOHEMATOLOGY

Name ______________________________

Laboratory Section ______________________ Date ____________

1. On each of the three slides shown in Figure Q47-1 the left side contains anti-A serum, the middle anti-B serum, and the right side anti-Rh_0(D) serum. One drop of an unknown blood sample was added to the antisera on each slide. On the basis of the reactions shown, determine the major blood type of each unknown blood sample and whether or not it contained the Rh_0 factor.

   a. Major blood factor ____________ Rh_0 factor ____________

   b. Major blood factor ____________ Rh_0 factor ____________

   c. Major blood factor ____________ Rh_0 factor ____________

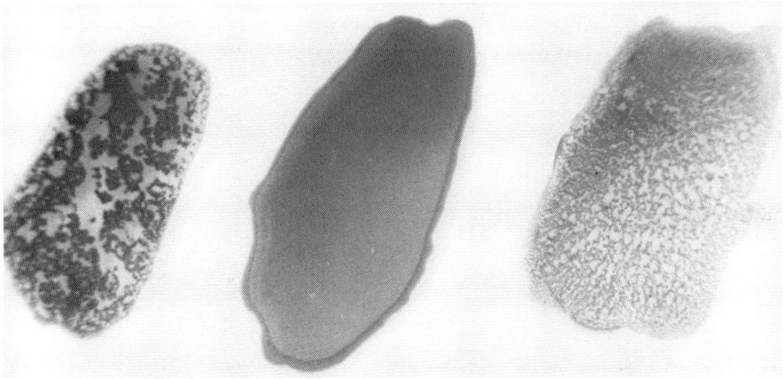

(a)

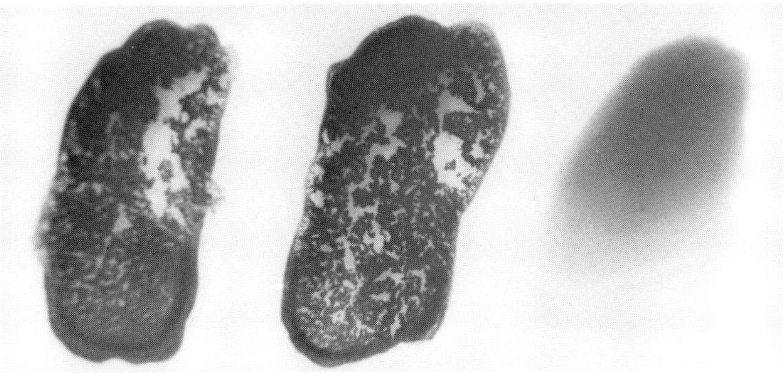

(b)

Figure Q47–1a,b

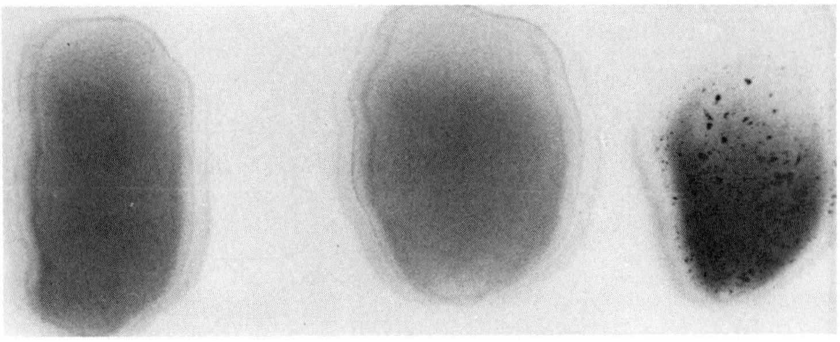

(c)

Figure Q47–1c

2. Define or explain the following.

   a. Cross-matching ______________________________

   ____

   b. Coombs' test ______________________________

   c. Hemolytic disease of newborn (Rh baby) ______________________________

   ____

   ____

   d. ABO incompatibility ______________________________

   ____

   e. Rh factor ______________________________

   f. Agglutination ______________________________

   ____

   g. Hemolysis ______________________________

   ____

3. List five major blood types other than the ABO and Rh groups.

   a. ______________ d. ______________

   b. ______________ e. ______________

   c. ______________

SECTION XIII

THE RESPIRATORY SYSTEM

When the term respiration is used alone, it generally refers to the overall metabolic activity that occurs within cells. When respiration is used in association with the respiratory system it refers to the process of gaseous exchange between the atmosphere, the blood, and the cells.

During the exchange of gases several processes are operating simultaneously. The movement of air between the atmosphere and the lungs is called *ventilation*. It is achieved by *breathing*, a method of creating differences in pressure in the chest cavity. These pressure changes are a direct consequence of a coordinated contraction of muscles, especially the intercostals and those of the diaphragm and abdomen. *External respiration* is the gaseous exchange of carbon dioxide, oxygen, and some nitrogen that occurs between the lung cells and the lung capillary blood. *Internal respiration* is a similar exchange between the blood and cells throughout the body.

In addition to its major role in the process of gaseous exchange, the respiratory system plays a minor role in temperature regulation and water balance, based on the amount of water vapor lost during exhalation. It also participates along with the kidney, liver, and gastrointestinal tract in the overall regulation of blood pH.

The average normal individual has a lung capacity of about 6000 ml. This total is made up of several respiratory air components including the expiratory reserve volume, inspiratory reserve volume, residual volume, and tidal volume (Figure XIII-1). A variety of pulmonary function tests are available to determine if lung tissues are functioning properly. In diseases such as carcinoma, emphysema, fibrosis, pneumonia, and pulmonary congestion there is a substantial reduction in normally functioning lung tissue. Several pulmonary function procedures are quite complex or require elaborate equipment. In Exercise 49 several of the more common tests are described and performed.

GENERAL REFERENCES

Lehrer, S. *Understanding Lung Sounds* (with audiocassette). Philadelphia: Saunders, 1984.

Levitzky, M. G. *Pulmonary Physiology*. New York: McGraw-Hill, 1982.

Tortora, G. J., and N. P. Anagnostakos, *Principles of Anatomy and Physiology*, 6th ed. New York: Harper & Row, 1990.

Witherspoon, J. D. *Human Physiology*. New York: Harper & Row, 1984.

SPECIFIC REFERENCES

Laszlo, G., and Sudlow, M. F. (Eds.). *Measurement in Clinical Respiratory Physiology*. New York: Academic Press, 1983.

Naeye, R. L. "Sudden Infant Death," *Scientific American*, 242:56–62, 1980.

Nero, A. V. Jr., "Controlling Indoor Air Pollution," *Scientific American*, 258: 42–48, 1988.

Schlesinger, R. B. "Defense Mechanisms of the Respiratory System," *BioScience*, 32: 45–50, 1982.

Weibel, E. R. "How Does Lung Structure Affect Gas Exchange?" *Chest*, 83:657–665, 1983.

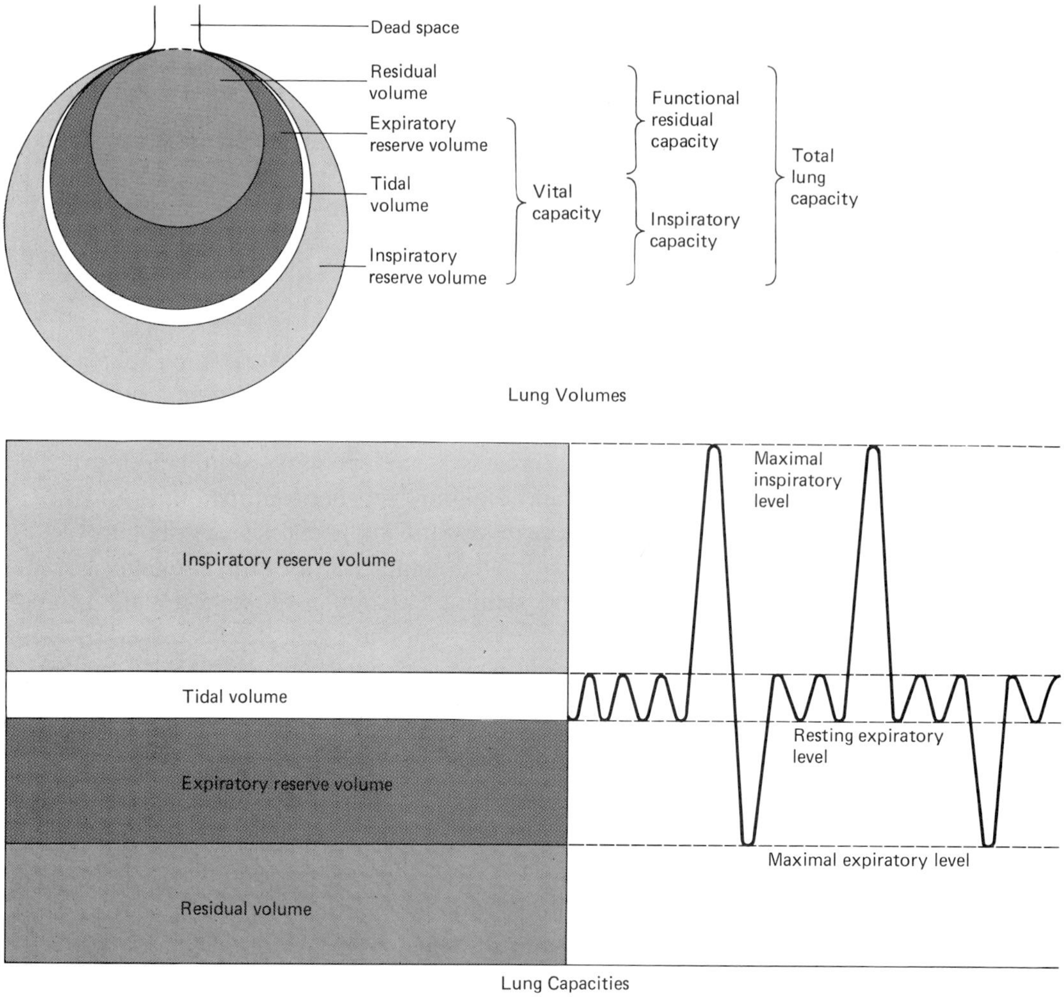

Figure XIII-1 Diagrammatic representation of respiratory air components.

EXERCISE 48

THE ANATOMY OF THE RESPIRATORY SYSTEM

Materials

The following materials should be provided for class use.

1. Cat dissection specimens
2. Dissecting tools
3. Model of the human respiratory tract
4. Charts of the human respiratory system
5. Demonstration of a mammalian head cut in sagittal section

ORGANS OF THE CAT RESPIRATORY SYSTEM

Locate the paired *external nares* of the cat. Air is taken into the nares, passes through the *nasal chamber* and the *internal nares,* and enters the *pharynx.* The pharynx is divided into three regions: (1) *nasopharynx,* (2) *oropharynx,* and (3) *laryngopharynx.* From the laryngopharynx the air passes through an opening called the *glottis* and enters the *larynx.* Beyond the larynx is the *trachea,* which divides into right and left *bronchi.* The bronchi penetrate the *lungs* and subdivide many times into *bronchioles,* finally ending in blind sacs called *alveoli,* which are in contact with the blood. The gases O_2 and CO_2 are exchanged between an alveolus and the blood (Figure 48-1).

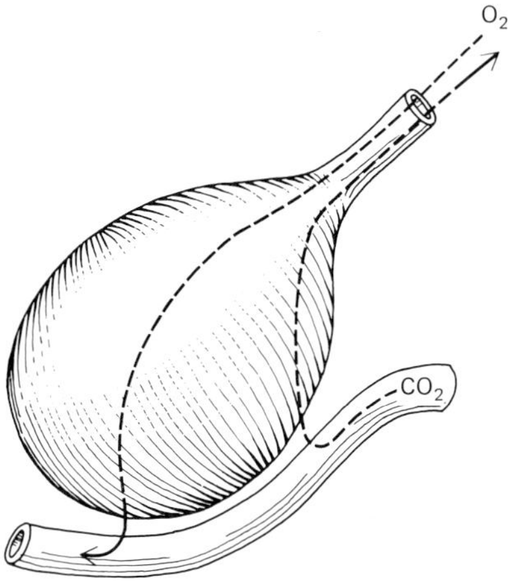

Figure 48-1 Exchange of gases between alveolus and capillary.

The mouth of the cat should be open. Observe as many organs as possible within the mouth cavity such as teeth and tongue. Then turn to the thoracic cavity and neck. Locate the *larynx* and observe once again the muscles that move its cartilages. Make a midsagittal cut into the ventral surface of the larynx and part of the trachea. Spread the two sides of the organ and observe the narrow *vocal cords* within it. If there is dried blood in the larynx, remove it with a wet paper towel or cloth. Posterior to the larynx is the *trachea,* which runs the length of the neck and into the thoracic cavity between the lungs. This tube never flattens because there is a series of cartilaginous rings that keep it open. At the lung the trachea bifurcates to form paired *bronchi,* which then penetrate the lung tissue and subdivide many times.

Locate the *right* and *left lungs.* The right lung contains three lobes: anterior, medial, and posterior. The left lung also has three lobes: anterior, medial, and posterior. A seventh lobe, the *mediastinal lobe,* is identified by some. The **mediastinal cavity** between the lungs contains the *heart, esophagus, thymus gland, trachea,* and major blood vessels. Each lung is enclosed by membranes. The *inner pleural membrane* is the **visceral peritoneum** and adheres closely to the lung. The *outer pleural membrane is the* **parietal peritoneum.** Between the two is a potential space called the **pleural cavity.** The membranes move with the lung as it expands and contracts during inhalation and exhalation. (Review Figure 48-2.)

THE HUMAN RESPIRATORY SYSTEM

LUNG

The human lung has five lobes:

Right superior lobe
Right medial lobe
Right inferior lobe
Left superior lobe
Left inferior lobe

The relationship of the pleura to the lungs and the thoracic cavity is the same as it is in the cat. There are two pleural membranes (inner and outer) and there is a potential space between them, the *pleural cavity.*

NASAL CAVITY

The nostrils, or *external nares,* open into a large bony cavity of the skull below the cranium and above the mouth (Figure 48-3). The bones involved are the *ethmoid, maxillae, palatines, vomer,* and the *inferior nasal conchae.* Review these bones. The ethmoid bone (which contains the superior and middle nasal conchae) forms the lateral walls and roof of the nasal cavity along with the inferior nasal conchae, which are separate bones. The floor of the nasal cavity is formed by the maxillae and the palatine bones. These latter four bones also form the **hard palate.** The cartilaginous part of the *nasal septum* that divides the nasal cavity into right and left sides is contiguous with the **perpendicular plate** of the *ethmoid* and the *vomer* bone. In this cavity air is warmed, filtered, and moistened. Sensory olfactory nerve endings are found in parts of the epithelial lining, which is the *pseudostratified ciliated type.*

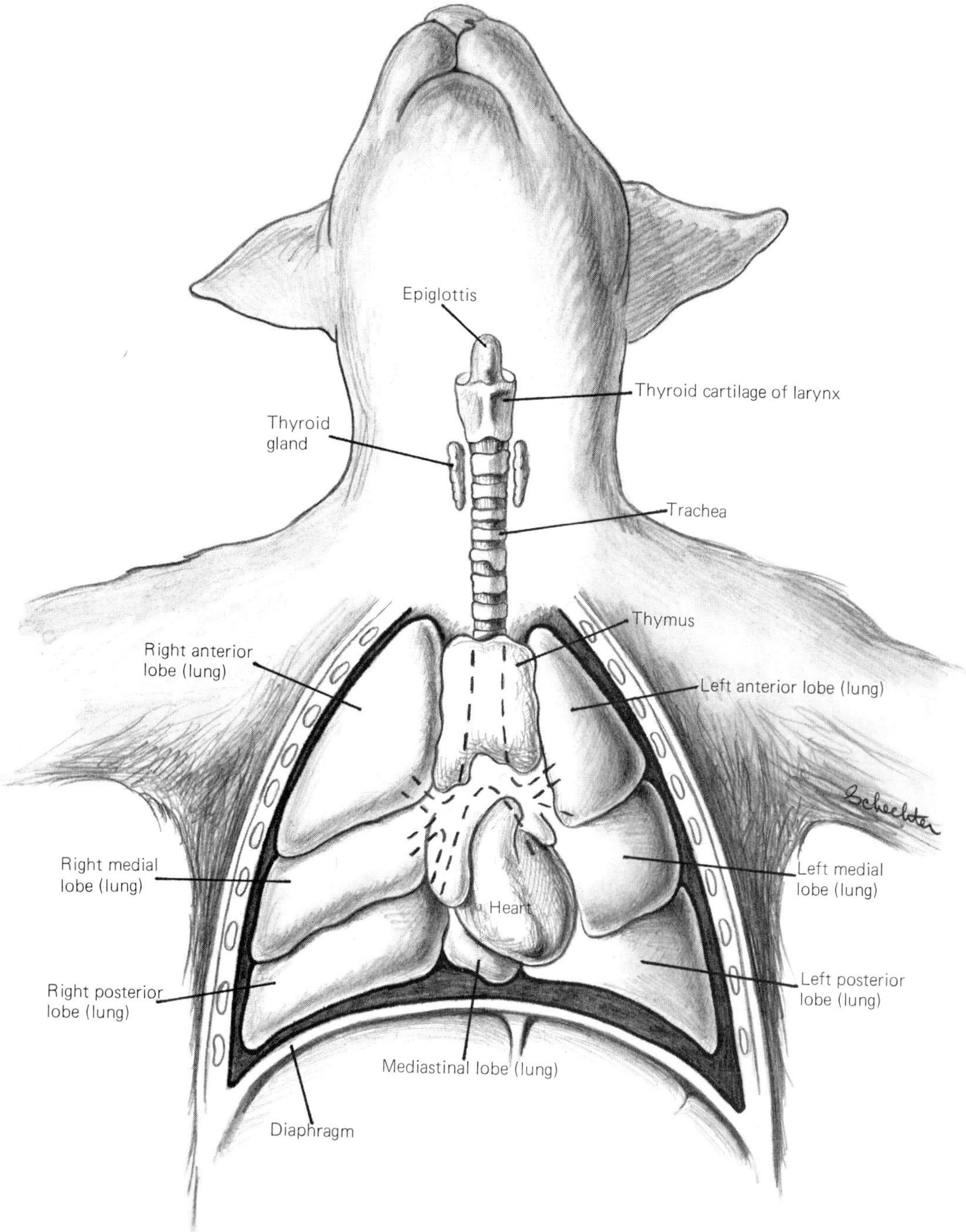
Epiglottis
Thyroid cartilage of larynx
Thyroid gland
Trachea
Thymus
Right anterior lobe (lung)
Left anterior lobe (lung)
Right medial lobe (lung)
Left medial lobe (lung)
Heart
Right posterior lobe (lung)
Left posterior lobe (lung)
Mediastinal lobe (lung)
Diaphragm

Figure 48-2

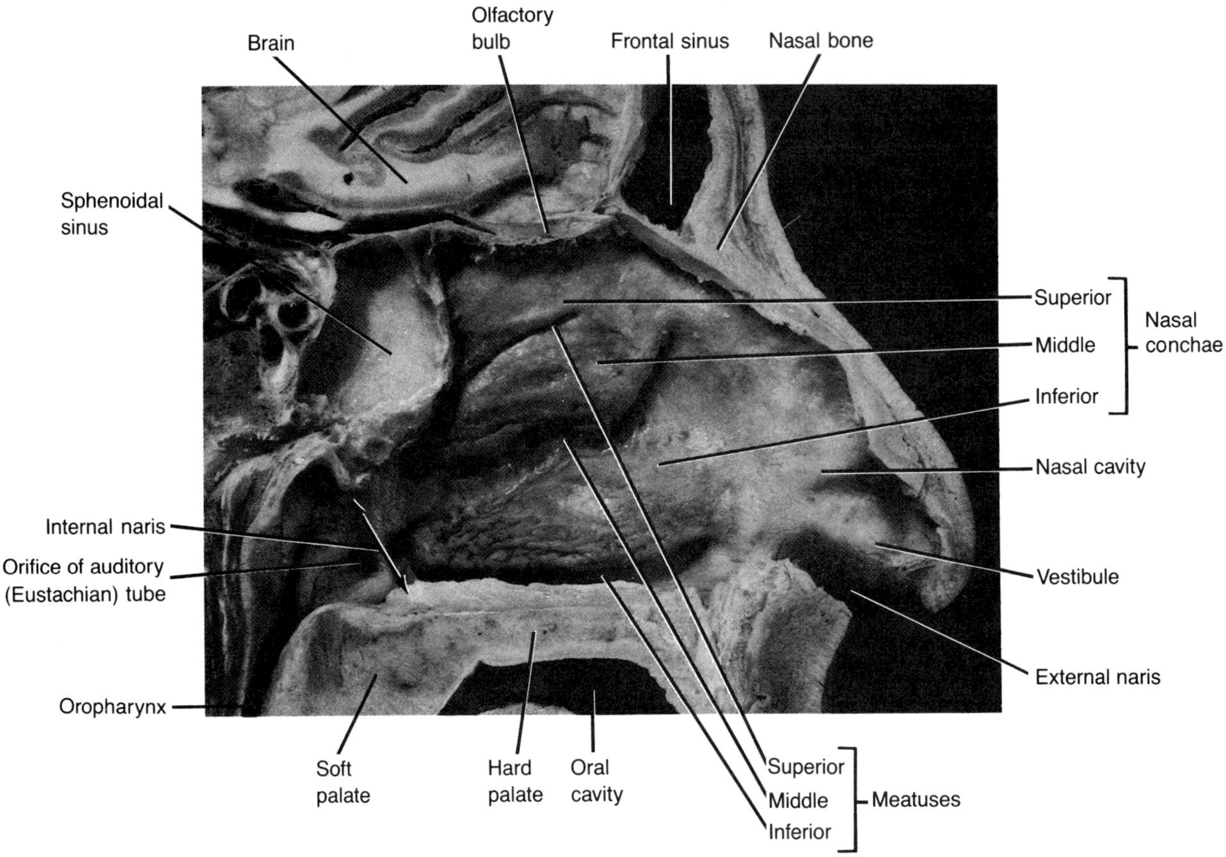

Figure 48-3 Midsagittal section of the human head showing upper respiratory tract. (Courtesy of J. A. Gosling, P. F. Harris, et al., *Atlas of Human Anatomy,* Gower Medical Publishing Ltd. 1985.)

PHARYNX

Air from the nasal cavity passes through the *internal nares* to enter a large cavity at the back of the mouth (the pharynx). It is a common passageway for both food and air and is highly muscular. The cavity is divided into three regions:

Nasopharynx
Oropharynx
Laryngopharynx

Lymphoid tissue, (tonsils), is found in the pharynx. Two types, the *palatine tonsils* and the *lingual tonsils,* are located in the oropharynx, and the *pharyngeal tonsils* (adenoids) are located in the nasopharynx.

LARYNX

The opening from the pharynx into the larynx between the vocal cords is called the *glottis.* Guarding this opening is a large laryngeal cartilage, the *epiglottis.* When one is swallowing, the epiglottis covers the opening to the glottis and the food enters the esophagus. When inhaling or talking, the glottis is open to permit air to flow through the larynx on its way to the lungs. Study Figure 48-4.

The larynx contains the vocal cords, which vibrate as air passes over them. These cords are capable of producing sounds. The framework of the larynx consists of nine cartilages that

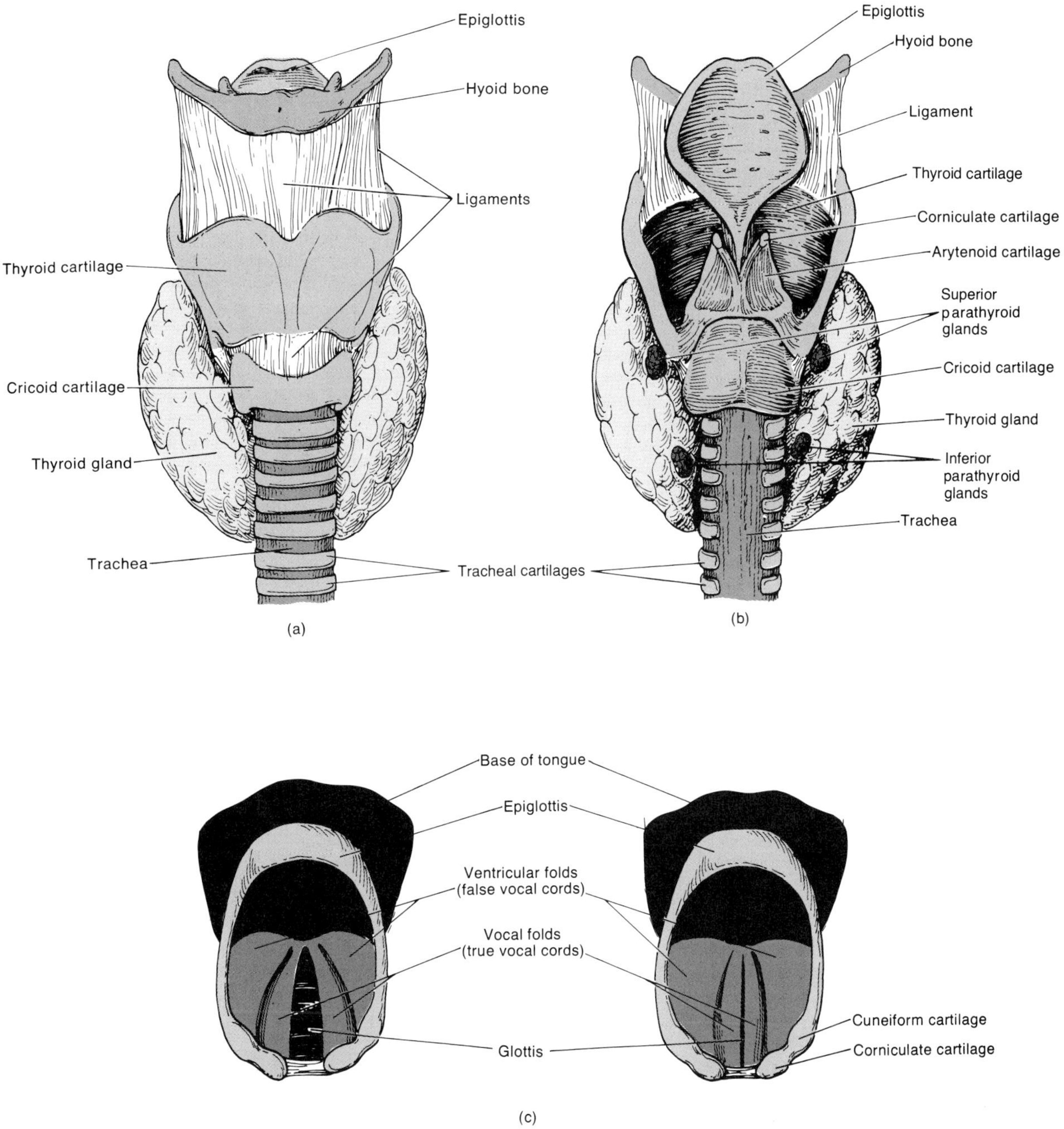

Figure 48-4 Larynx. (a) Anterior view. (b) Posterior view. (c) Viewed from above. In the figure on the left, the true vocal cords are relaxed. In the figure on the right, the true vocal cords are pulled taut.

are moved by muscles, many of which you have already observed in the cat. The laryngeal cartilages are:

Thyroid cartilage (forms the Adam's apple)
Epiglottis
Cricoid cartilage
Paired aretynoid cartilages
Paired corniculate cartilages
Paired cuneiform cartilages

TRACHEA, BRONCHI, AND BRONCHIOLES

The *trachea* is an air-conducting tube that is supported by cartilagenous rings. It begins at the larynx and terminates at the lungs in the mediastinal cavity where it divides into right and left bronchi. It is lined by pseudostratified ciliated epithelium and there are many mucus-producing goblet cells present. The *bronchi* and bronchioles branch about 24 times and end in air sacs, called *alveoli,* through which gases are exchanged between the lung and the blood.

Label Figure 48-5.

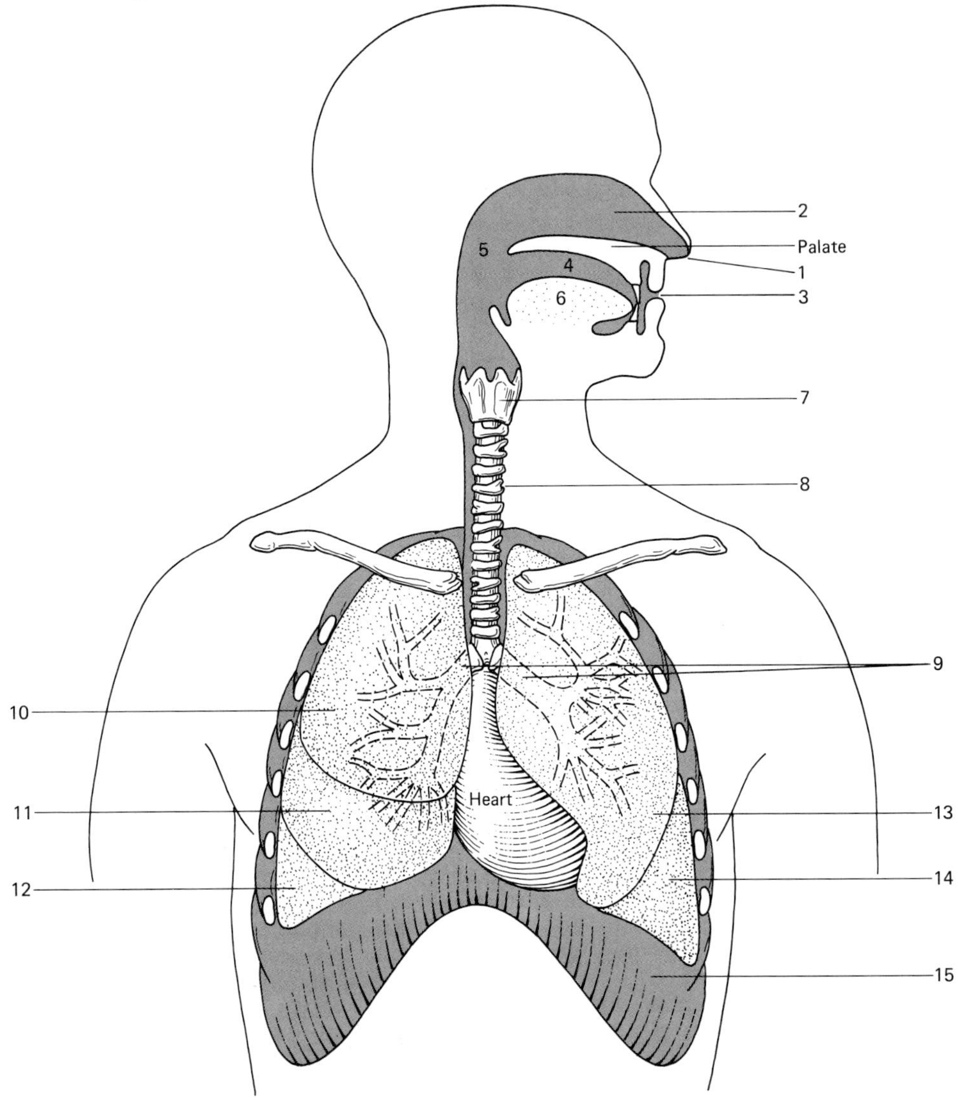

Figure 48-5 Human respiratory tract.

1. ______________
2. ______________
3. ______________
4. ______________
5. ______________
6. ______________
7. ______________
8. ______________
9. ______________
10. ______________
11. ______________
12. ______________
13. ______________
14. ______________
15. ______________

Questions for Review

EXERCISE **48**

THE ANATOMY OF THE RESPIRATORY SYSTEM

Name ______________________________

Laboratory Section ______________________ Date ____________

1. Match the phrases on the left with the correct term on the right.

a. ____________	Cavity between membranes surrounding the lung	1. Bronchioles
b. ____________	Name for two branches of the trachea	2. Pharynx
c. ____________	Respiratory passage that contains the vocal cords	3. Maxillary
d. ____________	Terminal part of respiratory duct system in contact with blood	4. 5
e. ____________	Opening into the larynx from the pharynx	5. Mediastinum
f. ____________	Number of lobes in human lung	6. Palatine
g. ____________	Name for the membranes surrounding the lungs	7. Ethmoid
h. ____________	A bone that makes up the hard palate	8. Larynx
i. ____________	Another bone that makes up the hard palate	9. Epiglottis
j. ____________	A prominent bone within the nasal chamber	10. Pleural
		11. Glottis
		12. 4
		13. Bronchi
		14. Alveoli
		15. Pericardial
		16. 6
		17. Mandible

2. Trace a molecule of oxygen from the air until it reaches the blood.

1. Air	8. ____________
2. ____________	9. ____________
3. Nasal chamber	10. Bronchioles
4. ____________	11. ____________
5. ____________	12. ____________
6. Glottis	13. Blood capillaries
7. ____________	

EXERCISE 49

PULMONARY FUNCTION TESTS

The respiratory system consists of a number of anatomic passageways through which the air we breathe travels to the delicately structured lungs. During *inspiration*, the movement of air from the atmosphere into the lungs, air is moved through the nose, pharynx, larynx, trachea, and successively smaller divisions of the bronchial tree until it reaches the numerous tiny air sacs, the pulmonary alveoli. Conditions at the alveoli are ideal for the physiological exchanges of gases. During *expiration*, the movement of air from the lungs into the atmosphere, the path is reversed. Air flows through these passageways during respiration because of a pressure gradient involving three important pressures:

1. *Atmospheric pressure*—the pressure exerted by the air surrounding the body, which at sea level is equal to 760 mm of mercury
2. *Intrapulmonary pressure*—the pressure within the respiratory space of the lungs (bronchi and alveoli)
3. *Intrapleural pressure*—the pressure within the pleural space between the lungs and the walls of the thoracic cage

Various factors are known to regulate respiration. For example, the temperature of blood flowing through the respiratory center may influence breathing. Above normal body temperature, as during fever or severe muscular exercise, results in an increase in the respiratory rate (number of breaths per minute). Subnormal temperature, as during hypothermia, results in a decreased respiratory rate. Other respiration regulating factors include pain and emotional state.

Before birth the lungs contain no air, but after birth, once they have been filled with air, they are never completely empty. No expiration ever completely empties the alveoli; neither are they completely filled. The ability of an individual's respiratory system to exchange air with the outside environment is important in evaluating normal function or the presence of respiratory disease. A number of different measurements can be made in studying the various properties of the respiratory system. These are described in Table 49-1. The instrument used to measure lung capacities, volumes of air exchanged, and rates of respiration is the *spirometer*. The spirometer can play an important role in the detection of restrictive or obstructive lung defects. This exercise describes procedures for making some of the respiratory measurements.

Materials

The following materials should be provided for class use.

1. Spirometers, such as the Collins Survey Spirometer, the Breon Spirometer, or other appropriate device (with their operator's manual, if possible)
2. Breathing tube attachment
3. Disposable mouthpieces (adult)

Table 49-1 Respiratory Volume Measurements

Measurement	Brief Description	Normal Values for Adults
Tidal volume (TV)	Quantity of air moved into and out of the lungs during a normal breath	500 ml
Respiratory rate (RR)	Number of breaths per minute	12–15
Minute volume (MV)	Quantity of air moved into and out of the lungs in 1 minute (MV = TV × RR)	6,000 ml/min
Inspiratory reserve volume (IR)	Quantity of air above tidal volume that can be forcefully inspired after a normal inhalation	3,000 ml
Expiratory reserve volume (ERV)	Quantity of air above tidal volume that can be forcefully expired after a normal inhalation	1,200 ml
Vital capacity (VC)	Maximum quantity of air that can be moved into and out of the lungs VC = IR + TV + ER)	4,500–5,500 ml
Residual air (RA)	Quantity of air left in the lungs after maximum expiration	1,200 ml
Total lung capacity (TLC)	Quantity of air the lungs can hold (TLC = VC + RA)	5,900 ml
Dead space	Air which is not available for gas exchange	100 ml (female) 150 ml (male)

4. Nose clips
5. Disinfectant and sterile gauze pads
6. Recording charts or paper
7. Recording pens and ink

Procedure 1: Identification of Spirometer Parts and Accessories

This procedure is to be performed by students individually.

1. Examine the spirometer provided.
2. With the aid of Figure 49-1 and the operator's manual, identify the typical parts and accessories of the instrument.
3. Your instructor will demonstrate the correct usage of the spirometer.

Procedure 2: Respiratory Rate

This procedure is to be performed by students in pairs.

1. Have your partner sit quietly for 5 minutes.
2. After the rest period, count the number of complete respiratory cycles (inspiration and expiration) in 60 seconds. Make two additional counts while your partner is still seated.
3. Make three counts with your partner standing and three counts in the prone position.
4. Determine the average respiratory rate in each position by dividing the number of times counts were taken into the number of complete cycles. Enter your findings in the results and observations section.
5. Change places with your partner and repeat steps 1–4.

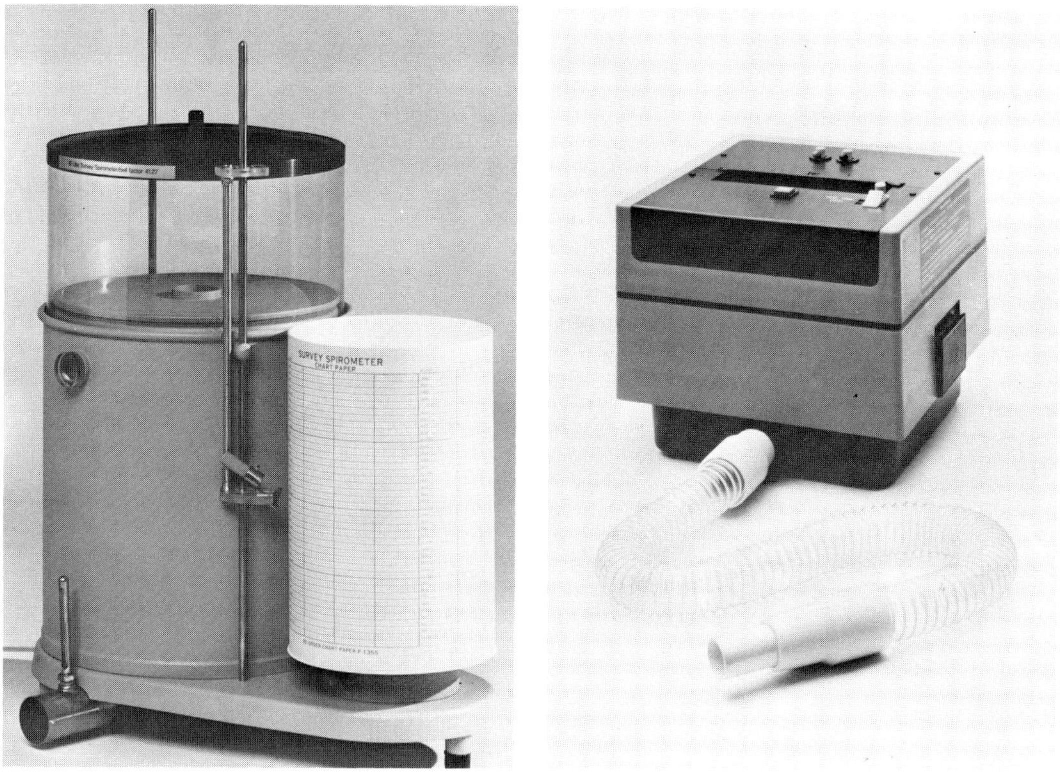

(a) (b)

Figure 49-1 Two instruments used for the measurement of lung volumes and capacities (spirometry). Note the various components of each. (a) Collins Respirometer. (Courtesy, Warren E. Collins, Inc., Braintree, MA.) (b) Breon Spirometer. (Courtesy, Breon Laboratories, Inc., New York, NY.)

Procedure 3: Tidal Volume

This procedure is to be performed by students in pairs.

1. Have your partner sit quietly for 5 minutes.
2. Disinfect a nose clip. Place it on your partner's nose so that the nasal passages are closed.
3. Place the disposable mouthpiece into your partner's mouth.
4. Start the recording system and have your partner breathe normally into the spirometer system for five complete respiratory cycles.
5. Record the time (in seconds) for each inspiration and the volume of air inspired. The average value for the inspiratory volume is the *tidal volume* at rest.
6. Repeat this procedure with your partner in the standing position and in the prone position.
7. Change places with your partner and repeat steps 1–6.
8. Record the measurements in the results and observations section.

Procedure 4: Expiratory Reserve Volume (ERV)

This procedure is to be performed by students in pairs.

1. Have your partner sit quietly and breathe normally.
2. Attach a nose clip as described in procedure 3.
3. Turn on the recording system.

4. Instruct your partner to breathe normally for four respiratory cycles and then on the fifth normal expiration to exhale as much as possible into the spirometer mouthpiece. (Help your partner by counting the inspirations and expirations of the complete cycles.)
5. Record the volume.
6. Repeat steps 4 and 5 two more times and calculate an average value.
7. Repeat the procedure with your partner in the standing position and in the prone position.
8. Change places with your partner and repeat steps 1–7.
9. Record the measurements in the results and observation section.

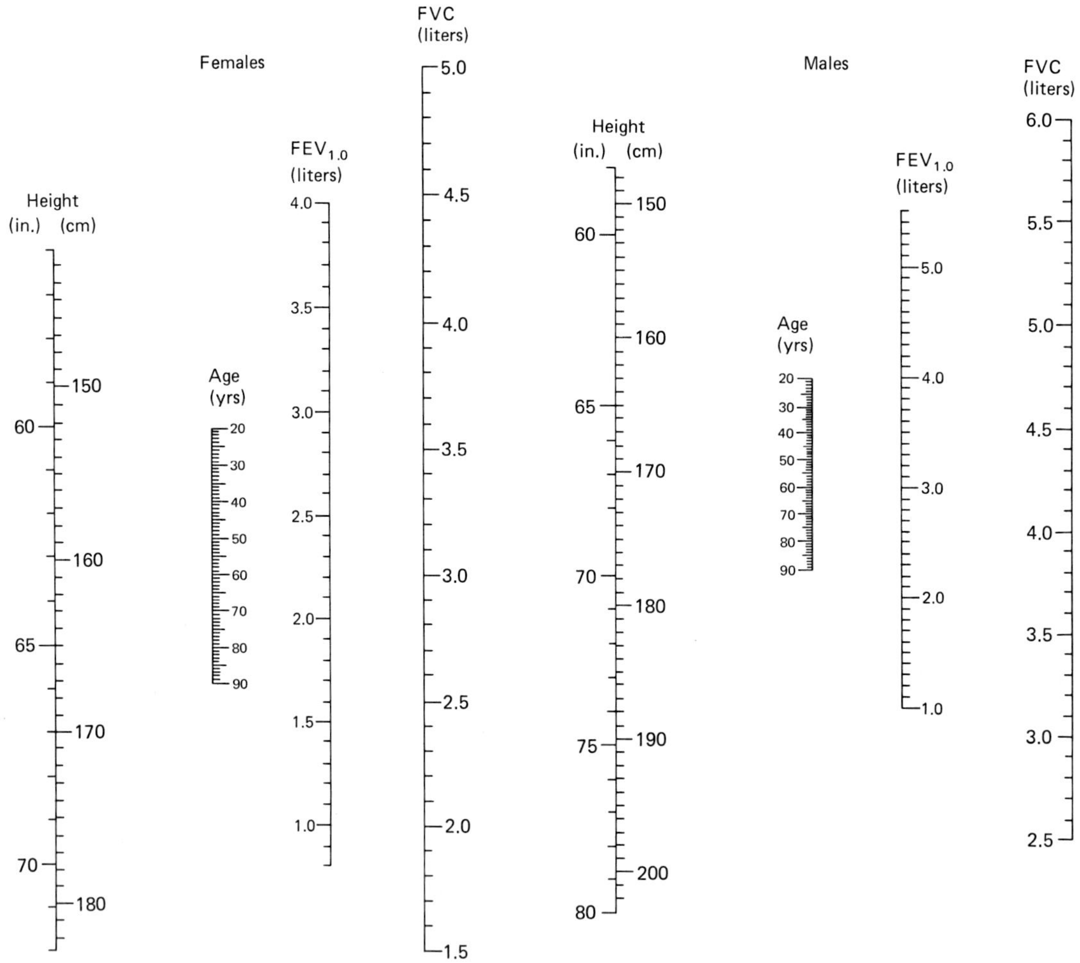

Figure 49-2 Scales for the determination of normal spirometric values for females and males. These values are expressed in terms of a normal range for age, height, and sex. To use these scales place the edge of a ruler at the correct point for the subject on the height scale and at the correct point on the age scale. The points at which the ruler intersects the FEV 1.0 (forced expiratory volume in one second) and FVC (forced vital capacity) scales give the normal values for those measurements. (Modified from T. L. Petty, *Office Spirometry.* New York: Breon Laboratories, 1980.)

Procedure 5: Forced Expiratory Vital Capacity (FVC)

This procedure is to be performed by students in pairs.

1. Have your partner sit quietly and breathe normally.
2. Attach a nose clip, and turn on the recording system.
3. Instruct your partner to breathe normally for four complete respiratory cycles and then on the fifth cycle to inhale as deeply as possible and to exhale as forcefully and steadily as possible into the spirometer mouthpiece.
4. Record the volume and time the expiratory portion of the last cycle. Determine the average FVC value.
5. Repeat the procedure two more times.
6. Change places with your partner and repeat steps 1–5.
7. Find the normal values for FVC on Figure 49-2.
8. Record the measurements in the results and observations section.

RESULTS AND OBSERVATIONS

1. Enter your respiratory rate determinations in Table 49-2 (A = average).

Table 49-2

	Sitting				Standing				Prone			
Student	1	2	3	A	1	2	3	A	1	2	3	A

2. Enter your tidal volume determinations in Table 49-3 (T = Time, V = volume, and A = average).

Table 49-3

	Sitting								Standing								Prone							
	1		2		3		A		1		2		3		A		1		2		3		A	
Student	T	V	T	V	T	V	T	V	T	V	T	V	T	V	T	V	T	V	T	V	T	V	T	V

3. Enter your expiratory reserve volume determinations in Table 49-4. (A = average.)

Table 49-4

	Sitting				Standing				Prone			
Student	1	2	3	A	1	2	3	A	1	2	3	A

4. Enter your forced expiratory vital capacity determinations in Table 49-5 (V = volume and T = time for forced expiration).

Table 49-5

	1		2		3		Average	
Student	V	T	V	T	V	T	V	T

Questions for Review

EXERCISE **49**

PULMONARY FUNCTION TESTS

Name ______________________________

Laboratory Section ______________________ Date __________

1. Distinguish between inspiration and expiration. ______________________________

____

____

2. Complete Table Q49-1 by briefly defining or explaining the terms listed and giving the normal values for each.

Table Q49-1

Measurement	Definition	Normal Value
Tidal volume	a.	b.
Minute volume	c.	d.
Inspiratory reserve	e.	f.
Expiratory reserve	g.	h.
Vital capacity	i.	j.
Residual air	k.	l.
Total lung capacity	m.	n.
Dead space	o.	p.
Respiratory rate	q.	r.

3. Define or explain the following.

   a. Atmospheric pressure ______________________________

   ____

   b. Intrapulmonary pressure ______________________________

   ____

   c. Intrapleural pressure ______________________________

   ____

4. Indicate whether the measurement values in Table Q49-2 are normal or abnormal for adults.

Table Q49-2

Measurement	Value(s)	Normal/Abnormal
Tidal volume	200 ml	a.
	500 ml	b.
Inspiratory reserve	1000 ml	c.
	2000 ml	d.
	3000 ml	e.
Vital capacity	2000 ml	f.
	4500 ml	g.
	5000 ml	h.
	5500 ml	i.
Total lung capacity	3500 ml	j.
	6000 ml	k.

5. List three major physiological functions of the respiratory system.

   a. ______________________________

   b. ______________________________

   c. ______________________________

6. How would the respiratory rate be affected by the following situations?

   a. High fever ______

   b. Running ______

   c. Hypothermia ______

7. What is hypoxia? ______

______

8. Define of explain hyperventilation. ______

______

9. List four factors that regulate respiration.

   a. ______ c. ______

   b. ______ d. ______

10. With the aid of your textbook or other reference complete the description of respiratory disorders in Table Q49-3.

Table Q49-3

Disorder	Description
a.	Collapse of lung tissue caused by extrinsic pressure or intrinsic pressure in pulmonary airways
Bronchitis	b.
Tuberculosis	c.
d.	Allergic reaction to proteins in foreign substances localized in respiratory membranes
Emphysema	e.
Pneumonia	f.
Pneumocystis infection	g.

SECTION XIV

THE DIGESTIVE SYSTEM

As a group, nutrients such as carbohydrates, lipids, proteins, vitamins, and minerals are obtained from foods. Three processes take place in the body to convert the food consumed into usable forms. These processes are digestion, absorption, and metabolism. Exercises in this section will consider selected components of the digestive system and emphasize the major types of normal as well as some abnormal activities associated with each one. Attention also is given to the role of micro-organisms in tooth decay (caries).

Enzymes and the effects of certain factors such as pH and temperature on enzymatic activity are examined. Tests used in the identification of proteins, carbohydrates, lipids, and other chemicals of importance in foods are demonstrated as well.

GENERAL REFERENCES

Davenport, H. W. *Physiology of the Digestive Tract,* 5th ed. Chicago: Year Book Medical Publishers, 1982.

Netler, F. H. *The CIBA Collection of Medical Illustrations,* Vol. 3, *Part I, Upper Digestive System.* Summit, NJ: CIBA Pharmaceutical Products, 1959.

Sleisenger, M. H., and J. S. Fordtran (Eds.). *Gastrointestinal Disease: Pathophysiology, Diagnosis, Management,* 3rd ed., Vols. 1 and 2. Philadelphia: Saunders, 1984.

Tortora, G. J., and N. P. Anagnostakos. *Principles of Anatomy and Physiology,* 6th ed. New York: Harper & Row, 1990.

Witherspoon, J. D. *Human Physiology.* New York: Harper & Row, 1984.

SPECIFIC REFERENCES

Davenport, H. W. "Why the Stomach Does Not Digest Itself," *Scientific American,* 226:86–93, 1972.

Moog, F. "The Lining of the Small Intestine," *Scientific American,* 245:154–176, 1981.

Peters, M. N., and C. T. Richardson. "Stressful Life Events, Acid Hypersecretion and Ulcer Disease," *Gastroenterology,* 84:114–119, 1983.

Track, N. S. "The Gastrointestinal Endocrine System," *Canadian Medical Association Journal,* 122:287–292, 1980.

EXERCISE 50

THE ORAL CAVITY

The floor of the human mouth lies in a horseshoe around the tongue and is continuous laterally with the gum, or *gingiva,* and centrally with the tongue. Near the front end are the openings of the submaxillary and sublingual salivary glands (Figure 50-1). These glands empty their secretions into the mouth. Here the fluids aid in swallowing, chewing, and digestion.

The tongue is made up of skeletal muscle. In addition to its role in chewing and swallowing, the tongue is important to humans because it assists in speech and contains receptors for taste. The rough surface of the tongue is due to several types of *papillae* (elevations) that are heavily supplied with taste buds (Color Plate 28). Taste buds are accumulations of taste cells that carry stimuli to the central nervous system. The sense of taste appears to be limited to four basic tastes—bitter, salty, sour, and sweet. These tastes are localized in various areas of the tongue and are demonstrated in Exercise 35.

The human is supplied with two sets of teeth, 20 deciduous teeth (baby or milk teeth) and 32 permanent teeth. Teeth are categorized into four groups. *Incisors* are used for cutting food, *canines* for tearing food, and *premolars* (bicuspids) and *molars* for grinding (Figure 50-2a). The parts of each tooth are the *crown,* the portion above the gum; the *root,* the

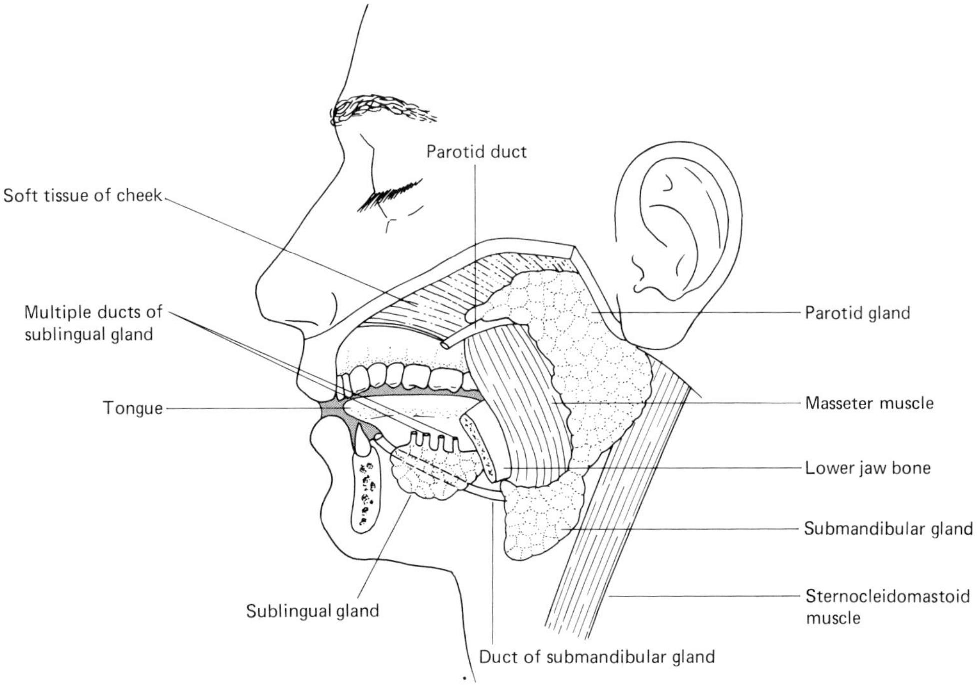

Figure 50-1 General organization of the human mouth.

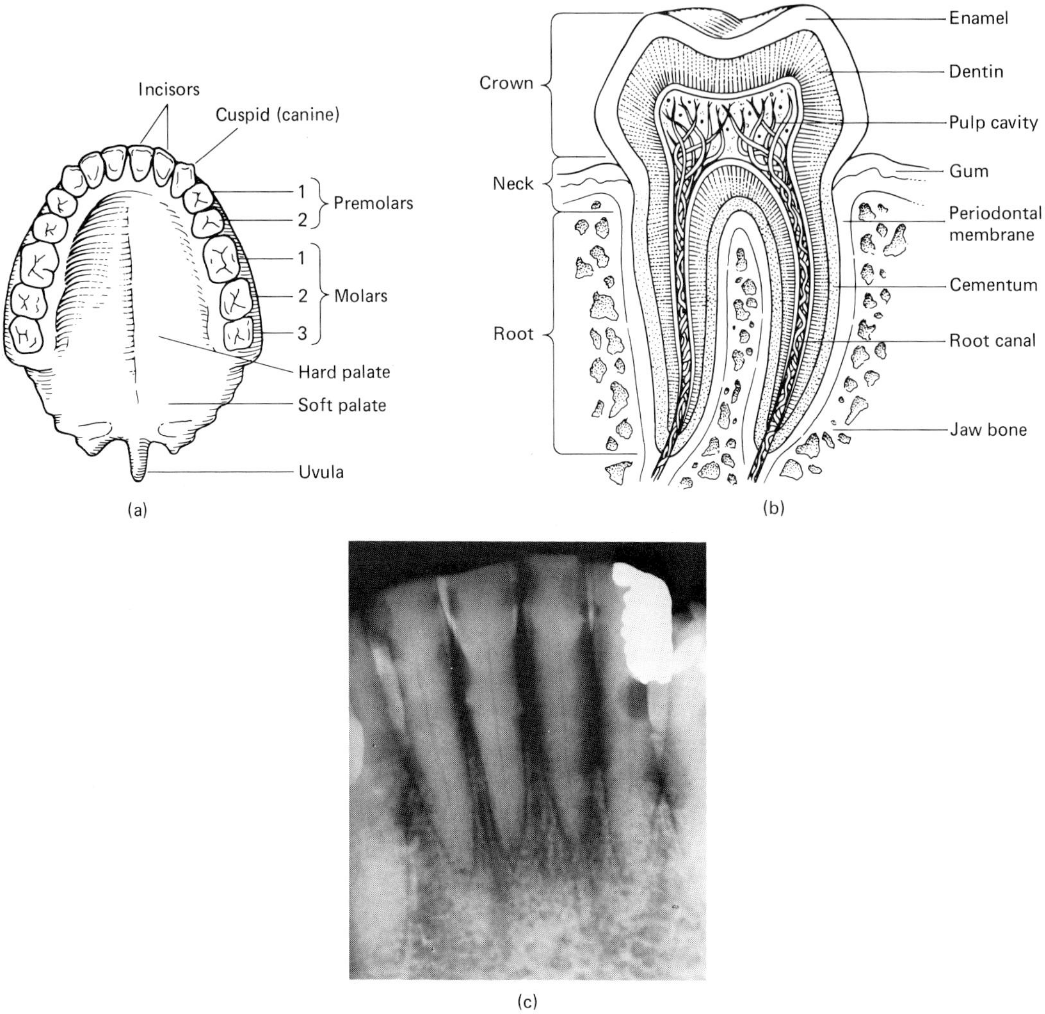

Figure 50-2 Types and structure of human teeth. (a) Major types of permanent teeth. (b) Structure of a tooth. (c) X ray of a tooth. Note the appearance of the filling.

structure embedded in the jaw; and the *neck*, the constricted region between the crown and the root (Figure 50-2b).

The mouth, being warm and moist and having a regular supply of fresh food, makes an ideal growth environment for various types of microorganisms. These organisms, which include bacteria, yeast (fungi), protozoa, and viruses, comprise the oral flora. The types of organisms in the oral flora are relatively stable. Although others may be introduced, perhaps with food, they are usually transients and seldom take up permanent residence.

Teeth tend to be covered with a gummy accumulation of salivary mucin and bacteria, called *dental plaque*. Such accumulations cause inflammations of the gum and are associated with tooth decay or dental caries.

In this exercise attention is directed to the categories of human teeth and the cat's oral cavity.

A. ORAL CAVITY OF THE CAT

Procedure

1. Examine the cat's mouth. Locate the parts of the mouth in Figure 50-3. The mouth was opened for you before the cat was killed to facilitate study of this area. If it is to be opened further in order to see posterior parts, bone forceps will be needed to cut back at the angles of the jaw. Follow directions given by your instructor.
2. Locate the papillae on the tongue. In which direction are the points of these structures directed?
3. Find the transverse ridges that form the animal's hard palate.
4. Locate the parotid, submaxillary, and sublingual glands and associated ducts.

The cat's mouth cavity (oral cavity) is divided into the *vestibule* and the mouth cavity proper by the gums and teeth (Figure 50-3). The vestibule is surrounded by lips and cheeks. Within the mouth cavity proper are the tongue and the teeth. Posterior to the mouth cavity is the pharynx. Observe that the tongue is tied to the floor of the mouth by a fold of tissue, the *frenulum linguae*. The tongue is a muscular organ (striated skeletal muscle) covered with a mucous membrane. On its rough outer surface are numerous papillae, many of which are horny and toothlike. Cats groom and clean themselves with their tongues. The *hard palate* forms seven or eight transverse ridges.

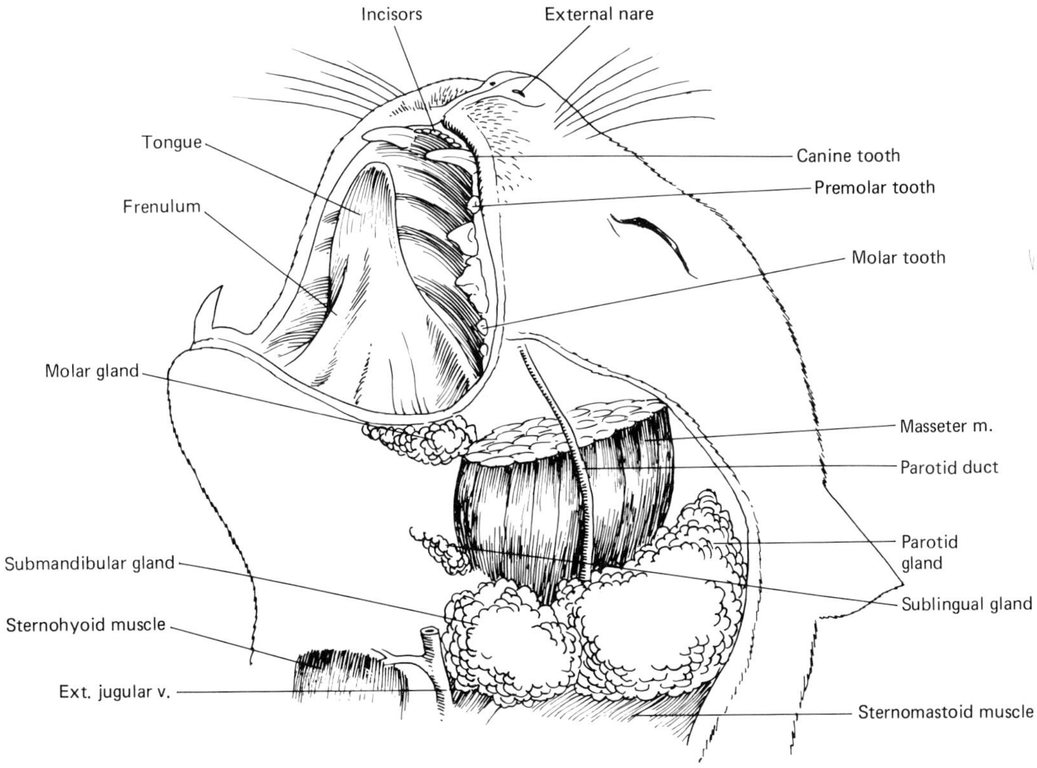

Figure 50-3 Oral cavity of the cat.

The cat has two sets of teeth in its lifetime. At birth, the kitten is toothless. The first or *deciduous* set of teeth are 26 in number. They are replaced by 30 *permanent* teeth, 14 in the lower jaw and 16 in the upper jaw. The teeth are of four types (Figure 50-3): incisors (12), canines (4), premolars (10), and molars (4).

There are five pairs of salivary glands in the cat. Three of the more prominent and accessible pairs were discussed in Exercise 19. Refer to Figures 19-2a and 19-2b.

1. *Parotid gland* and *parotid duct,* which open on the inside of the cheek opposite the last premolar tooth
2. *Submandibular gland* and *submandibular duct,* which open into the mouth cavity in front of the parotid duct
3. *Sublingual gland* and *sublingual duct,* which open into the mouth cavity in almost the same place but slightly anterior to the submaxillary duct
4. *Molar glands,* which are located under the mucosa of the lower lip and extend from the *masseter* muscle to the first canine tooth. Several small ducts open into the mouth cavity.
5. *Infraorbital glands,* which lie in the ventrolateral floor of the orbits. A single duct from each gland opens into the mouth cavity near the molar tooth. (Not shown in Figure 50-3.)

B. IDENTIFICATION OF SELECTED HUMAN MOUTH PARTS

Materials

The following materials should be provided for class use.

1. Models of human adult skull with teeth
2. Hand mirrors
3. Wooden tongue depressors

Procedure

This procedure is to be performed by students in pairs.

1. Examine a human skull model and with the aid of Figure 50-2a identify the major categories of adult teeth.
2. Use a hand mirror to examine your own mouth. Identify the major categories of teeth present.
3. Examine your laboratory partner's mouth and identify the base of the tongue, gum tissue, visible parts of teeth, and the major categories of teeth. Use Figure 50-2 for reference.

Questions for Review

EXERCISE **50**

THE ORAL CAVITY

Name ______________________________

Laboratory Section ______________________ Date __________

1. Identify the types and/or categories of teeth shown in Figure Q50-1.

a. ______________________ g. ______________________

b. ______________________ h. ______________________

c. ______________________ i. ______________________

d. ______________________ j. ______________________

e. ______________________ k. ______________________

f. ______________________

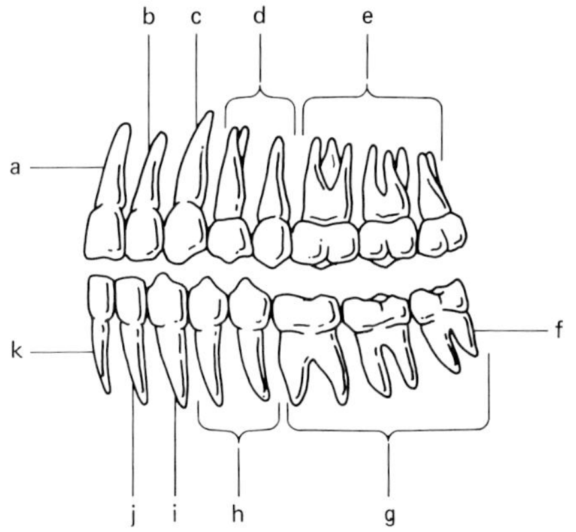

Figure Q50-1

2. Define or explain the following.

a. Oral flora ______________________________

____

b. Dental caries ______________________________

____

c. Dental plaque ______________________________

____

3. What are taste buds? ________________________________

____

4. List the categories of teeth found in the human, and give their respective functions. __

____

____

____

EXERCISE 51

THE ORGANS OF THE DIGESTIVE SYSTEM

Materials

The following materials should be provided for class use.

1. Cat dissection specimens
2. Dissecting tools
3. Model of the human torso
4. Chart of the human digestive system

INTRODUCTION

DIGESTION

Most foods we consume cannot be utilized directly and must be digested or chemically broken down into smaller components that can be absorbed into the blood and cells. This process is the major function of the *digestive system*. The chemical processes involved in the splitting of complex chemicals into simpler ones are *hydrolytic activity* (splitting in the presence of water molecules) and *enzymatic activity*. Hydrolysis is the basic mechanism for carbohydrate, protein, and fat digestion (the three major components of human diets). Specific enzymes are involved for each food substance. Digestion includes all the processes by which food is converted into end products that may be absorbed from the lumen of the intestine into the blood. Digestion begins in the mouth and continues in the stomach and small intestine. Elimination is the main function of the large intestine.

Many cell types contribute to digestion. Some produce mucus, while others secrete various chemicals including enzymes.

THE MAMMALIAN DIGESTIVE TRACT

Digestion and the subsequent absorption and elimination of nonabsorbed foodstuffs involve a series of organs that form a long digestive (alimentary or gastrointestinal) tract. This is the series of organs through which the food passes. Each organ in it is specifically constructed and adapted for different stages in the digestive process.

THE DIGESTIVE SYSTEM

The digestive system includes the digestive tract and associated organs such as the salivary glands, liver, pancreas, and gallbladder. The salivary glands, liver, and pancreas are glandular and produce secretions. The gallbladder is a storage sac for bile.

ORGANS OF THE DIGESTIVE SYSTEM OF THE CAT

Trace the esophagus of the cat from the thoracic cavity where it can be seen lying dorsal to the trachea. Follow it to and through the diaphragm. Below the diaphragm the digestive tract enlarges to form the stomach.

Study Figures 51-1 and 51-2 and identify the different parts of the *stomach.* The pouch that rises above the entrance of the *esophagus* is the *fundus.* The central area is called the body of the stomach. To the right of the body is the *pyloric region,* which ends at the *pylorus* (a sphincter muscle). The left surface, called the *greater curvature,* is convex and is the embryonic dorsal surface. The right surface, the *lesser curvature,* is concave and is the embryonic ventral surface. The stomach is actually lying on its right side. Note at this time the *gastrosplenic ligament* (a mesentery connecting the stomach and the spleen). The greater omentum is the fold of peritoneum attached to the stomach and spleen and extending over most of the organs in the abdominal cavity.

Make a longitudinal slit in the stomach to observe the inner wall of the organ. If it contains food, remove it with a spatula or a spoon. Observe that the inner stomach wall is thrown into prominent longitudinal folds. The folds *(rugae)* are characteristic of the stomach lining and will be seen in all stomachs except those extremely distended with food. The pyloric region of the stomach opens into the first part of the small intestine, the *duodenum* at the pyloric sphincter. The duodenum courses posteriorly for about 3 inches (the *descending portion of the duodenum),* then doubles back for another 2 inches (the *ascending portion of the duodenum),* at which point it terminates in a sharp backward bend and becomes known as the *jejunum.* The rest of the small intestine consists of jejunum and *ileum.* It is not possible to distinguish between them on gross examination, but they differ microscopically. The small intestine terminates at the end of the ileum in the ileo-cecal sphincter.

Cut longitudinally into the small intestine and feel the inner lining, which is velvety to the touch. The lining of the mammalian small intestine is thrown into millions of tiny folds that can be felt but not seen. The folds, called *villi,* serve to increase the surface area of the small intestine, which facilitates the absorption of digested foods.

The *mesentery proper* or *mesointestine,* which supports the small intestine, is attached to the dorsal body wall over a small area in the region of the duodenum, but it is extensive within the intestinal area connecting parts of the jejunum and ileum. Within it is a large lymph node called the *pancreas of Aselli.* Locate it.

The *large intestine,* or *colon,* begins at the ileocecal valve and is considerably larger in diameter than the small intestine. A blind sac, the *cecum,* marks the beginning of the colon. It is short and vestigial and does not give rise to an *appendix* as it does in the human. The *ascending colon* and *transverse colon* are short. Trace them to the *descending colon,* which merges imperceptibly with the *rectum,* which ends at the *anus.* The mesentery attaching the colon to the dorsal body wall is the *mesocolon.*

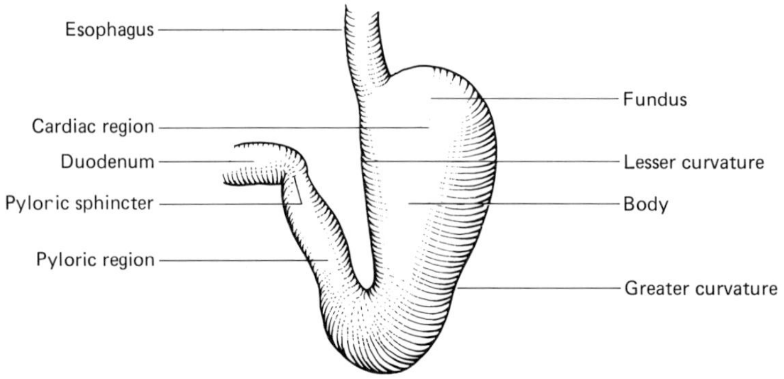

Figure 51-1 Stomach of the cat.

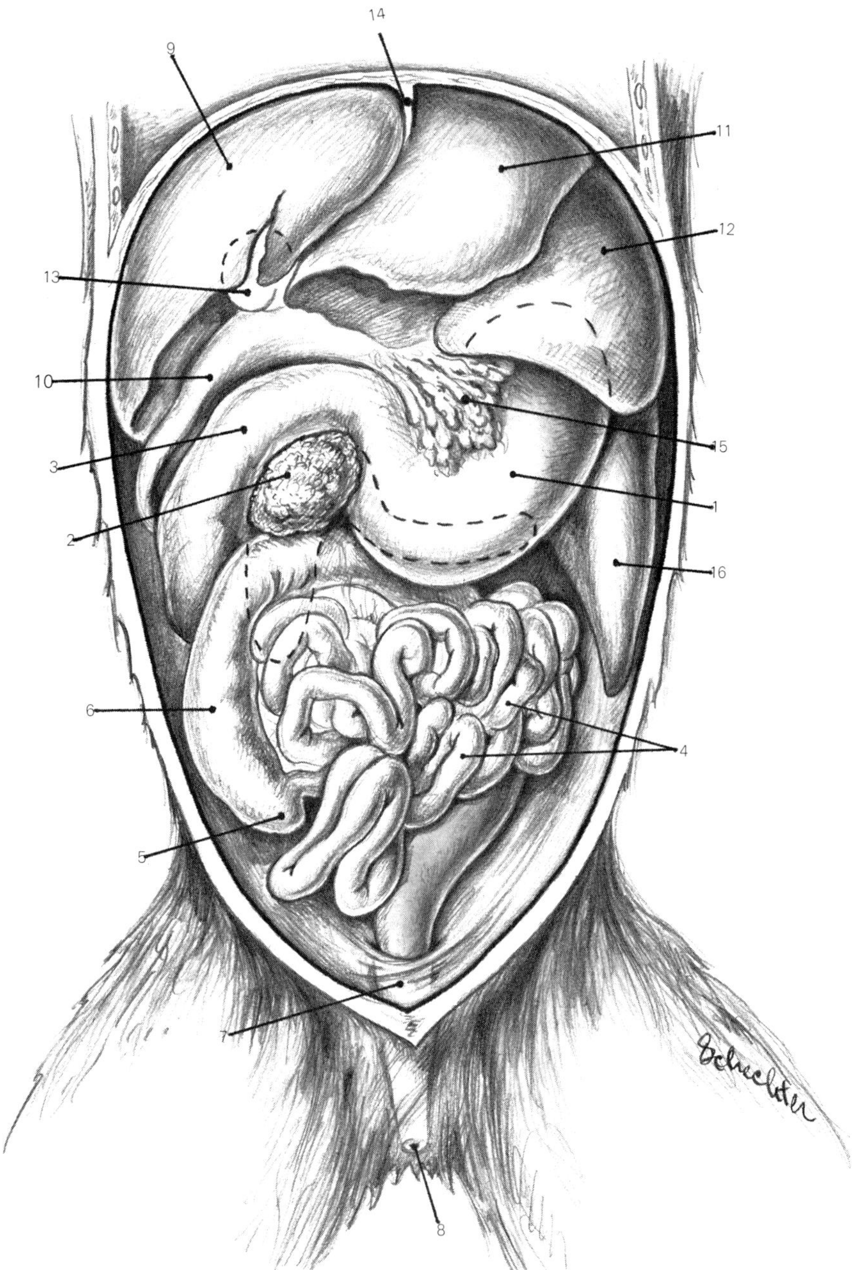

Figure 51-2 Digestive system of the cat.

1. ______________________ 9. ______________________

2. ______________________ 10. ______________________

3. ______________________ 11. ______________________

4. ______________________ 12. ______________________

5. ______________________ 13. ______________________

6. ______________________ 14. ______________________

7. ______________________ 15. ______________________

8. ______________________ 16. ______________________

ASSOCIATED DIGESTIVE ORGANS OF THE CAT

The *salivary glands* (parotid, submaxillary, and sublingual) have already been observed during the dissection of the muscles of the neck and jaw (see Figure 19-2a, b). Review their anatomy and their ducts. Saliva acts on food as it passes from the *mouth* to the *pharynx, esophagus,* and *stomach.*

The *pancreas* in the cat consists of two regions. Lying beside the *descending duodenum* is the *head of the pancreas.* Continuous with it, passing to the left side of the body, dorsal to the stomach, is the *tail of the pancreas.* The two *pancreatic ducts* are the *duct of Wirsung,* within the head, and the *duct of Santorini,* within the tail. The duct of Wirsung joins the *common bile duct* from the *liver* and opens into the duodenum near the *pylorus* at an internal swelling called the *ampulla of Vater.* Expose the pancreatic duct by removing parts of the pancreatic tissue from the dorsal surface gently with a dissecting needle at the place where the common bile duct joins the small intestine on the dorsal surface. Trace this pancreatic duct posteriad in the pancreas tissue. An accessory pancreatic duct, the duct of Santorini, usually joins the duct of Wirsung before entering the duodenum (Figure 51-3).

The *liver* is divided into five lobes (described in Exercise 37). Locate again the *right medial lobe* (the gallbladder lies within it), the *right lateral lobe,* and the *right caudal lobe.* On the left side of the cat are the *left medial lobe* and the *left lateral lobe.* The mesentery lying between the liver and the duodenum is the *lesser omentum.* In it is the long greenish

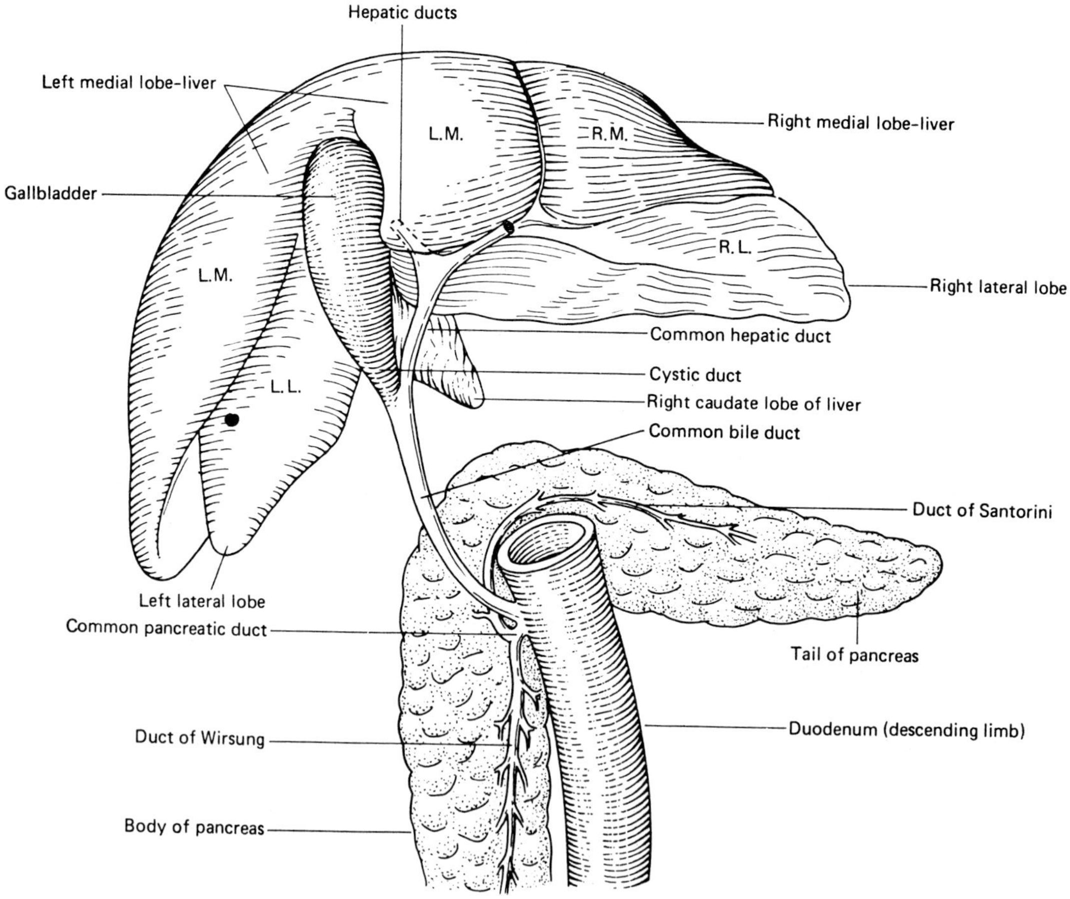

Figure 51-3 Digestive glands and their ducts. Liver raised to reveal gallbladder and ducts.

common bile duct along with several blood vessels. The *common bile* duct is formed by the confluence of *hepatic ducts* from each liver lobe and the *cystic duct* from the gallbladder. Find as many of these ducts as possible.

Bile from liver and gallbladder and the pancreatic juice from the pancreas flow together and enter the duodenum at the ampulla of Vater.

Be able to demonstrate all organs, mesenteries, and ducts mentioned in this exercise.

THE GREATER OMENTUM AND OTHER MESENTERIES OF THE CAT

The *greater omentum,* in which large quantities of fat are deposited, was noted in Exercise 37 and has been reflected back over the abdominal cavity to the left side of the cat. It is actually a large membranous sac containing a cavity called the *lesser peritoneal cavity.* (The *abdominal cavity* is known as the *greater peritoneal cavity.*) It has become greatly displaced due to many changes in position of the stomach during its embryonic development. The opening into this sac, the *epiploic foramen,* or *foramen of Winslow,* can be located by inserting a probe into an oval entrance located dorsal to the stomach and medial to the right lateral lobe of the liver. Place the probe or your little finger into this opening and it will be completely enclosed by the greater omentum. Demonstrate this to your instructor.

All mesenteries discussed thus far have been dorsal mesenteries. Most ventral mesenteries disappear during embryonic development, but remnants of two of them remain. Observe the small ventral mesentery between the liver below the diaphragm and the midventral body wall. This remnant of one of the many embryonic ventral mesenteries is called the *falciform ligament.* Another ventral mesentery connects the *urinary bladder* and the ventral body wall. Locate each of these mesenteries.

THE HUMAN DIGESTIVE SYSTEM

The human digestive system is much like that of the cat, but there are differences.

1. There are three pairs of salivary glands in the human but five pairs in the cat.
2. There are four human liver lobes but five in the cat.
3. The human pancreas is smaller and restricted to the region between the stomach and the duodenum. It is about 6 inches long and 1 inch wide. The cat has a bilobed pancreas.
4. The human ducts from liver and gallbladder have the same relationship to the duodenum as they do in the cat.
5. The single pancreas has but one duct in most humans, the duct of Wirsung, which opens into the common bile duct. An accessory pancreatic duct (duct of Santorini) occurs in a small percentage of people and opens separately into the duodenum. Cats have two ducts that drain two pancreatic masses.
6. The jejunum and the ileum together are approximately 20 feet long in the human. This area is proportionately shorter in the cat, reflecting its carnivorous diet.
7. The appendix arises from the cecum in the human but plays no part in digestion. It is absent in the cat.
8. The human ascending, transverse, and descending colons are defined more sharply than in the cat.
9. The sigmoid colon of the human is absent in the cat.
10. The longitudinal muscular layer of the human colon is arranged into three bands *(taeniae coli)* which, when contracted, gather the colon into a series of pouches called *haustra.* These are absent in the cat.

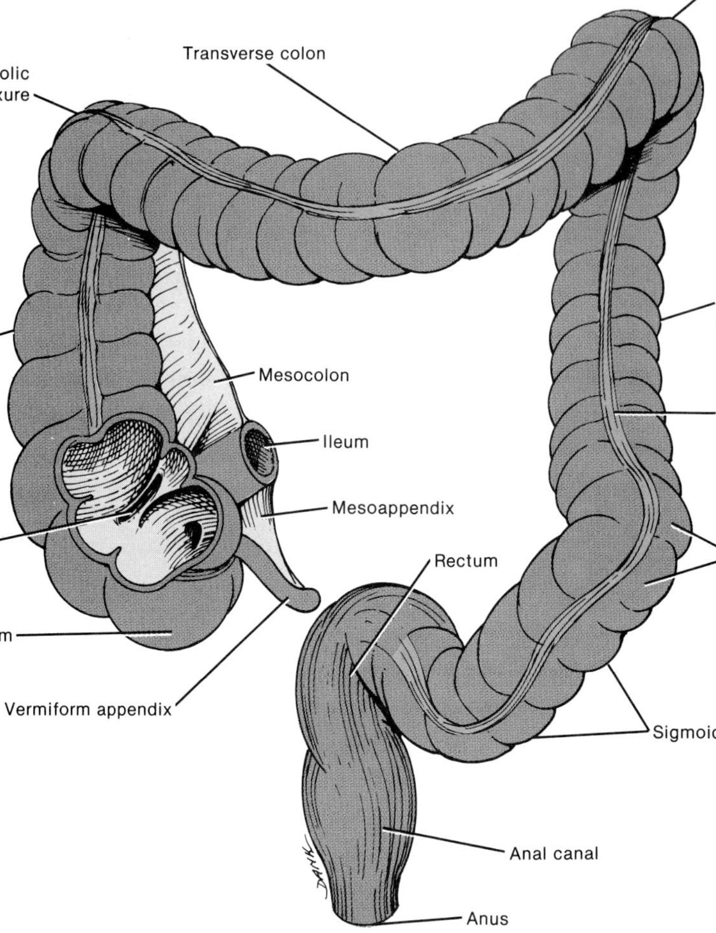

Figure 51-4 Colon of the human.

Study the drawing (Figure 51-4) of the human colon and become familiar with all its parts. Compare its length and shape with that of the cat. Label (Figure 51-5) and compare all parts in the human digestive system to those of the cat as to relative size, shape, and number of parts. This will serve as a review of both animal systems.

List all organs of the large intestine:

1. ______________________

2. ______________________

3. ______________________

4. ______________________

5. ______________________

6. ______________________

7. ______________________

8. ______________________

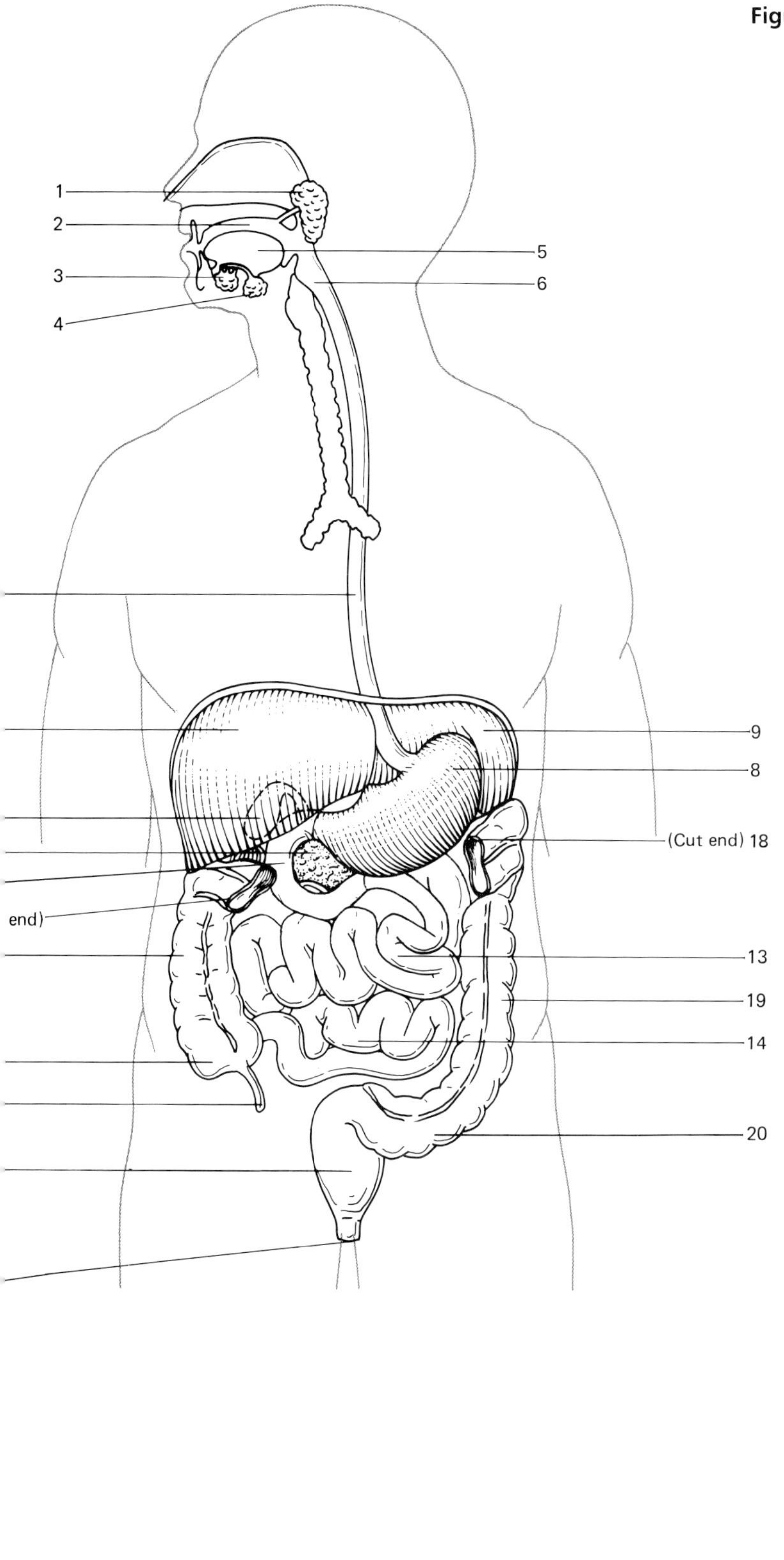

Figure 51-5 Digestive system of the human.

1. ______
2. ______
3. ______
4. ______
5. ______
6. ______
7. ______
8. ______
9. ______
10. ______
11. ______
12. ______
13. ______
14. ______
15. ______
16. ______
17. ______
18. ______
19. ______
20. ______
21. ______
22. ______

Questions for Review

EXERCISE **51**

THE ORGANS OF THE DIGESTIVE SYSTEM

Name ____

Laboratory Section ______________________________ Date ____________

1. Match the definitions on the left with the parts of the digestive system.

a. ____________ A salivary gland

b. ____________ Organ that stores bile

c. ____________ Duct from liver that opens into duodenum

d. ____________ Duct leading from gallbladder

e. ____________ Ventral mesentery supporting liver

f. ____________ Mesentery supporting large intestine

g. ____________ Tube that transports food to stomach

h. ____________ Part of digestive tract that opens into large intestine

i. ____________ Pouches within large intestine

j. ____________ Part of digestive tract through which food passes from mouth

k. ____________ Lymph gland located in mesentery of small intestine

l. ____________ Middle part of small intestine

m. ____________ Vestigial structure attached to cecum

n. ____________ Four-lobed organ ventral to stomach

o. ____________ Striated lining of stomach

1. Pancreatic duct
2. Falciform ligament
3. Esophagus
4. Haustra
5. Parotid
6. Rugae
7. Ileum
8. Gallbladder
9. Pancreas
10. Cystic duct
11. Appendix
12. Hepatic duct
13. Gastroplenic ligament
14. Liver
15. Pancreas of Aselli
16. Jejunum
17. Pharynx
18. Taeniae coli
19. Mesocolon
20. Common bile duct

EXERCISE 52

METABOLISM: ENZYME ACTION

The carbohydrates, fats, and proteins ingested as food must undergo specific changes before they can be absorbed from the digestive tract. Other substances such as water, vitamins, and minerals are ready for immediate absorption. Complex foods are prepared for absorption by *mechanical* and *chemical* activities. Mechanical activities include chewing, swallowing, and several different types of peristaltic contractions that break down large particles, mix the food with mucus and enzymes, and propel it through the digestive tract. Chemical activities break down the large molecules we eat into molecules that are small enough so that they can be absorbed through the walls of the digestive tract, transported through the body, and eventually used by body cells.

This exercise is concerned with chemical actions associated with enzymes. Several different enzymes and the factors that affect their activities will be considered. The procedures include specific tests used in the detection and identification of carbohydrates, lipids, and proteins.

This exercise also contains a demonstration of the Gastro-Test, a diagnostic tool designed to provide immediate determination of stomach pH without gastric incubation (insertion of a tube) or chemical analysis. This device (Figure 52-1) can be used to detect esophageal and gastric bleeding and various conditions with the absence of free hydrochloric acid in the stomach (achlorhydria). Achlorhydria-associated conditions include gastric cancer, thyroid

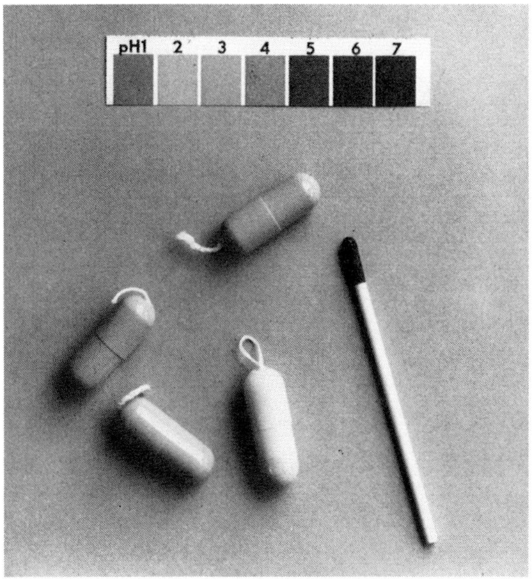

Figure 52-1 The components of the diagnostic tool Gastro-Test. The items shown include the weighted gelatin capsule containing a highly absorbant cotton floss (the floss loop can be seen protruding from the capsule), surface-marking pH stick, and a pH color chart.

goiter, pernicious anemia, and iron deficiency anemia. The Gastro-Test device consists of a weighted gelatin capsule with 70 cm of a highly absorbant cotton floss coiled within. One end of the floss protrudes through a hole in the cap while the other is loosely attached to the capsule.

A. HISTOLOGY OF THE ORGANS AND STRUCTURES ASSOCIATED WITH DIGESTION

Materials

The following materials should be provided for class use.

1. Prepared slides of
   a. Salivary glands
   b. Stomach, showing mucosa and chief parietal and mucous cells
   c. Small intestine, showing epithelium and goblet cells
   d. Large intestine, showing mucosa
   e. Pancreas
2. Anatomical charts of digestive system components

Procedure

This procedure should be performed by students individually.

1. Examine the prepared histological slides.
2. Sketch and label a representative field showing the characteristic cells of each organ or related structure in the results and observations section. Use Color Plates 22 and 41–43 as guides.
3. Answer the questions in the results and observations section.

B. THE GASTRO-TEST

Materials

The following materials should be provided for an instructor's demonstration.

1. Several Gastro-Test gelatin capsules (HDE Corporation, Mountain View, Calif.)
2. One surface marking pH stick
3. One pH color chart
4. A sheet of plain white paper
5. One glass of cold drinking water
6. Kim wipes
7. Table or flat horizontal surface

This procedure will be demonstrated by the instructor.

1. Have a student volunteer sit comfortably on a chair.
2. Obtain a glass of cold water. Have the student take a swallow of the water to moisten his or her mouth and throat.
3. Insert the Gastro-Test capsule into the mouth while holding the free end of the cotton floss outside the mouth.
4. Have the student take more water to swallow the capsule.
5. Instruct the student to lie on a laboratory table or flat surface for 10 minutes. *Still hold on to the floss.*

6. Have the student now sit up, open the mouth, and raise his or her chin.
7. Gently pull the floss from the mouth and place it on the sheet of white paper provided.
8. Touch the pH stick to the floss at one end and move it across to the opposite end.
9. Note the resulting colors for the appropriate regions corresponding to the mouth, esophagus, and stomach and compare them to the pH chart. The mouth and esophagus are normally pH 7, while the stomach may vary from pH 1 to slightly over 7. The finding on any part of the distal half of the floss of a pH 3 or lower is evidence that the stomach is able to secrete acid normally.
10. Enter your findings and answer the questions in the results and observations section.

C. DEMONSTRATION OF EXTRACELLULAR DIGESTION

Microorganisms are used here to show the extracellular nature of digestive processes.

Materials

The following materials should be provided for class demonstration.

1. One uninoculated starch agar plate
2. One starch agar plate centrally inoculated with the bacterium *Bacillus subtilis,* but not incubated
3. One starch agar plate inoculated with *B. subtilis* and incubated for 24 hours
4. Lugol's iodine solution in a container with dropper
5. Container with disinfectant for disposal of cultures

Procedure

This procedure is to be demonstrated by the instructor.

1. Examine the starch agar plates. Sketch the appearance and indicate the color of each on one side of the appropriate circle in the results and observations section.
2. Your instructor will flood the plates with an iodine solution.
3. After 15 seconds examine the plates. Sketch their appearance and indicate the colors formed on the other side of the appropriate circles in the results and observations section.
4. Answer the questions in the results and observations section.

D. SALIVARY AMYLASE ACTIVITY

Starch is one of the major carbohydrates in the human diet. To use starch, the body must first break down the large starch molecule into smaller glucose molecules. The process by which starch is hydrolyzed into smaller units begins in the mouth by the action of an enzyme *amylase.* In the next set of experiments the progress of this enzymatically catalyzed reaction and the various factors that can effect it will be followed by observing the disappearance of the reacting substrate (starch) using the iodine test.

Materials

The following materials should be provided for class use.

1. Three stock starch solutions, 5 percent, 1 percent, and 0.5 percent
2. Three buffer solutions, with pHs of 4, 7, and 9
3. Lugol's iodine or other similar iodine reagent in a dropper bottle
4. Pieces of paraffin or sugarless gum

5. Pyrex test tubes
6. Test tube racks
7. 12-ml Erlenmeyer flasks
8. Porcelain spot plates
9. Gauze
10. Glass or plastic funnels
11. 50-ml graduated cylinders
12. Water baths, 4°C, 37°C, and 70°C
13. Distilled water

Procedure 1: Salivary Amylase Stock Preparation

This procedure is to be performed by students in groups of four.

1. Two students should produce and collect 20 ml of saliva each. The sample should be measured and pooled in a clean Erlenmeyer flask. (Note that saliva excretion can be stimulated by drinking a glass of water and then chewing on a small piece of paraffin or sugarless gum.)
2. Add an equal amount of room temperature distilled water and mix the contents of the flask. Pass the mixture through a gauze filter system to remove most of the mucus. The resulting preparation is your amylase stock solution for the remaining procedures in this exercise. (If additional enzyme material is needed, repeat this procedure.)

Procedure 2: Effects of Temperature on Amylase Activity

This procedure is to be performed by students in pairs.

1. Label four test tubes numbers 1–4 and place them in a rack.
2. Add 5 ml of amylase stock solution to tubes 1, 2, and 3.
3. Place the tubes in the temperature environment specified below. Keep the tubes at these temperatures for at least 10 minutes (to allow all the solution to reach the desired temperature).

   Tube 1: 4°C
   Tube 2: 37°C
   Tube 3: 70°C
   Tube 4 (control): room temperature (25°C). This is the control tube for the starch–iodine reaction.

4. Add 5 ml of the 1 percent starch solution to each tube, shake gently, and return the mixtures to their water baths to incubate for 5 minutes.
5. Remove a small sample from each tube, place each into a separate depression on a spot plate, and test immediately for starch hydrolysis by adding 2 drops of the iodine reagent.
6. Enter your findings in the results and observations section.
7. Wash and rinse the tubes used.

Procedure 3: Effects of pH on Amylase Activity

This procedure is to be performed by students in pairs.

1. Number four test tubes and place them in a rack.
2. Add 3 ml of the amylase stock solution to each tube.
3. Add the following substances to the tubes as indicated:

   Tube 1: 3 ml buffer solution, pH 4
   Tube 2: 3 ml buffer solution, pH 7

Tube 3: 3 ml buffer solution, pH 9
Tube 4: 3 ml distilled water, pH 7

4. Shake and mix the contents and place them in a 37°C water bath for 10 minutes.
5. Add 3 ml of 1 percent starch solution to each tube, mix, and continue to incubate at 37°C.
6. Almost immediately test the contents of each tube for starch hydrolysis as described in procedure 2, step 5.
7. Continue to incubate the preparations at 37°C and take additional samples at 15 and 30 minutes. Test for starch hydrolysis as in step 6.
8. Enter your findings in the results and observations section.
9. Wash and rinse the tubes used.

Procedure 4: Effect of Substrate Concentration on Amylase Activity

This procedure is to be performed by students in pairs.

1. Number 12 test tubes and place them in a rack in two rows of six tubes each.
2. Add 1 ml of the amylase stock solution to each tube.
3. Add the following substances to the tubes as specified:

Row I		Row II	
Tube 1:	1 ml 1 percent starch solution	Tube 1:	1 ml 5 percent starch solution
Tube 2:	2 ml 1 percent starch solution	Tube 2:	2 ml 5 percent starch solution
Tube 3:	3 ml 1 percent starch solution	Tube 3:	3 ml 5 percent starch solution
Tube 4:	5 ml 1 percent starch solution	Tube 4:	5 ml 5 percent starch solution
Tube 5:	10 ml 1 percent starch solution	Tube 5:	10 ml 5 percent starch solution
Tube 6:	5 ml distilled water	Tube 6:	5 ml distilled water

4. Shake the tubes gently to mix the contents and place them in a 37°C water bath for 10 minutes.
5. Take samples and test for starch hydrolysis at 1 minute, 15 minutes, 30 minutes, and 60 minutes as described in procedure 2, step 5.
6. Enter your findings in the results and observations section.
7. Wash and rinse the tubes used.

DIGESTION OF FATS

Materials

The following materials should be provided for class use.

1. The following substances in appropriate containers
   a. Safflower oil
   b. Distilled water
   c. Ethyl alcohol
   d. Xylene
   e. Bile salts (5 percent solution)
2. 5-ml pipettes with bulbs
3. Test tubes
4. Test tube racks

Procedure

This procedure is to be performed by students in pairs.

1. Number five test tubes and place them in a rack.
2. Add 2 ml of safflower oil to each tube.
3. Add the following substances to the tubes as specified:

   Tube 1: 5 ml distilled water
   Tube 2: 5 ml ethyl alcohol
   Tube 3: 3 ml xylene
   Tube 4: 3 ml bile salts
   Tube 5: no additions

4. Shake the tubes gently to mix the contents and examine all tubes immediately. Look for a complete mixing of the contents of the tubes.
5. Repeat your examinations at 15 and 30 minutes.
6. Enter your findings in the results and observations section.

F. DIGESTION OF PROTEIN

Materials

The following materials should be provided for class use.

1. Small pieces of coagulated egg albumin
2. The following solutions in appropriate containers
   a. 1 percent trypsin
   b. 1 percent pepsin
   c. 0.5 percent sodium carbonate
   d. 0.8 percent hydrochloric acid
3. Distilled water
4. Biuret test reagents
5. Test tubes
6. Test tube racks
7. 5-ml pipettes with bulbs
8. 37°C water bath

Procedure: The Actions of Pepsin and Trypsin on Egg Albumin

This procedure is to be performed by students in pairs.

1. Prepare a test tube rack containing two rows of six tubes each. Number the tubes accordingly.
2. Add the following substances to the tubes as specified:

Row I		Row II	
Tube 1:	5 ml pepsin + 5 ml distilled water	Tube 1:	5 ml trypsin + 5 ml distilled water
Tube 2:	5 ml pepsin + 5 ml sodium carbonate	Tube 2:	5 ml trypsin + 5 ml sodium carbonate
Tube 3:	5 ml pepsin + 5 ml hydrochloric acid	Tube 3:	5 ml trypsin + 5 ml hydrochloric acid

Tube 4: 5 ml distilled water + 5 ml sodium carbonate
Tube 5: 5 ml distilled water + 5 ml hydrochloric acid
Tube 6: 5 ml distilled water

Tube 4: 5 ml distilled water + 5 ml sodium carbonate
Tube 5: 5 ml distilled water + 5 ml hydrochloric acid
Tube 6: 5 ml distilled water

3. Add several pieces of coagulated egg albumin to each tube, shake gently to mix, and place all tubes in a 37°C water bath.
4. Examine each tube at 15-minute intervals during a 90-minute period for evidence of the digestion of the egg albumin. This type of reaction is indicated by a disintegration of the protein and a general clearing of the protein material.
5. At the end of 90 minutes add 4 drops of the Biuret solution to each tube to test for the presence of the end products of protein digestion. The formation of a pink to violet color indicates the presence of polypeptides.
6. Record your findings in the results and observations section.
7. Wash and rinse the tubes used.

RESULTS AND OBSERVATIONS

HISTOLOGY OF THE ORGANS AND STRUCTURES ASSOCIATED WITH DIGESTION

1. Sketch and label representative portions of the tissues indicated.

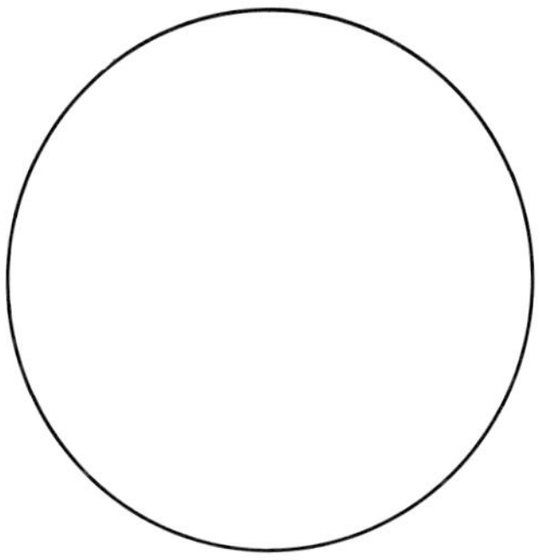

Salivary glands

Stomach

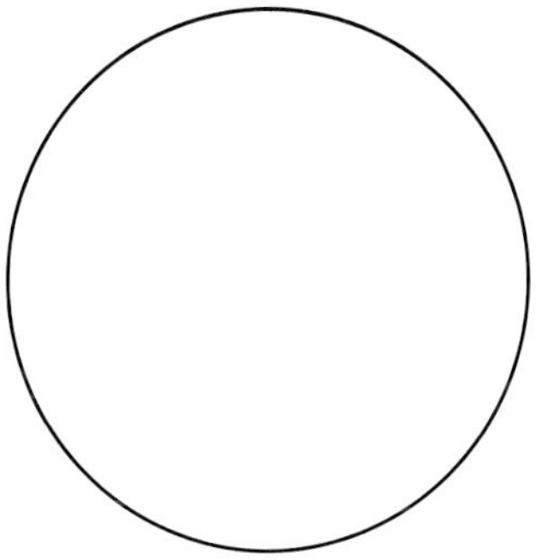

Small intestine

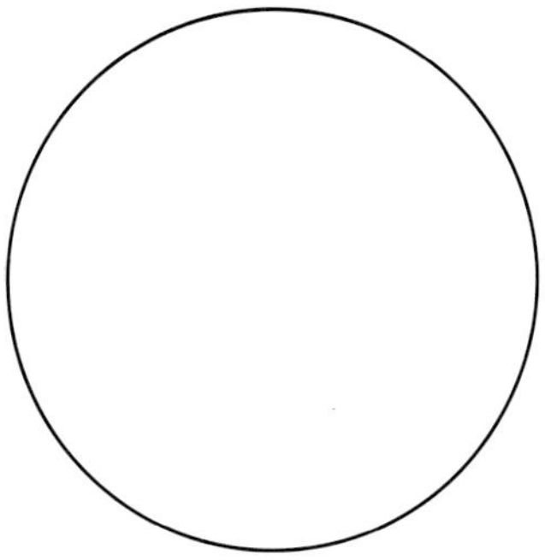

Large intestine

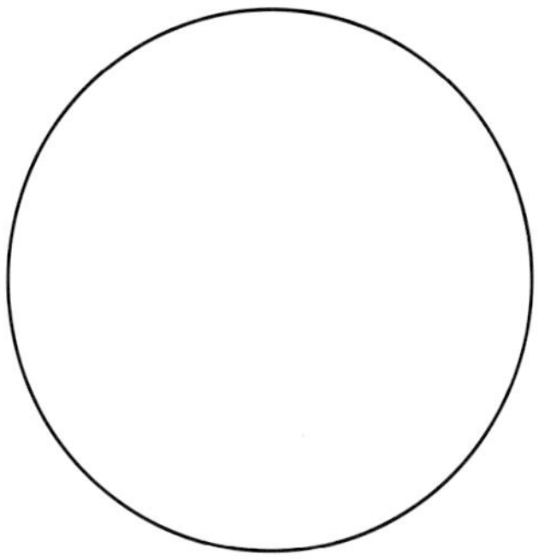

Pancreas

2. Indicate the specific product(s) of the structures or cells listed.

a. Salivary glands ____________

b. Chief cells ____________

c. Parietal cells ____________

d. Mucous cells ____________

e. Brunner's glands ____________

f. Goblet cells (small intestine) ____________

g. Pancreas ____________________________

____

THE GASTRO-TEST

1. Enter your findings in Table 52-1.

Table 52-1

Region	Color Observed	pH Reading
Mouth		
Esophagus		
Stomach		

2. Was the Gastro-Test capsule easily administered?

____

3. What types of conditions can this device be used to detect?

____

____

DEMONSTRATION OF EXTRACELLULAR DIGESTION

1. Sketch the appearance and indicate the color of the starch agar plates before and after the application of the iodine reagent.

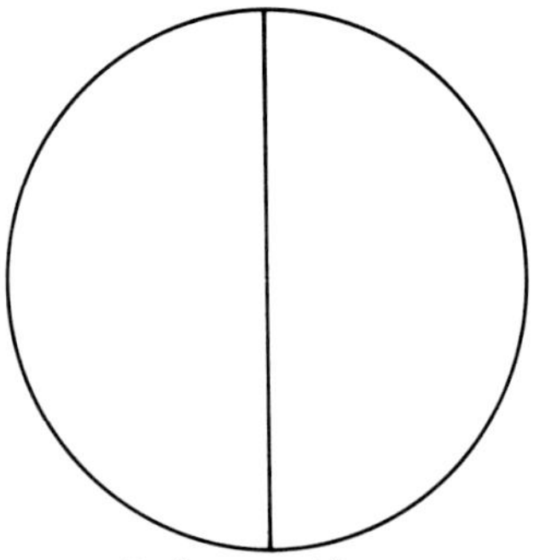

Before After

Uninoculated plate

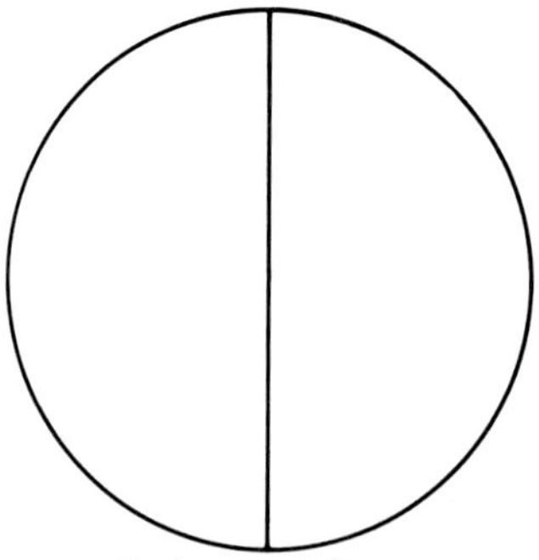

Before After

Inoculated, but not incubated plate

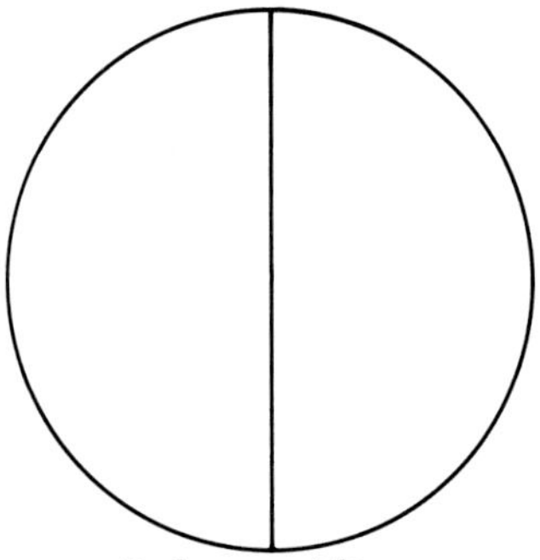

Before After

Inoculated and incubated plate

2. In which of the plates was starch hydrolysis observed? ______________

3. Why must an inoculated plate be incubated to produce a positive reaction? ________

____

4. What color develops after the addition of iodine in the cases of:

   a. Starch hydrolysis ______________ b. Presence of starch ______________

SALIVARY AMYLASE ACTIVITY

1. Enter in Table 52-2 the reaction to iodine at different temperatures and indicate whether or not starch hydrolysis occurred. Indicate a positive reaction with a "+" and a negative reaction with a "−".

Table 52-2

Tube	Temperature	Color After Addition of Iodine	Starch Hydrolysis
1	4°C		
2	37°C		
3	70°C		
4	25°C		

2. Enter in Table 52-3 the results of the tests for starch hydrolysis on substances at different pHs. Indicate a positive reaction with a "+" and a negative reaction with a "−".

Table 52-3

Tube	Substance	Immediate	15 min	30 min
1	Buffer, pH 4			
2	Buffer, pH 7			
3	Buffer, pH 9			
4	Distilled water, pH 7			

3. Enter in Table 52-4 the results of the tests for starch hydrolysis on substances of different substrate concentration. Indicate a positive reaction with a "+" and a negative reaction with a "–".

Table 52-4

		Concentration					
Row	Tube	Amount (ml)	Solution	1 min	15 min	30 min	60 min
I	1	1	1% starch				
I	2	2	1% starch				
I	3	3	1% starch				
I	4	5	1% starch				
I	5	10	1% starch				
I	6	5	distilled water				
II	1	1	5% starch				
II	2	2	5% starch				
II	3	3	5% starch				
II	4	5	5% starch				
II	5	10	5% starch				
II	6	5	distilled water				

4. Summarize your conclusions regarding the influence of temperature, pH, and substrate concentration on enzyme activity. ______________________________

____

____

____

5. What would have happened in the tests of amylase activity if alcohol had been used as the diluting fluid instead of distilled water?

____

____

DIGESTION OF FATS

1. Enter in Table 52-5 your observations of the degree to which the contents of the tubes mixed.

Table 52-5

Tube	Substance Mixed with Oil	Immediate	15 min	30 min
1	Distilled water			
2	Ethyl alcohol			
3	Xylene			
4	Bile salts			
5	None			

2. If there were any observable differences in any of the tubes at 15 or 30 minutes, explain why. ____

____

3. Where is bile produced in the body? ______________

4. What is the function of bile? ____________________________________

____

DIGESTION OF PROTEIN

1. Enter your observations and test results for protein digestion in Table 52-6. Indicate evidence of protein and a positive test reaction with a "+" and no evidence and a negative test reaction with a "−".

Table 52-6

			15 min		30 min		45 min		60 min		75 min		90 min	
Row	Tube	Solution	P	T	P	T	P	T	P	T	P	T	P	T
I	1	Distilled water												
I	2	Sodium carbonate												
I	3	HCl												
I	4	Sodium carbonate												
I	5	HCl												
I	6	______												
II	1	Distilled water												
II	2	Sodium carbonate												
II	3	HCl												
II	4	Sodium carbonate												
II	5	HCl												
II	6	______												

2. Which of the preparations produced a positive Biuret reaction?

____

3. What is digestion? ____

____

Questions for Review

EXERCISE **52**

METABOLISM: ENZYME ACTION

Name ____

Laboratory Section ______________________ Date ____________

1. Are the pH optimums the same for most enzymes? Explain. ______________________

____

____

2. What is the "lock-and-key" model of enzyme action? Use labeled diagrams to explain your answer.

3. List three major factors that affect enzyme action. Briefly explain how each affects enzymatic activity.

   a. ____

   ____

   b. ____

   ____

   c. ____

   ____

4. How are enzymes named? ____

____

____

5. Define or explain the following.

   a. Enzyme ______________________________

   b. Substrate ______________________________

   c. Optimum ______________________________

   d. Suffix ______________________________

   e. Active site ______________________________

   f. Achlorhydria ______________________________

EXERCISE 53

DIGESTION CHEMISTRY

Several tests and procedures, which are used in the identification of carbohydrates, lipids, proteins, and other substances of dietary importance, are demonstrated in this exercise. Be sure to handle all chemicals carefully. Observe safety rules.

A. CARBOHYDRATE IDENTIFICATION

Materials

The following materials should be provided for class use.

1. 1 percent known solutions in dropper bottles
   a. Glucose
   b. Lactose
   c. Maltose
   d. Albumin
   e. Starch
   f. Gelatin
2. Test materials in dropper bottles
   a. Honey
   b. Sugar-free drink
   c. Cola drink
   d. Distilled water
   e. Syrup
   f. Milk
   g. Saccharin
   h. Orange juice
3. Test reagents
   a. Benedict's reagent
   b. Lugol's iodine
   c. Molisch test reagents—5 percent alpha-naphthol alcoholic solution and concentrated sulfuric acid
4. Pyrex test tubes, test tube racks, and test tube holders
5. 1-ml and 10-ml pipettes with bulbs
6. Boiling water bath
7. Wax pencils or other marking devices

Procedure 1: Molisch Reaction

The Molisch reaction is a general test for carbohydrates. The formation of a purple or pink color upon the addition of the Molisch reagents indicates the presence of a carbohydrate.

This procedure is to be performed by students in pairs.

1. Label a test tube for each of the 1 percent known solutions. Place the tubes in a rack.
2. Introduce 10 drops of each known solution into its labeled tube. Use a separate pipette for each. Examine each preparation for any distinctive color, smell, or other noticeable property. Enter your findings in the results and observations section.
3. Add 2 drops of the 5 percent alpha-naphthol solution to each tube. Mix the contents by gently tapping the tube.
4. Tilt each tube and carefully add 5 ml of the concentrated sulfuric acid (H_2SO_4) so that it runs slowly down the inside surface. The acid is heavier and should form a layer beneath the test solutions.
5. Examine each tube for the formation of a pink or purple color at the boundary between the test solution and sulfuric acid. Record your findings in the results and observations section.
6. Dispose of the chemical solutions as directed by your instructor.
7. Repeat steps 1–6 with the test materials.

Procedure 2: Reducing Sugars Test

Benedict's reagent is used to detect the presence of a reducing sugar. When the cupric ion (Cu^{2+}) in the reagent reacts with a reducing sugar, a red precipitate of cuprous oxide forms.

This procedure is to be performed by students in pairs.

1. Label a test tube for each of the 1 percent known solutions. Place the tubes in a rack.
2. Introduce 10 drops of each known solution into its labeled tube. Use a separate pipette for each.
3. Using a 10-ml pipette, add 5 ml of Benedict's reagent to each test tube. Mix the contents by gently tapping the tubes.
4. Place all the tubes in a boiling water bath for 5 minutes.
5. Examine each tube for evidence of the presence of a reducing sugar (red precipitate) (see Color Plate 42). Enter your findings in the results and observations section.
6. Dispose of the chemical solutions as directed by your instructor.
7. Repeat steps 1–6 with the test materials.

Procedure 3: Starch Test

The presence of starch generally is indicated by the formation of a blue-black color upon the addition of an iodine reagent.

This procedure is to be performed by students in pairs.

1. Label a test tube for each of the 1 percent known solutions. Place the tubes in a rack.
2. Introduce 10 drops of each known solution into its labeled tube. Use a separate pipette for each.
3. Add 3 drops of the iodine reagent to each tube and mix.
4. Examine each tube for the formation of a blue-black color. Enter your findings in the results and observations section.
5. Dispose of the chemical solutions as directed by your instructor.
6. Repeat steps 1–5 with the test materials.

B. PROTEIN IDENTIFICATION

Materials

The following materials should be provided for class use.

1. 1 percent known solutions
   a. Albumin
   b. Cysteine
   c. Glucose
   d. Glycine
   e. Phenylalanine
   f. Starch
   g. Tryptophan
2. Test Materials
   a. Casein
   b. Hair
   c. Gelatin
3. Test reagents
   a. Biuret reagents—6 molar (M) sodium hydroxide (NaOH) and 0.5 percent copper sulfate ($CuSO_4$)
   b. 1 molar (M) lead acetate ($PbAc_2$) solution
   c. Concentrated nitric acid (HNO_3)
   d. Glacial acetic acid
   e. Litmus paper
4. 1-ml and 5-ml pipettes with bulbs
5. Pyrex test tubes, test tube racks, and test tube holders
6. Boiling water bath
7. Wax pencils or other marking devices

Procedure 1: Biuret Test

The Biuret is named for the purple compound that forms when a dilute alkaline solution of copper sulfate reacts with substances containing two or more peptide bonds.

This procedure is to be performed by students in pairs.

1. Label a test tube for each of the 1 percent known solutions and test materials. Place the tubes in a rack.
2. Introduce 1 ml of each solution into its labeled tube. Use a separate pipette for each.
3. Add 2 ml of the sodium hydroxide solution and 10 drops of the copper sulfate to each tube. Shake the tube gently to mix the contents.
4. Look for the presence of a pink to purple color (Color Plate 1). Enter your findings in the results and observations section.
5. Dispose of the chemical solutions as directed by your instructor.

Procedure 2: Sulfur Test

The sulfur test is used to detect the presence of amino acids in proteins that contain sulfur. When lead acetate is added to compounds containing sulfur, lead sulfide is formed. Lead sulfide is black.

This procedure is to be performed by students in pairs.

1. Label a test tube for each of the 1 percent known solutions. Put the tubes in a rack.
2. Introduce 1 ml of each known solution into its labeled tube. Use a separate pipette for each.
3. Add 2 ml of sodium hydroxide to each tube and place the tubes in a boiling water bath for 60 minutes.
4. Remove the tubes.

5. Neutralize the contents of each tube by carefully adding glacial acetic acid until blue litmus paper (pH indicator) turns red.
6. Add 10 drops of lead acetate.
7. Look for a black color. Enter your findings in the results and observations section.
8. Dispose of the chemical solutions as directed by your instructor.

Procedure 3: Xanthoproteic Test

This test detects the presence of benzene rings in amino acids. Such compounds produce yellow to red colors when exposed to nitric acid.

This procedure is to be performed by students in pairs.

1. Label a test tube for each of the 1 percent known solutions and the test materials. Place the tubes in a rack.
2. Introduce 1 ml of each known solution into its labeled tube. Use a separate pipette for each.
3. Carefully add 1 ml of concentrated nitric acid by letting the acid run down the side of each tube.
4. Place the tubes in a boiling water bath for 10 seconds.
5. Remove the tubes and observe the reactions. Enter your findings in the results and observations section.
6. Dispose of the chemical solutions as directed by your instructor.

Procedure 4: Denaturation by Heat

Certain proteins become denatured (permanently altered) and precipitate out of solution when exposed to boiling temperatures.

This procedure is to be performed by students in pairs.

1. Label a test tube for each of the 1 percent known solutions. Place the tubes in a rack.
2. Introduce 5 ml of each known solution into its labeled tube. Use a separate pipette for each.
3. Place the tubes in a boiling water bath for 3 minutes.
4. Examine each tube for the presence of a solid mass or precipitate. Enter your findings in the results and observations section.
5. Dispose of the chemical solutions as directed by your instructor.

C. PROPERTIES OF FATS

Materials

The following materials should be provided for class use.

1. Solvents
   a. Distilled water
   b. Ethyl alcohol
   c. 0.55 percent sodium chloride
   d. Xylene
2. Test solutions in dropper bottles
   a. 1 percent albumin
   b. Cottonseed or immersion oil
   c. Gelatin

   d. 2 percent glucose
   e. 2 percent starch
3. Test tubes and test tube racks
4. 5-ml pipettes with bulbs
5. Brown paper strips (7.5 × 5 cm)
6. Wooden applicator sticks
7. Wax pencils or other marking devices

Procedure 1: Solubility Testing

This procedure is to be performed by students in pairs. *Note:* Make certain that there are no open flames in the area where the following tests are being performed.

1. Label a test tube for each of the solvents. Place the tubes in a rack.
2. Introduce 5 ml of each solvent into its labeled tube. Use a separate pipette for each.
3. Add 6 drops of the test material albumin to each tube and shake.
4. Examine each tube immediately and after 15 minutes to see if the test material dissolved. Enter your findings in the results and observations section.
5. Dispose of the chemical solutions as directed by your instructor.
6. Repeat steps 1–5 with each of the other test materials.

Procedure 2: Grease Spot Test

This procedure is to be performed by students in pairs.

1. On a brown paper strip label a small area for each of the test materials.
2. Using a separate applicator stick for each, apply a small amount of each test material to its labeled area.
3. Examine each immediately and after 30 minutes for the presence of a grease spot. Enter your findings in the results and observations section.

D. IDENTIFICATION OF CONTENTS OF UNKNOWN SAMPLE

Materials

The following materials should be provided for class use.

1. Test reagents provided earlier for carbohydrate, lipid, and protein tests
2. Pyrex test tubes, test tube racks, and test tube holders
3. 1-ml and 10-ml pipettes with bulbs
4. Boiling water bath
5. One unknown sample (10 ml) per student

Procedure

This procedure is to be performed by students individually.

1. Obtain an unknown from your instructor.
2. Examine the unknown specimen for color, smell, and other distinctive properties.
3. Divide your specimen into several samples and identify its contents by performing whichever of the preceding procedures you think are necessary.
4. Enter your findings in the results and observations section.
5. Dispose of all chemical solutions as directed by your instructor.

RESULTS AND OBSERVATIONS

CARBOHYDRATE IDENTIFICATION

1. Enter in Table 53-1 observations of physical properties of the known solutions and test materials.

Table 53-1

Materials	Color	Smell	Other Properties
Glucose			
Lactose			
Maltose			
Albumin			
Starch			
Gelatin			
Honey			
Sugar-free drink			
Cola drink			
Distilled water			
Syrup			
Milk			
Saccharin			
Orange juice			

2. Enter in Table 53-2 the chemical test results.

Table 53-2

Materials	Carbohydrate Test	Reducing Sugars Test		Starch Test	
	Molisch Reaction	Reaction with Benedict's Reagent	Reducing Sugar Present	Reaction with Iodine	Starch Present
Glucose					
Lactose					
Maltose					
Albumin					
Starch					
Gelatin					
Honey					
Sugar-free drink					
Cola drink					
Distilled water					
Syrup					
Milk					
Saccharin					
Orange juice					

3. Which of the materials tested showed the presence of a reducing sugar? ____________

____

PROTEIN IDENTIFICATION

1. Enter your observations of the reactions to the chemical tests in Table 53-3.

Table 53-3

Material	Biuret Reaction	Sulfur Reaction	Xanthoproteic Reaction
Albumin			
Cysteine			
Glucose			
Glycine			
Phenylalanine			
Starch			
Tryptophan			
Casein			
Hair			
Gelatin			

2. Which of the materials tested contain peptide bonds?

____

3. Which of the amino acids tested contain a benzene ring?

____

4. How many of the materials tested showed the presence of sulfur? Which ones? ______

____

5. Which of the materials tested solidified when heated?

____

PROPERTIES OF FATS

1. Indicate in Table 53-4 the results of the fat solubility test.

Table 53-4

Solvent	Albumin		Cottonseed oil		Gelatin		Glucose		Starch	
	0 time*	After 15 min	0 time*	After 15 min	0 time*	After 15 min	0 time*	After 15 min	0 time*	After 15 min
Distilled water										
Ethyl alcohol										
Sodium chloride										
Xylene										

*Immediate reaction.

2. Which of the substances tested produced an obvious grease spot on the brown paper strip?

____

3. What other types of reactions besides those of solubility did you find? ______________

____

IDENTIFICATION OF CONTENTS OF UNKNOWN SAMPLE

1. List the tests used and enter your findings in Table 53-5.

Table 53-5

Test or Procedure	Description of Results
General appearance	

2. On the basis of the tests performed and results obtained, the unknown sample________

______________ contains ____

____

Questions for Review

EXERCISE **53**

DIGESTION CHEMISTRY

Name ____

Laboratory Section ______________________________ Date ____________

1. Complete Table Q53-1.

Table Q53-1

Chemical Test Reagent	Test Substance	Reaction Produced
Iodine	Glucose	a.
Iodine	Starch	b.
Benedict's solution	c.	Red precipitate
Benedict's solution	Galactose	d.
Molisch reagents (alpha-naphthol and concentrated sulfuric acid)	Albumin	e.
Molisch reagents	Sucrose	f.
Ninhydrin	g.	Purple color
Ninhydrin	Starch	h.
Biuret test reagent	Starch	i.
Biuret test reagent	Albumin	j.
Iodine	Orange juice and starch solution	k.

2. Define or explain the following.

   a. Reducing sugar ____

   ____

   b. Carbohydrate ____

   ____

   c. Vitamin ____

   ____

d. Hydrolysis ____________________

____

e. Lipase ____________________

____

3. Differentiate between the following.

a. Monosaccharide and polysaccharide ____________________

____

b. Amino acid and peptide ____________________

____

c. Exocrine and endocrine ____________________

____

4. What chemical elements are present in the compounds listed?

a. All carbohydrates ____________________

b. All proteins ____________________

c. All lipids ____________________

d. All nucleic acids ____________________

SECTION XV

THE URINARY AND REPRODUCTIVE SYSTEMS

The major function of the urinary system is to regulate the volume and composition of the extracellular fluid. The accumulation of unused or toxic by-products of metabolism and changes in the individual's environment, food ingestion, or metabolic activities may upset the balance or constancy of the body's internal environment. The kidneys, major components of the urinary system, are essential homeostatic organs because of their ability to remove wastes and to alter the volume and composition of blood through the production and excretion of urine.

The composition of urine can vary greatly. Some variation in the amount of substances normally found in the urine may reflect altered body activity or food and fluid intake. Elevated levels of these common components or the presence of substances normally not found in urine may indicate disturbed metabolic functions or abnormal renal activity.

Disorders and diseases of the kidneys may result in severe impairment of function. If the kidneys cannot adequately regulate the composition of the blood, an artificial device must be used in order to maintain life. The so-called artificial kidney uses a dialysis mechanism involving a membrane through which substances from the blood may diffuse.

Because of the importance of the kidneys and related structures, the functioning of the urinary system is routinely tested. Evaluation of kidney function usually involves determinations of the composition and volume of urine. Such tests also may reveal the presence of microbial infection.

This section considers certain physiological principles of renal function, the normal and abnormal components of urine, and the principle of hemodialysis. It examines the procedures involved in routine urine analysis tests and detection of bacterial infection. It also demonstrates the pH value and buffering capacity of several substances involved in the acid–base balance of body fluids.

The reproductive system consists of those organs necessary to produce, nourish, and protect sex (germ) cells—ova in the female and spermatozoa in the male. Exercises in this section describe the structures and functions associated with sex cell production and the stages of pregnancy and birth.

All eukaryotic cells exhibit the complex mechanism for the even distribution of hereditary material called *mitosis*. This process is quite efficient, and errors rarely occur. Mitosis results in the formation of two new cells, each containing the same number and types of chromosomes as in the original cell. The chromosome number normally remains constant from one generation of body, or somatic, cells to the next. In *meiosis*, or reduction division, the chromosome number is reduced by one-half. The process is always associated with the sexual process of *fertilization*. In fertilization the nuclei of two different cells, each containing one-half the normal number of chromosomes (haploid), fuse and establish the normal number of chromosomes (diploid). Errors during the stages of chromosome separation can result in genetic abnormalities. Karyotyping, a laboratory procedure, can be used to detect a number of chromosomal abnormalities. Exercises in this section also deal with meiosis, stages of early development, pregnancy testing, and karyotyping.

GENERAL REFERENCES

Davenport, H. W. *The ABC of Acid–Base Chemistry,* 6th ed. Chicago: University of Chicago Press, 1974.

Greep, R. O. (Ed.). *Reproductive Physiology IV.* Baltimore: University Park Press, 1983.

Lenin, A. *The Cycling Female.* San Francisco: Freeman, 1979.

Stamey, T. A., and R. W. Kindrachuk. *Urinary Sediment and Urinalysis.* Philadelphia: Saunders, 1984.

Sullivan, L. D., and J. J. Grantham. *Physiology of the Kidney,* 2nd ed. Philadelphia: Lea & Febiger, 1982.

Tortora, G. J., and N. P. Anagnostakos. *Principles of Anatomy and Physiology,* 6th ed. New York: Harper & Row, 1990.

Witherspoon, J. D. *Human Physiology.* New York: Harper & Row, 1984.

SPECIFIC REFERENCES

Asman, J. (Ed.). *Infertility: Diagnosis and Management.* New York: Springer-Verlag, 1984.

Beaconsfield, P., G. Birdwood, and R. Beaconsfield. "The Placenta," *Scientific American,* 243:94–102, 1980.

Fried, T. A., and J. H. Stein. "Glomerular Dynamics," *Archives of International Medicine,* 143: 729–787, 1983.

Fuchs, F., and A. Klopper (Eds). *Endocrinology of Pregnancy,* 3d ed. Philadelphia: Harper & Row, 1983.

Patterson, D. "The Causes of Down Syndrome," *Scientific American,* 257:52–60, 1987.

EXERCISE 54

ANATOMY OF THE UROGENITAL SYSTEM

Materials

The following materials should be provided for class use.

1. Cat dissection specimens
2. Dissecting tools
3. Model of human kidney
4. Charts of human kidney
5. Bone forceps
6. Models of human male and female reproductive organs
7. Charts of reproductive organs

The organs of the urogenital system (the urinary system plus the genital system) are closely related during embryonic development. In the lower chordates the urinary, reproductive, and digestive systems have a common outlet from the body, the *cloaca*. During the course of its development the human embryo passes through the cloacal stage until a subdivision of this region takes place. The result is a separate opening for fecal wastes (the anus) and for genital and urinary products (the urogenital sinus). In the human female there has been a further division of the urogenital region resulting in separate openings of the urinary tract and the genital tract. There are three separate openings in the human female. In the male, reproductive products and urine share the same duct and open together, as they do in both sexes of the cat.

DISSECTION OF THE CAT URINARY SYSTEM

If you have not already done so, remove the *parietal peritoneum,* which covers the kidney. Also, carefully remove any fat around the kidney. Since the kidneys are not actually inside the abdominal cavity, they are said to be located *retroperitoneally.* The right kidney is more anterior than the left. This situation is reversed in the human where the left kidney is superior to the right.

Identify once again the *renal artery* and *renal vein* associated with the concave medial surface of the kidney, which is called the *hilum.* Emerging from this same region is the broad yellow *ureter.* Follow this to the *urinary bladder.*

The urinary bladder collects urine produced by each kidney and stores it. When the urine is expelled, it leaves the urinary bladder through the *urethra,* which opens to the outside of the body. The urethra will be exposed in the dissection of the reproductive organs.

Just above and medial to the kidney is the hard, somewhat oval *adrenal gland* (an

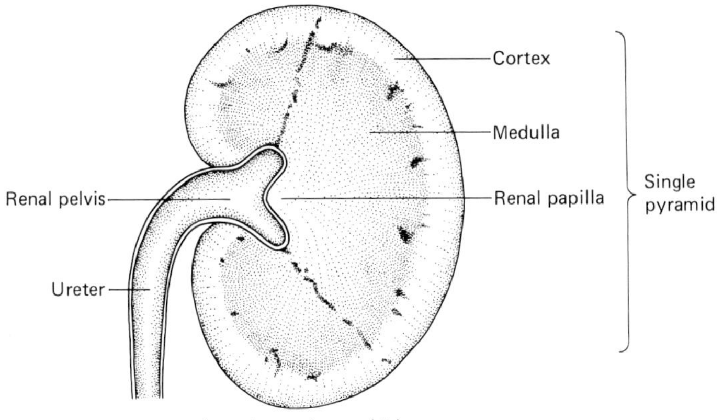

Figure 54-1 Section through cat kidney.

endocrine organ). In the cat the adrenal gland is not located directly above the kidney as it is in the human where it is called the *suprarenal gland.* Crossing the adrenal gland is the prominent *adrenolumbar vein.*

Without removing the kidney from the body or severing its attachment to the ureter, cut through it longitudinally with a sharp knife midway between the dorsal and ventral sides. Open it to see the *cortex, medulla, pelvis,* and its single *pyramid* (Figure 54-1).

THE HUMAN URINARY SYSTEM

The same organs you located in the cat are found in the human. There are some differences to note.

1. In the cat the right kidney is anterior to the left. In the human this is reversed with the left kidney superior to the right (Figure 54-2).

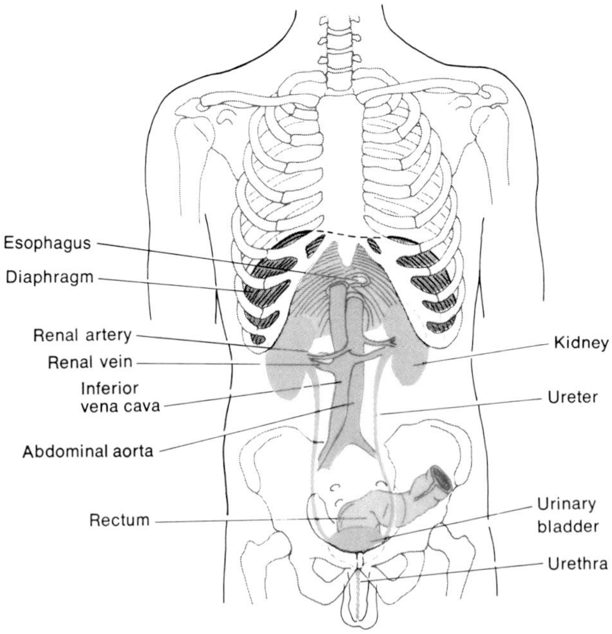

Figure 54-2a Locations of human urinary organs.

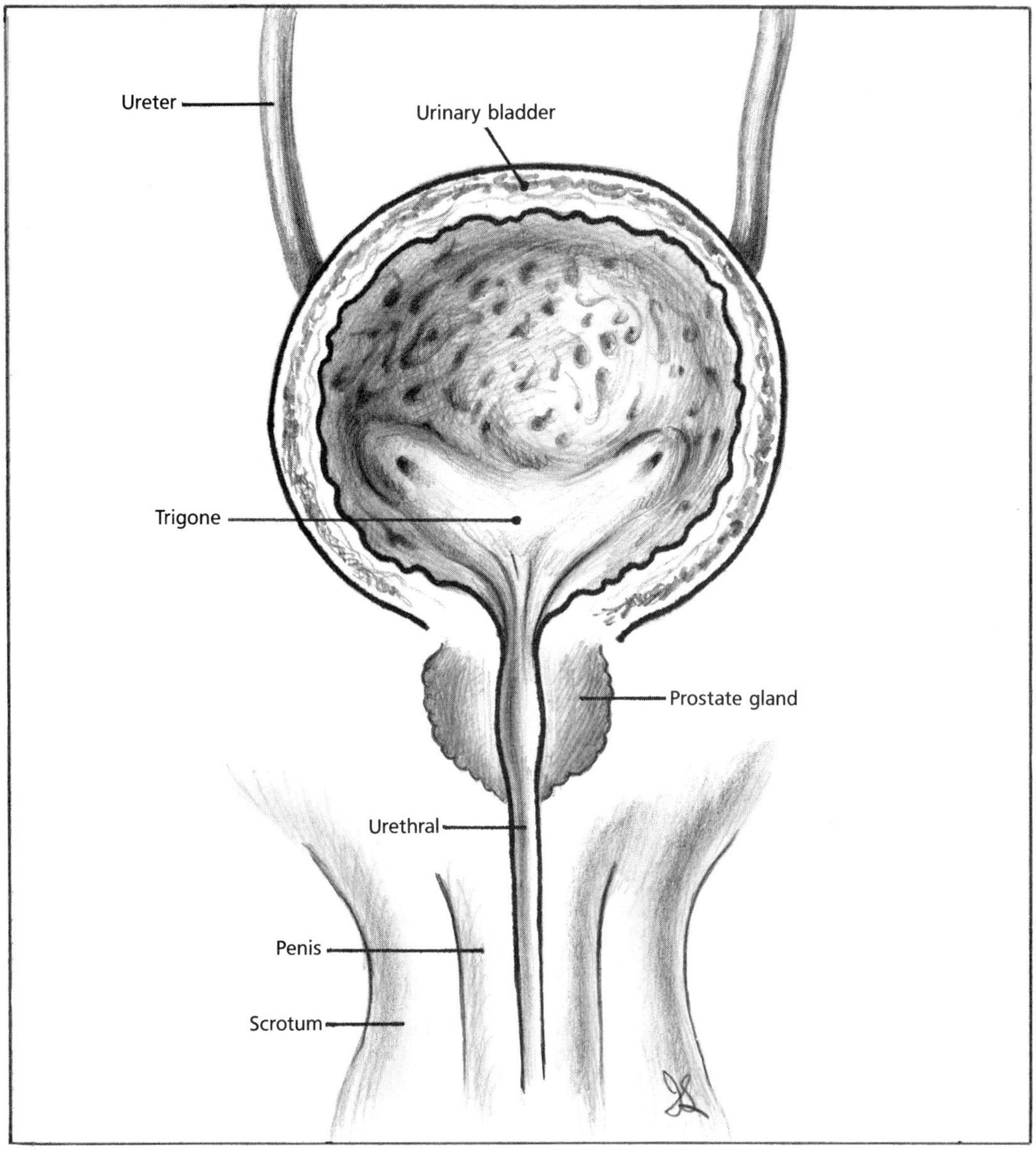

Figure 54-2b Human urinary bladder.

2. The cat has only one pyramid emptying into the renal pelvis, whereas the human has nine to twelve (Figure 54-3). Between pyramids are the renal columns.
3. The urinary bladder of the cat has a short neck between it and the urethra. The human bladder has no neck.
4. The human female has separate openings for the genital and urinary tracts. In the cat the urine opens into the small urogenital sinus, and ova and urine exit through the urogenital aperture.

Urine is produced by millions of tiny microscopic units called *nephrons* (Figure 54-4). Most of each nephron is located within the *cortex* area of the kidney and has an intimate association with the vascular system. Wastes are taken from the blood at the *glomerulus* into *Bowman's capsule* of the nephron. They pass through the *proximal tubule, loop of Henle, distal tubule,* and *collecting duct.* The collecting duct empties at the *renal papilla* of the *renal pyramids.* Wastes eliminated by the nephron are then transported as urine through a

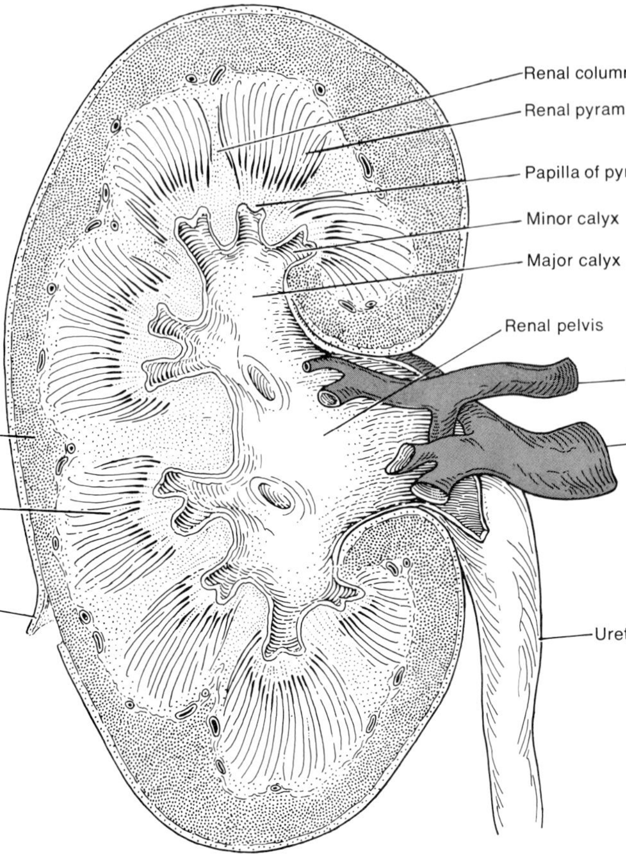

Figure 54-3 Coronal section through human kidney.

duct system of *minor calyces, major calyces, renal pelvis,* and *ureter.* The ureter carries urine to the *urinary bladder* where it is temporarily stored. Periodically urine is voided through the *urethra.* The collecting ducts pass through the medulla. Note the striations in this region. With your probe, locate the collapsed pelvis. The kidney is covered by a connective tissue capsule. Find all parts on Figures 54-3, 54-4, and on the kidney model.

THE REPRODUCTIVE SYSTEM OF THE MALE CAT

If the genital area of the cat has not yet been skinned, do it now. Remove the skin covering the *testes* and the *urogenital sinus* area. The skin covering the testis forms the *scrotal sac* or *scrotum.* During embryonic development the testes descend from the region in the abdominal cavity where they originally develop, through the abdominal wall and into the scrotal sac. The main reproductive duct leading away from the testis is the *vas deferens* and it should be traced (Figure 54-5). Together with the *spermatic artery* and the *spermatic vein* the vas deferens leaves the scrotum as the *spermatic cord* and enters the body cavity at the *inguinal canal.* Inside the body cavity the vas deferens leaves the spermatic cord, loops over the *ureter,* and enters the *urethra* below the *urinary bladder.* The spermatic blood vessels continue craniad to their place of origin, where they connect with the *descending abdominal aorta,* the *inferior vena cava,* and the *renal vein.* Follow the path of the vas deferens.

Return to the testis. To see it more clearly remove the tough layer of tissue, called the *tunica albuginea,* that surrounds it. Lying on the side of the testis, running the length of the organ, is a small, highly convoluted tubule, the *epididymis.* Other ducts for the transport of spermatozoa cannot be seen without further dissection and the use of the microscope.

Remove one of the testes and cut it in half longitudinally. Many *septa* create chambers

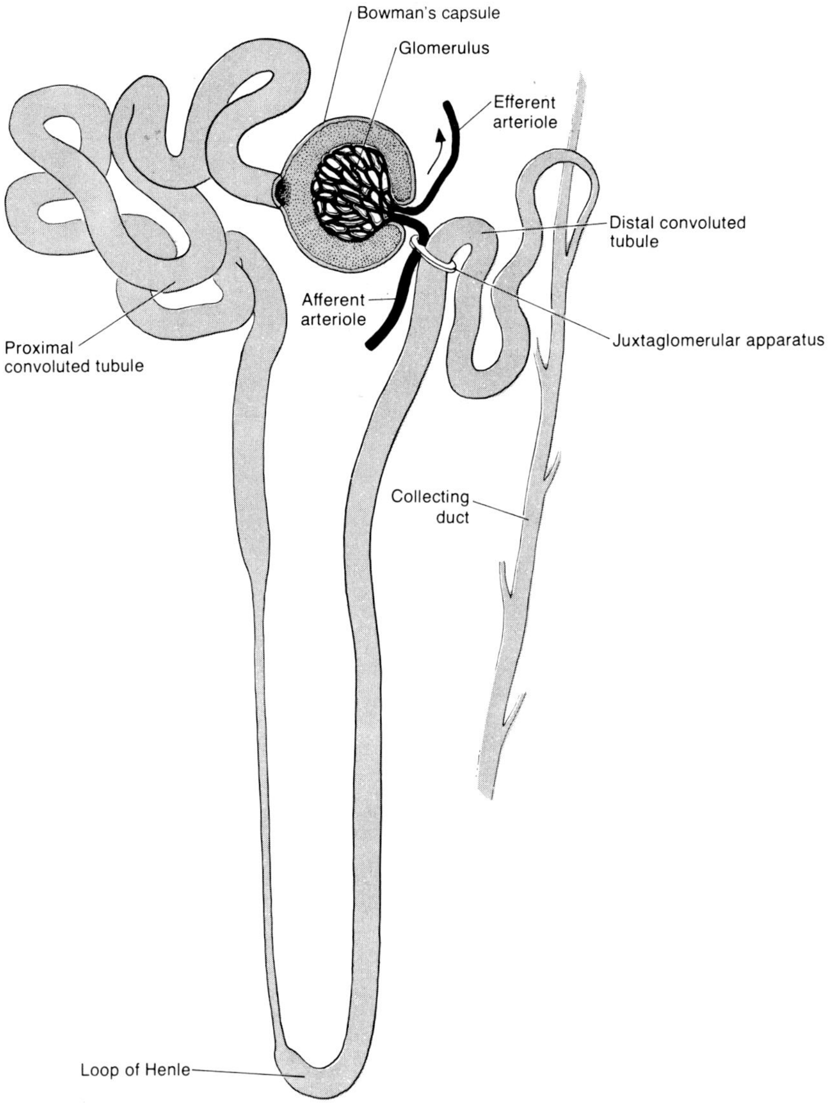

Figure 54-4 The nephron.

that contain the *seminiferous tubules* within which the *spermatozoa* are formed. Spermatozoa leave the testis through the *rete testes* and *vasa efferentia,* which then join the epididymis.

The posterior part of the reproductive system can be seen only if the *pelvic girdle* is split at the *pubic symphysis.* Using a nonrazor scalpel, cut through the muscle in the midline between the legs. If you are exactly at the midpoint, your scalpel blade will sink into the cartilage of the symphysis. Cut through the symphysis enough to weaken it, then place your hands on both legs and force them apart, breaking the joint and revealing the area posterior to the urinary bladder (Figures 54-5 and 54-6). Clear away all connective tissue and trace the entire length of the urethra. Posteriorly it enters the *penis* through which it exits the body.

At the point of entry of the vas deferens into the urethra, there is a large, hard mass of tissue surrounding the urethra. This is the *prostate gland.* About an inch further down the urethra there are paired hard masses, the *bulbourethral glands* or *Cowper's glands.* The prostate and bulbourethral glands secrete *semen* directly into the urethra. The cat does not have *seminal vesicles,* which are a third set of associated reproductive glands present in the human.

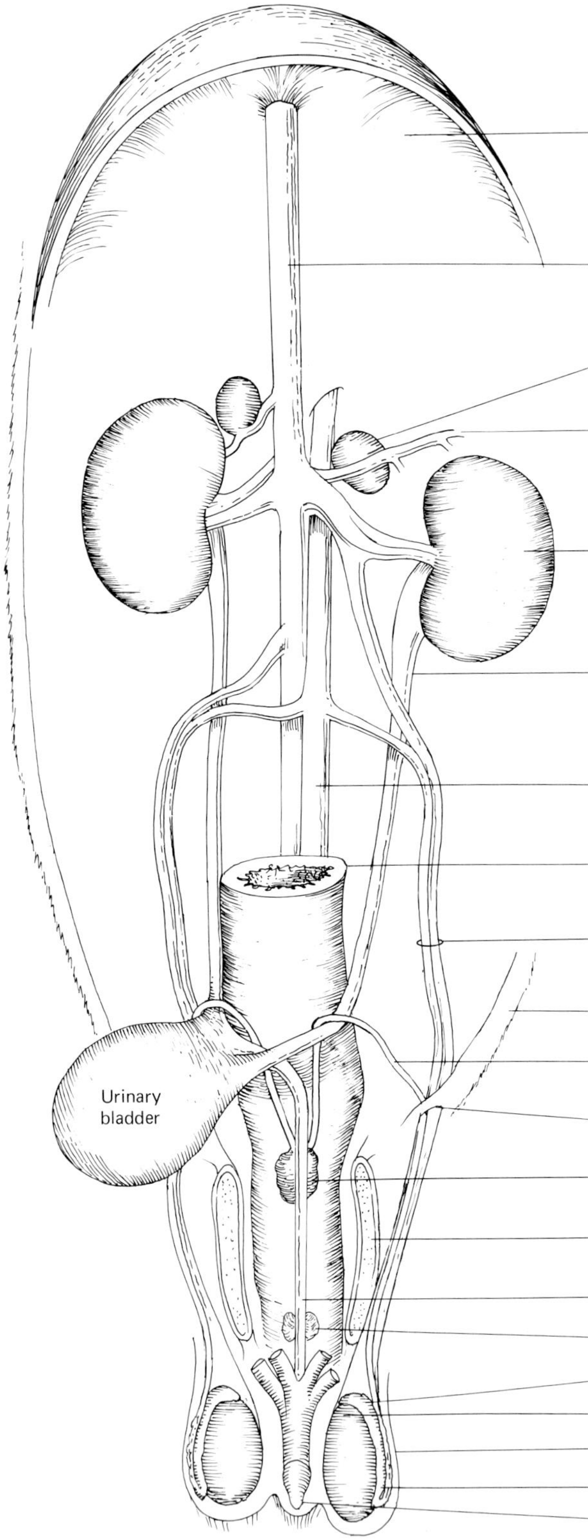

Figure 54-5 Urogenital system of the male cat.

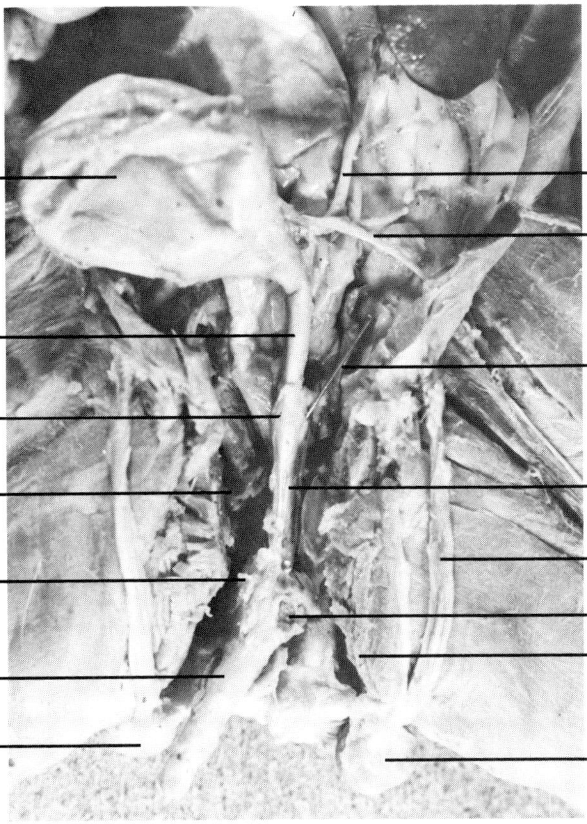

Figure 54-6 Male cat with a split pelvis.

1. Testis
2. Spermatic cord
3. Penis
4. Urethra
5. Bulbourethral gland
6. Prostate gland
7. Split pubic bone
8. Vas deferens
9. Ureter
10. Urinary bladder

THE REPRODUCTIVE SYSTEM OF THE FEMALE CAT

The uterus is the *bicornuate type,* meaning it has two large extensions (horns) from a central portion or body. The body of the uterus is located dorsal to the urinary bladder (Figure 54-7).

From the body of the uterus trace the paired *uterine horns* anteriorly. The uterine horns are very small in an immature female, large in a mature female, and enormous in a pregnant cat. In the pregnant cat the horns will contain *embryos* (Figures 54-8a, 54-8b, and 54-9).

At the anterior end of each uterine horn is the small oval *ovary,* surrounded by the expanded end of the *oviduct,* the *ostium.* The coiled oviduct may be seen lateral to the ovary. It connects with the *uterine horns.*

To observe the posterior end of the reproductive and urinary systems, follow the instructions given for the male reproductive tract and split the pubic symphysis (Figure 54-7). Trace the body of the uterus posteriorly as it passes dorsal to the *urethra.* It terminates

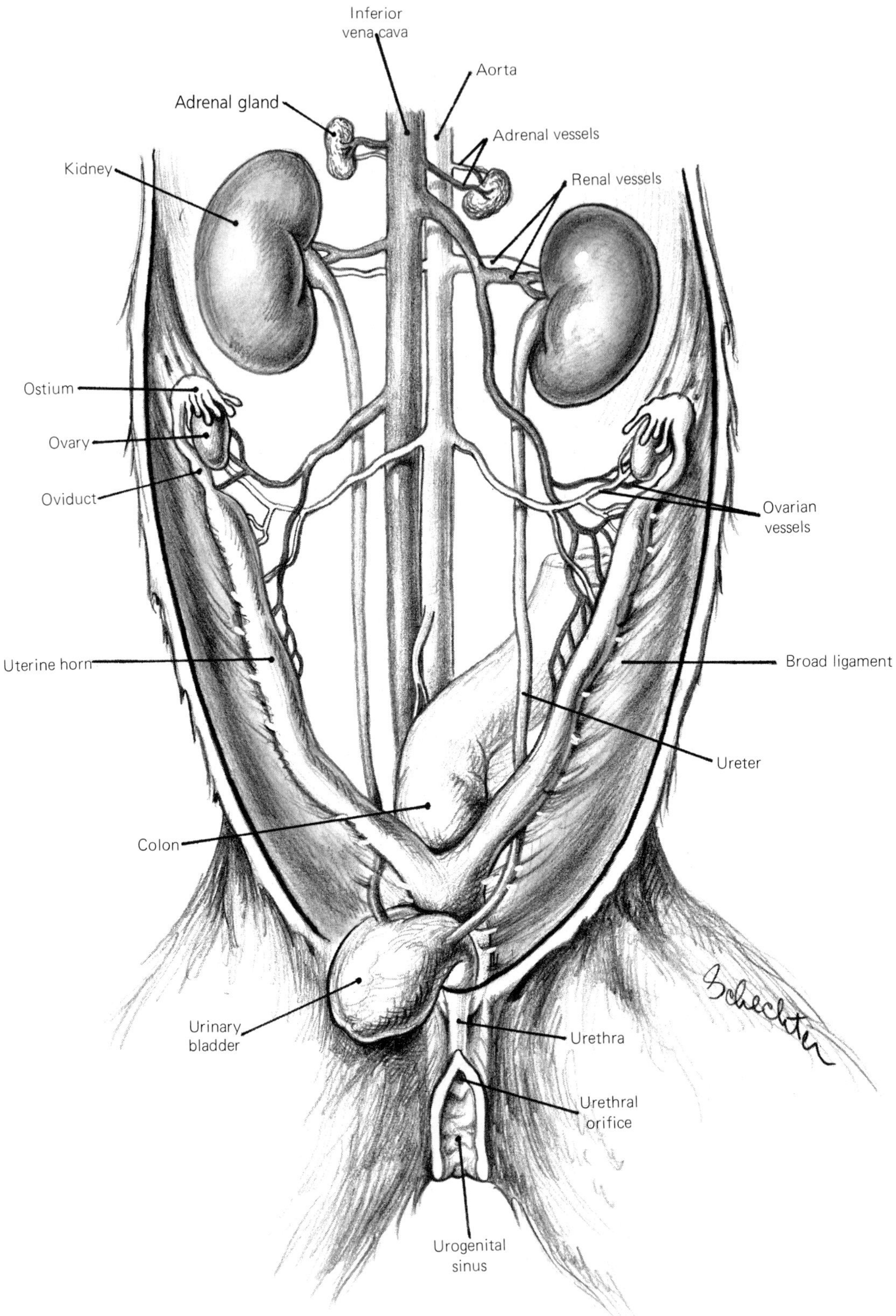

Figure 54-7 Reproductive system of a female cat.

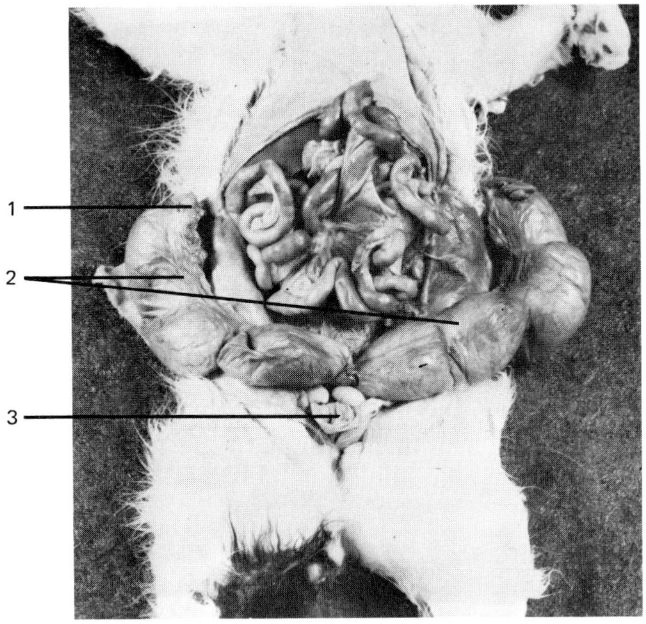

(a)

Figure 54-8a Pregnant cat showing uterine horns in a late stage of pregnancy.

1. Ovary
2. Uterine horns
3. Urinary bladder

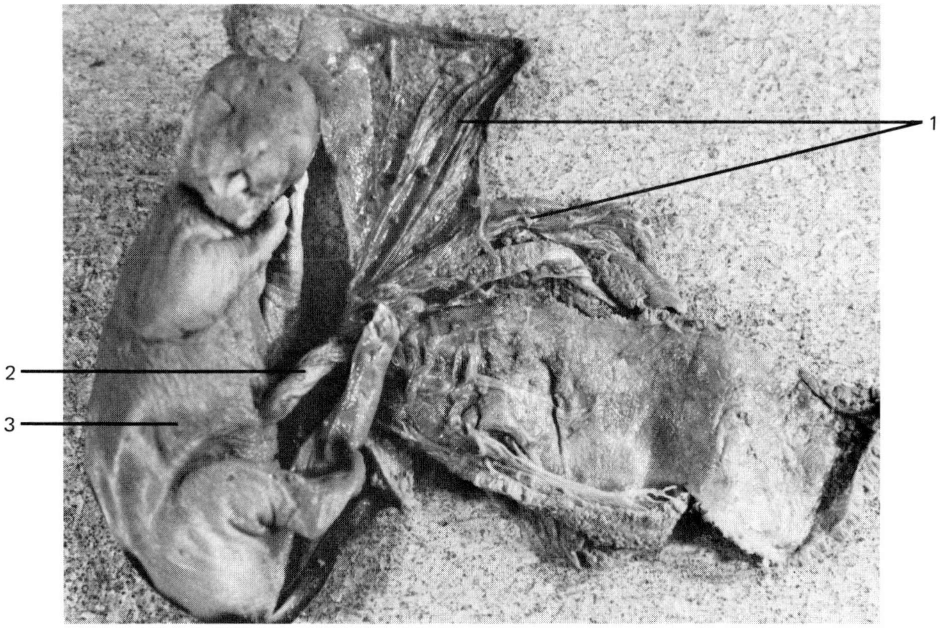

(b)

Figure 54-8b Cat embryo and placenta.

1. Placenta
2. Umbilical cord
3. Embryo

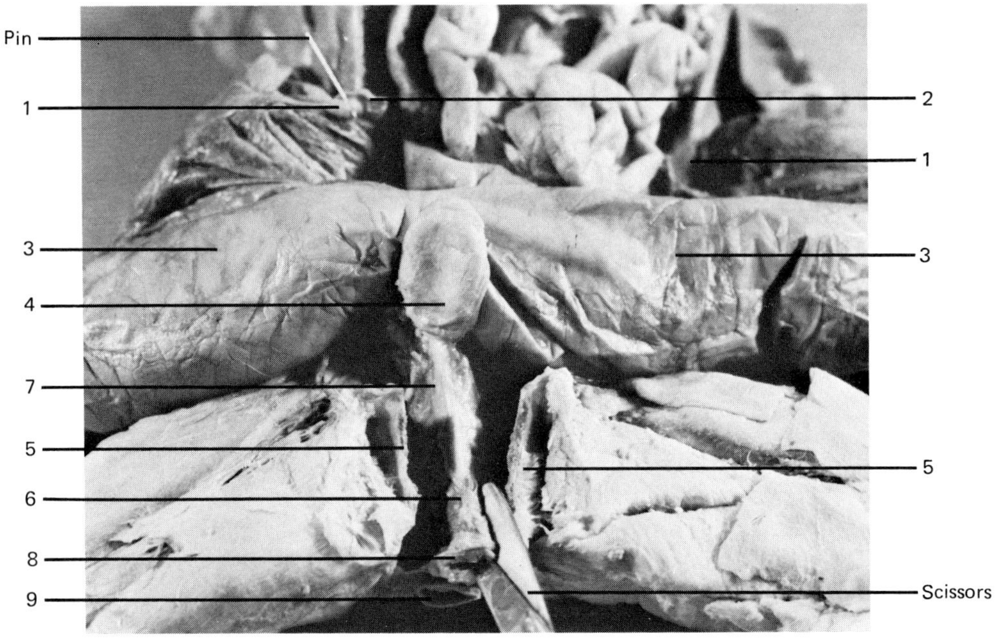

Figure 54-9 Pregnant female cat with a split pelvis. Scissors are shown inserted in the vagina.
1. Ovary
2. Oviduct
3. Uterine horn
4. Urinary bladder
5. Pubic symphysis (split)
6. Vagina
7. Body of uterus
8. Urogenital sinus opening
9. Anal opening

in the vagina. A short urogenital sinus is common to both ova and urine and opens distally as the urogenital aperture, at the end of the vagina. If the cat has large uterine horns or is pregnant, cut into the vagina from the urogenital sinus (aperture) craniad until you reach the *cervix of the uterus.* Make your cut to one side so as not to damage the urethra. Note the opening of the uterine cervix into the vagina.

THE HUMAN REPRODUCTIVE SYSTEM

The anatomy of the cat reproductive system is almost identical to the human. See Figure 54-10 for a diagram of the human male reproductive tract shown in lateral view from a midsagittal section. The presence of the *seminal vesicles,* a third pair of reproductive glands in addition to the *prostate gland* and the *bulbourethral glands,* is the only major difference. Find all parts of the cat reproductive tract and compare them with Figure 54-10.

The anatomy of the female reproductive tracts of the cat and the human are also similar (Figure 54-11). The main difference is in the *shape* of the uterus. The human has a *simplex uterus,* which usually accommodates one fetus at a time. Note its shape in Figure 54-11. The cat has a *bicornuate uterus* with horns capable of accommodating several fetuses of the typical multiple birth. The oviduct is usually called the *fallopian tube* or the *uterine tube* in human females where it is more prominent. Study Figure 54-11 and compare the sizes and shapes of the various organs in the two species.

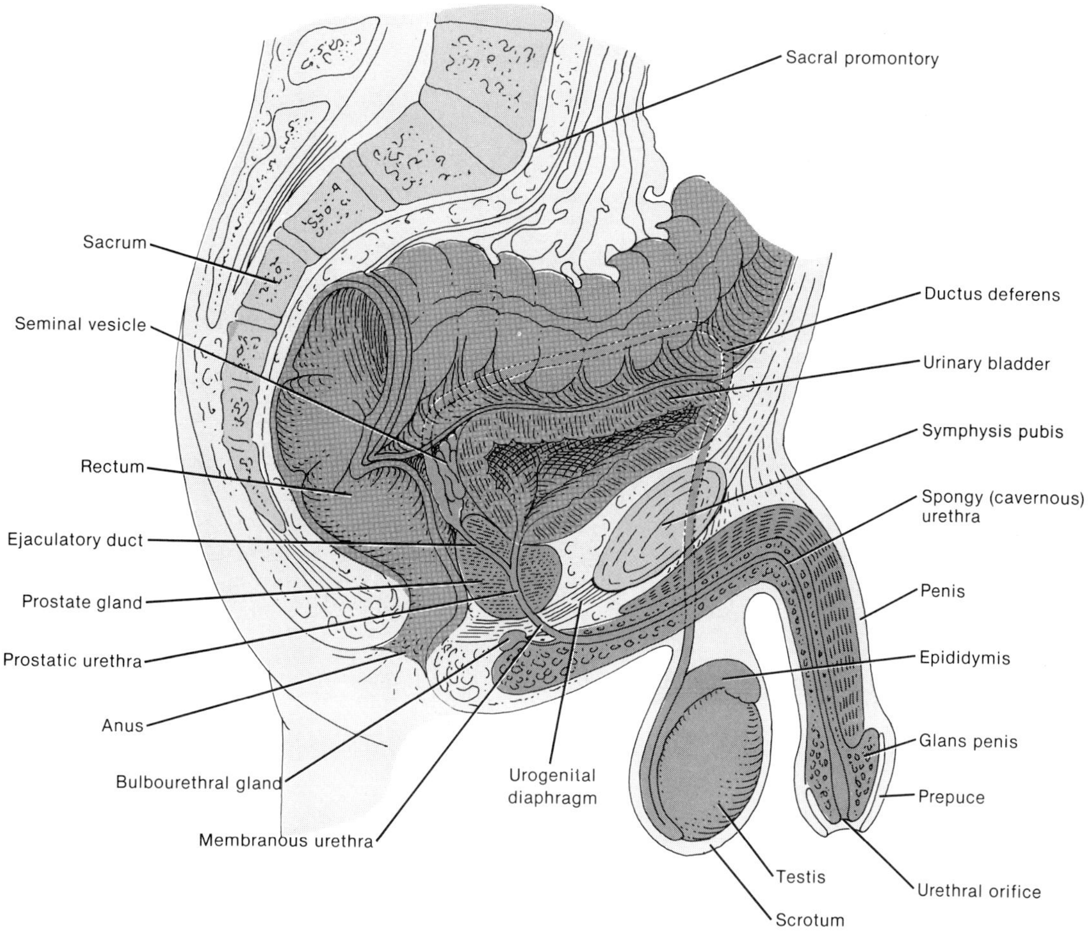

Figure 54-10 Human male reproductive system.

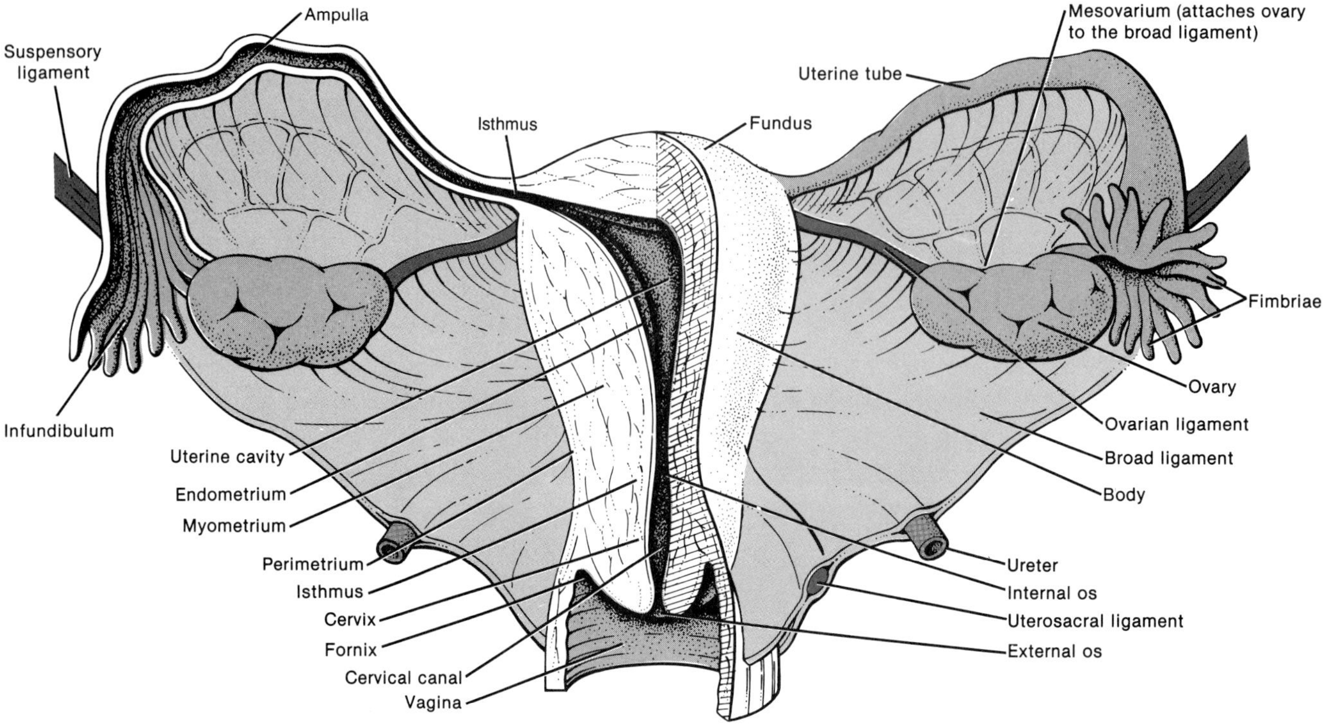

Figure 54-11 Human female reproductive system.

Questions for Review

EXERCISE **54**

ANATOMY OF THE UROGENITAL SYSTEM

Name ____________________

Laboratory Section ____________________ Date ____________

1. Trace the pathway of a waste product from the blood stream through the kidney to its elimination from the body. Some of the steps are provided. You fill in the rest.

   a. Afferent arteriole

   b. ____________________

   c. ____________________

   d. Proximal convoluted tubule

   e. ____________________

   f. ____________________

   g. Collecting duct

   h. ____________________

   i. ____________________

   j. Renal pelvis

   k. ____________________

   l. ____________________

   m. Urethra

2. Identify the parts of the human kidney in Figure Q54-1.

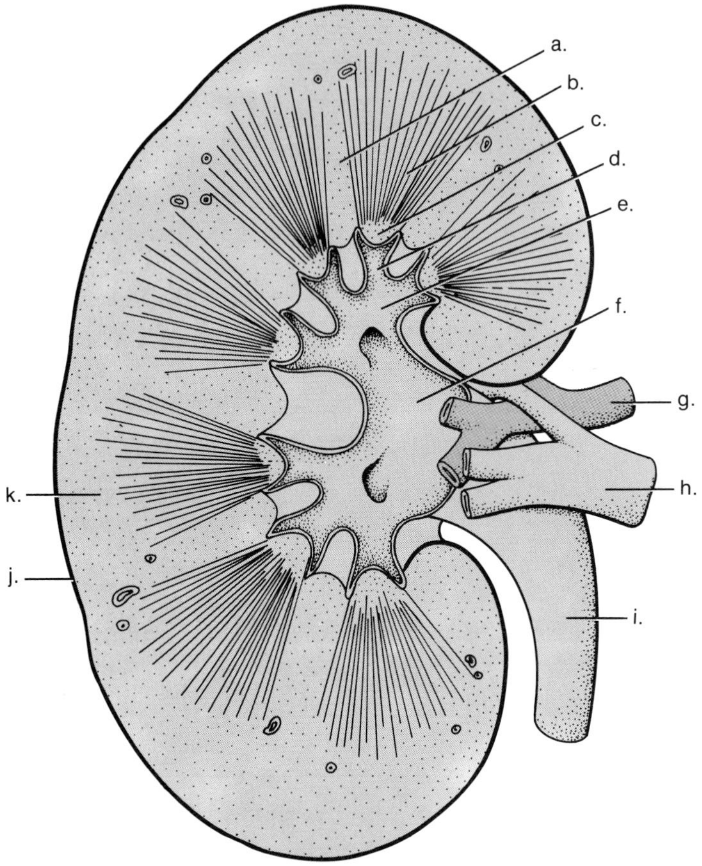

Figure Q54-1

a. ______________________________

b. ______________________________

c. ______________________________

d. ______________________________

e. ______________________________

f. ______________________________

g. ______________________________

h. ______________________________

i. ______________________________

j. ______________________________

k. ______________________________

3. What anatomical feature of the human female reproductive tract makes it difficult for her to have multiple births? Explain why this is so.

____

____

____

4. The anatomy of the male reproductive tract in both the cat and the human is still tied to the urinary system, while in the female the two systems have separated completely. Explain the relationship of the reproductive system and the urinary system in both male and female.

____

____

____

____

____

____

____

5. Trace the pathway of a human spermatozoan from its place of origin to the site of fertilization of an ovum. Some of the answers are provided. You fill in the rest.

a. Seminiferous tubules
b. ______________
c. ______________
d. Urethra
e. Vagina
f. ______________
g. ______________
h. Fallopian tube or oviduct

EXERCISE 55

FLUID AND ELECTROLYTE BALANCE AND A DEMONSTRATION OF DIALYSIS

Urine is an extremely complex water solution of various inorganic and organic substances that are the waste products of metabolism or are derived directly from ingested foods and fluids. The most important inorganic constituents of urine include ammonia, chlorides, phosphates, and sulfates. The chief organic components are creatinine, urea, and uric acid. Urine is a valuable indicator of various normal and abnormal situations, since most pathological conditions originating in the kidneys and various diseases involving other body organs alter its composition. Certain chemical substances and large numbers of microorganisms appear in urine only when pathological conditions are present. (Exercise 56 describes representative disease conditions and the substances that are associated with them.)

Urine formation is a process essential to the body's capability to rid itself of metabolic waste products. The kidneys are among the organs that assume a major responsibility for the regulation of the volume, pH, and chemistry of body fluids. Functional kidney performance depends on its basic units, the *nephrons* (Figure 55-1). These delicate, microscopic structures vary the concentration, volume, and composition of urine according to the needs of the body. Each nephron consists of a glomerulus (a tuft of capillaries enclosed by a double-walled cup called the renal or Malpighian corpuscle), the proximal convoluted tubule, the loop of Henle, the distal convoluted tubule, and the collecting duct, which leads to the renal pelvis. Each kidney contains approximately 1 million nephrons.

Urine formation begins in the glomerulus where water and small molecules are filtered from the blood. The blood supply to this filtering unit is provided by the afferent arteriole, which in turn is supplied by a renal artery. The resulting filtrate passes through long tubules. Physiologically important substances (such as glucose, water, electrolytes, including sodium, chloride, and phosphates, and some ions) are selectively reabsorbed, while other substances (such as ammonia, potassium, hydrogen ions, and foreign materials such as penicillin) are secreted into the urine. The final tubule fluid, urine, that enters the bladder is quite different in composition from the glomerular filtrate. The continuous adjustment of urine volume and electrolyte content within tubules is controlled by several substances and conditions including antidiuretic hormone (ADH) from the neurohypophysis, the sodium-retaining hormone, aldosterone, from the adrenal cortex blood flow, and the acid-base balance.

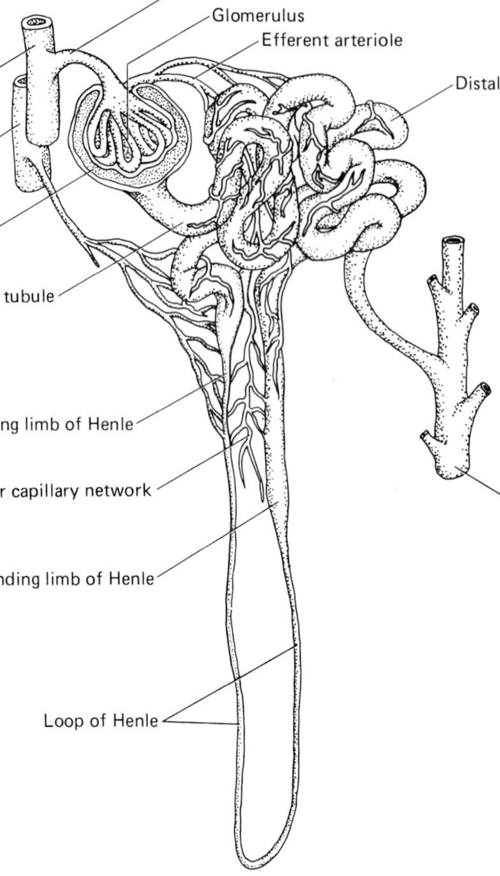

Figure 55-1 Microscopic features of a nephron.

ADH is regulated by osmoreceptors in the hypothalamus. When the body is dehydrated, the concentration of water in the blood falls below normal levels and consequently stimulates these receptors to bring about the release of ADH into the bloodstream and its transportation to the kidneys. The kidneys respond by decreasing the urine output, thereby conserving water. If, on the other hand, the blood contains higher than normal water concentrations, the osmoreceptors stop ADH secretion. The kidneys then release large quantities of urine and the volume of water in body fluids is brought back to normal. Secretion of ADH also can be influenced by various chemical factors, such as alcohol, nicotine, and other drugs, and physical and mental factors, such as pain, trauma, and stress.

Aldosterone may be stimulated by an increase in potassium ions (K^+), a decrease in sodium ions (Na^+), or a decrease in blood volume. The hormone acts on the tubule cells of the kidney and thereby causes an increase in the reabsorption of sodium ions and water. Sodium ions are thereby removed from urine and returned to the blood. This activity prevents the rapid loss of sodium and water from the body. Aldosterone also decreases the reabsorption of potassium ions, so large quantities of this ion are released into urine.

In the first part of this exercise the effects of salt, sugar, grapefruit juice (a source of potassium), and coffee or tea (sources of caffeine) on the volume, pH, specific gravity and chloride content of urine are considered. Students who are diabetic or are under medical treatment for a kidney or related problem should not participate in the first portion of this exercise. The principle of dialysis and factors that can affect the process will be considered in the second portion of this exercise.

Dialysis is a process by which certain substances in solution can be separated from one another by virtue of the difference in their rates of diffusion through a semipermeable membrane. *Hemodialysis* represents one type of dialysis used to correct a variety of conditions in the body that affect renal excretion. In situations such as drug overdose, temporary stoppage of urine formation, and extensive kidney destruction, hemodialysis may mean the difference between life and death.

Two types of commercial hemodialysis systems, frequently referred to as artificial kidneys, are in general use. Both operate on the same principle. Blood from the patient is circulated by a pump past one side of a semipermeable membrane, called a *dialyzing membrane* and usually made of cellophane. Circulating on the other side of the membrane is a *dialyzing solution* containing electrolytes in the same concentration as in normal plasma. As the blood and solution pass on either side of the membrane, excess electrolytes in the blood diffuse through the membrane into the solution because of the concentration gradient.

In one type of hemodialysis sytem, the *coil type* the dialyzing membranes are single or parallel flat cellophane tubes wrapped around a central core supported by plastic screening. The dialyzing or bathing solution is pumped through the concentric layers and over the flattened tubing of the system. The coil assembly system is presterilized and is generally discarded after use.

The second type of system, the *plate* or *layer dialyzer,* contains cellophane sheets instead of tubes (Figure 55–2). The sheets are held apart within plastic frames to allow passage of the fluid to be dialyzed between them. The dialyzing solution flows on the outside of the cellophane sheets. All parts of this system are sterilized before use.

The dialyzing solution is temperature controlled and has a composition that chemically approximates normal extracellular fluid. Specific needs of individuals on hemodialysis may require alterations in the extracellular fluid. Dialysis begins when blood from one of the patient's arteries is pumped through the system. After removal of the unwanted substances, the blood passing through the dialysis system is returned to the patient's circulation by a vein.

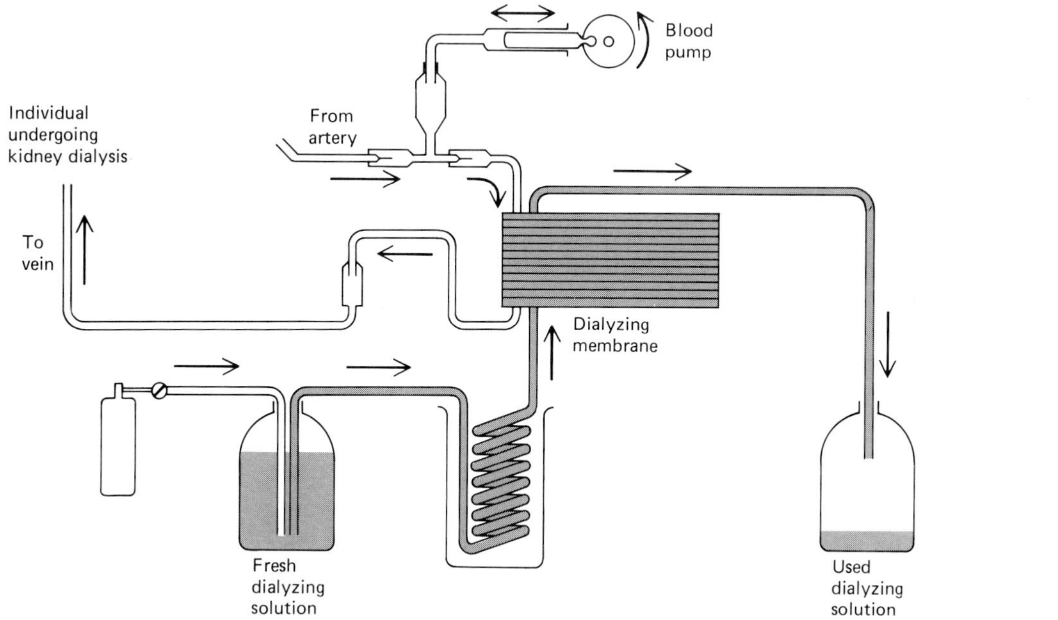

Figure 55-2 Schematic diagram of an artificial kidney system. This particular system shows the features of a plate or layer dialyzer.

A. EFFECTS OF SODIUM, POTASSIUM, CAFFEINE, AND WATER ON URINE COMPOSITION

Materials

1. The following materials should be provided for class use.
   a. Urinometers (cylinder and hydrometer float device)
   b. pH indicator paper strips (pHydrion or nitrazine test papers)
   c. Thermometers
   d. 250-ml beakers
   e. Filter paper
   f. Test Tubes
   g. Eyedroppers
   h. 20 percent potassium chromate solution in dropper bottle
   i. 2.9 percent silver nitrate solution
   j. Disposable drinking cups
   k. 1-ml pipettes, graduated in 0.1 ml, with bulbs
   l. Test tube racks
   m. Liquid soap and brushes for washing containers
2. The following materials should be provided for students in groups of five.
   a. Five urine specimen containers
   b. Five 500-ml graduated cylinders
   c. Five 250-ml beakers
   d. Test solutions in the volumes indicated:
      Cola drink (nondietetic), 750 ml
      Distilled water, 1,250 ml (500 ml is to be used with a salt tablet)
      Unsweetened, strong coffee or tea, 750 ml
      Unsweetened grapefruit juice, 750 ml
   e. One salt tablet (4.5g)

Procedure 1: Volume, Color, Turbidity, and Odor

This procedure is to be performed by students individually, but they are to work in groups of five. The instructor will indicate the facilities to use in order to obtain urine samples.

1. Obtain a urine specimen container and empty your bladder completely. Note the time and enter it in the appropriate space of the results and observations section. This sample is your normal urine specimen.
2. Use one of the 500-ml graduated cylinders to determine the volume of the normal specimen. Record the information in the results and observations section. The normal volume per 24 hours is 1 to 2 liters.
3. Note the color, the presence of turbidity (cloudiness), and any odor. Record your findings in Table 55–3 of the results and observations section. Normal urine is generally yellow to amber, is clear, and does not have a disagreeable odor.

Procedure 2: Specific Gravity and pH

This procedure is to be performed by students individually, but they are to work in groups of five.

1. Pour about 30 ml of the urine sample into a small beaker. Carefully pour the urine from the beaker into a clean urinometer cylinder until it is three-fourths full. Obtain more of the specimen if necessary. If foam forms on the surface, remove it with a piece of filter paper. Discard the paper as indicated by your instructor.

2. Obtain a thermometer and check the reading on it. If the reading is not 37°C, (carefully) shake it down to that level. Insert the thermometer in the urinometer and read the temperature after 3 minutes. The temperature of the sample is used in the determination of specific gravity in the next step. Clean the thermometer as indicated by your instructor.
3. Float the hydrometer in the cylinder. Make certain that it is not touching the bottom or sides of the cylinder. Adjust the value of the hydrometer reading according to the temperature to obtain the specific gravity. For each 3° above 25°C, add 0.001 to the hydrometer reading; for each 3° below 25°C, subtract 0.001. Record your findings in the results and observations section. The specific gravity of normal urine is 1.002 to 1.030. Remove the hydrometer and clean it as indicated by your instructor.
4. Dip a pH indicator test strip paper in the urinometer cylinder three consecutive times. Remove excess urine by touching the paper to the inside surface of the cylinder. After 1 minute compare the color of the strip with the color guide on the strip container or other available guide. Normal urine has an average pH of 6, but ranges from 4.8 to 7.5. Record your findings in the results and observations section.
5. Clean the urinometer as indicated by your instructor.

Procedure 3: Chloride Concentration Determination

This procedure is to be performed by students individually, but they are to work in groups of five.

1. Place two test tubes in a rack.
2. Using a 1-ml pipette, introduce 0.5 ml of the urine sample into each tube.
3. Add 1 drop of 20 percent potassium chromate to each tube.
4. Gently shake the tubes. The solution in the tubes should be yellow.
5. With an eyedropper, add 1 drop at a time of the silver nitrate solution provided to the first tube. Gently shake the tube after each drop. Count the number of drops required to change the color of the urine solution from yellow to brown. *Note:* One drop of the silver nitrate solution added is equivalent to 61 mg of chloride ion (CL^-) per 100 ml of urine.
6. Calculate the chloride concentration of the urine sample by multiplying the number of silver nitrate drops added by 61. The value should be reported in milligrams/100 ml of urine.
7. Repeat steps 5 and 6 with the second tube. Average your findings and enter them in Table 55–3 of the results and observations section.
8. Empty the tubes as indicated by your instructor and clean all glassware and other materials.

Procedure 4: Effects of Various Test Solutions on Urine Composition

This procedure is to be performed by students in groups of five.

1. Determine within your group the individuals who will drink a particular test solution or material. The test solutions include distilled water, cola drink, unsweetened grapefruit juice, and unsweetened strong coffee or tea. A salt tablet with water also will be used.
2. All members of the group should obtain the necessary volume of their test solution; 750 ml should be consumed. The individual selecting the salt tablet should swallow it with 500 ml of distilled water.
3. Drink the test solution as quickly as possible. If an individual is unable to consume the entire volume designated, the quantity consumed should be recorded in Table 55–2. Your instructor will indicate the pH, specific gravity, and chloride concentration of each test solution. Enter this information in Table 55–2.

4. Begin timing as soon as you have consumed the test solution entirely. Be prepared to obtain a urine sample at four 30-minute intervals. A total of four urine specimens will be collected.
5. Perform procedures 1, 2, and 3 with each urine specimen collected. Enter the findings obtained with your test solution in Table 55–3 of the results and observations section.
6. Compare your findings with those obtained by other members of your group, and answer the questions in the results and observations section.

RESULTS AND OBSERVATIONS

EFFECTS OF VARIOUS TEST SOLUTIONS ON URINE COMPOSITION

1. List the food and drink items you consumed approximately 2 hours before starting the experiment in Table 55–1. Indicate the approximate quantity of each item.

Table 55-1

Food or Drink Item	Quantity	Food or Drink Item	Quantity

2. a. Approximate time of urinary bladder emptying prior to the laboratory exercise:

   ____

   b. Time at which normal urine specimen was obtained at the beginning of the laboratory exercise: ____________________

3. a. Enter the pH, odor, specific gravity, and relative chloride content of each test solution provided for your group in Table 55–2.

Table 55–2

Test Solution	pH	Odor	Specific Gravity	Relative Chloride Content	Volume Consumed
Distilled water					
Unsweetened cola drink					
Unsweetened coffee or tea					
Unsweetened grapefruit juice					
Salt tablet and water					

   b. Which test solution did you use? ____________________

Table 55–3

Property	Normal Sample	Experimental Samples			
		30 Min	60 Min	90 Min	120 Min
Volume (in ml)					
Color					
Odor					
Turbidity (clouding)					
pH					
Specific gravity					
Relative chloride content (in mg/ml)					
Rate of urine formation (in ml/min)					

4. Enter the findings obtained with your urine samples in Table 55–3.
   a. To calculate your rate of urine production for the normal sample, find the difference in time between emptying the bladder before the exercise and taking the normal urine sample. Then calculate the number of milliliters produced per minute. Enter the result in Table 55–3.
   b. Calculate the rate of urine formed for each interval after drinking the test solution in ml/min. Enter the results in Table 55–3.

5. a. Were any of the properties observed with the normal urine sample changed by the test solution consumed? If so, which ones? ______________________________

   ____

   b. Offer an explanation for any changes observed. ______________________________

   ____

6. a. Were any of the properties observed with the normal urine samples of your group changed by the test solutions they consumed? If so, which ones? ______________________________

   ____

   b. Offer an explanation for any changes observed. ______________________________

   ____

B. DEMONSTRATION OF DIALYSIS

Materials

1. The following materials should be provided per four students.
   a. Two pieces of dialysis tubing, each 20 cm (about 8 inches) long
   b. One 50-ml beaker and two 250-ml beakers
   c. One small funnel
   d. Four Pyrex test tubes (10–15 ml capacity)
   e. One 100-ml graduated cylinder

2. The following materials should be provided for class use.
   a. 2 percent glucose solution
   b. 5 percent starch solution
   c. 10 percent sodium chloride (NaCl) solution
   d. 10 percent egg albumin solution
   e. Benedict's reagent (for glucose determinations)
   f. Silver nitrate reagent (for Cl^- determinations)
   g. Iodine reagent (for starch determinations)
   h. Sodium hydroxide (NaOH) and copper sulfate ($CuSO_4$) reagents (for protein determinations)
   i. Distilled water, kept at 4°C, room temperature (25°C), and 37°C
   j. String for dialysis bags
   k. Boiling water bath

Procedure 1: Preparation of Dialysis Bags

This procedure is to be performed by students in groups of four.

1. Soak the dialysis tubing in distilled water (25°C) until soft.
2. While the dialysis tubing is softening, mix in a 50-ml beaker 10 ml each of the albumin, glucose, sodium chloride, and starch solutions.
3. When the tubing material is soft, prepare a dialysis bag by tying one end of the tube with a small piece of string. Make certain that the knot is tight enough so that the system will not leak when filled.
4. Have a laboratory partner hold a small funnel in the open end of the dialysis bag. Fill the bag with the prepared chemical mixture. Leave only enough space at the top of the bag so that the tube can be tied again to seal the bag.
5. Close the open end of the dialysis bag by tying a knot with another piece of string.
6. Carefully rinse the filled dialysis bag with distilled water (25°C).
7. Place the bag in a 250-ml beaker containing distilled water at 4°C. The bag should be completely submerged.
8. Prepare another dialysis bag by repeating steps 1–6. Submerge the second bag in a beaker containing distilled water at 37°C.

Procedure 2: Testing the Dialysate

This procedure is to be performed by students in groups of four. The instructor will demonstrate the testing procedures for the *dialysate* (the chemicals passing out of the dialysis bag) before the students perform the procedure.

1. Immediately after submerging the dialysis bags in the beakers, carefully pour about 20 ml of the water from each beaker into separate graduated cylinders.
2. Divide the sample of the 4°C water into four separate test tubes (about 5 ml each). To determine which of the substances added to the dialysis mixture are present in the dialysate perform the following tests, using one test tube for each test.

a. The *Biuret* or *protein (egg albumin) test.* Add approximately 5 ml of the sodium hydroxide and 2 ml of the copper sulfate reagents. The formation of a pink to violet color is a positive reaction. Refer to Color Plate 1.
b. *Chloride ion (Cl^-)test.* Add 8 drops of the silver nitrate solution. The formation of a white precipitate is a positive result.
c. *Glucose test.* Add 10 ml of the Benedict's reagent and place tube in a boiling water bath for about 5 minutes. The formation of an orange color is a positive result (color Photograph 42).
d. *Starch test.* Add 3 drops of the iodine reagent. The formation of a purple to black color is a positive result.

3. Remove 20-ml samples at 15 minutes, 30 minutes, 1 hour, 2 hours, and 48 hours (or during the next laboratory period).
4. With each sample perform the tests listed in step 2.
5. Enter your findings in the results and observations section.
6. Repeat steps 2–5 with the second dialysis system (37°C water).
7. Answer the questions in the results and observations section.

RESULTS AND OBSERVATIONS

1. Describe the test results that would indicate the presence or absence of the substances listed in Table 55-4.

Table 55-4

Substance	Positive Result	Negative Result
Chloride ion		
Egg albumin		
Glucose		
Starch		

2. Enter in Table 55-5 the results of the tests performed. Indicate the presence of a substance with a "+" and its absence with a "−".

Table 55-5

Time of Sampling	Chloride Ion		Egg Albumin		Glucose		Starch	
	4°C	37°C	4°C	37°C	4°C	37°C	4°C	37°C
Immediate								
15 min								
30 min								
1 hr								
2 hr								
48 hr								

3. What other methods could be used to test for the following substances?

   a. Glucose ______________________________

   b. Albumin ______________________________

4. Briefly explain the relationship between dialysis and kidney function. ______________

   ____

   ____

5. Briefly describe the appearance of the dialysis bags at the beginning and at the end of the experiment.

   ____

   ____

6. What general conclusions can you draw from the results of this experiment?__________

   ____

   ____

Questions for Review

EXERCISE **55**

THE KIDNEY AND FLUID AND ELECTROLYTE BALANCE

Name ______________________________

Laboratory Section ______________________ Date ____________

1. Briefly describe the general events involved with urine formation. ______________________________

____

____

____

2. List the three distinct processes involved with urine formation.

   a. ____________________ c. ____________________

   b. ____________________

3. Define or explain:

   a. Glomerulus ______________________________

   ____

   b. Nephron ______________________________

   ____

   c. Antidiuretic hormone ______________________________

   ____

   d. Aldosterone ______________________________

   ____

4. Give the function of the following structures in urine formation.

   a. Glomerulus ______________________________

   ____

   b. Proximal tubule ______________________________

   ____

DEMONSTRATION OF DIALYSIS

5. Define dialysis. ______________________________

____

____

6. What is hemodialysis? ______________________________

____

7. List and briefly describe two types of systems used for hemodialysis.

   a. ______________________________

   ____

   b. ______________________________

   ____

8. List three conditions in which hemodialysis is necessary.

   a. ______________ c. ______________

   b. ______________

9. Define or explain the following.

   a. Electrolyte ______________________________

   ____

   b. Osmotic pressure ______________________________

   ____

   c. Artificial kidney ______________________________

   ____

   d. Reagent ______________________________

   ____

EXERCISE 56

ROUTINE URINALYSIS

Urine is a watery solution containing nitrogenous waste and inorganic salts that are removed from blood plasma and eliminated by the kidneys. Table 56-1 lists the general properties of normal urine. Table 56-2 lists the average amounts of constituents of normal urine.

Table 56-1 Physical Properties of Urine

Property	Normal Feature	Abnormal Features
Color	Generally light yellow to amber	Dark red or brown
Odor	Aromatic	Fishy, ammonialike
Turbidity (cloudiness)	Clear	Hazy, may contain obvious shreds of mucus
Specific gravity (at 25°C)	1.002–1.030	Any major change from normal range
pH (hydrogen ion concentration)	6.0 average, but ranges from 4.8 to 7.5	Any major variation from normal range
Volume	1–2 liters per 24 hr	Any major variation from normal range

Urine and blood are analyzed more frequently than any other body fluid. As with blood, the composition of urine is a functional indicator of the state of body health at any given period of time. Some abnormal constituents and the types of conditions responsible for their appearance in urine are listed in Table 56-3. The detection of several such abnormal constituents can be achieved by tests that are based on color changes caused by enzymes or chemical reagents. These tests may be qualitative (showing the presence of an abnormal constituent) or quantitative (showing the concentration of the constituent). For most of these tests (Table 56-4) the color reacting substance has been impregnated on plastic strips so that a urine sample can be easily applied. An example of a diagnostic urinary test strip demonstrating the features described here is Phenistix. It is used for the detection of the hereditary metabolic defect, phenylketonuria (PKU). A baby born with this defect is unable to synthesize adequate amounts of the enzyme phenylalanine hydroxylase. As a consequence, the amino acid phenylalanine is not converted to the compound tyrosine, but to phenylpyruvate. Excessive amounts of phenylpyruvate in the baby's body can damage its central nervous system. If PKU is detected early, foods that contain phenylalanine can be eliminated from the diet and brain damage can be avoided. The Phenistix is based on a reaction between a ferric ion reagent and phenylpyruvic acid to produce a gray-green color. The diagnostic reagent strip with a test patch containing the reagent is immersed in the urine specimen, tapped to remove excess urine, and in 30 seconds compared to a color chart on the strip container. As little as 8 mg of phenylpyruvate per deciliter can be detected. A blood test generally is recommended with positive Phenistix results. Table 56-4 describes several commercially available urine diagnostic test strips. Tests using reagent strips will be performed in this exercise.

Table 56-2 Solid Constituents of Normal Urine

Constituent	Daily Normal Values
Bilirubin	0.02 mg
Calcium	30–75 mg
Creatinine	1 g
Glucose	130 mg
Sodium	125–260 meq (milliequivalents)
Urea nitrogen	6–17 g
Uric acid	250–750 mg
Average amount of solids	60 g

Table 56-3 Abnormal Constituents of Urine and Associated Causes

Abnormal Constituent and Term Used to Indicate Its Presence	Associated Causes
Acetone bodies or ketone bodies (acetonuria or ketosis)	Diabetes mellitus Starvation Diarrhea in children Pernicious vomiting of pregnancy
Albumin (albuminuria)	Damaged kidney tubules Chemical poisoning Prolonged exercise Cold baths Altered glomeruli
Bile	Obstructive jaundice (blockage of bile duct resulting in the yellowing of tissues)
Bilirubin	Liver disorders Obstructive jaundice
Glucose (glycosuria)	Diabetes mellitus
Hemoglobin	Conditions involving marked blood destruction, such as: Chemical poisoning Poisonous snake bites Malaria Transfusion with incompatible blood
Leucocytes and other components of pus (pyuria)	Cystitis (infection of the urinary bladder) Infection of the kidney Urethritis (infection of the urethra) Tuberculosis of the kidney
Red blood cells (hematuria)	Infection of glomeruli Tumors of the urinary tract Kidney and bladder stones Tuberculosis Circulatory system disorders
Sugars, galactose, and fructose	Severe liver disease Inherited metabolic defect

Table 56-4 Examples of Commercially Available Urinary Test Strips

Product Name and Manufacturer	Tests for the Presence of	Color Reaction for Definite Positive Reaction	Significance of Results	Limitations
Chemstrip (Bio-Dynamics)	Leukocytes Nitrates	Purple Red-violet	Both tests used in combination detect possible urinary tract infection	Certain antibiotic drugs interfere Large amounts of ascorbic acid (orange juice, vitamin C tablets) decrease sensitivity
Clinistix (Ames)	Glucose	Purple-blue	Results can be used to adjust diet and/or medication for diabetics and others	Vitamin C, aspirin, and some drugs may produce a false negative result
Combistix* (Ames)	pH Glucose Protein	Orange-yellow, and Green-blue Green-brown Green-blue	Test results may provide information concerning kidney function, acid–base balance, and carbohydrate metabolism	False positive reactions may occur with alkaline, highly buffered specimens
Diastix (Ames)	Glucose	Green-brown	Highly specific for glucose	Generally none
Ketostix (Ames)	Urinary ketones (acetoacetates and acetone)	Lavender-purple	Detects ketoacidosis	Phthalein medications may produce false positive reactions

(continued)

Table 56-4 *(Continued)*

Multistix (Ames)	pH	Broad range of colors depending on pH	Provides information on acid–base balance	Contamination (runoff) from other reaction patches causes color changes
	Glucose	Green-brown	Provides information on carbohydrate metabolism	Vitamin C and aspirin may cause false reactions
	Protein	Green-blue	Provides information on kidney function	Highly alkaline urine and some drugs interfere with test results
	Bilirubin	Various shades of tan	Provides information on liver disorders and diseases	Various drugs and vitamin C may cause false reactions
	Ketone	Maroon	Detects ketoacidosis, indication of carbohydrate metabolism disorders	Certain drugs interfere with test results
	Nitrite	Pink	Detection of urinary tract infection	Drugs and vitamin C may cause false results
	Urobilinogen	Brown-orange	Sensitive to elevated urobilinogen secretion	Certain drugs may produce false reactions
	Blood	Green spots	Highly sensitive to the presence of hemoglobin and indicates possible infection	Vitamin C, microorganisms, and elevated protein may produce false reactions
Phenistix (Ames)	Urinary phenylketones	Blue-gray	Test used for phenylketonuria (PKU) screening	Aspirin, excess bilirubin, and phenothiazines interfere with the test

*A variation of this product, Hema-Combistix, incorporates a test for the detection of occult blood in urine.

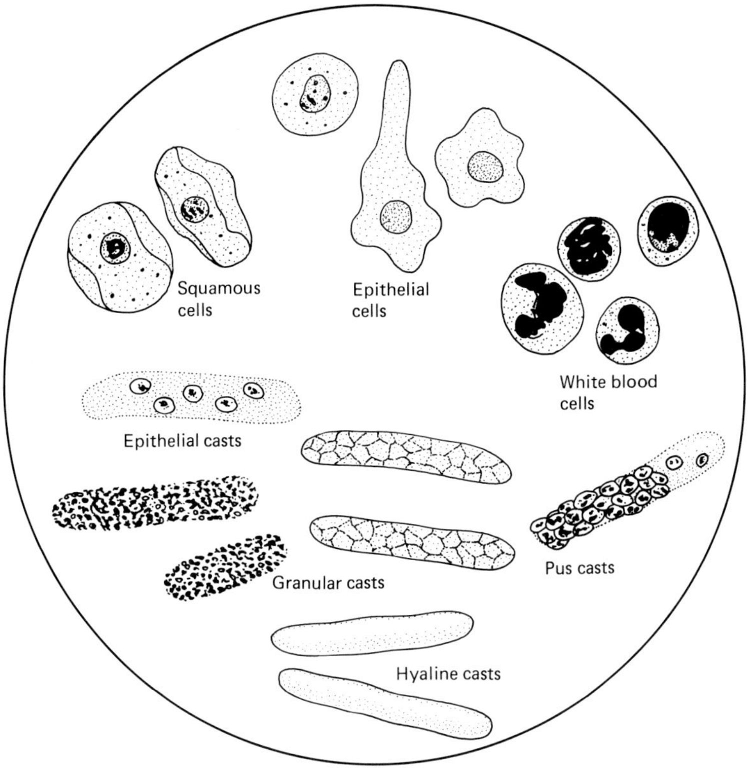

Cells and Casts

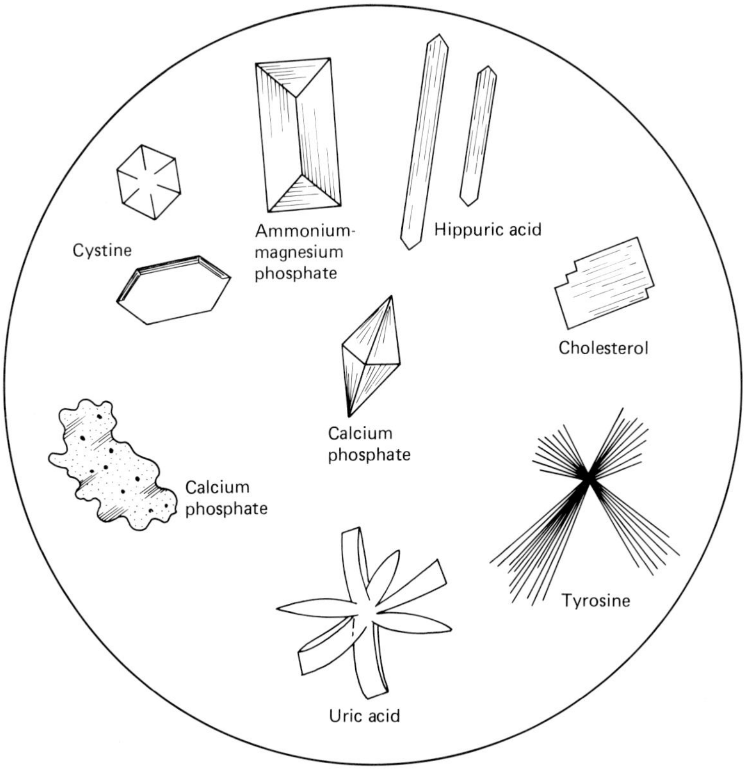

Crystals

Figure 56-1 Elements found in urine sediments.

Microscopic examination of a drop of urine that has been centrifuged may show the abnormal presence of blood and other cells, casts, and renal calculi (Figure 56-1). Casts are small masses of material that have hardened in the lumens of tubules and assumed the shape of the lumen. Renal calculi are tiny stones formed within the kidney.

Normal urine is generally sterile or contains 1000 or fewer bacteria per milliliter because of the microorganisms making up the normal urethral flora. The presence of a microbial infection in any portion of the urinary system usually produces significant changes in urine specimens. The presence of large numbers and kinds of bacteria in urine, or *bacteriuria*, can be determined by specific bacteriological techniques and the use of commercially produced dipstick or related devices (see Figure 56-2).

To be of value a urine analysis must be done on a representative specimen, which should be examined within 30 minutes after it is passed or be stored under refrigeration. Because an early morning specimen is concentrated, it is more likely to reveal abnormalities than urine collected at some other time. In some laboratories standard determinations are semi-automated.

Routine analysis of a urine specimen involves a description of physical properties (color, turbidity, odor, specific gravity, and pH) and determination of the presence or absence of abnormal constituents. A microscopic examination of urinary sediment and tests for the detection and identification of microorganisms such as bacteria also may be required. Several of these aspects of urinalysis are considered in this exercise.

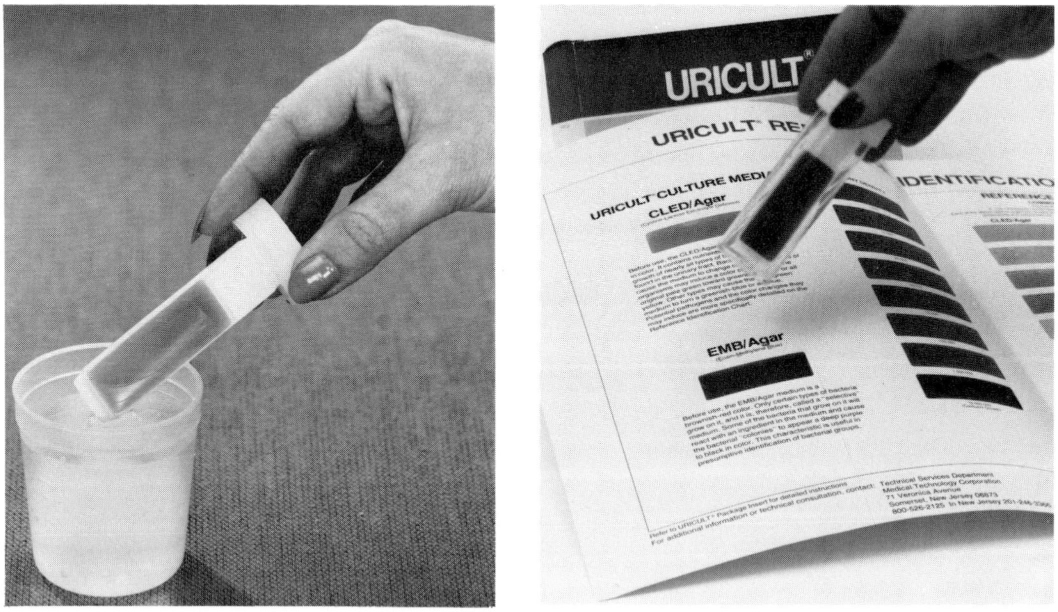

(a) (b)

Figure 56-2 One approach to testing urine for the presence of an infectious disease caused by bacteria. (a) This dip paddle device contains nutrients on which bacteria can grow. It is dipped into a urine specimen to collect the microorganisms. (b) After an incubation period evidence of an infection is clear. Each of the small clumps on the dipstick is a bacterial colony containing 1000 organisms. (Courtesy, Medical Technology Corporation.)

A. DESCRIPTION OF PHYSICAL PROPERTIES

Materials

The following materials should be provided for class use.

1. Urinometer (cylinder and hydrometer float device)
2. pH indicator paper strips (pHydrion or nitrazine test papers)
3. Two freshly voided urine samples
4. Thermometers
5. 250-ml beakers
6. Filter paper
7. Liquid soap and brushes for washing of containers

Procedure

This procedure is to be performed by students individually.

1. Pour about 30 ml of one of the samples into a beaker. Carefully pour the urine from the beaker into a clean urinometer cylinder until it is three-fourths full. Obtain more of the specimen if necessary. If foam forms on the surface, remove it with a piece of filter paper. Discard the paper as directed by your instructor.
2. Note the color, odor, and turbidity of the sample. Refer to Table 56-1. Record your findings in the results and observations section.
3. Check the reading on a thermometer. If it is not at 37°C, shake it down to that level. Insert the thermometer in the urinometer and read the temperature after 3 minutes. The temperature of the sample is used in the determination of specific gravity in the next step. Clean the thermometer as directed by your instructor.
4. Float the hydrometer in the cylinder. Make certain that it is not touching the bottom or the sides of the cylinder. Adjust the value of the hydrometer reading according to temperature to obtain the specific gravity. For each 3° above 25°C, add 0.001 to the hydrometer reading; for each 3° below 25°C, subtract 0.001. Record your findings in the results and observations section. Remove the hydrometer and clean it as directed by your instructor.
5. Dip a pH indicator test strip paper in the urinometer cylinder three consecutive times. Remove excess urine by touching the paper to the inside surface of the cylinder. After 1 minute compare the color of the strip with the color guide on the strip container or other available guide. Compare your findings with the values in Table 56-1. Record your findings in the results and observations section.
6. Clean the urinometer as directed by your instructor.
7. Repeat steps 1–6 with the second urine sample.

B. DETERMINATION OF PRESENCE OF ABNORMAL CONSTITUENTS

Materials

The following materials should be provided for class use.

1. N-Multistix (Ames) and Phenistix (Ames) reagent strips with their respective containers showing test results or individual or other combined test reagent strips for urinalysis, such as *Albustix, Clinistix, Ketostix, Hemastix,* and *Bili-Labstix.*
2. Two freshly voided urine samples
3. 250-ml Pyrex beakers
4. 10-ml pipettes with bulbs

5. Benedict's reagent in dropper bottles
6. Beta-phenylpyruvic acid test solution in a dropper bottle
7. Pyrex test tubes
8. Electric hot plate or other heating system
9. Liquid soap and brushes
10. Pasteur pipettes with bulbs

Procedure 1: N-Multistix Reagent Strips

This procedure is to be performed by students individually.

1. Pour about 125 ml of one of the urine samples into a beaker.
2. Remove one N-Multistix reagent strip from its container. Be careful not to handle the specific color squares. Examine the strip and the identification or test interpretation portion of the label on the container. Note that this strip contains material for determinations of pH, protein, glucose, ketones, bilirubin, blood, nitrite, and urobilinogen.
3. Dip the test strip into the urine sample and tap the excess liquid off on the inside top surface of the beaker. Wait 15–30 seconds and compare the color reactions with the test interpretation label. (Refer to Color Plate 45.) Record your findings in the results and observations section. Discard the test strip as directed by your instructor.
4. Repeat steps 1–3 with the second urine sample.

Procedure 2: Phenistix Reagent Strips

This procedure is to be performed by students individually.

1. Remove one Phenistix reagent strip from its container. Be careful not to handle the specific color squares. Examine the strip and the identification or test interpretation portion of the label on the container. Note that this strip contains material for the determination of phenylpyruvate.
2. Place 2 drops of the beta-phenylpyruvic acid test solution on the strip reaction side. Wait 30 seconds and compare the results with the color chart on the container. Dispose of the strip as indicated by your instructor.
3. Pour about 25 ml of one of the urine samples into a clean beaker.
4. Obtain a fresh Phenistix strip and dip it into the urine sample. Tap the excess liquid off on the inside top surface of the beaker. Wait 30 seconds and compare the color reactions with the test interpretation label. Record your findings in the results and observations section. Discard the test strip as directed by your instructor.
5. Repeat steps 3 and 4 with the second urine sample.

Procedure 3: Quantitative Test for Glucose

This procedure is to be performed by students individually.

1. Pipette 5 ml of Benedict's reagent into two separate Pyrex test tubes.
2. Add 10 drops of the urine sample to each tube and mix by tapping the tubes. Place the tubes in a beaker of water on a hot plate or other heating system and boil for 5 minutes. Allow the tubes to cool and determine the quantity of glucose according to Table 56-5 (See Color Plate 44). Record your findings in the results and observations section.
3. Clean the equipment used as directed by your instructor.
4. Repeat steps 1–3 with the second urine sample.

Table 56-5 Benedict's Reagent Test Results

Color	Glucose	Result Reported
Clear, white turbidity or green turbidity	Zero	Negative
Yellow-green	0.5–1.5	Weak positive
Yellow-orange	1.5–4.0	Moderately positive
Red	4.10 and over	Strongly positive

C. URINARY SEDIMENT EXAMINATION

Materials

The following materials should be provided for class use.

1. Sedi-stain (concentrated stain preparation for urinary sediment examination) with dropper
2. Centrifuge and centrifuge tubes
3. Urine disposal container
4. Two freshly voided urine samples
5. Glass slides and cover slips
6. Microscopes
7. 100-ml capacity graduated cylinders
8. Disposable Pasteur pipettes with bulbs

Procedure

This procedure is to be performed by students individually.

1. Use a graduated cylinder to transfer a urine sample to a centrifuge tube.
2. Centrifuge your sample as directed by your instructor. Usually a speed of 1500 revolutions per minute (rpm) for 5 minutes is sufficient.
3. Without disturbing the sediment in the tube, pour off the fluid into the urine disposal container or as directed by your instructor.
4. Add 2 drops of the Sedi-stain preparation to the urine sediment, and mix by tapping the tube with your index finger.
5. Transfer 1 drop of the stained material to a glass slide. Cover the specimen with a cover slip.
6. Examine the preparation under low and high power. With the aid of Figure 56-1 identify the various items in the sediment. Sketch a representative field in the results and observations section.
7. Repeat steps 1–6 with the second urine sample.

D. DETECTION OF BACTERIURIA

Materials

The following materials should be provided for class use.

1. Bacturcult tubes with counting strip (one per student)
2. Urine specimen with low bacterial count
3. Urine specimen with high bacterial count
4. Sterile small beakers
5. Urine disposal container

Procedure

This procedure is to be performed by students in pairs.

1. Pour a small amount of one of the urine specimens into a beaker.
2. Open one Bacturcult tube and pour the urine specimen into it.
3. *Immediately* pour the urine back into the beaker and allow all the fluid to drain completely from the Bacturcult tube for a few seconds.
4. Replace the screw cap snugly on the tube, loosen the cap about a half turn, and incubate the tube in your laboratory drawer for at least 48 hours.
5. Dispose of the urine specimen as directed by your instructor.
6. Have your laboratory partner repeat steps 1–5 with the other urine specimen.
7. After incubation, place the counting strip around an area of the tube where there appears to be an even distribution of bacterial colonies. Such colonies appear as varying sized specks on the sides of the tube. (Refer to Color Plate 46.)
8. Count the number of colonies within the circle and enter your findings in the results and observations section. If the number of colonies is less than 25, no urinary infection exists. A count of greater than 50 is a definite indication of a urinary tract infection.

E. UNKNOWN SPECIMENS

Materials

The following materials should be provided for class use.

1. All materials listed for the other procedures
2. Containers for urine specimen collection (one per student)
3. Unknown urine specimen (one per student)

Procedure

This procedure is to be performed by students individually.

1. Collect a sample of your own urine in the container provided by your instructor.
2. Obtain an unknown specimen from your instructor.
3. Perform the following procedures and tests on both specimens and time yourself while carrying out the entire analysis. Enter your findings and indicate your time in the results and observations section.
   a. Description of physical properties (color, odor, turbidity, pH, and specific gravity)
   b. Chemical determinations for bilirubin, blood, glucose (qualitative and quantitative), ketones, nitrite, and urobilinogen
   c. Bacteriuria determination
   d. Sediment examination
4. Discard the specimens and materials used as directed by your instructor. Clean all glassware.

RESULTS AND OBSERVATIONS

DESCRIPTION OF PHYSICAL PROPERTIES

1. Enter your findings in Table 56-6.

Table 56-6

Property	Urine Sample 1	Urine Sample 2
Color		
Odor		
Turbidity		
Specific gravity		
pH		

2. Were there any unusual properties associated with the urine samples? If so list them.

____

____

____

DETERMINATION OF PRESENCE OF ABNORMAL CONSTITUENTS

1. Enter your findings in Table 56-7. Indicate the following items: the color (C) of the reagent test strip; the units (U) represented by the color reaction if appropriate; the color (NC) of a normal result; and the units (NU) that correspond to the normal result if appropriate.

Table 56-7

	Urine Sample 1				Urine Sample 2			
Test	C	U	NC	NU	C	U	NC	NU
Protein								
Glucose								
Ketones								
Bilirubin								
Blood								
Nitrite								
Phenylpyruvic acid								
Urobilinogen								
Benedict's								

2. Were any abnormal results observed with the urine specimens? Explain. ______________________________

3. Why should urine specimens be refrigerated if tests cannot be performed soon after collection? ______________________________

URINARY SEDIMENT EXAMINATION

1. Sketch the results of your urinary sediment examination.

Specimen 1 Specimen 2

2. What did your urine specimens contain? ______________________________

DETECTION OF BACTERIURIA

1. Indicate the bacterial counts for each of the urine specimens tested.

   a. Specimen 1 ____________

   b. Specimen 2 ____________

2. Which of the specimens showed signs of an infection? ____________

3. Was this procedure a simple one to perform? ______________________________

UNKNOWN SPECIMENS

1. Enter your findings in Table 56-8.

Table 56-8

Procedure or Test	Results for Student Specimen	Results for Unknown Specimen #____
Color		
Turbidity		
pH		
Bilirubin*		
Blood*		
Glucose (qualitative)		
Glucose (quantitative)		
Ketone*		
Nitrite*		
Phenylpyruvic acid*		
Protein*		
Urobilinogen*		
Bacterial count†		
Urinary sediment examination		

*These determinations are to be made with reagent strips.
†This result can be determined only after 24 hours.

2. Were any of the results obtained with the unknown specimen abnormally high? If so, indicate which ones.

____

____

3. How long did it take you to perform the urinalysis? ____________

4. How did your time compare with others in your laboratory?

____

____

Questions for Review

EXERCISE **56**

ROUTINE URINALYSIS

Name ______________________________

Laboratory Section ______________________ Date ____________

1. In Table Q56-1, distinguish between diabetes mellitus and diabetes insipidus as to the properties of urine.

Table Q56-1

Property of Urine	Diabetes Mellitus	Diabetes Insipidus
Volume	a.	b.
Specific gravity (low, normal, high)	c.	d.
Presence of sugar	e.	f.
Color	g.	h.

2. What would a 200,000 bacterial colony count on a dipstick device indicate? ________

____

3. List five abnormal constituents of urine and indicate the possible cause for their appearance.

Abnormal constituent	Possible cause
a. ____________	____________
b. ____________	____________
c. ____________	____________
d. ____________	____________
e. ____________	____________

4. Define or explain the following.

   a. Albuminuria ______________________________

   b. Casts ______________________________

   ____

   c. Urea ______________________________

d. Uric acid ______________________________

e. Creatine ______________________________

f. Glycosuria ______________________________

g. Jaundice ______________________________

h. Phenylketonuria ______________________________

5. List three microbial diseases of the urinary system.

a. ____________________ c. ____________________

b. ____________________

EXERCISE 57

ACID–BASE BALANCE

The body must maintain a fairly constant balance of acids and bases because the functions of cells are modified greatly by any departure from narrow limits of such substances. An acid is any substance that dissociates into hydrogen ions (H^+), while a base dissociates into hydroxyl ions (OH^-). The pH of a solution is a measure of its degree of acidity or alkalinity (basicity). In a healthy individual the pH of body fluids is stabilized between 7.35 and 7.45. The acid–base balance is maintained by mechanisms that control the concentration of H^+ ions of body fluids.

Respiration functions in maintaining acid–base balance because CO_2 concentration affects pH. An increase in CO_2 lowers pH, while decrease raises it. *Kidney excretion* contributes to homeostasis of pH by the secretion and reabsorption of H^+ ions, NH_4^+ (ammonium) ions, and $NaHCO_3$ (sodium bicarbonate).

This exercise considers pH determinations and the buffering capacity of body fluids and other substances.

A. pH MEASUREMENTS

Materials

The following materials should be provided for class use.

1. 0.85 percent NaCl solution (saline)
2. 5 percent albumin
3. Milk
4. Distilled water
5. Orange juice
6. Lysol or other disinfectant solution
7. pH meter
8. 50-ml beakers
9. 10-ml pipettes with bulbs
10. pH 7 buffer solution
11. Distilled water wash bottle
12. pH paper (range 1–11)

Procedure

This procedure is to be performed by students in pairs. Since different types of pH meters are operated differently, your instructor will demonstrate and give specific directions for the pH meter you will use.

1. Examine the pH meter provided for class use. Locate the following parts: electrodes, pH scale or digital display, temperature control, and calibration control.
2. Observe the demonstration of the use of the pH meter by your instructor.
3. Using a 10-ml pipette, introduce a sufficient amount of the pH 7 buffer into a beaker or similar container to cover the tips of the pH meter electrodes.
4. Turn on the meter and allow it to warm up for at least 15 minutes or as directed by your instructor.
5. Adjust the pH meter with appropriate controls so that the meter will show a pH value of 7. Once this is accomplished the meter has been calibrated.
6. Using separate pipettes for the introduction of samples, determine the pH of each specimen provided. Rinse the electrodes between measurements.
7. Determine the pH of each of the specimens with the pH paper. Simply immerse a small piece of the paper in the specimen and compare the resulting color with the scale on the side of the paper dispenser.
8. Enter all findings and answer the questions in the results and observations section.

B. BUFFERING CAPACITY OF BODY FLUIDS AND OTHER SUBSTANCES

Materials

1. The following materials should be provided for class use.
   a. 0.85 percent NaCl solution (saline)
   b. 5 percent albumin solution
   c. 1 percent glucose solution
   d. Blood plasma
   e. Distilled water
   f. 0.1 molar sodium bicarbonate ($NaHCO_3$) solution
   g. 0.01 normal hydrochloric acid (HCl) solution (reagent)
   h. 0.2 percent phenol red solution in dropper bottle (pH indicator)
   i. pH meter
   j. 1-ml and 10-ml pipettes with bulbs
   k. 50-ml beakers
   l. Pipette disposal container
   m. Distilled water wash bottle
2. The following specimens should be provided by students for testing purposes.
   a. Freshly voided urine
   b. Saliva
   c. Containers for specimens

Procedure

This procedure is to be performed by students in groups of four.

1. Using a separate 10-ml pipette for each, introduce the different specimens into separate beakers. There should be enough in each beaker to cover the tips of the pH meter electrodes.
2. Add 4–5 drops of the phenol red pH indicator to each specimen. Mix by swirling the contents of the beaker. (Phenol red is red in alkaline solutions, orange in neutral solutions, and yellow in acidic solutions.) Record the color of the original solution and the changes in color that occur upon the addition of acid in the results and observations section.
3. With the pH meter determine the pH of each specimen.

4. Use a 1-ml pipette to add 4 drops of the hydrochloric acid reagent to each specimen. Note the color changes, if any, and determine the pH.
5. Working with one specimen at a time, add hydrochloric acid drop by drop, measuring the pH after each addition, to determine how many drops of the acid are necessary to bring about a 0.1 pH unit change and a 1.0 pH unit change. Be certain to mix the solutions carefully after each addition of acid.
6. Record all of your findings and answer the questions in the results and observations section.

C. EFFECTS OF EXERCISE ON THE RATE OF CARBON DIOXIDE (CO_2) PRODUCTION

Materials

The following materials should be provided for class use.

1. Distilled water
2. Plastic straws
3. 0.2 percent phenol red solution (pH indicator)
4. Timer
5. 100-ml graduated cylinders
6. 10-ml pipettes with bulbs
7. 0.1 normal sodium hydroxide (NaOH) solution

Procedure

This procedure is to be performed by students in pairs.

1. Fill a beaker with 100 ml of distilled water. Add 5 ml of 0.10 normal NaOH and 5 drops of phenol red indicator (red in alkaline, orange in neutral, and yellow in acidic solutions).
2. Sit quietly and exhale through a plastic straw into the solution. Have your partner determine the time it takes to turn the solution from orange to yellow.
3. Repeat steps 1 and 2 after vigorous exercise such as running up and down stairs or jogging in place for 5 minutes.
4. Change places with your partner and repeat steps 1–3.
5. Record all your findings and those of at least three other laboratory members in the results and observations section.

RESULTS AND OBSERVATIONS

pH MEASUREMENTS

1. Record the pH values for each solution tested in Table 57-1.

Table 57-1

Specimen	Value with pH Meter	Value with pH Paper
Saline		
Albumin		
Milk		
Distilled water		
Orange juice		
Lysol		
*		
*		

*Other specimen provided.

2. What factors affect the pH of body fluids? Explain. ______________________________

____

____

____

BUFFERING CAPACITY OF BODY FLUIDS AND OTHER SUBSTANCES

1. Record in Table 57-2 the pH values for each solution tested.

Table 57-2

Specimen	Initial Color	Initial pH	Color After 4 Drops of HCl	pH After 4 Drops of HCl	Number of Drops of HCl to Produce Changes of:	
					0.1 pH Unit	1.0 pH Unit
Saline						
Albumin						
Glucose						
Blood plasma						
Distilled water						
Sodium bicarbonate						
Urine						
Saliva						
*						

*Other specimen provided.

2. Which of the specimens tested appeared to have the greatest buffering capacity?

____

3. Which of the specimens tested appeared to have the least buffering capacity?

____

4. What factors influence buffering capacity? ______________________________

____

EFFECTS OF EXERCISE ON THE RATE OF CARBON DIOXIDE (CO_2) PRODUCTION

1. Record in Table 57-3 the times required to produce an indicator color change.

Table 57-3

Student	Time at Rest	Time After Exercise

2. Was there a significant difference in the times at rest and after exercise? Explain. ____

____

Questions for Review

EXERCISE **57**

ACID–BASE BALANCE

Name ____

Laboratory Section ______________________ Date ____________

1. Define or explain acid–base balance. ______________________________

____

2. a. What is a chemical buffer? ______________________________

____

b. List two types of buffer pairs used to maintain acid–base balance. ____________

____

____

c. What is the function of a buffer system? ______________________________

____

3. List three mechanisms that maintain the critical acid–base balance in the body.

a. ____

b. ____

c. ____

4. What is pH? ____

____

5. Give the normal pH values for the following substances.

a. Blood ____________________

b. Gastric juice____________________

c. Urine ____________________

6. a. What is metabolic acidosis? ______________________________

____

b. List two situations that produce metabolic acidosis.

____

____

c. What happens to an individual if this form of acidosis is not corrected? ______________________________

____

____

7. What is respiratory acidosis? ______________________________

____

____

8. a. Distinguish between metabolic and respiratory alkalosis.

____

____

b. List two conditions that can result in metabolic alkalosis.

____

____

c. What will happen if metabolic alkalosis is not corrected?

____

____

EXERCISE 58

GAMETE FORMATION, STAGES OF EARLY DEVELOPMENT, AND KARYOTYPING

After completing this exercise, you should be able to:

1. Distinguish between mitosis and meiosis
2. Describe and identify the stages of meiosis
3. Recognize characteristic structures of ovarian tissue and the general appearance of spermatozoa
4. Identify early stages of division of the zygote
5. Prepare a karyotype from a metaphase spread of chromosomes
6. List and describe four structural changes of chromosomes
7. Recognize normal and selected abnormal karyotypes

As indicated in Exercise 4, chromosomes carry the hereditary chemical material, deoxyribonucleic acid (DNA). All cells of the body, except gametes (the sex cells), contain two complete chromosome sets. For humans the total cellular number is 46 or 23 pairs. Twenty-two pairs (44) are referred to as the *autosomes,* and the remaining pair as the sex chromosomes. In the female one chromosome, the X chromosome, is duplicated, while in the male, one X chromosome is replaced by a smaller one, the Y chromosome. Thus, under normal conditions, human body cells contain 46 chromosomes, including either two X chromosomes for a female or one X and one Y for a male.

Meiosis, or reduction division, is a process by which the chromosome number is reduced by half. While mitosis involves one division of the nucleus following a single duplication of the normal or diploid number of chromosomes, meiosis consists of two distinct nuclear divisions resulting in four new nuclei. Meiosis is associated with the production of sex cells. Figure 58–1 diagrammatically shows meiosis in the formation of female sex cells (oogenesis) and male sex cells (spermatogenesis). Only six chromosomes are used in this figure to show how they are distributed during the process.

In organisms that reproduce sexually, the union of the haploid sperm produced by a male parent and the haploid ovum produced by the female parent forms the diploid fertilized egg or *zygote.* The zygote undergoes a type of mitotic division known as *cleavage.* The cell division is so rapid that there is insufficient time for a cell to enlarge between divisions. The number of cells increases, but the cells become smaller and smaller so that the mass of cells that develops is about the same size as the original zygote known as the *blastula* (Figure 58–2). The blastula generally consists of cells organized in a hollow sphere that surrounds a fluid-filled cavity, the *blastocoel* (Figure 58–2b). Processes collectively known as *gastrulation,* convert the single layered blastula into a multilayered structure with functionally

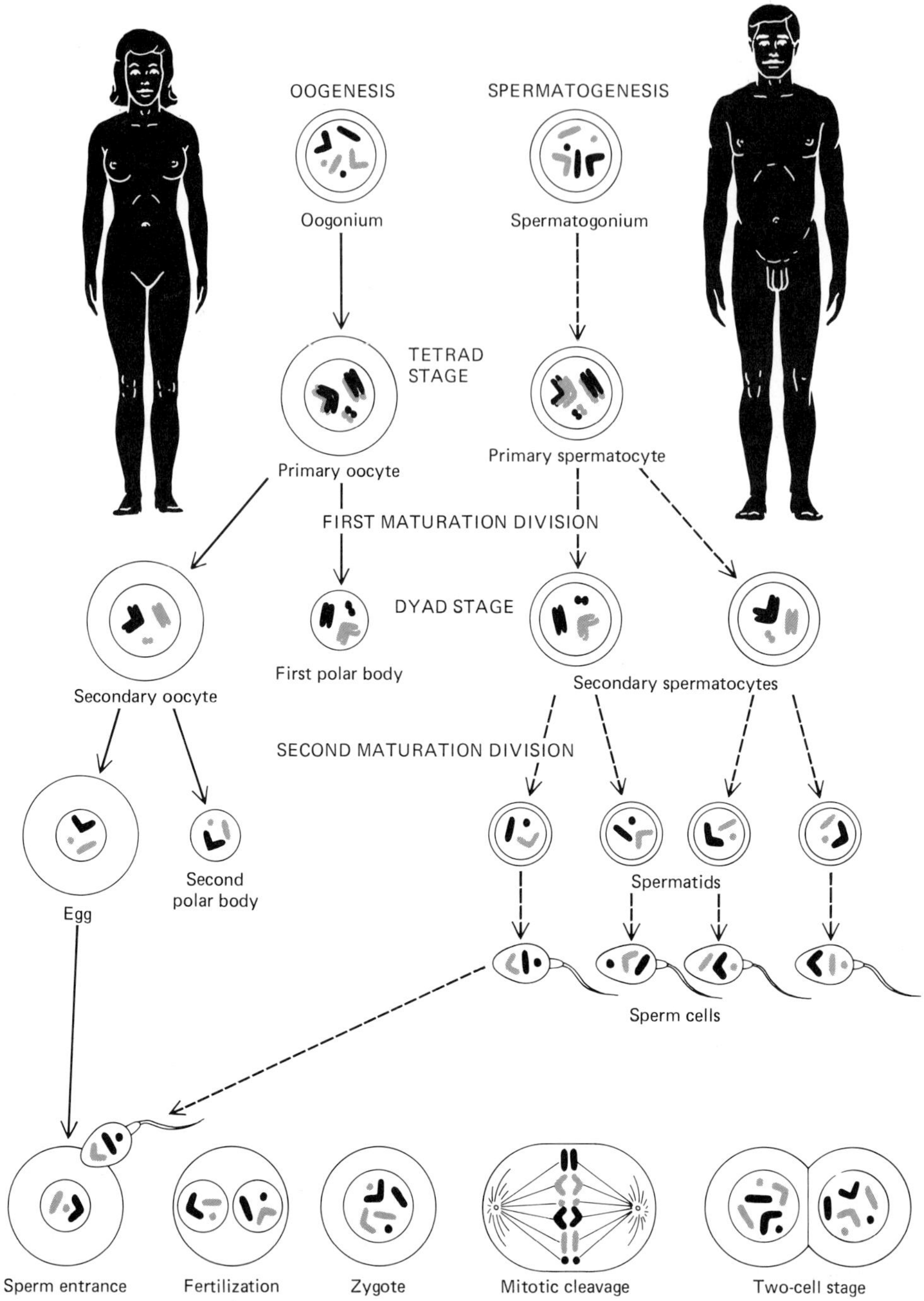

Figure 58-1 Meiosis in female sex cell production (oogenesis) and male sex cell production (spermatogenesis). Each diploid spermatogonium produces four haploid sperm cells. Meiosis is essentially the same in the female, except that the division of the cytoplasm is unequal, and only one functional haploid egg is produced from each diploid oogonium.

specialized areas. These changes in the developing form or embryo depend on a series of cellular movements and the segregation of specific groups of cells. Demonstration slides will be used in this exercise to show these early stages of development.

Since the discovery that human somatic (body) cells contain 46 chromosomes, a number of diseases have been found to be related to abnormalities in chromosome structure and number. Detailed cytogenetic procedures have been established for studying chromosomes. *Karyotyping* is one method commonly used to demonstrate many chromosomal aberrations

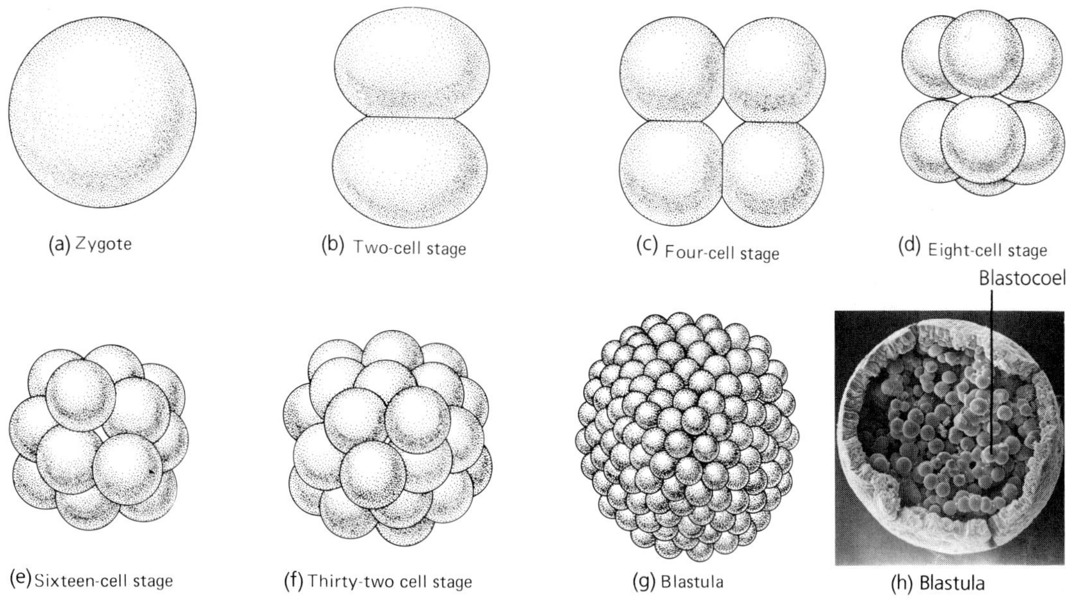

Figure 58-2 Early stages of development. *a* through *g*. The repeated mitotic divisions until the formation of the blastula are known as cleavage. (*h*) A scanning micrograph of the internal features of a blastula (From Komazaki, S. *Cell Differentiation,* 24:25–32, 1988).

(deviations from the normal). The method (Figure 58–3) involves growing cells such as leukocytes or skin cells in tissue culture for a defined period, stopping cell division during metaphase by the use of a drug such as colchicine, staining the chromosomes, and photographing the metaphase spread of chromosomes (Figure 58–4a). Once the photograph is made, the individual chromosomes are cut out and placed in order according to size, shape, and staining pattern. The longest chromosome is 7 to 8 μm, while the shortest one is about 1 to 2 μm. In a routine analysis a chromosomal count is done. There may be more than the normal number or a rearrangement of those present. Differences from the normal number of chromosome arrangement are usually indicative of clinically important genetic defects.

When paired, numbered, and grouped in a standard way, the pattern formed by the chromosomes is called a *karyotype* (Figure 58–4b and Table 58–1). The grouping system makes use of the location of the centromere associated with the chromosome. A chromosome with a centromere in the middle, which divides it into two equal arms, is called *metacentric.* If the centromere is somewhat nearer one end, creating a distinct long and short arm, the chromosome is *submetacentric.* When the centromere is very near to one end, so that the chromosome has a very short arm and longer long arm, the chromosome is called *acrocentric.*

Staining methods called *banding techniques* are used in cytogenetics to identify chromosomes and abnormalities. Banding refers to the horizontal, alternating light and dark crossbands that result from the staining. Different staining techniques are used for different purposes. Some identify and differentiate the chromosomal pairs. Such techniques routinely are used for detecting and defining chromosomal aberrations. Other procedures stain specific chromosome areas to identify special problem situations. The most popular banding technique in G (Giemsa) banding. Other techniques are Q (quinacrine dye) bands and C (centromere staining) bands. Figure 58-5 is a diagrammatic representation of banding patterns revealed by staining and the system used to specify location on the chromosome. The chromosomes are identified by number. The short and long arms are labeled *p* and *q*, respectively. The bands are specified by a two-digit numbering system. Thus, if the chromosome shown in Figure 58–5 were chromosome 1, the light band at the tip of the

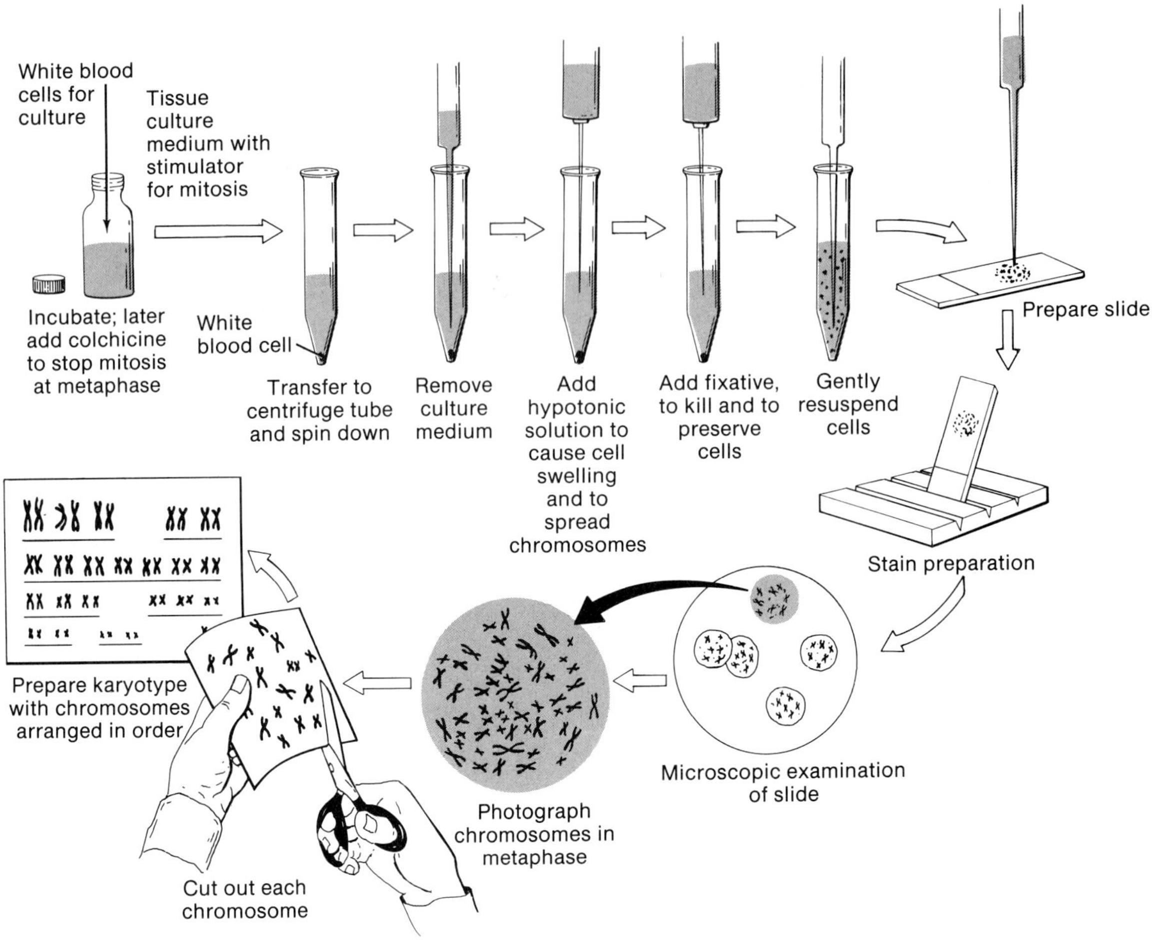

Figure 58-3 The steps in a karyotyping procedure.

short arm would be designated 1 *p*36, and the light band at the bottom end of the long arm would be 1 *q*44.

Chromosomal aberrations or abnormalities are of two types: numerical changes and structural changes. Numerical changes are deviations from the total complement number of 46. Such changes may result from a misdivision during metaphase called *nondisjunction.* In this situation the sister chromatids fail to separate, resulting in one daughter cell with an extra chromosome and one with a missing chromosome. One extra chromosome at a pair location is called *trisomy;* one missing is referred to as *monosomy.* The consequences depend on the chromosomes involved. An example of a trisomy involving an autosome is Down's syndrome or trisomy 21, that is, an extra chromosome at pair 21 (Figure 58–6). Abnormalities involving sex chromosomes include Klinefelter's syndrome, an XXY combination in a male, and Turner's syndrome, a single X chromosome in a female (Table 58–2).

In a karyotype description numerical changes are indicated by a "+" or a "−" placed before a chromosome number to indicate an extra or missing whole chromosome (see the caption for Figure 58–6).

There are several types of structural changes in chromosomes. A *deletion* (del) occurs when a chromosome is broken and a particular fragment is lost. This form of aberration usually results in severe malformations in the individual.

A *translocation* (t) involves a break in a chromosome and the transfer of part or all of one

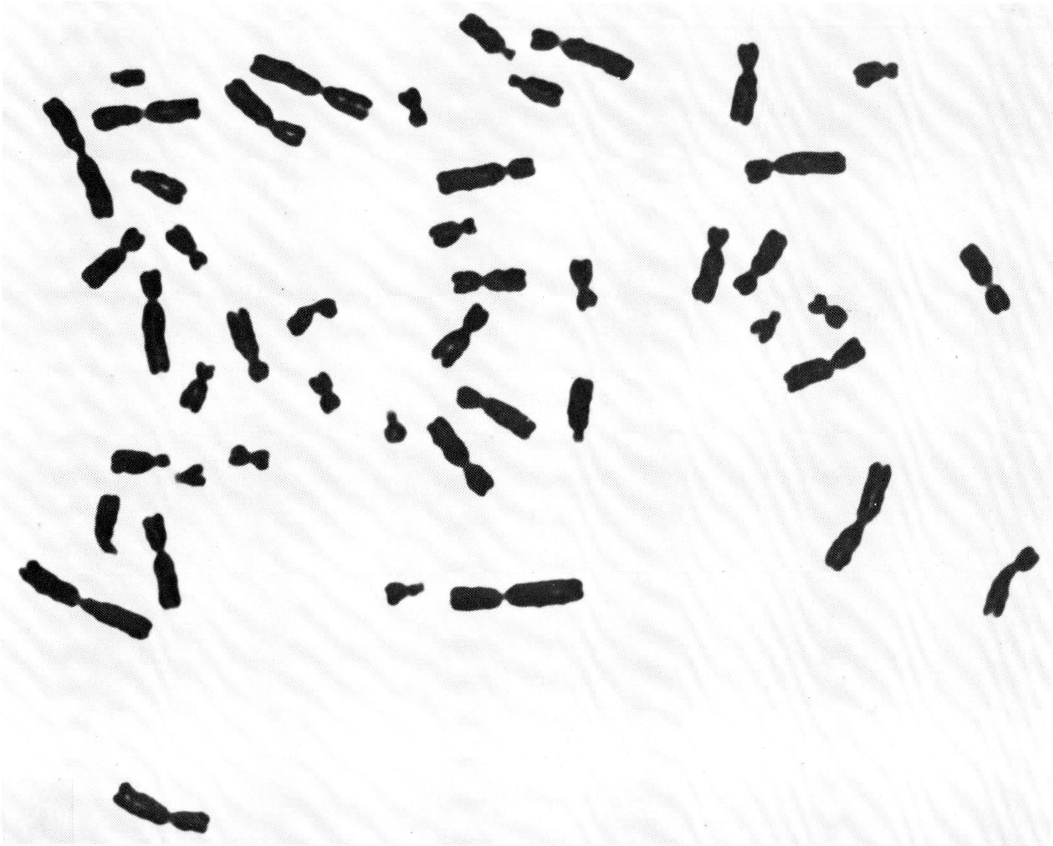

(a)

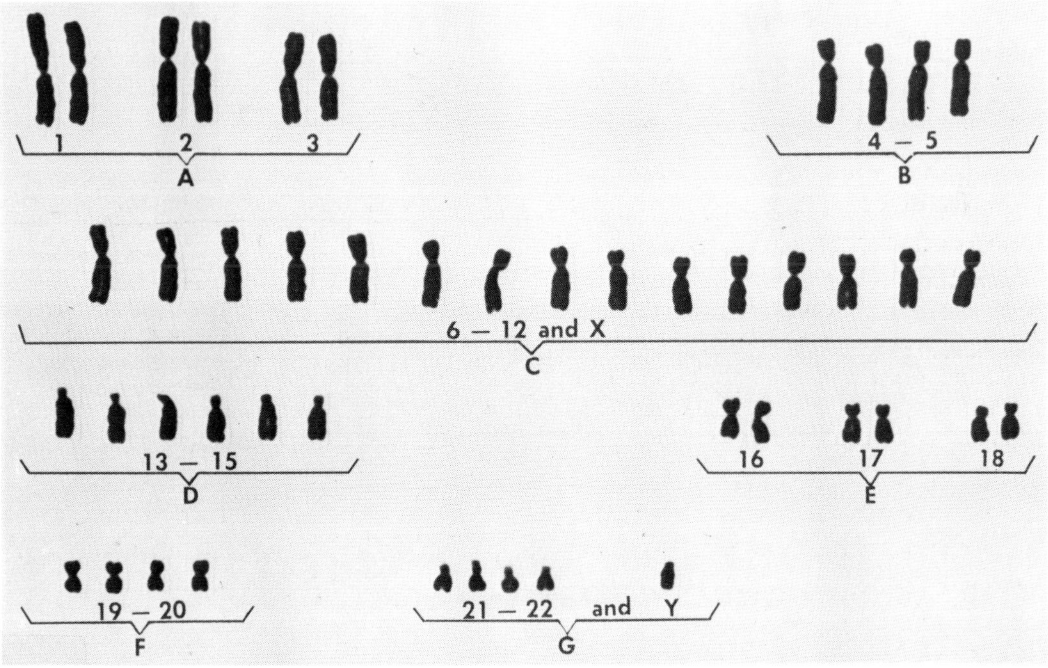

(b)

Figure 58-4 Karyotyping. (a) A metaphase spread from a normal male cell. (b) A karyotype of a normal male. The chromosomes are grouped and arranged according to their shapes and sizes. The X sex chromosome is located at the right of group C. The Y chromosome is located at the right of group G. (From L. Larson, *Amer. J. Med. Technol., 49:*687-697, 1983.)

Table 58-1 The Grouping System for Chromosomes in a Karyotyping Procedure

Group	Description
Group 1–3 or A	The three largest pairs of chromosomes, all of which are metacentric
Group 4–5 or B	Two slightly smaller submetacentric pairs
Group 6–12 and X or C	Different size autosomes, some metacentric and some submetacentric, and X chromosome (sex chromosomes are sometimes placed separately at the lower right of a karyotype as in Figure 58-5)
Group 13–15 or D	Three pairs, nearly as large as those in group 6–12, which are submetacentric and have satellites (tiny masses of chromatic attached to the short arm)
Group 16–18 or E	Three smaller pairs, which are metacentric or submetacentric and lack satellites
Group 19–20 of F	The two smallest metacentric pairs
Group 21–22 and Y or G	The two smallest acrocentric autosomes, which have satellites, and the Y chromosome, which is also acrocentric and typically larger than the others, but the size varies

Table 58-2 Human Chromosomal Aberrations

Abnormallity	Type of Chromosome	Total number of Chromosomes and Abnormal Pair	Resulting Conditions
Down's syndrome	Autosome	47, trisomy 21	Growth stunted: characteristic facial changes; mental retardation; increased susceptibility to infection; affects both sexes
Klinefelter's syndrome	Sex	47,XXY	Affects males; atrophic (abnormally small testes; enlarged breasts; secondary sex characteristics underdeveloped or lacking; usually sterile
Turner's syndomre	Sex	45, X–	Affects females; small, webbed neck; secondary sex characteristics lacking; ovaries underdeveloped; absence of menstruation
XXX female (superfemale)	Sex	47,XXX	Generally appears normal; slightly depressed mental ability; can produce normal children

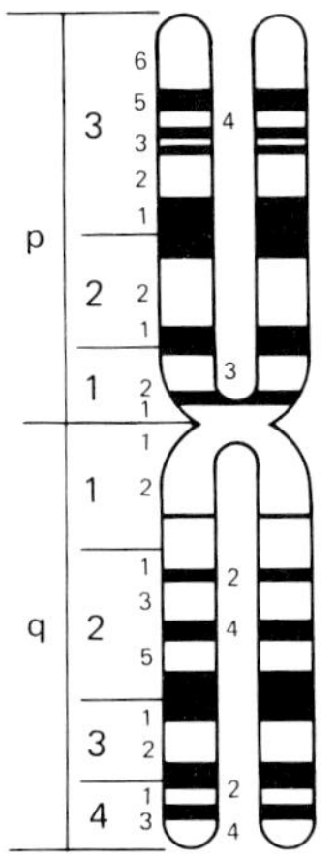

Figure 58-5 A diagrammatic representation of the human chromosome's banding patterns revealed by the different kinds of staining techniques. The letter *p* represents the short arm, *q* the long arm. Two numbers specify the bands.

chromosome to another. In a karyotype description translocations are indicated by a "t" placed before the numbers of chromosomes involved given in parenthesis. The break points are also indicated. For example, t(2:10) (*p*24;*q*26) specifies a translocation involving chromosomes 2 and 10, breaks occurring on the short arm (*p*) at band 24 and on the long arm (*q*) at band 26.

An *inversion* (inv) results when a section breaks off a chromosome and reattaches in an inverted position. The section may be one end of a chromosome or a fragment created by two breaks within the chromosome.

An *isochromosome* (i) is caused by an unequal division of the centromere.

A *ring chromosome* is formed when the ends of both the long and short arms of a chromosome break off and then reattach to each other to form a ringlike structure.

This exercise includes the preparation of karyotypes and identification of chromosomal parts and abnormalities.

A. DEMONSTRATIONS OF MEIOSIS, SEX CELLS AND CLEAVAGE STAGES

Materials

The following materials should be provided for class use.

1. Prepared slides of roundworm (*Ascaris equorum*) eggs showing meiosis
2. Models or charts showing stages of meiosis
3. Prepared slides of human spermatozoa
4. Prepared slides of ovarian tissue showing Graafian follicles
5. Prepared slides of starfish cleavage stages
6. Microscopes for demonstration slides
7. Charts and diagrams showing cleavage stages and general appearance of sex cells

Procedure 1: Meiosis

This procedure is to be performed by students individually.

1. Examine a prepared slide of portions of the oviducts of an *Ascaris*. The preparation should show eggs undergoing meiosis.
2. With the aid of Figure 58-1 and other material showing the stages of meiosis, identify metaphase, anaphase, and telophase of the first meiotic division and anaphase and telophase of the second meiotic division.

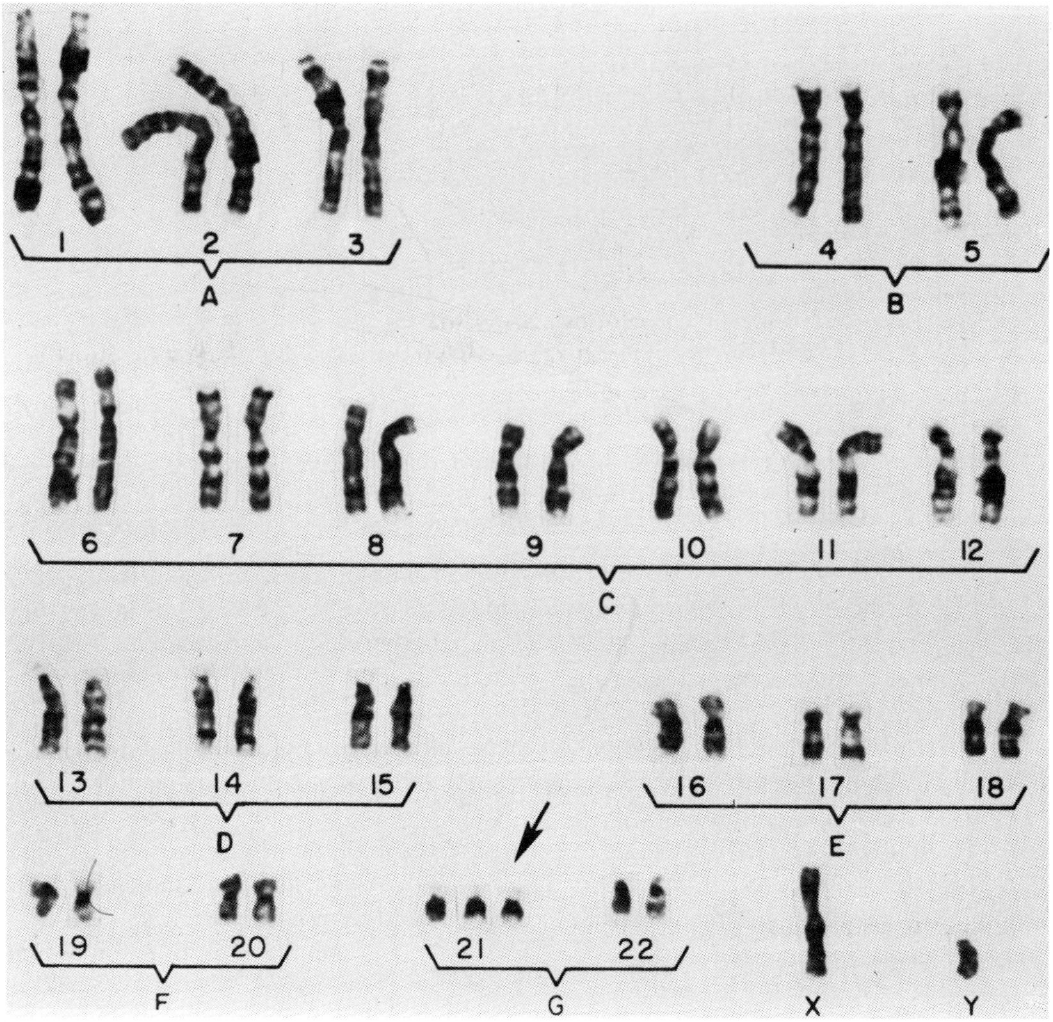

(a)

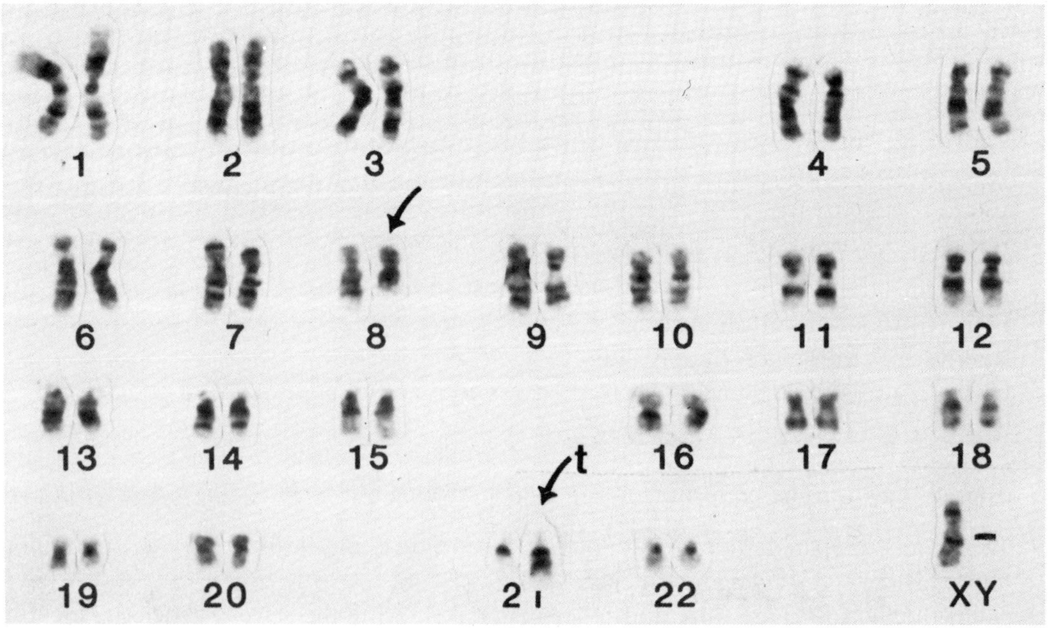

(b)

Figure 58-6 Karyotypes. (a) A karyotype of a patient with trisomy 21 (arrow). G (Giemsa) bands can be seen on the individual chromosomes. (b) A G-stained karyotype of a patient exhibiting a translocation (t) involving pairs 8 and 21. Note also that the Y chromosome is missing (− Y) and therefore the individual has only 45 chromosomes. This condition is found in individuals with a form of acute leukemia. The karyotype is expressed as 45, X − Y, t (8:21). (From A. Cork. *Amer. J. Med. Technol., 49:*703-714, 1983.)

3. Sketch and label these stages in the results and observations section.
4. Count the number of chromosomes (as best you can) in the anaphase of both meiotic divisions.
5. Answer the questions in the results and observations section.

Procedure 2: Sex Cell Demonstration

This procedure is to be performed by students individually.

1. Examine the demonstration slide of ovarian tissue.
2. With the aid of Color Plate 47 and other references locate the following structures: Graafian or vesicular follicle, the fluid-filled cavity of the follicle, and a primary oocyte within a follicle.
3. Sketch and label a representative view in the results and observations section.
4. Examine the slide containing spermatozoa.
5. With the aid of Color Plate 48 and other references identify the head and thinner tail portions. Look for abnormalities such as two heads or no tail. Sketch and label a normal and an abnormal spermatozoon in the results and observations section.
6. Answer the questions in the results and observations section.

Procedure 3: Cleavage Stages Demonstration

This procedure is to be performed by students individually.

1. Examine the demonstration slide showing cleavage stages in the starfish.
2. Identify the following stages: zygote, two-cell, four-cell, eight-cell, sixteen-cell, and blastula. (See Figure 58–2.)
3. Sketch and label the stages as indicated in the results and observations section.
4. Answer the questions in the results and observations section.

B. KARYOTYPING

Materials

The following materials should be provided per two students.

1. Two copies of Figure 58–7 (both removed from students' manuals) or similar reproductions of photographs of metaphase chromosome spreads
2. One pair of scissors
3. One millimeter ruler
4. Rubber cement or similar material with which to paste

Procedure

This procedure is to be performed by students individually.

1. Examine the chromosome spread in Figures 58–7a.
2. Carefully cut out the chromosomes and arrange them in pairs according to approximate length and general appearance. Refer to Figure 58–4b.
3. Paste the chromosomes on one of the karyotype analysis sheets provided in the results and observations section. Use Figure 58–4b as a guide.
4. Compare your karyotype with Figures 58–b and determine the number of chromosomes; if the karyotype is for a female or male, and if the karyotype is an abnormal one for a human.
5. Record your findings in the results and observations section.
6. Repeat this entire procedure with the chromosome spread in Figure 58–7b.

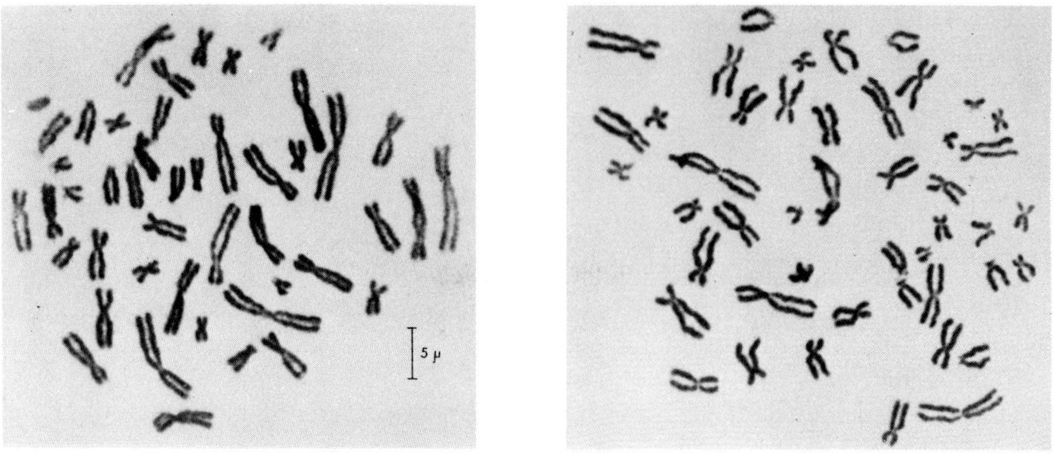

(a) (b)

Figure 58-7 Metaphase chromosome spreads.

RESULTS AND OBSERVATIONS

MEIOSIS

1. Sketch and label stages of meiotic division.

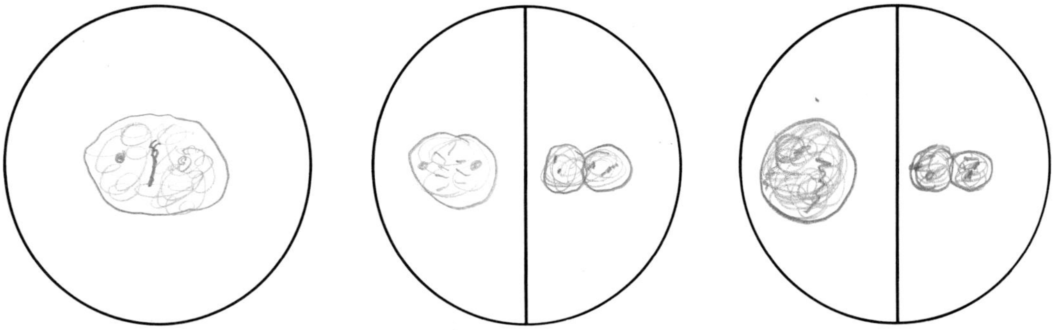

Metaphase, Meiosis I | Anaphase, Telophase, Meiosis I | Anaphase, Telophase Meiosis II

2. Were the same numbers of chromosomes in anaphase, in meiosis I and meiosis II? If not, explain.

No, because they break up + divide

SEX CELLS

1. Sketch representative views of a Graafian follicle and representative spermatozoa.

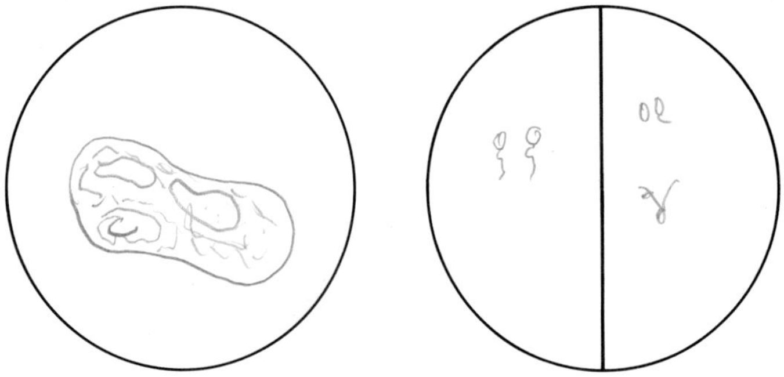

Graafian follicle | Normal | Abnormal

2. Were abnormal spermatozoa present? If so, what percentage of the cells observed were abnormal? 1/2

3. What factors are responsible for abnormal spermatozoa formation? genetics some medication

4. Complete the following diagram of possible meiotic divisions.

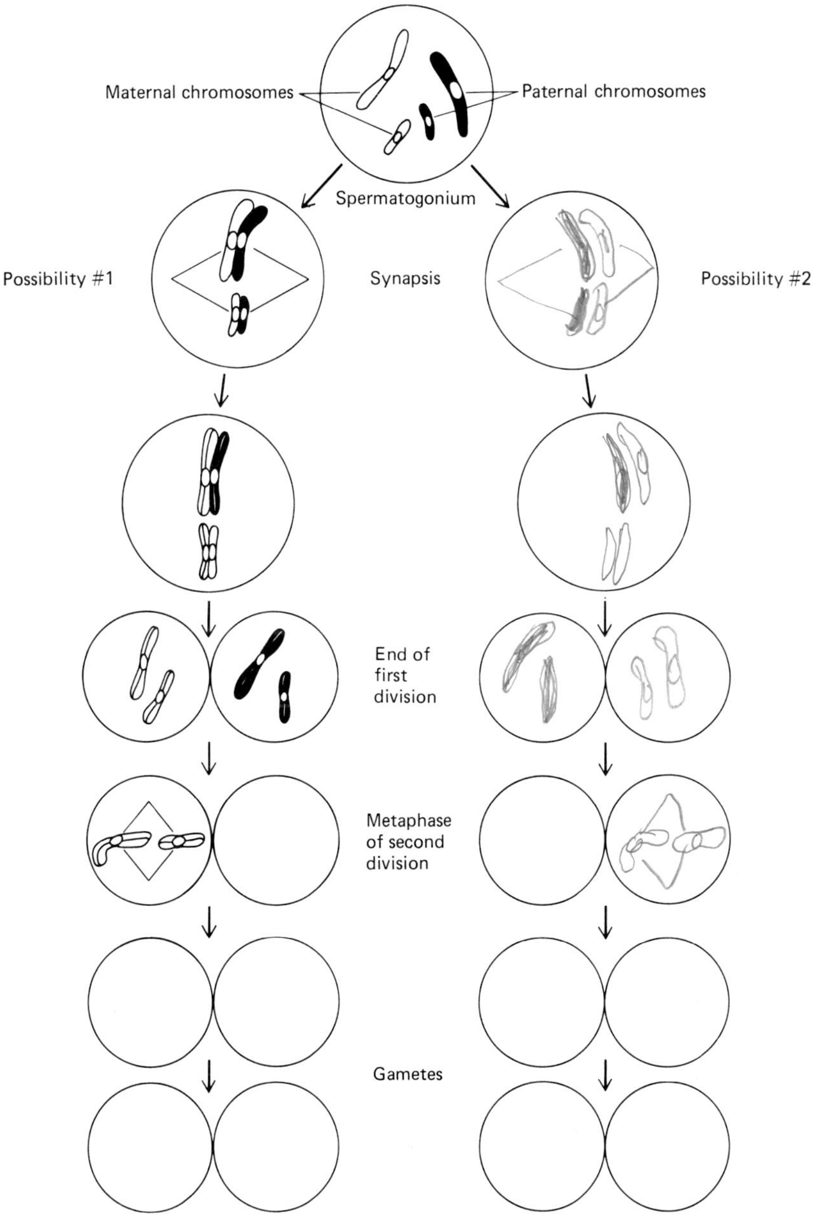

Figure 58-8 Possible meiotic divisions in spermatogenesis. Only two pairs of chromosomes are shown.

CLEAVAGE STAGES

1. Sketch stages of cleavage as indicated.

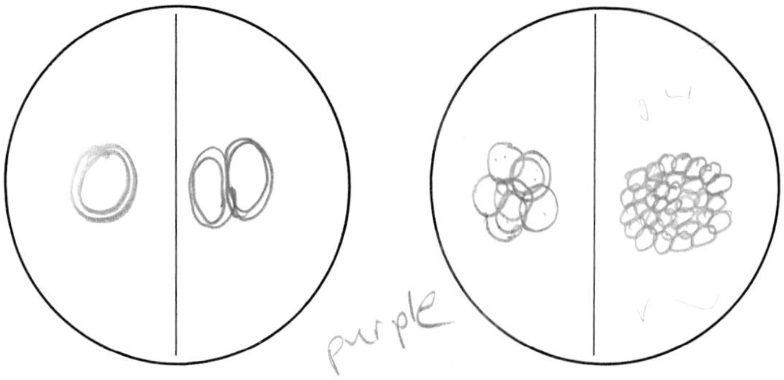

Zygote Two-cell stage Eight-cell stage Blastula

2. Compare the zygote and blastula stages. Which one is larger? Blastula

3. Are each of the cells in the eight-cell and blastula stages the same size? no

KARYOTYPING

1. Examine the karyotypes you prepared from Figures 58–7a and 58–7b.

A

B

1 2 3 4 5

C

6 7 8 9 10 11 12

D

E

13 14 15 16 17 18

F

G

19 20 21 22 X X X Y

Karotype analysis sheet for Figure 58-7a.

2. Which is the karyotype of a human female? ______________ Which is the karyotype of a human male? ______________

3. Are the karyotypes normal? If not list the aberrations. ______________

____

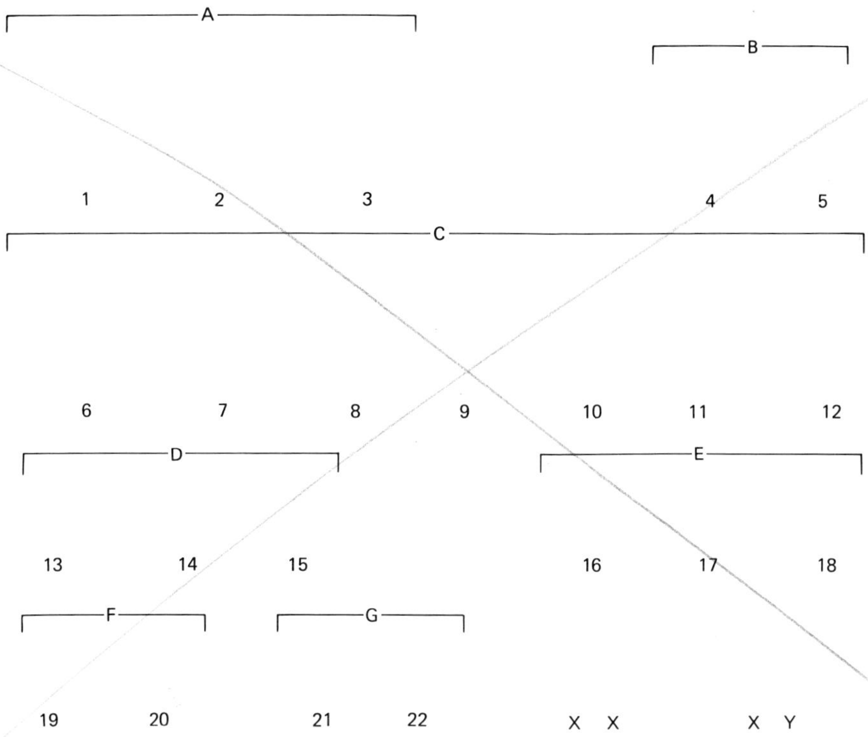

Karotype analysis sheet for Figure 58-7b.

Questions for Review

EXERCISE **58**

GAMETE FORMATION, STAGES OF EARLY DEVELOPMENT, AND KARYOTYPING

Name ______________________________

Laboratory Section ______________________ Date ____________

1. List three features of meiosis that distinguish it from mitosis.

   a. ______________________________

   b. ______________________________

   c. ______________________________

2. Define or explain:

   a. Chromosome ______________________________

   ____

   b. Chromatid ______________________________

   ____

   c. Hapliod ______________________________

   d. Diploid ______________________________

   e. Meiosis ______________________________

   ____

   f. Down's syndrome ______________________________

   g. Nondisjunction ______________________________

3. List and describe four specific structural chromosomal aberrations.

   a. ______________________________

   b. ______________________________

   c. ______________________________

   d. ______________________________

4. Examine the karyotype in Figure Q58-1.

   a. Is the arrangement of chromosomes correct for a normal human? ____________

   b. Is the karyotype typical of a normal human? ____________

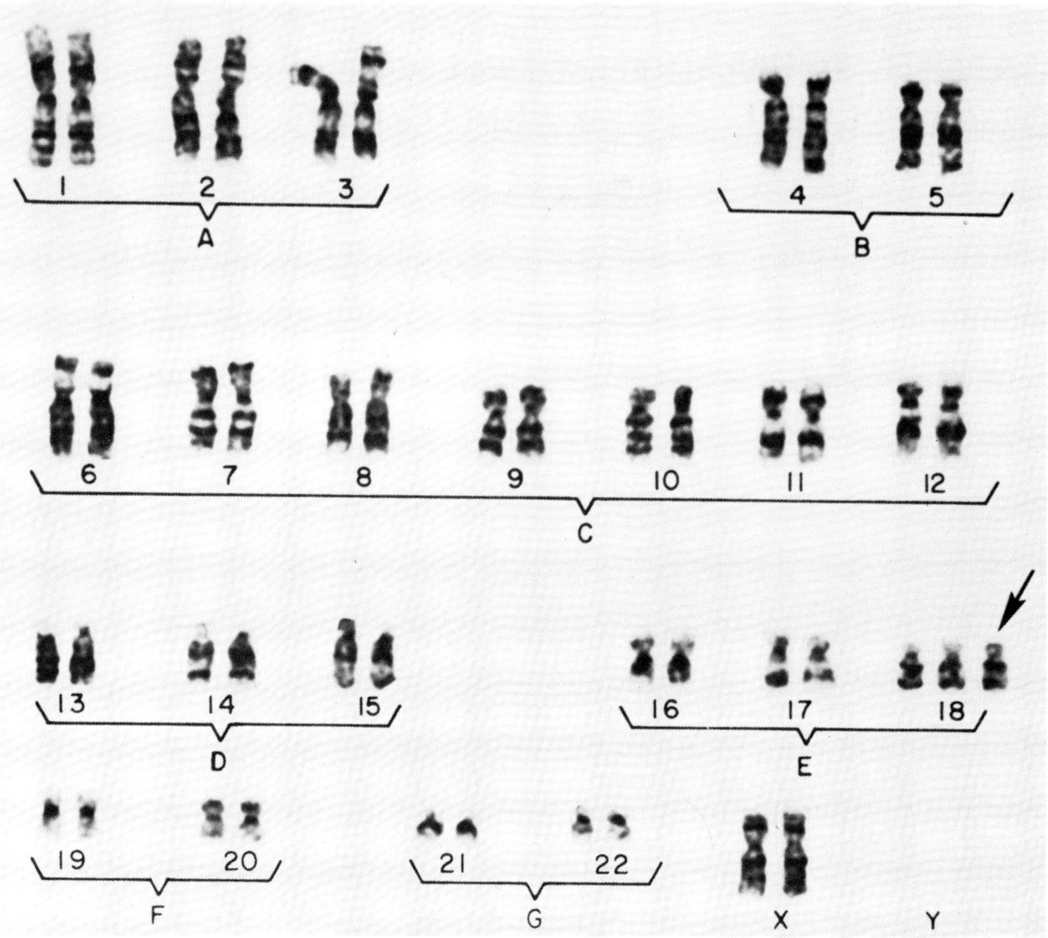

Figure Q58-1 (From L. Larson, *Amer. J. Med. Technol., 49:*687-697, 1983.)

c. What sex is the individual? ____________

d. Describe any observable aberrations. ______________________________

____

EXERCISE 59

HUMAN CHORIONIC GONADOTROPIN AND PREGNANCY TESTING

Human chorionic gonadotropin (HCG) appears in the urine and serum of pregnant women in large concentrations. The hormone is a glycoprotein produced by placenta trophoblastic cells and plays a major role in maintaining the developing corpus luteum. The corpus luteum maintains the state of pregnancy by the production of the hormones progesterone and estradiol until the placenta and fetus begin the secretion of steroid and protein hormones. HCG levels increase rapidly in urine following implantation of a fertilized ovum and may reach detectable levels shortly after conception. Laboratory test demonstration of the hormone's presence is generally indicative of pregnancy. However, false positive results may be obtained if the urine and serum contain high levels of luteinizing hormone (LH), if the individual is menopausal, or if certain cancerous states exist, such as choriocarcinoma and chorioepithelioma, or in situations of a spontaneous abortion. In the latter situation, results remain positive for approximately one week.

The immunoenzymatic test for pregnancy considered in this exercise is an accurate, rapid, and simple procedure. It clearly replaces the various tests that require specialized equipment, radioactive reagents, or the use of animals such as frogs, rats, mice, or rabbits.

The Pregnospia test is an immunoenzymatic assay. It uses a revolutionary type of diagnostic tool, the monoclonal immunoglobulin (antibody). The production of pure or monoclonal (single-type) antibodies is achieved by an in vitro hybridization process in which an antibody-producing cell is fused with a myeloma (particular cancer) cell under special experimental conditions. In this process the genetic materials of both cells are combined to create a hybrid cell that contains desirable characteristics from both of the parent cells.

The Prognosia test is an agglutination immunoassay. The monoclonal antibodies mentioned earlier are coated onto special gold particles, which have an intense reddish-purple color in solution. These antibodies are noted for their specificity for special sites on the intact HCG molecule and one of its polypeptide subunits. When the HCG antibody preparation reacts with HCG present in a urine specimen, connecting molecular bridges are formed, binding the gold particles together and thereby causing them to clump or agglutinate. The resulting agglutination is evidenced by a distinct color change versus the original reddish-purple color of the HCG antibody preparation. If there is no HCG present in the urine sample, the agglutination reaction will not change color. This result is interpreted as a negative reaction.

Materials

The following materials should be provided per two students. (All reagents are available in the commercial kit.)

1. Four tubes of Organon Accusphere reagent (gold sol particles bonded with monoclonal antibodies against HCG)
2. Buffer (2 ml)
3. One 5-ml pipette, calibrated in 0.1-ml units, and a rubber bulb
4. Three 1-ml pipettes, calibrated in 0.1-ml units, and rubber bulbs
5. One known positive urine control
6. One known negative urine control
7. Two unknown specimens (one per student)
8. One marking pen or similar device
9. Container for disposing of pipettes

Procedure

This procedure is to be performed by students in pairs.

1. Obtain four tubes containing the Accusphere reagent.
2. Number the tubes accordingly, 1 (known positive), 2 (known negative), 3 (unknown), and 4 (unknown).
3. Remove the rubber stoppers from the four tubes.
4. With the 5-ml pipette, introduce 0.4 ml of the buffer provided to each tube.
5. Replace the stoppers.
6. Gently shake each tube for approximately 15–30 seconds until the Accusphere reagent is completely resuspended.
7. Using separate 1-ml pipettes, add 0.1 ml of the positive, negative, and the unknown samples to their respectively marked tubes.
8. Gently shake the tubes for about 30 seconds to achieve thorough mixing.
9. Place the tubes in a rack or on a flat surface and wait 5 minutes.
10. Examine the tubes for a color change. If a change does not occur within this time period, continue to observe the tubes for another 25 minutes.
11. Compare the unknown tubes with the results shown in the positive and negative samples (tubes 1 and 2).
12. Interpret your findings and enter them in the results and observation section.
13. Check your results with the instructor.
14. Answer any questions pertaining to this exercise in the questions for review section.

RESULTS AND OBSERVATIONS

1. Enter your findings in Table 59-1.

Table 59-1

Specimen	Agglutination	HCG Present	Indicative of Pregnancy
Known positive			
Known negative			
Unknown #____			
Unknown #____			

Questions for Review

EXERCISE **59**

HUMAN CHORIONIC GONADOTROPIN AND PREGNANCY TESTING

Name ______________________________

Laboratory Section ______________________ Date ____________

1. What is HCG? ______________________________

2. Briefly explain the principle involved in the urinary chorionic gonadotropin in vitro test.

3. List three advantages of the in vitro immunoenzymatic test for the detection of pregnancy when compared to the use of laboratory animals for the same purpose.

   a. ______________________________

   b. ______________________________

   c. ______________________________

4. What is the appearance of the results of the UCG (urinary chorionic gonadotropin) test or comparable test?

   a. Negative result ______________________________

   b. Positive result ______________________________

5. What conditions produce a false positive result? ____________________

____

____

6. List two limitations of the *in vitro* test.

a. ____________________

b. ____________________

Index

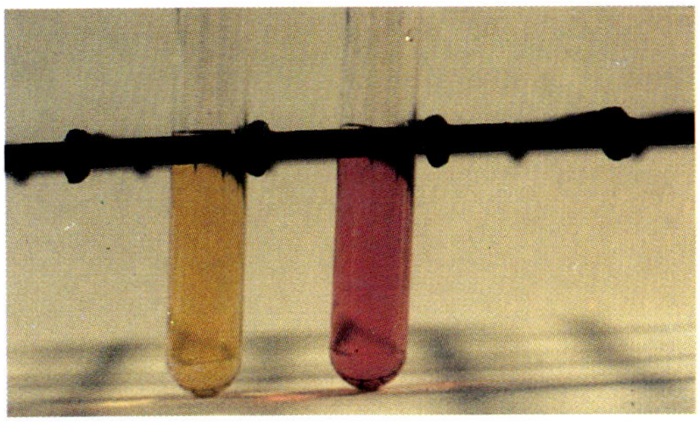

Plate 1 The appearance of a positive (purple) Biuret test.

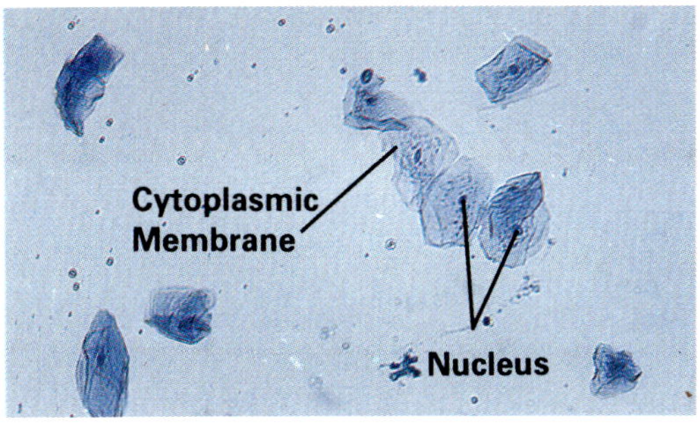

Plate 3 Cheek cells. Note the presence of a nucleus and a cytoplasmic membrane in these cells.

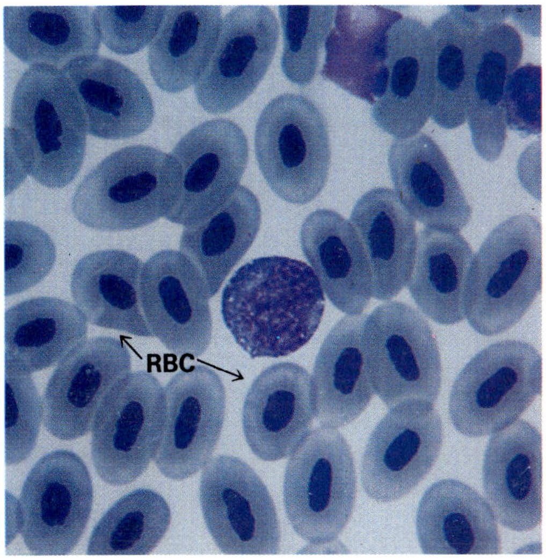

Plate 5 Frog blood. Here both red (RBC) and white blood cells are nucleated. Compare this photo with that of human blood cells.

Plate 2 The letter "e" under low power.

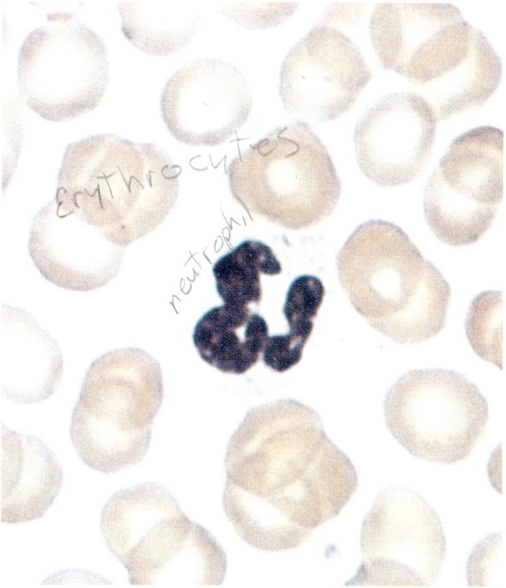

Plate 4 A human blood smear showing erythrocytes (red blood cells) and a neutrophil.

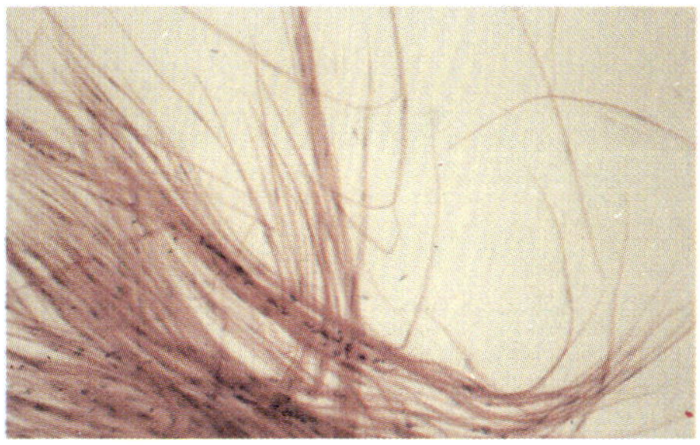

Plate 6 A preparation of smooth muscle cells. (Courtesy Carolina Biological Supply Company.)

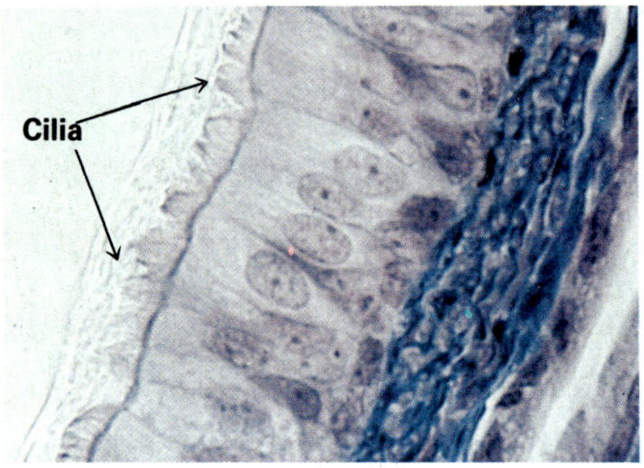

Plate 7 A view of tracheal epithelium. Pseudostratified epithelium bearing cilia and underlying connective tissue are shown in this micrograph. (Courtesy Carolina Biological Supply Company.)

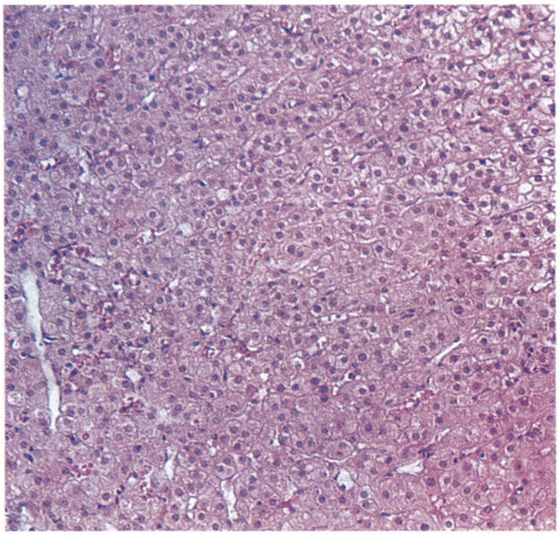

Plate 9 The liver.

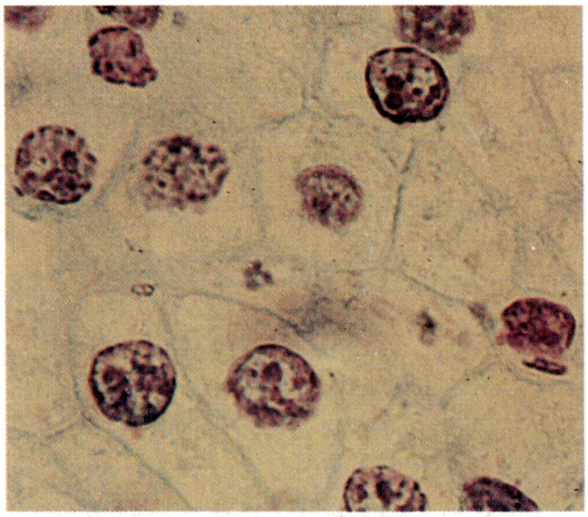

Plate 11 Feulgen reaction showing the purple appearance of DNA in cell nuclei.

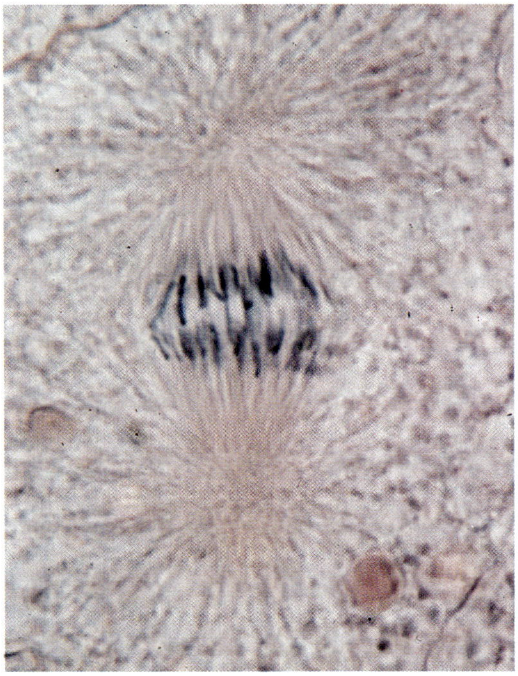

Plate 8 The appearance of an aster. Note the strand composition of the cellular component.

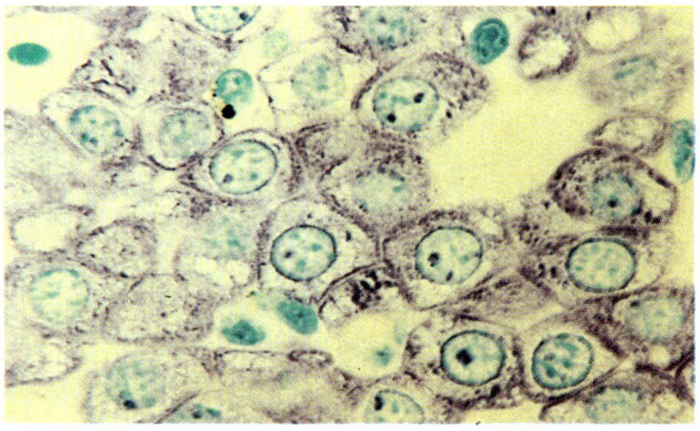

Plate 10 DNA and RNA in animal cells, stained with Azure B. The DNA in the nucleus appears bluish-green and the RNA in the smaller, intranuclear nucleolus appears purple. (Courtesy Ward's Natural Science Establishment, Inc.)

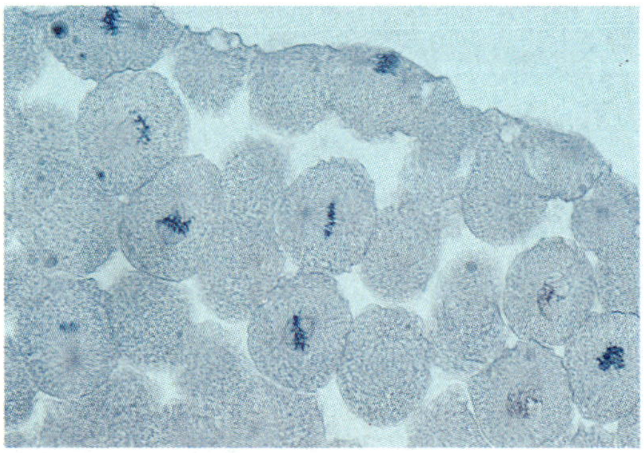

Plate 12 Mitosis in whitefish cells. Note the various stages of mitosis and the appearance of chromosomes.

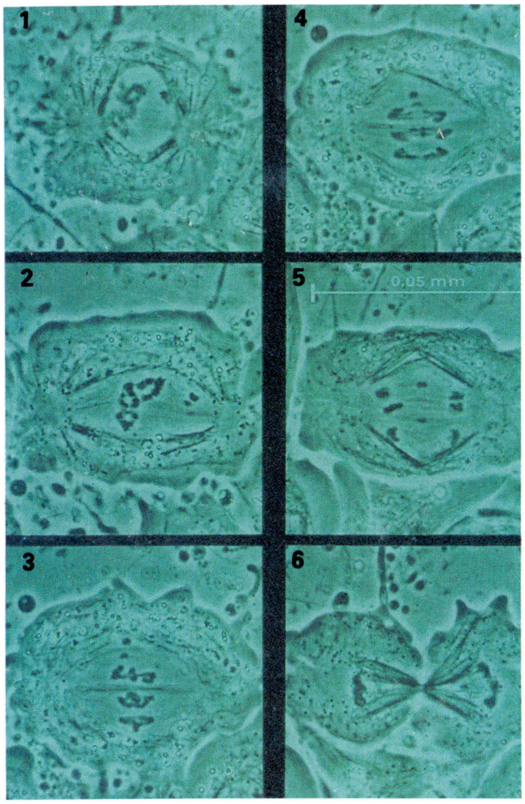

Plate 13 A living cell undergoing mitosis as seen under phase contrast microscopy.

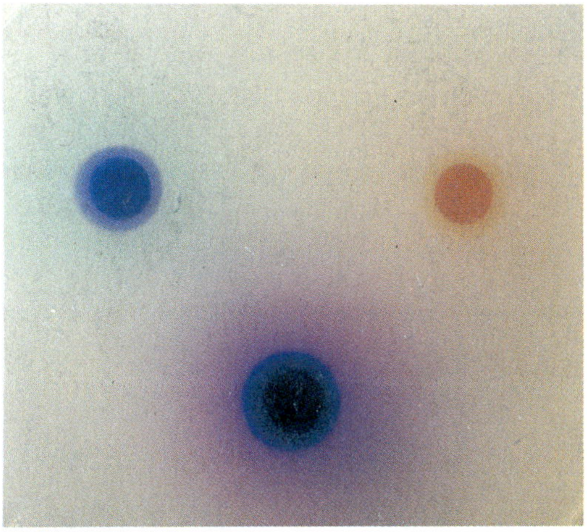

Plate 14 Dye diffusion zones: crystal violet (CR), methylen blue (M13), and safranin (S).

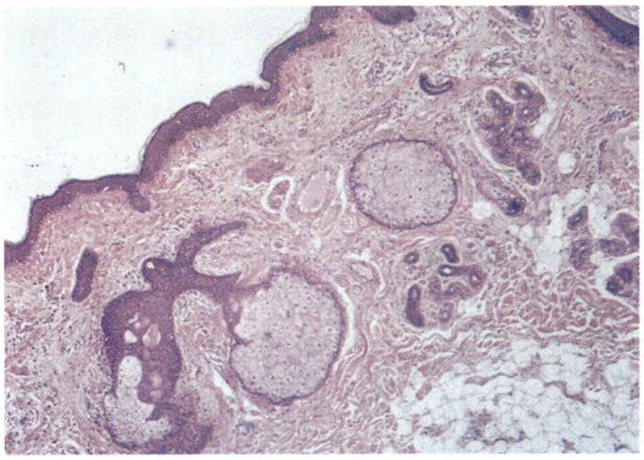

Plate 15 The simple branched acinar sebaceous glands. Sudoriferous or sweat glands are also present.

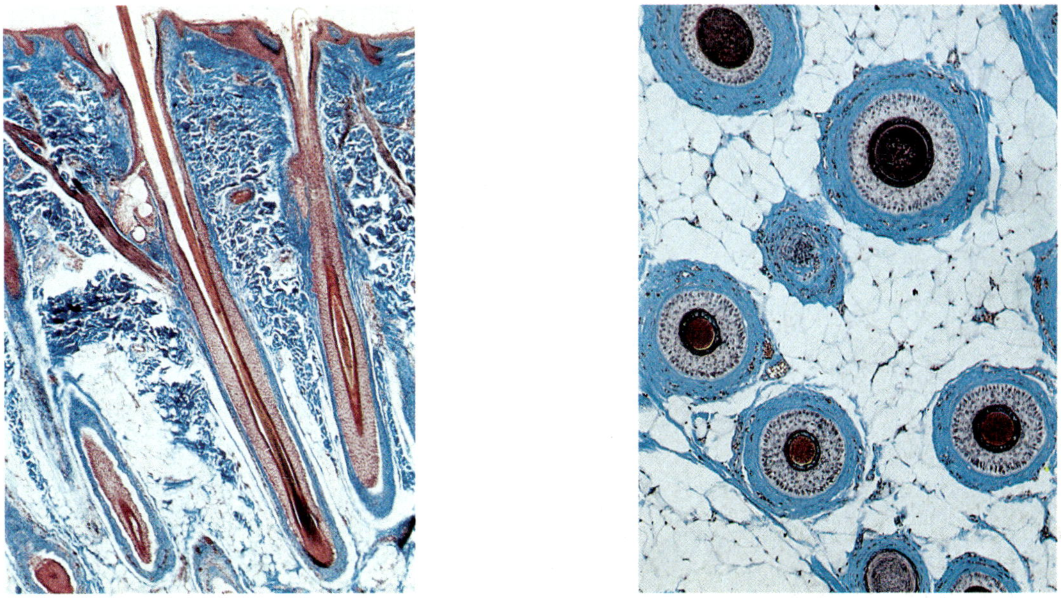

Plates 16a, b Human scalp showing the features of hair follicles. (a) Longitudinal section. Compare this view with Figure 10-10a. Particular structures included here are the hair follicle, sebaceous glands, and arrector pili. (b) This transverse section shows the central cortex, the surrounding cuticle layer, internal root sheath, external root sheath, and the connective tissue sheath. (Courtesy Fisher Scientific Company.)

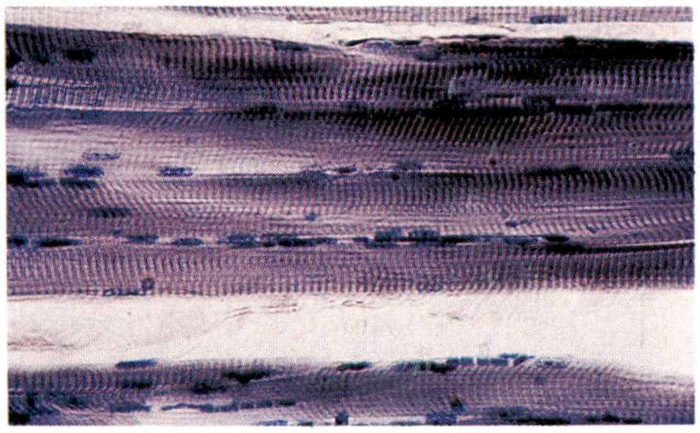

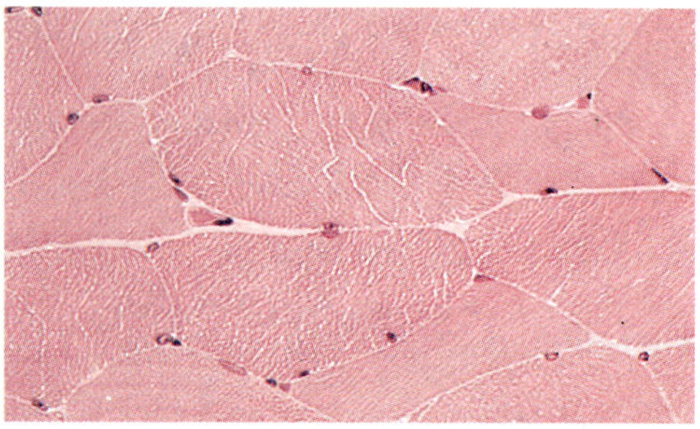

Plates 17a,b Skeletal muscle. (a) A longitudinal section. (b) A cross section. (Courtesy Carolina Biological Supply Company.)

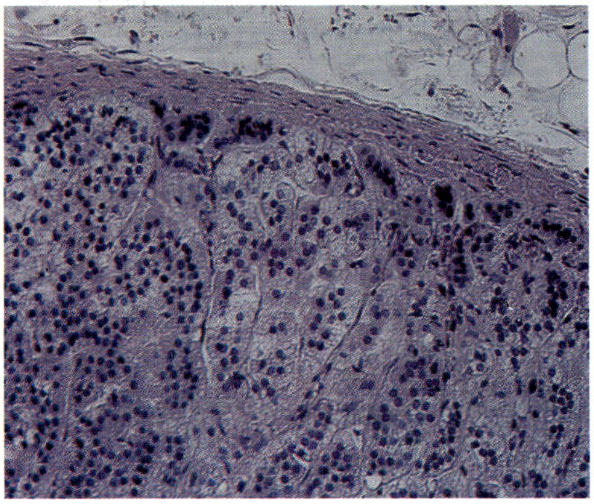

Plate 18 The adrenal gland. Capsule and the zone glomerulosa. This zone contains secretory cells arranged in irregular ovoid clumps separated by connective tissue strands.

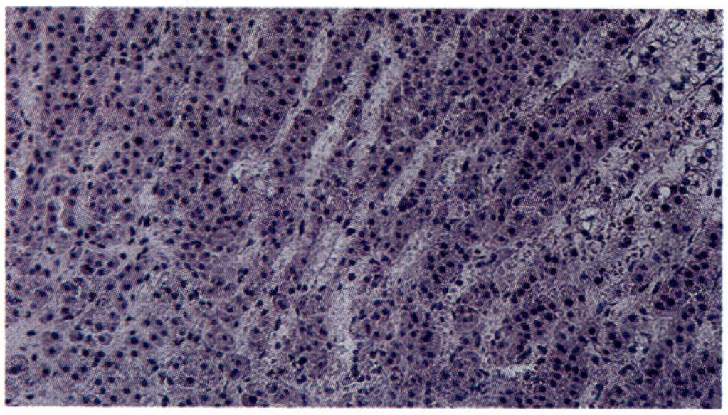

Plate 19 The zona fasciculata, the intermediate and broadest of the three zones of the adrenal cortex. This zone consists of narrow cords of large secretory cells, with lightly staining cytoplasm.

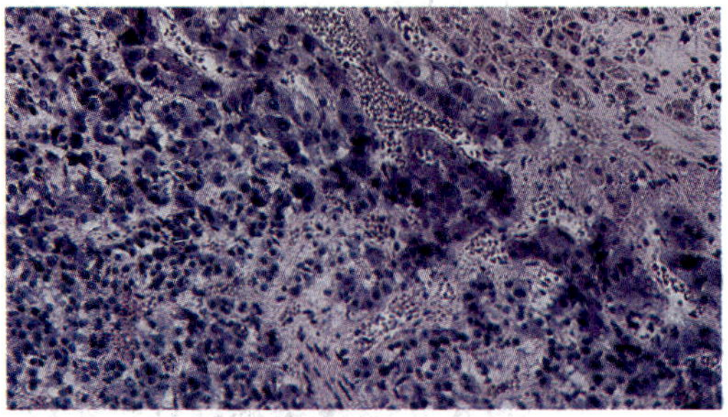

Plate 20 The zona reticularis, the thin, innermost zone of the adrenal cortex. This zone contains an irregular network of bunching cords and clumps of small glandular cells with darkly staining cytoplasm.

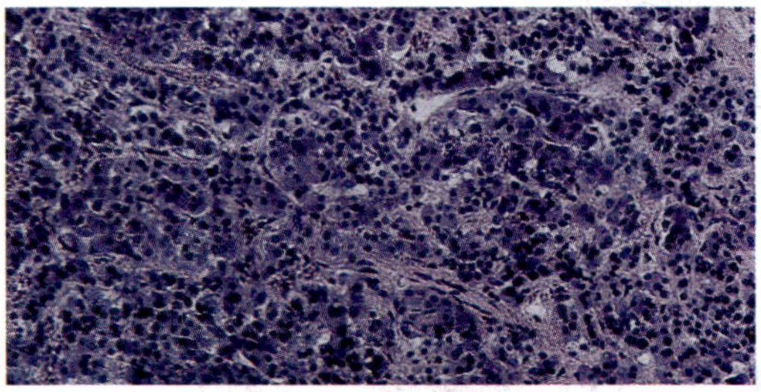

Plate 21 The adrenal medulla consists of closely packed, irregular clumps of secretory cells. These cells have large granular nuclei and dark blue to purple staining cytoplasm.

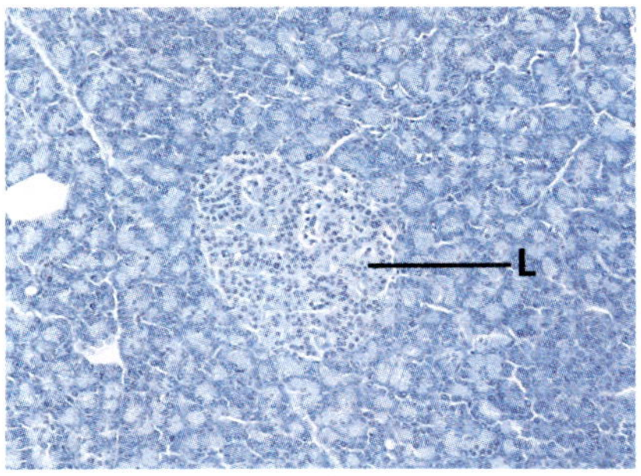

Plate 22 The pancreas. The centrally located islets of Langerhans (L) and the surrounding acini are shown.

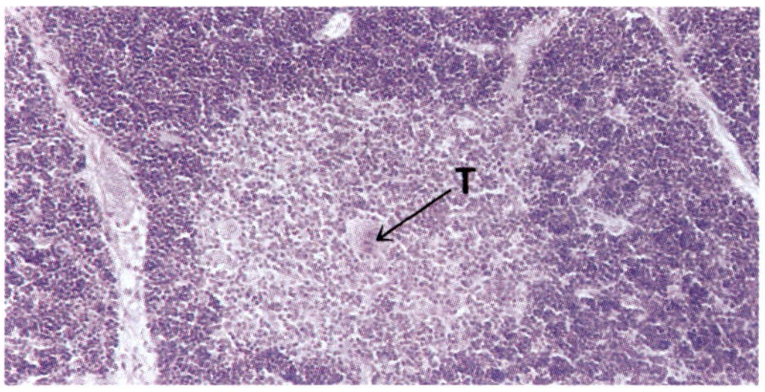

Plate 23 The thymus gland. The lighter, inner medulla region, and the outer, cortex region of a lobule are shown. A centrally located thymic (T) corpuscle also is present.

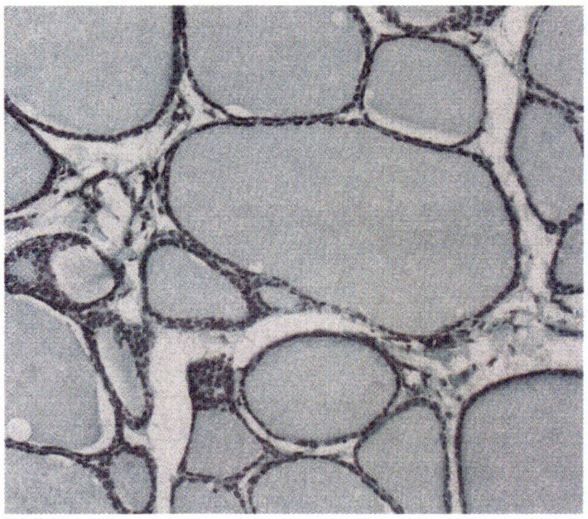

Plate 24 Appearance of a normal thyroid gland.

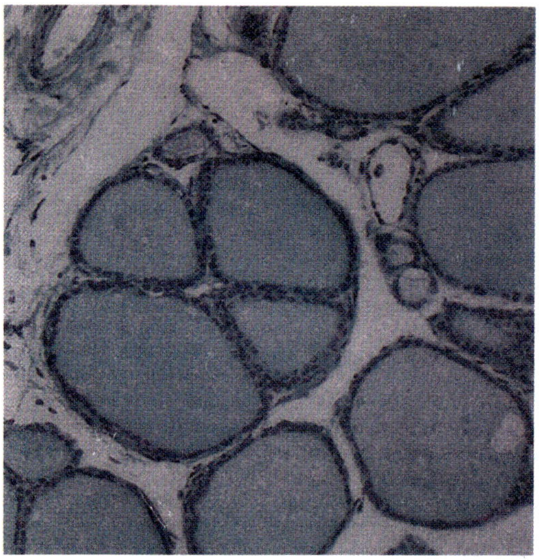

Plate 25 Thyroid preparation from a goiter situation.

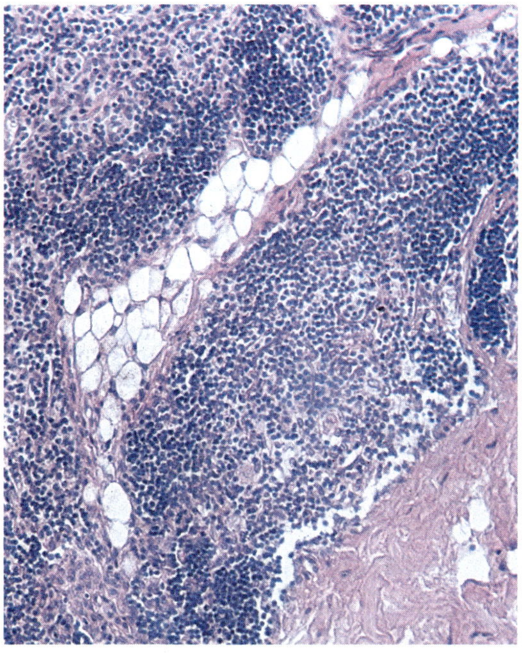

Plate 26 A portion of the parathyroid gland.

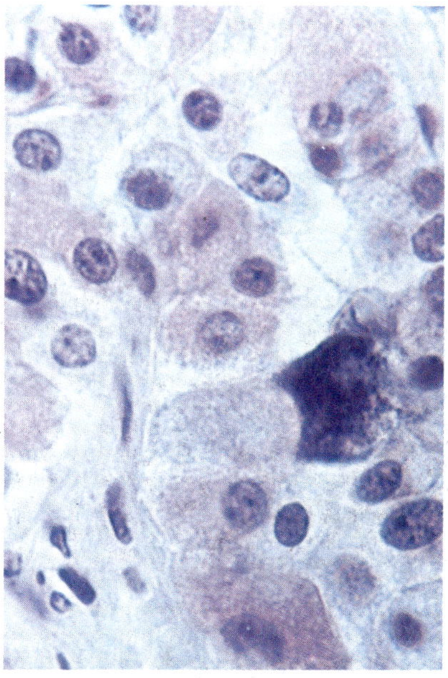

Plate 27 A higher magnification of the parathyroid.

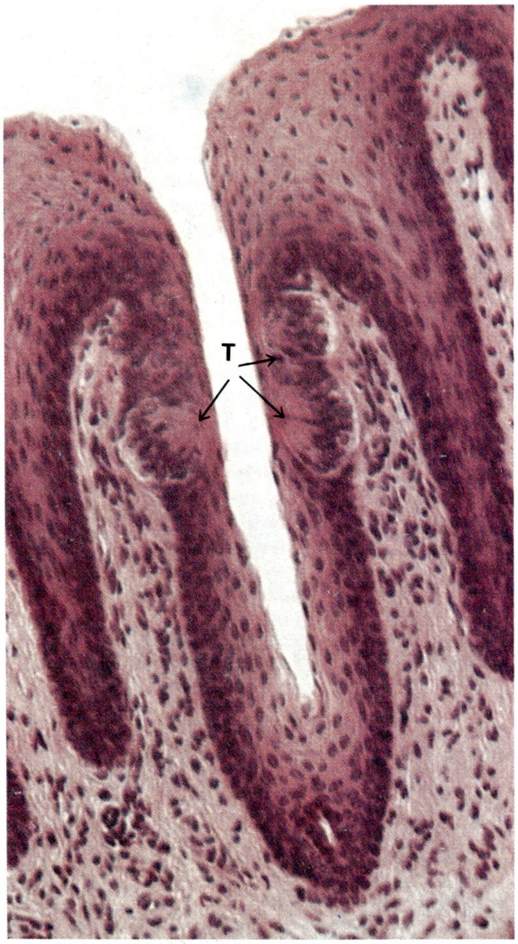

Plate 28 Taste buds.

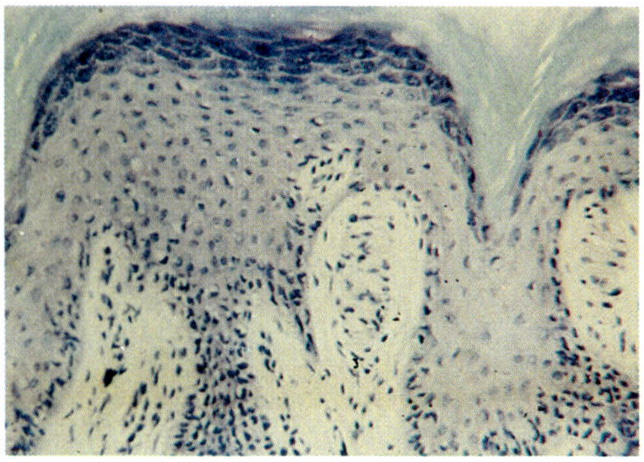

Plate 29 Meisner's corpuscles are small oval, encapsulated sensory organs located in the dermal papillae immediately beneath the skin epithelium. (Courtesy Ward's Natural Science Establishment, Inc.)

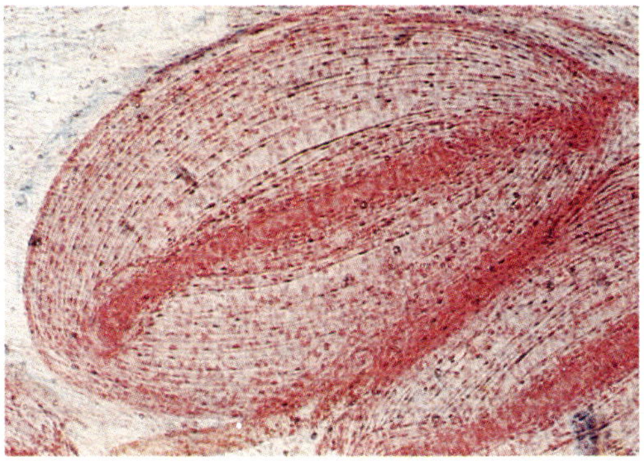

Plate 30 The Pacinian corpuscle, an encapsulated nerve ending associated with the sensations of touch and pressure. Dendritic endings are enclosed within many onionlike layers of connective tissue. Such layers absorb some of the pressure when the stimulus is applied and maintained.

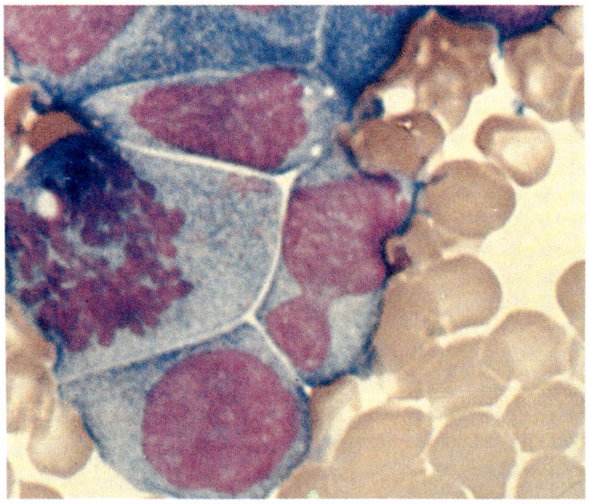

Plate 31 A bone marrow preparation, showing the features of an acute monocytic leukemia. A cell in mitosis can be seen in this smear. Bone marrow is usually obtained by aspiration, which involves inserting a needle into the marrow cavity and withdrawing a sample. The sternum, ilium, iliac crest and vertebral spinous process are the most common sites used for aspiration. (Courtesy Dr. I. Mirchandani, Wayne State University.)

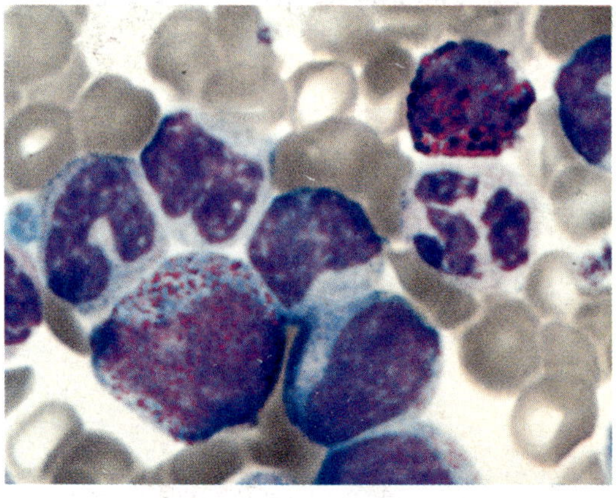

Plate 32 Chronic granulocytic leukemia. (Courtesy Dr. I. Mirchandani, Wayne State University.)

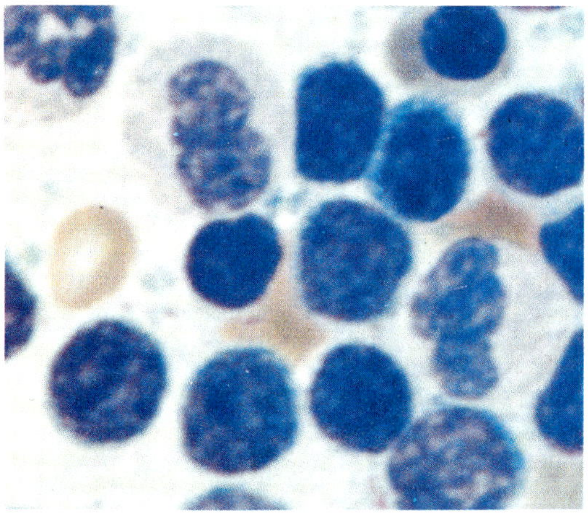

Plate 33 Chronic lymphocytic leukemia. (Courtesy Dr. I. Mirchandani, Wayne State University.)

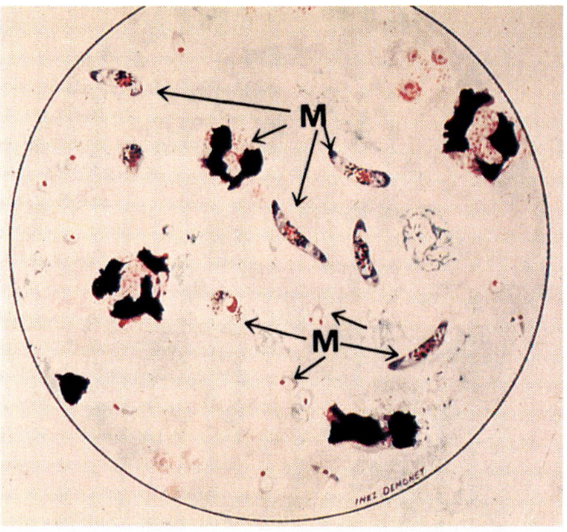

Plate 35 Stages of the malarial (M) parasite *Plasmodium falciparum.*

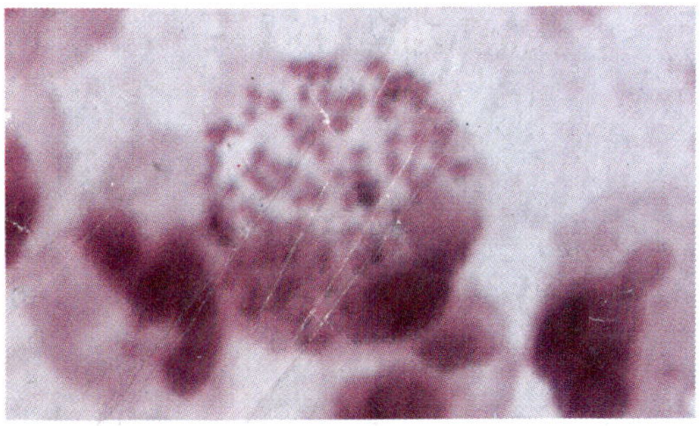

Plate 37 The intracellular location of *Neisseria gonorrhoeae,* the causative agent of the sexually transmitted disease gonorrhea.

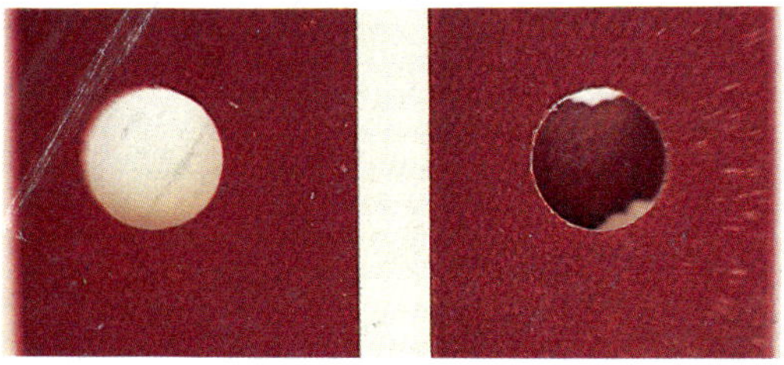

Plate 38 A Tallquist reaction showing a normal hemoglobin content.

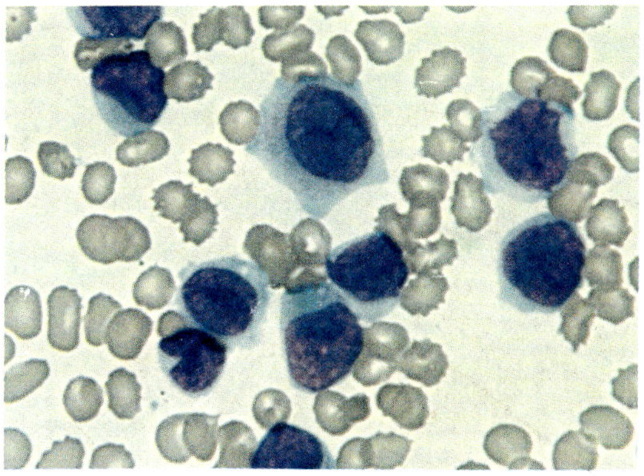

Plate 34 Acute myelomonocytic leukemia. (Courtesy Dr. I. Mirchandani, Wayne State University.)

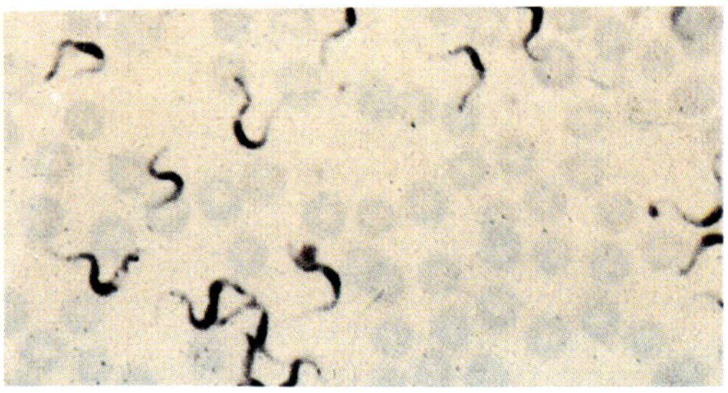

Plate 36 The causative agents of African sleeping sickness.

Plate 39 Fecal occult blood test. The slightest blue color is a positive test.

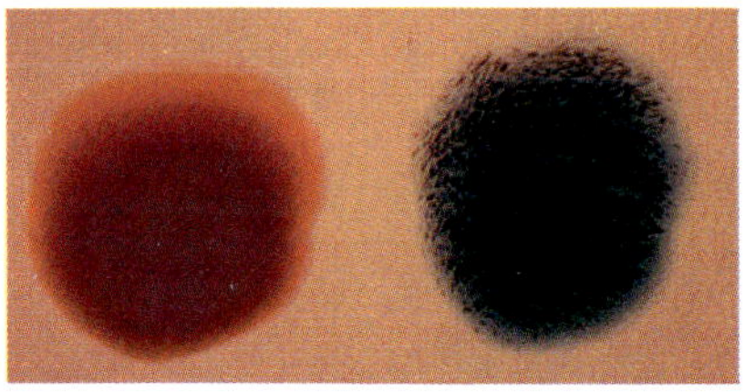

Plate 40 A typical agglutination reaction. The small clumps of red cells are quite obvious.

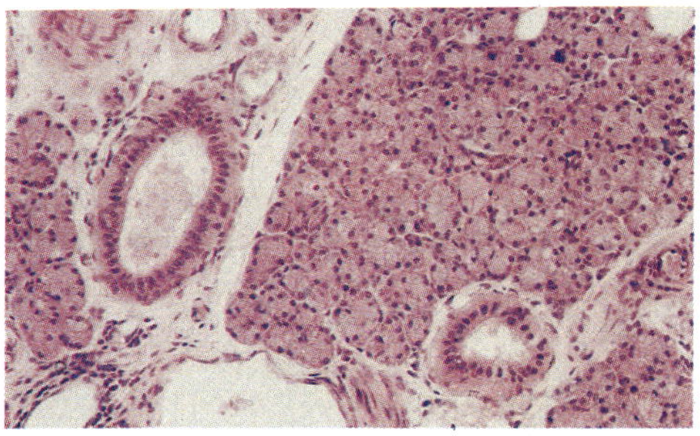

Plate 41 The submandibular salivary gland.

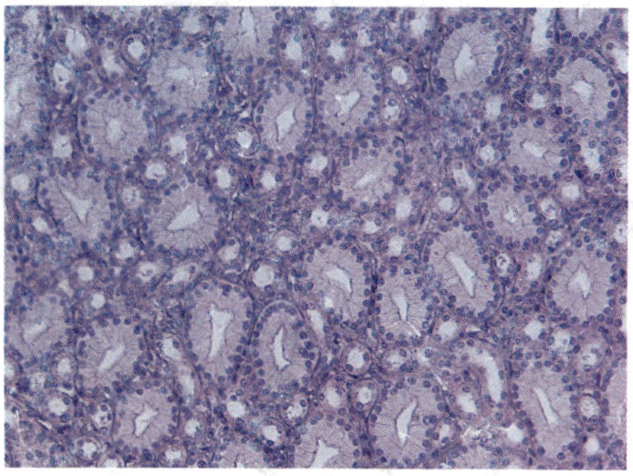

Plate 42 The parotid gland.

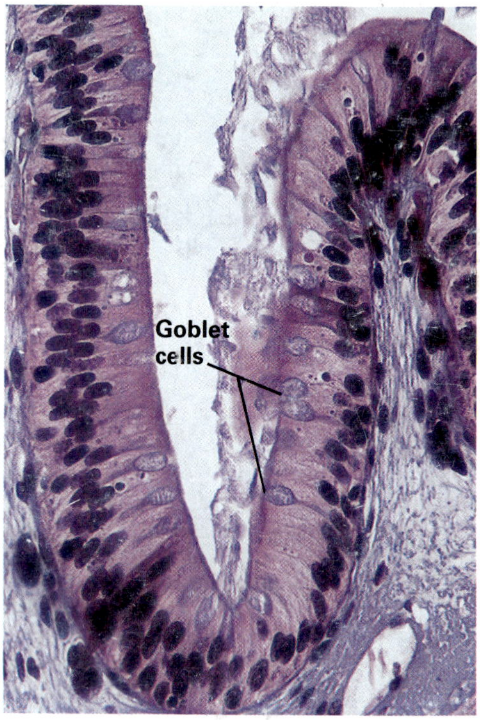

Plate 43 A portion of intestine showing secretory cells (goblet cells) and columnar cells).

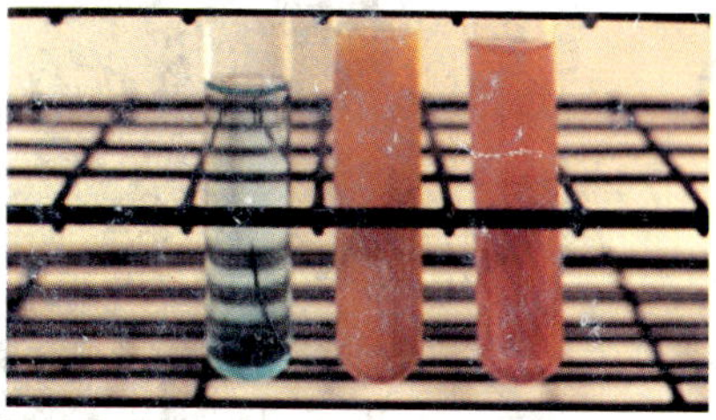

Plate 44 The results of a Benedict's test. Left tube shows no glucose, while the one in the center and on the right clearly contain glucose.

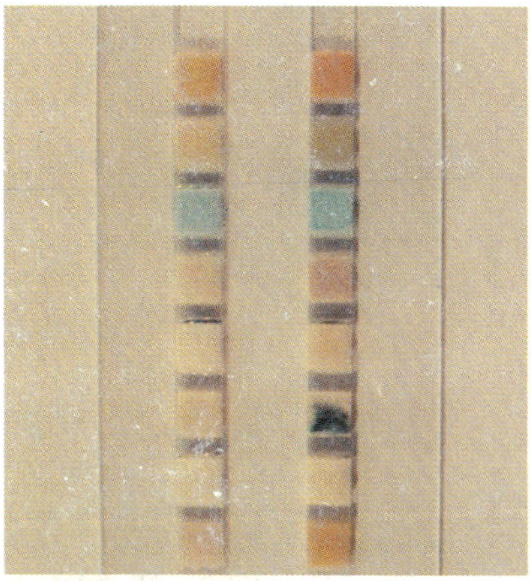

Plate 45 Multistix reactions. The strip on the left is normal.

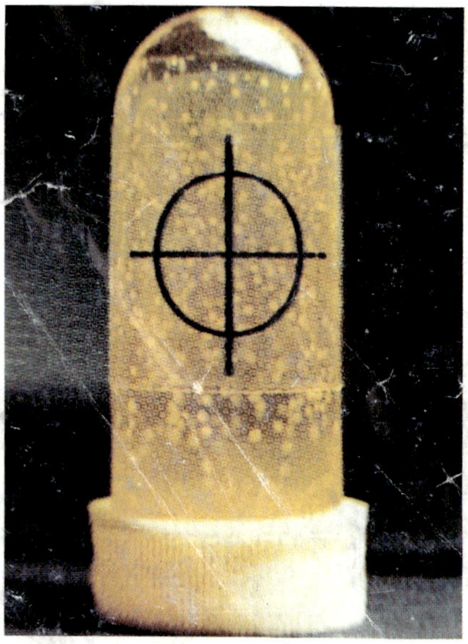

Plate 46 Bacturcult results showing a bacterial count under 100,000/ml. Note the appearance of the counting strip.

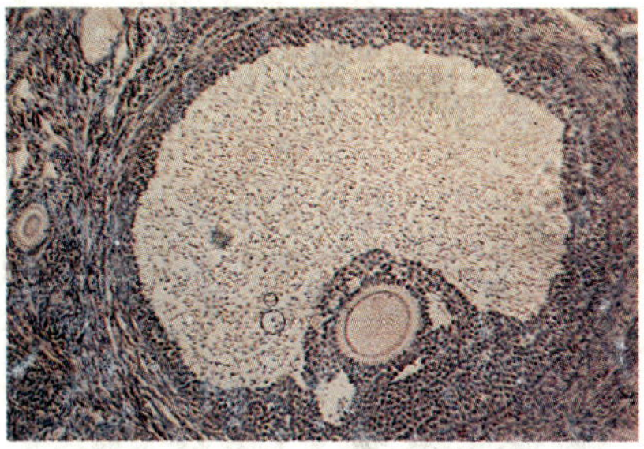

Plate 47 Human ovary showing a follicle in the first stages of growth. (Courtesy Carolina Biological Supply Company.)

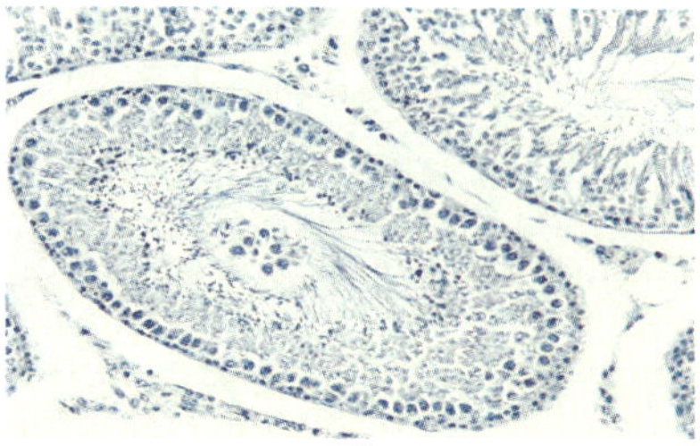

Plate 48 Microscopic view of human testes.